T0188846

Handbook of North European Garden Plants
With Keys to Families and Genera

This book provides a means for the accurate identification of over 190 families and 2220 genera of flowering plants cultivated out-of-doors in gardens in north-west Europe and in other geographical regions with a similar climate, including parts of North America. The text is an abridged version of *The European Garden Flora* (published in six volumes) and constitutes a handy single volume digest that covers families and genera, but not species. A key to all the families is provided and, for each of the families, a key to the genera within it. Scientifically rigorous descriptions of families and genera follow, including information on the number of constituent genera or species, and details of geographical distribution. Illustrations of genera from most of the major families are included to aid accurate identification.

JAMES CULLEN is Director of the Stanley Smith (UK) Horticultural Trust. He has a wealth of horticultural experience, having spent 17 years as Assistant Regius Keeper at the Royal Botanic Garden, Edinburgh, UK, and 24 years as chief editor of *The European Garden Flora*.

HANDBOOK OF
NORTH
EUROPEAN
GARDEN
PLANTS

With Keys to Families and Genera

EDITED BY JAMES CULLEN

CAMBRIDGE
UNIVERSITY PRESS

CAMBRIDGE
UNIVERSITY PRESS

University Printing House, Cambridge CB2 8BS, United Kingdom

Cambridge University Press is part of the University of Cambridge.

It furthers the University's mission by disseminating knowledge in the pursuit of education, learning and research at the highest international levels of excellence.

www.cambridge.org
Information on this title: www.cambridge.org/9780521004114

© Cambridge University Press 2001

First published 2001

A catalogue record for this publication is available from the British Library

ISBN 978-0-521-65183-7 Hardback
ISBN 978-0-521-00411-4 Paperback

CONTENTS

ILLUSTRATIONS

INTRODUCTION and ACKNOWLEDGEMENTS

This book provides a means for the accurate identification of the families and genera of flowering plants cultivated out-of-doors in gardens in north-western Europe -- north-western and northern France, the British Isles, Ireland, Holland, Belgium, northern Germany, Denmark, Norway, Sweden, Finland and Iceland. The selection of families and genera has been inclusive rather than restrictive, so that those genera which can be grown only in places with an unusually mild climate within the overall area (e.g. parts of western France, western Ireland, the Scilly Isles, etc.) are not excluded. This selection also covers most of the families and genera widely cultivated in eastern North America.

The keys and descriptions are, with minor modifications, taken from the much more comprehensive *The European Garden Flora* volumes I--VI (Cambridge University Press, 1984--2000), which, of course, includes plants cultivated in southern Europe and in glasshouses elsewhere. The book is thus a one-volume digest of the much larger Flora, and provides an up-to-date equivalent of J.W.C. Kirk's extremely useful *A British Garden Flora* (Edward Arnold, London, 1927). In order to keep the text as short as possible, notes on the various families and genera have been reduced to a minimum, and, for each genus, the number of species it is thought to contain, and the number of these in cultivation are given as a fraction immediately following the generic name. Thus '**27. Hemerocallis** Linnaeus. 15/8 (and many hybrids and cultivars)' indicates that the genus *Hemerocallis* is generally accepted as containing 15 species, of which 8, represented by the species themselves, and by many hybrids and cultivars, are likely to be found in cultivation.

In order to make the task of identification more precise and accurate, black and white line illustrations (essentially diagrammatic and not to scale) are provided showing diagnostic features of various genera from the major families, or (for those in the glossary), illustrating the usage of terms. Some of these illustrations are modified from those already published in *The European Garden Flora*, but most of them are taken, with the author's and artist's permission from Hickey & King, *100 Families of Flowering Plants*, Cambridge University Press, edn 1, 1981, edn 2, 2000. The re-use of these excellent illustrations is a particular pleasure for the editor, and thanks are due to Michael Hickey and Clive King for allowing it. The editor's thanks are also due to Professor John Parker and the staff of Cambridge University Botanic Garden for encouragement and facilities, and especially to Dr Mark Winfield for help with the computer-reorganisation of some of the illustrations. Finally, an acknowledgement is due to the authors of accounts of various groups in the original Flora; they are too numerous to list here (there are 175 of them), but this book would not have been possible without their work.

Key to Subclasses

1a. Cotyledon 1, terminal; leaves usually with parallel veins, sometimes these connected by cross-veinlets; leaves without stipules, opposite only in some aquatic plants; flowers generally with parts in 3s; mature root-system wholly adventitious   *MONOCOTYLEDONS*

  b. Cotyledons usually 2, lateral; leaves usually net-veined, with or without stipules, alternate, opposite or whorled; flowers with parts in 2s, 4s or 5s, or parts numerous; primary root-system (taproot) usually persistent, branched   *DICOTYLEDONS* (see p. 106)

MONOCOTYLEDONS

Plants herbaceous or woody, frequently with bulbs, corms or rhizomes. Primary root usually quickly lost, the mature root system wholly adventitious, fibrous. Seedling leaf (cotyledon) 1. Leaves usually alternate or all basal, usually with several parallel veins, more rarely with a distinct midrib giving off parallel lateral veins, very rarely the veins forming a network. Parts of the flower usually in 3s or multiples of 3.

Key to Families

1a. Ovary superior or flowers completely without perianth (including all aquatics with totally submerged flowers)   2

  b. Ovary inferior or partly so (if aquatic, flowers borne at or above water-level)   23

2a. Trees, shrubs or prickly scramblers with large, pleated, usually palmately or pinnately divided leaves; flowers more or less stalkless in fleshy spikes or panicles with large basal bracts (spathes)   **17. PALMAE**

  b. Plants without the above combination of characters   3

3a. Perianth entirely translucent (sometimes brown) or reduced to bristles, hairs, narrow scales, or absent   4

  b. Perianth well-developed, though sometimes small, never entirely translucent or shiny   9

4a. Flowers in small, 2-sided or cylindric spikelets provided with overlapping bracts (spikelets sometimes 1-flowered)   5

  b. Flowers arranged in heads, superposed spikes, racemes, panicles or cymes, never in spikelets as above   6

5a. Leaves alternate, in 2 ranks, on a stem which is usually hollow and with cylindric internodes; leaf-sheath usually with free margins, at least in the upper part; flowers arranged in 2-sided spikelets (sometimes 1-flowered) each usually subtended at the base by 2 sterile bracts (glumes); each flower usually enclosed between a lower lemma and an upper palea (sometimes absent); perianth of 2 or 3 concealed scales (lodicules), more rarely 6 or absent; styles generally 2, feathery   **16. GRAMINEAE**

  b. Leaves usually spirally arranged on 3 sides of the cylindric or more usually 3-angled stems which usually have solid internodes; young leaf-sheaths closed, though sometimes opening later; flowers arranged in 2-sided or cylindric spikelets often with a 2-keeled or 2-lobed glume at the base; each flower subtended only by a glume; perianth of several bristles, hairs or scales, or absent; style 1 with 2 or 3 papillose stigmas   **21. CYPERACEAE**

6a. Inflorescence a simple fleshy spike (spadix) of inconspicuous flowers subtended by or rarely joined to a large bract (spathe); leaves often net-veined or lobed (plant rarely a small, evergreen, floating aquatic)

18. ARACEAE

b. Plant without the above combination of characters   7

7a. Flowers bisexual; perianth-segments 6, translucent or shiny; ovary with
3--many ovules   **13. JUNCACEAE**

b. Flowers unisexual; perianth-segments a few threads or scales; ovary with
1 ovule   8

8a. Flowers in 2 superposed, elongate, brownish or silvery spikes; ovary borne
on a stalk with hair-like branches   **20. TYPHACEAE**

b. Flowers in spherical heads; ovary not stalked   **19. SPARGANIACEAE**

9a. Carpels free or slightly united at base   10

b. Carpels united for most of their length, though the styles may be free, or
carpel solitary   12

10a. Inflorescence a spike, sometimes bifid; perianth-segments 1--4

4. POTAMOGETONACEAE

b. Inflorescence not a spike; perianth-segments 6   11

11a. Placentation parietal or marginal   **2. BUTOMACEAE**

b. Placentation basal   **1. ALISMATACEAE**

12a. All perianth-segments similar   13

b. Perianth-segments of the outer and inner whorls conspicuously different,
the former usually sepal-like, the latter petal-like   21

13a. Inflorescence subtended by an entire, spathe-like sheath; plants aquatic

11. PONTEDERIACEAE

b. Inflorescence not as above; plants terrestrial   14

14a. Plant woody, or not woody but bearing rosettes of long-lived, fleshy or
leathery leaves at or near ground-level   15

b. Plant herbaceous, leaves usually not long-lived and in rosettes, if so, then
deciduous and not fleshy   19

15a. Leaf-stalk bearing 2 tendrils; leaves net-veined   **5. LILIACEAE**

b. Leaf-stalk without tendrils; leaves parallel-veined   16

16a. Leaves very small, scale-like or spiny, their function taken over by flattened
stems (cladodes) on which the inflorescences are usually borne

5. LILIACEAE

b. Plant with true leaves; cladodes absent   17

17a. Shrubs or woody climbers with scattered stem-leaves; flowers solitary,
usually large and hanging; placentation mostly parietal   **5. LILIACEAE**

b. Plant without the above combination of characters   18

18a. Leaves leathery and more or less thin, if succulent, then with a spine-like or
cylindric tip; flowers usually green or whitish, bell- or cup-shaped or with a
narrow tube and spreading lobes, often more than 1 to each bract

6. AGAVACEAE

b. Leaves succulent, usually without a spine-like or cylindric tip; flower
usually red, yellow or orange, tubular, the lobes not spreading, always 1
per bract   **5. LILIACEAE**

19a. Leaves very small, scale-like or spiny, their function taken over by flattened
or needle-like stems (cladodes) on which the inflorescences are usually
borne   **5. LILIACEAE**

b. Plant with true leaves, cladodes absent   20

20a. Leaves evergreen, clearly stalked; flowers more than 1 to each bract, with a narrow tube as long as or longer than the spreading lobes **6. AGAVACEAE**

    b. Leaves deciduous, usually without distinct stalks; flowers of various shapes, rarely as above, always 1 to each bract   **5. LILIACEAE**

21a. Flowers solitary or in umbels; leaves broad, opposite or in a single whorl near the top of the stem   **5. LILIACEAE**

    b. Flowers in spikes, heads, cymes or panicles; leaves not as above   22

22a. Stamens 6, or 5--3 with 1--3 staminodes; anthers basifixed; leaves usually borne on the stems, often with closed sheaths, never grey with scales; bracts neither overlapping nor conspicuously coloured

                                               15. COMMELINACEAE

    b. Stamens 6, staminodes 0; anthers dorsifixed; leaves mostly in basal rosettes, often rigid and spiny-margined, when on the stem usually grey with scales; bracts usually overlapping and conspicuously coloured

                                             14. BROMELIACEAE

23a. Flowers radially symmetric or weakly bilaterally symmetric; stamens 6, 4, 3 or rarely many   24

    b. Flowers strongly bilaterally symmetric or asymmetric; stamens usually 5, 2 or 1 (very rarely 6)   33

24a. Unisexual climbers with heart-shaped or very divided leaves; rootstock tuberous or woody   **10. DIOSCOREACEAE**

    b. Plants without the above combination of characters   25

25a. Rooted or floating aquatics; stamens 2--12; ovules distributed all over the carpel-walls (diffuse-parietal placentation)   **3. HYDROCHARITACEAE**

    b. Terrestrial or marsh plants, or epiphytes; stamens 3 or 6, rarely many; placentation axile or parietal (ovules restricted to a few rows on the carpel-walls)   26

26a. Stamens 3, staminodes absent; leaves often sharply folded, their bases overlapping; style-branches often divided   **12. IRIDACEAE**

    b. Stamens 6, or 3 plus 3 staminodes; leaves not usually as above; style-branches not divided   27

27a. Perianth consisting of an outer, calyx-like whorl and an inner, corolla-like whorl; bracts usually overlapping and conspicuously coloured

                                             14. BROMELIACEAE

    b. Segments of the perianth not in 2 dissimilar whorls; bracts not as above   28

28a. Ovary half-inferior   29

    b. Ovary fully inferior   30

29a. Anthers opening by pores   **8. TECOPHILAEACEAE**

    b. Anthers opening by slits   **5. LILIACEAE**

30a. Leaves long-persistent, evergreen   **6. AGAVACEAE**

    b. Leaves dying down annually   31

31a. Flowers in a spike; leaves fleshy, often spotted with brown, the margins more or less rolled around each other in bud   **6. AGAVACEAE**

    b. Flowers in umbels or solitary; leaves not usually fleshy or spotted with brown, but flat, pleated or with the margins folded outwards in bud   32

32a. Leaves mostly basal, densely hairy, pleated or with prominent veins

                                             9. HYPOXIDACEAE

    b. Leaves various, not usually densely hairy, pleated or with prominent veins, basal or not   **7. AMARYLLIDACEAE**

33a. Fertile stamens 6; perianth-segments all similar, tube curved and unevenly swollen; stems below ground, fleshy   **6. AGAVACEAE**

  b. Fertile stamens 5, 2 or 1, very rarely 6; staminodes, which may be petal-like, often present; perianth-segments usually differing among themselves; fleshy underground stems rare   34

34a. Fertile stamens 2 or 1, united with the style to form a column; pollen usually borne in masses (pollinia); leaf-veins, when visible, all parallel to margins   **25. ORCHIDACEAE**

  b. Fertile stamens 5 or 1, rarely 6, not united to the style; pollen granular; leaves with distinct midrib more or less parallel to margins, the secondary veins parallel, running from midrib to margins   35

35a. Fertile stamen with normal structure, not petal-like   **22. ZINGIBERACEAE**

  b. Fertile stamen in part petal-like, and with only 1 pollen-bearing anther-lobe   36

36a. Leaf-stalk with a swollen band (pulvinus) at the junction with the blade; ovary smooth, with 1--3 ovules   **24. MARANTACEAE**

  b. Leaf-stalk without a pulvinus at the junction with the blade; ovary usually warty, with numerous ovules   **23. CANNACEAE**

1. ALISMATACEAE

Aquatic or marsh herbs, usually perennial. Leaves entire, alternate or basal. Flowers bisexual or unisexual, radially symmetric, borne in umbels, racemes or panicles. Sepals 3, free. Petals 3, free, often falling early. Stamens 3--many; anthers opening by slits. Ovary superior; carpels 3--many, free or united at the base; ovules usually 1 per carpel, with basal placentation. Fruit a group of achenes. See Figure 1, p. 8.

There are 13 genera from all parts of the world, and 90 species, some of which are cultivated in and around ornamental pools or in aquaria.

  1a. Leaves sagittate   **1. Sagittaria**

    b. Leaves not sagittate   2

  2a. Some flowers unisexual; achenes winged   **1. Sagittaria**

    b. Flowers all bisexual; achenes ribbed or furrowed but not winged   3

  3a. Achenes with 3 ribs on the back and a double inner rib, spirally arranged in a spherical head   **2. Baldellia**

    b. Achenes otherwise ribbed, in a single, sometimes irregular whorl   4

  4a. Achenes 6--15, oblong-ovoid, in a hemispherical head; style apical   **3. Luronium**

    b. Achenes 11--28, laterally compressed, in a more or less flat head; style lateral   **4. Alisma**

1. Sagittaria Linnaeus. 20/few. Rootstock often with stolons and tubers. Leaves aerial, floating or submerged. Some flowers unisexual, usually in whorls of 3, forming racemes or panicles with female or bisexual flowers at the base and male flowers above, or occasionally with the flowers all male or female. Petals white. Stamens usually numerous. Carpels numerous, each with 1 ovule, spirally arranged on the receptacle, developing into a head of laterally compressed, winged, beaked achenes. *Mostly America.*

5

2. Baldellia Parlatore. 2/1. Sometimes with stolons. Leaves elliptic to lanceolate or linear-lanceolate. Flowers bisexual in 1--3 whorls in umbels or racemes. Stamens 6. Carpels numerous, free, spirally arranged in a spherical head. Achenes ovoid, longitudinally 5-ribbed (3 ribs on the back and an inner double rib), each with a short apical beak. *Europe, North Africa.*

3. Luronium Rafinesque. 1/1. Stems to 50 cm or more, floating or creeping and rooting at the nodes. Floating leaves long-stalked, blades to 4 × 1.5 cm, elliptic to ovate; submerged leaves linear. Flowers bisexual, 1.2--1.5 cm in diameter, long-stalked. Petals white, each with a yellow blotch at the base. Stamens 6. Carpels 6--15 in an irregular whorl. Achenes oblong-ovoid, 2.5 mm, each with 12--15 longitudinal ribs and a short, apical beak. *Europe.*

Formerly placed in the genus *Elisma* Buchenau.

4. Alisma Linnaeus. 9/2--3. Leaves aerial, floating or submerged. Flowers bisexual, in panicles or occasionally in racemes or umbels. Stamens 6. Carpels 11--28 in a single whorl, free. Achenes laterally compressed, each with a short beak. *Almost cosmopolitan, mostly northern hemisphere.* Figure 1, p. 8.

2. BUTOMACEAE

Annual or perennial aquatic herbs with rhizomes. Leaves basal or alternate, not submerged. Flowers bisexual, in axillary clusters or umbels which are long-stalked; more rarely flowers solitary. Sepals 3, sometimes petaloid. Petals 3, larger than the sepals. Stamens 6--many, anthers opening by longitudinal slits. Carpels 6--many, free or united at the base, with many ovules with marginal or parietal placentation. Fruit a whorl of follicles.

4 genera and 13 species with a scattered distribution..

1. Butomus Linnaeus. 1/1. Perennial herbs with thick, creeping rhizomes. Leaves linear, triangular in cross-section in the lower part. Umbels many-flowered; flower-stalks 5--10 cm. Sepals petaloid, pink. Stamens 6--9. *Eurasia.*

Species of **Hydrocleis** Richard, which have leaves with distinct blades and flowers with yellow petals, may be grown in very sheltered areas.

3. HYDROCHARITACEAE

Submerged or floating, annual or perennial herbs. Leaves alternate, opposite or whorled, sometimes with minute scales at the nodes. Flowers bisexual or unisexual, 1 or more together subtended by spathes formed from 1 or 2 united or free bracts. Sepals 3. Petals 3 or absent. Stamens 2--20, anthers usually opening by longitudinal slits. Ovary inferior, of 3--6 united carpels; ovules numerous on intrusive parietal placentas; styles 3--15. Fruit berry-like.

A cosmopolitan but mainly tropical family. Species of some 9 genera are cultivated.

1a. Leaves with distinct stalks, in basal rosettes   **4. Hydrocharis**
  b. Leaves arranged along the stems, or, if in basal rosettes, then stalkless   2
2a. Leaves in a basal rosette   3
  b. Leaves distributed along the stems   4
3a. Leaves rigid, margins with prominent spiny teeth; plant Aloe-like
     6. Stratiotes

b. Leaves not rigid but ribbon-like, margins without prominent, spiny teeth
7. Vallisneria
4a. All leaves spirally arranged   **5. Lagarosiphon**
  b. Leaves opposite or whorled, at least in the upper part of the stem   5
5a. Flowers conspicuous, held above the water-surface, with nectaries
1. Egeria
  b. Flowers minute, submerged or borne at the water-surface, without
    nectaries   7
6a. Leaves in whorls of 3--8, with fringed nodal scales   **3. Hydrilla**
  b. Leaves in whorls of 3--4, with inconspicuous, unfringed nodal scales
2. Elodea

1. Egeria Planchon. 2/2. Dioecious or monoecious aquatic perennial herbs. Leaves submerged, whorled, stalkless, linear. Flowers conspicuous, with nectaries, held above the water-surface. Male spathes 2--4-flowered; petals of male flowers much longer than sepals; stamens 9. Ovary stalkless, styles 3, free, bifid or occasionally 3-fid. *Warm temperate South America; naturalised in Europe.*
Sometimes listed under *Elodea*.

2. Elodea Michaux. 15/3. Dioecious or monoecious, aquatic perennial herbs, often with stolons. Leaves submerged, whorled or sometimes opposite, stalkless, usually linear, with 2 minute scales at each node. Flowers inconspicuous, without nectaries, borne and pollinated at the water-surface. Spathes axillary, stalkless, with 1 or rarely 3 male flowers, or 1 female flower, or rarely 1 bisexual flower. Sepals 3. Petals 3. Stamens 3--9. Ovary stalkless; styles 3, free, bifid or occasionally entire. *America; now widely naturalised elsewhere.*

3. Hydrilla Richard. 1/1. Dioecious, submerged, branched herb with stolons, overwintering as turions. Leaves in whorls of 3--8 or the lowest opposite, lanceolate, each with a single mid-vein, stalkless, minutely toothed, with 2 brown, fringed scales in each axil. Flowers borne and pollinated at the water-surface, solitary, emerging from a tubular spathe. Sepals 3. Petals 3. Male flowers breaking free and floating on the water-surface at maturity; stamens 3. Female flowers stalkless, with an elongate perianth-tube; sepals smaller than petals; staminodes 3, styles not apparent, stigmas 3 (rarely to 5), ovary red. Fruit cylindric. *Europe, east Asia, east Africa, Australia.*

4. Hydrocharis Linnaeus. 2/1. Aquatic perennial herbs. Leaves stalked, all basal. Flowers unisexual, held above the water. Male spathes stalked, with 1--4 flowers; female spathes stalkless, 1-flowered. Petals much longer than sepals. Stamens 9--12, the outer sometimes sterile and reduced to staminodes. Styles 6, bifid, free. *Europe, western Asia, North Africa.*
One species of **Ottelia** Persoon, from the warmer areas of the world, may be cultivated in very mild areas; it is rooted in the substrate, without stolons and the spathes are tubular, ribbed or winged.

5. Lagarosiphon Harvey. 15/1. Dioecious, submerged plants. Leaves stalkless, each with 2 minute scales at the node. Flowers submerged until open, pollinated at the water-surface. Male spathe with many flowers: stamens 3, staminodes 3. Female

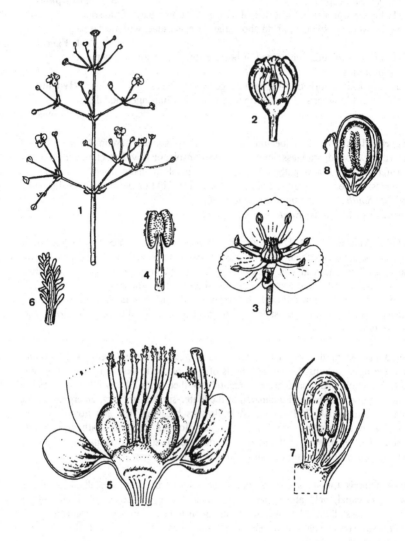

Figure 1. Alismataceae. *Alisma plantago-aquatica*. 1, Inflorescence. 2, Flower bud. 3, Flower. 4, Stamen. 5, Longitudinal section of flower. 6, Style-apex. 7, Longitudinal section of a carpel still attached to the receptacle. 8, Longitudinal section of mature fruit.

8

spathes usually each with 1 flower; ovary stalkless; styles 3, free, bifid. Sepals and petals 3 each. *Africa, Madagascar, India.*

6. Stratiotes Linnaeus. 1/1. Dioecious, submerged, Aloe-like perennial herb with stolons. Leaves all basal, stalkless. Flowers held above the water-surface. Male spathes with several flowers, female spathes each with a single flower. Petals larger than sepals. Stamens 12, surrounded by numerous staminodes. Ovary stalkless, styles 6, bifid, free. *Eurasia.*

Plants are mostly submerged, but rise to the surface at flowering time.

7. Vallisneria Linnaeus. 10/3. Dioecious, submerged plants with stolons. Leaves all basal, alternate, stalkless. Flowers pollinated at the water-surface. Male spathes almost stalkless or stalked, many-flowered, female spathes each long-stalked and 1-flowered; ovary linear; styles 3, free, bifid. Sepals and petals 3 each. *Cosmopolitan.*

4. POTAMOGETONACEAE

Submerged or floating, annual or perennial aquatic herbs. Stems elongate, flexible, erect. Rhizomes floating or creeping, slender, elongate, often producing specialised winter-buds (turions). Leaves opposite or alternate, rarely in whorls of 3, sheathing at their bases. Flowers bisexual, in axillary or terminal bractless spikes, inconspicuous, radially symmetric. Perianth-segments 4. Stamens 4. Ovary superior; carpels usually 4 (rarely 1--3), free or shortly joined at the base; ovules 1 per carpel. Style short. Fruit a small green or brown achene or drupe.

A family of 2 genera, distributed throughout the world. Only *Potamogeton* is cultivated.

1. Potamogeton Linnaeus. 100/2. Rhizomes branching, sometimes producing turions. Leaves alternate, submerged or partially floating; submerged leaves thin, translucent, linear; floating leaves leathery, opaque; stipule-like organs present, free or attached to the leaf-bases. Flowers wind- or water-pollinated. Fruit a drupe. *Cosmopolitan.*

5. LILIACEAE

Perennial herbs or shrubs, sometimes climbing or scrambling, with rhizomes, fleshy or fibrous roots, bulbs or corms. Leaves very variable, borne on the stems or all basal, usually alternate, more rarely opposite or whorled, very rarely very reduced, their functions taken over by needle-like, thread-like or flattened cladodes; veins usually parallel, more rarely net-like. Flowers usually with bracts, solitary or in racemes, spikes, panicles, clusters or umbels which are sometimes subtended by spathes. Perianth usually of 6 free or united segments, more rarely the segments 4 or more than 6. Stamens usually 6, more rarely 3 or 4 or more than 6, filaments free or united, borne on the perianth or not; anthers basifixed or dorsifixed and often versatile, usually opening by slits. Ovary usually superior, usually 3-celled with 2--many ovules per cell with axile placentation, more rarely 1-celled with numerous ovules with parietal placentation. Fruit a berry or capsule. Seeds 1--many. See Figures 2, p. 26, 3, p. 28 & 4, p. 36.

As treated here in the traditional sense, a family of 220 genera and 3500 species from all over the world. Many of the genera are cultivated.

There are varying views as to how the family should be defined and classified; the book *Monocotyledons: a comparative survey*, by R. Dahlgren and H.T. Clifford

9

(1982) should be consulted for further details. In summary, genera **1--11** & **93** are now often placed in *Melanthiaceae*; **12, 13, 74--84** & **92** in *Convallariaceae*; **20--22** & **29** in *Anthericaceae*; **14**, together with *Phormium*, here included in the *Agavaceae*, in *Phormiaceae*; **15--19** & **30** in *Asphodelaceae*; **23, 24** & **45--60** in *Hyacinthaceae*; **25** in *Aphyllanthaceae*; **26** in *Hostaceae*; **27** in *Hemerocallidaceae*; **28** in *Blandfordiaceae*; **31--33** in *Colchicaceae*; **34--43** in the *Liliaceae* in the restricted sense; **44** in *Alstroemeriaceae*; **61--73** in *Alliaceae*; **85--88** in *Trilliaceae*; **89** in *Asparagaceae*; **90** & **91** in *Ruscaceae*; **94--97** in *Philesiaceae*; and **98** in *Smilacaceae*.

Key to groups

1a. Stem woody, persistent through the winter   *Group 7* (p. 16)
  b. Stem not woody and persistent through the winter, though leaves sometimes evergreen   2
2a. Ovaries below ground, hidden in the bulbs or corms   *Group 5* (p. 15)
  b. Ovaries exposed, above ground   3
3a. Ovary partly or fully inferior   *Group 6* (p. 16)
  b. Ovary completely superior   4
4a. Flowers in umbels which are subtended by spathes (rarely flower solitary and subtended by 2 united spathes)   *Group 4* (p. 15)
  b. Flowers in inflorescences of various kinds, or solitary, not subtended by spathes   5
5a. Styles 3 (rarely 4), separate and distinct, or stigmas 3, separate, borne directly on the ovary   *Group 1*
  b. Style 1, sometimes 3-branched at the apex, or style absent, stigmas united, borne directly on the ovary   6
6a. Perianth-segments free at the base, or united for less than one-tenth of their total length   *Group 2* (p. 11)
  b. Perianth united at the base for more than one-tenth of its length   *Group 3* (p. 13)

Group 1

1a. Plant a climber   **98. Smilax**
  b. Plant not a climber   2
2a. Leaves 2, opposite or in 1 or 2 whorls on the flowering stem, never long and grass-like   3
  b. Leaves several to many, all basal or borne on the flowering stem, if 2 and opposite then grass-like, at least 8 times longer than broad   6
3a. Flower solitary, terminal   4
  b. Flowers in umbels   5
4a. Leaves 4 or more in a whorl; parts of the flowers in 4s   **87. Paris**
  b. Leaves in whorls of 3; parts of the flowers in 3s   **88. Trillium**
5a. Leaves 2 at the base of each flowering stem; stamens 3   **85. Scoliopus**
  b. Leaves in 2 whorls on each flowering stem; stamens 6   **86. Medeola**
6a. Some part of the plant obviously hairy or covered with scales   7
  b. No part of the plant obviously hairy or covered in scales   9
7a. Perianth-segments conspicuously clawed   **10. Melanthium**
  b. Perianth-segments not clawed   8
8a. Leaves broadly ovate to elliptic with many conspicuous veins, narrowed towards the stalk-like, sheathing bases   **11. Veratrum**

b. Leaves narrow, without conspicuous veins, not narrowing towards the
   sheathing bases   **9. Zigadenus**

9a. Plant with a bulb or corm   **8. Stenanthium**
  b. Plant with a rhizome or fleshy roots   10
10a. Perianth-segments very unequal, the 2 or 3 lower small and thread-like
                               7. Chionographis
  b. Perianth-segments all more or less equal   11
11a. Leaves narrowed towards their bases, shortly stalked   12
  b. Leaves not narrowed towards their bases, not stalked   14
12a. Perianth-segments somewhat swollen or inflated towards the base; fruit a
    berry   **83. Disporum**
  b. Perianth-segments not as above; fruit a capsule   13
13a. Flowers unisexual   **5. Chamaelirion**
  b. Flowers bisexual   **4. Helonias**
14a. Stem-leaves numerous; flower-stalks very long   **3. Xerophyllum**
  b. Stem-leaves few; flower-stalks very short or absent   **1. Tofieldia**

Group 2

1a. Plants with bulbs or corms   2
  b. Plants with rhizomes or fleshy or fibrous roots   20
2a. Leaves borne on the flowering stem   3
  b. Leaves borne at the base of the scape   13
3a. Perianth-segments without nectaries at their bases   4
  b. Perianth-segments with nectaries at their bases   5
4a. Flowers in panicles   **34. Gagea**
  b. Flowers solitary or in racemes or umbels   **38. Tulipa**
5a. Style absent, stigmas borne directly on the ovary   **36. Calochortus**
  b. Style present   6
6a. Anthers basifixed   7
  b. Anthers dorsifixed, often versatile   9
7a. Perianth-segments reflexed from the base or very near it
                         37. Erythronium
  b. Perianth-segments not reflexed from near their bases, reflexed, if at all,
    at their tips only   8
8a. Perianth-segments in 2 distinct whorls; bulb-tunics usually absent, thin and
    white if present   **39. Fritillaria**
  b. Perianth-segments not in 2 distinct whorls; bulb-tunics always brown
                     35. Lloydia
9a. Perianth-segments twisting together after flowering; flower-stalks jointed
                     24. Chlorogalum
  b. Perianth-segments not twisting together after flowering; flower-stalks not
    jointed   10
10a. Perianth-whorls differing markedly in size or colour and markings;
    filaments swollen at their bases; nectary usually formed from a number of
    plates arranged in a fan shape   **42. Nomocharis**
  b. Perianth-whorls similar, or segments of the inner whorl slightly narrower
    than those of the outer; nectary various, not as above   11

11a. Bulb persistent after flowering, over-wintering; leaves various, but neither long-stalked and heart-shaped nor linear, and very rarely produced in autumn or winter **40. Lilium**
  b. Bulb dying after flowering, the plant persisting by offsets; leaves long-stalked and heart-shaped or linear and produced in autumn or winter 12
12a. Leaves long-stalked and heart-shaped **43. Cardiocrinum**
  b. Leaves not stalked, linear **41. Notholirion**
13a. Perianth-segments reflexed from their bases or almost so
  37. Erythronium
  b. Perianth-segments not at all reflexed or, if so, at their apices only 14
14a. Inner perianth-segments erect, usually hooded or each with a swelling at the tip, outer segments spreading **51. Albuca**
  b. All perianth-segments similar, not as above 15
15a. Filaments flattened, usually broadest at the base, occasionally winged and toothed; flowers not blue **49. Ornithogalum**
  b. Filaments thread-like, or if flattened then lanceolate to elliptic, narrow at the base; flowers often blue 16
16a. Perianth withering and falling after flowering **52. Urginea**
  b. Perianth withering but not falling after flowering, persisting around or below the capsule 17
17a. Ovules 2 per cell; perianth-segments to 6 mm **23. Schoenolirion**
  b. Ovules several per cell; perianth-segments more than 6 mm 18
18a. Each perianth-segment 3-veined **48. Camassia**
  b. Each perianth-segment 1-veined 19
19a. Bracts 2 to each flower **46. Hyacinthoides**
  b. Bract 1 to each flower, or bracts absent **45. Scilla**
20a. Leaves borne on the flowering stem 21
  b. Leaves at the base of the stem only 32
21a. Fruit a berry 22
  b. Fruit a capsule 27
22a. Anthers opening by pores **14. Dianella**
  b. Anthers opening by slits 23
23a. Flowering stem leafy over most of its length below the inflorescence 24
  b. Flowering stem with only 2 or 3 leaves along its length or with several leaves concentrated towards the base 26
24a. Flowers solitary or in pairs, their stalks fused to the stem for some distance so that they appear to be borne well above their bract, the stalk with a downward bend at about the middle of the free part **82. Streptopus**
  b. Flowers and flower-stalks not as above 25
25a. Leafy part of the stem unbranched; perianth-segments neither swollen nor slightly inflated above the base **80. Smilacina**
  b. Leafy part of the stem branched; perianth-segments often swollen or slightly inflated above the base **83. Disporum**
26a. Flowering stem bearing 2 or 3 heart-shaped leaves; flowers with parts in 4s
  81. Maianthemum
  b. Flowering stem with a few leaves which are not heart-shaped; flowers with parts in 3s **75. Speirantha**

12

27a. Flowers solitary or in a head subtended by several bracts
    25. Aphyllanthes
  b. Flowers not as above   28
28a. Leaves evergreen; seeds with linear or thread-like appendages at both ends
    6. Heloniopsis
  b. Leaves not evergreen; seeds not as above   29
29a. Flowers in long racemes   30
  b. Flowers in clusters, panicles or false umbels, or solitary   31
30a. Leaves very numerous, not obviously in 2 ranks; stamens unequal,
    filaments hairless   **16. Asphodeline**
  b. Leaves few, obviously in 2 ranks; stamens equal, filaments woolly
    2. Narthecium
31a. Perianth pure yellow, not spotted; flowers pendent; capsule splitting
    between the septa   **12. Uvularia**
  b. Perianth spotted; flowers ascending or erect; capsule splitting along the
    septa   **13. Tricyrtis**
32a. Fruit a berry   33
  b. Fruit a capsule   34
33a. Perianth-segments hairy outside   **79. Clintonia**
  b. Perianth-segments not hairy outside   **75. Speirantha**
34a. Stamens 3   **85. Scoliopus**
  b. Stamens 6   35
35a. At least 3 of the filaments hairy   36
  b. None of the filaments hairy   37
36a. Perianth whitish inside, purplish outside   **19. Simethis**
  b. Perianth yellowish or brownish inside and out   **18. Bulbine**
37a. Ovules 2 in each cell of the ovary   **15. Asphodelus**
  b. Ovules 4--8 in each cell of the ovary   38
38a. Anthers dorsifixed: raceme 1-sided   **21. Paradisea**
  b. Anthers basifixed; raceme (or panicle) not 1-sided   39
39a. Flowers 6--10 in a raceme; scape 20--70 cm; perianth-segments white,
    each with 3 veins   **20. Anthericum**
  b. Flowers 50--800 in a raceme; scape to 2.4 m, if less than 70 cm then each
    perianth-segment with 1 vein   **17. Eremurus**

Group 3
  1a. Flowering stem leafy or bearing cladodes   2
  b. Flowering stem leafless, leaves all basal   7
  2a. True leaves small, scale-like, their function taken over by numerous
    cladodes which are borne singly or in clusters in the axils of the scale-
    leaves   **89. Asparagus**
  b. True leaves present, green; cladodes absent   3
  3a. Leaves distinctly narrowed to stalks at their bases; racemes more or less
    1-sided   4
  b. Leaves not narrowed to the stalks; racemes not 1-sided   5
  4a. Perianth spherical to bell-shaped, the tube two-thirds of the total length;
    fruit a berry   **74. Convallaria**

b. Perianth narrowly funnel-shaped, the tube less than half the total length;
   fruit a capsule   **26. Hosta**
5a. Flowers bilaterally symmetric; stamens deflexed   **16. Asphodeline**
  b. Flowers radially symmetric; stamens not deflexed   6
6a. Leaves mostly towards the base of the stem, none with flowers in their
   axils   **93. Aletris**
  b. Leaves spread along the stem, the uppermost with flowers or flower-
   clusters in their axils   **84. Polygonatum**
7a. Plant with a bulb   8
  b. Plant with a rhizome or fleshy or fibrous roots   18
8a. Perianth-tube longer than lobes   9
  b. Perianth-tube shorter than to as long as the lobes   14
9a. Leaves 2, ovate, oblong or almost circular, spreading horizontally along
   the ground; scape absent or very short; perianth-lobes reflexed, almost
   as long as tube   **60. Massonia**
  b. Leaves various, not as above; perianth-lobes much shorter than the tube,
   not usually reflexed   10
10a. Bracts obvious, as long or almost as long as the flower-stalks
                    57. Brimeura
  b. Bracts minute or absent   11
11a. Perianth constricted at the mouth of the tube   **58. Muscari**
  b. Perianth not constricted at the mouth of the tube   12
12a. Capsule 3-angled   **55. Bellevalia**
  b. Capsule rounded, not angled   13
13a. Perianth 4--9 mm   **56. Hyacinthella**
  b. Perianth 1--3.5 cm   **54. Hyacinthus**
14a. Perianth disc-shaped, the lobes spreading from a short, obconical tube   15
  b. Perianth cylindric or bell-shaped, the lobes spreading or not   16
15a. Filaments united into a corona with projecting lobes   **53. Puschkinia**
  b. Filaments flat, not united   **47. Chionodoxa**
16a. Plant to 1 m; flowers 2.5 cm or more, white or greenish   **60. Galtonia**
  b. Plant much smaller; flowers less than 2.5 cm, variously coloured   17
17a. Perianth blue or bluish purple   **55. Bellevalia**
  b. Perianth white, greenish or brownish orange or pink   **50. Dipcadi**
18a. Leaves evergreen, not stalked, often thick, fleshy or leathery   19
  b. Leaves thin, usually deciduous, if evergreen then stalked   20
19a. Scape flattened; leaves thin   **22. Comospermum**
  b. Scape not flattened; leaves thick, fleshy or leathery   **77. Rohdea**
20a. Leaves distinctly stalked   21
  b. Leaves not stalked   22
21a. Perianth funnel-shaped with 6 lobes; flowers in racemes borne above
   ground-level   **26. Hosta**
  b. Perianth not funnel-shaped, with 6--8 lobes; flowers solitary, at ground-
   level or almost so   **78. Aspidistra**
22a. Flowers stalkless   23
  b. Flowers stalked   24
23a. Fruit a berry; perianth-lobes longer than the tube   **76. Reineckea**
  b. Fruit a capsule; perianth-tube longer than lobes   **93. Aletris**

24a. Perianth funnel-shaped, the lobes longer than the tube   **27. Hemerocallis**
  b. Perianth cylindric or bell-shaped, the tube longer than the lobes   25
25a. Ovary stalked; filaments attached near the middle of the perianth
                                                28. Blandfordia
  b. Ovary not stalked; filaments attached at the base of the perianth
                                                30. Kniphofia

Group 4

1a. Plants with rhizomes or fleshy roots, bulbs absent   2
  b. Plants with bulbs (which sometimes have a rhizome beneath them)   3
2a. Perianth with 6 outgrowths forming a corona   **62. Tulbaghia**
  b. Perianth without outgrowths   **61. Agapanthus**
3a. Perianth with free segments or with a very short tube   4
  b. Perianth with a distinct tube   10
4a. Plant smelling of onion or garlic   5
  b. Plant not smelling of onion or garlic   6
5a. Outer perianth-segments with 3--7 veins; flower-stalks swollen below the
    flower; ovary with many ovules   **65. Nectaroscordum**
  b. Outer perianth-segments with 1 vein; flower-stalks not swollen below the
    flowers; ovary with 1--8 ovules per cell   **63. Allium**
6a. Spathe 1   **68. Caloscordum**
  b. Spathes 2 or more   7
7a. Spathes green, leaf-like; plant less than 20 cm   **34. Gagea**
  b. Spathes papery; plant 20 cm or more   8
8a. Spathes more than 2   **67. Muilla**
  b. Spathes 2   9
9a. Flower-stalks jointed at the top; filaments united at the base into a cup-like
    structure   **66. Bloomeria**
  b. Flower-stalks not jointed at the top; filaments free   **64. Nothoscordum**
10a. Flower solitary (rarely 2 flowers together), subtended by 2 spathes which
    are united into a tube below   **70. Ipheion**
  b. Flowers more than 2, in umbels subtended by 2 or 4 free spathes   11
11a. Stamens enclosed in the perianth-tube   **69. Leucocoryne**
  b. Stamens projecting from the perianth-tube   12
12a. Leaves not keeled beneath; veins weak or not visible   **71. Brodiaea**
  b. Leaves keeled beneath; veins distinct   13
13a. Anthers versatile; ovary shortly stalked   **72. Triteleia**
  b. Anthers basifixed; ovary stalkless   **73. Dichelostemma**

Group 5

1a. Style 1, branched above   2
  b. Styles 3, free   3
2a. Perianth with a tube below   **29. Leucocrinum**
  b. Perianth without a tube, the segments clawed, with auricles at the junction
    of the blade and claw which cohere, making the perianth appear super-
    ficially as though it has a tube   **32. Bulbocodium**
3a. Perianth with a tube below   **31. Colchicum**

b. Perianth-segments free to the base, though the bases of the segments are close and superficially appear to be joined   **33. Merendera**

Group 6

1a. Flowers in panicles; styles 3, free   **9. Zigadenus**
  b. Flowers in umbels, spikes or racemes or solitary; style 1   2
2a. Stem-leaves twisted at the base through 180°; flowers in umbels
    44. Alstroemeria
  b. Stem-leaves not twisted at their bases; flowers in racemes or spikes
    92. Ophiopogon

Group 7

1a. Leaves scale-like, their function taken over by cladodes which are cylindric, thread-like or flattened, borne singly or in clusters in the axils of the scale-leaves   2
  b. Leaves normal, cladodes absent   4
2a. Stamens free; scale-leaves with thickened, often spine-like or projecting bases   **89. Asparagus**
  b. Stamens united into a tube; scale-leaves not as above   3
3a. Flowers in racemes of 5--8 at the tips of the branches, separate from the cladodes   **90. Danaë**
  b. Flowers in clusters on the surfaces of the cladodes   **91. Ruscus**
4a. Leaves with net-veins, their stalks bearing 2 tendrils   **98. Smilax**
  b. Leaves with parallel veins, or veins not visible, without tendrils   5
5a. Ovary 3-celled; flower less than 4 cm   **94. Luzuriaga**
  b. Ovary 1-celled; flowers more than 4 cm   6
6a. Plant a small, upright shrub   **95. Philesia**
  b. Plant climbing or twining   7
7a. Perianth-segments all equal in length   **97. Lapageria**
  b. Perianth-segments unequal, the outer somewhat shorter than the inner
    96. × Philageria

1. Tofieldia Hudson. 18/2. Perennial herbs with rhizomes. Leaves narrowly sword-shaped, densely tufted in 2 opposite ranks. Flowering stems erect, with or without leaves. Flowers small, in terminal racemes or heads. Flower-stalks short, each bearing a bract at the base and sometimes a bracteole just below the flower. Perianth persistent; segments 6, free, narrow, ascending or spreading. Stamens free, filaments hairless; anthers broadly ovate, basifixed. Ovary superior, carpels 3, free above, each gradually narrowed into a short, persistent style. Fruit a spherical to ellipsoid capsule; seeds very small, numerous, ellipsoid to narrowly oblong. *North temperate areas.*

2. Narthecium Hudson. 8/1. Perennial herbs with rhizomes and fibrous roots. Leaves basal and also borne on the stem. Flowers in a usually dense raceme; bracts about as long as the flower-stalks. Perianth-segments 6, free, spreading in flower, erect and persistent in fruit. Stamens 6, anthers versatile, opening to the outside of the flower; filaments woolly. Ovary superior, 3-celled, 3-sided, ovules numerous in each cell. Style simple. Fruit a capsule splitting between the septa. Seeds numerous, tailed. *North temperate areas.*

3. Xerophyllum Michaux. 2/2. Perennial herbs with stout, woody, stem-like rhizomes. Leaves numerous, densely tufted, glaucous on the backs, mostly basal, linear, slightly broader at the base and finely tapered to the tip; margins very hard and rough or finely toothed. Flowering stem stout, unbranched, erect, bearing smaller and shorter stem-leaves with membranous margins. Flowers white or yellowish white, numerous in long, dense terminal racemes. Perianth-segments 6 in 2 whorls, all similar, free, spreading, ovate, persistent. Stamens 6, free. Styles 3, thread-like, erect, becoming curled. Ovary superior. Fruit a 3-lobed, 3-celled capsule with 2--4 wedge-shaped seeds per cell. *North America.*

4. Helonias Linnaeus. 1/1. Evergreen perennial producing leaves and inflorescences from a horizontally branched, tuberous rhizome. Leaves shortly stalked, lanceolate, in basal rosettes. Racemes dense, spike-like, conical; scapes stout, hollow, with bracts. Flowers small, bisexual, star-shaped, sweetly scented. Perianth of 6 pink, persistent segments. Stamens 6, with long filaments and projecting, blue-grey anthers. Ovary 3-celled; styles 3, divided to the base. Fruit a 3-celled capsule; seeds many, each with a white appendage. *Eastern North America.*

5. Chamaelirion Willdenow. 1/1. Dioecious herb with a bitter, tuberous rootstock. Stems unbranched, erect, hairless. Lower leaves in loose rosettes, obovate to spathulate, stalked; stem-leaves smaller, the uppermost linear to lanceolate, stalkless. Raceme dense, cylindric, the male often drooping at the tip at first, the female more slender; bracts absent. Perianth-segments 6, free, white, linear-spathulate. Stamens 6, female flowers usually with 6 staminodes. Ovary with 3 short styles, stigmas decurrent. Fruit a capsule. Seeds numerous, linear-oblong, winged. *Eastern North America.*

6. Heloniopsis Gray. 4/1. Evergreen herb with short rhizomes. Leaves in a basal rosette. Scape erect, bracts leaf-like but small, lanceolate. Racemes few-flowered, often condensed and umbel-like, flowers drooping. Perianth-segments 6, free, spreading, equal. Stamens 6, filaments subulate. Ovary 3-celled, style simple with a capitate stigma. Fruit a 3-lobed capsule; seeds numerous, linear, each with a thread-like appendage at each end. *Japan, Korea, Taiwan.*

7. Chionographis Maximowicz. 7/1. Hairless herbs with short, thick rhizomes. Basal leaves elliptic to oblanceolate or ovate, often long-stalked; stem-leaves smaller, stalkless. Flowers many in a terminal spike, white, bisexual, bilaterally symmetric. Perianth-segments 3--6, unequal, the upper 3 or 4 petal-like, linear to thread-like, the lower 2 or 3 small or absent. Stamens 6, filaments very short or absent. Ovary 3-celled, styles 3, free. Fruit a capsule, seeds 2 per cell. *China, Japan, Korea.*

8. Stenanthium (Gray) Kunth. 5/2. Perennial herbs with small bulbs. Basal leaves 2--4, long and narrow, bright green, arched above, hairless, acute. Flowering stems slender, erect, stem-leaves few, bract-like above. Flowers bisexual or unisexual in dense or loose terminal racemes or panicles. Perianth with 6 linear, curved lobes and a tube below which is fused to the base of the ovary. Stamens 6, short, borne at the bases of the perianth-lobes. Styles 3, free. Fruit a 3-beaked capsule with many narrow, oblong seeds. *North America, eastern Asia.*

9. Zigadenus Michaux. 18/7. Perennial herbs with rhizomes or bulbs. Leaves usually narrow, linear, hairless. Stems simple, leafy. Flowers in terminal panicles or racemes, greenish to yellowish white. Perianth-segments 6, free, spreading, persistent, ovate to lanceolate, each with 1 or 2 greenish glands near the base. Ovary superior or partly inferior, 3-celled, with several ovules to each cell; styles 3, persistent. Fruit a 3-lobed capsule which splits between the septa to the base. *North America, 1 species in northern Asia.*

10. Melanthium Linnaeus. 5/2. Perennial herbs with rhizomes and fibrous roots. Leaves basal and on the stem. Inflorescence a panicle with its axis, branches and the backs of the flowers roughly hairy. Flowers bisexual and male in the same inflorescence. Perianth of 6 free, widely spreading segments, each with a conspicuous narrow claw and an expanded blade which has 2 glands (appearing as black spots) at the base. Stamens 6, borne on the claws of the perianth-segments. Ovary superior, cylindric, 3-celled with several ovules in each cell; styles 3, free, divergent. Fruit an inflated capsule, 3-lobed in section, 3-beaked at apex. Seeds flat, broadly winged. *North America.*

11. Veratrum Linnaeus. 45/few. Robust perennial herbs with rhizomes. Stems simple, leafy, usually hairless below, downy or coarsely hairy above. Leaves alternate, broadly ovate to elliptic, many-veined, pleated and more or less erect and overlapping, narrowed into long, sheathing bases. Flowers numerous in a terminal panicle or raceme, usually bisexual, occasionally male only. Perianth-segments 6, free, widely spreading, elliptic or lanceolate, persistent in fruit, white, green, reddish brown or almost black. Stamens 6, free. Ovary superior, 3-celled, ovules numerous. Fruit a 3-celled capsule containing numerous seeds. *Northern hemisphere.*

12. Uvularia Linnaeus. 5/5. Rootstock a rhizome. Stems simple or branched, leafy. Leaves alternate, stalkless, sometimes clasping the stems. Flowers drooping, yellow, narrowly bell-shaped, the perianth-segments 6, free. Stamens 6, much shorter than perianth-segments, anthers opening by slits. Ovary superior, 3-celled, style with 3 branches. Fruit a 3-lobed or 3-winged capsule which splits between the septa. *North America.*

13. Tricyrtis Wallich. 16/9. Perennial herbs. Stems simple or little-branched, leafy. Leaves alternate, sometimes in 2 rows, ovate to lanceolate, stalkless, lightly pleated, sometimes spotted with dark green. Flowers bisexual. Perianth-segments 6, free, erect at base, spotted, 3--6 times as long as broad, the 3 outer each with a conspicuous, sometimes bilobed, spherical or ovoid nectarial pouch. Stamens 6, curved outwards towards the top. Ovary 3-celled, narrowly oblong; style single with 3 recurved, bifid stigmas. Fruit a furrowed capsule opening along the septa. Seeds flat. *Eastern Asia.*

14. Dianella Lamarck. 30/7. Rhizomatous perennials. Leaves in 2 ranks crowded at the base of the stem, linear or lanceolate, keeled beneath. Flowers borne in a panicle. Perianth-segments free, spreading, each with 3--8 veins. Filaments thickened; anthers basifixed, opening by terminal pores. Ovary superior, style 1. Fruit a berry. *Tropical Asia, Australia, New Zealand, Polynesia.*

18

15. Asphodelus Linnaeus. 12/5. Annuals, or perennials with rhizomes. Leaves basal, linear, with membranous, sheathing bases. Inflorescence a dense raceme or panicle. Bracts membranous. Flower-stalks jointed. Perianth-segments free or united at the extreme base, 1-veined, spreading. Stamens shorter than the perianth-segments, free, anthers dorsifixed. Ovary superior, 3-celled. Fruit a capsule; seeds 6, winged. *Mediterranean area to the Himalaya.*

16. Asphodeline Reichenbach. 19/6. Biennial or perennial herbs with clusters of fleshy roots. Leaves numerous, linear, with wide, membranous, sheathing bases and often rough margins. Inflorescence branched or unbranched with the flowers in racemes. Bracts membranous. Flower-stalks jointed. Flowers more or less bilaterally symmetric. Perianth united at the base, each lobe with 3 central veins (superficially appearing as 1 vein), the outer 3 lobes narrower than the inner 3. Stamens unequal; stamens and style curved downwards; anthers dorsifixed. Ovary superior, 3-celled. Fruit a capsule with 6 unwinged, sharply angled seeds. *Mediterranean area to the Caucasus.*

17. Eremurus Bieberstein. 40--50/12. Perennials with thick rhizomes, the rhizome-neck bearing fibrous or membranous remains of old leaves. Leaves basal, forming a tuft or rosette, usually narrow and often keeled beneath. Flowers many, borne in a raceme on an unbranched stem; bracts membranous. Perianth-segments 6, almost free, each with 1, 3 or 5 dark central veins, the inner segments often broader than the outer. Anthers basifixed. Ovary superior; style 1. Fruit a 3-celled, usually spherical capsule. Seeds usually winged. *Western & central Asia.*

18. Bulbine Wolf. 30/4. Herbaceous perennials with somewhat to very swollen, tuber- or bulb-like bases, usually stemless. Leaves in rosettes, fleshy, terete or flat, occasionally subterranean. Scape well-developed. Flowers few to many in terminal racemes. Perianth of 6, free, yellow or brownish, spreading segments. Stamens 6, free, the filaments of all, or of 3 only, each with a patch of hairs. Ovary superior, 3-celled with several ovules in each cell. Fruit a capsule containing few seeds. *Africa, a few species from Australia.*

Species of **Bulbinella** Kunth, with mostly unisexual flowers and filaments without hairs may be grown in very mild areas.

19. Simethis Kunth. 1/1. Roots fleshy, from a short rhizome. Flowering stem 12--40 cm, hairless. Leaves mostly basal, 15--60 cm × 2.5--7 mm, linear with sheathing bases. Flowers borne in a terminal panicle, on non-jointed stalks. Perianth-segments free, 8--11 mm, white above, purplish beneath. Filaments white, hairy; anthers yellow, dorsifixed. Style single. straight. Ovary superior, ovules 2 per cell. Fruit a capsule. Seeds 3--6, black, each with a whitish swelling at one end. *West Mediterranean area.*

20. Anthericum Linnaeus. 30/2. Perennial hairless herbs with rhizomes and somewhat fleshy roots. Leaves basal, linear with membranous sheathing bases. Flowers in a loose raceme or panicle. Bracts usually small and membranous. Flower-stalks usually jointed. Perianth-segments spreading, free or shortly joined at the extreme base, each with 3 central veins. Stamens equal, straight; filaments hairless; anthers basifixed. Ovary superior, 3-celled, with 4--8 ovules in each cell. Fruit a capsule. *Mainly tropical and southern Africa, but also in Europe, America and eastern Asia.*

21. Paradisea Mazzucato. 2/2. Similar to *Anthericum* but flowers always borne in a raceme, perianth trumpet-shaped, the segments with a claw, style and stamens curved upwards and anthers dorsifixed. *South Europe.*

22. Comospermum Rauschert (*Alectorurus* Makino). 1/1. Rootstock a short rhizome. Leaves evergreen, basal, in 2 ranks, narrowly strap-shaped, curved, 10--50 cm. Flowering stem 15--40 cm, flattened with a slightly winged margin, bearing a many-flowered panicle. Flowers white tinged with lilac, to pale pink, of 2 forms which occur on different plants: one variant has the stamens projecting, the other has the stamens equal in length to the perianth-segments. Flower-stalks jointed. Perianth joined at the base. Anthers dorsifixed. Style 1, with a slightly thickened stigma. Ovary superior. Ovules 2 per cell. Seeds 3-angled, brown with a tuft of white hairs at one end. *Japan.*

23. Schoenolirion Durand. 5/1. Bulbs with membranous or fibrous tunics. Leaves basal, linear. Flowers in a raceme which is sometimes branched at the base. Flower-stalks jointed. Perianth-segments free, becoming papery with age. Stamens 6, joined to base of perianth-segments; anthers versatile. Style 1, short. Ovary superior, with 2 ovules in each cell. Capsule ovoid, with black seeds. *Southwest & southeast USA.*

24. Chlorogalum (Lindley) Kunth. 5/2. Bulbs with fibrous or membranous tunics. Leaves mostly basal, linear; stem-leaves much reduced. Flowers borne on jointed stalks, in a terminal panicle. Perianth-segments free, with 3 central veins, twisting together over the ovary after pollination. Stamens with versatile anthers. Ovary superior. Style 1, slightly 3-lobed at the apex. Capsule with 1 or 2 black seeds in each cell. *Western coastal USA extending into Mexico.*

25. Aphyllanthes Linnaeus. 1/1. Perennial with a dense, fibrous rootstock. Leaves reduced to reddish brown sheaths, which surround the lower part of the numerous slender, wiry, glaucous stems 15--40 cm high. Flowers terminal, solitary or in compact groups of 2 or 3, subtended by about 5 reddish brown, scarious bracts. Perianth-segments free, erect in the lower half, spreading above. Anthers dorsifixed, opening inwards. Ovary superior; style 1. Stigma 3-lobed. Fruit a 3-seeded capsule enclosed by the bracts. *Southwest Europe, Morocco.*

26. Hosta Trattinick. *c.* 50/30 (and many hybrids). Perennial herbs with short, thick, fleshy rhizomes. Leaves mostly basal; stalks fleshy, erect or ascending, channelled, often winged, sometimes purple-spotted; blades elliptic or oblong to ovate or almost circular, acuminate, acute or obtuse, with tapered, rounded, truncate or cordate bases, often decurrent, membranous to leathery, flat or with marginal undulations or wrinkles, yellowish or mid-green to dark green (with many yellow and variegated varieties and cultivars), dull, shiny or glaucous, with 2--14 pairs of side-veins. Flowering stems erect or ascending, almost always unbranched, occasionally with a few leaves or large leaf-like bracts below; upper bracts erect or spreading, often channelled, persistent or withering at flowering. Flowers few to many in a loose to dense head, bell- or trumpet-shaped, white or various shades of bluish purple: perianth fused below into a narrow tube, which broadens upwards; lobes 6, spreading or recurved. Stamens 6, not or slightly projecting, usually free from perianth-tube, curved upward just below anthers; anthers versatile, white, yellow or purplish. Ovary superior, 3-celled; style

projecting beyond stamens, curved upwards. Seeds many, membranous, winged. *Mostly Japan, a few species in Korea and China.*
Known as *Funkia* Sprengel until early in the twentieth century.

27. Hemerocallis Linnaeus. 15/8 (and many hybrids and cultivars). Herbaceous perennials with rhizomes, roots fibrous, often swollen and fleshy at their ends. Leaves in 2 ranks, deciduous or persisting well into winter, linear, tapered gradually to the apex, usually recurved, flat or folded at the base. Flowers in a raceme or panicle borne on a scape, bracts small or large. Perianth united into a tube at the base, more or less funnel-shaped or openly funnel-shaped, 6-lobed, yellow or orange, rarely reddish. Stamens 6, deflexed downwards, attached to the top of the perianth-tube. Ovary superior, 3-celled, with many ovules, style thread-like. Fruit a 3-angled or 3-winged capsule. *Eastern Asia.*

28. Blandfordia J.E. Smith. 4/1. Deciduous, hairless perennial herbs with fibrous roots and often with rhizomes. Leaves sheathing, linear, triangular in cross-section, often 2-ranked. Perianth tubular to bell-shaped with 6 equal lobes, varying in colour from red to orange and yellow. Scapes solitary, bearing 3--20 bisexual flowers with bracts, in terminal racemes. Stamens 6, included in perianth and attached at or below the middle of the tube. Ovary superior, 3-celled with many ovules in each cell; style 1. Fruit a stalked, 3-celled, ovoid, capsule. Seeds many. *Australasia.*

29. Leucocrinum Gray. 1/1. Rootstock a cluster of short, fleshy roots. Leaves narrowly linear, forming a tuft, 10--20 cm × 2--6 mm, surrounded at the base by membranous bracts. Stem absent to very short, below ground-level, flowers borne on subterranean stalks 5--30 mm, appearing at ground-level, white, fragrant. Perianth joined into a slender tube 5--12 cm, the free lobes 1.4--2 cm, narrowly oblong. Stamens 6, borne near the top of the perianth-tube; anthers 4--6 mm. Style 1, slender, stigma slightly lobed. Ovary subterranean with several black, angled seeds in each cell. *Western USA.*

30. Kniphofia Moench. 70/11 (and many hybrids). Plants with thick rhizomes, rarely with aerial stems. Leaves usually basal, in several ranks, linear, tapering gradually to the apex, usually keeled beneath, margins smooth or very finely toothed. Scape simple, erect, with a few small, sterile bracts beneath the raceme. Raceme usually dense, flowers usually opening from the bottom upwards, occasionally from the top downwards, spreading or drooping, white or red, the buds nearly always red: bracts scarious, exceeding the flower-stalks. Perianth with a cylindric or narrowly bell-shaped, often curved tube to 4.5 cm and 6 small lobes. Stamens 6, often becoming spirally twisted and withdrawn after shedding pollen, filaments attached to the perianth near the base. Ovary superior, 3-celled with numerous ovules in each cell. Style usually projecting. Capsule spherical or ovoid, sometimes 3-angled. *Mainly eastern and southern Africa, Madagascar, southern Arabia.*

Species of **Aloe** Linnaeus, with fleshy leaves, flowers in racemes or panicles, the perianth in 2 distinct whorls, segment of the outer whorl free or united, those of the inner whorl variably joined to those of the outer, may be grown in very favoured areas. This is sometimes placed in the family *Aloaceae*.

31. Colchicum Linnaeus. 45/30. Perennial stemless herbs with corms; tunics membranous, papery or leathery, frequently extended into a tubular, persistent false-stem or neck. Leaves basal, partially developed at flowering or developing after flowering (occasionally developing as the flowers fade). Flowers solitary or in clusters, each subtended by a small bract and very shortly stalked, the stalks elongating as the capsule ripens. Perianth bell-, funnel- or star-shaped, purple, pink or white, sometimes tessellated, with a tube at the base and 6 lobes in 2 equal or almost equal whorls, bases of the lobes occasionally with auricles, the throat of the perianth sometimes ridged on either side of the filament-bases to form a short 'filament-channel', the ridges hairless or downy. Stamens borne near the bases of the perianth-lobes, in 1 or 2 series; filaments slender, sometimes thickened at the base; anthers dorsifixed and versatile, or basifixed and rigid. Styles 3, free, stigmas point-like or unilaterally decurrent along the style; ovary subterranean. Capsule 3-celled, splitting along the septa. Seeds numerous, spherical or almost so. *Europe, North Africa, western Asia to India and China.*

32. Bulbocodium Linnaeus. 2/2. Perennials with corms. Leaves linear to linear-lanceolate, obtuse. Perianth of 6 free segments, each divided into a blade and a long, narrow claw; claws held together by teeth or auricles at the base of the blade, thus forming a tube. Stamens 6, with slender filaments borne at the base of the blade. Ovary superior, but below ground, 3-celled, containing many ovules. Style 3-fid above, undivided below. Seeds spherical. *South & east Europe.*

Sometimes considered to be part of *Colchicum.*

33. Merendera Ramond. 10/6. Perennial, stemless herbs. Corms oblong-ovoid, occasionally narrow and more or less horizontal, each enclosed by a membranous or leathery tunic which is usually extended into a short neck. Developing leaves and inflorescences enclosed within a membranous, cylindric sheath. Leaves basal, partially developed at flowering or developing after flowering. Flowers purple to pink or white, solitary or in clusters. Perianth-segments free, each with a long, narrow claw and a broader, linear, narrowly elliptic to narrowly obovate blade, with or without 2 auricles at the base. Stamens borne close to the base of the blade. Anthers versatile or basifixed, pollen yellow. Styles 3, free to the base; stigmas point-like. Ovary subterranean. Capsule opening along the septa with 3 abruptly pointed flaps, maturing at or just above ground-level by the elongation of the flower-stalk. Seeds numerous, spherical or almost so. *South Europe, North Africa, western Asia, Ethiopia.*

34. Gagea Salisbury. 50/few. Perennial herbs with small bulbs which are sometimes closely entwined with persistent, thickened roots. Basal leaves 1 or 2, arising from the bulb, linear or linear-lanceolate, hollow, solid or flat. Flowers usually yellow, rarely white, either in an umbel subtended by an opposite pair of leaf-like spathes on an otherwise leafless stem, or in a few-flowered panicle with at least 1 leaf on the stem below the point of branching. Perianth-segments free, equal, without nectaries. Anthers basifixed. Ovary superior, style 1. Seeds flat or pear-shaped. *Europe to central Asia.*

35. Lloydia Reichenbach. 12/1--2. Perennial hairless herbs possessing bulbs with brown, fibrous, membranous tunics. Leaves basal and on the stem, linear. Flowers 1 or 2 at the top of the stem. Perianth-segments free, each usually with a gland at the base.

Stamens borne at the base of the perianth-segments; anthers basifixed. Ovary superior, 3-celled. Fruit a capsule with 3 ribs. *Temperate northern hemisphere.*

36. Calochortus Pursh. 57/10. Perennial herbs with ovoid bulbs with membranous or fibrous coats. Stems leafy or sometimes scape-like, often branched, frequently bearing bulbils. Leaves usually linear, the basal one large and conspicuous at flowering, or withering before flowering. Inflorescence cymose or more or less umbellate. Flowers conspicuous, often marked with colour contrasts. Perianth of 2 distinct whorls. Sepals usually lanceolate, hairless. Petals obovate to oblanceolate, usually more or less hairy on the inner surface and with a depression or gland near the base. Stamens 6, attached to the bases of the perianth-segments, the filaments broadened at the base. Ovary superior, 3-celled with numerous ovules in each cell, triangular or 3-winged in section. Fruit spherical to linear, 3-angled or 3-winged in section. *North & central America (mainly California).*

37. Erythronium Linnaeus. 25/12. Perennial herbs with membranous-coated corms. Leaves usually 2, basal or almost so, often mottled. Flowers nodding, solitary or several in a raceme, rarely in an umbel whose stem is below ground-level. Perianth-segments free, usually recurved, those of one or both whorls usually with 2 or 4 inflated appendages near the base. Stamens shorter than the perianth-segments; anthers basifixed. Stigma entire to 3-lobed. Ovary superior. Fruit a 3-celled capsule. *Mainly temperate North America, 1 species in Eurasia.*

38. Tulipa Linnaeus. 100/many. Bulbous perennials. Bulb-tunics of various textures, hairless inside or lined with hairs. Leaves few, rather fleshy, alternate, decreasing in size up the stem. Flowers usually solitary, rarely to 12. Perianth-segments free. Nectaries absent. Filaments broadened towards the base, hairy or not. Anthers basifixed, opening inwards. Ovary superior. Style very short or absent. Stigma 3-lobed. Fruit a spherical or ellipsoid capsule containing numerous, flat seeds. *North temperate Old World, especially Central Asia.*

39. Fritillaria Linnaeus. 100/50. Spring ephemerals, dormant in summer. Bulbs of 2 or more thick, closely wrapped scales, usually spherical or spindle-shaped, rarely with a thin, white, papery tunic, or bulbs of thick, separate scales, often surrounded at the base with numerous, white, rice-like bulbils. Basal leaves (on non-flowering shoots) 1, lanceolate or rarely linear. Stem-leaves usually alternate, more rarely opposite or whorled. Bracts similar to the leaves but smaller, alternate or whorled. Flowers usually nodding, bell-shaped to tubular, rarely saucer-shaped or conical. Perianth-segments free, in 2 whorls, those of the inner whorl wider than those of the outer, each with a usually conspicuous nectary. Style 1, entire or 3-fid. Capsule erect (rarely nodding), flat-topped, sometimes 6-winged, the style falling. Seeds flat (rarely spherical). *Temperate areas of the northern hemisphere (except eastern North America).*

40. Lilium Linnaeus. 100/50 and many hybrids. Rootstock a bulb composed of fleshy overlapping scales, sometimes with stolons. Stem erect, unbranched, leafy. Leaves scattered or in whorls, usually linear or lanceolate, sometimes with bulbils in the axils. Flowers in a terminal raceme or umbel, or sometimes solitary, funnel-, cup-, bowl- or bell-shaped, with the perianth-segments usually spreading or reflexed to a greater or lesser extent, or of turk's-cap type. Perianth-segments free, each with a nectar-bearing furrow or gland at the base, papillose or sometimes hairy at the apex. Stamens borne at the base of the perianth-segments, usually free from one another; anthers versatile. Ovary superior; style 1. Fruit a capsule with 3 cells. Seeds many, flat, 2 rows in each cell. *Temperate northern hemisphere.* Figure 2, p. 26.

41. Notholirion Boissier. 6/4. Differs from *Lilium* in flowering once and then dying (the plant perennates by offsets) and having bulbs with few scales covered by a ribbed tunic; long basal leaves produced in autumn and winter; style with 3 narrow branches. *Afghanistan to western China.*

42. Nomocharis Franchet. 7/7. Similar to *Lilium* but with the flowers opening flat or sometimes shallowly cup-shaped, and with the inner perianth-segments entire to fringed and bearing basal nectaries which usually have ridges of tissue arranged in a fan shape. Filaments swollen, with a needle-like appendage (aciculus) at the top. *Western China, Burma, northeast India.*

43. Cardiocrinum (Endlicher) Lindley. 3/3. Differs from *Lilium* in the monocarpic bulbs which die after flowering (the plant perennates by offsets), the large, long-stalked, heart-shaped leaves and the capsule which splits into 3 valves which are fringed with teeth. *Himalaya, China, Japan.*

44. Alstroemeria Linnaeus. 50/10. Rootstock fibrous, with clusters of tubers or creeping rhizomes. Leaves alternate or scattered up the stem, larger on the sterile shoots, with short stalks which are twisted through 180°, many-veined. Flowers usually bilaterally symmetric, borne usually in a terminal, simple or compound umbel, rarely solitary. Perianth-segments free, clawed at the base, almost equal or the outer 3 broader and shorter than the inner. Stamens often unequal, usually curved downwards then upwards at the tips; anthers basifixed. Style slender, curved downwards, with a 3-fid stigma. Ovary inferior, 3-celled with numerous ovules in each cell. Fruit a capsule containing numerous seeds. *South America.*

Some species of **Bomarea** Mirbel, which are similar, but climbers with smaller, drooping, radially symmetric flowers, may be grown out-of-doors in very mild areas.

45. Scilla Linnaeus. 90/12. Perennial herbs with ovoid to spherical bulbs of numerous free scales which are progressively renewed annually. Leaves few to several, all basal, linear to elliptic, sometimes channelled. usually hairless, appearing before, with or after the flowers. Scapes hairless, erect, usually 1--4 per bulb. Flowers few to many, in terminal racemes. Bracts absent or 1 subtending each flower. Perianth-segments 6, free, occasionally close at the base giving the appearance of a short tube, blue, purple, pink or white, often each with a darker midrib, spreading. Filaments 6, free, equal, inserted at the base of the perianth; anthers dorsifixed. Ovary superior, almost spherical to ovoid, 3-celled, with 1--12 ovules per cell, stalkless; style 1, straight, stigma small, truncate. Fruit a capsule, seeds spherical or oblong, sometimes angled, pale brown to black, each occasionally bearing an appendage. *Europe, western Asia, southern Africa.*

A few species of **Ledebouria** Roth, with the perianth-segments curved at the base to make a cup-shaped flower and a stalked ovary, may be grown in very mild areas.

46. Hyacinthoides Medikus. 5/5. Perennial herbs. Bulbs with tubular, coalescent scales, completely renewed each year. Flowers in a raceme, each subtended by 2 bluish, linear-lanceolate bracts. Perianth-segments free to the base, erect or spreading, usually blue. Filaments free, attached to the perianth-segments; anthers versatile. Ovary superior, 3-celled. Style simple, stigma capitate. Fruit a capsule. Seeds spherical, black. *Western Europe, northern Asia.* Figure 3, p. 28.

47. Chionodoxa Boissier. 6/6 and some hybrids. Perennial herbs with small bulbs, tunics brown. Leaves usually 2. Scapes with 1--15 flowers in a loose raceme. Flowers blue or pinkish, often with a white central zone. Perianth with a short, obconical tube at the base, the lobes spreading. Stamens borne at the apex of the perianth-tube; filaments of unequal length, flattened. Anthers dorsifixed. Ovary superior, style 1. Fruit an almost spherical capsule. *Western Turkey, Cyprus, Crete.*

Hybrids between *Chionodoxa* and *Scilla* are frequent, grown as × **Chionoscilla** Nicholson.

48. Camassia Lindley. 9/few. Perennial herbs. Bulbs usually large, ovoid or spherical, with brown or black tunics. Leaves basal, linear, keeled, sheathing at the base, hairless, with entire margins. Scape terete, bearing a dense or loose raceme of flowers. Flower-stalks in axils of lanceolate-acuminate bracts. Perianth-segments spreading, persistent, blue, purple or white. Stamens 6, attached at the base of the segments; filaments thread-like; anthers versatile. Ovary 3-celled; style thread-like; stigma 3-lobed. Fruit a 3-lobed capsule, opening by 3 slits, with several black seeds in each cell. *North America.*

49. Ornithogalum Linnaeus. 80/*c.* 15. Plants with bulbs which are usually sub-terranean with whitish or brownish papery tunics, more rarely partly exposed and green and fleshy. Leaves in a rosette, linear to lanceolate or obovate, margins ciliate or not, sometimes marked with a whitish line above. Raceme corymbose, pyramidal, almost spherical or cylindric, with 2--very many flowers. Bracts usually conspicuous. Perianth-segments 6, equal, usually white, more rarely orange, yellow or red, each usually marked with a green stripe on the outside, usually widely spreading, more rarely erect. Stamens 6, filaments flattened, often broadened to the base, sometimes winged or abruptly widened towards the base. Ovary superior, cylindric to spherical, yellow, green or purplish black, 3-celled. Ovules numerous. Style terminal, long or short. Fruit a capsule with many seeds. *Mediterranean area, southern Africa.*

50. Dipcadi Medikus. 55/1. Hairless perennial herbs with bulbs. Leaves linear, often sheathing at the base. Flowers bisexual in loose racemes. Perianth tubular to bell-shaped, united in the lower third, the inner 3 lobes held erect, the outer 3 spreading. Stamens 6, included in perianth. Ovary superior, style 1. Fruit a 3-celled capsule. *Southern and tropical Africa, southwest Europe, East Indies.*

51. Albuca Linnaeus. 30/10. Bulbous perennials. Leaves basal, linear or lanceolate, usually hairless. Flowers in loose, terminal racemes. Perianth-segments free, often with a central band of a contrasting colour, the outer spreading, the inner erect, hooded at the tips. Outer whorl of stamens sometimes sterile. Ovary superior; style 1. Fruit a capsule; seeds black, flattened. *Tropical and southern Africa.*

Figure 2. Liliaceae. *Lilium.* Range of perianth shapes.

52. Urginea Steinheil. 100/1. Perennial herbs. Bulbs with numerous free scales. Leaves narrow, deciduous. Flowers stalked, in long, erect racemes, the perianth falling after flowering, the stalks with small, often persistent bracts. Perianth-segments 6, free, spreading. Filaments thread-like, sometimes inflated at the base, inserted at the base of the perianth-segments. Style .equalling or exceeding the stamens. Fruit a 3-angled capsule, seeds many, flattened and winged. *Mostly Africa, some in the Mediterranean area.*

53. Puschkinia Adams. 1/1. Perennial herbs with small bulbs, tunics brown. Leaves usually 2. Scapes 5--20 cm, leafless, with 4--10 flowers in a loose raceme. Flowers almost stalkless or the lowest with stalks to 10 mm, pale blue with darker stripes, rarely white or greenish. Perianth 7--10 mm, with a short tube at the base, lobes erect or slightly spreading, not widely open; throat with a 6-lobed corona, its lobes alternating and projecting between the anthers. Anthers almost stalkless, borne on the corona, dorsifixed. Ovary superior, style 1. Fruit an almost spherical capsule. *Turkey, north Iran, north Iraq, Lebanon.*

54. Hyacinthus Linnaeus. 3/3. Perennial herbs with bulbs. Leaves present in spring, 2 or more on each flowering bulb, with smooth or rough margins. Flowers in a raceme arising between the leaves. Bracts small, 2-lobed. Flower-stalks much shorter than perianth. Flowers bisexual. Perianth with a narrow tube for about half to two-thirds of its length, its lobes divergent, spreading or recurved. Stamens attached to perianth at or below the middle of the tube, not exceeding perianth-lobes. Ovary superior, 3-celled, with 2--several ovules in each cell; style much shorter than perianth-tube. Capsule nearly spherical. Seeds black, wrinkled. *Western & central Asia.*

55. Bellevalia Lapeyrouse. 50/10. Perennial herbs with bulbs. Leaves 2--several on each flowering bulb, margins cartilaginous and sometimes ciliate, appearing in autumn or spring. Flowers in a raceme; bracts usually minute, 2-lobed. Flowers bisexual, radially or slightly bilaterally symmetric, nodding or horizontal. Perianth with a tube for one-quarter of its length or usually more, violet-blue to brownish, forming a 6-lobed bell which is not constricted at the mouth though (rarely) the lobes close the mouth. Stamens attached to perianth at bases of lobes, rarely lower, not projecting beyond them; filaments flattened, broadly to narrowly triangular, shortly united with each other at base. Ovary 3-celled, with 2--6 ovules in each cell; style 1. Fruit a capsule, obtuse at apex, the cells in cross-section forming prominent, acutely to acuminately angled lobes. Seeds rounded or occasionally pear-shaped, smooth, dull, black, but sometimes with a bluish bloom. *Mediterranean and Black Sea areas.*

Species of **Lachenalia** Murray, with rounded capsules and flowers with the 3 outer perianth-lobes shorter than the 3 inner, all in bright colours, may occasionally be grown out-of-doors in mild areas.

56. Hyacinthella Schur. 16/5. Small perennial herbs with bulbs, leafless in summer and winter. Bulb-scales coated with powdery crystals. Leaves 2 or sometimes 3 on each (undivided) flowering bulb, the second (or third if there are 3) narrower than the first. Flowers in a raceme on a stalk arising between the leaves and overtopping them. Bracts minute. Flower-stalks not longer than flowers. Flowers not more than 1.2 cm, bisexual, radially symmetric or nearly so, more or less bell-shaped, blue to whitish or pinkish. Perianth-tube longer than the lobes, persistent in fruit. Stamens (in ours)

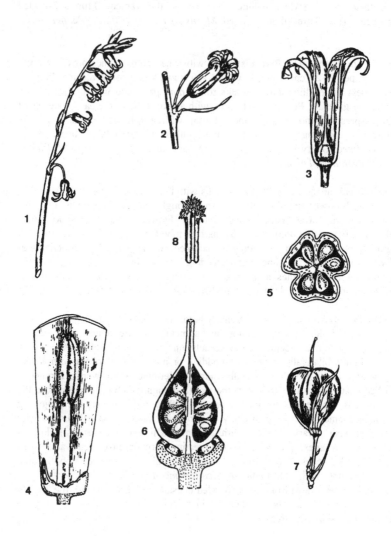

Figure 3. Liliaceae. *Hyacinthoides non-scriptus*. 1, Inflorescence. 2, Flower. 3, Longitudinal section of flower. 4, Young inner stamen. 5, Transverse section of ovary. 6, Longitudinal section of ovary. 7, Fruit. 8, Stigma.

attached just below the perianth-lobes and with filaments shorter than to slightly longer than the anthers. Ovary superior, 3-celled with 2--4 ovules in each cell. Style simple. Fruit a rounded capsule to 5 mm wide with the cells rounded on the back. Seeds black, seed-coat wrinkled. *East & southeast Europe, western Asia.*

57. Brimeura Salisbury. 2/2. Small hairless perennial herbs with bulbs, leafless in summer and winter (?). Leaves several on each flowering bulb. Flowers in a raceme arising between the leaves. Bracts about as long as flower-stalks or longer. Flowers bisexual, radially symmetric. Perianth with a long tube and 6 shorter lobes. Stamens concealed or exposed. Ovary superior, 3-celled with 2--4 ovules in each cell. Style simple. Fruit a capsule, to 5 mm. Seeds black, shining, finely wrinkled. *South Europe.*

58. Muscari Miller. 36/12. Herbaceous perennials with bulbs, with or without offsets. Leaves 1--8, basal. Scape present, flowers in racemes with minute bracts. Perianth united for most of its length into a cylindric, tubular, bell-shaped or urceolate tube, the 6 lobes small, blue, brownish, yellow, whitish or white, not changing as the flower dies, often of a different colour from the tube. Stamens 6, anthers not projecting from perianth-tube. The raceme often has sterile flowers near the apex, which may differ in colour or tone from the fertile, sometimes forming a tuft (comus). Ovary superior, 3-celled, with 2 ovules in each cell. Capsule strongly angled, with 2 seeds per cell. Seeds black, often shiny, often minutely wrinkled. *Mediterranean area, southwest Asia.*

The genus is variable, and is sometimes split into 4 smaller genera: **Muscari** in the strict sense (*Muscarimia* Kosteletzky) -- perianth of strongly scented fertile flowers strongly constricted towards apex, yellow or greyish white, the 'shoulders' forming a 6-lobed crown below the normal lobes; sterile flowers few or none; **Leopoldia** Parlatore -- perianth of fertile flowers strongly constricted, but without a crown, yellow or greenish; sterile flowers usually numerous forming a blue, violet or pink terminal tuft; **Botryanthus** Kunth -- perianth of fertile flowers constricted towards apex, blue, violet or blackish (rarely white); sterile flowers few or none; and **Pseudomuscari** Garbari & Greuter -- perianth of fertile flowers not constricted above.

59. Galtonia Decaisne. 3/3. Perennial herbs; bulbs with membranous tunics. Leaves few, basal, flaccid, flat. Scape terete, bearing a loose raceme of white or green flowers. Flower-stalks more or less spreading at flowering time, curving upwards in fruit, sub-tended by large bracts. Flowers bell-shaped, horizontal or slightly nodding. Perianth persistent, the tube slightly to much shorter than the obovate or oblong lobes. Stamens 6, attached just below the mouth or at about the middle of the perianth-tube, shorter than the lobes. Ovary oblong, 3-angled. Fruit a 3-celled capsule. Seeds numerous, angled, brownish or black. *South Africa.*

60. Massonia Houttuyn. 8/1. Perennial herbs; bulbs with whitish scales surrounded by brown membranous scales. Leaves 2, ovate, oblong or almost circular, spreading horizontally on the ground. Scape absent or very short. Inflorescence usually with large bracts. Flowers stalked, usually pleasantly scented. Perianth-lobes fused in the lower part, usually spreading or reflexed above. Stamens attached at the mouth of the perianth-tube. Ovary superior. Style usually slightly longer than the stamens. Fruit a winged or deeply lobed capsule, opening by slits. Seeds numerous, black. *South Africa.*

61. Agapanthus L'Héritier. 10/5 (and some hybrids). Perennial herbs with fleshy roots arising from a tuberous rootstock. Leaves usually basal, linear to strap-shaped with entire margins. Flowers few to many in an umbel on a usually stout scape 30--150 cm. Perianth tubular to spreading, united towards the base, ranging from dark violet or deep blue to white. Stamens 6, inserted on the perianth-tube. Ovary of 3 cells, superior, developing into a many-seeded capsule. Seeds flat and winged, black. *Mainly South Africa.*

62. Tulbaghia Linnaeus. 21/5. Rootstock a bulb or rhizome. Leaves 4--8, strap-shaped to linear, basal. Flowering stem solitary, erect, bearing an umbel of 6--40 flowers which is subtended by 2 bracts. Perianth united into a tube for about half its length; mouth of tube with a fleshy corona which is cylindric or composed of 3 free scales. Anthers stalkless on the perianth-tube, in 2 whorls one above the other. Style short. Stigma capitate. Capsule 3-celled, containing triangular black seeds. *Temperate & tropical Africa.*

63. Allium Linnaeus. 690/over 100. Perennial herbs, mostly smelling of onion, usually with well-formed bulbs, arising in some species from a short rhizome. Leaves linear to elliptic, basal or sheathing the stem, flat to channelled, terete or semiterete, solid or hollow. Flowers usually many (unless replaced by bulbils) in an umbel on a solid or hollow stem and enclosed at first within a spathe consisting of 1 or more bracts. Perianth star-shaped, cup-shaped or bell-shaped, with 6 usually free segments. Stamens 6; filaments free or united at base, simple or variously toothed; anthers opening by longitudinal slits. Ovary 3-celled, each cell with 1--8 ovules, usually 2; style slender; stigma usually simple but 3-lobed in a few species. Fruit a 3-celled capsule splitting between the septa. Seeds angled or rounded, blackish, in a few species with a spongy appendage. *Europe, Asia, America, extending to Sri Lanka and Mexico.*

64. Nothoscordum Kunth. 20/1. Perennial herbs with bulbs. Leaves linear, basal. Scape erect, bearing an umbel of 4--15 flowers which is subtended by 2 papery bracts (spathe-valves). Flower-stalks unequal. Perianth shortly united at the base. Ovary superior, 3-celled. Capsule 3-celled, containing angular black seeds. *North & South America.*

65. Nectaroscordum Lindley. 3/1. Differs from *Allium* in that the outer perianth-segments have 3--7 veins, the flower-stalks are swollen below the flowers and the ovary contains many ovules. *Southeast Europe, western Asia.*

A species of **Bessera** Schultes, which has drooping scarlet flowers with stamens projecting and a staminal tube which is toothed at the top, may occasionally be grown in very mild areas.

66. Bloomeria Kellogg. 3/1. Differs from *Nothoscordum* in having the flower-stalks jointed at the top, and the filaments broadened at the base into a cup-like structure with 2 short teeth at the top. *Southwest USA, northern Mexico.*

67. Muilla Watson. 5/1. Differs from *Nothoscordum* in having almost terete leaves, the umbel subtended by more than one spathe and the ovary with many ovules. *Southwest USA, northern Mexico.*

68. Caloscordum Herbert. 1/1. Similar to *Nothoscordum* but umbel subtended by only one bract. Bulb with whitish papery tunic. Leaves 2--6, very narrowly linear, channelled above, shorter than flowering stem, often somewhat withered at flowering time. Stem 10--25 cm, bearing an umbel of 10--20 flowers which is subtended by a single bract. Flower-stalks unequal. Perianth bright pink, 6--8 mm, joined in the lower one-third, lobes spreading-reflexed, each with a darker central vein, perianth 5--7 mm across the lobes. Stamens borne on the perianth-tube, Ovary superior. *Northeast Asia.*

69. Leucocoryne Lindley. 5/3. Herbaceous perennials. Bulbs to 2 cm, spherical to ovoid, tunic dark brown. Leaves 2--5, deciduous, basal, narrowly linear, usually channelled, 15--25 cm × 1--3 mm, appearing after the flowers. Scapes to 50 cm, bearing umbels of 2--12 flowers subtended by 2 spathes. Perianth funnel-shaped to very widely funnel-shaped, white to purple or bluish violet. Stamens 3 with dorsifixed anthers. Staminodes 3. Stigmas capitate. Ovary superior, 3-celled. Ovules many. *Chile.*

One species of **Milla** Cavanilles, with spathes 4, stamens projecting from the perianth, and flowers stalkless but appearing stalked because of the long perianth-tube, may be grown in very mild areas.

70. Ipheion Rafinesque. 10/2. Perennials with small bulbs and fleshy roots. Leaves basal, linear. Flowers solitary (rarely in umbels of 2), with 2 partly united bracts on the scape some distance below the flower. Perianth with a narrowly bell-shaped tube and spreading lobes. Stamens within the perianth-tube: anthers dorsifixed, opening inwards. Ovary superior. Stigma small, obscurely 3--4-lobed. Fruit a many-seeded capsule. *Temperate South America.*

71. Brodiaea J.E. Smith. 15/5. Perennial herbs arising from corms which are more or less spherical and have dark brown fibrous coats. Leaves linear, crescent-shaped in section, without a keel, with vein impressions weak or lacking. Flowers in an umbel subtended by scarious bracts. Perianth united into a tube below, lobes of the outer whorl narrower than those of the inner. Stamens 3, opposite the inner perianth-lobes, usually alternating with 3 staminodes; filaments joined to the perianth-tube for most of their lengths, lacking appendages; anthers basifixed. Ovary superior, stalkless, 3-celled with several ovules in each cell: stigma with 3 spreading, winged lobes. Seeds flattened, angled. *Western North America.*

In the past, *Brodiaea* has included *Triteleia, Dichelostemma* and *Ipheion.*

72. Triteleia Lindley. 17/7. Like *Brodiaea*, but corms flattened and with pale brown coats. Leaves distinctly keeled beneath and with evident veins. Staminodes 0, fertile stamens 6, anthers versatile; ovary shortly stalked, seeds more or less spherical, angled. *Western North America.*

73. Dichelostemma Kunth. 7/4. Like *Brodiaea* but leaves distinctly keeled beneath and with obvious veins; bracts coloured; fertile stamens sometimes 6; stigma with 3 short lobes. *Western North America.*

74. Convallaria Linnaeus. 1/1. Hairless perennial herb, with creeping rhizomes. Stems erect, with several green or violet scales below and 1--4 leaves above. Leaf-blades ovate-lanceolate to elliptic, acute to acuminate. Scapes solitary, sharply angled, arising from scale-axils, shorter than the leaves. Inflorescence a 1-sided raceme. Flowers

5--13, nodding; flower-stalks usually curved downwards and exceeding the ovate-lanceolate bracts. Perianth spherical-bell-shaped, white, united for half to two-thirds of its length, lobes 6, reflexed at tips. Ovary superior, 3-celled, with 4--8 ovules in each cell. Style simple; stigma capitate. Fruit a red berry. *North temperate areas.*

75. Speirantha Baker. 1/1. Stemless perennial herbs with thick rhizomes and numerous stolons. Leaves erect, 6--8 in a rosette, oblanceolate, tapered slightly to the base but not stalked, apex acute. Scape with a few, small, sheathing, scale-leaves, to 15 cm. Flowers in a loose raceme, with bracts. Perianth-segments 6, free, spreading, white. Stamens 6, free. Ovary superior, 3-celled, with 3 or 4 ovules in each cell. Style 1; stigma 3-grooved. Fruit a berry. *China.*

76. Reineckea Kunth. 1/1. Perennial herbs. Leaves persistent at the apex of prostrate, rhizome-like stems, linear-lanceolate, somewhat narrowed to the base. Flowers stalkless in a terminal spike on a scape 6--10 cm tall, bracteoles triangular-ovate. Flowers pale to deep pink, perianth becoming reflexed, united at the base to form a tube. Stamens and style equal. Ovary superior. Fruit a spherical red berry, seldom formed in cultivation. *China, Japan.*

77. Rohdea Roth. 1/1 with many cultivars. Perennial herbs with rhizomes and fibrous roots. Leaves in a basal rosette, leathery, thick, variable in shape, to 45 cm. Flowers in short, dense spikes borne among the leaves and shorter than them. Perianth greenish yellow, bell-shaped. united into a tube for most of its length. Stamens 6, borne on the perianth-tube, filaments very short or absent. Ovary superior, 3-celled, with 1 ovule in each cell. Stigma peltate, style absent. Fruit a spherical, red or yellow, usually 1-seeded berry. *Southeast China, Japan.*

78. Aspidistra Ker Gawler. 8/3. Evergreen, stemless, perennial herbs with thick rhizomes and roots. Leaves all basal, arising singly or in groups at intervals along the rhizome. Leaf-blade leathery and glossy, lanceolate to elliptic, wedge-shaped at base, acute, with sunken midrib. Leaf-stalk winged, one-third to half as long as blade, subtended by a papery sheath. Flowers bisexual, with bracts, borne individually at ground-level on short stalks, rather inconspicuous and usually obscured by the foliage. Perianth united to form a spherical, bell-shaped or urceolate flower, lobes 6--8, usually dull greenish brown, often with purple spotting. Stamens 6 or 8, included in perianth-tube. Stigma peltate. Ovary 3- or 4-celled, superior. Fruit a berry. *Himalaya to Japan.*

79. Clintonia Rafinesque. 5/5. Rootstock a short rhizome. Leaves basal, sheathing at the base, entire. Flowering stem erect, unbranched, leafless. Flowers borne in an umbel or raceme, occasionally solitary, bell-shaped to star-shaped. Perianth-segments free, downy outside at least at the base. Stamens borne at the bases of the perianth-segments. Ovary superior. Stigma with 2 or 3 lobes. Fruit a berry with 2 or 3 cells. *Himalaya, eastern Asia, North America.*

80. Smilacina Desfontaines. 25/4. Perennial herbs with creeping rhizomes. Stems unbranched, leafy. Leaves alternate, stalkless or with short stalks. Flowers borne in terminal racemes or panicles. Flower-stalks jointed. Perianth-segments free. Stamens borne at the bases of the perianth-segments. Ovary superior. Fruit a few-seeded berry. *North America, eastern Asia, Himalaya.*

81. Maianthemum Weber. 3/1. Perennial herbs, with slender creeping rhizomes, bearing solitary, long-stalked leaves. Flowering stems erect, with scales at base and 2 or 3 alternate leaves above. Flowers white, in terminal racemes. Perianth-segments 4, free, spreading or reflexed. Flower-stalks 1--several at a node, slender, with bracts. Stamens 4, shorter than perianth. Ovary superior, 2-celled, with 2 ovules in each cell. Style simple, short: stigma slightly 2-lobed. Fruit a 1- or 2-seeded berry. *North temperate areas.*

82. Streptopus Richard. 10/4. Herbaceous perennials with rhizomes. Stems erect, leafy. Leaves alternate, stalkless or clasping. Flowers solitary or in pairs, nodding, borne on a stalk which is fused to the stem for some distance. Flower-stalks with a downward bend at about the middle. Perianth bell-shaped, joined at the extreme base. Stamens borne at base of perianth-lobes. Ovary superior. Style usually with 3 branches. Fruit a many-seeded berry. *Temperate northern hemisphere.*

83. Disporum Don. 10/7. Perennial herbs with creeping rhizomes. Stems erect, leafy, branched. Leaves alternate, stalkless or shortly stalked. Flowers terminal, solitary or in a few-flowered cluster, usually drooping. Perianth-segments free, often slightly inflated or swollen at the base. Stamens borne at the base of perianth-segments; anthers opening by slits; filaments hairless. Ovary superior, 3-celled. Style 1, with an entire or 3-branched stigma, or styles 3, free. Fruit a berry. *North America, Himalaya, eastern Asia, Malaysia.*

84. Polygonatum Miller. 50/12. Perennial herbs with horizontal creeping rhizomes. Stems unbranched, solitary, erect at least in lower part, leafy in upper part. Leaves all borne on the stem, stalkless, alternate, opposite or whorled, often glaucous beneath. Flowers solitary or in clusters, borne in the leaf-axils. Bracts persistent or deciduous, usually small and membranous, rarely large and leafy. Perianth united into a tube, usually for more than half its length; corona absent. Stamens borne on the tube; anthers included, 2-lobed at the base. Ovary superior, 3-celled. Style slender, included. Stigma 3-lobed. Fruit a spherical berry, red, orange or blue-black when ripe. *North temperate areas.*

85. Scoliopus Torrey. 2/2. Perennial, hairless herb with a short, slender rootstock. Leaves usually 2. Flowers borne in an umbel at the top of the short, underground stem, each flower thus appearing to be solitary. Perianth-segments 6, free, in 2 dissimilar whorls. Sepals (outer segments) broad, spreading to reflexed. Petals (inner segments) narrow, erect. Stamens 3, attached to the base of the sepals. Ovary superior. Style 1, very short. Stigmas 3, linear. Ovary 3-angled, 1-celled. Fruit a capsule. *Western USA.*

86. Medeola Linnaeus. 1/1. Perennial herbs with horizontal, white tubers to 8 cm, smelling of cucumber. Stem simple, 20--90 cm, hairy at first, eventually becoming hairless. Leaves in 2 whorls, the lower whorl about halfway up the stem with 5--11 lanceolate to obovate, shortly acuminate leaves, the upper whorl just below the flowers, with 3, occasionally 4 or 5 smaller, ovate leaves. Flowers 3--9, nodding, in an umbel at the top of the stem, flower-stalks to 2.5 cm, spreading or recurved, becoming more or less erect at fruiting-time. Perianth of 6 free, similar segments, 7--8 mm, pale greenish yellow, recurved. Stamens 6, filaments longer than anthers. Ovary superior. Styles 3, linear, recurved. Fruit a spherical, dark purple or red berry with 3 cells, 4--8 mm across. *North America*.

87. Paris Linnaeus. 20/1. Perennials with creeping rhizomes. Leaves 4 or more in a whorl near the top of the unbranched stem. Flowers solitary, terminal. Outer perianth-segments 4--6, green, inner 4--6, yellow, very narrow. Stamens 4--10 with short, fat filaments and basifixed anthers. Ovary superior. Styles free, usually 4. Fruit a fleshy, berry-like capsule. *Temperate Eurasia*.

88. Trillium Linnaeus. 30/30. Perennial herbs with short, thick rhizomes. Stem simple, usually erect and hairless. Leaves 3 in a whorl at the top of the stem, with netted veins. Flower solitary, stalked or not. Perianth of 2 dissimilar whorls of free segments. Sepals (outer whorl) 3, usually green. Petals (inner whorl) 3, varying in colour, rarely green. Stamens 6, with basifixed anthers. Styles 3. Ovary superior with 3 or 6 angles or wings. Fruit a 3-celled berry, usually spherical. *Mainly North America, also western Himalaya, northeast Asia*.

89. Asparagus Linnaeus. 60/1. Perennial herbs, shrubs or climbers, with rhizomes and usually with tubers. Leaves reduced to small scales at the nodes, their bases hardened and often projecting as spines. The function of the leaves is taken over by cladodes (leaf-like stems) which are either borne singly or in groups of 3--50 at the nodes in the axils of the scale-leaves; the cladodes may be flattened, leaf-like, awl-like or thread-like. Flowers bisexual or functionally unisexual, borne singly or in few- to many-flowered clusters among the cladodes, more rarely in terminal, umbel-like clusters, or in racemes borne on the older shoots; flower-stalks distinctly jointed. Perianth shortly united at the base, bell-shaped or with the 6 lobes widely spreading. Stamens 6. Ovary superior, 3-celled. Fruit usually a spherical berry containing 1--6 seeds. *Old World, absent from Australasia*.

Some 24 species, most requiring glasshouse protection, are grown for the sake of their feathery shoots; some, wrongly called Smilax, are in demand as background material for bouquets, and 1 species is widely grown as a luxury vegetable.

90. Danaë Medikus. 1/1. Evergreen shrubs with short rhizomes; stems branched. Leaves papery, those of main stems mostly about 1.5--2.5 cm, ovate-lanceolate, readily shed, those of axillary branches about 2 mm, persistent, 5 or 6 on each branch, subtending leaf-like stems (cladodes). Cladodes mostly 3--7 cm, ovate-lanceolate, asymmetric. Flowers bisexual, nearly spherical, in racemes of 5--8 at tips of branches. Perianth cream-coloured; segments joined for most of their length; mouth much narrower than the diameter of the flower. Stamens united into a tube enclosed within the perianth; anthers 6. Ovary superior. Style slender, with 3 stigmatic lobes, reaching the mouth of the staminal tube. Fruit an orange-red berry with 1 or rarely 2 large seeds. *Western Asia*. Figure 4, p. 36.

91. Ruscus Linnaeus. 6/6. Evergreen herbs with short rhizomes and unbranched stems, or shrubs with branched stems. Leaves reduced to papery scales subtending branches or leaf-like stems (cladodes). Flowers unisexual, inconspicuous, borne in succession in clusters which are subtended by a bract and are usually solitary on the face of the cladode. Perianth-segments 6, free, the inner smaller than the outer, green, tinged or peppered with purple. Stamens united into a fleshy tube which is present in the flowers of both sexes; in the male flowers the tube is topped by stalkless anthers which are represented in the female flower by minute, papery flanges. Ovary filling the space within the staminal tube, vestigial in male flowers. Stigma more or less entire, protruding from the neck of the staminal tube. Fruit a red berry with thin flesh and 1--4 large seeds. *Azores to the south Caspian area.* Figure 4, p. 36.

92. Ophiopogon Ker Gawler. 20/3. Herbaceous perennials with short rhizomes and sometimes stolons as well. Roots fine or thickened, sometimes swollen and tuberous. Leaves all basal, in tufts, linear, leathery. Flowers in a terminal raceme or spike on a 3-angled or flattened scape. Perianth bell-shaped, of 6 free, equal, drooping, pale purple to white segments. Stamens 6, the filaments very short, the anthers lanceolate, acute, more or less united about the style. Ovary partly inferior, 3-celled. Style simple, straight or slightly curved, stigma small. Seeds fleshy, berry-like, exposed early, spherical or oblong, usually blue and often persistent. *India, China, Japan, south to west Malaysia, Borneo & the Philippines.*

93. Aletris Linnaeus. 10/2. Perennial herbs with rhizomes and fibrous roots. Leaves mostly in a basal rosette. Flowering stem erect, bearing a terminal, spike-like raceme. Flowers almost stalkless, held ascending-erect. Perianth united into a tube for most of its length, the outside of the tube warty and 6-ridged. Stamens 6. Ovary superior, 3-celled, each cell with several ovules. Fruit a beaked capsule, surrounded by the persistent but dried and shrunken perianth. *Eastern North America, eastern Asia.*

94. Luzuriaga Ruiz & Pavón. 4/2. Slightly woody trailing plants with rooting stems, sometimes climbing trees, producing simple or branched lateral leafy shoots. Lowest leaves of lateral shoots often scale-leaves. Main leaves jointed to the stems, very shortly stalked; blades with parallel veins and a few cross-veins, lower surface dark green but facing upwards, either because the shoot is pendent or because the leaf-stalk is twisted, upper surface (facing downwards) strongly glaucous except for green stripes following the veins. Flowers usually solitary, axillary, their stalks about as long as the perianth. Perianth-segments 6, free, white. Stamens 6; anthers basifixed or dorsifixed and versatile. Ovary superior, 3-celled; style simple. Floral parts becoming spotted with orange-brown when dry. Ovules few in each cell. Fruit a berry. *Peru, Chile, Falkland Islands, New Zealand.*

95. Philesia Jussieu. 1/1. Shrub, 15--30 cm or, in nature, to 120 cm, with underground stolons. Main stem bearing scale-leaves and branches. Branches angled, with scale-leaves at base and shortly stalked main leaves above. Leaf-blades pinnately veined, hard in texture, dark green above, glaucous beneath, recurved at the edges. Flowers solitary or few at the ends of the leafy branches, nodding, subtended by several pale green scale-leaves overlapping each other and the base of the flower. Perianth of 3 outer sepal-like segments and 3 inner petal-like segments, the inner slightly united at the base and each with a basal nectarial pouch. Stamens with filaments slightly united to perianth at base and united to each other for one-quarter to two-thirds of their length: anthers basifixed, opening by slits. Ovary 1-celled with 3 parietal placentas; style club-shaped and slightly lobed at apex. Ovules many. Fruit a berry. Perianth becoming spotted with orange-brown when dry. *Chile.*

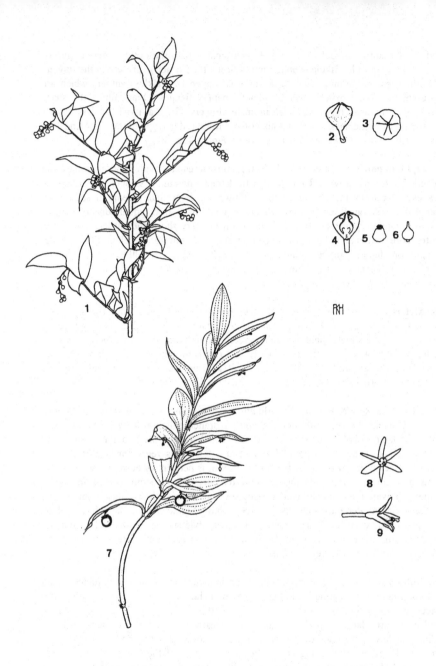

Figure 4. Liliaceae. 1–6, *Danaë racemosa*. 1, Flowering shoot. 2, Flower from the side. 3, Flower from above. 4, Longitudinal section of flower. 5, Staminal tube with anthers. 6, Ovary. 7–9, *Ruscus colchicus*. 7, Flowering shoot. 8, Female flower, partly closed, from the side. 9, Male flower from above.

96. × **Philageria** Masters. 1/1. Similar to *Lapageria* in habit but much less vigorous. Leaf-blades 3-veined. Flowers arranged as in *Philesia*, similar to those of *Lapageria* but outer perianth-segments not more than two-thirds the length of inner; stamens united to each other and to the perianth for *c*. 4 mm. *Garden Origin.*
An artificial hybrid between species of *Lapageria* and *Philesia*.

97. Lapageria Ruiz & Pavón. 1/1. Woody climber, spreading strongly by underground stolons. Stems thick, twining, branched, to *c*. 10 m. Leaves alternate, with stalks to 1 cm; blades leathery, with parallel main veins and net-veins between them. Flowers 1--3 on short scaly shoots in axils of upper leaves, pendent. Perianth-segments 6, entire, each with a pouched nectary at the base. Stamens 6, free from one another or very slightly joined at the base; filaments subulate; anthers basifixed, opening by slits. Ovary superior, 1-celled with 3 parietal placentas; style club-shaped and slightly lobed at apex. Ovules many. Fruit a berry. Seeds pale yellow or brownish. *Chile.*

98. Smilax Linnaeus. 200/13. Woody or herbaceous, evergreen or deciduous, perennial climbers with rhizomes or tubers. Stems terete or angled, usually spiny at least below, branched. Leaves alternate, simple, sometimes shallowly lobed, sometimes spiny on the margins, with 3--9 prominent veins interconnected by net-veins; lower leaves reduced to scales; leaf-stalks often very short, bearing a pair of tendrils (modified stipules) near the base, and sometimes a pair of stipule-like auricles. Flowers white to pale green, yellow or brown, lateral, in clusters, racemes, umbels, racemes of umbels, or solitary. Male and female flowers on separate plants. Perianth-segments free: male flowers with 6 free stamens borne at the base of the perianth-segments; female flowers with up to 6 staminodes and a superior ovary on which 1--3 stigmas are borne directly (styles absent). Fruit a spherical to ovoid, red, blue or black berry with 1--6 seeds. *Tropics, some in temperate Eurasia and North America.*

6. AGAVACEAE

Plants of very variable form, trees with trunks, shrubs with thin, woody stems, herbs, or stemless rosette-plants, often monocarpic. Leaves in rosettes, either borne directly on the rootstocks or at the ends of the stems or branches, more rarely distributed along the stems, usually thickened or fleshy, mostly hard, leathery and persisting for several years, bases thickened and sheathing, margins often spiny or horn-like, apices often spine-tipped (the apex formed by the upcurving of the margins which coalesce and become a solid, hard or soft spine or rarely a cylindric projection), usually very fibrous, venation often indistinct. Flowers in terminal panicles or racemes, usually bisexual, not always produced each year. Perianth usually united into a long or short tube at the base, the lobes 6, more or less equal. Stamens 6, usually borne at the top of the perianth-tube, more rarely attached to the perianth. Ovary superior or inferior, usually 3-celled, ovules usually many per cell, more rarely 1 per cell or the ovary 1-celled with 1--3 ovules. Fruit a capsule or berry, or indehiscent, dry and spongy.

A very troublesome family from the point of view of identification. Its separation from the *Liliaceae* (p. 9) and the *Amaryllidaceae* (p. 39) is based, at least in part, on cytological, chemical and anatomical characters, and this makes a clear diagnosis of the family difficult to prepare.

The family is very variable, including species that seem to have little in common with each other, except that they are either woody, or have thick, fibrous leaves, or both. However, the family is recognised in most recent studies of flowering plant families, and so is included here; its content and circumscription follows the system of Hutchinson (*Families of Flowering Plants*, 662--5, 1959) in including *Phormium*, rather than that of the *Syllabus der Pflanzenfamilien*, edn 12, in which this genus is treated as a member of the *Liliaceae*. In more recent classifications, this genus, together with *Dianella* (here treated as part of the *Liliaceae*), and some others, are placed in the *Phormiaceae*.

The genera in the family are relatively easily defined on floral characters, though, so defined, they are vegetatively very variable. Unfortunately, flowers are usually not available in cultivation -- many of the species flower only at long intervals (Agaves are known colloquially as Century Plants), or take many years to attain the flowering condition and then die; others are often grown as house or office plants and generally die or are discarded before they attain sufficient size for flowering to take place.

1a. Ovary inferior **2. Agave**
  b. Ovary superior 2
2a. Leaves folded towards the base, equitant; flowers bilaterally symmetric
4. Phormium
  b. Leaves neither folded towards the base nor equitant; flowers radially
    symmetric 3
3a. Perianth-lobes 2.5 cm or more; filaments thickened towards their apices
1. Yucca
  b. Perianth-lobes always less than 2.2 cm; filaments not thickened towards
    their apices **3. Cordyline**

1. Yucca Linnaeus. 30/1--2. Plants stemless or with short stems or tall, woody trunks. Leaves borne in rosettes at the ends of the stems or branches, persistent, stiff or flexible, fleshy, margins entire, finely toothed or splitting off as threads, bases expanded and somewhat sheathing. Inflorescence a much-branched, erect or rarely hanging panicle, stalk long and conspicuous or short and hidden by the leaves. Flowers bisexual, subtended by bracts. Perianth bell-shaped, hemispherical or spherical, opening more widely at night, radially symmetric, of 6 lobes which are united at their bases for a short distance. Stamens 6, the filaments free from the perianth, or, more rarely, united to it for some distance, usually curved out towards their apices, which are swollen and broader than the anthers. Ovary superior, 3-celled, each cell with many ovules, usually tapering into the 3- or 6-lobed stigma (more rarely the style thread-like). Fruit a capsule or indehiscent and dry or fleshy. Seeds winged or wingless. *USA, Mexico, Guatemala, West Indies.*

Some 20 species cultivated, only 1 or 2 reliably hardy.

2. Agave Linnaeus. 100/1--2. Monocarpic or perennial leaf-succulents without fleshy underground stems. Stems absent or short, sometimes forming trunks. Leaves in rosettes, persisting for many years, small to very large, often very fleshy, usually hard and leathery, usually with spines or teeth on the margins (more rarely the margins unrolling as fine threads) and with a conspicuous terminal spine. Inflorescence a large spike or panicle with many bisexual, radially symmetric flowers. Perianth tubular, of 6 lobes which are united for most of their length. Stamens 6, anthers projecting from the perianth. Ovary inferior, 3-celled, with many ovules in each cell, Fruit a capsule. Seeds flattened, black. *Central America, West Indies, southern USA; naturalised elsewhere.*

Over 100 species. Very many species are recorded as having been in cultivation, but their identification is dubious, and there is much uncertainty. Most are not hardy in N Europe. One species of **Manfreda** Salisbury (a genus often included in *Agave*) may be grown in very mild areas: it is a perennial herb with succulent roots. Also, one species of **Polianthes** Linnaeus, which is also herbaceous, with sweetly scented, bilaterally symmetric flowers, may also be grown in similar areas.

3. Cordyline R. Brown. 20/10. Plants woody, tufted or tree-like, branching usually falsely dichotomous. Leaves crowded in rosettes at the ends of the branches, persistent, usually lanceolate or sword-shaped, stalked, the lateral veins diverging from the conspicuous midrib at an acute angle. Flowers in dense, terminal panicles, which often appear lateral due to the growth of a new vegetative shoot from a nearby leaf-axil. Flowers bisexual, bracteate (often with 2 bracts to each flower-stalk), stalked, radially symmetric. Perianth of 6 lobes united into a short tube at the base, the lobes usually reflexed. Stamens 6, anthers versatile, filaments flattened, attached to the top of the perianth-tube. Ovary superior, 3-celled, with several ovules in each cell. Fruit a small berry which dries with age; seeds several, curved, black. *Southeast Asia, Australasia, Polynesia, Hawaii.*

A few species of **Dracaena** Linnaeus (now often placed in the *Dracaenaceae*, as is *Cordyline*) may also be grown in very mild areas; they often have variegated leaves, and there is 1 ovule in each cell of the ovary.

4. Phormium Forster & Forster. 2/2. Rhizomatous perennial plants, ultimately becoming woody at the base. Leaves persistent, linear, folded towards the base and equitant, keeled, marked with many fine, close, longitudinal stripes. Flowering axis long, bearing alternate, deciduous bracts which are scarious, the upper bracts subtending and entirely enclosing the short, alternately branched flowering branches. Flower-stalks jointed near the apex. Flowers bisexual, bilaterally symmetric. Perianth-lobes 6, more or less equal, erect, united into a tube at the base. Stamens 6, projecting from the perianth. Ovary superior, 3-celled, elongate, ovules many in each cell. Fruit a long, many-seeded capsule, the seeds flattened and almost winged, black and shiny. *New Zealand.*

7. AMARYLLIDACEAE

Perennial herbs, usually with bulbs, more rarely with rhizomes. Leaf-bases sometimes forming a false-stem or neck above the bulb. Leaves often in 2 ranks, usually flat or with the margins bent outwards when young, usually basal, more rarely borne on the stem. Inflorescence an umbel or reduced to a solitary terminal flower, subtended by usually 2 (rarely 1 or more than 2) spathes (bracts) which enclose the whole inflorescence in bud and usually persist beneath it in flower; individual flowers often with bracts which are smaller than the spathes. Perianth radially or bilaterally symmetric, either of 6 free segments, or joined into a tube at the base and with 6 lobes which are all alike or, more rarely, those of the outer whorl different from those of the inner. Stamens 6, usually borne on the top of the perianth-tube, often deflexed and curved upwards at their tips; anthers usually dorsifixed and versatile, sometimes basifixed, opening by slits or more rarely by pores. A corona often present, joining the bases of the stamens or as outgrowths of the perianth; it is funnel-shaped, cup-shaped, cylindric or disc-like (more rarely a low ridge); scales (perianth-scales) or tufts of hairs sometimes present at the bases of the filaments. Ovary completely inferior, 3-celled with 2--many ovules in each cell, placentation axile or basal. Fruit a capsule or berry with few to many seeds. See Figure 5, p. 44.

A family of 65--70 genera and about 850 species, found in most parts of the world. About half the genera are cultivated, several of them requiring glasshouse protection in northern Europe. The limits of the family are controversial, and parts of the *Agavaceae* and *Liliaceae* (subfamilies *Allioideae* and *Alstroemenoideae*) are sometimes included within it. Genus **1** is sometimes treated as forming the family *Ixioliriaceae*. The delimitation of genera within the family is also difficult, frequently based on the presence or absence of a corona (and, if present, its nature). Different authors define the corona in different ways; in this account the small appendages (perianth-scales) or tufts of hairs found at the attachment of stamens to perianth in some genera (*Hippeastrum, Zephyranthes*) are not regarded as forming a corona.

1a. Plant with a flowering stem leafy in the lower part   **1. Ixiolirion**
  b. Plant with a leafless scape, the leaves all basal   2
2a. Corona present, joining the bases of the filaments so that they are
      separately borne on a tube, cup or ridge   3
  b. Corona absent, or, if present, the filaments borne free from it; perianth-
      scales or tufts of hairs may be present at the bases of the filaments   4
3a. Ovules borne in 1 or 2 vertical rows in each cell of the ovary
                                         13. Pancratium
  b. Ovules borne side-by-side in each cell of the ovary   **12. Vagaria**
4a. Anthers basifixed, opening by pores   5
  b. Anthers dorsifixed, usually versatile, opening by slits   7
5a. Flowers erect; filaments almost as long as anthers   **10. Lapiedra**
  b. Flowers pendent on arching stalks; filaments much shorter than anthers   6
6a. Segments of the inner whorl of the perianth different from those of the
      outer whorl   **9. Galanthus**
  b. All perianth-segments similar   **11. Leucojum**
7a. Spathes 1 or 2, united below to form a tube around the flower-stalks   8
  b. Spathes 1--several (usually 2), not united below to form a tube   11
8a. Corona distinct, trumpet-, funnel- or cup-shaped, more rarely a disc or low
      ridge, arising at the junction of the perianth tube and lobes   **6. Narcissus**
  b. Corona absent; perianth-scales or tufts of hairs sometimes present at the
      bases of the filaments   9
9a. Perianth-lobes at most 1 cm   **7. Tapeinanthus**
  b. Perianth-lobes 2 cm or more   10
10a. Stigma simple, capitate or minutely 3-toothed; perianth usually yellow
      inside and out, rarely white   **3. Sternbergia**
  b. Stigma distinctly 3-fid; perianth usually not yellow, occasionally yellow
      inside   **2. Habranthus**
11a. Ovules 1--3 in each cell of the ovary   **4. Lycoris**
  b. Ovules 4 or more in each cell of the ovary   12
12a. Perianth-tube at most 5 mm, the lobes curving away from it, separate and
      distinct, margin often wavy   **8. Nerine**
  b. Perianth-tube at least 1 cm, the lobes forming a funnel shape, overlapping
      or almost so, margins not wavy   **5. Amaryllis**

1. Ixiolirion (Fischer) Herbert. 4/2. Perennial herbs with bulbs. Leaves mostly in a basal rosette, usually winter-persistent, also present on the stems. Flowers in a terminal umbel with or without several racemosely arranged flowers below it. Perianth radially symmetric, either of 6 free segments or united into a tube below and 6-lobed. Stamens 6, attached to the perianth-tube (if present) or to the bases of the perianth-segments or -lobes. Ovary 3-celled; ovules numerous in each cell; placentation axile. Fruit a capsule. Seeds numerous, black. *Southwest & central Asia.*

2. Habranthus Herbert. 20/4. Perennial herbs with bulbs. Leaves linear. Flowers solitary, terminal (in a reduced, 1-flowered umbel), subtended by a spathe (a pair of fused bracts) which is tubular in its lower half and usually somewhat bifid above. Perianth slightly bilaterally symmetric, funnel-shaped, with a short or very short tube below. Corona absent. Stamens 6, deflexed, then curving upwards towards the apex, of 2 different lengths. Style deflexed downwards then curving upwards, exceeding the stamens; stigma 3-fid. Fruit a top-shaped capsule containing numerous seeds. *Temperate South America.*

One species of **Zephyranthes** Herbert, with stamens of equal lengths, more or less straight and spreading, may occasionally be grown in very mild areas.

3. Sternbergia Waldstein & Kitaibel. 8/6. Perennial herbs with bulbs; tunics membranous, brown or black. Leaves basal, linear or strap-shaped to narrowly lanceolate, flat or shallowly channelled above, sometimes keeled beneath, often twisted. Flowering stems 1--several, sometimes below ground at flowering, elongating and often arching in fruit. Flowers solitary, usually yellow, rarely white, appearing in spring or autumn. Spathe membranous, tubular below, split above. Perianth radially symmetric; tube cylindric, narrow; lobes 6, in 2 similar whorls, oblanceolate to obovate. Corona absent. Stamens 6, in 2 unequal whorls, borne at top of perianth-tube, shorter than perianth-lobes. Style 1, bearing an entire or minutely 3-toothed capitate stigma. Fruit a cylindric to spherical capsule containing numerous large, dark seeds, often with fleshy appendages. *From Spain to India (Kashmir).*

4. Lycoris Herbert. 12/5. Herbaceous perennials with ovoid bulbs. Leaves basal, linear or strap-shaped, usually appearing after the inflorescences. Flowers in a terminal umbel borne on a solid scape, the umbel subtended by 2 free spathes. Perianth bilaterally or radially symmetric, with a usually very short tube at the base, lobes equal, spreading or strongly recurved; perianth-scales minute. Stamens 6, erect or ascending, borne on the throat of the perianth-tube, deflexed, anthers versatile. Ovary 3-celled, with 2 or 3 ovules per cell. Style thread-like, stigma minute. Capsule almost spherical to ovoid. Seeds black-brown. *China, Japan.*

5. Amaryllis Linnaeus. 1/1. Perennial herb with bulbs. Leaves appearing after the flowers, oblong, somewhat narrowed to the base. Scape stout, solid. Flowers 5--many in an umbel subtended by 2 large, equal, free spathes which enclose the whole umbel in bud. Flowers ascending to erect. Perianth somewhat bilaterally symmetric, with a short tube and 6 oblanceolate lobes which are somewhat clawed, acute, the 3 outer lobes each with a small, hairy, inwardly pointing appendage just below the apex. Stamens 6, deflexed and then curving upwards towards the apex. Ovary 3-celled, with few ovules in each cell. Style deflexed and then curving upwards, stigma capitate. Fruit a capsule, splitting irregularly. Seeds few. *South Africa.*

41

The one species has been hybridised with *Crinum moorei* to produce a range of hybrids bearing the name × **Amarcrinum** Coutts (× *Crinodonna* Anon.), and also with species of *Nerine*. A few species of **Hippeastrum** Herbert, which have hollow scapes, may occasionally be grown in very mild areas, as may one species of **Cyrtanthus** Linnaeus, which has narrowly tubular, sweetly-scented flowers.

6. Narcissus Linnaeus. 70/25 and many hybrids and cultivars. Perennial herbs with bulbs. Leaves 1--several from each bulb, basal, erect, spreading or prostrate, almost cylindric in section to flat and broad. Scape present. Inflorescence an umbel of 2--20 flowers or flowers solitary, subtended by a usually scarious bract (spathe). Perianth with a tube below, 6-lobed. Corona almost always present, usually conspicuous, free from the stamens. Stamens 6, usually in 2 whorls, more rarely in 1. Ovary 3-celled with many ovules. Capsule ellipsoid to almost spherical, many-seeded. Seeds sometimes with an appendage. *Europe, Mediterranean area.*

7. Tapeinanthus Herbert. 1/1. Very like *Narcissus*. Leaves absent at flowering. Flowers with perianth-tube extremely short, corona rudimentary or absent. *Spain, North Africa.*

Perhaps not distinct from *Narcissus*.

8. Nerine Herbert. 30/12. Perennial herbs with bulbs, the bulb sometimes with an elongate neck. Leaves present or absent at flowering. Scape usually exceeding the leaves. Umbel with 4--20 flowers, subtended by 2 persistent spathes. Perianth usually bilaterally symmetric, more rarely radially symmetric, with a very short tube and 6 arching and spreading lobes which often have crisped margins. Stamens 6, attached to the perianth-tube at the base, usually deflexed, more rarely erect, sometimes with appendages at the bases of the filaments. Ovary inferior, 3-celled and -lobed, with usually 4 ovules in each cell. Fruit a membranous capsule. *South Africa.*

9. Galanthus Linnaeus. 12/12. Perennial herbs with bulbs. Leaves basal, 2, linear, strap-shaped or oblanceolate, enclosed in a tubular, membranous sheath at the base. Spathe of 2 fused bracts. Flowers pendent. Perianth-segments free, the outer 3 acute to more or less obtuse, spathulate or oblanceolate to narrowly obovate, shortly clawed, erect-spreading, the inner half to two-thirds as long as the outer, usually notched, oblong, spathulate or oblanceolate, tapered to the base, erect, with a green patch around the notch and sometimes also at the base. Stamens inserted at the base of the perianth, shorter than the inner segments. Filaments much shorter than the anthers. Anthers basifixed, opening by terminal pores. Style slender, exceeding the anthers; stigma capitate. Capsule ellipsoid or almost spherical, opening by 3 flaps. Seeds light brown, each with an appendage. *Europe, Turkey, Iran, the Caucasus.* Figure 5, p. 44.

10. Lapiedra Lagasca. 2/1. Perennial herb with a bulb. Leaves few, basal. Flowers erect in an umbel of 4--9; spathe of 2 bracts. Perianth-segments all similar, free, spreading. Filaments as long as anthers; anthers basifixed, opening by pores. Capsule flattened-spherical, 3-lobed, surrounded by the persistent, withered perianth. Seeds few. *Mediterranean area.*

11. Leucojum Linnaeus. 10/7. Perennial herbs with bulbs. Flowers bell-shaped, nodding, white or pink, solitary or in umbels of up to 5 (rarely 7), subtended by a spathe of 1 or 2 bracts. Perianth-tube and corona absent. Perianth-segments free, all more or less alike. Anthers blunt at the tip, not pointed as in *Galanthus*, opening by pores. Capsule erect, more or less spherical; seeds numerous. *South, west & central Europe, North Africa, southwest Asia.*

12. Vagaria Herbert. 4/1. Perennial herbs with white fleshy roots and narrow, flask-shaped bulbs that are clustered together, each covered to the neck with a papery brown tunic, often sheathing the scape and leaf-bases for up to 15 cm. Leaves 4--7, linear, obtuse and with a conspicuous median white band. Scape somewhat flattened and slightly 4-angled, solid. Bracts 2--4, papery. Flowers bisexual, 6--9 per umbel, stalks extending after flowering. Perianth-tube *c.* 1 cm, lobes 6, each narrowly elliptic to lanceolate with hooded apex, white with a conspicuous green midrib, somewhat keeled beneath. Corona arising at throat of perianth, consisting of 12 linear, acute teeth, occurring in pairs at the bases of the filaments. Anthers 6, versatile. Ovary 3-celled, ovules 2 or 3 per cell, placentation axile; stigma capitate. Fruit a capsule containing several 3-sided, glossy black seeds. *North America, Middle East.*

13. Pancratium Linnaeus. 16/4. Perennial herbs with large bulbs and several broadly linear to strap-shaped basal leaves which are more or less in 2 ranks. Scape bearing an umbel of 3--15 flowers (rarely a solitary flower), subtended by usually 2 scarious spathes. Flowers large, white, fragrant. Perianth-tube present, perianth-lobes linear-lanceolate, spreading or almost erect; corona conspicuous, united to the lower part of the filaments, which thus appear to be inserted on its margin. Anthers dorsifixed. Stigma capitate. Ovary 3-celled with ovules in 1 or 2 vertical rows in each cell. Fruit a many-seeded capsule; seeds black, angular, dry. *Canary Islands, Mediterranean area to tropical Asia & Africa.*

8. TECOPHILAEACEAE
Herbs with corms or tubers. Leaves basal, linear or ovate, with parallel veins. Flowers bisexual, radially symmetric, solitary or in a raceme or panicle. Perianth-segments 3, united at base into a short tube, or not united. Fertile stamens 6, or 3 with 3 infertile staminodes. Ovary half-inferior, 3-celled, ovules numerous. Fruit a capsule.

A family of 6 genera and about 20 species native in California, South & Central America, and South Africa. Only 1 genus is in general cultivation.

1. Tecophilaea Colla. 2/1. Corms with fibrous tunics. Leaves 2 or 3, linear, enclosed in a sheath when young, to 10 cm long, green, hairless. Perianth blue to purple with darker veins and white throat, 4 cm long. *Chile.*

9. HYPOXIDACEAE
Perennial herbs with rhizomes or corms, and fleshy as well as fibrous roots. Leaves often pleated, usually hairy, stalked or not, usually all basal. Flowers solitary or in the axils of bracts in racemes, occasionally contracted into a dense head, rarely borne at ground-level. Perianth of 6 lobes usually united into a (sometimes short) tube below, the lobes spreading, usually at least the outer 3 hairy on their backs. Stamens 6, borne on the perianth-tube. Ovary inferior, 3-celled, with numerous, axile ovules in each cell (the ovary is occasionally subterranean, borne between the leaf-sheaths, and the perianth is raised into the air by a beak on top of the ovary). Fruit a capsule opening by

43

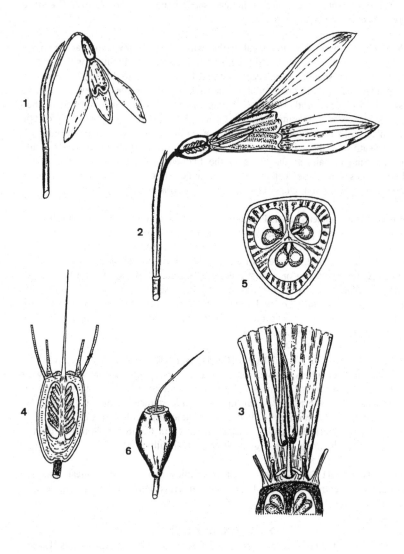

Figure 5. Amaryllidaceae. *Galanthus nivalis.* 1, Flower. 2, Longitudinal section of flower. 3, Stamen and adjacent perianth-segment. 4, Longitudinal section of ovary. 5, Transverse section of ovary. 6, Developing fruit.

44

a lid, a berry or, when the ovary is subterranean, capsule-like but thin-walled and breaking up irregularly. Seeds usually black.

A family of 3--8 genera (different authors vary in the number of genera recognised) from most of the warmer parts of the world. Its classification is confused and identification can be difficult.

    1a. Flowers yellow or orange, rarely white when each perianth-lobe has a dark
        purple blotch at the base   **1. Hypoxis**
     b. Flowers red, pink or white, perianth-lobes without dark blotches at the
        base   **2. Rhodohypoxis**

1. Hypoxis Linnaeus. 150/7. Plants with corms and thick, fleshy roots. Leaves mostly basal, sometimes borne on the stem, ridged or not, often hairy, their bases sheathing and overlapping. Scape present, bearing a solitary flower or a 2--12-flowered raceme. Perianth-lobes 6, the outer 3 slightly broader than the inner and greenish and usually hairy outside. Stamens 6. Ovary inferior, 3-celled, with several ovules in each cell. Style present or very short, stigma 3-lobed. Capsule containing several seeds, opening by means of a lid formed from the ovary wall. Seeds black or brown, spherical or elongate. *North America, Africa, tropical Asia, Australia.*

2. Rhodohypoxis Nel. 6/1. Perennial herbs with fleshy and fibrous roots. True stem absent. Leaves all basal, usually hairy. Scape present or absent (when absent the ovary is within the leaf-sheaths at or below ground-level). Flowers solitary or few. Perianth of 6 lobes united into a tube below, the lobes spreading, the 3 inner conspicuously clawed and each with a pronounced, inflexed curve at the junction of claw and blade, which close the mouth of the tube. Stamens 6, borne on the perianth-tube at 2 levels. Ovary 3-celled, with numerous ovules in each cell; when borne at ground-level, the ovary has a beak which raises the perianth into the air. Fruit a capsule opening by a lid or, if borne at ground-level, breaking up irregularly to release the seeds. Seeds black. *Southern Africa.*

10. DIOSCOREACEAE

Herbaceous or slightly woody climbers with slender, twining stems arising from tubers or rhizomes. Aerial tubers sometimes present. Leaves usually alternate, sometimes opposite, entire, usually heart-shaped, sometimes palmately lobed, net-veined with 3--11 primary veins extending upwards from the junction with the leaf-stalk. Flowers small, greenish, usually unisexual (the plants then dioecious), in axillary spikes, racemes or panicles. Perianth of 6 lobes in 2 whorls, usually united at the base. Male flowers with 6 stamens in 2 whorls, sometimes 1 whorl reduced to staminodes or absent, anthers opening by slits. Female flowers with inferior ovaries of 3 united carpels; placentation usually axile, ovules usually 2 in each cell; styles 3 or 1 with 3 stigmas. Fruit a capsule, often 3-winged, or a berry. Seeds usually winged.

A family of 6 genera and 630 species, mainly from tropical and warm temperate areas. In these areas several species of *Dioscorea* are grown for their edible tubers (yams) but in Europe the genus is grown only for ornament.

    1a. Fruit a berry   **1. Tamus**
     b. Fruit a capsule   **2. Dioscorea**

1. Tamus Linnaeus. 5/1. Perennial, dioecious herbs with large, cylindric to ovoid tubers. Leaves alternate, heart-shaped. Flowers in axillary racemes. Perianth bell-shaped. Stamens 6, rudimentary in the female flower. Style with 3 bilobed stigmas. Fruit a berry with few, unwinged, spherical seeds. *Atlantic Islands, Europe, North Africa, southwest Asia.*

2. Dioscorea Linnaeus. 600/few. Dioecious climbers with twining stems arising from tuberous roots, sometimes bearing aerial tubers in the leaf-axils. Perianth of male flowers bell-shaped. Perianth of female flowers deeply 6-lobed. Fruit a 3-angled or 3-winged capsule; seeds flat, usually winged. *Tropics and subtropics.*

11. PONTEDERIACEAE

Annual or perennial aquatic herbs with rhizomes or stolons. Leaves alternate, linear or with a distinct stalk and blade. Flowers bisexual, solitary or paired or in spikes or racemes which are subtended by a spathe, radially or bilaterally symmetric. Perianth-segments usually 6. Stamens 3 or 6 or occasionally only 1. Ovary superior, of 3 united carpels; ovules 3--many, placentation axile or parietal. Style 1. Fruit a capsule or nutlet.

There are 9 genera, all native in tropical or subtropical aquatic habitats. None of the species is native in Europe, although 4 are now naturalised; 4 genera are represented in cultivation.

1. Pontederia Linnaeus. 5/1. Perennial herbs with submerged, floating or creeping stems. Leaves submerged, floating or emergent, linear or stalked. Flowers in a spike, blue. Perianth tubular, hairy outside, 2-lipped, each lip with 3 lobes. Stamens 6, anthers equal. Fruit a nutlet with 1 seed. *America.*

One species of **Heteranthera** Ruiz & Pavón, which has 3 stamens and stalkless flowers, may be grown in very mild areas.

12. IRIDACEAE

Herbs (rarely shrubby) with corms or rhizomes. Leaves basal and borne on the stem, often erect and equitant, usually in 2 ranks. Flowers solitary or borne in spikes or panicles (see figure 6, p. 48), each subtended by a bract and sometimes a bracteole, occasionally more than 1 flower to a bract. Large bracts (spathes) may subtend the whole or part of the inflorescence. Perianth radially or bilaterally symmetric, with a tube at the base or the segments completely free. Stamens usually 3. Ovary inferior, usually 3-celled, occasionally 1-celled; ovules numerous, placentation axile. Style usually 3-branched above, the branches sometimes further divided or broad and petaloid. Fruit a capsule splitting between the septa. See figures 6, p. 48, 7, p. 50, 8, p. 52 & 9, p. 54.

A family of about 70 genera and 1500 species from most parts of the world; the cultivated species are mainly from the Mediterranean area and South Africa. The genera are often difficult to distinguish, and care must be taken using the key below.

    1a. Perianth united into a tube at the base, which may be very short (some genera with extremely short perianth-tubes are keyed out under 1b as well as here)   2
    b. Perianth-segments completely free (but including some genera with extremely short, inconspicuous perianth-tubes)   17

2a. Leaves neither arranged in 2 ranks nor equitant   3
  b. Leaves arranged in 2 ranks, usually equitant   6
3a. Leaves each with a white stripe on the upper surface   **9. Crocus**
  b. Leaves without such a stripe   4
4a. Styles petaloid   **1. Iris**
  b. Styles not petaloid   5
5a. Filaments united, at least at the base   **10. Galaxia**
  b. Filaments free   **11. Romulea**
6a. Perianth bilaterally symmetric (sometimes due only to the curved tube)   7
  b. Perianth radially symmetric (tube always straight)   8
7a. Perianth-tube slender throughout or with a very slender basal part,
    abruptly expanded above   **21. Crocosmia**
  b. Perianth-tube gradually widening from the base   **22. Gladiolus**
8a. Perianth-lobes distinctly in 2 whorls, the lobes of the outer whorl differing
    from those of the inner in shape and size   9
  b. Perianth-lobes all similar, apparently in 1 whorl, or if obviously in 2 whorls
    then the lobes all similar in size and appearance   10
9a. Rootstock consisting of finger-like tubers; inner perianth-lobes greenish
    yellow, blades of outer lobes dark purple-brown; ovary 1-celled
         4. Hermodactylus
  b. Rootstock a bulb or rhizome; flowers never with the above colour
    combination; ovary 3-celled   **1. Iris**
10a. Filaments united   11
  b. Filaments free   13
11a. Style not branched, stigma entire or 3-toothed   **15. Solenomelus**
  b. Style 3-branched, stigmas 3   12
12a. Flowers stalkless   **20. Ixia**
  b. Flowers with stalks (which may be concealed within bracts)
         14. Sisyrinchium
13a. Plant woody   **16. Nivenia**
  b. Plant not woody   14
14a. Stems arching over; bracts scarious, conspicuous; flowers pendent
         19. Dierama
  b. Stems not arching over; bracts inconspicuous, green or scarious; flowers
    not pendent   15
15a. Plants with rhizomes   **17. Schizostylis**
  b. Plants with corms   16
16a. Corm-tunics hard and woody   **18. Geissorhiza**
  b. Corm-tunics papery or fibrous   **20. Ixia**
17a. Style-branches petaloid   18
  b. Style-branches not petaloid   22
18a. Flower with an apparent perianth-tube which is, in fact, a beak on top of
    the ovary   **3. Gynandriris**
  b. Flower without a beak on top of the ovary   19
19a. Flowering stem with many long branches; claw of outer perianth-segments
    marked with transverse bands   **2. Pardanthopsis**
  b. Flowering stem unbranched or with a few short branches; claw of outer
    perianth-segments not marked with transverse bands   20

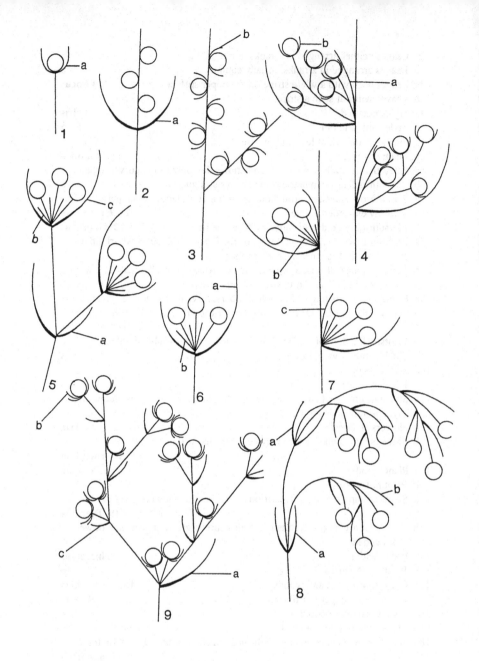

Figure 6. Iridaceae. Inflorescence types (a, spathe; b, bract; c, secondary spathe; flowers indicated by open circles). See text for references to individual diagrams.

20a. Filaments united, at least at the base; rootstock a corm   **5. Moraea**
  b. Filaments usually free; rootstock a rhizome or bulb   21
21a. Leaves usually all basal: perianth-tube present, though sometimes
     extremely short   **1. Iris**
  b. Leaves basal and also borne on the stem; perianth-tube absent   **6. Dietes**
22a. Filaments free   23
  b. Filaments united, at least at the base   25
23a. Perianth-segments distinctly in 2 whorls, the segments of the outer whorl
     differing from those of the inner in shape or size   **7. Trimezia**
  b. Perianth-segments all similar, in 1 or 2 whorls   24
24a. Perianth-segments twisting spirally after flowering   **13. Belamcanda**
  b. Perianth-segments not twisting spirally after flowering   **18. Geissorhiza**
25a. Rootstock a corm   **8. Homeria**
  b. Rootstock a rhizome or just a cluster of fibrous roots   26
26a. Segments of the outer whorl of the perianth smaller than those of the inner
     whorl   **12. Libertia**
  b. Perianth-segments all of the same size   **14. Sisyrinchium**

1. Iris Linnaeus. 250/90 and many hybrids. Rhizomatous or bulbous perennial herbs. Leaves usually basal and in 2 ranks, flat, channelled, 4-angled or nearly cylindric. Flowers 1--several, borne within 2 spathes. Perianth radially symmetric, united at the base into a tube. Falls (outer lobes) narrowed towards the base into a claw (haft), the blade sometimes bearded. Standards (inner lobes) usually erect or arching, sometimes horizontal or deflexed, narrowed towards the base into a claw, rarely bearded or very reduced. Stamens 3, filaments free, borne at the base of the falls. Style-branches 3, each 2-lobed beyond the stigma, coloured and petaloid, each covering a stamen. Capsule cylindric to ellipsoid, more or less round to triangular in cross-section, often with 3 or 6 ribs; seeds numerous, sometimes each bearing a fleshy appendage (aril). *Northern hemisphere.* Figures 7, p. 50 & 8, p. 52.

2. Pardanthopsis Lenz. 1/1. Related to and resembling *Iris.* Stems much-branched with many small, short-lived flowers. Perianth-segments free, each fall with transverse bands on the claw. Perianth-segments twisting spirally after flowering. *Siberia, north China, Mongolia.*
  Until recently included in *Iris.*

3. Gyanandriris Parlatore. 9/3. Rootstock a corm, surrounded by fibrous tunics. Leaves 1 or 2, linear, channelled. Flowers in terminal and axillary cymes, short-lived, like those of *Iris* or *Moraea* in appearance with petaloid style-branches, and free perianth-segments differentiated into falls (outer) and standards (inner). Stamens partly united, closely adherent to the underside of the style-branches. Ovary extended upwards as a slender, sterile beak on which the perianth is directly inserted; perianth-tube absent. Capsule with many seeds. *Southern Africa, Mediterranean area.*

4. Hermodactylus Miller. 1/1. Rootstock of 2--4 palmately branched tubers. Leaves in 2 ranks, narrowly linear, 4-angled, longer than the stem. Flowering stem to 40 cm. Flowers radially symmetric, terminal, solitary, partly enclosed by a pale green bract, yellowish green except for the blade of the falls which is very dark purple-brown to

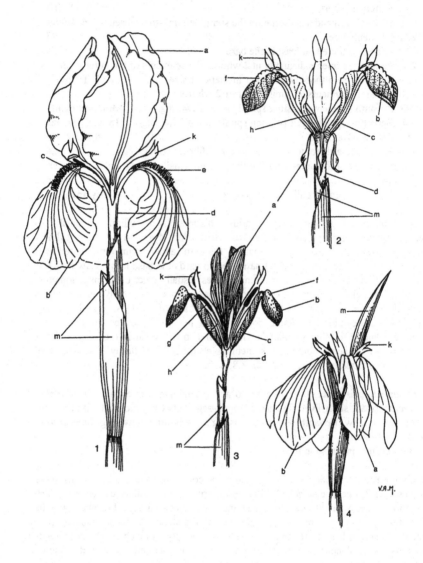

Figure 7. Iridaceae. Flower shapes and structures in *Iris* (a, standard; b, blade of fall; c, claw of fall; d, perianth-tube; e, beard; f, ridge; g, anther; h, style-branch; k, style-lobe; m, bract).

50

almost black. Falls 4--5 cm; standards 2--2.5 cm. Perianth-tube short, funnel-shaped. Ovary with a short, slender sterile beak at the apex. *Mediterranean area.*

5. Moraea Miller. 100/10. Rootstock a corm with membranous, fibrous or netted tunics. Leaves 1--several, usually linear. Stem with reduced bract-like sheathing leaves above. Flowers 1--several, borne within bracts (spathes) which may be green or dry and papery. Flowers often short-lived. Perianth radially symmetric, segments free. Falls (outer segments) larger than standards (inner segments), reflexed or spreading, each bearing a conspicuous nectar guide. Standards reflexed to erect, entire or 3-lobed. Stamens opposite outer perianth segments, adpressed to style-branches. Filaments completely joined, to joined at the base only. Style-branches petaloid with 2 lobes (crests). Fruit a capsule containing angled seeds. *Africa south of the Sahara.*

6. Dietes Klatt. 6/4. Rhizome thick, creeping. Leaves several, in 2 ranks, leathery, linear to sword-shaped. Stem erect, hairless, branched above, bearing leaves at the lower nodes and bracts at the upper nodes. Flowers short-lived, in groups, each group subtended by a spathe-bract. Perianth radially symmetric, segments free, the outer larger than the inner, with an ascending claw and spreading blade and with a nectar guide at the base of the blade. Filaments usually free. Style short, divided into 3 large, flattened, petaloid branches which divide at the tips into paired style-crests; stigma-lobes borne on the lower surface. Fruit a 3-celled capsule with many seeds. *East, central & southeast tropical Asia, coastal southern Africa, Lord Howe Island.*

One species of the very similar genus **Cypella** Herbert, with pleated leaves, may occasionally be grown in very mild areas.

7. Trimezia Herbert. 5/1. Rhizome bulb-like, covered with the fibrous remains of old leaf-bases. Leaves linear to linear-lanceolate. Perianth radially symmetric, segments free, in 2 dissimilar whorls, shortly clawed at the base. Stamens borne opposite the style-branches; filaments free. Fruit a dry capsule on a long stalk. Seeds black, angled. *Central & tropical South America, West Indies.*

There are several other genera from California and South America which may occasionally be grown in very mild areas: **Tigridia** Jussieu (flowers 3--15 cm across, stigmas on radii alternating with anthers), **Rigidella** Lindley (perianth-segments red or orange, coiling spirally after flowering), **Herbertia** Sweet (perianth-segments purple or bluish, hairless, anther-connective linear) and **Alophia** Herbert (perianth-segments purple or bluish, each with a line of hairs, anther-connective broad, fiddle-shaped). Similarly, the one species of the tropical African genus **Ferraria** Burman (with conspicuously fringed style-branches), may occasionally be found.

8. Homeria Ventenat. 31/5. Corm covered with netted tunics. Leaves solitary or several, linear. Stem erect, hairless, usually branched. Inflorescence with several flowers, each flower subtended by 2 leafy bracts. Perianth radially symmetric, segments free, almost equal or the inner ones slightly smaller, usually with a basal claw, the blade with a basal nectary which may extend down the claw. Stamens opposite the outer perianth-segments, filaments united. Style-branches 3, entire, flattened, with or without crests. Fruit a cylindric capsule, with a flat top or beaked. *Southern Africa.*

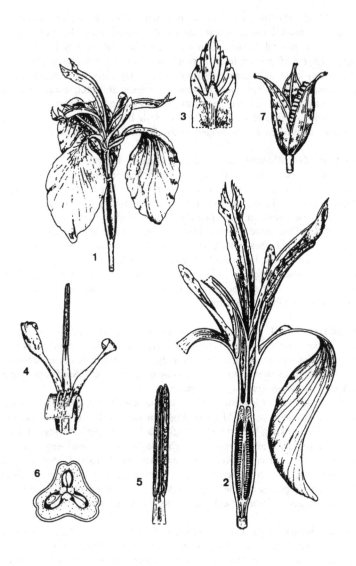

Figure 8. Iridaceae. *Iris pseudacorus*. 1, Flower. 2, Longitudinal section of flower. 3, Tip of style-branch. 4, Stamen and two inner perianth-segments. 5, Anther. 6, Transverse section of ovary. 7, Mature fruit.

9. Crocus Linnaeus. 80/35 and hybrids. Corm enclosed by several fibrous, papery or leathery tunics. Sheathing leaves up to 5, enclosing the aerial shoot. Leaves all basal, linear, not in 2 ranks, present at flowering time or absent and appearing later; upper surface green or grey-green with a central whitish stripe. Flowers 1--several, produced in autumn, winter or spring, each carried on a short subterranean flower-stalk which is sometimes subtended by a membranous spathe (prophyll -- the 'basal spathe' of various authors). Bracts 1 or 2, membranous, subtending the ovary and sheathing the perianth-tube (the 'floral spathe' of various authors). Perianth radially symmetric with a long narrow tube; lobes 6, usually equal or inner whorl sometimes smaller. Style with 3 or more branches. Ovary subterranean. Capsule cylindric or ellipsoid, maturing above ground level by elongation of the flower-stalk; seeds numerous. *Europe, west & central Asia.* Figure 9, p. 54.

One species of **Syringodea** Hooker, with leaves without a white stripe above and perianth-tube longer than the lobes, may be grown in very favoured areas.

10. Galaxia Thunberg. 12/3. Corms with fibrous tunics which are often vertically ribbed. Stem underground or slightly above ground-level, elongating in fruit. Leaves borne at the top of the stem, not 2-ranked. Perianth with a long tube which raises the flower above the leaves; lobes more or less equal, spreading. Filaments partly to completely joined, inserted at the throat of the perianth-tube. Stigmas entire or fringed. Ovary more or less stalkless, ovoid. Fruit a capsule containing many small, angled seeds. *South Africa.*

11. Romulea Maratti. 80/15. Perennial herbs with corms. Corms with hard, brown tunics, usually asymmetric at the base. Basal leaves usually 2, rarely 1 or up to 6, usually 4-grooved; stem-leaves 1--6; all leaves linear, usually hairless and 4-grooved, not in 2 ranks. Flowering stems erect or recurved (especially in fruit), rarely absent. Flowers upright, surrounded by an outer bract and an inner bracteole. Perianth united into a short tube below, lobes all similar. Stamens 3, exceeding or shorter than the style, sometimes aborted. Style 1 with 3 bifid stigmas, the branches thread-like. Capsule borne on an elongate stem. Seeds spherical. *Mediterranean area* (10 species), *South Africa* (about 70 species)

The single species of **Gelasine** Herbert, which has blue flowers and membranous corm-tunics, may be grown in very mild areas.

12. Libertia Sprengel. 20/5. Perennials with short, creeping rhizomes and fibrous roots. Leaves numerous, mostly basal, equitant, linear, overlapping, mostly flat and rigid. Flowering stems erect, bearing a few reduced leaves. Flowers radially symmetric, in loose clusters or panicles borne terminally and in the axils of the stem-leaves. Perianth-segments entirely free, spreading, the inner segments usually longer than the outer. Stamens 3; filaments slightly fused at base. Style with 3 entire, linear, spreading branches; fruit a many-seeded, 3-celled capsule. *Australasia, South America.*

13. Belamcanda Adanson. 1--2/1. Rhizomatous perennial with branched stems and leaves borne in fans. Perianth-segments more or less alike, not differentiated into outer and inner, free, twisting spirally after flowering. Stamens free, not held against style. Styles 3, slender, not expanded and petaloid. Capsule splitting into 3 segments which reflex, exposing many shiny black seeds on the central axis. *India to eastern Asia.*

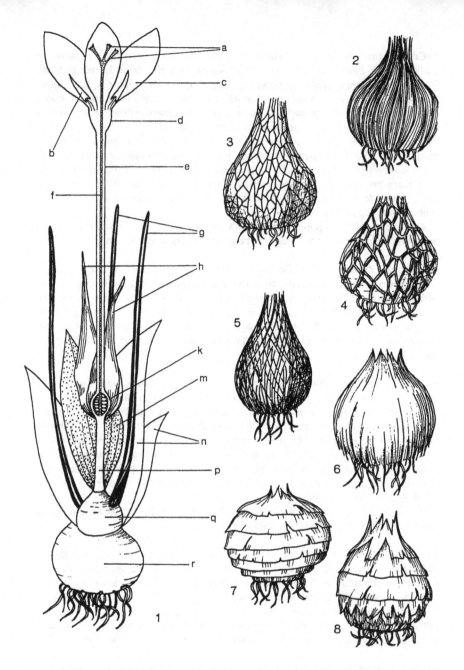

Figure 9. Iridaceae. Diagnostic features of *Crocus*. 1, Parts of the plant (a, style-branch; b, anther; c, perianth-lobe, d, throat; e, perianth-tube; f, style; g, leaf; h, bract; k, ovary; m, spathe; n, sheathing leaf; p, scape; q, new corm; r, old corm). 2--8, Different types of corm-tunic.

14. Sisyrinchium Linnaeus. Up to 200/15. Rhizome absent or very short; roots fleshy or fibrous. Stem erect, simple or branched, terete or flattened, often winged. Leaves mostly basal, 2-ranked, linear to strap-shaped or very narrowly lanceolate, acute, pale to dark or bluish green, often blackening on drying. Stem-leaves broader and shorter. Spathe-bracts usually in pairs, lanceolate, acute, often with white translucent margins. In branched species a lower (primary) pair of spathe-bracts gives rise to inflorescence-stalks, on the latter are further (secondary) pairs of spathe-bracts, from which the flower-stalks arise. In unbranched species, a single pair of spathe-bracts gives rise directly to the flower-stalks. Flower-stalks terete or flattened, often arched, lengthening in fruit. Flowers solitary or in clusters. Perianth radially symmetric, very shortly fused at base, lobes oblong, spreading, often notched with a short point in the notch. Stamens arising from base of perianth; filaments united into a cylindric or flask-shaped tube for some or all of their length. Ovary 3-celled, spherical to ovoid; ovules many. Style mostly hidden within filament-tube, with 3 acute or capitate, entire, stigmatic branches. *North & South America.* See figure 6, p. 48.

15. Solenomelus Miers. 2/2. Herbs with short rhizomes. Leaves 2-ranked, mostly basal or clustered near stem-base, a few higher up. Flowers many, on short stalks, in clusters enclosed by a pair of spathe-bracts. Perianth yellow or blue, radially symmetric, with a slender cylindric tube, lobes spreading, similar. Filaments attached at throat, completely fused into a tube; anthers oblong. Style simple; stigma entire or with 3 very short teeth. *Temperate South America.*

16. Nivenia Ventenat. 8/2. Evergreen shrubs. Stems branching or rarely unbranched, woody below. Leaves linear, 2-ranked, acute, overlapping at base. Flowering stem flat and 2-edged, ending in a spike, corymb or umbel subtended by a single spathe. Each flower or pair of flowers subtended by 3 bracts. Flowers radially symmetric, stalkless, falling entire on fading. Perianth with a distinct tube and 6 similar lobes. Flowers of 2 types: either with long stamens and short styles or with long styles and short stamens. In some species both types occur, but on separate plants. Filaments free, borne at top of perianth-tube. Style-branches small, entire. Capsule spherical to cylindric. *South Africa.* See figure 6, p. 48.

17. Schizostylis Backhouse & Harvey. 1/1. Rootstock short, rhizome-like (rarely producing a corm); roots fibrous. Basal leaves 2--4, 2-ranked, 30--45 cm × 6--12 mm, linear, keeled; midrib distinct. Stem 30--90 cm, slender, terete, erect, bearing *c.* 3 reduced leaves. Flowers 4--14 in a 2-sided spike, radially symmetric, each with an unequal pair of green lanceolate bracts. Perianth-tube 2.5--3 cm, narrowly conical, yellowish or reddish green; perianth-lobes *c.* 2.5 cm, similar, ovate to oblong, acute, overlapping, scarlet to crimson. Stamens 3; filaments slender, free, attached to top of perianth-tube; anthers bright yellow. Style equalling perianth-tube, divided into 3, entire, slender, red branches *c.* 1.8 cm long. *South Africa.*

18. Geissorhiza Ker Gawler. 70/6. Corm ovoid, asymmetric, flat-based; tunics of hard, woody, overlapping layers, splitting longitudinally. Basal leaves lanceolate, 2-ranked, linear or thread-like, often curled, with margin or mid-vein often thickened. Stem simple or branched, erect, bearing a few reduced leaves. Flowers radially symmetric in a 1- or 2-sided spike, each with a pair of leaf-like bracts. Perianth-tube short, erect, widening above; lobes 6, in 2 similar whorls, spreading. Stamens borne at lobe-bases; filaments free. Style slender, longer than perianth-tube; branches short, entire, recurved. *Southern Africa*.

A few species of **Hesperantha** Ker Gawler, with the style shorter than the perianth-tube, its branches long and ascending, may be grown in very mild areas.

19. Dierama Koch. *c.* 20/3. Rootstock a corm. Leaves mostly basal, linear, stiff, 2-ranked. Stems exceeding the leaves, arching over towards the tip, each ending in a loose, 1-sided panicle with numerous, very slender primary branches, each of which terminates in a pendulous, crowded, spike-like raceme of 3--6 radially symmetric flowers; bracts scarious. Perianth bell-shaped, its lobes about twice as long as the tube. Filaments free. Stigma 3-lobed, lobes spathulate. Capsule spherical, with 6 seeds. *Southern and east tropical Africa*. See figure 6, p. 48.

20. Ixia Linnaeus. 45/8. Perennial herbs with corms with papery or fibrous tunics. Stems simple or little-branched. Leaves few, mostly basal, 2-ranked, linear, lanceolate or thread-like, usually with a false midrib. Inflorescence a spike or a panicle of spikes with few, usually erect branches; spikes with few to many flowers borne in 2 ranks or spirally arranged. Bracts short but exceeding the ovaries, usually 3-toothed at their apices. Perianth of 6 equal or almost equal lobes, joined below into a short or long tube. Stamens attached at the throat of the tube or within it, filaments free or united into a short tube. Style thread-like, exceeding the stamens, stigma 3-fid. Fruit a thin-walled capsule opening by 3 flaps and containing many angular seeds. *South Africa*.

21. Crocosmia Planchon. 7/4 and several hybrids. Perennial herbs with small, flattened corms and creeping stolons which readily produce new corms. Leaves mostly basal, 2-ranked, conspicuously ribbed, sometimes pleated. Inflorescence usually a panicle of spikes, more rarely a simple spike, overtopping the leaves. Bracts small, scarcely exceeding the ovaries. Perianth with a curved tube, slender throughout or slender at the base and then abruptly expanded, the lobes spreading, all similar or the uppermost slightly longer than the others. Stamens 3, free, borne on the perianth-tube, arching to the upper side of the flower, the anthers either just projecting from the mouth of the perianth-tube, or with very long filaments, the anthers borne at the level of the tips of the perianth-lobes or beyond this. Style somewhat longer than the stamens, 3-fid at the apex. Fruit a capsule opening by 3 splits and containing many seeds. *Southern Africa*.

22. Gladiolus Linnaeus. 180/20 with many hybrids and selections. Herbaceous, usually with corms, often producing cormlets at base or on stolons. Lowest leaves reduced to sheaths, main leaves 2-ranked, usually equitant, linear to narrowly lanceolate, upper leaves much shorter, stem-clasping. Flowering stem terminating the leafy shoot or arising separately, simple or sparingly branched; flowers bisexual, stalkless, each with a bract and bracteole. Perianth in cultivated species bilaterally symmetric, with a tube at the base. Stamens arising on the tube, arching to the upper side of perianth. Style arching like the stamens and as long, 3-branched near the apex. *Africa, Madagascar, Mediterranean area and northern Europe*.

Species of several other South African genera with more or less bilaterally symmetric perianths may be cultivated in mild areas: **Tritonia** Ker Gawler (style exceeding the stamens), **Sparaxis** Ker (perianth-tube funnel-shaped, bracts with raggedly toothed apices), **Synnotia** Sweet (3 lower lobes of the perianth united beyond the mouth of the tube, perianth lavender to deep purple with yellow stripes), **Chasmanthe** N.E. Brown (like *Synnotia*, but perianth red or orange and yellow) and **Anomalesia** N.E. Brown (3 lower perianth-lobes very inconspicuous).

13. JUNCACEAE

Annual or perennial herbs with erect or horizontal rhizomes. Leaves basal, long and narrow with sheathing bases, or reduced to scales. Flowers bisexual, crowded into heads or panicles. Perianth segments 6, in 2 whorls. Stamens 3 or 6. Ovary superior with 1--3 cells; stigmas 3, ovules 3 or numerous. Fruit a dehiscent capsule. See figure 10, p. 58.

A family of 9 genera distributed throughout the world, but absent from the tropical lowlands. In Europe, *Juncus* and *Luzula* are native, and these are also the only cultivated genera.

1a. Leaves hairless; capsule with numerous seeds	**1. Juncus**
b. Leaves hairy; capsule with 3 seeds	**2. Luzula**

1. Juncus Linnaeus. 225/1. Hairless perennial herbs with rhizomes. Stamens 6 or fewer, opposite the perianth-segments and attached to their bases. Capsule with numerous seeds. *Cosmopolitan.*

2. Luzula de Candolle. 30/2. Hairy perennial herbs with rhizomes and stolons. Leaves flat; leaf-sheath closed. Flowers in cymes, sometimes in dense heads. Capsule with 3 seeds. *North temperate areas.* Figures 10, p. 58 & 11, p. 60.

14. BROMELIACEAE

Perennial herbs, rarely shrubby; terrestrial, growing on rocks or epiphytic. Roots usually present, but acting as holdfasts in all but terrestrial species. Leaves usually in rosettes, sheathing at the base, simple, entire or variously toothed or spiny, usually bearing peltate scales (modified hairs). Inflorescence terminal or lateral, stalkless or borne on a scape which may bear primary bracts, simple or compound, composed of racemes, spikes, panicles, heads or solitary flowers, usually with brightly coloured floral bracts. Sepals 3, free or united. Petals 3, free or united, sometimes bearing appendages (basal scales) near the base. Stamens 6 in two whorls. Ovary superior to inferior. Fruit a capsule or berry.

A large and mostly tropical family of 46 genera and 2100 species. All but one of the species (a *Pitcairnia* found in West Africa) are natives of the New World. Most require glasshouse protection in northern Europe.

1a. Stem absent or short and not immediately visible	**2. Fascicularia**
b. Stem well-developed and evident	**1. Puya**

1. Puya Molina. 170/few. Perennial, terrestrial herbs, usually with stems. Leaves very leathery, in dense rosettes, with distinct sheaths and spiny-margined blades which are not narrowed towards the base. Inflorescence usually a panicle. Sepals much shorter than the petals. Petals with distinct claws and usually broad blades, the blades coiled together spirally after flowering. Stamens usually somewhat shorter than the petals. Ovary superior or almost entirely so. Fruit a capsule. *South America.*

57

Figure 10. Diagnostic features of Juncaceae and Cyperaceae. 1, Inflorescence of *Luzula*. 2, Flower of *Luzula*. 3, Flower of *Scirpus* (a, perianth-bristles). 4, Inflorescence of *Cyperus*. 5, Spikelet of *Cyperus*. 6, Flower of *Cyperus* (b, glume). 7, Inflorescence of *Carex* (c, male spike; d, female spike). 8, 9, portions of stem and leaf-base of *Carex* (e, leaf-sheath; f, ligule). 10, Male flower of *Carex* (g, glume). 11, Female flower of *Carex*.

2. Fascicularia Mez. 9/2. Terrestrial, stemless or shortly stemmed epiphytes. Leaves forming very dense rosettes, but without a central cup; blades linear, rigid, strongly spiny-margined. Inflorescence simple, dense, sunk in the centre of the rosette. Flowers stalkless or shortly stalked. Sepals free, keeled, scaly, especially toward the apex. Petals free, obtuse, fleshy, blue, with 2 small scales near the base. Stamens not projecting from the corolla. Ovary completely inferior. *Chile.*

Nine species, 2 cultivated, not reliably hardy.

15. COMMELINACEAE

Perennial or annual herbs, often fleshy. Stems often swollen at the nodes. Leaves simple, entire, with tubular sheathing bases, margins of young leaves inrolled. Flowers radially or bilaterally symmetric, bisexual, or rarely with bisexual and unisexual flowers on the same plant, in terminal or axillary 1- to many-flowered, 1-sided, or coiled cymes (cincinni), often aggregated or fused in pairs. Sepals 3. Petals 3, rarely 1 of them much reduced or absent. Stamens 6 or fewer, free or borne on the petals, all similar or variously differentiated, or some reduced to staminodes. Ovary superior, 3- or rarely 2-celled with several to 2, or rarely 1, ovules per cell. Fruit a capsule, dehiscent or sometimes indehiscent, dry or fleshy; seeds few, with endosperm, position of embryo marked by a thickening.

A mainly tropical family of about 35 genera and 600 species, with species of horticultural value in *Tradescantia* and a few other genera. The flowers, which open in strict succession along the cymes (cincinni), last a few hours only, the petals then deliquescing into a mass with the stamens and ovary which makes dissection and study well-nigh impracticable. The cymes are also often so condensed and umbel-like as to make their construction difficult to discern. The diagnostic feature of *Tradescantia* and allied genera, however, is the fusion of the cymes in pairs to make a 2-sided unit, recognisable in *Tradescantia* itself by the accompanying paired bracts.

1a. Flowers bilaterally symmetric, the stamens of 2 or more types
1. Commelina
  b. Flowers radially symmetric, stamens all similar   **2. Tradescantia**

1. Commelina Linnaeus. 200/few. Usually perennial, often tuberous-rooted. Flowers variously coloured, bisexual or some unisexual, bilaterally symmetric, in single or paired cymes more or less enclosed by a folded, keeled, spathe-like bract, a few species also with subterranean, cleistogamous flowers. Sepals 3, free, unequal, the outermost hooded; petals 3, free, usually the upper 2 clawed, the lower reduced, more rarely all equal; fertile stamens 3, together on upper side of the flower, often of 2 kinds, filaments naked; staminodes 3 or 2, with sterile anthers, filaments naked. Ovary 3 or 2-celled, with 1--2 ovules per cell. *Tropics.*

One species of **Tinantia** Scheidweiler, with bracts not folded, may be grown in very mild areas.

2. Tradescantia Linnaeus. 69/7. Perennials, habit diverse, roots fibrous or tuberous. Inflorescences mostly terminal, the cymes fused in pairs and subtended by paired, boat-shaped bracts similar to or differentiated from the leaves. Flowers bisexual, radially symmetric. Sepals 3, usually free. Petals 3, usually free or sometimes united at base, blue-violet, purple, pink or white. Stamens 6, all fertile, similar and more or less

Figure 11. Juncaceae. *Luzula campestris.* 1, Whole plant. 2, Single flower at late bud stage. 3, Open flower. 4, Longitudinal section of part of flower. 5, Stigma. 6, Longitudinal section of ovary. 7, Transverse section of ovary. 8, Opening fruit.

60

equal; anthers versatile with broad connectives. Ovary 3-celled with 2 ovules (rarely 1) in each cell; stigma knob-like. Capsule dehiscent. *Northern South America.*

16. GRAMINEAE (*Poaceae*)

Annual or perennial herbs, rarely woody, often with rhizomes or stolons. Flowering stems (culms) usually cylindric, with hollow internodes and solid nodes, rarely solid throughout, with a growing point at the base of each internode. Leaves alternate, 2-ranked, consisting of sheath, ligule and blade. Sheaths surrounding the stem, with free, overlapping or united margins, sometimes with auricles at the mouth. Ligule situated at the junction of sheath and blade, usually membranous, sometimes a row of hairs or rarely absent. Blade linear to thread-like, rarely lanceolate to ovate, usually with more or less prominent, parallel ribs, generally constricted at the junction with the sheath, rarely shortly stalked. Flowers usually bisexual, consisting of 1--3 (rarely to 6 or more) stamens, and an ovary with 2 (rarely 3) styles; 2 (rarely 0, 3 or 6) small, translucent scales (lodicules) present near base of filaments; each flower enclosed in 2 bracts (a lower lemma and an upper palea). Flowers 1--many, inserted alternately on 2 sides of slender, jointed axis (rhachilla) and subtended by usually 2 (rarely 0, 1 or 3) bracts (glumes), the whole forming a spikelet. Glumes sometimes represented by 1 or more bristles forming an involucre. Lemma often with a bristle (awn); palea usually membranous or transparent, 2-keeled, sometimes very small or absent. Lemma with a thickened, sometimes elongate and pointed base (callus). Ovary 1-celled, superior; ovule 1, attached on the inner side of the cell, to a point or line visible in fruit as the hilum. Fruit a grain (caryopsis) or rarely with a free, membranous pericarp. See figures 12, p. 62 & 13, p. 74.

A large cosmopolitan family of about 660 genera and 9000 species. Over 500 species are mentioned in horticultural literature, but these are mainly employed in lawns, sports grounds and landscaping, or offered as novelties of special interest to flower arrangers. Relatively few of them can fairly lay claim to a place in the garden as ornamentals. Many are important cereal and fodder crops.

The bamboos (woody grasses) are placed here at the end of the family, with a separate general description and key to the genera.

Key to groups

    1a. Culms not woody   *Group 1* (Herbaceous grasses)

      b. Culms woody   *Group 2* (Bamboos, p. 78)

Group 1 (Herbaceous grasses)

    1a. Spikelets unisexual, male spikelets different in appearance from female
        spikelets   2

      b. All or some of the spikelets bisexual, rarely unisexual but then male
        spikelets similar to female spikelets: 3

    2a. Inflorescences bisexual, female below, male above, projecting
                                          14. Tripsacum

      b. Inflorescences unisexual, the female enclosed   **15. Zea**

    3a. Spikelets 2-flowered, falling entire at maturity, with the upper flower
        bisexual and the lower male or sterile; spikelets usually flattened from top
        to bottom   4

      b. Spikelets with 1--many flowers, breaking up at maturity above the more or
        less persistent glumes, or if falling entire then not with 2 flowers with the

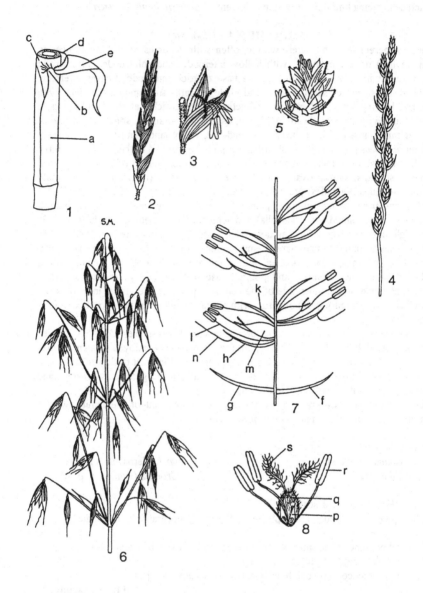

Figure 12. Gramineae. 1, Portion of a culm of a bamboo (a, culm-sheath; b, auricle, c, bristle; d, ligule; e, blade of leaf). 2, Spikelet of a bamboo. 3, Flower of a bamboo. 4, Spike of stalkless spikelets. 5, A stalkless spikelet. 6, A panicle of stalked spikelets. 7, Exploded diagram of a spikelet (f, lower glume; g, upper glume; h, lemma; k, palea; l, stamen; m, ovary; n, awn). 8, Flower of *Avena sativa* (p, lodicule; q, ovary; r, anther; s, stigma).

upper flower bisexual and the lower male or sterile; spikelets usually
    flattened from side to side or terete   15

4a. Upper lemma firmer in texture than the lower glume; spikelets seldom
    paired   5

  b. Upper lemma thinner in texture than the lower glume; spikelets paired   10

5a. Spikelets subtended by an involucre of 1 or more bristles   6

  b. Spikelets not subtended by bristles   7

6a. Bristles retained on the branches after the spikelets have fallen
                                   26. Setaria

  b. Bristles falling with the spikelet   **27. Pennisetum**

7a. Inflorescence a panicle   **25. Panicum**

  b. Inflorescence composed of racemes   8

8a. Spikelet with a bead-like swelling at its base   **23. Eriochloa**

  b. Spikelet passing smoothly into its stalk without a bead-like swelling   9

9a. Upper lemma cartilaginous with flat, thin margins covering most of the
    palea and often overlapping   **22. Digitaria**

  b. Upper lemma leathery to hard and somewhat brittle with narrow, inrolled
    margins clasping only the edges of the palea   **24. Echinochloa**

10a. Spikelets in each pair alike   11

  b. Spikelets in each pair different   12

11a. Raceme axis tough   **17. Miscanthus**

  b. Raceme axis fragile   **16. Spodiopogon**

12a. Internodes and stalk of the raceme with a translucent median line
                                     20. Bothriochloa

  b. Internodes and stalk of the raceme not as above   13

13a. Racemes borne in a panicle   14

  b. Racemes paired   **21. Andropogon**

14a. Spikelets flattened from side to side   **19. Chrysopogon**

  b. Spikelets flattened from top to bottom   **18. Sorghastrum**

15a. Inflorescence composed of spikes or racemes   16

  b. Inflorescence a panicle   26

16a. Spikes or racemes several in a group   17

  b. Spikes or racemes single   18

17a. Spikelets falling entire at maturity   **11. Spartina**

  b. Spikelets breaking up at maturity   **12. Bouteloua**

18a. Spikelets 1 at each node of the spike   19

  b. Spikelets 2--4 at each node of the spike   22

19a. Inflorescence axis tough   20

  b. Inflorescence axis fragile   21

20a. Spikelets 1-flowered   **51. Mibora**

  b. Spikelets 2- or more-flowered   **28. Brachypodium**

21a. Spikelets with the side of the lemma towards the axis   **29. Aegilops**

  b. Spikelets with the back of the lemma towards the axis   **42. Gaudinia**

22a. Spikelets with 1 flower   **34. Hordeum**

  b. Spikelets with 2 or more flowers   23

23a. Inflorescence axis fragile   **31. Sitanion**

  b. Inflorescence axis tough   24

24a. Glumes absent or reduced to 2 short bristles   **32. Hystrix**

b. Glumes well developed   25
25a. Spikelets with 2--6 fertile flowers   **30. Elymus**
    b. Spikelets with 1 fertile and 1 sterile flower   **33. Taeniatherum**
26a. Spikelets with 1 flower   27
    b. Spikelets with 2 or more flowers, or of 2 kinds   38
27a. Glumes feathery; panicle spherical to oblong   **56. Lagurus**
    b. Glumes hairless to shortly hairy   28
28a. Lemma surrounded by fine white hairs from the base, the hairs one-third
    as long as to much longer than the lemma   29
    b. Lemma hairless at base or with only a tuft of very short hairs there   30
29a. Lemma about half the length of the glumes   **54. Calamagrostis**
    b. Lemmas as long as the glumes   **10. Muhlenbergia**
30a. Spikelets falling entire   31
    b. Spikelets breaking up at maturity   33
31a. Glumes conspicuously awned   **53. Polypogon**
    b. Glumes not or shortly awned   32
32a. Lemma awned   **49. Alopecurus**
    b. Lemma awnless   **68. Phaenosperma**
33a. Inflorescence a cylindric, spike-like panicle   **50. Phleum**
    b. Inflorescence an open or contracted panicle   34
34a. Lemma not strongly hardened and shiny when mature   35
    b. Lemma strongly hardened and shiny when mature   36
35a. Lemma awnless or with a sharply bent awn   **52. Agrostis**
    b. Lemma with a long straight awn   **55. Apera**
36a. Lemma awnless   **47. Milium**
    b. Lemma awned   37
37a. Lemma hairless; awn often falling off   **67. Piptatherum**
    b. Lemma hairy; awn persistent   **66. Stipa**
38a. Spikelets of 2 different kinds, mostly in groups of 3--7 with 1 central
    bisexual spikelet and 2--6 male or sterile spikelets surrounding it   39
    b. Spikelets not obviously of different kinds   41
39a. Male or sterile spikelets persistent   **59. Cynosurus**
    b. Sterile and fertile spikelets falling in groups   40
40a. Lemma of fertile spikelets awned   **58. Lamarckia**
    b. Lemma of fertile spikelets awnless   **48. Phalaris**
41a. Inflorescence a dense, ovoid to spherical head of stalkless spikelets
    57. Echinaria
    b. Inflorescence not a dense head of stalkless spikelets   42
42a. Ligule a row of hairs, sometimes surmounting a short membrane   43
    b. Ligule membranous   50
43a. Lowest flower male or sterile   44
    b. Lowest flower bisexual   45
44a. Inflorescence feathery   **6. Phragmites**
    b. Inflorescence not feathery   **2. Uniola**
45a. Lemmas, at least of the female flowers, hairy   46
    b. Lemmas hairless   49
46a. Awn of lemma short, scarcely exceeding the teeth   **5. Arundo**
    b. Awn of lemma conspicuous   47

68a. Ovary crowned by a fleshy hairy cap, the styles arising from beneath it
35. Bromus

   b. Ovary hairless or hairy, the styles arising from its tip   **63. Festuca**

69a. Spikelets with 1--3 flowers   **65. Melica**

   b. Spikelets with 3--20 flowers   **64. Glyceria**

1. Ehrharta Thunberg. 30/1. Annuals or perennials. Leaves flat or rolled; ligule membranous. Inflorescence a panicle or raceme. Spikelets stalked, flattened from side to side, with 3 flowers, the 2 lowest reduced to the lemma, the uppermost complete, glumes persistent, membranous, shorter than the flowers: lower lemmas tough, often hairy, awned or awnless; fertile lemma smaller and thinner. *South Africa; a few naturalised in North America and Australia.*

2. Uniola Linnaeus. 4/1. Perennials, tufted or rhizomatous, with slender to stout stems. Leaf-blades linear to narrowly lanceolate, flat or rolled. Inflorescence a loose or contracted and dense panicle. Spikelets with few to many flowers, strongly compressed from side to side, often large, falling entire. Glumes persistent, keeled, rigid. Lemmas compressed and keeled, firm to tough, many-veined, the lower 1--6 sterile. *North & South America.*

3. Chasmanthium Link. 5/1. Tufted to loosely rhizomatous perennials of diffuse habit. Leaf-blades linear to narrowly lanceolate, flat. Inflorescence an open or contracted panicle. Spikelets with few to many flowers, strongly compressed from side to side. Glumes shorter than lemmas, 3--7-veined, acute to acuminate. Lemmas 5--15-veined, the keels toothed or ciliate, acuminate, entire or bifid, the lower 1--4 sterile. *North America.*

4. Cortaderia Stapf. 15/3. Large tufted perennials with separate male and female plants in some species and in others female and bisexual plants. Leaves flat; ligule a line of hairs. Inflorescence a large panicle. Spikelets compressed from side to side with 2--7 (rarely 1) flowers. Glumes unequal, membranous, 1-veined. Lemma membranous, with shaggy hairs, 3-veined, with a terminal awn. Rachilla with shaggy hairs, breaking above the glumes and above each flower. *Temperate South America, New Zealand.*

5. Arundo Linnaeus. 3/1. Large rhizomatous perennials. Leaves flat. Inflorescence a loose panicle. Spikelets compressed from side to side, with few, mostly bisexual flowers. Glumes almost equal, membranous, equalling the flowers, with 3--9 veins, persistent. Lemma with 3--9 veins and with long soft hairs on the lower half of the back, not keeled. Rachilla hairless. *Mediterranean area to China and Japan.*

6. Phragmites Adanson. 4/1. Rhizomatous perennials, with erect, stout or robust stems. Leaf-blades linear to narrowly lanceolate, flat, deciduous. Panicle large, dense, profusely branched, silky-hairy. Spikelets with 3--11 flowers, the lowest male or sterile, the middle bisexual, the uppermost more or less reduced; rachilla bearded above with long silky hairs; glumes with 3--5 veins: lemmas narrow, rounded or slightly keeled on the back, hairless; lowest lemma much longer than the glumes, with 3--7 veins, more or less persistent; fertile lemmas acuminate with 1--3 veins. *Cosmopolitan.*

7. Hakonechloa Makino. 1/1. Perennial with long scaly rhizomes, moderately slender stems, linear-lanceolate, finely pointed leaf-blades and minutely ciliate ligules. Inflorescence a loose panicle. Spikelets stalked, slightly compressed from side to side, with 3--9 flowers; joints of the axis bearded at the tip; glumes persistent, pointed, membranous, with 3--9 veins, unequal, the lower shorter; lemmas much exceeding the glumes, overlapping, linear-lanceolate in side view, with a short straight awn from the tip, rounded on the back, membranous, with 3 veins, narrowed at the base into a bearded stalk. *Japan.*

8. Chionochloa Zotov. 20/1. Coarse perennials, forming tussocks from 20 cm to 2.5 m tall. Leaves mostly deeply grooved. Inflorescence a panicle. Spikelets with several flowers. Glumes shorter than spikelets. Lemmas mostly with 7--9 veins, distinctly lobed, the awn a conspicuous prolongation of the middle vein, mostly bent or twisted at the base. *New Zealand, southern Australia.*

9. Molinia Schrank. 2/1. Tufted perennials. Leaves flat, ligule a row of hairs. Inflorescence a panicle. Spikelets compressed from side to side with 1--4 flowers. Glumes almost equal, membranous, shorter than the flowers, the lower with 1 vein or vein absent, the upper with 1--3 veins. Lemma membranous, with 3 veins, rounded on the back. Palea about equalling lemma. *Europe, north & southwest Asia.*

10. Muhlenbergia Schreber. 130/1. Annuals or perennials, with mostly slender stems. Leaf-blades thread-like or linear, flat or rolled; ligules membranous. Inflorescence a loose or contracted and dense panicle. Spikelets stalked, small, with 1 flower, awned or awnless. Glumes persistent, thin, with 1--3 veins, shorter than or as long as the lemma. Lemma narrow, membranous to firm, usually with 3 veins, awned or awnless, with the awn straight or flexuous. *Warm temperate & tropical America.*

11. Spartina Schreber. 14/1. Rhizomatous perennials. Leaves flat or rolled. Inflorescence a number of spikes arranged in a raceme. Spikelets strongly compressed from side to side, in 2 rows, closely adpressed to one face of the triangular spike-axis, each with 1 or rarely 2 flowers. Glumes unequal, papery, the lower 1-veined. the upper with 1--3 (rarely to 6) veins, about as long as lemma. Lemma 1--6-veined, leathery, with a wide membranous margin. Palea slightly shorter than lemma. *Western & southern Europe, North & South Africa, America, South Atlantic Islands.*

12. Bouteloua Lagasca. 45/2. Annuals or perennials, tufted or with stolons, with stiff, slender stems and narrow, flat or rolled leaves. Inflorescence of 1--many, dense, 1-sided spikes or racemes borne on a common axis. Spikelets stalkless or nearly so, in 2 rows on one side of the axis, with 1 bisexual flower, and with 1 or more rudimentary flowers above it; lemmas 3-veined, the fertile one short-awned or mucronate, the rudimentary usually 3-awned. *Warm temperate & tropical America.*

One species of **Chloris** Swartz, which has all its racemes arising at the top of the stem, may be cultivated in very mild areas.

13. Eragrostis Wolf. 300/4. Annuals or perennials. Leaves flat or rolled, narrow; ligule a ring of hairs (rarely membranous). Inflorescence a panicle. Spikelets compressed from side to side, with 2--many flowers. Glumes unequal to almost equal, membranous, shorter than the flowers, 1-veined. Lemma 3-veined, membranous, usually awnless, falling separately or with the palea and grain. Palea transparent. 2-keeled, often persistent. Stamens 3 or 2. *Tropics & subtropics.*

14. Tripsacum Linnaeus. 13/1. Perennial, often broad-leaved and robust, usually rhizomatous. Inflorescences terminal and axillary, of palmately arranged racemes, each raceme female below, with fragile, swollen internodes, male above, with tough, narrow internodes. Female spikelets single without a trace of pairing, deeply sunk in the internodes, the callus transversely truncate with central ridge or peg; lower glume hard and somewhat brittle, smooth, slightly winged at tip; lower flower sterile, without palea. Male spikelets paired, both stalkless or 1 raised on a free stalk; glumes papery; both flowers male. *Central America.*

15. Zea Linnaeus. 4/1. Annual. Leaves flat. Male inflorescence a terminal panicle of spike-like racemes; female inflorescence axillary, of numerous spikelets arranged in longitudinal rows on a thickened axis, the whole enclosed in leaf-sheaths. Male spikelets in pairs, 1 almost stalkless, the other stalked; flowers 2; glumes equal, membranous; lemma and palea transparent; stamens 3, lodicules 2. Female spikelets with 2 flowers, the lower sterile; glumes wider than long, fleshy below, transparent above; lemmas short, transparent; lodicules absent; style long, shortly bifid at apex, hairy throughout its length. *Central America.*

One species of **Coix** Linnaeus, with the female inflorescence enclosed in a hard, shiny jacket, may be grown in very mild areas.

16. Spodiopogon Trinius. 5/1. Perennial or annual, stems slender to stout; ligules membranous. Leaf-blades flat, linear to narrowly lanceolate. Inflorescence a loose or contracted panicle of spike-like racemes. Spikelets in pairs, 1 stalkless, the other stalked, in few-jointed, fragile racemes, all alike, falling entire at maturity, with 2 flowers, the lower flower male and the upper bisexual; glumes firm, hairy: lemmas transparent, enclosed by the glumes, the upper awned. *Temperate & tropical Asia.*

17. Miscanthus Andersson. 12/2. Tufted or rhizomatous, usually with tall, stout, reed-like stems, short membranous ligules and long, narrow, flat or folded leaf-blades. Inflorescence an oblong-elliptic or fan-shaped panicle of slender, hairy, spike-like racemes, the latter with a continuous axis. Spikelets surrounded with long hairs from the base, falling entire at maturity, paired, those of each pair unequally stalked, with 2 flowers but with only the upper flower bisexual; glumes firm, awnless, enclosing the lemmas, the latter delicate, transparent, and the upper usually awned. *Central, eastern and southeastern Asia, Polynesia.*

One species of **Imperata** Cirillo, which has a spike-like, cylindric inflorescence, may be grown in very mild areas.

18. Sorghastrum Nash. 20/1. Annual or perennial; stems tufted, slender to stout, erect; ligules scarious. Leaf-blades narrow, flat or rolled. Inflorescence a loose or dense, narrow panicle, its branches terminating in few-jointed racemes. Spikelets solitary and stalkless at each raceme-node, accompanied by the hairy stalk of a suppressed spikelet, falling entire at maturity, with 2 flowers, the upper flower alone fertile; glumes leathery, the lower rounded or flattened on the back, the upper boat-shaped; lemmas transparent, enclosed by the glumes, the upper with a bent and twisted awn. *Tropical Africa, warm temperate & tropical America.*

19. Chrysopogon Trinius. 24/1. Perennial, mostly tufted. Leaf-blades linear, often harsh and glaucous. Inflorescence a terminal panicle with whorls of slender, persistent branches bearing terminal racemes, these reduced to a triad of 1 stalkless and 2 stalked spikelets. Stalkless spikelet compressed from side to side. Lower glume rounded on back. Inner flower reduced to a transparent lemma; upper lemma 2-toothed or entire, with a sharply bent awn. *Tropical & warm temperate areas.*

20. Bothriochloa Kuntze. 30/2. Annuals or perennials. Inflorescence of 1--many fragile racemes bearing paired, dissimilar spikelets, 1 stalkless, the other stalked; internodes of raceme-axis and spikelet-stalks with a translucent median line. Stalkless spikelet with 2 flowers, flattened from top to bottom; lower flower reduced to a transparent lemma; upper flower bisexual with entire lemma passing into a hairless, sharply bent, awn. Stalked spikelet male or sterile, unawned. *Warm temperate areas.*

One species of **Saccharum** Linnaeus, with the lower glume membranous to almost leathery, its veins not raised, may be grown in very mild areas.

21. Andropogon Linnaeus. 120/1. Annuals or perennials. Inflorescence of paired or palmately arranged, fragile racemes bearing paired, dissimilar spikelets, 1 stalked, the other stalkless. Stalkless spikelet with 2 flowers, the callus inserted in the cup-like apex of the internode; lower glume with 2 keels; lower flower reduced to a transparent lemma; upper flower bisexual with a 2-toothed lemma passing into a hairless, sharply bent awn. Stalked spikelet male or sterile, awnless or with a very fine hair at the apex. *Warm temperate & tropical areas.*

22. Digitaria Haller. 250/1. Annuals or perennials. Leaves flat; ligule membranous. Inflorescence of 1-sided racemes arranged palmately or upon a short central axis and bearing the spikelets in adpressed groups of 1--5 or more. Spikelets flattened from top to bottom, with 2 flowers; lower glume small or absent; upper glume as long as or much shorter than the spikelet; lower flower sterile, represented only by a lemma which is usually as long as the spikelet; upper flower bisexual, the lemma cartilaginous, with its transparent margins enfolding and concealing most of the palea. *Tropics & warm temperate areas.*

23. Eriochloa Kunth. 30/1. Annuals or perennials. Ligule a line of hairs. Inflorescence of 1-sided racemes arranged along a central axis. Spikelets flattened from top to bottom, with 2 flowers; lower glume vestigial, attached to the swollen, bead-like, lowest rachilla-internode; upper glume almost as long as the spikelet; lower flower male or empty, the lemma resembling the upper glume; upper flower bisexual, the lemma leathery, clasping only the margins of the palea. *Tropics.*

24. Echinochloa Palisot de Beauvois. 25/1. Annuals or perennials. Ligule absent or a line of hairs. Inflorescence of 1-sided racemes arranged along a central axis. Spikelets in several rows, convex beneath, flattened above, toothed or awned, with 2 flowers. Glumes unequal, the lower shorter than, the upper as long as the spikelet. Lower flower male or sterile, the lemma resembling the upper glume; upper flower bisexual, the lemma hard and rather brittle, smooth, clasping only the margins of the palea. *Warm temperate areas.*

One species of **Paspalum** Linnaeus, with spikelets with a single glume, may be grown in very mild areas.

25. Panicum Linnaeus. 400/5. Annuals or perennials. Inflorescence a panicle. Spikelets more or less flattened from top to bottom, unawned, with 2 flowers; lower glume shorter than, upper glume as long as the spikelet; lower flower male or sterile, its lemma resembling the upper glume, with or without a palea; upper flower bisexual, the lemma hard and rather brittle, clasping only the margins of the palea. *Widespread.*

A species of **Rhynchelytrum** Nees, which has the upper lemma cartilaginous and the upper flower flattened from side to side, may be occasionally grown in very mild areas.

26. Setaria Palisot de Beauvois. 100/5. Annuals or perennials. Inflorescence a panicle, usually spike-like, the spikelets subtended by bristles which persist on the axis after the spikelets fall. Spikelets oblong to ovate; glumes unequal; lower flower male or sterile, as long as the spikelet; upper flower bisexual, its lemma hard and rather brittle, boat-shaped, often with a wrinkled surface. *Tropics, warm temperate areas.*

27. Pennisetum Richard. 120/5. Annuals or perennials. Inflorescence spike-like, each spikelet or cluster of spikelets enclosed by an involucre of slender bristles which are free throughout and fall with the spikelets. Spikelets lanceolate to oblong with 2 flowers; lower glume often minute, upper glume very small to as long as the spikelet; lower flower male or sterile, its lemma as long as the spikelet or much reduced; upper flower bisexual, its lemma membranous to thinly leathery. *Tropics, warm temperate areas.*

One species of **Cenchrus** Linnaeus, which has the involucral bristles flattened into a little disc below, may be grown in very mild areas.

28. Brachypodium Palisot de Beauvois. 16/1. Perennials or annuals, rarely woody below, often with extensively branched rhizomes. Inflorescence a raceme or alternate, shortly stalked spikelets inserted in 2 ranks, with the backs of the lemmas towards the axis. Spikelets 1 (rarely to 3) at each node, usually with numerous flowers. Glumes unequal, shorter than the lowest flower. Glumes and lemmas acuminate, mucronate or with a straight or flexuous, apical awn. Palea equalling or a little shorter than the lemma, notched or truncate, the keels ciliate or rough with short, stiff hairs. Ovary hairy at the apex. Grain narrowly elliptic to oblanceolate; hilum linear. *Temperate parts of the northern hemisphere.*

Species of the genus **Lolium** Linnaeus are frequently grown as lawn, turf or forage grasses. They will key out to *Brachypodium* in the key to genera (p. 63), but can be distinguished by their spikelets, in which the lower glume is present, the upper absent.

29. Aegilops Linnaeus. 20/1. Annuals. Leaves usually flat. Inflorescence a spike. Spikelets solitary at the nodes of the axis, all bisexual or the upper sterile and the lower vestigial. Flowers 2--8, the upper usually male or vestigial. Glumes equal, leathery, truncate, often with 1 or more teeth or awns, usually rounded on the back. Lemma thin below, leathery, and strongly veined towards the toothed or awned apex. Palea with 2 keels. *Mediterranean area.*

30. Elymus Linnaeus. 100/2. Tufted or rhizomatous perennials. Leaves flat or more or less rolled; ligule short, membranous. Inflorescence a spike, axis usually tough. Spikelets solitary or in groups of 2 or 3 at each node, almost stalkless, usually overlapping, with 2--11 flowers; rachilla separating above the glumes and beneath each

70

flower or sometimes the spikelets falling entire at maturity. Glumes with 1--11 veins, awnless or with short awns. Lemma lanceolate, with 5 veins. Palea 2-keeled. *North temperate areas.*

31. Sitanion Rafinesque. 2/1. Tufted perennials with slender stems; ligules membranous. Leaf-blades flat or rolled. Inflorescence a very bristly spike. Spikelets with 2--few flowers usually in pairs, alternating on opposite sides at each node of the fragile main axis of the spike, the axis breaking at the base of each joint at maturity; glumes bristle-like, with 1 or 2 veins, extending into 1--3 or more awns; lemmas rounded on the back, 5-veined, firm, slightly 2-toothed at the apex, the central vein extending into a long slender spreading awn, sometimes with 1 or more lateral awns. *Western North America.*

32. Hystrix Moench. 6/1. Erect perennials, with flat leaf-blades. Inflorescence a loose, bristly spike. Spikelets with 2--4 flowers, 1--4 at each node of a continuous, flattened axis, horizontally spreading or ascending at maturity. Glumes reduced to short or minute awns, the first usually rudimentary, both often absent from upper spikelets. Lemmas convex, rigid, tapering into long awns, with 5 veins, the veins obscure except towards the tip. Palea about as long as the body of the lemma. *North America, north India, China, New Zealand.*

33. Taeniatherum Nevski. 2/1. Annuals. Leaves flat or rolled. Inflorescence a dense spike with spikelets in 2 ranks, separating at the nodes below the spikelets at maturity. Spikelets arranged in pairs at the nodes of the axis; flowers 2, the lower bisexual, the upper rudimentary and sterile. Glumes joined at the base, narrowly subulate, rigid. Lemmas lanceolate, each with a flattened callus and a long, flexuous, rough awn. *South Europe, Middle East.*

34. Hordeum Linnaeus. 25/2. Annuals or perennials. Leaves usually flat. Inflorescence a compressed, linear to oblong spike. Spikelets 3 at each node, dispersed together at maturity, the triplets arranged in 2 longitudinal rows, the fertile flowers in 2, 4 or 6 longitudinal rows, each triplet with a central bisexual spikelet and 2 lateral bisexual, male or sterile spikelets. Flowers 1 per spikelet, rarely 2. Glumes linear-subulate to lanceolate and awned, free to the base. Lemma 5-veined, ovate. Palea narrowly ovate, keeled. Rachilla usually prolonged in central spikelets. *Temperate northern hemisphere, South America.*

35. Bromus Linnaeus. 100 or more/6. Annuals, biennials or perennials. Leaves flat or somewhat rolled; ligule membranous, often jagged. Inflorescence a panicle. Spikelets with 1--many flowers; glumes with 1--9 veins, persistent at maturity, the upper usually the larger; lemma with few to many veins, awnless or with an almost terminal awn. Ovary crowned by a fleshy, hairy cap, the styles emerging from beneath it. Grain normally adherent to the palea and tightly enclosed within the lemma; hilum narrow, elongate. *North temperate areas, South America, South Africa and on mountains in the tropics.*

36. Aira Linnaeus. 9/2. Annuals. Leaves often rolled. Inflorescence a panicle. Spikelets compressed from side to side, with 2 bisexual flowers, the upper stalkless or almost so. Glumes about equal, membranous, equalling or longer than the flowers, with 1--3 veins. Lemma with 5 veins in the basal half, bifid at apex, with an awn arising from below the middle on the back; sometimes the lower or both flowers awnless. Palea shorter than lemma. Rhachilla not prolonged. Grain more or less fusiform, longitudinally furrowed, hairless. *Eurasia.*

37. Deschampsia Palisot de Beauvois. 40/2. Perennials, usually tufted. Leaves flat or bristle-like. Inflorescence a panicle. Spikelets compressed from side to side, with 2 (rarely 3) bisexual flowers. Glumes more or less equal, membranous, somewhat shorter than or equalling the flowers, with 1--3 veins. Lemma obscurely 5-veined, truncate and irregularly toothed at apex, with short hairs at the base; awn arising on the back, straight or sharply bent. Palea as long as the lemma, with 2 rough keels. Rachilla prolonged. *Temperate areas, on mountains in the tropics.*

38. Holcus Linnaeus. 8/1. Perennials or annuals. Leaves flat. Inflorescence a panicle. Spikelets compressed from side to side, with 2--3 flowers, the lower bisexual, the upper usually male. Glumes almost equal, membranous, longer than the flowers, strongly keeled, the lower with 1 vein, the upper with 3 veins. Lemma obscurely 3--5-veined, leathery, shiny, the upper or both usually with an awn on the back. Palea membranous, slightly shorter than the lemma. Rachilla shortly prolonged. *Europe, temperate Asia, North & South Africa.*

39. Avena Linnaeus. 15/3. Annuals. Leaves flat; ligule membranous. Inflorescence a panicle. Spikelets large, with 2--5 (rarely 1, or as many as 7) flowers. Glumes lanceolate, acuminate, usually about equal, thinly papery, with a shiny scarious margin and several veins. Lemma usually leathery, 2-toothed or with 2 fine hairs (rarely almost entire), generally with an awn on the back; awn usually with a thick, twisted lower part, and a thinner, straight upper part, usually sharply bent. Palea tough, shorter than the lemma, with translucent margin and 2 ciliate keels. *Temperate areas.* Figure 12, p. 62.

40. Helictotrichon Besser. 60/2. Tufted perennials. Leaves distinctly ribbed above; ligule of stem-leaves usually truncate or toothed; sheaths of basal leaves usually open to the base. Inflorescence a compound panicle. Spikelets usually numerous, erect or spreading, slightly compressed from side to side, with 2--4 fertile and 1 (or rarely 2) sterile flowers; fertile flowers usually awned. Glumes lanceolate, unequal, membranous, acute, the lower with 1 vein, the upper with 3 veins. Lemma lanceolate, with 5--7 veins, hairless, slightly 2-toothed; awn sharply bent, arising at about the middle of the lemma, more or less terete in the lower part. Palea more or less bifid at apex, shortly ciliate on keels. Ovary densely hairy. *North temperate areas, South Africa; on mountains in the tropics.*

41. Arrhenatherum Palisot de Beauvois. 6/1. Perennials; basal internodes often swollen and more or less spherical. Leaves flat, rolled when young, not strongly ribbed above. Inflorescence a panicle. Spikelets slightly compressed from side to side, usually with 3 flowers, 2 fertile and 1 rudimentary. Lower flower male, with a sharply bent awn arising from the lower third of the back of the lemma, the upper female or bisexual, usually with a short, slender, straight bristle, rarely both flowers bisexual and awned. Glumes unequal, translucent, the lower with 1--3 veins, the upper about as long as the flowers, with 3 (rarely 5) veins. Lemma with 5--9 veins. Palea shorter than lemma, shortly 2-toothed at apex, ciliate on keels. Lodicules 2. Stamens 3. Flowers falling together at maturity. Grain more or less terete, with hilum extending for half to two-thirds of its length. *Europe, North Africa, northern & western Asia.* Figure 13, p. 74.

42. Gaudinia Palisot de Beauvois. 3/1. Annuals or biennials. Leaves flat. Inflorescence a spike with spikelets in 2 ranks; axis fragile, breaking up above the insertion of the spikelet. Spikelets compressed from side to side, stalkless, more or less adpressed to the concave axis, with 3--11 flowers. Glumes unequal, shorter or equalling the spikelet, the lower with 3 (rarely 5), the upper with 4--7 (rarely up to 11) strong veins. Lemma obscurely 7--9-veined, leathery, often with an awn on the back. Palea shorter than lemma. *South Europe, Mediterranean area.*

43. Anthoxanthum Linnaeus. 15/1. Annuals or perennials, often smelling of coumarin (new-mown hay). Leaves flat. Inflorescence a dense panicle. Spikelets compressed from side to side, of 2 sterile flowers and a terminal, bisexual flower. Glumes membranous, very unequal, the lower with 1 vein, the upper with 3 veins. Lemma of sterile flowers membranous, toothed or with 2 oblong, obtuse lobes at apex, with 3 veins and an awn on the back. Lemma of fertile flower somewhat hardened, shorter than the sterile ones, with 5--7 veins. Stamens 2. Lodicules absent. *Europe, north Asia, North & South America.*

44. Hierochloe R. Brown. 15/2. Perennials, smelling of coumarin (new-mown hay). Leaves flat or more or less rolled, those on the flowering stem with short blades; ligule truncate, obtuse or acute. Inflorescence a panicle. Spikelets compressed from side to side, with 2 male flowers and 1 terminal, bisexual flower. Glumes membranous, almost equal, ovate, about as long as flowers, obscurely 3-veined. Lemmas of male flowers membranous, obscurely 3--5-veined, obtuse, mucronate or awned. Palea shorter than lemma, membranous, with 2 veins. Lemma of bisexual flower obscurely 5-veined, hard and shiny in fruit, hairy towards apex. Palea shorter than lemma, membranous, with 1 vein. Stamens 3 in male flowers, 2 in bisexual flower. *Temperate & arctic areas.*

45. Koeleria Persoon. 30/1. Tufted perennials. Leaves flat or rolled. Inflorescence a spike-like or loose panicle. Spikelets compressed from side to side, with 2--5 flowers (rarely only 1). Glumes equal or unequal, the upper longer and usually equalling the first flower, the lower about two-thirds as long as the first flower, distinctly keeled, with 1--5 veins. Lemma lanceolate to ovate-lanceolate, keeled, mostly longer than the glumes, usually obtuse, with or without an awn. Palea as long as or shorter than the lemma, 2-keeled, bifid at the apex. *Temperate areas, on mountains in Africa.*

46. Rostraria Trinius. 10/1. Annuals. Inflorescence a panicle. Spikelets laterally compressed, with 2--5 (rarely only 1, or as many as 10) flowers. Glumes keeled, cartilaginous at the base. Lemma bifid, usually prominently 5-veined, keeled, with a straight or slightly curved awn inserted near the apex. Palea 2-keeled, bifid. Rachilla and callus hairless or with short hairs. *Temperate Eurasia, North Africa.*

47. Milium Linnaeus. 6/1. Annuals or perennials. Leaves flat. Inflorescence a panicle. Spikelets slightly flattened from top to bottom, with 1 flower. Glumes equal, membranous, longer than the flower, with 3 veins. Lemma with 5 veins, leathery and

73

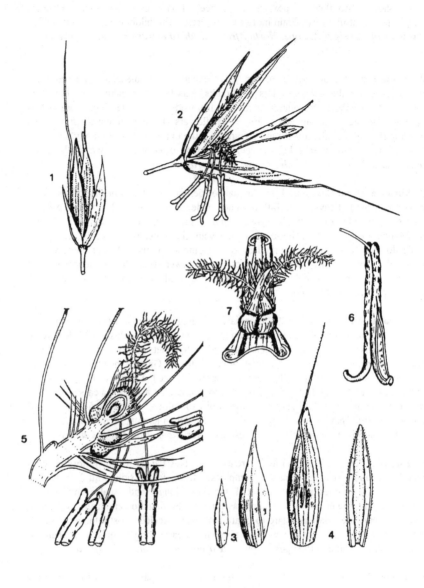

Figure 13. Gramineae. *Arrhenatherum elatius.* 1, Single spikelet. 2, Open spikelet. 3, Glumes. 4, Lemma and palea. 5, Longitudinal section of flowering spikelet. 6, Dehisced anther. 7, Flower with stamens removed showing styles and lodicules.

74

shiny in fruit. Palea as long as the lemma, leathery. *Temperate North America, Europe, Asia.*

48. Phalaris Linnaeus. 15/4. Annuals or perennials. Leaves flat. Inflorescence an ovoid or cylindric panicle, usually dense and unlobed. Spikelets strongly compressed from side to side, with 2 or 3 flowers, the lower 1 or 2 reduced to lemmas, the uppermost bisexual. Glumes almost equal, papery, longer than the flowers, with 3--5 veins, often winged on the keel. Lower lemmas small, linear to lanceolate; upper lemma with 5 veins, leathery. Palea leathery. *Temperate areas, mainly Mediterranean.*

49. Alopecurus Linnaeus. 25/1. Usually almost hairless annuals or perennials. Inflorescence a spike-like panicle. Spikelets compressed from side to side, with 1 flower. Glumes almost equal, often united below, with 3 veins. Lemma translucent, with 3 veins, usually awned from the back; margins often united below. Palea usually absent, or if present, small. Ovary hairless; styles usually united below. Rachilla breaking up below the glumes. *North temperate areas.*

50. Phleum Linnaeus. 15/1. Annuals or perennials. Leaves flat. Inflorescence a dense, spike-like, ovoid or cylindric panicle. Spikelets strongly compressed from side to side with 1 flower. Glumes membranous, almost equal, longer than the lemma, keeled, shortly awned, with 3 veins, with the margins overlapping along most of their length. Lemma membranous, truncate or obtuse, with 1--7 veins, unawned, hairless to densely ciliate. Palea equalling or almost equalling the lemma, ovate-lanceolate, obtuse, with 2 veins. Stamens 2--3. *Temperate areas.*

51. Mibora Adanson. 1/1. Annuals. Leaves flat. Inflorescence spike-like, 1-sided. Spikelets compressed from side to side, with 1 flower. Glumes more or less equal, membranous, longer than the flower, with 1 vein, persistent. Lemma with 5 veins, truncate, densely hairy, thinner than the glumes. Palea as long as lemma, hairy. *Europe, Mediterranean area.*

52. Agrostis Linnaeus. 120/1. Perennials or annuals. Leaves flat or bristle-like. Inflorescence a panicle. Spikelets compressed from side to side, with 1 flower. Glumes more or less equal to somewhat unequal, membranous, longer than the flower, usually with 1 vein. Lemma membranous or scarious, truncate, with 3--5 veins, often with the lateral veins slightly projecting from the margin, sometimes with a sharply bent awn on the back. Palea shorter than the lemma, often very small. Rachilla prolonged or not. Callus usually hairy. *Widespread.*

53. Polypogon Desfontaines. 10/1. Annuals or perennials. Leaves flat. Inflorescence a dense panicle. Spikelets somewhat compressed from side to side, with 1 flower. Glumes more or less equal, with 1 vein, papery, rough with small, stiff hairs, longer than flower, usually with an apical awn. Lemma with 5 veins, transparent, truncate. Palea transparent. Rachilla not prolonged. *Warm temperate areas.*

54. Calamagrostis Adanson. 250/2. Perennials. Leaves flat or rolled; ligule membranous. Spikelets narrow, with 1 flower, in dense to rather loose panicles, Lemma membranous, with 3--5 veins, shorter than the glumes, surrounded usually by long hairs from the base, with an awn arising from the back or apex. Rachilla separating above the glumes. *Temperate areas.*

55. Apera Adanson. 3/1. Annuals. Leaves flat or rolled. Inflorescence a panicle. Spikelets compressed from side to side, with 1 flower (rarely more). Glumes unequal, membranous, the lower with 1 vein, the upper with 3 veins and about as long as the lemma. Lemma obscurely 5-veined, papery, rounded on the back, with a long awn near the apex. Palea about equalling lemma. Rhachilla shortly prolonged. *Eurasia.*

56. Lagurus Linnaeus. 1/1. Annual. Leaves flat. Inflorescence a dense, ovoid or cylindric to almost spherical panicle. Spikelets compressed from side to side, with 1 flower. Glumes almost equal, linear, membranous, with 1 vein and a long, densely ciliate, bristle-like apex. Lemma with 5 veins, membranous, hairy at the base, with 2 long apical bristles and a sharply bent awn on the back. Palea somewhat shorter than lemma. Rachilla prolonged, ciliate. *Mediterranean area.*

57. Echinaria Desfontaines. 1/1. Annual. Leaves usually flat. Inflorescence a dense, ovoid or spherical, prickly head. Spikelets almost stalkless, somewhat compressed from side to side, with 3 or 4 (rarely 1) flowers. Glumes more or less equal, membranous, the lower with 2 (rarely 5) strong veins which run out at the margins, the upper with a midrib which runs out at the apex. Lemma leathery, with 5 (rarely 7) very strong veins prolonged as flattened awns, which are deflexed at maturity. Palea as long as the lemma, the 2 (rarely 5) veins prolonged as flattened awns. *Mediterranean area.*

58. Lamarckia Moench. 1/1. Annual. Leaves flat. Inflorescence a rather condensed, 1-sided panicle. Spikelets of 2 kinds, all somewhat compressed from side to side, falling in groups of 3--5; fertile spikelets with 1 rudimentary and 1 fertile flower, surrounded by 2--4 sterile spikelets with many flowers. Glumes thin, almost equal, acute. Lemma of fertile spikelets papery, awned, of the sterile spikelets membranous, unawned. Palea 2-keeled. *Mediterranean area.*

59. Cynosurus Linnaeus. 5/1. Perennials or annuals. Leaves flat. Inflorescence a more or less 1-sided, dense panicle. Spikelets of 2 different kinds, the fertile compressed from side to side, with 1--5 flowers, the sterile below the fertile and consisting of narrow, rigid glumes and lemmas. Glumes of fertile spikelets thin, almost equal, acute. Lemma papery, with an apical awn. Palea 2-keeled, shortly bifid at apex, about as long as lemma. *Mediterranean area, Europe.*

60. Dactylis Linnaeus. 5/1. Perennials. Leaves flat or rolled. Inflorescence a panicle with spikelets in dense clusters on the spreading or erect branches. Spikelets compressed from side to side with 2--5 flowers. Glumes almost equal, keeled, somewhat curved, the lower with 1 vein, the upper with 3 veins. Lemma with 5 veins, papery, mucronate or shortly awned at apex. Palea about as long as lemma. *Europe, temperate Asia, Mediterranean area.*

61. Briza Linnaeus. 12/3. Almost hairless annuals or perennials. Inflorescence usually much-branched. Spikelets ovoid or broadly triangular, compressed from side to side; flowers 4--20. Glumes ovate-circular, heart-shaped at the base, with 3--9 veins and with wide scarious margins. Lemma almost circular, heart-shaped at the base, with

7--9 veins, closely overlapping. Palea ovate to obovate, obtuse, shorter than to almost equalling the lemma, with 2 narrowly winged keels. Ovary hairless; styles terminal. Grain flattened on one side. *Northern hemisphere.*

62. Poa Linnaeus. 250/few. Annuals or perennials. Inflorescence a panicle. Spikelets compressed from side to side, with 2--10 flowers (rarely only 1). Glumes keeled. membranous, usually with 3 veins or the lower with 1 vein. Lemma with 5 veins, keeled, membranous, awnless or rarely with a short terminal awn. Palea 2-keeled, the keels with very small teeth or ciliate. Grain ellipsoid; hilum basal. *Temperate areas.*

63. Festuca Linnaeus. 300/6. Tufted or rhizomatous perennials. Leaves folded, rolled or flat; sheaths open or closed. Inflorescence a panicle. Spikelets stalked, with 4 or more flowers (rarely only 2), compressed from side to side. Glumes 2, the lower usually with 1 vein, the upper wider, usually with 3 veins. Lemma with rounded back, not keeled, with or without a terminal or almost terminal awn. Palea 2-keeled, scarious. *Temperate & subtropical areas.*

64. Glyceria R. Brown. 20/1. Hairless perennials. Leaves flat; sheaths often entire; ligule membranous. Inflorescence a panicle with a 3-sided main axis. Spikelets ovate to linear in outline, with 3--many flowers. Glumes unequal, transparent, with 1 vein, shorter than the lowest lemma. Lemma awnless, ovate to oblong-lanceolate, rounded on the back, with 5--11 veins. Palea about as long as lemma. *Northern hemisphere.*

65. Melica Linnaeus. 25/2. Perennials. Leaves flat or rolled; sheaths entire. Inflorescence usually a rather loose, often simple panicle. Spikelets slightly compressed from side to side, with 1--several bisexual flowers, and 2 or 3 sterile lemmas forming a terminal, club-shaped structure. Glumes more or less unequal, firmly membranous, with 3--5 veins. Fertile lemma rounded on the back, leathery in fruit, the veins 5--9 (rarely to 13). *Europe, temperate Asia, North Africa, temperate America.*

66. Stipa Linnaeus. 300/5. Tufted perennials, rarely annuals. Leaves usually pleated or rolled, at least when dry, with prominent ribs on the upper surface. Inflorescence a panicle. Spikelets somewhat compressed from side to side, with 1 flower. Glumes usually more or less equal, translucent or membranous, much longer than the lemma, with 1--3 veins; lemma usually leathery, rolled or terete, entire or shortly bifid at apex, hairy, awned from apex or sinus; hairs usually confined to the veins and margins, those of the overlapping margins forming a single line. Callus bearded, usually long and pointed; awn usually with a thick twisted lower part (column) and a thinner, straight upper part (limb) hairless or hairy, usually twice bent. Palea translucent, 2-veined, usually enclosed by lemma. Rhachilla not prolonged. *Tropical & temperate areas.*

67. Piptatherum Palisot de Beauvois. 25/2. Perennials. Leaves flat or rolled. Inflorescence a loose panicle. Spikelets somewhat compressed from side to side, with 1 flower. Glumes unequal, the lower with 5 veins, the upper with 3--5 veins. Lemma obscurely 3--5-veined, leathery, with a terminal awn that is eventually shed. Palea about as long as lemma, leathery. Callus very small. Rhachilla not prolonged. *Old World subtropics.*

68. Phaenosperma Bentham. 1/1. Tufted perennial with slender to stout stems, scarious ligules and flat, narrowly lanceolate leaf-blades which have numerous cross-veins. Inflorescence a loose panicle with clustered branches. Spikelets falling at maturity, elliptic-oblong, at length widely gaping, awnless, with 1 flower; glumes thinly membranous, the lower about half the length of the upper, which is as long as the spikelet; lemma as long as the spikelet, rounded on the back, becoming firmly papery, with 3--5 veins, hairless; grain spherical or almost so, with longitudinal grooves, and a free pericarp. *Northeast India, Burma, China, Taiwan, Korea & Japan.*

Group 2 (Bamboos). Woody, evergreen, perennial grasses, Rhizomes sometimes thick and short, sometimes thin and far-running. Culms emerging in spring, summer or autumn, as thick, pointed, asparagus-like shoots, growing rapidly; mature culms 20 cm to 46 m, 3 mm to 65 cm in circumference, usually hollow except at the nodes, which are marked by a more or less prominent line; below this there is usually another line which is the raised scar left by a sheathing leaf (culm sheath): these are either pushed off by the developing branches or are persistent; the culm sheaths are alternate, on 2 sides of the culm, smooth and shining inside, and usually have a stalkless, short-lived blade which often bears auricles and bristles at its junction with the sheath; a ligule is also present. Crowns sometimes pendent with the weight of the leaves and branches. Branches single at each node (except for *Chusquea)* but the branch itself is branched very near its base, so that many branches appear to arise at each node. Foliage leaves largest on new culms, with basal persistent sheaths, frequently with auricles and bristles at the sheath apex and always with a ligule; blades stalked on the sheaths, deciduous at a precise line or joint, with at least one margin with fine, forwardly pointing teeth, the veins parallel but joined by many transverse veinlets, the venation often showing as a distinct mosaic of tessellation. Inflorescence of panicles, spikes or racemes, sporadically produced in Europe. Spikelets various, with 0--4 (usually 2) glumes, and a varying number of flowers (sometimes partly abortive) which may be in part unisexual or otherwise imperfect. Lodicules usually 3. Stamens 3--6 in cultivated species. Style 1, stigmas 2 or 3.

Because bamboos flower and fruit uncertainly in Europe the keys and descriptions given here are based almost entirely on vegetative characters. Identification using these is not entirely easy, and there is much confusion in the naming of the cultivated species. In order to use the account properly a hand lens ($\times$ 10 or $\times$ 15) is necessary, and the plants should be examined whole, if possible, at the most suitable time of the year, that is, soon after the culms have grown to their full height but while they retain their sheaths, which should be examined at the middle of the culm rather than above or below on mature plants. The bristles which occur on the culm and leaf-sheaths are important for identification: they may be absent, or present, when their colour and surface are important. The surface may be entirely smooth, or roughened with small projections which can be seen with the hand lens. In a few cases the bristles are rough near the base and smooth above. The presence or absence of vein tessellation in the leaves may be seen by holding them up to the light and examining them through the lens.

Bamboos form most of the subfamily *Bambusoideae* of the *Gramineae*; this subfamily is native over wide areas of eastern Asia, central and southern Africa and America from Maryland south to Argentina. Because of uncertainties in classification, it is not possible to state the number of genera and species of bamboos, and there is much still to be learned about their classification.

1a. Culm always solid when mature   2
  b. Culms sooner or later hollow (except at the nodes)   4
2a. Culms D-shaped in section   **70. Shibataea**
  b. Culms terete in section   3
3a. Culms to 40 cm; branches 1--3 at each node   **75. Pleioblastus**
  b. Culms to 6 or rarely to 10 m; branches very numerous at each node
     82. Chusquea
4a. Culms not terete, at least above   5
  b. Culms terete throughout   8
5a. Culms obtusely 4-cornered in section   **76. Chimonobambusa**
  b. Culms at least in part with an obvious groove   6
6a. Culms grooved throughout, or at least between those nodes where
    branches are borne   **69. Phyllostachys**
  b. Culms mainly terete below, grooved only above   7
7a. Culms with a large hollow, held stiffly erect; leaves 1.6 cm or more broad
     74. Semiarundinaria
  b. Culms ultimately bending with the weight of the crown, and with a small
    hollow; leaves at most 6 mm broad   **72. Otatea**
8a. Branches always 1 at each node   9
  b. Branches mostly more than 1 at each node   13
9a. Culms ascending or flopping; sheath-bristles, if present, rough
    throughout   10
  b. Culms erect; sheath-bristles absent, or, if present, entirely smooth or
    rough only near the base   11
10a. Culms persistently ascending; leaves 4--11 on each branch, at most 40 cm
     77. Sasa
  b. Culms ultimately flopping; leaves 1 or 2 on each branch, to 60 cm
     78. Indocalamus
11a. Sheath-bristles rough near their bases; culms to 1.5 cm in circumference
     79. Sasaella
  b. Sheath-bristles smooth or absent; culms more than 3 cm in circumference
     12
12a. Culms to 3 m, without white waxy bloom below the nodes
     80. Sasamorpha
  b. Culms to 6 m, with white waxy bloom below the nodes   **81. Pseudosasa**
13a. Leaves without visible tessellation   14
  b. Leaves with visible tessellation   15
14a. New shoots appearing in late summer or autumn   **76. Chimonobambusa**
  b. New shoots appearing in spring   **73. Arundinaria**
15a. Culm-sheaths soon falling, though often hanging from a small attachment
    at the base for some time; branches forming on the culms from the base
    upwards   **71. Semiarundinaria**
  b. Culm-sheaths persistent, not hanging as above; branches forming on the
    culms from the top downwards   16
16a. Bristles rough   **73. Arundinaria**
  b. Bristles smooth   17
17a. Bristles and ligule dark-coloured   **74. Sinarundinaria**
  b. Bristles and ligules whitish   18

18a. Culms finally with a small hollow, emerging in late summer; culm-
sheaths marbled, blades minute   **76. Chimonobambusa**
   b. Culms with a wide hollow, emerging in spring; culm-sheaths not marbled,
blade obvious   **75. Pleioblastus**

69. Phyllostachys Siebold & Zuccarini. 60/10. Rhizomes short or elongate. Culms
close or spaced, with hollow internodes which are grooved on alternate sides above
each node (except sometimes for those below the lowest branches); nodes prominent.
Branches developing in sequence from below upwards, 3 at each node, the outer
unequal, the central small and depauperate, sometimes absent; branchlets numerous.
Sheaths soon deciduous (those at the base sometimes more persistent), bristles rough
or absent. Leaves 6--20 cm, visibly tessellate, hairless above, paler and sometimes
downy near the base beneath; bristles rough but soon deciduous. Inflorescences spicate
or racemose. Spikelets with usually 2 glumes and 5--13 flowers. Lodicules 3, ciliate.
Stamens 3. Stigmas 3. *China; introduced elsewhere.*

70. Shibataea Nakai. 5/1. Usually tufted (though rhizomes occasionally running).
Culms D-shaped in section, slender, at most 6 mm in circumference, solid, with
prominent nodes. Sheaths fairly persistent, hairless, without auricles or bristles.
Branches 2 or 3 at each node, short, all more or less equal. Leaves visibly tessellate,
broadly lanceolate, without bristles. Inflorescence a panicle. Spikelets with 2--3 glumes
and 1 or 2 flowers. Lodicules 3. Stamens 3. Stigmas 3. *Japan, China.*

71. Semiarundinaria Nakai. 20/1. Culms usually tightly clumped (but capable of
spreading), hollow, terete or with the upper internodes grooved, held stiffly erect.
Sheaths soon falling, except on late shoots, but often hanging by their bases for some
time, the inner surface highly polished, usually purplish red, without bristles. Branches
erect, 3--8 at each node, developing from the base of the culm upwards, the lower
nodes without branches. Leaves to 20 cm, visibly tessellated and with rough bristles.
Inflorescence a number of spikes. Spikelets with glumes absent or rarely 1 present,
with 3--6 flowers of which the upper 1--3 are male only. Lodicules 3, ciliate. Stamens
3. Stigmas 3. *Eastern Asia.*

72. Otatea (McClure & Smith) Calderon & Söderstrom. 2/1. Rhizome short or
elongate. Young plants tufted, becoming more open with age. Culms slender, solid at
first, later hollow, the upper internodes each with a shallow groove on one side, the
whole culm ultimately arching with the weight of the crown. Sheaths soon deciduous,
without bristles. Branches 3 at each node initially, later more numerous. Leaves
without visible tessellation, 3--6 mm broad, the sheaths mostly without auricles or
bristles. Inflorescence a panicle. Spikelets with 2 glumes and 3--7 flowers. Lodicules 3.
Stamens 3. Stigmas 2. *Mexico to Honduras.*

73. Arundinaria Michaux. Several/8. Rhizomes slender and elongate, or compact and
much-branched. Culms erect, hollow, terete; new shoots emerging in spring. Sheaths
persistent with well-developed blades and dark, rough bristles. Branches several at
each node, foliage developing from the top downwards. Leaves variably tessellate or
not; sheaths bearing dark, tough bristles (rarely bristles absent); ligule often
dark-coloured. Inflorescence racemose or paniculate. Spikelets with 2 or rarely 3
glumes and several flowers of which the upper may be unisexual. Lodicules 3 or rarely

80

6. Stamens usually 3 or 6. Stigmas 3. *Mainly Himalaya & China, also America & South Africa.*

74. Sinarundinaria Nakai. 1/1. Like *Arundinaria*, but with a much-branched rhizome, habit tufted, branches almost equal, arising horizontally, leaf-bristles smooth and white. Inflorescence a small panicle. Spikelets with 2 glumes and 2--4 flowers. Lodicules 3. Stamens 3. Stigmas 2--3. *Northern China.*

75. Pleioblastus Nakai. 20/8. Rhizomes mostly elongate, slender. Culms erect, almost always hollow but thick-walled, terete. New shoots arising in spring, leafing from the top downwards. Sheaths persistent, bearing smooth, whitish bristles. Branches usually 3--7 at each node or sometimes branches only 1 or 2, all in the lower part of the culm. Leaves with visible tessellation and with smooth, whitish bristles which are occasionally absent. Inflorescence a spike or raceme. Spikelets with 2 glumes and 5--13 flowers. Lodicules 3, ciliate. Stamens 3. Stigmas 3. *China, Japan.*
    Formerly considered to be part of *Arundinaria.*

76. Chimonobambusa Makino. Several/4. Rhizomes slender, mostly shortly running. Culms hollow, terete or 4-cornered: new shoots appearing in late summer or autumn. Sheaths without bristles, variably persistent. Branches 3--many, developing from the top downwards during the following year. Leaves visibly tessellate or not, with smooth bristles or bristles absent. Inflorescence a panicle. Spikelets with 0--2 glumes and several flowers. Lodicules 3. Stamens 3. Stigmas 2. *Himalaya, China, Japan.*
    Species of **Bambusa** Schrader (culms to 15 m, leaves and sheaths silver-hairy) and **Dendrocalamus** Nees (culms more than 15 m, more than 10 cm in circumference), may occasionally be grown out-of-doors in very mild areas.

77. Sasa Makino & Shibata. 30/5. Rhizomes slender. Culms distant, ascending, hollow, terete, 2--3 m tall, 3--6 cm in circumference, unbranched below. Sheaths persistent, mostly shorter than the internodes, bristles spreading and rough or rarely absent. Nodes rather prominent, with white waxy bloom below. Branches 1 at each node. Leaves 4--11 per branch, thick and hairless, visibly tessellate; bristles rough, whitish, rigid, borne at right-angles to the branch, sometimes absent; ligule downy. Inflorescence a panicle. Spikelets with 2 glumes and 4--10 flowers. Lodicules 3. Stamens 6. Stigmas 3. *Japan, Korea.*

78. Indocalamus Nakai. 20/1. Differs from *Sasa* essentially in floral characters. The one cultivated species has less prominent nodes without white, waxy bloom, short internodes which are exceeded by the sheaths and very large leaves (the largest of any bamboo hardy in Europe), which are borne 1 or 2 to a branch; bristles, when present, dark. Stamens 3. Styles 2. *Malaysia, China.*

79. Sasaella Makino. 12/1. Like *Sasa*, but culms erect, to 15 m tall and to 1.5 cm in circumference, bristles not spreading, rough below only, smooth above. Leaves 5--8 per branch. Inflorescence a panicle. Spikelets with 2 glumes and 5--10 flowers. Lodicules 3. Stamens 6. Stigmas 3. *Japan.*

80. Sasamorpha Nakai. 4/1. Like *Sasa* but culms erect to 3 m tall and *c.* 3 cm in circumference, sheaths without any bristles, longer than the internodes. Leaves 2--5 per branch, without bristles. Inflorescence a panicle. Spikelets with 2 glumes and 4--8 flowers. Lodicules 3. Stamens 6. Stigmas 3. *Eastern Asia.*

81. Pseudosasa Nakai. 6/1. Rhizomes slender, usually clump-forming, occasionally running. Culms erect, to 6 m. Sheaths persistent, longer than the internodes, bristles usually absent, if present then smooth and white. Branches 1 at each node, the lower nodes without branches. Leaves 4--7 per branch, without auricles or bristles. Inflorescence a panicle. Spikelets with 2 glumes and 3--8 flowers. Lodicules 3. Stamens 3 or rarely 4. Stigmas 3. *Eastern Asia.*

82. Chusquea Kunth. 90/2. Rhizomes elongate, slender. Culms solid, terete, more than 1 cm in circumference. Sheaths persistent, without bristles. Branches numerous at each node. Leaves narrow, without bristles. Inflorescence usually a panicle. Spikelets with usually 4 glumes and with 1 (rarely 2) flower(s). Lodicules 3. Stamens 3. Stigmas usually 2, occasionally with 1 branched again. *Mexico to southern Argentina.*

17. PALMAE (*Arecaceae*)

Evergreen shrub- or tree-like plants or woody climbers, sometimes spiny. Stems solitary or clustered, or colonial from long rhizomes, rarely branched aerially, erect or prostrate, sometimes apparently stemless, smooth, spiny or covered in leaf-sheaths. Leaves alternate; sheaths tubular at first, sometimes forming a column (crownshaft) above the stem or splitting, often toothed or spiny; stalk usually present. Leaf-blade pinnate, bipinnate, palmate, costapalmate (i.e. palmate but with a short extension of the stalk between each pair of leaflets), or undivided with palmately or pinnately arranged folds; sometimes a protrusion (hastula) present on upper surface of leaf-stalk where the leaflets are attached. Leaflets (segments) with 1--several ribs, acute, bifid, truncate, or irregularly toothed. Inflorescences among or below the leaves or aggregated into a terminal mass (the stem then dying after fruiting), 1 (rarely more) per node, paniculate to spicate, the stalk bearing a basal bract with 2 keels, and upper bracts few to many (rarely none), all persistent or deciduous; inflorescence branches (where present) subtended by bracts, branching occurring at up to 4 orders. Flowers unisexual (when plant monoecious or dioecious) or bisexual, stalkless or stalked, or sunk into pits, borne singly, in clusters or in 3s (triads) of 2 lateral males and a central female, or in pairs of bisexual and male, or female and neuter, or of male flowers only. Sepals and petals usually 3 in each whorl, free or united, or perianth of one undifferentiated series. Stamens usually 6, sometimes from 3 to almost 1000. Ovary superior, with carpels free or united, usually 3, sometimes 1, 2 or more than 3; ovule 1 only in each carpel, variously attached; carpels sometimes infertile and fused into a pistillode. Fruit fleshy or dry, sometimes covered in scales, hairs or spines, sometimes with bony endocarp; seeds 1--3 or more, sometimes with a fleshy seed-coat (sarcotesta); endosperm smooth or furrowed, the furrows penetrated by folds of the seed-coat. See figure 14, p. 84.

A large, almost entirely tropical and subtropical family of 212 genera and about 2700 species.

It is difficult to ascertain which palms are grown in Europe; the selection presented here is based on what might be expected to grow in northern parts. The account does not deal with juveniles of many species which enter the pot-plant trade as ephemeral houseplants with little chance of survival.

82

1a. Leaves pinnate or pinnately divided   **4. Phoenix**
  b. Leaves palmate or palmately divided   2
2a. Stem covered with a skirt of dead leaves; inflorescence-bracts woody
                                         3. Washingtonia
  b. Stem with or without persistent leaf-bases but not with a skirt; bracts thin
    or leathery   3
3a. Dwarf, clustering palms; leaf-stalks with conspicuous spines
                                       2. Chamaerops
  b.Usually erect, solitary palms; leaf-stalks without conspicuous spines
                                       1. Trachycarpus

1. Trachycarpus Wendland. 6/6. Small to moderate, solitary or clustering, dioecious palms, rarely monoecious or with unisexual and bisexual flowers on the same plant; trunks often covered with persistent, fibrous leaf-sheaths. Leaves palmate, divided into 1-ribbed segments. Inflorescences among the leaves, with several thin bracts on the inflorescence-stalk, branching to 3 orders. Flowers in clusters of 2--4, with 3 basally overlapping sepals and 3 overlapping petals. Male flowers with 6 stamens and 3 small pistillodes. Female flowers with 6 staminodes and 3 carpels. Bisexual flowers with 6 stamens and 3 carpels. Fruit spherical to kidney-shaped or oblong-ovoid with remains of stigma at apex. Seed with smooth endosperm and lateral embryo. *Subtropical Asia.*

2. Chamaerops Linnaeus. 1/1. Clustering, shrubby palms, mostly dioecious or rarely with a few flowers of the opposite sex in each inflorescence. Stems suckering and forming low clumps but occasionally producing an erect trunk to 5 m, covered in more or less persistent, fibrous leaf-sheaths. Leaves palmate; stalks with short spines pointing towards the apex; blade held stiffly and deeply divided into 1-ribbed segments. Inflorescences among the leaves, with a single large bract at the base and smaller bracts subtending the few branches. Flowers borne singly, but crowded together, golden-yellow. Calyx a low, 3-toothed cup. Petals 3, overlapping. Stamens or staminodes 6. Carpels 3, free or reduced to pistillodes. Fruit rounded, reddish brown, fleshy, with a rancid smell. *South Europe, North Africa.*

3. Washingtonia Wendland. 2/2. Robust, solitary, bisexual palms, with trunks (in nature) clothed in persistent dead leaves. Leaves costapalmate with 1-ribbed segments; stalks with spines along margins. Inflorescences borne among leaves, exceeding them in length and bearing several empty bracts on the stalk, and woody, open, sword-shaped bracts each subtending a hanging inflorescence-branch which bears slender, flower-bearing branches. Flowers cream, with tubular calyx; petals 3, deciduous; stamens 6; carpels 3, free, except at the very tip, where united into a common style. Fruit with 1 seed with smooth endosperm. *USA, northwest Mexico.*

4. Phoenix Linnaeus. 17/7. Solitary or clustering, dwarf to robust, dioecious palms. Leaves pinnate with acute segments which are V-shaped in section, the lowermost modified as spines. Inflorescence among the leaves, subtended by a single, usually deciduous bract, the stalk compressed and bearing single or clustered branches at the tip. Flowers creamy yellow, borne singly. Male flowers with sepals joined into a 3-lobed cup and 3 petals which are edge-to-edge, stamens usually 6. Female flowers with 3 sepals joined into a cup, and 3 overlapping petals; carpels 3, free. Fruit usually developing from only 1 carpel, cylindric to spherical with dry to fleshy mesocarp. Seed

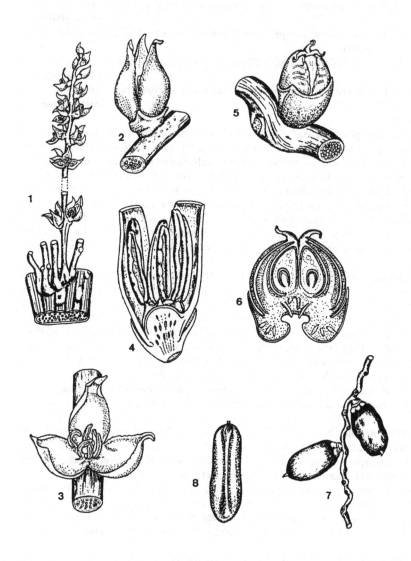

Figure 14. Palmae. *Phoenix sylvestris.* 1, Flattened branch from a male inflorescence. 2, Male flower just opening. 3, Open male flower. 4, Longitudinal section of a male flower. 5, Female flower just opening. 6, Longitudinal section of a female flower. 7, Fruits. 8, A single seed.

single, longitudinally grooved, with smooth endosperm and a basal or lateral embryo. *Old World tropics & subtropics.* Figure 14, p. 84.

18. ARACEAE

Plants perennial, herbaceous or slightly woody, usually hairless, sometimes with a distinct juvenile phase. Leaves alternate or basal, usually stalked and with a broad blade. Scale-leaves present at least at some stage. Leaf-stalk usually with a definite sheath; sheath usually embracing the principal stem but if not, the leaves are interspersed with scale-leaves. Leaf-blade simple, lobed or compound, often with netted veins; primary lateral veins sometimes alternating with smaller straight or sinuous veins (the interprimary veins) which originate from the midrib or are formed by amalgamation of branches from the primary laterals; 1 or more marginal veins often present near the edge. Plants bisexual or, rarely, unisexual. Inflorescence consisting of a stalk and a spadix with its spathe. Spadix sometimes more or less united to spathe, sometimes with a stalk (stipe) between spathe attachment and lowest flowers, sometimes with a sterile terminal appendix. Fertile flowers bisexual or unisexual. Unisexual flowers arranged in zones on spadix, the female usually below the male, sometimes few in number, the female occasionally solitary. Sterile flowers sometimes present, usually either as filaments or flattish scales, occurring between male and female zones or above male zone. Perianth-segments 0, 4, 6 or occasionally more. Stamens 0, 1, 4 or 6 (rarely more), free or (in male flowers) sometimes joined; filaments usually short or none. Staminodes sometimes present in female flowers. Ovary with 1--4 cells, sometimes with incomplete partitions. Style usually short or absent. Ovules 1--many. Fruit usually a berry. See figure 15, p. 88.

A family of about 115 genera and 2000 species, mainly tropical but with some subtropical or temperate aquatic and some temperate tuberous representatives. Many tropical Araceae are woody climbers with aerial roots.

1a. Waterside plant with iris-like leaves fragrant when crushed; spadix
    protruding sideways from an apparent leaf   **1. Acorus**
  b. Plant not as above   2
2a. Stipe several times longer than spadix, part of it wrapped in the sheathing
    part of the spathe   3
  b. Stipe not longer than spadix   4
3a. Spathe reaching fertile part of spadix, its limb well-developed, boat-shaped;
    marsh plants with large leaves   **2. Lysichiton**
  b. Spathe not reaching fertile part of spadix, its limb like a small bract;
    aquatic plant with submerged floating and emergent leaves   **4. Orontium**
4a. Spathe in the form of an upwardly open funnel or chalice, white, yellow,
    pink or, occasionally, green; spadix without appendage   **6. Zantedeschia**
  b. Spathe not shaped as above; if upwardly open, then dark red, or spadix
    with an appendage, or both conditions apply   5
5a. Spadix uniform, flowers all or mostly bisexual   **5. Calla**
  b. Spadix with zones bearing female, male and sometimes sterile flowers
    exclusively, sometimes with 1 or more non-flowering zones   6
6a. Lower part of the spathe in the form of a closed tube, formed by fusion
    of the margins   7
  b. Lower part of spathe with free but overlapping margins   9
7a. Leaves pedate; spadix appendage *c.* 30 cm   **10. Sauromatum**

b. Leaves simple; spadix appendage much less than 30 cm   **8**

8a. Spadix with the zones of male and female flowers adjacent; leaf-blade heart-shaped or sagittate   **12. Arisarum**

b. Spadix with a gap which sometimes bears sterile flowers between the zones of male and female flowers; leaf-blade tapered or truncate at base   **11. Biarum**

9a. Flowers all bisexual; spadix nearly as broad as long; temperate marsh plants flowering early in spring, before their leaves appear   **3. Symplocarpus**

b. Flowers mostly unisexual; spadix not nearly as broad as long; plants not as above   **10**

10a. Female part of spadix completely joined to spathe   **14. Pinellia**

b. Female part of spadix completely free from spathe   **11**

11a. Leaves compound or deeply lobed   **12**

b. Leaves simple   **13**

12a. Spadix unisexual or with male flowers not touching or (usually) showing both these conditions   **13. Arisaema**

b. Spadix bisexual; male flowers more or less touching   **9. Dracunculus**

13a. Sterile flowers between male and female zones of spadix, each in the form of a thread or prong, often with a swollen base   **8. Arum**

b. Sterile flowers each in the form of a flat plane, occasionally with a rudimentary ovary forming a boss in the middle, or sterile flowers absent   **7. Peltandra**

1. Acorus Linnaeus. 2/2. Rhizomes branched, bearing linear leaves and stems which are leafless except for the spathe. Leaves equitant, without distinct stalks. Spathe similar to the leaves (except for the basal part, which is narrow and stem-like), continuing the line of the inflorescence-stalk, so that the spadix appears lateral. Spadix stalkless, without appendix. Flowers very numerous and densely packed, bisexual, with 6 obovate perianth-segments, 6 stamens, and a broadly ovoid, 2- or 3-celled ovary. Fruits not formed in Europe. *South & east Asia, southern North America.*

Sometimes placed in the separate family *Acoraceae.*

2. Lysichiton Schott. 2/2. Robust, rhizomatous, stemless herbs, forming large clumps. Leaves ovate-oblong, truncate at the base, shortly stalked, without a sheath, appearing about the same time as the flowers. Spathe arising from ground-level, withering before the fruit is ripe, the lower part consisting of a narrow sheath enclosing the very long stipe of the spadix, but with free margins, the upper part erect, elliptic, strongly concave. Spadix cylindric, on a long stipe, without appendix. Flowers crowded, bisexual; perianth-segments 4, often unequal; stamens 4; ovary 2-celled. Berries partly embedded in the spadix; seeds 2. *Western North America.*

3. Symplocarpus Salisbury. 2/1. Foetid, stemless herbs with stout, vertical rhizome. Leaves numerous, forming large clumps; stalk channelled, without a sheath; blade ovate, cordate, acute, entire. Spathes borne at ground-level, very strongly concave, somewhat suggestive of a shell. Spadix with a short stipe enclosed within the spathe, shortly cylindric or almost spherical. Flowers bisexual; perianth-segments 4; stamens 4; ovary 1-celled. Fruits very deeply embedded in the spadix so as to form a compound fruit, succulent on the surface, dry and spongy within. *Temperate eastern Asia, eastern North America.*

4. Orontium Linnaeus. 1/1. Plant aquatic, rhizomatous. Leaves all basal, with long stalks, clustered, mixed with scale-leaves; blade ovate, entire, submerged, floating or emerging. Inflorescence-stalk slender, about as long as leaves, white above, terminating in a slender yellow spadix. Spathe much reduced, consisting of a sheath surrounding the lower part of the very long stipe and bearing a small, bract-like green blade which withers early. Flowers numerous, crowded, bisexual, with 4--6 perianth-segments, 4--6 stamens and a 1-celled ovary. Fruit green, deeply embedded in the spadix, containing a single seed. *Eastern USA.*

5. Calla Linnaeus. 1/1. Rhizomatous aquatic plants. Leaves all basal, mixed with scale-leaves, long-stalked, broadly ovate to kidney-shaped, shortly pointed, cordate; sheathing bases of the leaf-stalks persisting on the rhizome. Spathe similar to the leaf-blades in shape but smaller, erect, concave but scarcely enfolding the spadix even at its base, greenish outside, white on inner surface during flowering period, persistent in fruit; 2 or 3 spathes per spadix are sometimes developed. Spadix green, shortly cylindric, with a short stipe and without appendix. Flowers crowded, without perianths, mostly bisexual, but the uppermost usually male; stamens 6 or more; ovary 1-celled. Berry red, with 4--10 seeds. *Eurasia, North America.*

6. Zantedeschia Sprengel. 6/6. Plants with oblique, tuberous, underground rhizomes, which bear apical clusters of leaves or leaves and inflorescences. Scale-leaves absent. Leaf-stalk spongy. Leaf-blade wavy, lobed at the base or entire; main lateral veins joining a marginal vein, minor ones forming a network. Inflorescence as tall as the leaves or taller, fragrant; stalk spongy. Spathe normally white or brightly coloured, obliquely funnel-shaped, withering but persistent in fruit. Spadix much shorter than spathe, yellowish. Flowers without perianths, unisexual. Zones of male and female flowers adjacent. Male flowers not individually recognisable. Female flowers with or without staminodes. Ovary usually with 3 cells, each with 1--8 ovules. Style short; stigma entire. Fruit a berry. Seed leathery. *Southern Africa.*

7. Peltandra Rafinesque. 4/2. Plants with horizontal rhizomes bearing rosettes of scale-leaves and long-stalked foliage leaves. Leaf-blade sagittate or hastate, with 2 or 3 marginal veins; lateral veins mostly parallel but the smallest joined by cross-veins; each basal lobe with a distinct midrib. Inflorescence-stalk as long as or slightly longer than leaf-stalks. Spathe with margins overlapping in the zone of female flowers, open above: upper part decaying in fruit. Flowers without perianths, unisexual. Male flowers of 4 or 5 stamens united to form a flat plate with anthers around the edge, opening by pores at the top. Female flowers with a cup-like structure round the ovary. Ovary 1-celled; ovules 1 or few; style present. *Eastern North America.*

8. Arum Linnaeus. 12/8. Tuberous herbs. Leaves all basal; stalk shortly sheathing at base, usually about twice as long as blade; blade hastate to sagittate, entire. Inflorescence-stalks shorter than leaf-stalks or nearly equalling them, usually visible above the ground. Lower part of spathe narrow, with overlapping margins, enclosing the floral region of the spadix and usually the lower part of its appendix; limb usually lanceolate, concave, longer than the spadix. Flowers without perianths, unisexual. Spadix with zones of male and female flowers usually separated by a zone of sterile flowers; sterile flowers usually present also above the male. Male flowers with 3 or 4

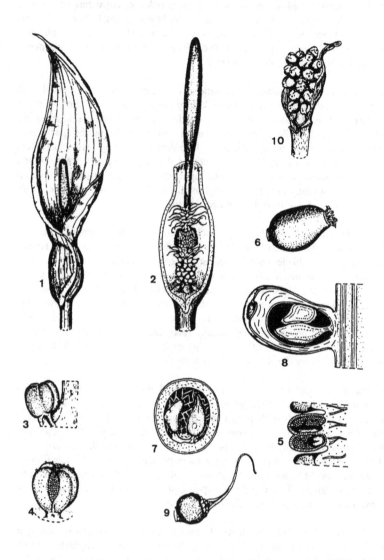

Figure 15. Araceae. *Arum maculatum.* 1, Flowering shoot. 2, Flowering shoot with part of the spathe cut away. 3, A single fertile male flower just before anther-dehiscence. 4, Male flower at anther-dehiscence. 5, Longitudinal section of part of the male portion of the inflorescence. 6, Fertile female flower. 7, Transverse section of ovary. 8, Longitudinal section of ovary. 9, Sterile female flower. 10, Fruiting inflorescence.

stamens; ovary 1-celled. Berry red; seeds 1--6. *Canary Islands to Iran*. Figure 15, p. 88.

9. Dracunculus Schott. 3/3. Tuberous plants *c.* 70--180 cm, with several large leaves. Leaves basal, their stalks with long, overlapping sheaths which together resemble a stem. Leaf-blade pedately divided or deeply pedately lobed. Lower part of spathe with overlapping margins enclosing lower part of spadix, limb large, widely open. Flowers without perianths, unisexual. Spadix with zones of male and female flowers adjacent or separated; appendix long. Male flowers with 2--4 stamens. Ovary 1-celled. Berry green; seeds 6. *Canary Islands, Mediterranean area*.

10. Sauromatum Schott. 2/2. Tuberous herbs, bearing one leaf at a time. Leaf-blade pedately cut or divided. Inflorescence appearing before the leaves; stalk short, so that the spathe may rest on the ground. Spathe with margins fused below to form a tube; limb lanceolate; both parts withering after flowering. Flowers without perianths, unisexual. Spadix with zones of male and female flowers about equal, with a long sterile zone between them which bears club-shaped sterile flowers below, and a long curved appendix. Male flowers indistinct. Ovary 1-celled, with 2--4 ovules. *East & west Africa, southern Asia*.

11. Biarum Schott. 12/few. Plants with small flattened-spherical tubers producing roots above. Scale-leaves conspicuous around bases of foliage leaves and inflorescences. Leaf-blade entire, not lobed. Inflorescence-stalk concealed below ground or slightly exposed in fruit. Spathe with margins of lower part united to form a tube, withering before fruit is ripe. Flowers unisexual, without perianths. Spadix with a wide space between zones of male and female flowers on which are usually borne some slender sterile flowers; appendix as long as remainder of spadix or longer. Male flowers with 1 or 2 stamens. Ovary 1-celled, with 1 ovule. Fruiting spadix approximately spherical. Berry white or pale green, sometimes striped with purple. *Mediterranean area, western Asia*.

12. Arisarum Targioni-Tozzetti. 2/2. Rootstock a tuber or rhizome. Leaves long-stalked, sagittate, entire. Spathe withering before fruit is ripe, the lower part with margins united so as to form a closed tube, the upper part open in front but curved forwards at the tip so as largely to conceal the spadix. Spadix shorter than or equalling the spathe, with a short floral zone at the base and a long appendix. Flowers without perianths, unisexual, the female few, the male situated immediately above the female. Stamen 1; ovary 1-celled. Berry greenish, with about 6 seeds. *Mediterranean area, Atlantic Islands*.

13. Arisaema Martius. 150/10. Plants with tubers (often with offsets or stolons), or rarely rhizomes. Leaves solitary or few, compound, long-stalked, all basal, though their sheathing bases sometimes conceal the lower part of the inflorescence-stalk so as to make the leaves appear to be borne on a stem. Two or more whitish, papery, narrow-oblong sheathing scale-leaves are usually present, enfolding the bases of the leaf-stalks and the inflorescence-stalk. Lower part of spathe narrow, with overlapping margins, surrounding the floral part of the spadix and sometimes much of its appendix as well; upper part concave or nearly flat, erect or bent forwards at the tip, but seldom completely concealing the spadix. Spadix with a long appendix. Flowers without perianths, unisexual, the male usually not touching; borne, in monoecious plants, immediately above the female. Stamens 1--5, united when more than 1. Ovary 1-celled; seeds few. Most species are normally dioecious but occasionally bear some abortive male flowers or filamentous sterile flowers above the female. *Mainly eastern Asia & the Himalaya, a few species in North America & east Africa.*

14. Pinellia Tenore. 6/2. Slender plants growing from tubers. Leaves basal, their blades simple, lobed or compound. Inflorescence arising from tuber separately from the leaves. Spathe narrow, with margins overlapping below and with a flat or channelled limb above. Flowers without perianths, unisexual. Spadix with zone of female flowers entirely united with the spathe; male zone free; appendix slender. Male flowers with 1 or 2 stamens. Ovary 1-celled. Berry with a single seed. *Eastern Asia.*

19. SPARGANIACEAE

Aquatic or semi-terrestrial, monoecious, perennial herbs. Stems floating or erect, from a creeping rhizome. Leaves linear, in 2 ranks, hairless, sheathing the stem at the base. Flowers in globular heads (capitula), of which the upper are male and the lower female. Perianth-scales brown, 3 or 4 in female flowers, 1--6 in male flowers. Stamens 1--8. Ovary superior with 1 cell. Fruit indehiscent, with 1 seed.

A family of a single genus.

1. Sparganium Linnaeus. 20/3. Description as for the family. *North temperate areas.*

20. TYPHACEAE

Perennial, monoecious herbs, aquatic or semi-terrestrial. Flowers in cylindric spikes (partial inflorescences), male above female. Male flowers with stamens in clusters of 1--3, or rarely as many as 8; filaments irregularly fused. Female flowers on stalks which bear hair-like branches; ovary superior with 1 cell; style persistent. Fruit dry, usually dehiscent, with 1 seed. See figure 16, p. 92.

A family of a single genus, distributed throughout the world in suitable freshwater habitats, but generally absent from the tropics.

1. Typha Linnaeus. 15/4. Hairless perennial herbs, with stout, creeping rhizomes. Stems erect. Leaves mostly basal, in 2 ranks, erect, with sheath closely enclosing stem. *Cosmopolitan.* Figure 16, p. 92.

21. CYPERACEAE

Usually perennial herbs, with solid stems which are often triangular in section, and linear leaves sometimes reduced to sheaths. Florets inconspicuous, wind-pollinated, bisexual or unisexual, subtended by a small bract (glume) and arranged in spikes. Perianth absent or represented by bristles persisting round the base of the 1-seeded indehiscent fruit (nut). See figures 10, p. 58, & 17, p. 94.

A large cosmopolitan family with about 100 genera and 4000 species, mostly growing in wet places, and of little horticultural importance. In addition to the 4 genera treated here, species of **Eriophorum** Linnaeus are sometimes grown for their decorative, cottony fruiting heads; they would perhaps be more widely cultivated if they did not require wet, peaty soils.

90

1a. Flowers all unisexual; female flowers enclosed in a sac (utricle) which
    persists round the fruit   **4. Carex**
  b. Flowers mostly bisexual; utricle absent   2
2a. Inflorescence a simple terminal spike   3
  b. Inflorescence of several (often many) spikes   4
3a. Fruit with persistent style-base spike without a basal bract   **2. Eleocharis**
  b. Fruit without a persistent style-base; spike with a more or less leafy basal
    bract   **1. Scirpus**
4a. Spikes more or less cylindric; glumes and flowers not in 2 ranks
    1. Scirpus
  b. Spikes compressed, glumes and flowers in 2 ranks   **3. Cyperus**

1. Scirpus Linnaeus. Many/few. Rhizomatous or tufted, mostly perennial herbs of varied habit, sometimes more or less leafless. Inflorescence usually compound, with bisexual flowers arranged spirally in spikes. Perianth represented by bristles at base of ovary, sometimes absent. Stamens and stigmas 3 or 2. Fruit a nut. *Cosmopolitan.* Figure 10, p. 58.

2. Eleocharis R. Brown. Many/1. Tufted or rhizomatous, leafless herbs, with cylindric or angled stems sheathed at base and bearing solitary, bractless spikes. Flowers as in *Scirpus.* Fruit a nut with persistent style-base. *Cosmopolitan.*

3. Cyperus Linnaeus. 500/6. Tufted or rhizomatous herbs with terminal, umbellate or capitate inflorescences. Spikes many-flowered, flattened; flowers bisexual, in 2 ranks. Perianth absent; stamens usually 3. Fruit a nut. *Tropics & subtropics.* Figure 10, p. 58.

4. Carex Linnaeus. 1500/20. Rhizomatous or tufted perennials, with grass-like leaves in 3 rows, and stems often triangular in section. Inflorescence compound, terminal, containing (in all cultivated species) both male and female flowers grouped in spikes. Perianth absent; male flowers consisting of 2 or 3 stamens, female consisting of an ovary with 2 or 3 stigmas and enclosed in a sac or utricle. Fruit a small nut enclosed in the persistent utricle. *Mainly north temperate areas.* Figures 10, p. 58 & 17, p. 94.

22. ZINGIBERACEAE

Perennial herbs, often highly aromatic; rootstock usually fleshy. Leaves arranged in 2 ranks or in a spiral, leaf-sheaths encircling the stem, each commonly with a membranous outgrowth (ligule) at the junction of sheath and leaf. Inflorescence terminal on a tall or short leafy shoot, or from a separate, leafless shoot. Flowers usually arising in the axils of bracts, each flower often subtended by a bracteole. Calyx and corolla tubular, petals 3. Functional stamen 1, sometimes with a crest at apex. the remainder of the staminal whorl modified into 1 usually showy staminode (lip) and, in most cases, 2 variously formed lateral staminodes. Ovary inferior, commonly 3-celled with axile placentation. Fruit a dehiscent capsule or fleshy berry.

    A family with over 40 genera and about 1000 species, found mainly in the tropics of the Old World but with some representatives in the New World and in subtropical Asia. It is remarkably uniform in vegetative structure. Tropical members yield the spices ginger, turmeric and cardamom. Only 1 genus has species generally hardy in Europe.

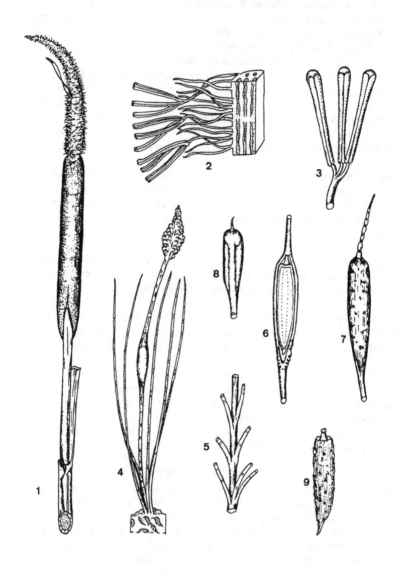

Figure 16. Typhaceae. *Typha latifolia*. 1, Complete inflorescence, male above, female below. 2, Part of main axis with three male flowers. 3, A single male flower. 4, Part of axis with a single female flower. 5, Base of the stalk of the ovary showing bristles. 6, Longitudinal section of ovary. 7, Young fruit. 8, Abortive ovary. 9, Seed.

1. Roscoea J.E. Smith. 17/6. Inflorescence terminal on the leafy stem, sometimes long-stalked. Bracts overlapping, each subtending a single flower. Bracteoles absent. Perianth usually purple. Upper petal hooded. Lip entire or 2-lobed, sometimes each lobe notched. Staminodes petaloid. Anther versatile, spurred at the base. Fruit elongate, not splitting readily. *China, Himalaya.*

Only 1 species hardy as far north as Britain. One species of **Cautleya** Hooker, which has yellow perianths and fruits splitting readily, may be grown in mild areas.

23. CANNACEAE

Perennial rhizomatous herbs with generally unbranched, erect, leafy shoots. Leaves large, alternate, pinnately veined, with sheathing bases. Inflorescence a spike, raceme or panicle, commonly with 2 flowers in the axil of each bract. Flowers bisexual, asymmetric. Calyx of 3 usually free sepals. Corolla of 3 petals, united below into a tube of varying length. Stamens forming the showy part of the flower; in the cultivated species represented by 4 petaloid staminodes (one of which, the lip, is usually smaller and reflexed) and a fertile, petaloid stamen bearing a 1-celled anther on its margin, all united below to form a tube or a short united region. Ovary inferior, 3-celled, with many ovules, papillose or warty; style single, flat or club-shaped, conspicuous in the centre of the flower. Fruit a 3-valved capsule bearing the persistent sepals at its apex.

A family consisting of a single genus with about 50 species, native in the tropics, subtropics and warm temperate regions of the New World.

1. Canna Linnaeus. ?50/several. Description as for family. *Southern USA, tropical America.*

Several species are grown as ornamentals (often planted out in summer, protected in a cool house in winter) and numerous hybrids have been produced. The spectacular flowers and foliage of the hybrid Cannas make them important as summer bedding plants in the warmer parts of Europe. Most of them are complex hybrids.

24. MARANTACEAE

Perennial, rhizomatous herbs. Leaves generally in 2 ranks, winged sheath usually well-developed; blade pinnately veined, often noticeably asymmetric, with a pulvinus at its junction with the petiole (the leaf-stalk consists of 3 parts: the winged sheath, the unwinged petiole and the pulvinus). Inflorescence a spike, raceme or head made up of 2-ranked or spiral bracts with paired flowers, or groups of paired flowers in their axils; rarely a more diffuse panicle. Flowers bisexual, asymmetric, often in mirror-image pairs. Calyx of 3 nearly free sepals. Corolla of 3 petals united below into a tube of variable length. Stamens borne on the corolla, basically in 2 whorls, the outer whorl represented by 1 or 2 petaloid staminodes (rarely absent), the inner by a fertile stamen with a 1-celled anther, often bearing a petaloid structure in addition, and 1 or 2 petaloid staminodes of specialised structure. Ovary inferior, 3-celled or 1-celled by abortion, with a single ovule in each cell. Fruit a capsule, berry or nut.

A family of 32 genera and about 350 species occurring in the tropics and subtropics of both the New and Old Worlds, but with its principal development in the former. Many are grown under glass, mainly for the decorative, coloured and patterned foliage, and some flower only rarely or never. One species is occasionally grown out-of-doors in northern Europe.

93

Figure 17. Cyperaceae. *Carex acutiformis.* 1, Inflorescence with one terminal and one lateral male spike. 2, Female spikes (borne below the male). 3, Young male flower. 4, Mature male flower. 5, Female flower. 6, Longitudinal section of fruit showing utricle. 7, Transverse section of ovary. 8, Transverse section of mature fruit. 9, Mature fruit.

1. Thalia Linnaeus. 12/2. Tall marsh plants with leaves basal, with long stalks. Inflorescence a rather loose panicle with many flowers, borne on a long scape. Bracts deciduous. Corolla-tube short. Outer staminode 1, petaloid. Ovary 1-celled due to abortion of the other cells, with 1 ovule. Stigma 2-lipped. Fruit indehiscent. *Tropical & warm temperate America, Africa.*

25. ORCHIDACEAE

Plants usually with rhizomes, sometimes with root-tubers, growing in soil, on rocks or as epiphytes, some (not ours) without chlorophyll. Aerial roots frequent. Stems usually present, often fleshy and swollen, when referred to as pseudobulbs. Leaves deciduous or persistent, borne on the stems (more rarely directly on the rhizomes) in 2 ranks or spirally arranged, rolled or folded when young, often hard and leathery. Flowers bilaterally symmetric or rarely asymmetric, solitary or in racemes or panicles borne terminally or laterally, usually on scapes which bear bracts; usually the flowers are turned upside down by a twist of 180° in the flower-stalk and ovary (flowers resupinate). Sepals 3, free or united. Petals 3, 2 of them usually similar to each other and to the sepals, the other (the lowermost in resupinate flowers) usually different in shape, size and colour, and referred to as the lip. Stamen(s), style and stigmas united into a solid structure known as the column. Anthers 2 or more frequently 1, borne at or near the apex of the column. Pollen aggregated into powdery or waxy masses known as pollinia, of which there may be 2, 4, 6, 8 or more. Pollinia usually borne on a sticky pad (the viscidium), or more rarely on 2 viscidia, to which they may be attached by pollinial tissue, or by a structure of different hardness and texture known as the stipe. Stigmas 3, the lateral 2 receptive and usually joined, forming a hollow on the inner surface of the column below the anther, or rarely borne on stalks; the median stigma not receptive, variously modified, borne just above the fertile 2 and beneath the anther, forming a structure known as the rostellum which is sometimes conspicuous, and contributes material to the viscidium. Ovary inferior, usually 1-celled, more rarely 3-celled, placentation parietal. Ovules numerous. Fruit a capsule. Seeds numerous, very small and dust-like. See figures 18, p. 96, 19, p. 98 & 20, p. 104.

The Orchidaceae, with about 20,000 species, is probably the largest flowering plant family. It is extremely variable both vegetatively and florally, and the flowers show numerous adaptations to insect pollination. They vary from the extremely beautiful to the bizarre, and the plants hold a rather special place in gardening, with many enthusiasts growing them (and little else) on a large scale. The literature on the family is extensive, including many illustrated works, taxonomic accounts and horticultural guides, and there are several journals devoted solely to the family (see below).

Most orchids are of tropical origin, and require greenhouse protection throughout the whole of Europe; about 25 genera contain species which are hardy at least in parts of northern Europe.

The identification of orchids can be difficult, particularly at the genus level, and because the plants are so variable and so different from those of most other plant families, a specialised terminology has been developed. The important terms used in this account are briefly defined below, and some indication is given of how the various characters are to be observed.

Growth habit. Orchids grow either monopodially (i.e. with a growing point that continues vegetative growth from year to year, the inflorescences being borne laterally), or sympodially (with a growing point that ceases growth after some time, generally by flowering, further growth being continued by a lateral bud formed on the

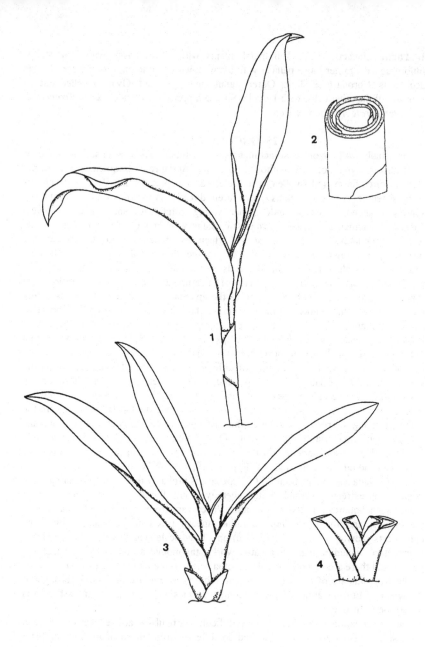

Figure 18. Orchidaceae. 1, Stem with rolled leaves. 2, Transverse section of the bases of two rolled leaves. 3, Stem with folded leaves. 4, Transverse section of the bases of several folded leaves.

96

older growth). This feature is often quite difficult to appreciate in individual specimens, and is not used, as such, in the key to genera. It is important, however, in that monopodial orchids do not produce pseudobulbs, whereas sympodial orchids may do so.

Pseudobulbs. A pseudobulb is a swollen fleshy stem; whether or not a particular plant has pseudobulbs or more normal stems is a very important feature in its generic identification. In a few species and genera there is no sharp dividing line between a 'normal' stem and a pseudobulb; the keys to the genera have been constructed, as far as possible, to allow for these to be identified whether they have been considered as having a stem or a pseudobulb. The pseudobulbs, if present, are borne on the rhizomes, and may consist of several internodes, when they will bear leaves or leaf-scars along their lengths; such pseudobulbs are described here as 'compound'. Alternatively, they may consist of a single internode, when they bear leaves or leaf-scars only at the apex; such pseudobulbs are described here as 'simple'. The pseudobulbs may be close together or distant or, in a few genera, they form chains built up of several pseudobulbs borne one on top of the other.

Leaves. The manner in which the leaves are packed when young is an important character in identification of the genus. The young leaves may be rolled, so that one margin overlaps the other and the back of the young leaf is rounded, or they may be folded once longitudinally. This character is generally easy to see in most plants when young leaves are present, and it is usually quite easy to see in mature leaves, as the rolling or folding persists at the base. In some genera which bear few leaves it may be difficult to decide on this character when the leaves are mature; the following features help in making a decision. Folded leaves generally retain a longitudinal line or small fold down the middle over most of their length, and usually have a definite keel on the outside towards the base; very hard, leathery leaves are almost all folded. Rolled leaves are often pleated or have prominent veins, do not usually have a single line or fold down the middle, and are rounded on their backs towards the base. See figure 18, p. 96.

Inflorescences. These may be terminal on the stem or pseudobulb, or lateral; in the latter case the inflorescence can arise in a leaf-axil or a leaf-scar axil on the pseudobulb or stem, or from the rhizome at the bases of the pseudobulbs.

Flowers. In all the keys and descriptions the flowers are considered to be resupinate (i.e. with their stalks and ovaries twisted through 180° so that the lip petal is lowermost with the column above it) unless the contrary is stated. In species with arching or hanging inflorescences the degree of resupination of the individual flowers may vary, those flowers towards the hanging apex being non-resupinate (they are effectively turned upside down by the hanging of the inflorescence), while those at the more erect base may be properly resupinate. However, in a few genera with hanging inflorescences the flowers are all strictly resupinate.

In many genera the base of the column is prolonged at an angle to the rest, forming a column-foot. In all genera with a foot the lip is attached at or near the end of it. In many of these genera the lateral sepals are also borne on the sides of the foot, which results in a humped, angular, chin-like projection being visible on the outside of the back of the flower. This projection is known as a mentum, and is of some importance in identification. See figure 19, p. 98.

Pollinia. Much of the classification of the family is based on the structure of the pollinia and their associated organs. In this account we have attempted to use these characters as little as possible, but the pollinia cannot be completely ignored,

97

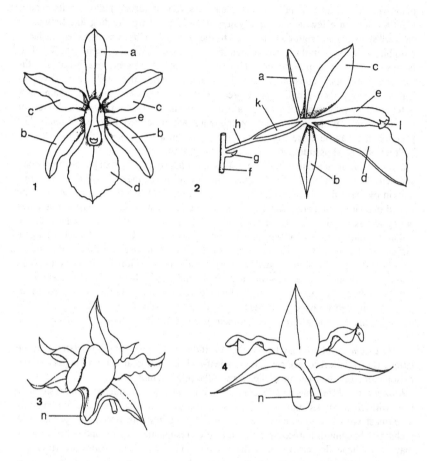

Figure 19. Orchidaceae. 1, A generalised orchid flower (a, upper sepal; b, lateral sepal; c, petal; d, lip; e, column). 2, Longitudinal section of such a flower (a–e as above; f, inflorescence-stalk; g, bract; h, flower-stalk; k, ovary; l, anther). 3, 4, Front and back views of a flower with a pronounced mentum (n, mentum).

98

particularly their number. They can be extracted from the flower by sliding a needle slowly upwards along the inner side of the column, when they will attach themselves to the needle, and can be examined through a hand lens (a magnification of ×10 or ×15 is usually sufficient). The number of pollinia can usually easily be seen, though care must be taken, as the pollinia are occasionally not all extracted at once.

A few species have 2 deeply bilobed pollinia, which can look like 4; again, other species may have extra, infertile pollinia which are shrunken and colourless. In most of the tropical species in cultivation the pollinia are waxy and of well-defined shape. In some hardy species they are granular or powdery and in a few others they are each made up of several to many individual packets (known technically as massulae) which separate very easily from each other.

The column. This is very variable in shape, size and details of structure: a selection is shown in the illustration.

1a. Annual growth persisting throughout the year   2
  b. Annual aerial growth dying off completely each autumn or winter, leaving no part above ground except, occasionally, a rosette of small leaves or a solitary leaf formed after the current year's growth has died down   3
2a. Plant with distinct pseudobulbs or stems swollen above or below, not usually hairy   **20. Liparis**
  b. Plant without pseudobulbs or swollen stems; stems densely hairy   **6. Goodyera**
3a. Lip slipper-like   4
  b. Lip not slipper-like   5
4a. Stem with 2 or more separated nodes and 2 or more leaves; lateral sepals pointing downwards, usually fused   **1. Cypripedium**
  b. Stem with 1 node and 1 leaf; lateral sepals pointing upwards, free   **21. Calypso**
5a. A spur, or rarely 2 spurs, present, arising from the base of the lip or the base of the central sepal   6
  b. Spur(s) absent   17
6a. Flower with the lip, which is similar to the petals, uppermost (i.e. flower not resupinate); spur 2 mm or less   **15. Nigritella**
  b. Flower with the lip, which is different from the petals, lowermost (i.e. flower resupinate); spur more than 2 mm   7
7a. Spur 2.5 cm or more   8
  b. Spur less than 2.5 cm   10
8a. Lip strap-shaped, unlobed   **18. Platanthera**
  b. Lip not strap-shaped, variously lobed and dissected   9
9a. Uppermost leaf tightly sheathing the stem; column with 1 stigmatic area   **13. Orchis**
  b. Uppermost leaf not tightly sheathing the stem; column with 2 stalked stigmatic areas   **19. Habenaria**
10a. Spur 10 mm or more, slender   11
  b. Spur less than 10 mm, slender or not   13
11a. Spur cylindric, obtuse; flowers usually yellow   **14. Dactylorhiza**
  b. Spur thread-like, acute; flowers purple, pink or white   12
12a. Raceme conical; flowers with a foxy smell, lip deeply 3-lobed with 2 basal ridges   **12. Anacamptis**

b. Raceme cylindric; flowers fragrant; lip shallowly 3-lobed, without basal ridges **16. Gymnadenia**

13a. Lip spirally twisted or its lateral lobes with wavy margins 14

b. Lip not spirally twisted, margins not wavy 15

14a. Central lobe of lip 2 or more times longer than the lateral lobes, strap-shaped, spirally twisted; bracts shorter than flowers **10. Himantoglossum**

b. Central lobe of lip less than 2 times longer than the lateral lobes, oblong, not spirally twisted; bracts longer than the flowers **11. Barlia**

15a. Lip strap-shaped, terminating in 3 short teeth; flowers greenish **17. Coeloglossum**

b. Lip not strap-shaped, variously lobed; flowers not greenish 16

16a. Root-tubers ovoid; leaves mostly in a basal rosette, the uppermost usually completely sheathing the stem; all bracts membranous, shorter than the flowers **13. Orchis**

b. Root-tubers lobed; leaves not in a basal rosette; upper leaves sheathing the stem at their base only, often transitional to the leaf-like bracts; bracts usually as long as or longer than the flowers **14. Dactylorhiza**

17a. Stem with 2 more or less opposite leaves; lip strap-shaped, 2-lobed at the apex **4. Listera**

b. Stem with more than 2 leaves which are distributed along it or are in a basal rosette; lip not as above 18

18a. Flowers arranged in a tight spiral **5. Spiranthes**

b. Flowers not arranged in an obvious spiral 19

19a. Lip divided into 2 parts by a constriction in the middle 20

b. Lip not divided into 2 parts 22

20a. Roots tuberous; leaves not pleated; bracts large, coloured like the sepals, sheathing the basal parts of the flowers **8. Serapias**

b. Roots fibrous, tubers absent; leaves pleated; bracts green and leaf-like, not sheathing the basal parts of the flowers 21

21a. Flowers shortly stalked, borne mostly on 1 side of the axis and horizontal or downwardly pointing; sepals and petals spreading **2. Epipactis**

b. Flowers not stalked, borne all round the axis, pointing upwards; sepals and petals not widely spreading **3. Cephalanthera**

22a. Lip velvety or covered with short hairs or with a geometric pattern or with a shining area or any combination of these, if lobed the lobes not long and slender **7. Ophrys**

b. Lip without any of the above characteristics, the lateral and central lobes slender, like the arms and legs of a man **9. Aceras**

1. Cypripedium Linnaeus. 35/13. Terrestrial, aerial parts dying away in winter. Leaves deciduous, usually pleated, rolled when young, spirally arranged or in a single, almost opposite pair. Flowers persistent, solitary or up to 12 per raceme; bracts leaf-like. Sepals edge-to-edge in bud. Petals and upper sepal free, lateral sepals usually fused into a single lower sepal below the lip. Lip mostly extended into a sac-like pouch with an inturned rim, the opening partially blocked by a staminode. Stigmatic surface, 2 fertile anthers and staminode borne on the column. Rostellum absent. Pollen in exposed granular masses. Seeds fusiform in a single-celled ovary. *Northern hemisphere.*

2. Epipactis Zinn. 20/5. Perennial herbs with creeping or vertical rhizomes and numerous fleshy but not tuberous roots. Aerial growth dying off in winter. Leaves spaced along the stem, spirally arranged or in 2 opposite rows, strongly veined, pleated. Flowers held horizontally or hanging, often only on 1 side of the axis. Bracts green and leaf-like, not sheathing the basal parts of the flowers. Sepals and petals spreading or sometimes scarcely opening. Lip in 2 distinct parts, the basal cup-like, the apical flat and variously shaped. Spur absent. Pollinia rapidly breaking up. *North temperate areas.*

3. Cephalanthera Richard. 14/3. Perennial herbs with erect or short, creeping rhizomes. Roots fibrous. Aerial growth dying down in winter. Leaves evenly spaced along stem, pleated. Flowers stalkless or very shortly stalked in a loose terminal spike, pointing upwards; bracts green and leaf-like, not sheathing the basal parts of the flowers. Sepals and petals hooded, scarcely opening. Lip constricted at the middle into 2 distinct parts, the base concave and clasping the base of the column, the apex with several ridges. Spur absent. Pollinia 2, club-shaped, each longitudinally divided, powdery. *North temperate areas.*

4. Listera R. Brown. 10/2. Herbaceous perennials with short rhizomes. Aerial growth dying down in winter. Leaves 2 (rarely 3--4), ovate, almost opposite, arising at about the middle of the stem or below it. Flowers in a spike-like raceme. Sepals and petals more or less equal in size. Lip longer than the sepals and petals, strap-shaped, its apex 2-lobed. Spur absent. Pollinia 2, club-shaped, each longitudinally divided and made up of easily separable masses. *Temperate areas.*

5. Spiranthes Richard. 30/4. Perennial herbs with slender or sturdy, fusiform tubers. Aerial growth dying off in winter. Leaves all basal in a rosette or evenly distributed on the stem, ovate-elliptic or lanceolate, rolled when young. Inflorescence elongate, twisted, with many flowers which are close, spirally arranged and often scented. Sepals and petals almost equal, incurved, the tips free. Lip entire, its apical margins variously frilled or not. Spur absent. Pollinia 2, each made up of easily separable masses. *Temperate areas, mainly North America.*

6. Goodyera R. Brown. 80/5. Terrestrial herbs, rhizomes creeping, sometimes above ground, aerial growth persistent through the winter. Leaves membranous or fleshy, often asymmetric, mostly basal in rosettes, more rarely scattered on erect stems, often with differently coloured veins, rolled when young. Inflorescence erect, flowers few to many, arranged spirally or on 1 side, or uniformly in a cylindric spike or raceme. Sepals directed forwards or spreading. Petals directed forwards over the column. Lip saccate, the sac usually with hairs inside, entire, the tip usually pointed and reflexed. Column short, blunt or with 2 long or short teeth on the rostellum. Pollinia 2, club-shaped, each longitudinally divided and made up of easily separable masses. *North temperate areas, southeast Asia, Australasia.*

7. Ophrys Linnaeus. 50/10. Perennial herbs. Tubers 2 (occasionally 3), spherical, fleshy. Leaves lanceolate, oblong or ovate, rolled when young, mostly in a basal rosette, sometimes spaced along the stem when usually sheathing, usually appearing in autumn after the current season's growth has died back. Flowers few to *c.* 15 in a short or long, loose or rarely dense spike. Bracts leaf-like, often inrolled. Sepals spreading, usually greenish or pinkish, oblong or ovate, obtuse, all equal in length, the upper concave, erect or curved over the column. Petals usually smaller and narrower than the sepals, often hairy or velvety. Lip spurless, hairy, complex in structure, often like a bee, wasp or other insect, oblong, square, rounded or diamond-shaped, usually marked with a coloured, hairless, often complex, patterned area (the speculum), lobed or not, with or without humps or horn-like basal protuberances; apex often with a short, tooth-like, deflexed or upcurved appendage. Column erect or curved forwards; anther-connective beak-like, its apex obtuse or pointed. Pollinia 2, club-like, made up of small, easily separable masses. *Europe, North Africa, southwest Asia.*

8. Serapias Linnaeus. 6/3. Terrestrial herbs. Roots tuberous, tubers 2--4, occasionally 5, stalkless and in some species also produced at the end of root-like stolons, fleshy, ovoid. Stems erect. Leaves lanceolate, mostly basal or spaced along the stem, rolled when young, later often folded. Aerial growth dying off during the current year. Spike loose- or dense-flowered; bracts very conspicuous, boat-shaped, usually purplish or glaucous. Sepals and petals forming a pointed hood, usually coloured like the bracts, veins distinct; sepals lanceolate; petals ovate at base, tapering to a long point. Lip 3-lobed, composed of a basal and an apical part, usually hairy on the upper surface, lateral lobes upturned, central lobe usually abruptly bent downwards, tongue-like, the basal part with a swollen area or a double ridge at the base. Pollinia 2, club-shaped, consisting of easily separable masses. Column apex beak-like. *Azores, Mediterranean area to the Caucasus.*

9. Aceras R. Brown. 1/1. Herbaceous perennial. Aerial growth usually dying down in winter or plant sometimes overwintering as a basal rosette. Stems to *c.* 50 cm arising from 2 ovoid tubers. Leaves unspotted, the lower crowded, spreading, lanceolate, obtuse, the upper smaller, erect, sheathing the stem. Spike to 20 cm, bearing 50 or more flowers. Sepals obovate-lanceolate, petals linear, slightly shorter, all yellowish or greenish yellow, hooded over the column. Lip *c.* 1.2 cm, shaped like a man, the lateral lobes being the arms and the divided central lobe the legs. Spur absent. Pollinia 2, club-shaped, composed of easily separable masses. *West Europe, Mediterranean area.*

One species of the Australian genus **Pterostylis** R. Brown may be grown in areas with a very mild climate; it has flowers with a lip which is sensitive to touch, but small and hidden in a hood formed by the sepals and petals.

10. Himantoglossum Koch. 9/1. Herbaceous perennials with 2 ovoid tubers. Aerial growth dying down in winter. Leaves in a basal rosette and arranged along the stem. Spike elongate, bracts equalling the flowers. Sepals and petals free, forming a hood over the column. Lip long, strap-shaped, 3-lobed, the central lobe bifid and spirally twisted, 2 or more times longer than the lateral lobes. A single spur present, arising from the base of the lip. Pollinia 2, club-shaped, composed of easily separable masses, attached to a single viscidium. *Europe, southwest Asia.*

11. Barlia Parlatore. 1/1. Very similar to *Himantoglossum* but bracts as long as or longer than the flowers, the sepals erect, not hooded, and the lip much shorter, its central lobe not spirally twisted. *Canary Islands, Mediterranean area.*

12. Anacamptis Richard. 1/1. Perennial herbs. Tubers spherical or ovoid. Stems to 75 cm with up to 8 leaves. Leaves mostly basal, unspotted, lanceolate, upper leaves shorter, rolled when young. Aerial growth dying down in winter. Spike to 8 cm, conical, becoming shortly cylindric, dense. Bracts linear, slightly longer than ovary. Flowers pink, red or purple, occasionally white, emitting a faint, fox-like smell. Upper sepal and petals incurved, lateral sepals spreading. Lip deeply 3-lobed, as broad as long or broader than long, lobes oblong, apex obtuse or truncate; base of lip with 2 erect ridges leading towards the mouth of the spur. Spur 1--1.4 cm, slender, downward-pointing. Pollinia 2, club-like, made up of easily separated masses. *Europe, southwest Asia.*

13. Orchis Linnaeus. 35/10. Perennial herbs. Aerial growth dying down in winter. Tubers 2 or 3, spherical or ovoid. Leaves mostly in a basal rosette, spotted or unspotted, stem-leaves usually sheath-like. Flowers in a short or long, dense- or loose-flowered spike. Bracts membranous, Sepals and petals almost equal, free, all curved forward over the column, or the lateral sepals spreading and the upper sepal and petals incurved and hooded over the column. Lip mostly 3-lobed, sometimes unlobed. Spur 8--25 mm, arising from the base of the lip. Pollinia 2, club-like, powdery. *Europe, North Africa, Asia.*

Dactylorhiza is often included under *Orchis* and most nurserymen still offer plants under the latter name.

14. Dactylorhiza Nevski. 30/6 and several hybrids. Terrestrial. Root-tubers palmately lobed or divided into finger-like sections. Aerial parts dying down in winter. Leaves 3--15, spirally arranged on flowering stem, not forming a basal rosette at flowering time, sometimes spotted or blotched with brownish purple; upper leaves often small and bract-like. Flowers several to many, in compact, spherical, conical or cylindric racemes. Bracts leaf-like, green or purplish green. Sepals and petals free; lateral sepals spreading or erect: upper sepal and petals curved forwards forming a hood over the column. Lip simple or 3-lobed, spurred, usually marked with lines, dots or streaks. Column short, rostellum 3-lobed. *Europe, North Africa, North America.* Figure 20, p. 104.

15. Nigritella Richard. 1/1. Herbaceous, terrestrial perennial. Tubers 2, palmately lobed. Stems to 28 cm. Leaves 6--11, crowded, linear to narrowly lanceolate, channelled, mostly basal, unspotted. Aerial growth dying down in winter. Spike conical becoming ovoid or elongate, dense. Bracts slender, as long as or longer than flowers. Flowers *c.* 5 mm, blackish crimson, red, yellowish or rarely white, vanilla-scented; sepals and petals linear to lanceolate; tips acute. Lip uppermost, triangular to lanceolate-ovate, entire. Spur *c.* 2 mm, conical, shorter than ovary. Pollinia 2, granular. *Scandinavia to northern Spain & Greece.*

One species of the genus **Satyrium** Swartz may be cultivated in areas with a very mild climate; it has non-resupinate flowers with 2 backwardly-pointing spurs.

16. Gymnadenia R. Brown. 10/2. Herbaceous, terrestrial perennials. Tubers palmately lobed. Leaves linear to lanceolate, arranged along lower part of stem, unspotted. Aerial growth dying down in winter. Flowers in a cylindric spike, sweetly scented. Upper sepal and petals incurved to form a hood over the column; lateral sepals spreading almost horizontally. Lip shallowly 3-lobed. Spur slender, shorter or longer than ovary. Pollinia 2, club-like, powdery. *Europe, temperate Asia.*

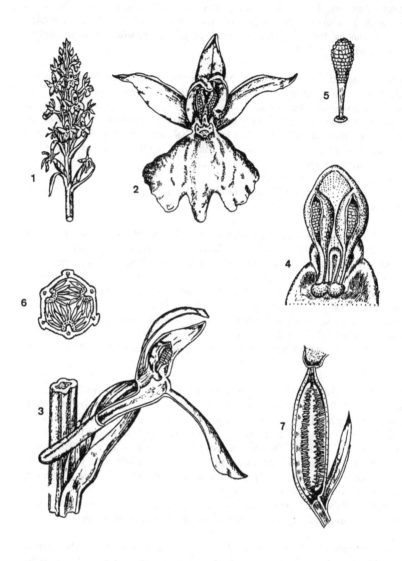

Figure 20. Orchidaceae. *Dactylorhiza fuchsii.* 1, Inflorescence. 2, Flower from the front. 3, Side view of flower with part of the perianth removed. 4, Upper part of the column showing stamens and pollinia. 5, Single pollinium. 6, Transverse section of the ovary. 7, Longitudinal section of the ovary.

17. Coeloglossum Hartmann. 2/2. Herbaceous, terrestrial perennials with 2 ovoid tubers. Leaves unspotted, the lower ovate or oblong, obtuse, the upper smaller, lanceolate and pointed. Aerial growth dying down in winter. Flowers slightly scented, 5--25 in a loose cylindric spike. Sepals and petals hooded over the column, greenish or yellow-green often tinged reddish, sepals ovate, petals relatively narrower, almost linear. Lip strap-shaped, the apex shortly 2-lobed with or without a short middle tooth, with a short, rounded spur. Pollinia 2, club-shaped, attached to separate viscidia. *North temperate areas.*

18. Platanthera Richard. 100/2. Herbaceous, terrestrial perennials. Tubers 2 or more, spindle-shaped, elongate. Stems erect, unbranched. Leaves basal or on the stem. Aerial growth dying down in winter. Raceme usually with many small or medium-sized, white or greenish flowers. Sepals and petals free, upper sepal and petals usually incurved over the column, lateral sepals spreading or recurved. Lip entire with the base extended into a spur. Anthers distinctly separated. Pollinia club-shaped, granular. *North temperate areas.*

19. Habenaria Willdenow. 800/5. Terrestrial herbs. Roots tuberous, fleshy, elongate or ovoid. Stems erect. Leaves spaced along stem or sometimes basal, linear to lanceolate, sheathing at base, rolled when young. Aerial growth dying down in winter. Raceme few- to many-flowered. Upper sepal often forming a hood with the petals; lateral sepals spreading (in one species the basal parts of the sepals, petal lobes, column and lip are united). Lip variously lobed or fringed, base spurred, spur slender, long. Column short or long, slender or thick. Pollinia 2, club-like, breaking up into easily separable masses. Stigmas 2, borne on club-like arms, or stalkless. *Temperate North America, eastern Asia, tropics & subtropics.*

20. Liparis Richard. 200/4. Perennial herbs, terrestrial, growing on rocks or epiphytic. Rhizome short. Stems fleshy, thicker at base, sometimes pseudobulbous. Leaves 1--7, membranous or leathery, jointed at the base or not, rolled or folded when young. Inflorescence lateral, racemose with few to many flowers. Sepals and petals oblong, reflexed, the petals usually narrower and a little shorter than the sepals. Lip broadly oblong to rounded, sharply bent backwards at or about the middle. Column long and slender, slightly winged. Pollinia 4, in 2 pairs, waxy. *Mostly Old World tropics.*

Species of two other genera which have pseudobulbs may be grown in areas with a very mild climate; they are: **Bletilla** Reichenbach (racemes terminal, pollinia 8) and **Pleione** D. Don (racemes lateral, lip not bent backwards, rolled and concealing the base of the column, pollinia 4).

21. Calypso Salisbury. 1/1. Terrestrial, 5--20 cm. Rootstock a corm producing a solitary leaf which dies down in winter. Leaf 3--10 cm, broadly ovate, pleated, rolled when young. Scape bearing a terminal flower. Sepals and petals lanceolate, spreading and ascending, often twisted, purplish pink. Lip 1.5--2.5 cm, inflated and saccate, with 2 small horns at the base, whitish to pale pink, marked with purple and with white or yellow hairs twoards the base. Pollinia waxy, flat, in 2 pairs fixed to a detachable viscidium. *North temperate areas.*

DICOTYLEDONS

Plants herbaceous or woody. Primary root (taproot) often persisting, enlarging and branched. Seedling leaves (cotyledons) usually 2, lateral. Leaves alternate, opposite or whorled, rarely absent, veins usually forming a branched network. Parts of the flower usually in 2s, 4s or 5s, or numerous.

Key to Families

Key to groups

1a. Petals present, free from each other at their bases (rarely united
    above the base), usually falling as individual petals, or petals absent   2
  b. Petals present and all united at the base into a longer or shorter tube,
    usually falling as a complete corolla   9
2a. Flowers unisexual and without petals, at least the males borne in catkins
    which are usually deciduous; plants always woody   *Group 1*
  b. Flowers with or without petals, unisexual or bisexual, never in catkins;
    plants woody or not   3
3a. Ovary consisting of 2 or more carpels which are completely free from
    each other, their styles also completely free from each other
                                  Group 2 (p. 107)
  b. Ovary consisting of a single carpel or of 2 or more carpels which are
    united to each other wholly or in the greater part, rarely the bodies of
    the carpels free but the styles completely united   4
4a. Perianth of 2 or more whorls, more or less clearly differentiated into
    calyx and corolla (calyx rarely small and obscure; excluding aquatic
    plants with minute, quickly deciduous petals and branch-parasites with
    opposite, leathery leaves)   5
  b. Perianth of a single whorl (which may be corolla-like) or perianth
    completely absent, more rarely the perianth of 2 or more whorls but
    the segments not or scarcely differing from whorl to whorl   8
5a. Stamens more than twice as many as the petals   *Group 3* (p. 108)
  b. Stamens twice as many as the petals or fewer   6
6a. Ovary partly or fully inferior   *Group 4* (p. 111)
  b. Ovary completely superior   7
7a. Placentation axile, apical, basal, marginal or free-central   *Group 5* (p. 112)
  b. Placentation parietal   *Group 6* (p. 115)
8a. Stamens borne on the perianth or ovary inferior (perianth of female
    flowers sometimes very small)   *Group 7* (p. 115)
  b. Stamens free from the perianth, ovary superior or naked (i.e. not
    surrounded by a perianth)   *Group 8* (p. 117)
9a. Ovary partly or fully inferior   *Group 9* (p. 119)
  b. Ovary completely superior   10
10a. Corolla radially symmetric   *Group 10* (p. 120)
  b. Corolla bilaterally symmetric   *Group 11* (p. 123)

Group 1

1a. Leaves pinnate   2
  b. Leaves simple and entire, toothed or lobed (sometimes deeply so)   3
2a. Leaves without stipules; fruit a nut   **27. JUGLANDACEAE**

    b. Leaves with stipules; fruit a legume   **87. LEGUMINOSAE**
3a. Leaves opposite, evergreen, entire; fruit berry-like   **145. GARRYACEAE**
    b. Leaves alternate, deciduous or evergreen; fruit not berry-like   4
4a. Ovules many, parietal; seeds many, cottony-hairy; male catkin erect with
    the stamens projecting between the bracts, or hanging and with
    fringed bracts   **28. SALICACEAE**
    b. Ovules solitary or few, not parietal; seeds few, not cottony-hairy; male
    catkins not as above   5
5a. Leaves dotted with aromatic glands   **26. MYRICACEAE**
    b. Leaves not dotted with aromatic glands   6
6a. Styles 3, each often branched; fruit splitting into 3 mericarps; seeds with
    appendages   **94. EUPHORBIACEAE**
    b. Styles 1--6, not branched; fruit and seeds not as above   7
7a. Plants with milky sap   **33. MORACEAE**
    b. Plants with clear sap   8
8a. Male catkins compound, i.e., each bract with 2 or 3 flowers attached to it;
    styles 2   **29. BETULACEAE**
    b. Male catkins simple, i.e. each bract with a single flower attached to it;
    styles 1 or 3--6   **30. FAGACEAE**

Group 2
    1a. Trees with bark peeling off in plates; leaves palmately lobed; flowers
    unisexual in hanging, spherical heads   **80. PLATANACEAE**
    b. Combination of characters not as above   2
    2a. Perianth-segments and stamens borne independently below the ovary, or
    perianth absent (i.e. ovary superior or naked)   3
    b. Perianth-segments and stamens borne on a rim or cup which is itself borne
    below the ovary (i.e. ovary superior, perianth and stamens perigynous) 21
    3a. Aquatic plants with peltate leaves and flowers with 3 sepals
        64. NYMPHAEACEAE
    b. Terrestrial plants, or, if aquatic, then without peltate leaves and flowers
    with more than 3 sepals   4
    4a. Herbs, succulent shrubs or shrubs with yellow wood, or climbers with
    bisexual flowers and opposite leaves   5
    b. Trees or shrubs which are neither succulent nor with yellow wood, if
    climbers then with unisexual flowers and alternate leaves   10
    5a. Perianth absent   **66. SAURURACEAE**
    b. Perianth present   6
    6a. Leaves succulent; stamens in 1 or 2 whorls   **82. CRASSULACEAE**
    b. Leaves not succulent; stamens spirally arranged, not obviously in whorls 7
    7a. Petals fringed; fruits formed from each carpel borne on a common stalk
    (gynophore)   **79. RESEDACEAE**
    b. Petals (when present) not fringed, but sometimes modified for nectar-
    secretion; fruits formed from each carpel not borne on a common stalk 8
    8a. Leaves opposite or whorled; flowers small, stalkless, in axillary clusters;
    ovule 1, placentation basal   **39. PHYTOLACCACEAE**
    b. Combination of characters not as above   9

9a. Sepals differing among themselves, green; stamens ripening from the
    inside of the flower outwards, borne on a nectar-secreting disc
                                               69. PAEONIACEAE
  b. Sepals all similar, green or petal-like; stamens ripening from the outside
     of the flower inwards; disc absent   **60. RANUNCULACEAE**
10a. Leaves simple   11
  b. Leaves compound   20
11a. Sepals and petals 5 each   12
  b. Sepals and petals not 5 each   13
12a. Leaves opposite; stamens 5--10   **101. CORIARIACEAE**
  b. Leaves alternate; stamens more than 10   **68. DILLENIACEAE**
13a. Unisexual climbers   14
  b. Erect trees or shrubs, usually flowers bisexual   15
14a. Carpels many; seeds not U-shaped   **51. SCHISANDRACEAE**
  b. Carpels 3 or 6; seeds usually U-shaped **63. MENISPERMACEAE**
15a. Stamens each with a truncate connective which overtops the anther; fruit
    usually fleshy, formed from the closely contiguous products of several
    carpels (i.e. a syncarp); endosperm convoluted   **50. ANNONACEAE**
  b. Connectives of stamens not as above; fruit not as above; endosperm not
    convoluted   16
16a. Carpels spirally arranged on an elongate receptacle; stipules large, united,
    early deciduous leaving a ring-like scar   **48. MAGNOLIACEAE**
  b. Carpels in 1 whorl; stipules absent, minute, or united to the leaf-stalk, not
    leaving a ring-like scar when fallen   17
17a. Petals present   18
  b. Petals absent   19
18a. Sepals free, overlapping, more than 6; ovule 1 in each carpel
                                        52. ILLICIACEAE
  b. Sepals 2--6, united or if free, then edge-to-edge in bud; ovules more than
    1 in each carpel   **49. WINTERACEAE**
19a. Leaves in whorls; flowers bisexual; sepals minute or absent
                                      58. EUPTELEACEAE
  b. Leaves opposite or alternate; flowers unisexual; sepals 4
                                59. CERCIDIPHYLLACEAE
20a. Unisexual climbers, or erect shrubs with blue fruits; perianth parts in 3s
                                   62. LARDIZABALACEAE
  b. Erect shrubs, fruits not blue; perianth parts not in 3s **69. PAEONIACEAE**
21a. Flowers unisexual; leaves evergreen   **53. MONIMIACEAE**
  b. Flower bisexual; leaves usually deciduous   22
22a. Inner stamens sterile; perianth of many segments; leaves usually opposite
                             54. CALYCANTHACEAE
  b. Stamens all fertile; perianth of 4--9 segments; leaves usually alternate
                                    86. ROSACEAE

Group 3
  1a. Herbaceous climber; leaves palmately divided into stalked leaflets;
     petals 2, stamens 8   **91. TROPAEOLACEAE**
    b. Combination of characters not as above   2
  2a. Perianth and stamens hypogynous, borne independently below the superior

ovary   3

  b. Perianth and stamens either perigynous, borne on the edge of a rim or cup which itself is borne below the superior ovary, or epigynous, borne on the top or the sides of the partly or fully inferior ovary   29

3a. Placentation axile or free-central   4

  b. Placentation marginal or parietal   19

4a. Placentation free-central; sepals 2   **42. PORTULACACEAE**

  b. Placentation axile; sepals usually more than 2   5

5a. Leaves all basal, tubular, forming insect-trapping pitchers; style peltately dilated   **74. SARRACENIACEAE**

  b. Leaves not as above; style not peltately dilated   6

6a. Leaves alternate   7

  b. Leaves opposite or rarely whorled   17

7a. Anthers opening by terminal pores   8

  b. Anthers opening by slits   9

8a. Shrubs with simple leaves without stipules, often covered with stellate hairs; stamens inflexed in bud; fruit a berry   **71. ACTINIDIACEAE**

  b. Combination of characters not as above   **116. ELAEOCARPACEAE**

9a. Perianth-segments of inner whorl tubular or bifid, nectar-secreting; fruit a group of partly to fully coalescent follicles   **60. RANUNCULACEAE**

  b. Combination of characters not as above   10

10a. Leaves with translucent aromatic glands   **96. RUTACEAE**

  b. Leaves without such glands   11

11a. Sap milky; flowers unisexual   **94. EUPHORBIACEAE**

  b. Sap not milky; flowers bisexual   12

12a. Succulent herb with spines; bark hard and resinous; stamens 15 in groups of 3 of which the central is the largest   **90. GERANIACEAE**

  b. Combination of characters not as above   13

13a. Stipules absent; leaves evergreen   **72. THEACEAE**

  b. Stipules present; leaves usually deciduous   14

14a. Filaments free; anthers 2-celled   15

  b. Filaments united into a tube, at least around the ovary, often also around the style; anthers often 1-celled   16

15a. Nectar-secreting disc absent; stamens more than 15; leaves simple   **117. TILIACEAE**

  b. Nectar-secreting disc present, conspicuous; stamens 15; leaves dissected   **92. ZYGOPHYLLACEAE**

16a. Styles divided above, several; stipules often persistent; carpels 5 or more   **118. MALVACEAE**

  b. Style 1, stigma head-like or lobed, 1--several; stipules usually deciduous; carpels 2--5   **119. STERCULIACEAE**

17a. Sepals united, falling as a unit; fruit separating into boat-shaped units   **70. EUCRYPHIACEAE**

  b. Sepals and fruit not as above   18

18a. Leaves simple, without stipules, often with translucent glands; stamens often united in bundles   **73. GUTTIFERAE**

  b. Leaves pinnate, without translucent glands; stamens not united in bundles   **92. ZYGOPHYLLACEAE**

19a. Aquatic plants with cordate leaves; style and stigmas forming a disc on
   top of the ovary   **64. NYMPHAEACEAE**
   b. Combination of characters not as above   20
20a. Carpel 1 with marginal placentation   21
   b. Carpels 2 or more, placentation parietal   22
21a. Leaves bipinnate or modified into phyllodes (flattened leaf-stalks), with
   stipules   **87. LEGUMINOSAE**
   b. Leaves various but not as above, without stipules   **60. RANUNCULACEAE**
22a. Leaves modified into active insect-traps, the 2 halves of the blade fringed
   and closing rapidly when stimulated   **75. DROSERACEAE**
   b. Combination of characters not as above   23
23a. Leaves opposite   24
   b. Leaves alternate   26
24a. Styles numerous; floral parts in 3s   **76. PAPAVERACEAE**
   b. Styles 1--5; floral parts in 4s or 5s   25
25a. Style 1; stamens not united in bundles; leaves without translucent glands
                                                       126. CISTACEAE
   b. Styles 3--5, free or variously united below; stamens united in bundles
   (sometimes apparently free); leaves with translucent or blackish glands
                                                       73. GUTTIFERAE
26a. Sepals 2 or rarely 3, quickly deciduous   **76. PAPAVERACEAE**
   b. Sepals 4--8, persistent in flower   27
27a. Leaves scale-like; styles 5   **127. TAMARICACEAE**
   b. Leaves not as above; styles 1, 2 or 3   28
28a. Ovary closed at the apex, borne on a stalk (gynophore); none of the petals
   fringed   **77. CAPPARIDACEAE**
   b. Ovary open at apex, not borne on a stalk; at least some of the petals
   fringed   **79. RESEDACEAE**
29a. Flowers unisexual; leaf-bases oblique   **132. BEGONIACEAE**
   b. Flowers bisexual; leaf-bases not oblique   30
30a. Placentation free-central; ovary partly inferior   **42. PORTULACACEAE**
   b. Placentation not free-central; ovary either completely superior or
   completely inferior   31
31a. Aquatic plants with cordate leaves   **64. NYMPHAEACEAE**
   b. Terrestrial plants; leaves various   32
32a. Stamens united into bundles on the same radii as the petals; staminodes
   often present; plants usually rough with stinging hairs   **130. LOASACEAE**
   b. Combination of characters not as above   33
33a. Sepals 2, united, falling as a unit as the flower opens; plants herbaceous
                                                       76. PAPAVERACEAE
   b. Sepals 4 or 5, usually free, not falling as a unit; mostly trees or shrubs   34
34a. Carpels 8--12, superposed   **137. PUNICACEAE**
   b. Carpels fewer, side-by-side   35
35a. Leaves with stipules   36
   b. Leaves without stipules   37
36a. Leaves opposite or in whorls; plants woody   **84. CUNONIACEAE**
   b. Leaves alternate; plants woody or herbaceous   **86. ROSACEAE**
37a. Leaves with translucent aromatic glands; style 1   **136. MYRTACEAE**

  b Leaves without such glands; styles usually more than 1

83. SAXIFRAGACEAE

Group 4

  1a. Petals and stamens numerous; plant succulent   2
    b. Petals and stamens each 10 or fewer; plants usually not succulent   3
  2a. Stems succulent, often very spiny; leaves usually absent, very reduced or
       falling early   **47. CACTACEAE**
    b. Stems and leaves succulent, spines usually absent   **41. AIZOACEAE**
  3a. Placentation parietal, placentas sometimes intrusive   4
    b. Placentation axile, basal, apical or free-central   5
  4a. Climbing herbs with tendrils; flowers unisexual   **133. CUCURBITACEAE**
    b. Erect herbs or shrubs, if climbing, then without tendrils; flowers usually
       bisexual   **83. SAXIFRAGACEAE**
  5a. Stamens as many as and on the same radii as the petals; trees or shrubs
       with simple leaves   **114. RHAMNACEAE**
    b. Stamens more numerous than petals or, if as numerous, then not on the
       same radii as them; plants herbaceous or woody, leaves simple or
       compound   6
  6a. Flowers borne in umbels, these sometimes condensed into heads or in
       superposed whorls; leaves usually compound   7
    b. Flowers not in umbels; leaves usually simple   8
  7a. Fruit splitting into 2 mericarps; flowers usually bisexual; petals overlapping
       in bud; usually herbs without stellate hairs   **147. UMBELLIFERAE**
    b. Fruit a berry; flowers often unisexual; petals edge-to-edge in bud; plants
       mostly woody, often with stellate hairs   **146. ARALIACEAE**
  8a. Style 1   9
    b. Styles more than 1, often 2 and divergent   14
  9a. Floating aquatic herb; leaf-stalks inflated   **135. TRAPACEAE**
    b. Terrestrial herbs, trees or shrubs; leaf-stalks not inflated   10
  10a. Ovule 1 in each cell of the ovary   11
    b. Ovules 2--numerous in each cell of the ovary   13
  11a. Petals overlapping in bud; flowers often unisexual   **142. NYSSACEAE**
    b. Petals edge-to-edge in bud; flowers usually bisexual   12
  12a. Stamens with swollen, hairy filaments; petals recurved **141. ALANGIACEAE**
    b. Stamens without swollen, hairy filaments; petals not recurved

144. CORNACEAE

  13a. Sap milky; petals 5; ovary 3-celled   **193. CAMPANULACEAE**
    b. Sap watery; petals 2 or 4; ovary usually 4-celled   **138. ONAGRACEAE**
  14a. Stipules absent   **83. SAXIFRAGACEAE**
    b. Stipules present, though sometimes early deciduous   15
  15a. Ovary with half or more superior and half or less inferior; leaves opposite

84. CUNONIACEAE

    b. Ovary mostly or completely inferior; leaves alternate   16
  16a. Anthers opening by valves; cells of ovary as many as styles; stellate hairs
       frequent   **81. HAMAMELIDACEAE**
    b. Anthers opening by slits; cells of ovary ultimately twice as many as styles;
       stellate hairs absent   **86. ROSACEAE**

Group 5

1a. Perianth bilaterally symmetric   2
 b. Perianth radially symmetric (the stamens sometimes not so due to deflexion)   13
2a. Anthers cohering above the ovary like a cap   **108. BALSAMINACEAE**
 b. Anthers free, not as above   3
3a. Anthers opening by terminal pores   **100. POLYGALACEAE**
 b. Anthers opening by longitudinal slits or by valves   4
4a. Plants herbaceous   5
 b. Plants woody (shrubs, trees or climbers)   9
5a. Leaves with stipules   6
 b. Leaves without stipules   7
6a. Carpel 1; fruit a legume, sometimes 1-seeded   **87. LEGUMINOSAE**
 b. Carpels 5; fruit a capsule or berry, or splitting into mericarps   **90. GERANIACEAE**
7a. Sepals, petals and stamens borne independently below the ovary (rarely the petals and stamens somewhat united at the base); leaves peltate   **91. TROPAEOLACEAE**
 b. Sepals, petals and stamens borne on a rim, cup or tube which itself is borne below the ovary; leaves not peltate   8
8a. Leaves opposite   **134. LYTHRACEAE**
 b. Leaves alternate or all basal   **83. SAXIFRAGACEAE**
9a. Stamens as many as or fewer than petals, borne on the same radii as them   **106. SABIACEAE**
 b. Stamens more numerous than the petals or, if as many or fewer, not on the same radii as them   10
10a. Carpel 1   **87. LEGUMINOSAE**
 b. Carpels 2 or more   11
11a. Leaves opposite, palmate; sepals united at the base   **105. HIPPOCASTANACEAE**
 b. Leaves alternate, usually pinnate; sepals free   12
12a. Stipules large, borne between the bases of the leaf-stalks; plants not climbing   **107. MELIANTHACEAE**
 b. Stipules absent, or, if present, not borne as above and plants climbing   **104. SAPINDACEAE**
13a. Anthers opening by terminal pores   14
 b. Anthers opening by longitudinal slits or by valves   17
14a. Low shrubs with unisexual flowers; stamens 4, petals 4, each usually 2- or 3-lobed, rarely a few unlobed   **116. ELAEOCARPACEAE**
 b. Combination of characters not as above   15
15a. Carpels 3; style divided above into 3 stigmas; nectar-secreting disc absent   **149. CLETHRACEAE**
 b. Carpels usually 4 or 5, very rarely 3; style undivided or divided into 4 or 5 branches; nectar-secreting disc present around the base of the ovary   16
16a. Petals about as long as broad, not clawed; evergreen herbs or low shrubs; style divided above into 4--5 stigmas, rarely unlobed; seeds each with a wing-like projection at each end   **150. PYROLACEAE**

b. Petals usually longer than broad, somewhat clawed; evergreen or deciduous shrubs; style undivided, stigmas as many as carpels, borne within a cup-like sheath; seeds various, rarely as above   **151. ERICACEAE**

17a. Placentation free-central (ovary rarely with septa below) or basal   18

b. Placentation axile or apical   23

18a. Stamens as many as petals and on the same radii as them   19

b. Stamens more or fewer than petals, if as many then not on the same radii as them   22

19a. Small evergreen shrubs; leaves with translucent dots: flowers unisexual, sepals and petals 4; fruit a 1-seeded drupe   **154. MYRSINACEAE**

b. Combination of characters not as above   20

20a. Anthers opening by valves; stigma 1   **61. BERBERIDACEAE**

b. Anthers opening by longitudinal slits; stigmas more than 1   21

21a. Sepals 5; ovule 1, basal on a long, curved stalk   **156. PLUMBAGINACEAE**

b. Sepals 2 or rarely 3; ovules usually numerous, rarely 1 and then not on a long curved stalk   **42. PORTULACACEAE**

22a. Ovary lobed, consisting of several rounded humps, the style arising from a depression between them; leaves pinnatisect   **88. LIMNANTHACEAE**

b. Ovary not lobed, style terminal; leaves simple, entire   **44. CARYOPHYLLACEAE**

23a. Petals and stamens both numerous; plants with succulent leaves and stems   **41. AIZOACEAE**

b. Combination of characters not as above   24

24a. Small hairless annual herb growing in water or on wet mud; seeds pitted   **129. ELATINACEAE**

b. Combination of characters not as above   25

25a. Sepals, petals and stamens borne on a rim, cup or tube (perigynous zone) which itself is inserted below the ovary   26

b. Sepals, petals and stamens inserted individually below the ovary (rarely the stamens united to the bases of the petals)   32

26a. Stamens as many as the petals and borne on the same radii as them   **114. RHAMNACEAE**

b. Stamens more or fewer than the petals, or if of the same number, then not on the same radii as them   27

27a. Style 1   28

b. Styles more than 1, often 2 and divergent   29

28a. Perigynous zone not prominently ribbed; seeds with arils; mostly trees, shrubs or climbers   **111. CELASTRACEAE**

b. Perigynous zone prominently ribbed; seeds without arils; mostly herbs   **134. LYTHRACEAE**

29a. Fruit an inflated, membranous capsule; leaves mostly opposite, compound   **112. STAPHYLEACEAE**

b. Combination of characters not as above   30

30a. Leaves opposite, leathery   **84. CUNONIACEAE**

b. Leaves usually alternate or all basal, not leathery if opposite   31

31a. Trees or shrubs; hairs often stellate; anthers usually opening by valves; fruit a few-seeded, woody capsule   **81. HAMAMELIDACEAE**

b. Herbs or shrubs; hairs simple or absent; anthers opening by longitudinal slits; fruit a capsule, not woody   **83. SAXIFRAGACEAE**

32a. Leaves with translucent, aromatic glands   **96. RUTACEAE**
  b. Leaves without such glands   33
33a. Sap often milky; flowers unisexual; styles 3, often further divided
     94. EUPHORBIACEAE
  b. Combination of characters not as above   34
34a. Flower with a well-developed disc, usually nectar-secreting, below and
    around the ovary   35
  b. Flower without a disc, nectar secreted in other ways   42
35a. Stamens as many as and on the same radii as the petals   **115. VITACEAE**
  b. Stamens more or fewer than the petals, or if of the same number not on
    the same radii as them   36
36a. Resinous trees or shrubs   **102. ANACARDIACEAE**
  b. Herbs, shrubs or trees, not resinous, sometimes aromatic   37
37a. Plant herbaceous   **92. ZYGOPHYLLACEAE**
  b. Plant woody (tree, shrub or climber)   38
38a. Flowers, or at least some of them, functionally unisexual (i.e. anthers
    not producing pollen or ovary containing no ovules)   39
  b. Flowers functionally bisexual   40
39a. Leaves alternate; ovary with 2--5 carpels, not flattened
     98. SIMAROUBACEAE
  b. Leaves opposite; ovary with 2 (rarely 3) carpels, usually flattened
     103. ACERACEAE
40a. Leaves entire or toothed; stamens 4 or 5, emerging from the disc; seeds with
    arils   **111. CELASTRACEAE**
  b. Combination of characters not as above   41
41a. Leaves without stipules, not fleshy; filaments united into a tube
     99. MELIACEAE
  b. Leaves with stipules, fleshy; filaments free   **92. ZYGOPHYLLACEAE**
42a. Plant herbaceous   43
  b Plant woody (tree, shrub or climber)   45
43a. Leaves always simple; ovary 6--10-celled by the development of 3--5
    secondary septa during maturation of the flower   **93. LINACEAE**
  b. Leaves lobed or compound; secondary septa absent   44
44a. Leaves with stipules   **90. GERANIACEAE**
  b. Leaves without stipules   **89. OXALIDACEAE**
45a. Filaments of stamens united below   46
  b. Filaments of stamens entirely free from each other   47
46a. Stamens 2   **160. OLEACEAE**
  b Stamens 3 or more   **119. STERCULIACEAE**
47a. Stamens 8--10   **124. STACHYURACEAE**
  b. Stamens 3--6   48
48a. Staminodes present in flowers which also contain fertile stamens
     109. CYRILLACEAE
  b. Staminodes absent from flowers with fertile stamens, present only in
    female flowers   49
49a. Trees with opposite, pinnate leaves; twigs tipped with large, dark buds;
    fruit a samara   **160. OLEACEAE**
  b. Combination of characters not as above   50
50a. Sepals united at the base   **110. AQUIFOLIACEAE**

114

b. Sepals entirely free from one another   51
51a. Ovule 1 per cell; petals 3--4   **97. CNEORACEAE**
  b. Ovules many per cell; petals 5   **85. PITTOSPORACEAE**

Group 6

1a. Sepals, petals and stamens perigynous, borne on a rim or cup which
   itself is inserted below the ovary   2
  b. Sepals, petals and stamens hypogynous, inserted individually below
   the ovary   3
2a. Annual, aquatic herb; stamens 6   **78. CRUCIFERAE**
  b. Combination of characters not as above   **83. SAXIFRAGACEAE**
3a. Perianth bilaterally symmetric   4
  b. Perianth radially symmetric   7
4a. Ovary open at apex; some or all of the petals fringed **79. RESEDACEAE**
  b. Ovary closed at the apex; no petals fringed   5
5a. Petals and stamens 5; carpels 3   **123. VIOLACEAE**
  b. Petals and stamens 4 or 6; carpels 2   6
6a. Ovary borne on a stalk (gynophore); stamens projecting beyond petals
                             77. CAPPARIDACEAE
  b. Ovary not borne on a stalk; stamens not projecting beyond petals
                             76. PAPAVERACEAE
7a. Petals and stamens numerous   **41. AIZOACEAE**
  b. Petals and fertile stamens each fewer than 7   8
8a. Stamens alternating with much-divided staminodes   **83. SAXIFRAGACEAE**
  b. Stamens not alternating with much-divided staminodes   9
9a. Leaves insect-trapping and -digesting by means of stalked, glandular hairs
                             75. DROSERACEAE
  b. Leaves not as above   10
10a. Climbers with tendrils; ovary and stamens borne on a common stalk
   (androgynophore); corona present   **125. PASSIFLORACEAE**
  b. Combination of characters not as above   11
11a. Petals 4, the outer pair 3-fid; sepals 2   **76. PAPAVERACEAE**
  b. Petals not as above; sepals 4 or 5   12
12a. Stamens usually 6, 4 longer and 2 shorter, rarely reduced to 2; carpels 2;
   fruit usually with a secondary septum   **78. CRUCIFERAE**
  b. Stamens 4--10, all more or less equal; carpels 2--5, fruit without a
   secondary septum   13
13a. Petals each with a scale-like appendage at the base of the blade; leaves
   opposite   **128. FRANKENIACEAE**
  b. Petals without appendages; leaves alternate or all basal   14
14a. Stipules present   **123. VIOLACEAE**
  b. Stipules absent   15
15a. Leaves alternate, scale-like   **127. TAMARICACEAE**
  b. Leaves usually all basal, normally developed   **150. PYROLACEAE**

Group 7

1a. Aquatic plants or rhubarb-like marsh plants with cordate leaves   2
  b. Terrestrial plants, not as above   5

2a. Stamens 8, 4 or 2; leaves either deeply divided or cordate at the base
                                 139. HALORAGACEAE
  b. Stamens 6 or 1; leaves undivided, not cordate at the base   3
3a. Stamens 6; leaves all basal   **78. CRUCIFERAE**
  b. Stamen 1; leaves opposite or in whorls   4
4a. Leaves in whorls; fruits small, indehiscent, dry, 1-seeded, not lobed
                                140. HIPPURIDACEAE
  b. Leaves opposite; fruit 4-lobed with up to 4 seeds   **173. CALLITRICHACEAE**
5a. Trees or shrubs   6
  b. Herbs, climbers or parasites   20
6a. Flowers small, the central one bisexual, the rest male, in heads subtended
    by 2 large, white bracts   **143. DAVIDIACEAE**
  b. Inflorescence and flowers not as above   7
7a. Plant covered by scales; fruit enclosed in the berry-like, persistent, fleshy
    calyx   **121. ELAEAGNACEAE**
  b. Plant not covered by scales; fruit not as above   8
8a. Stamens as many as, and on radii alternating with, the perianth-segments
                                  114. RHAMNACEAE
  b. Stamens not arranged as above   9
9a. Stamens 4, situated at the tops of the spoon-shaped, petal-like perianth-
    segments   **35. PROTEACEAE**
  b. Combination of characters not as above   10
10a. Stipules present, sometimes falling early   11
  b. Stipules absent   15
11a. Ovary of a single carpel   12
  b. Ovary of 2--6 united carpels   13
12a. Stamens numerous, borne on the reflexed inner surface of the funnel-
    shaped perianth   **86. ROSACEAE**
  b. Stamens 10, perianth not as above, the stamens not borne on it
                                  87. LEGUMINOSAE
13a. Styles 3--6; fruit a nut surrounded by a scaly cupule   **30. FAGACEAE**
  b. Styles 2; fruit not as above   14
14a. Leaves alternate; stellate hairs usually present; fruit a woody capsule
                              81. HAMAMELIDACEAE
  b. Leaves opposite; stellate hairs absent; fruit a non-woody capsule
                              84. CUNONIACEAE
15a. Ovary of a single carpel   16
  b. Ovary of 2 or 3 united carpels   18
16a. Ovary inferior   **36. SANTALACEAE**
  b. Ovary superior   17
17a. Leaves with aromatic glands; stamens 8 or more, all borne at more or
    less the same level in the perigynous zone   **55. LAURACEAE**
  b. Leaves without aromatic glands; stamens 2 or 8--10, when borne at
    different levels in the perigynous zone   **120. THYMELEACEAE**
18a. Ovary superior   **103. ACERACEAE**
  b. Ovary inferior   19
19a. Placentation parietal   **83. SAXIFRAGACEAE**
  b. Placentation axile   **144. CORNACEAE**

20a. Branch-parasites with green, forked branches or flowers stalkless on
    branches of host   **37. LORANTHACEAE**
  b. Plants not parasitic as above   21
21a. Perianth absent; flowers in spikes   **66. SAURURACEAE**
  b. Perianth present; flowers not usually in spikes   22
22a. Leaf-base oblique; ovary inferior, 3-celled   **132. BEGONIACEAE**
  b. Leaf-base not oblique; ovary not as above   23
23a. Ovary superior   24
  b. Ovary inferior   29
24a. Carpel 1, containing a single apical ovule; perianth tubular
    120. THYMELEACEAE
  b. Combination of characters not as above   25
25a. Carpels 3 (rarely 2), ovule 1, basal; perianth persistent in fruit; leaves
    usually alternate, entire   26
  b. Combination of characters not as above   27
26a. Leaves without stipules; stamens 5   **43. BASELLACEAE**
  b. Leaves usually with stipules united into a sheath (ochrea); stamens usually
    6--9   **38. POLYGONACEAE**
27a. Leaves alternate, usually lobed or compound   **86. ROSACEAE**
  b Leaves opposite, usually entire   28
28a. Ovule 1, fruit a nut; stipules translucent and papery or rarely absent
    44. CARYOPHYLLACEAE
  b. Ovules numerous; fruit a capsule; stipules absent   **134. LYTHRACEAE**
29a. Leaves pinnate; ovary open at apex   **131. DATISCACEAE**
  b. Leaves not pinnate; ovary closed at apex   30
30a. Ovary 6-celled; perianth 3-lobed or tubular and bilaterally symmetric
    67. ARISTOLOCHIACEAE
  b. Combination of characters not as above   31
31a. Ovules 1--5; seed 1   32
  b. Ovules and seeds numerous   33
32a. Perianth-segments thickening in fruit; leaves alternate
    45. CHENOPODIACEAE
  b. Perianth-segments not as above; leaves opposite or alternate
    36. SANTALACEAE
33a. Styles 2; placentation parietal   **83. SAXIFRAGACEAE**
  b. Style 1; placentation axile   **138. ONAGRACEAE**

Group 8
  1a. Aquatic plants, either submerged or at least partially covered by flowing
    water; leaves whorled, much divided   **65. CERATOPHYLLACEAE**
   b. Terrestrial plants, not as above   2
  2a. Stipules present, sometimes falling early   3
   b. Stipules entirely absent   12
  3a. Ovary 1-celled, containing a single ovule   4
   b. Ovary 1--several-celled, containing more than a single ovule   7
  4a. Styles 2--4, usually 3, free   **38. POLYGONACEAE**
   b. Style 1, sometimes divided above into 2 stigmas   5
  5a. Ovule basal; herbs or shrubs, flowers never sunk in a fleshy receptacle
    34. URTICACEAE

   b. Ovule apical; trees, shrubs or woody climbers, if herbs then flowers sunk in a fleshy receptacle   6

6a. Sap watery; style 1; flowers often bisexual   **31. ULMACEAE**

   b. Sap milky; styles usually 2, rarely 1; flowers usually unisexual

                              33. MORACEAE

7a. Placentation parietal or free-central   8

   b. Placentation axile   9

8a. Shrubs or trees; leaves alternate; placentation parietal

                      122. FLACOURTIACEAE

   b. Herbs; leaves usually opposite; placentation free-central

                      44. CARYOPHYLLACEAE

9a. Sap milky; styles usually 3, often divided; ovules 1 or 2 per cell

                      94. EUPHORBIACEAE

   b. Combination of characters not as above (ovules occasionally 2 per cell) 10

10a. Stellate hairs usually present; stamens 5 or 10, filaments united below

                      119. STERCULIACEAE

   b. Stellate hairs absent; stamens not as above   11

11a. Style 1; trees or shrubs   **116. ELAEOCARPACEAE**

   b. Styles 3 or 4; herbs   **66. SAURURACEAE**

12a. Ovary 1-celled, containing a single ovule   13

   b. Ovary 1--several-celled, containing more than a single ovule   19

13a. Tree with milky sap; styles 2   **32. EUCOMMIACEAE**

   b. Combination of characters not as above   14

14a. Stamens with filaments united, at least at the base   **46. AMARANTHACEAE**

   b. Stamens with the filaments completely free   15

15a. Styles 2--5, completely free   16

   b. Style 1 or style absent, stigmas stalkless on the ovary   17

16a. Stamens 9; perianth petal-like, of 6 segments united below; flowers in an umbel or head partly enclosed in an involucre of 4--8-lobed bracts

                      38. POLYGONACEAE

   b. Stamens usually 5; perianth not as above, generally of 2--5, free, sepal-like segments; inflorescence not as above   **45. CHENOPODIACEAE**

17a. Leaves opposite or in whorls   **40. NYCTAGINACEAE**

   b. Leaves alternate   18

18a. Leaves with translucent, aromatic glands; stamens with enlarged connectives; fruit a syncarp   **50. ANNONACEAE**

   b. Leaves without translucent aromatic glands; stamens without enlarged connectives; fruit a berry   **39. PHYTOLACCACEAE**

19a. Styles 2 or more, free for all or most of their length   20

   b. Style 1, sometimes lobed above into 2 or more stigmas   24

20a. Ovary of 2 or 3 cells   21

   b. Ovary of 4 or more cells   22

21a. Ovary usually 3-celled (rarely 2-celled); ovules 1 or 2 per cell, all fertile

                      113. BUXACEAE

   b. Ovary 2-celled, though the septum is incomplete; ovules 2 per cell, but only 1 of the 4 developing into a seed   **95. DAPHNIPHYLLACEAE**

22a. Ovules 1 per cell; shrubs or large herbs   **39. PHYTOLACCACEAE**

   b. Ovules many per cell; trees   23

23a. Leaves alternate; carpels 4   **56. TETRACENTRACEAE**

b. Leaves in whorls; carpels 6--10   **57. TROCHODENDRACEAE**

24a. Placentation parietal or free-central   25

b. Placentation axile, apical or basal   26

25a. Placentation parietal; perianth-segments 2, free   **76. PAPAVERACEAE**

b. Placentation free-central; perianth-segments 5, united below   **155. PRIMULACEAE**

26a. Leaves modified into insect-trapping pitchers   **74. SARRACENIACEAE**

b. Leaves not modified into insect-trapping pitchers   27

27a. Heath-like shrublets with narrow leaves with their margins revolute; ovule 1 per cell   **152. EMPETRACEAE**

b. Plants not as above; ovules 2 or more per cell   28

28a. Leaves opposite; stamens 2   **160. OLEACEAE**

b. Leaves alternate; stamens 3 or more   29

29a. Resinous trees or shrubs; leaves simple or compound; fruit 1-seeded, drupe-like; ovule 1 per cell, apical or basal   **102. ANACARDIACEAE**

b. Non-resinous trees, shrubs or climbers; leaves usually compound; fruit not as above; ovules 2 per cell, axile   **104. SAPINDACEAE**

Group 9

1a. Inflorescence a head surrounded by an involucre of bracts; ovule always solitary   2

b. Inflorescence and ovules not as above   3

2a. Each flower with a cup-like involucel; stamens 4, free; ovule apical   **192. DIPSACACEAE**

b. Involucel absent; stamens 5, their anthers united into a tube; ovule basal   **197. COMPOSITAE**

3a. Leaves alternate or all basal   4

b. Leaves opposite or whorled   13

4a. Anthers opening by pores; fruit a berry or drupe   **151. ERICACEAE**

b. Anthers opening by slits; fruit various   5

5a. Evergreen trees or shrubs; corolla white, campanulate; ovary half-inferior; placentation free-central, ovules few   **154. MYRSINACEAE**

b. Combination of characters not as above   6

6a. Climbers with tendrils and unisexual flowers; stamens 1--5; placentation parietal; fruit berry-like   **133. CUCURBITACEAE**

b. Combination of characters not as above   7

7a. Stamens 10--many; plants woody   8

b. Stamens 4--5; plants woody or herbaceous   10

8a. Leaves gland-dotted, smelling of eucalyptus; corolla completely united, unlobed, falling as a whole   **136. MYRTACEAE**

b. Combination of characters not as above   9

9a. Hairs stellate or scale-like; stamens in 1 series, anthers linear   **158. STYRACACEAE**

b. Hairs absent or not as above; stamens in several series; anthers broad   **159. SYMPLOCACEAE**

10a. Stigmas surrounded by a sheath   **194. GOODENIACEAE**

b. Stigmas not surrounded by a sheath   11

11a. Stamens as many as and on the same radii as the petals   **155. PRIMULACEAE**

119

    b. Stamens not as above   12

12a. Stamens 2 or 4, borne on the corolla; sap not milky   **184. GESNERIACEAE**

    b. Stamens 5 or more, free from the corolla; sap usually milky

                                                 193. CAMPANULACEAE

13a. Placentation parietal; stamens 2, or 4 and paired   **184. GESNERIACEAE**

    b. Placentation axile or apical; stamens 1 or more, if 4 then not paired   14

14a. Stamens 1--3; ovary with 1 ovule   **191. VALERIANACEAE**

    b. Stamens 4 or more; ovary usually with 2 or more ovules   15

15a. Leaves divided into 3 leaflets; flowers few in a head   **190. ADOXACEAE**

    b. Leaves simple or rarely pinnate; inflorescence various, usually not as above   16

16a. Stipules usually borne between the bases of the leaf-stalks and sometimes looking like leaves; ovary usually 2-celled, more rarely 5-celled; flowers usually radially symmetric; fruit capsular, fleshy or schizocarpic

                                                 167. RUBIACEAE

    b. Stipules usually absent, when present not as above; ovary usually 3-celled (occasionally 2--5-celled), sometimes only 1 cell fertile; flowers often bilaterally symmetric; fruit a berry or drupe   **189. CAPRIFOLIACEAE**

Group 10

1a. Stamens 2   **160. OLEACEAE**

    b. Stamens more than 2   2

2a. Carpels several, free; leaves succulent   **82. CRASSULACEAE**

    b. Carpels united or, if the bodies of the carpels are free, then the styles united, rarely ovary of a single carpel; leaves usually not succulent   3

3a. Corolla papery, translucent, 4-lobed; stamens 4, projecting from the corolla; leaves with parallel veins, often all basal   **188. PLANTAGINACEAE**

    b. Combination of characters not as above   4

4a. Stamens more than twice as many as petals   5

    b. Stamens up to twice as many as petals   9

5a. Leaves with stipules; filaments of stamens united into a tube around the ovary and style   **118. MALVACEAE**

    b. Leaves without stipules; filaments free   6

6a. Anthers opening by pores   **71. ACTINIDIACEAE**

    b. Anthers opening by longitudinal slits   7

7a. Leaves with translucent, aromatic glands; calyx cup-like, unlobed

                                              96. RUTACEAE

    b. Leaves without such glands; calyx not as above   8

8a. Ovules 2 per cell; flowers usually unisexual   **157. EBENACEAE**

    b. Ovules many per cell; flowers bisexual   **72. THEACEAE**

9a. Stamens as many as petals and on the same radii as them   10

    b. Stamens more or fewer than petals, if as many then not on the same radii as them   14

10a. Placentation axile   **115. VITACEAE**

    b. Placentation basal or free-central   11

11a. Trees or shrubs; fruit a berry or drupe   **154. MYRSINACEAE**

    b. Herbs (occasionally woody at the extreme base); fruit a capsule or indehiscent   12

12a. Sepals 2, free   **42. PORTULACACEAE**

b. Sepals 4 or more, united   13

13a. Corolla persistent and papery in fruit; ovule 1 on a long stalk arising from the base of the ovary   **156. PLUMBAGINACEAE**

  b. Corolla not persistent and papery in fruit; ovules many, on a free-central placenta   **155. PRIMULACEAE**

14a. Flower compressed with 2 planes of symmetry; stamens united in 2 bundles   **76. PAPAVERACEAE**

  b. Combination of characters not as above   15

15a. Leaves bipinnate or replaced by phyllodes; carpel 1, fruit a legume   **87. LEGUMINOSAE**

  b. Combination of characters not as above   16

16a. Anthers opening by pores   17

  b. Anthers opening by longitudinal slits, or pollen in coherent masses (pollinia)   18

17a. Stamens free from corolla-tube, often twice as many as the corolla-lobes   **151. ERICACEAE**

  b. Stamens attached to the corolla-tube, as many as the corolla-lobes   **176. SOLANACEAE**

18a. Leaves alternate or all basal; carpels never 2 and free or almost so but with a single terminal style   19

  b. Leaves opposite or whorled, alternate only when carpels 2, free or almost so and style 1, terminal   33

19a. Plant woody, leaves usually evergreen, often spiny-margined; stigma not stalked, borne directly on the top of the ovary   **110. AQUIFOLIACEAE**

  b. Combination of characters not as above   20

20a. Procumbent herbs with milky sap and stamens free from the corolla-tube   **193. CAMPANULACEAE**

  b. Combination of characters not as above   21

21a. Ovary 5-celled   22

  b. Ovary 2-, 3- or 4-celled   23

22a. Leaves fleshy; anthers 2-celled; fruit often deeply lobed, schizocarpic   **175. NOLANACEAE**

  b. Leaves leathery; anthers 1-celled; fruit a capsule or berry   **153. EPACRIDACEAE**

23a. Ovary 3-celled   24

  b. Ovary 1-, 2- or 4-celled   25

24a. Dwarf evergreen shrublets; 5 staminodes usually present; petals imbricate in bud   **148. DIAPENSIACEAE**

  b. Herbs or climbers with tendrils; staminodes absent; petals each overlapping and overlapped by one other in bud   **168. POLEMONIACEAE**

25a. Flowers in spirally coiled cymes, or the calyx with appendages between the lobes; style terminal or arising from between the lobes of the ovary   26

  b. Flowers not in spirally coiled cymes, calyx without appendages; style terminal   27

26a. Style terminal; fruit a capsule, usually many-seeded   **170. HYDROPHYLLACEAE**

  b. Style arising from the depression between the 4 lobes of the ovary; fruit of up to 4 nutlets, or more rarely a 1--4-seeded drupe   **171. BORAGINACEAE**

27a. Placentation parietal   28
  b. Placentation axile   29
28a. Corolla-lobes edge-to-edge in bud; leaves simple and cordate or peltate, or of 3 leaflets, hairless; aquatic or marsh plants   **164. MENYANTHACEAE**
  b. Corolla-lobes variously overlapping in bud; leaves never as above; not aquatics or marsh plants   **184. GESNERIACEAE**
29a. Ovules 1 or 2 in each cell of the ovary   30
  b. Ovules 3--many in each cell of the ovary   32
30a. Arching shrubs with small purple flowers in clusters on the previous year's wood   **177. BUDDLEJACEAE**
  b. Combination of characters not as above   31
31a. Sepals free; corolla-lobes each overlapping and being overlapped by one other and infolded in bud; twiners, herbs or dwarf shrubs   **169. CONVOLVULACEAE**
  b. Sepals united; corolla-lobes not as above in bud; trees or shrubs   **171. BORAGINACEAE**
32a. Corolla-lobes folded, edge-to-edge or each overlapping and overlapped by one other in bud; septum of the ovary oblique, not in the horizontal plane   **176. SOLANACEAE**
  b. Corolla lobes variously overlapping but not as above in bud; septum of ovary in the horizontal plane   **178. SCROPHULARIACEAE**
33a. Trailing, heather-like shrublet   **151. ERICACEAE**
  b. Plant not as above   34
34a. Milky sap usually present; fruit usually of 2 almost free follicles united by a common style; seeds with silky appendages   35
  b. Milky sap absent; fruit a capsule or fleshy, carpels united; seeds without silky appendages   36
35a. Pollen granular; corona absent; corolla-lobes edge-to-edge in bud   **165. APOCYNACEAE**
  b. Pollen usually in coherent masses (pollinia); corona usually present; corolla-lobes edge-to-edge or overlapping in bud   **166. ASCLEPIADACEAE**
36a. Flowers in coiled cymes; usually herbs   **170. HYDROPHYLLACEAE**
  b. Flowers not in coiled cymes; herbs or shrubs   37
37a. Placentation parietal; carpels 2   38
  b. Placentation axile; carpels 2, 3 or 5   39
38a. Leaves compound; epicalyx present   **170. HYDROPHYLLACEAE**
  b. Leaves simple; epicalyx absent   **163. GENTIANACEAE**
39a. Stamens fewer than corolla-lobes   **172. VERBENACEAE**
  b. Stamens as many as corolla-lobes   40
40a. Carpels 5; shrubs with leaves with spiny margins   **162. DESFONTAINIACEAE**
  b. Carpels 2 or 3; herbs or shrubs, leaves not as above   41
41a. Leaves without stipules; carpels 3; corolla-lobes each overlapping and overlapped by 1 other in bud; herbs   **168. POLEMONIACEAE**
  b. Leaves with stipules (often reduced to a ridge between the leaf-bases); corolla-lobes variously overlapping, or edge-to-edge in bud; plants usually woody   42
42a. Corolla usually 5-lobed; stellate and/or glandular hairs absent   **161. LOGANIACEAE**

    b. Corolla 4-lobed; stellate and glandular hairs present   **177. BUDDLEJACEAE**

Group 11

1a. Stamens more numerous than the corolla-lobes, or anthers opening by pores   2
    b. Stamens as many as corolla-lobes or fewer, anthers not opening by pores   6

2a. Anthers opening by pores; leaves undivided; ovary of 2 or more united carpels   3
    b. Anthers opening by longitudinal slits; leaves dissected or compound; ovary of a single carpel   5

3a. The 2 lateral sepals petal-like; filaments united   **100. POLYGALACEAE**
    b. No sepals petal-like; filaments free   4

4a. Shrubs with alternate or apparently whorled leaves; stamens 4--25   **151. ERICACEAE**
    b. Herbs with opposite leaves; stamens 5   **163. GENTIANACEAE**

5a. Leaves pinnate, or of 3 leaflets; perianth not spurred   **87. LEGUMINOSAE**
    b. Leaves laciniate; upper petals spurred; upper sepal helmet-like or spurred   **60. RANUNCULACEAE**

6a. Stamens as many as corolla-lobes; corolla weakly bilaterally symmetric   7
    b. Stamens fewer than corolla-lobes; corolla strongly bilaterally symmetric   12

7a. Stamens on the same radii as the corolla-lobes; placentation free-central   **155. PRIMULACEAE**
    b. Stamens on different radii from the corolla-lobes; placentation axile   8

8a. Leaves of 3 leaflets, with translucent, aromatic glands; stamens 5, the upper 2 fertile, the lower 3 sterile   **96. RUTACEAE**
    b. Combination of characters not as above   9

9a. Ovary of 3 carpels; ovules many   **168. POLEMONIACEAE**
    b. Ovary of 2 carpels; ovules 4 or many   10

10a. Flowers in coiled cymes; fruit of up to four 1-seeded nutlets   **171. BORAGINACEAE**
    b. Flowers not in coiled cymes; fruit a many-seeded capsule   11

11a. Corolla-lobes each overlapping and overlapped by 1 other in bud; stamens 5, equal; leaves opposite; climber   **161. LOGANIACEAE**
    b. Corolla-lobes various overlapping in bud; stamens 4 or 5 and unequal; leaves usually alternate   **178. SCROPHULARIACEAE**

12a. Placentation axile; ovules 4 or many   13
    b. Placentation parietal, free-central, apical or basal; ovules many or 1 or 2   19

13a. Ovules numerous but not in vertical rows in each cell of the ovary   14
    b. Ovules 4, or more numerous but then in vertical rows in each cell of the ovary   16

14a. Seeds winged; mainly trees, shrubs or climbers with opposite, pinnate, digitate or rarely simple leaves   **180. BIGNONIACEAE**
    b. Seeds usually wingless; mainly herbs or shrubs with simple leaves   15

15a. Corolla-lobes variously overlapping in bud; septum of ovary in the horizontal plane; leaves opposite or alternate   **178. SCROPHULARIACEAE**

b. Corolla-lobes usually folded, edge-to-edge or overlapping and being overlapped by 1 other in bud; septum of ovary oblique, not in the horizontal plane; leaves alternate   **176. SOLANACEAE**

16a. Fruit a capsule; ovules 4--many, usually in vertical rows in each cell of the ovary   17

b. Fruit not a capsule; ovules 4, side-by-side   18

17a. Leaves all opposite, often prominently marked with cystoliths; flower-stalks without swollen glands at the base; capsule usually opening elastically, seeds usually on hooked stalks   **181. ACANTHACEAE**

b. Upper leaves alternate, cystoliths absent; flower-stalks with swollen glands at the base; capsule not elastic, seeds not on hooked stalks   **182. PEDALIACEAE**

18a. Style arising from the depression between the 4 lobes of the ovary, or if terminal then corolla with a reduced upper lip; fruit usually of four 1-seeded nutlets; calyx and corolla often 2-lipped   **174. LABIATAE**

b. Style terminal; corolla with well-developed upper lip; fruit usually a berry or drupe; calyx often more or less radially symmetric, not 2-lipped   **172. VERBENACEAE**

19a. Ovules 4--many; fruit a capsule, rarely a berry or drupe   20

b. Ovules 1 or 2; fruit indehiscent, often dispersed in the persistent calyx   24

20a. Ovary containing 4 ovules, side-by-side   **172. VERBENACEAE**

b. Ovary containing many ovules   21

21a. Placentation free-central; corolla spurred; leaves modified for trapping and digesting insects   **186. LENTIBULARIACEAE**

b. Placentation parietal or apical; corolla not spurred, rarely swollen at base; leaves not insectivorous   22

22a. Leaves scale-like, never green; root-parasites   **178. SCROPHULARIACEAE**

b. Leaves green, expanded; free-living plants   23

23a. Seeds winged; mainly climbers with opposite, pinnately divided leaves   **180. BIGNONIACEAE**

b. Combination of characters not as above   24

24a. Capsule with a long beak separating into 2 curved horns; plant sticky-velvety   **183. MARTYNIACEAE**

b. Capsule without beak or horns; plant velvety or variously hairy or hairless   **184. GESNERIACEAE**

24a. Flowers in heads surrounded by an involucre of bracts; ovule 1   **179. GLOBULARIACEAE**

b. Flowers not in heads, often in spikes; ovules 1 or 2   **178. SCROPHULARIACEAE**

26. MYRICACEAE

Dioecious or monoecious shrubs or trees. Leaves alternate, resinous gland-dotted, usually fragrant when crushed. Flowers wind-pollinated, usually in short catkins, usually unisexual, without sepals or petals. Male flowers with 2--16 stamens, though usually 4. Female flowers consisting of a 1-celled ovary containing 1 ovule; style short with 2 stigmatic branches. Fruit a small drupe, sometimes nut-like.

A small family of 3 genera mostly from temperate or subtropical regions of the Old and New Worlds. Their roots contain nitrogen-fixing bacteria in nodules. They are mainly cultivated for their fragrant foliage.

la. Leaves entire or toothed, not incised, without stipules   **1. Myrica**
  b. Leaves pinnatifid, with pointed, more or less heart-shaped stipules
                                                2. Comptonia

1. Myrica Linnaeus. 35/5. Deciduous or evergreen shrubs or small trees. Leaves entire or toothed, without stipules. Flowers unisexual. Male catkins ellipsoid-cylindric, rarely branched; filaments free or somewhat united at base. Female catkins usually ovoid, usually stalkless, ovary stalkless, with 2 or more rounded bractlets. Fruit spherical or ovoid, covered in wax in some species. *Widespread, absent from Australasia.*

2. Comptonia Aiton. 1/1. Deciduous shrubs to 1.5 m, dioecious or monoecious. Stems hairy when young. Leaves linear to narrowly lanceolate, 5--30 cm, deeply pinnatifid into broad, obliquely rounded lobes, hairy, stipulate. Flowers unisexual; male catkins flexuous-cylindric, 2--3 cm, with 3--6 stamens per flower; female catkins spherical, ovary with 8 linear bractlets. Fruiting catkins spherical, bur-like, 1--2.5 cm. with long, linear, persistent bracts. Nutlets ellipsoid, 4--5 mm. *East & southeast USA.*
  Grown for its sweetly fragrant foliage when brushed against or crushed.

27. JUGLANDACEAE

Deciduous trees or shrubs with unisexual flowers, often aromatic. Leaves (in our genera) pinnate with 5--21 opposite leaflets (reduced to 1 in some cultivars). Male flowers with or without an inconspicuous perianth, borne in catkins on the twigs of the previous year; female flowers in spikes, borne on the twigs of the current year; perianth present. Ovary inferior, 1-celled or incompletely 2- or 4-celled; ovule 1, erect from the base. Fruit a nut surrounded by a fleshy cover formed by the perianth and subtending bracts and bracteoles. Seed without endosperm; cotyledons often much contorted. See figure 21, p. 126.
  A family of 7--8 genera mainly from the northern hemisphere.

  1a. Pith continuous; catkins 3 or more together   2
    b. Pith septate; catkins solitary   3
  2a. Fruits more than 1 cm, few (1--20) on a pendent fruit-stalk   **2. Carya**
    b. Fruits less than 1 cm, many (*c.* 100) in an upright. cone-like structure
        bearing many persistent bracts   **1. Platycarya**
  3a. Buds stalkless, with scales; fruit a large nut, without a wing   **3. Juglans**
    b. Buds stalked, often naked; fruit a winged nut   **4. Pterocarya**

1. Platycarya Siebold & Zuccarini. 1/1. Trees to 15 m in the wild, usually shrubs in cultivation. Pith continuous. Leaflets 7--15, stalkless, lanceolate to narrowly ovate, doubly toothed, 4--10 × 1--3 cm. Male catkins 5--8 cm, erect, shortly stalked, few together below the female catkin but overtopping it; flowers without perianths, stamens 8--10. Female catkins solitary, erect, ovoid-oblong, 3--4 cm, brown, with overlapping, narrowly lanceolate, persistent bracts; flowers without perianths, bracteoles 2, united to the ovary; styles 5. Fruit a winged nut to 5 mm, many together in an upright, cone-like structure. *Japan, Korea, China, Taiwan.*

2. Carya Nuttall. 17/12. Trees with fissured or stripping bark and solid pith. Male flowers in drooping catkins, female flowers in 2--10-flowered terminal spikes. Fruit a large nut surrounded by a thick, green husk that splits to a varying degree into 4

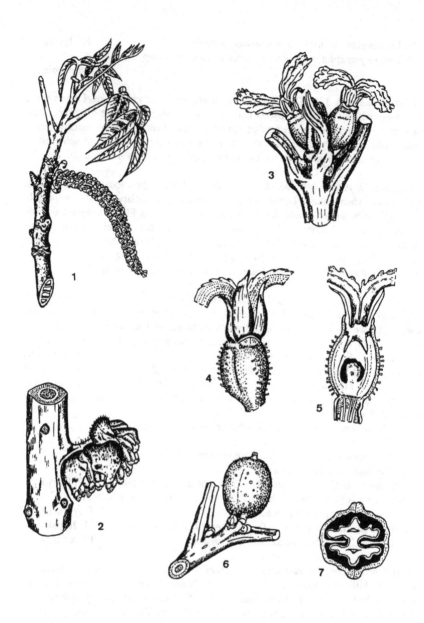

Figure 21. Juglandaceae. *Juglans regia*. 1, Branch-tip with pendulous male catkin and terminal female flowers. 2, A single male flower. 3. Two terminal female flowers. 4, Female flower. 5, Female flower in section. 6, Young fruit with glandular outer skin. 7, Section of mature fruit.

126

sections, borne 1--20 together on a pendent stalk. *North America, a few in Mexico, China & 'Indochina'.*

3. Juglans Linnaeus. 20/8. Trees (occasionally shrubby) with furrowed bark and septate pith. Buds with scales. Leaves pinnate (may be reduced to 1 in some cultivars). Male flowers in drooping catkins, female in short, few-flowered spikes. Fruits solitary or in racemes. Large nuts in fleshy, indehiscent or eventually bilobed husks. *North & South America, southeast Europe to eastern Asia.* Figure 21, p. 126.

4. Pterocarya Kunth. 10/5 and some hybrids. Trees with furrowed bark and septate pith. Leaves large, with 5--27 toothed leaflets. Male flowers in short catkins, female flowers in much longer catkins. The fruits are winged nuts in long, pendent spikes. *Asia.*

28. SALICACEAE

Dioecious, deciduous or rarely evergreen trees and shrubs. Leaves undivided, usually alternate and with stipules, commonly developing after the flowers. Flowers in spike- or raceme-like catkins, each flower subtended by a membranous catkin-scale; perianth reduced to a cup-shaped disk or to 1 or more nectary-scales, occasionally absent. Stamens 2--many; filaments sometimes united. Ovary 1-celled; stigmas 2--4; ovules numerous. Fruit a capsule opening by 2--4 segments. Seeds small, surrounded by a tuft of silky hairs; endosperm absent. See figure 22, p. 128.

There are 2 genera, mainly temperate and sparsely represented in the southern hemisphere.

1a. Catkin-scales entire; buds generally with 1 cap-shaped scale, very rarely
with 2 scales; catkins erect or spreading, rarely pendent   **1. Salix**
  b. Catkin-scales toothed or fringed; buds with several overlapping scales;
catkins pendent   **2. Populus**

1. Salix Linnaeus. 300/80 and many hybrids. Trees, shrubs and dwarf shrubs with sympodial branching. Buds generally cap-shaped (calyptrate) with 1 apparent scale, rarely with 2 free scales. Leaves mostly lanceolate, oblong or obovate, entire or shortly toothed; leaf-stalk generally short, not compressed laterally; stipules often conspicuous and persistent. Catkins mostly erect or spreading, rarely pendent, stalked or almost stalkless, with a varying number of more or less leaf-like basal bracts; catkin-scales entire, persistent or sometimes soon falling; flowers primarily insect-pollinated (rarely wind-pollinated) with 1 or more small nectary-scales. Stamens usually 2 (sometimes up to 12), with free or occasionally united filaments. Ovary usually flask-shaped, style single, short or long, sometimes almost absent, stigmas 2, sometimes bifid. Fruit a capsule opening by 2 segments, the segments usually recurving from the apex at maturity. *Temperate areas, mostly in the northern hemisphere.* Figure 22, p. 128.

2. Populus Linnaeus. 40/30. Trees, generally with monopodial branching. Buds with several overlapping scales, often sticky and smelling of balsam. Leaves ovate, triangular-ovate or diamond-shaped, sometimes cordate at the base, rarely narrow; stalks usually long, sometimes compressed laterally; stipules inconspicuous, soon falling. Catkins developing before the leaves, without any obvious bracts, the males (and often the females) pendent; catkin-scales toothed or fringed. Perianth an oblique,

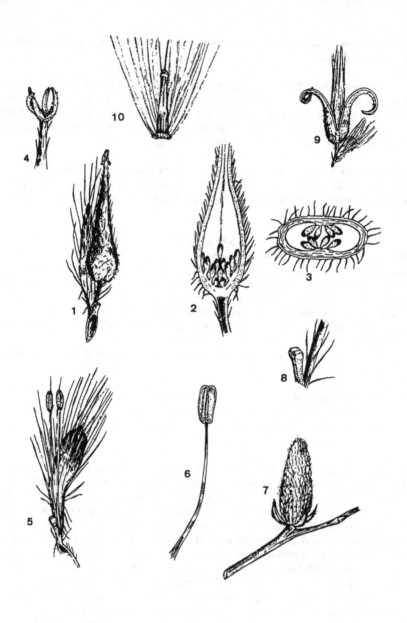

Figure 22. Salicaceae. *Salix caprea*. 1, Single female flower. 2, Longitudinal section of ovary. 3, Transverse section of ovary. 4, Tip of style showing stigmas. 5, Single male flower with two stamens. 6, Single stamen. 7, Young male catkin. 8, Base of filaments to show nectary. 9, Capsule opening to release plumed seeds. 10, A single seed.

128

cup-shaped disc. Stamens usually numerous, commonly with crimson anthers. Ovary broadly flask-shaped, style short or absent, stigmas 2--4. Fruit a capsule opening by 2--4 segments. Wind-pollinated. *North temperate areas.*

29. BETULACEAE

Monoecious, deciduous trees or shrubs. Leaves alternate, simple, toothed or shallowly lobed, with deciduous stipules. Flowers unisexual; male flowers in erect or pendent catkins, 3 to each bract, with perianth very small or absent and 2--15 stamens. Female flowers in usually erect catkins, clusters or short spikes, perianth present or absent, ovary inferior, 2-celled below, with 2 styles. Fruits (nuts or, if very small, sometimes called 'nutlets') in catkins, clusters or spikes, sometimes forming woody 'cones'; involucre made up of united bracts and bracteoles present or absent; fruits compressed and winged or not.

A small family of 6 genera. This is often divided into 2 families, *Betulaceae* in the strict sense, containing genera **1--2**, and *Corylaceae* (genera **3--6**). See figures 23, p. 130 & 24, p. 132.

1a. Nut small, compressed, winged, without an involucre; male flowers with 2--4 stamens, with a perianth   2
  b. Nut often large. not compressed or winged, with a leaf-like involucre made up of bracts and bracteoles; male flowers with 4--15 stamens, without a perianth   3
2a. Fruiting catkins ovoid, cone-like; fruiting catkin-scales 5-lobed, woody, persistent   **1. Alnus**
  b. Fruiting catkins cylindric to narrowly ovoid, breaking up at maturity; fruiting catkin-scales 3-lobed, falling with the fruit   **2. Betula**
3a. Female flowers in pendent catkins; anthers not hairy   4
  b. Female flowers in erect clusters or short spikes; anthers hairy at apex   5
4a. Involucre of fruit flat, 3-lobed   **3. Carpinus**
  b. Involucre of fruit bladder-like   **4. Ostrya**
5a. Flowers appearing before the leaves   **6. Corylus**
  b. Flowers appearing with the leaves   **5. Ostryopsis**

1. Alnus Miller. 35/15 and a few hybrids. Trees or shrubs. Male flowers with 4 stamens, anther-lobes separated by a forked connective, and a 4- or 5-partite perianth. Female flowers 2 to each bract. Fruiting catkin an ovoid 'cone' with 5-lobed, persistent, woody scales from which the ripe, winged nuts are released. *North temperate areas.*

2. Betula Linnaeus. 60/17. Trees or shrubs, usually with slender twigs and thin, often papery bark. Male flowers with 2 stamens, each bifid below the anthers, and a minute perianth. Female flowers 3 to each bract. Fruiting catkins ('cones') cylindric to narrowly ovoid, with 3-lobed scales falling when the winged nuts are ripe. *Northern hemisphere.* Figure 23, p. 130.

3. Carpinus Linnaeus. 35/9. Deciduous, much-branched trees generally without central stems, rarely shrubs. Bark grey, smooth or scaly. Branches slender, buds acute with many overlapping scales. Leaves alternate, arranged in 2 rows, margins toothed, veins 6--24 pairs. Flowers unisexual, appearing with the leaves. Male catkins pendulous, enclosed in buds in winter, borne on short shoots; flowers lacking perianth

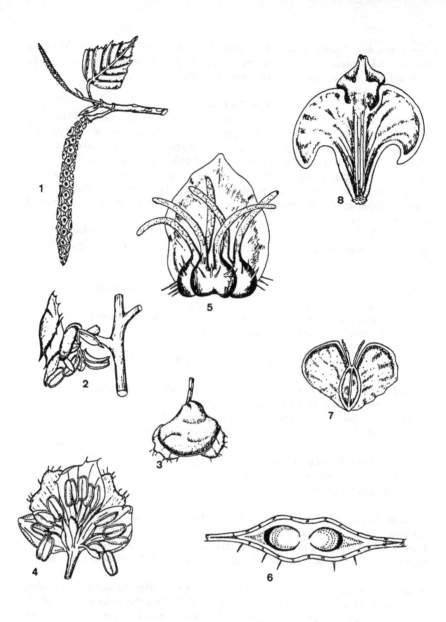

Figure 23. Betulaceae. *Betula pendula.* 1, Pendent male catkin. 2, Side view of male catkin-scale. 3, Male catkin-scale from outside. 4, Male catkin-scale from above showing the three male flowers. 5, Female catkin-scale from above showing three female flowers. 6, Transverse section of ovary. 7, A single fruit. 8, Scale from a female catkin in the fruiting stage.

or bracteoles, stamens 4--12 borne in the bract-axils, filaments bifid almost to the base. Female flowers in terminal catkins, 2 per bract, each with 6 bracteoles; perianth with 6--10 teeth, joined to the 2-celled ovary; styles short with 2 stigmas. Fruit a 1-seeded, ribbed nut with a large 3-lobed solitary bract, which continues to grow after fertilisation. *North temperate areas, chiefly eastern Asia & North America.*

4. Ostrya Scopoli. 10/3. Deciduous trees with rough scaly bark. Buds pointed, covered in overlapping scales. Leaves alternate, stalked, ovate to ovate-oblong, acuminate, margins sometimes with 2 rows of forward-pointing teeth; blades hairless to softly hairy, with 9--15 pairs of veins. Flowers unisexual, appearing in spring with the leaves. Male catkins pendulous, developing in autumn, flowers lacking perianth but with bracts surrounding 5--15 stamens. Female catkins erect, with 2 flowers in each bract-axil, calyx closely adpressed to ovary. Bract and 2 bracteoles joined to form a tubular involucre. Stigmas 2, linear. Fruit a ribbed nut, enclosed by the involucre, which expands considerably after fertilisation. *North America, Europe, Asia.*

5. Ostryopsis Decaisne. 2/1. Deciduous shrubs with hairy branches and pointed buds covered by overlapping scales. Leaves alternate, simple, with toothed margins. Flowers unisexual, appearing with the leaves. Male flowers in pendulous catkins, perianth absent, filaments divided into 2 at apex, anthers hairy at apex. Female flowers in very short spikes, each bract subtending 2 flowers, involucre with 3 teeth. Calyx united to ovary, ovule 1 per cell, style bifid. Fruit a nut enclosed in a tubular involucre. *China.*

6. Corylus Linnaeus. 15/12. Deciduous shrubs, more rarely trees. Winter buds broadly ovoid, obtuse, with many overlapping scales. Leaves broadly ovate, softly hairy, margins with 2 rows of forward-pointing teeth. Flowers unisexual, usually appearing before the leaves. Male catkins pendulous, flowers lacking perianths, each bract with 4--8 stamens, filaments bifid, anthers with long hairs at apex. Female flowers in capitate clusters, enclosed in small scaly buds, with only the styles projecting, ovaries with 1 or 2 ovules per cell, styles bifid to base, usually red. Fruit a spherical or ovoid nut, with woody pericarp, enclosed or surrounded by a leafy involucre; involucral bracts toothed or dissected, joining to form a tube. *North America, Europe, Asia.* Figure 24, p. 132.

30. FAGACEAE

Monoecious trees or shrubs. Buds with overlapping scales, small to large and conspicuous. Leaves alternate, sometimes 2-ranked, stalked, entire to toothed, scalloped or lobed, evergreen or deciduous. Stipules usually deciduous. Flowers unisexual; male flowers in erect or pendent catkins with a single flower to each bract, or in stalked heads or solitary or in 3-flowered cymes, perianth 4--7-lobed, stamens 8--40; female flowers solitary or in 3's enclosed by a cupule made up of overlapping and partly fused scales, these solitary or in spikes, rarely at the bases of the male catkins, ovary inferior, 3--6-celled, styles 3. Ovules 2 per cell. Fruit of 1--3 nuts enclosed in the enlarged cupule. See figure 25, p. 134.

A family of 8 genera and about 600 species found all over the world except in Africa south of the Sahara.

1a. Male flowers solitary or in 3s or in long-stalked heads   2

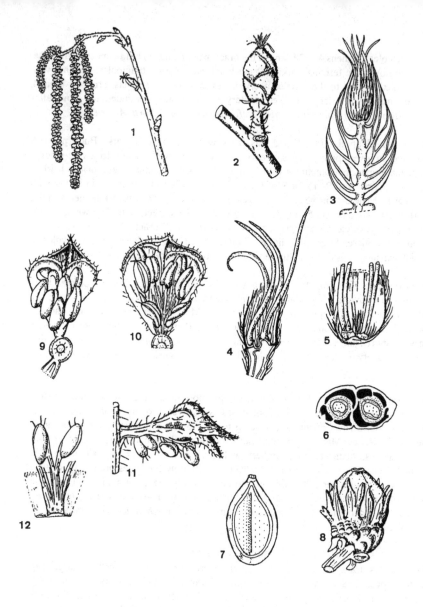

Figure 24. Betulaceae/Corylaceae. *Corylus avellana.* 1, Shoot with pendent male catkins, female inflorescence below. 2, Female inflorescence. 3, Longitudinal section of female inflorescence. 4, Longitudinal section of a pair of female flowers, only the styles distinguishable. 5, Basal parts of female flowers showing the small united bracteoles at base. 6, Transverse section of ovary. 7, Longitudinal section of seed. 8, Mature fruit with irregularly lobed cupule. 9, Male flower, anthers undehisced. 10. Male flower, anthers dehisced. 11, Male flower from the side. 12, A single stamen.

132

b. Male flowers in erect or pendent catkins   3
2a. Male and female flowers solitary or in 3s; cupule with 1--3 nuts
                                                            2. Nothofagus
   b. Male flowers in heads, female in pairs; cupule with 2 nuts   **1. Fagus**
3a. Male flowers in pendent catkins   **7. Quercus**
   b. Male flowers in stiffly erect catkins   4
4a. Leaves deciduous, toothed; ovary 6-celled   **3. Castanea**
   b. Leaves evergreen, usually not toothed; ovary 3-celled   5
5a. Nut solitary, clearly projecting from the cupule   **6. Lithocarpus**
   b. Nuts 1--3, more or less completely enclosed in the cupule   6
6a. Young shoots and undersides of the leaves covered in bright yellow hairs
   (which age to tawny)   **5. Chrysolepis**
   b. Young shoots and undersides of leaves hairy or not, hairs, if present, not
   bright yellow   **4. Castanopsis**

1. Fagus Linnaeus. 10/10. Usually large, monoecious trees with smooth, grey bark. Buds large, spindle-shaped. Leaves in 2 ranks, entire, toothed or scalloped (lobed in some cultivars), usually silky-hairy, at least when young. Male flowers in long-stalked, more or less spherical heads; perianth 4--6-lobed, stamens 8--12. Female flowers usually 2 together surrounded by a stalked, 4-lobed, scaly cupule made up of needle-like or somewhat flattened appendages; perianth 4--6-lobed. Ovary 3-celled. Fruit a nut. released on the splitting of the cupule into 4 segments. *Temperate parts of the northern hemisphere.* Figure 25, p. 134.

2. Nothofagus Blume. 20/8. Deciduous or evergreen, monoecious trees or sometimes shrubs. Leaves alternate, often asymmetric, pleated, flat or the margins rolled under in bud; buds small, ovoid. Male flowers solitary or in 3-flowered cymes with 4--6-lobed perianth and 8--40 stamens. Female flowers 1--3 in a stalkless or shortly stalked cupule which is 2--4-lobed and made up of entire, toothed or laciniate scales, often with glandular tips or appendages. Nuts 1--3 per cupule. *Temperate and tropical areas of the southern hemisphere.*

3. Castanea Miller. 20/6. Deciduous, monoecious trees or shrubs. Buds small, ovoid. Leaves in 2 ranks, toothed, often hairy and often with short glands which are translucent when young, yellowish or orange when mature and whitish when old, on the surface. Male catkins erect, often showy, with a 6-lobed perianth and 10--12 (rarely to 20) stamens. Female flowers borne towards the bases of the male catkins or on separate catkins, usually grouped in 3s, each group surrounded by a cupule (involucre) made up of spine-like, spreading or reflexed scales. Ovary 6-celled. Nuts large, 1--3 (rarely more) borne within the enlarged cupule which splits at maturity into 2--4 flaps or segments. *Temperate areas of the northern hemisphere.*

4. Castanopsis (D. Don) Spach. 100/4. Large, evergreen, monoecious trees with scaly bark. Buds small, ovoid. Leaves usually very leathery, usually toothed. Male catkins borne at the ends of branches; female catkins distinct. Male flowers with a 5--6-lobed perianth and 10--12 stamens. Female flowers usually 3 together within a common, often spiny cupule which ultimately splits into segments. Nuts ripening in their second year. *Southeast Asia.*

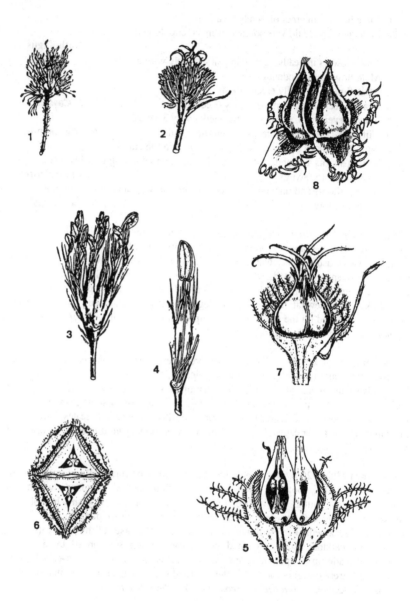

Figure 25. Fagaceae. *Fagus sylvatica.* 1, Male inforescence (pendent on tree). 2, Female inflorescence with two flowers. 3, A single male flower. 4, A male flower sectioned to show the attachment of the filaments. 5, Longitudinal section of two female flowers. 6, Transverse section of cupule showing sections of two ovaries. 7, Cupule with the front portion removed, showing the female flowers. 8, Cupule in fruit opened out to reveal two fruits.

5. Chrysolepis Hjelmqvist. 1--2/1. Similar to *Castanopsis* hut leaves with bright yellow hairs beneath (fading to tawny as the leaves age), the cupule made up from its inception of 7 free, spiny segments, 5 surrounding the 3 nuts, 2 others within and between the nuts; female flowers borne at the bases of the male catkins. *Western North America.*

6. Lithocarpus Blume. 100/few. Trees, like *Quercus* but leaves always evergreen, usually very leathery, entire or toothed. Male catkins upright, stiff. Stamens 10--12. *Tropics and subtropics.*

7. Quercus Linnaeus. 600/many and many hybrids. Monoecious trees or shrubs, often with stellate or tree-like hairs. Leaves evergreen, semi-evergreen (persisting from one spring to the next) or deciduous, alternate, entire, toothed or variously lobed. Male flowers in long, slender, hanging catkins, each flower with a 4--7-lobed perianth, 4--6, rarely to 12 stamens and. often, a rudimentary ovary in the centre. Female flowers solitary or a few together on a short or long common stalk; perianth inconspicuous, 6-lobed, ovary inferior, usually 3-celled, with 3 long or short styles; ovules 2 per cell. Fruit a nut (acorn) containing a single seed, surrounded at least at the base by a cup (cupule) made up of scales which are closely adpressed at least in their lower parts. *Mainly northern hemisphere, but reaching south of the equator in South America & southeast Asia.*

31. ULMACEAE

Trees or rarely shrubs, without latex. Leaves alternate, 2-ranked, simple, usually toothed, usually with the 2 halves unequal, with deciduous stipules. Hairs simple. Flowers bisexual or unisexual on monoecious plants. Perianth of 1 whorl of 3--8 segments. Stamens the same number as the perianth-segments or twice as many. Ovary superior, 1-celled, with 1 ovule attached at the top; style-branches 2. Fruit a nutlet or winged nutlet, or a drupe. See figure 26, p. 136.

A family of about 15 genera in both northern and southern hemispheres.

  la. Veins of leaves not running into teeth, looped and joining other veins
                                                                    1. Celtis
   b. Veins of leaves running direct to teeth, sometimes forking before entering
      them   2
  2a. Teeth of leaves uniform, mucronate; fruit a small drupe produced in autumn
                                                                  2. Zelkova
   b. Teeth of leaves usually of 2 sizes, the larger alternating with 1 or more
      smaller ones, not mucronate; fruit winged, produced in late spring (in
      ours)   **3. Ulmus**

1. Celtis Linnaeus. 70/2. Trees. Leaves simple, toothed and deciduous in our species, asymmetric, with veins looped near the margin, not running directly into teeth. Flowers on the current year's growth, appearing with the leaves, in clusters with male flowers below and bisexual flowers above. Sepals and stamens 4 or 5. Fruit a drupe with thin flesh, ripening in autumn. *North temperate areas, tropics.*

2. Zelkova Spach. 5/2. Trees or shrubs. Leaves deciduous, simple, toothed, with veins running direct to teeth. Flowers on current year's growth, solitary or few

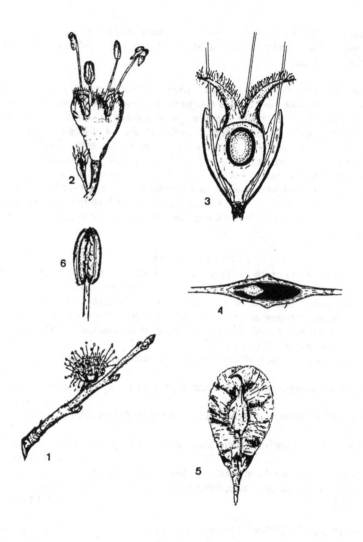

Figure 26. Ulmaceae. *Ulmus procera*. 1, Inflorescence. 2, A single flower. 3, Longitudinal section of a flower. 4, Transverse section of ovary. 5, Fruit. 6, Stamen.

136

together, male in lower axils of the shoot, bisexual higher up. Sepals and stamens 4 or 5. Fruit a green drupe, wider than long, almost stalkless, ripening in autumn. *Asia*.

3. Ulmus Linnaeus. ?40/10. Deciduous, rarely half-evergreen trees. Leaves shortly stalked; blades toothed, usually doubly (with small teeth on the sides of the major teeth), the 2 halves unequal in size and shape (at least in a proportion of the leaves). Flowers in stalkless clusters appearing in spring before the leaves or rarely in autumn. Perianth *c*. 3 mm, of 4--9 segments joined at the base. Stamens the same number as and longer than the perianth-lobes. Fruits flattened, circular to elliptic, with an apical notch, green during development but dry, papery and brownish when ripe, with a single seed occupying a swelling on the mid-line, ripening within a few weeks of flowering. *N America, Europe, Asia*. Figure 26, p. 136.

In much of Europe numerous vegetatively propagated clones occur, some of which produce little viable seed.

32. EUCOMMIACEAE

Deciduous, dioecious trees, stems hollow, pith septate. Leaves alternate, simple, without stipules. Flowers solitary, without petals or sepals, in the axils of bracts below the upper leafy portions of the shoots, the males crowded, the females separated. Fruit a samara.

A family of a single genus.

1. Eucommia Oliver. 1/1. Deciduous trees to 20 m, dioecious. Leaves stalked, simple, narrowly ovate to elliptic, 7--15 cm, acuminate, pinnately veined, hairy along the veins when young, margins with fine teeth pointing forwards. Male flowers stalked, consisting of 6--12 linear stamens *c*. 1 cm. Female flowers stalked, ovary long, *c*. 1 cm, 1-celled with 2 hanging ovules, style terminal, 2-lobed, reflexed. Fruit ellipsoid, 3--4 cm with narrow, longitudinal wings, terminally notched. *Central China*.

33. MORACEAE

Trees, shrubs, herbs or vines, evergreen or deciduous, dioecious or monoecious, most with milky sap containing latex. Leaves alternate or rarely opposite, simple, entire or lobed or toothed, with 2 (rarely 1) deciduous stipules. Flowers unisexual, reduced and small, radially symmetric, in globular heads or catkins, sometimes densely packed inside hollow receptacles. Calyx-lobes 4--5, sometimes absent. Petals absent. Stamens 1--5, opposite the calyx-lobes. Ovary usually 1-celled, ovules mostly pendulous. Styles simple or branched. Fruit an achene, nut or drupe, sometimes enclosed in the fleshy, hollow receptacle. Seed with usually curved embryo.

There are 50 genera and 1200--1500 species in this mainly tropical family. Two genera (6 & 7) that do not have milky sap and possess 5 (not 4) sepals and stamens are sometimes separated into a distinct family, *Cannabaceae* (*Cannabinaceae*, *Cannabidaceae*). See figure 27, p. 138.

1a. Annual or perennial herbs or vines | 2
b. Trees or shrubs, occasionally climbing | 3
2a. Vines, perennial or annual | **6. Humulus**
b. Erect herbs | **7. Cannabis**
3a. Trees with spines on the branches | 4
b. Trees without spines, or shrubs | 5

137

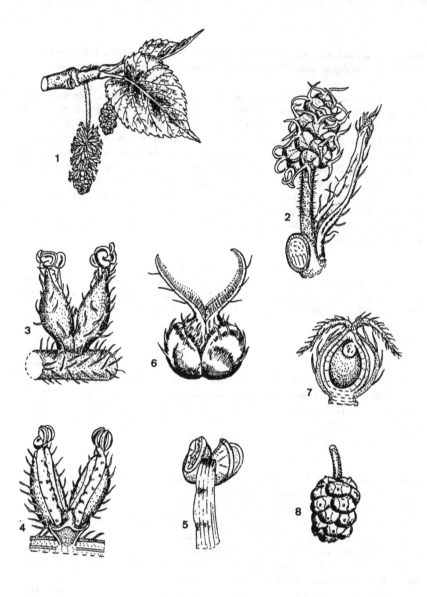

Figure 27. Moraceae. *Morus nigra.* 1, Two pendent male catkins. 2, Female inflorescence with several flowers. 3, Male flower from the side. 4, Longitudinal section of male flower. 5, Upper part of stamen showing dehisced anthers. 6, Longitudinal section of female flower. 7, Ripe compound fruit.

4a. Male flowers in a loose raceme; leaves always entire   **3. Maclura**
  b. Male flowers in spherical heads; leaves sometimes lobed   **4. Cudrania**
5a. Receptacle (fig) fleshy and urn-shaped with the flowers borne inside
                                                                 5. Ficus
  b. Fruit not as above   6
6a. Fruiting head succulent, juicy, female flowers with 4 distinct sepals
                                                                 1. Morus
  b. Fruiting head hairy, not as above, female flowers with a tubular calyx
                                                              2. Broussonetia

1. Morus Linnaeus. 12/5. Deciduous trees, monoecious or dioecious. Leaves alternate, simple, entire or lobed, toothed. Flowers in axillary spikes. Male flowers with 4 calyx-segments and 4 stamens. Female flowers with 4 calyx-segments, style with 2 branches. Fruits juicy, seeds enclosed by the enlarged, succulent calyx. *Temperate Asia & North America.* See Figure 27, p. 138.

2. Broussonetia Ventenat. 8/3. Trees or shrubs, monoecious or dioecious. Leaves alternate, toothed, sometimes 3-lobed. Inflorescence axillary. Male flowers in cylindric catkins, with 4 calyx-lobes and 4 stamens curved inwards in bud. Female flowers in spherical heads; calyx tubular or ovoid, with 3 or 4 teeth, persistent in fruit. Style simple. Fruiting heads small, spherical, hairy. Fruit fleshy, projecting from calyx. *East Asia to Polynesia.*

3. Maclura Nuttall. 1/1. Dioecious tree to 15 m; branches with spines (to 4 cm long). Leaves deciduous, alternate, simple, entire, 5--12 × 2--6 cm, ovate to oblong, acuminate, dark green and hairless above, paler and downy beneath. Flowers green, axillary, inconspicuous; males in a loose raceme. Fruit a spherical aggregate fruit to 14 cm in diameter, resembling an orange. *Eastern USA.*

4. Cudrania Trécul. 8/1. Small trees or climbing shrubs, dioecious. Branchlets often reduced to spines. Leaves alternate, entire or 3-lobed. Inflorescences usually spherical, small. Male flowers with 4 calyx-lobes, stamens 4, anthers erect; ovary absent or rudimentary. Female flowers with 4 calyx-lobes closely surrounding the ovary; style undivided or 2-branched. Fruit with enlarged bracts, fleshy. *Tropical & subtropical areas.*

5. Ficus Linnaeus. 700/50. Mostly evergreen trees, shrubs, trailers or climbers with milky sap. Leaves alternate (rarely opposite), usually stalked, often large and showy with prominent veins, membranous or papery to thinly or thickly leathery. sometimes abrasive, simple or lobed, with an entire, toothed or irregular margin. Stipules paired (very rarely solitary), enclosing the developing bud. Climbers usually seen in cultivation as the juvenile, non-reproductive phase with creeping stems bearing numerous, clinging, adventitious roots and 2 ranks of small, stalkless, often asymmetric leaves. Flowers minute, densely packed inside hollow receptacles (figs) inside which pollination and fruit development take place. Flowers of three types: male, functional female, and non-functional female (gall flowers), modified for the nourishment and reproduction of minute wasps intimately involved in pollination. In most species, plants are either male (bearing figs with male and gall flowers) or female (producing fertile figs), though some species are bisexual having all three types of flower within each fig. *Cosmopolitan, mainly subtropical & tropical.*

6. Humulus Linnaeus. 3/2. Monoecious vine-like herbs. Leaves opposite, with stalks, cordate at base. Flowers unisexual, inflorescences pendulous, male branched. Male flowers with 5-lobed calyx and 5 stamens. Female flowers in cone-like inflorescences; calyx membranous, ovary stalkless and style with 2 branches. *Temperate Eurasia.*

7. Cannabis Linnaeus. 1/1. Erect annual herb to 2.5 m, dioecious. Leaves alternate (lower leaves may be opposite), stalked, palmately divided, segments 3--9, linear-lanceolate, margins deeply toothed. Inflorescences erect, glandular; males branched. Male flowers with 5-lobed calyx, stamens 5; female inflorescence a raceme, female flowers with membranous calyx closely adhering to ovary; style with 2 branches. Embryo curved. *Asia.*

Sometimes placed with *Humulus* in a separate family (*Cannabinaceae*). Cultivated widely for its fibres (hemp) and (generally illegally) for its narcotic juices.

34. URTICACEAE

Dioecious or monoecious herbs, undershrubs or rarely trees with very soft wood. Leaves alternate or opposite, sometimes with stinging hairs; stipules present but often deciduous; cystoliths (usually spindle-shaped) abundant in stems and leaves. Flowers mostly unisexual, in cymes, sometimes crowded on a common enlarged receptacle. Male flowers with 4 or 5 perianth-lobes and 4 or 5 stamens inflexed in bud, exploding when ripe. Female flowers with 2--4-lobed perianth, often with staminodes, ovary superior, free from or fused to perianth, 1-celled. Fruit an achene or a fleshy berry. Seed with oily endosperm. See figure 28, p. 142.

About 50 genera and 2000 species from temperate and tropical regions throughout the world, but poorly represented in Australia.

1a. Flowers inconspicuous, solitary; stems creeping; leaves small, more or
    less circular; stipules absent; mat-forming herb   **3. Soleirolia**
  b. Flowers in conspicuous inflorescences; leaves and habit not as above   2
2a. Stinging hairs present; leaves opposite   **1. Urtica**
  b. Stinging hairs absent; leaves opposite or alternate   **2. Boehmeria**

1. Urtica Linnaeus. 100/3. Annual or perennial herbs, usually with stinging hairs; stems ridged or 4-angled. Leaves opposite, margins toothed, stipules free. Flowers unisexual, green, in axillary cymes. Perianth with 4 lobes, which are unequal in the female flowers. Stigma brush-like. Fruit a shining, flattened nut. *Temperate areas.* Figure 28, p. 142.

2. Boehmeria Jacquin. 100/4. Mostly perennial herbs in cultivation, monoecious or dioecious. Leaves alternate or opposite, toothed, with 3 veins, stipules mostly free. Flowers usually unisexual, in axillary clusters. Male flowers with 4-lobed perianth, stamens 4. Female flowers with tubular perianth with 2--4 lobes at the apex, opening often constricted. Stigma hairy, persistent. Ovary stalked. Fruits sometimes enlarged. with 2 acute wings, enclosed in the persistent perianth. *Subtropical and temperate eastern Asia.*

A few species of the genus **Pilea** Lindley, may be found in very sheltered areas. They are like *Boehmeria*, but have a brush-like stigma and 3-lobed, non-tubular perianth.

3. Soleirolia Gaudichaud-Beaupré. 1/1. Slender, creeping, hairy perennial herb, forming dense, evergreen mats. Stems to 20 × 0.5 mm, rooting at nodes. Leaves pale green, alternate, circular, entire, shortly stalked, 2--6 mm, without stipules. Flowers unisexual, axillary, solitary, inconspicuous. Female flowers enclosed in an involucre of 1 bract and 2 bracteoles, perianth tubular with 4 narrow lobes. Male flowers with 4 perianth-lobes. Fruit enclosed by persistent perianth and involucre. Achene ovoid, shining. *West Mediterranean area.*
Formerly known as *Helxine* Requien.

35. PROTEACEAE
Evergreen trees and shrubs, occasionally monoecious or dioecious, sometimes with woody tubers (lignotubers). Leaves alternate (rarely opposite or in whorls), simple or pinnate, often toughened (sclerified), toothed or spiny. Flowers unisexual or bisexual, radially symmetric or bilaterally symmetric, in complex inflorescences. Perianth-segments 4, petal-like, coloured. Stamens 4, anthers usually stalkless, rarely 1--3 infertile, 1--4 hypognous scales present or absent. Ovary superior; stigma 1, often flattened and modified as a pollen-presenter; carpel 1, ovules 1--many, marginal. Fruit a woody 'cone' or many-seeded follicle or 1-seeded achene, often fire-resistant.

About 75 genera and perhaps 1300 species, almost entirely confined to the southern hemisphere where they are becoming increasingly popular as ornamental garden plants. Species of most genera are not hardy in northern Europe, and only 2 are dealt with here.

1a. Flowers red, paired, hairless, in unbranched inflorescences **1. Embothrium**
  b. Flowers not as above, not paired or in branched inflorescences **2. Lomatia**

1. Embothrium Forster & Forster. 1/1. Trees with a single trunk and without suckers, or tall shrubs suckering to form thickets. Leaves leathery, alternate, lanceolate to ovate, to 15 × 4 cm, hairless, green to dark green above, pale beneath; stalk to 1.5 cm. Flowers in pairs arranged in loose, terminal racemes. Perianth to 3 cm, bilaterally symmetric, scarlet to orange-red, very rarely pale yellow or white; segments separating at maturity. Anthers 4, stalkless. Style slightly curved with conical pollen-presenter. Fruit a woody pod, greenish yellow becoming red-brown when mature, with persistent style, containing numerous winged seeds. *Temperate South America.*

2. Lomatia R. Brown. 12/7. Trees or shrubs. Leaves toughened, leathery, alternate (rarely opposite), entire or divided. Flowers white to cream, in pairs in axillary or terminal panicle-like inflorescences. Perianth bilaterally symmetric; anthers all fertile; hypogynous scales 3 (if 4 then the 4th scale minute). Fruit a woody follicle with numerous winged seeds. *Australia (Tasmania), South America.*

36. SANTALACEAE
Herbs, shrubs or trees, parasitic or partially parasitic on the roots of other plants. Leaves opposite or alternate, simple, entire (rarely scale-like and falling early), without stipules. Flowers in spikes, racemes, cymes, clusters or solitary, unisexual or bisexual, radially symmetric. Sepals 3--6, somctimes petal-like. Petals absent. Stamens 3--6.

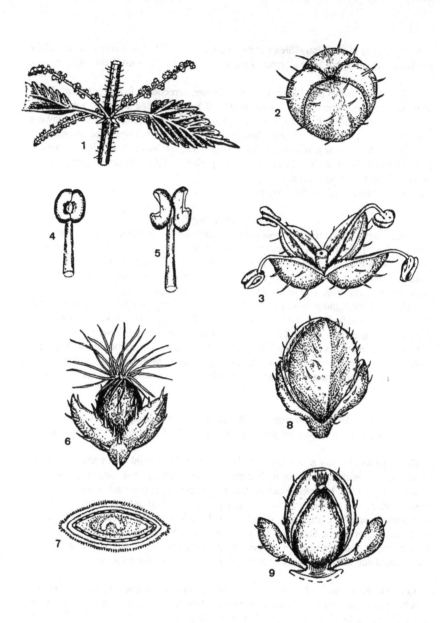

Figure 28. Urticaceae. *Urtica dioica.* 1, Part of stem with three inflorescences. 2, Male flower in bud. 3, Male flower. 4, Stamen, anthers undehisced. 5, Stamen, anthers dehisced. 6, Female flower. 7, Transverse section of female flower. 8, Perianth enclosing fruit. 9, As for 8, but one perianth-segment removed.

Ovary inferior or at least partly so (in all cultivated genera), 1-celled with 2 or 3 (rarely 1) ovules borne at the apex of a basal, central placenta; style absent or present, stigmas 2--6. Fruit a nut or drupe.

Thirty-five genera and about 400 species, mostly found in the tropics, though some extend into temperate regions. All are partially or completely parasitic, and few are found in gardens.

1a. Leaves opposite or almost so   **1. Buckleya**
   b. Leaves all alternate   2
2a. Plant with herbaceous stems woody only at or near ground-level
                                          4. Comandra
   b. Plant with persistent, aerial, woody branches   3
3a. Male and female flowers in spikes   **2. Pyrularia**
   b. Male flowers in axillary racemes; female flowers solitary or in clusters
     of up to 3   **3. Osyris**

1. Buckleya Torrey. 5/1. Shrubs. Leaves opposite, in 2 ranks, shortly stalked, lanceolate. Flowers unisexual, males and females on separate plants. Male flowers in umbel-like racemes, with 4 ovate sepals and 4 stamens, without patches of hairs on the sepals behind the stamens. Female flowers solitary, terminal or axillary, with 4 sepal-like bracts and 4 sepals; style 1, stigmas 4. Fruit a nut. *Eastern North America, China.*

2. Pyrularia Michaux. 3/1. Large shrubs, downy when young. Leaves alternate, shortly stalked. Flowers unisexual, males and females on separate plants. Male flowers with 5 sepals and 5 stamens, each sepal with a patch of hairs just behind the attachment of the stamen. Female flowers with 5 sepals and an ovary that tapers strikingly towards the base, making the flowers appear to be superficially stalked. Fruit a pear-shaped drupe. *North America, Himalaya.*

3. Osyris Linnaeus. 10/1. Shrubs. Leaves alternate, shortly stalked. Flowers unisexual; males and females on separate plants. Male flowers in axillary racemes with 3 or 4 sepals and 3 or 4 stamens, each sepal with a patch of hairs just behind the attachment of the stamen. Female flower solitary or in clusters of 2 or 3 in the leaf-axils; ovary tapering towards the base. Fruit a drupe. *Mediterranean area to China.*

4. Comandra Nuttall. 6/1. Plants with woody, creeping stocks at ground-level and erect, simple, herbaceous flowering stems. Leaves alternate, almost stalkless. Flowers bisexual, in terminal panicles. Sepals 5, petal-like, white. Stamens 5, each with a tuft of hairs behind its attachment. Style long. Fruit a nut. *Europe, North America.*

37. LORANTHACEAE

Evergreen shrubs parasitic on various trees. Leaves opposite or more rarely whorled, sometimes scale-like, entire, without stipules. Flowers small, unisexual (in all cultivated species), radially symmetric. Perianth of 1 or 2 whorls. Calyx 2--6-lobed. Petals 4--6 or absent. Stamens 2--6, attached to the petals, at least at their bases. Ovary inferior, 1-celled, the ovules not differentiated from the placenta. Fruit a dry or sticky berry; seed 1, embryos 1--3. See figure 29, p. 146.

A mainly tropical, parasitic family of about 40 genera and 1400 species. Only a few are grown for ornament. *Viscum* is sometimes placed in the separate family, *Viscaceae*.

1. Viscum Linnaeus. 65/1. Evergreen shrub with hard wood. Leaves well-developed, with parallel veins. Flowers crowded in cymes. Calyx 4-lobed or 4-toothed in the female flowers, absent in the male. Petals usually 4. Stamens 4, joined to the petals for most of their length, anthers opening by pores. Fruit a sticky, white or yellow berry. *Widespread*. Figure 29, p. 146.

Parasitic on various trees; commonly cultivated for ornament at especially at Christmas.

38. POLYGONACEAE

Annual to perennial herbs, shrubs or climbers, rarely trees. Leaves mostly alternate, simple, variable in outline; stipules often united into a membranous sheath (ochrea). Flowers bisexual or unisexual, radially symmetric, often in racemes or raceme-like inflorescences. Perianth-segments 3--6, sepal- or petal-like, free or united at the base, often enlarging in fruit and becoming hardened, papery or fleshy. Fruit a 3-angled or flattened and 2-angled nut, usually enclosed by the persistent perianth. See figure 30, p. 148.

A family of about 30 genera and 800 species, mostly in north temperate regions. Relatively few species are cultivated.

1a.	Erect, branched shrubs or trees	2
b.	Herbs, dwarf shrubs or woody climbers	3
2a.	Often spiny shrubs; leaves less than 3 cm	**5. Atraphaxis**
b.	Spineless trees or shrubs; leaves usually at least 3 cm	**2. Rumex**
3a.	Perianth-segments 4; nut flattened, 2-angled	**3. Oxyria**
b.	Perianth-segments 5 or 6; nut usually not flattened, usually 3-angled	4
4a.	Woody climbers	5
b.	Herbs or dwarf shrubs	6
5a.	Perianth not fleshy in fruit	**6. Polygonum**
b.	Perianth fleshy in fruit	**7. Muehlenbeckia**
6a.	Stem-leaves few or absent; stamens always 9	7
b.	Stem-leaves usually numerous; stamens not more than 8	8
7a.	Leaves palmately lobed; flowers in long panicles	**4. Rheum**
b.	Leaves not palmately lobed; flowers in compact heads surrounded by whorls of bracts	**1. Eriogonum**
8a.	Perianth white and fleshy in fruit	**7. Muehlenbeckia**
b.	Perianth usually not fleshy in fruit, if fleshy then reddish	9
9a.	Perianth-segments 5, white or pink, equal in fruit or the outer larger	**6. Polygonum**
b.	Perianth-segments 6, greenish, the inner much larger than the outer in fruit	**2. Rumex**

1. Eriogonum Michaux. 150/6. Annual to perennial herbs or small shrubs. Leaves mostly basal, small, alternate or in whorls, entire. Ochreae absent. Flowers bisexual, in compact heads or umbels partly enclosed by a tubular or bell-shaped involucre of 4--8-lobed bracts. Perianth-segments 6, petal-like, united in lower part, enlarged in fruit. Stamens 9. Stigmas 3. Fruit a 3-angled nut. *Mostly western North America*.

2. Rumex Linnaeus. 150/few. Annual or perennial herbs, rarely shrubs, hairless or inconspicuously hairy, usually with stout roots or rhizomes. Flowers bisexual or unisexual, in whorls arranged in panicles, green or reddish, wind-pollinated. Perianth-segments 6 in 2 whorls of 3, the outer whorl usually inconspicuous, the inner whorl enlarged in fruit and becoming hardened or papery, usually 1, 2 or all 3 segments with corky tubercles developing (the ripe inner perianth-segments are often known as 'valves'). Stamens 6. Stigmas 3 feathery. Fruit a 3-angled nut, enclosed by the enlarged inner perianth-segments. *Mainly north temperate areas.*

3. Oxyria Hill. 2/1. Hairless perennial herbs with stout, scaly rootstocks. Leaves almost all basal, with long stalks. Flowers in loose, leafless panicles. Perianth-segments sepal-like, in 2 pairs, those of the inner pair enlarged in fruit. Flowers bisexual. Stamens 6. Stigmas 2. Fruit lens-shaped, broadly winged. *Arctic & mountainous regions of the northern hemisphere.*

4. Rheum Linnaeus. 50/7. Robust perennial herbs with stout, woody rhizomes. Leaves mostly basal, large, palmately lobed. Ochreae loose, persistent. Flowers bisexual, in panicles, wind-pollinated. Perianth-segments 6, free, not enlarging in fruit. Stamens 9. Stigmas 3, almost stalkless. Fruit a 3-winged nut. *Temperate & subtropical Asia.*

5. Atraphaxis Linnaeus. 20/3. Dwarf, often spiny shrubs with spreading or twisted branches. Leaves small, deciduous. Ochreae cartilaginous, bifid. Flowers bisexual in short racemes. Perianth-segments 4 or 5, in 2 whorls, those of the inner whorl enlarged in fruit. Stamens 6 or 8. Stigmas 2 or 3. Nut 2- or 3-angled. *Near East & Asia.*

6. Polygonum Linnaeus. 300/30. Annual to perennial herbs, dwarf shrubs or climbers. Leaves usually distinctly longer than wide. Flowers bisexual, in spikes or panicles. Perianth-segments usually 5 (rarely 4), usually equal, free or united at the base, petal-like. Stamens 5--8. Stigmas 2 or 3. Nut 3-angled or 2-angled and flattened, enclosed by the persistent, sometimes winged, perianth-segments, or protruding by up to half its length. *Mostly north temperate areas.* Figure 30, p. 148.

Frequently now divided into a number of smaller genera: *Aconogonon* (Meissner) Reichenbach, *Bilderdykia* Dumortier, *Bistorta* (Linnaeus) Adanson, *Fagopyrum* Miller, *Fallopia* Adanson, *Persicaria* (Linnaeus) Miller, *Pleuropteropyrum* Grossheim, *Reynoutria* Houttuyn and *Tovara* Adanson.

7. Muehlenbeckia Meissner. 20/7. Shrubs or climbers, usually hairless, mostly deciduous. Ochreae flimsy, soon disappearing. Flowers unisexual, with males and females on separate plants, or with unisexual and bisexual flowers on the same plant. Perianth-segments 5, united at base, enlarged and fleshy in fruit, usually white. Stamens 8. Stigmas 3, more or less stalkless. Nut 3-angled, partly fused with the persistent perianth. *Australia, temperate South America.*

About 20 species, 7 of them cultivated.

39. PHYTOLACCACEAE

Trees, shrubs (some climbing) and herbs. Leaves alternate, simple, entire, stalked or not, usually without stipules, or if stipules present then very small. Flowers usually bisexual and radially symmetric, mostly in terminal or axillary racemes, often subtended by bracts and bracteoles. Perianth-segments 4 or 5, usually free and

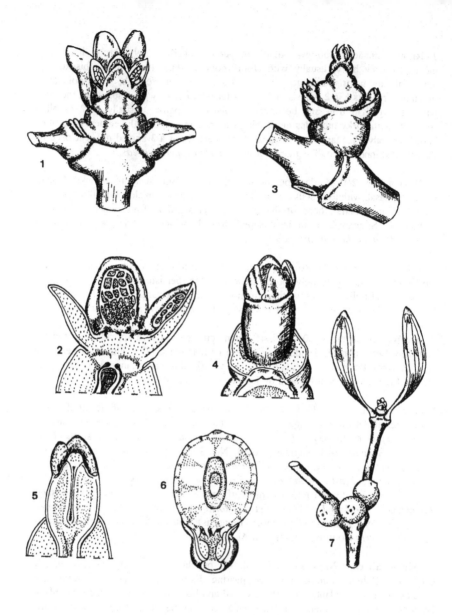

Figure 29. Loranthaceae. *Viscum album.* 1, Male inflorescence. 2, Longitudinal section of male flower showing the stamens attached to the perianth-segments. 3, Female inflorescence. 4, Female flower, receptacular tissue cut away. 5, Longitudinal section of female flower. 6, Longitudinal section of fruit. 7, Part of female plant showing three fruits.

146

persistent. Stamens as many as the perianth-segments or more numerous. Ovary usually superior; carpels 1--many, free or united, each containing a single ovule; styles short, as many as carpels, or absent. Fruit a fleshy berry, dry nut or rarely a capsule.

A family of about 20 genera and 100 species native to the warmer parts of the world, mainly America and Africa.

3a. Racemes axillary; leaf-blades not more than 7 cm   **1. Ercilla**
  b. Racemes terminal or opposite the leaves; leaf-blades usually more than
    7 cm   **2. Phytolacca**

1. Ercilla Jussieu. 2/1. Climbing evergreen shrubs to 20 m with slender, very leafy stems bearing aerial roots. Leaves 2.5--7 × 2--5.5 cm, ovate to oblong, blunt at apex, tapered to cordate at base, hairless; stalk 3--6 mm. Flowers in dense axillary racemes 1--5 cm, each flower 6--7 mm in diameter, shortly stalked and subtended by a bract and 2 bracteoles. Perianth-segments 5, elliptic, green with purplish margins. Stamens 6--10, white, protruding. Fruit a berry, rarely produced in cultivation. *Chile, Peru.*

2. Phytolacca Linnaeus. 25/8. Perennial herbs, shrubs (rarely climbing) or trees, hairless or nearly so, sometimes dioecious. Leaves ovate, elliptic or lanceolate, usually stalked. Racemes dense or open, erect or drooping, terminal or opposite the leaves. Flowers usually bisexual, sometimes unisexual; 1 bract and 2 bracteoles usually present. Perianth-segments 5, greenish to white or pink. Stamens 5--30, in 1 or 2 whorls, Carpels 5--16, free or united. Fruit a somewhat flattened spherical 'berry', usually glossy black with purple juice. *Mainly in the tropics, though extending into subtropical areas in both hemishperes.*

40. NYCTAGINACEAE

Annual or perennial herbs, woody vines, shrubs or trees. Leaves usually entire, alternate, opposite or whorled, without stipules. Inflorescence usually cymose. Flowers bisexual or unisexual, sometimes subtended by coloured bracts. Perianth usually 5-lobed, united below into a tube. Stamens 1--many, usually 5 and then alternate with the perianth-lobes, free or united at base and sometimes branched above. Ovary superior, 1-celled, containing a single basal ovule, and bearing a long style. Fruit an achene, enclosed in the persistent perianth.

A family of 30 genera and about 290 species from the tropics and subtropics, especially America.

1. Mirabilis Linnaeus. 60/2. Annual or perennial herbs, often with tuberous roots. Stems usually branched, hairless or glandular-hairy. Leaves opposite, the lower stalked, the upper almost stalkless. Flowers fragrant or scentless in axillary cymes, surrounded by a tubular or narrowly bell-shaped, calyx-like involucre of 5 bracts. Perianth funnel-shaped, with a long tube, contracted above the ovary, and a spreading, slightly 5-lobed limb. Stamens 3--5, filaments hairless. Fruit ellipsoid to spherical or ovoid, often angled or ribbed. *North, Central & South America.*

Species and hybrids of **Bougainvillea** Jussieu (spiny climbing shrubs, flowers subtended by large, coloured bracts) and species of **Abronia** Jussieu (leaves in pairs, fleshy, those in each pair of different sizes, bracts not united and involucre-like) are grown in glasshouses and may survive outside in extremely sheltered areas.

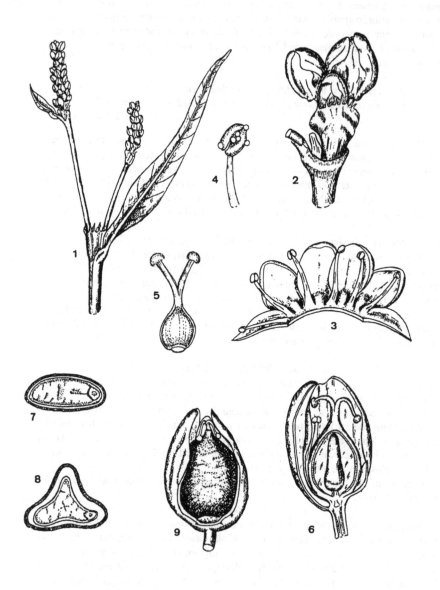

Figure 30. Polygonaceae. *Polygonum persicaria*. 1, Part of shoot showing two inflorescences and the stipular sheath (ochrea). 2, Three flowers from the base of the
inflorescence. 3, Perianth opened out to show the stamens. 4, Stamen. 5, Ovary with
two styles. 6, Longitudinal section of a flower showing the single, erect ovule. 7,
Transverse section of a bicarpellary ovary. 8, Transverse section of a tricarpellary
ovary. 9, Fruit, part of the persistent, enveloping perianth removed.

148

41. AIZOACEAE

Succulent annual or perennial herbs or small shrubs. Leaves usually opposite, fleshy, without stipules, often the leaves of each pair slightly to almost completely joined at the base (in some species the plant-body consists only of a single pair of closely adpressed or divergent leaves), leaves very variable in shape, often papillose, sometimes toothed, sometimes with translucent 'windows' at the broad, flat apex. Flowers bisexual. radially symmetric, axillary or terminal and solitary or in terminal cymes, often with fleshy bracts and bracteoles. Sepals 4--15, mostly free, sometimes united into a tube at the base. Petals very numerous in 1--several rows, mostly free, rarely joined into a short tube at the base. Stamens numerous; staminodes often present between the petals and stamens and transitional between the two. Nectary usually present, either as 5 or 6 distinct glands, or as a continuous, sometimes scalloped ring. Most or all of the ovary inferior, 4--20-celled; ovules usually numerous, usually parietal, rarely axile. Stigmas usually free, more rarely united into a columnar or disc-shaped structure. Fruit a capsule, usually opening when wetted and closing when dry, very complex in structure, more rarely breaking up irregularly or fleshy and indehiscent. Seeds numerous, usually with long funicles.

A large family, sometimes divided into 3 (*Aizoaceae* in the strict sense, *Molluginaceae* and *Mesembryanthemaceae*), with about 120 genera (most of them split off from the old genus *Mesembryanthemum* during the 20th century) and about 2500 species, mostly from South Africa, though with a few native in other arid parts of the world.

The classification of the family as currently understood is based largely on the structure of the fruits, which are in many ways remarkable and complex structures. In a few genera the fruits are fleshy and indehiscent or break up irregularly, but in most the capsules open when wetted, releasing the seeds gradually, and then close again when dry; the organs responsible for this opening and closing are unique to the family and are particularly important in classification at the genus level. Fortunately, only very few genera are hardy in northern Europe; the *European Garden Flora* (volume **3**: 133) should be consulted for details of the others.

1a. Perennials; fruit juicy, indehiscent   **2. Carpobrotus**
  b. Annuals; fruit dehiscent, opening when wetted   **1. Dorotheanthus**

1. Dorotheanthus Schwantes. 10/1. Much-branched annual plants, the branches often reddish and markedly papillose. Leaves at first in rosettes, later opposite or alternate, flat, narrowed to the base, entire, soft, usually spathulate. Flowers terminal or axillary, stalked, without bracts. Sepals 5, unequal, covered with clear papillae. Petals in 2 whorls. Anthers brown or black. Ovary 5-celled, placentation parietal; stigmas 5, hardened and persistent in fruit. Fruit a capsule, dehiscent when wetted. Seeds pear-shaped, smooth. *South Africa (Cape Province),*

2. Carpobrotus N.E. Brown. 23/3. Small perennial herbs with long, 2- or 3-angled, stout, trailing branches. Leaves opposite, thick, 3-angled in section, margins entire or toothed, smooth, grey-green with small translucent spots and often swollen below at the base. Flowers solitary, terminal, stalked. Sepals 5. Petals numerous, purple, red, pink, yellow or white. Ovary 10--16-celled, placentation parietal; stigmas 10--16. Fruit fleshy and indehiscent, without valves. Seeds slightly compressed. *South Africa; a few species naturalised in parts of Europe.*

42. PORTULACACEAE

Herbs or rarely shrubs or trees, often fleshy. Leaves alternate, opposite or all basal, entire, with or without stipules, usually somewhat fleshy (leaves sometimes very reduced and hidden between the enlarged, parchment-like stipules). Flowers bisexual, usually radially symmetric, in racemes or cymes (often condensed and cluster-like), rarely solitary; bracts often conspicuous. Sepals usually 2, free or united at base, persistent or deciduous. Petals 3--18, usually 5, free or rarely slightly united at the base. Stamens 3--many, anthers opening by slits. Ovary superior or at least partly inferior, of 2, 3 or rarely more carpels, 1-celled, ovules 1--many, placentation basal when ovule 1, free-central when more. Fruit usually a capsule opening by lobes at the apex or by a split developing along the circumference, the top falling like a lid. Seeds 1--many, with or without appendages. See figure 31, p. 152.

A cosmopolitan family of 19 genera and about 900 species, centred in South Africa and America.

1a. Ovary partly or wholly inferior   **1. Portulaca**
  b. Ovary fully superior   2
2a. Capsule splitting around its circumference, the top falling like a lid: petals usually more than 5   **2. Lewisia**
  b. Capsule splitting into lobes at the top; petals usually 5   3
3a. Stigmas 2; capsule opening by 2 lobes   **4. Spraguea**
  b. Stigmas 3; capsule opening by 3 lobes or by loss of the outer skin   4
4a. Ovules more than 6   **3. Calandrinia**
  b. Ovules 3--6   5
5a. Stem leaves several, alternate or opposite; basal rosette absent   **6. Montia**
  b. Stem leaves 2, opposite; basal rosette usually present   **5. Claytonia**

1. Portulaca Linnaeus. 100/2. Usually sprawling annual herbs. Leaves fleshy, flat or terete, alternate or some opposite. Flowers solitary, terminal, stalkless. Sepals 2. Petals 6 or 7. Stamens 7--many. Ovary at least partly inferior, 1-celled, with many ovules. Style 1, stigmas 3--9. Fruit a capsule opening by splitting around its circumference above the point of attachment of the sepals. Seeds many. *Mostly tropical.*

One species of **Portulacaria** Jacquin, which is a succulent shrub or small tree, may be found in the mildest areas.

2. Lewisia Pursh. 16/8 and many hybrids. Hairless perennial herbs with corms or fleshy roots and a crown which may be small or large. Leaves all or mostly basal, evergreen or deciduous. Scapes long or short, bearing 1--3 flowers or many-flowered panicles, bracts occasionally borne close to the sepals and similar to them. Sepals usually 2, occasionally to 8. Petals 4--18, often unequal in width, sometimes so in length. Stamens 5--many. Ovary 1-celled with 3--8 stigmas. Fruit a thin-walled capsule, splitting around its circumference near the base and also splitting into 3--8 flaps at the apex. Seeds numerous, sometimes with arils. *Western North America, adjacent Mexico.*

3. Calandrinia Humboldt et al. 150/3. Annual or perennial herbs, more or less succulent. Leaves alternate or all basal. Flowers in racemes, panicles or umbel-like clusters, remaining open for 1 day or less. Sepals 2, persistent. Petals usually 5, red or purple, rarely white. Stamens 3--14. Ovary superior, style short, stigmas 3. Fruit a

capsule opening by 5 flaps at the apex. Seeds numerous. *Mostly South America &*
Australia. Figure 31, p. 152.

Species of **Talinum** Adanson (herbs with deciduous sepals) and **Anacampseros**
Linnaeus (herbs with persistent sepals and 15--60 stamens) are grown in glasshouses in
northern Europe, and may be found outside in very mild areas.

4. Spraguea Torrey. Few/1. Plants mostly perennial from a deep, woody taproot.
Leaves mostly basal, in rosettes. Flowering stems very short. Inflorescence compact,
head-like, made up of umbels of cymes. Sepals 2, persistent, conspicuous, scarious.
Petals 4. Stamens 3. Ovary with 1--8 ovules; style 1, stigmas 2. Fruit a membranous
capsule opening by 2 lobes at the apex. *Western North America.*

5. Claytonia Linnaeus. 24/4. Annual or perennial herbs, usually with a basal rosette of
leaves and with 2 opposite leaves on the stem, these sometimes united, stem-leaves
otherwise absent. Flowers in long or contracted racemes or in panicles. Sepals 2.
Petals 5, white or pinkish, Stamens 5. Ovary containing 3--6 ovules. Style 3-lobed.
Fruit a capsule opening by 3 lobes at the apex. *North America, east Asia; a few
species naturalised in Europe.*

6. Montia Linnaeus. 37/1. Annual or perennial herbs similar to *Claytonia*, but stems
with numerous opposite or alternate leaves; ovules always 3. (There are also important
differences in the structure of the pollen but these are not recorded here.) *Western
North America, South America, Australia.*

43. BASELLACEAE

Perennial hairless herbs with slender, twining, climbing or decumbent stems growing
from tuberous rootstocks. Leaves alternate, entire, fleshy. Flowers radially symmetric,
bisexual or unisexual, in axillary spikes, racemes or panicles. Bracteoles 2, often united
with the perianth. Perianth-segments 5, free or basally united. Stamens 5, opposite the
perianth-segments. Ovary superior, 1-celled. ovule 1. Style branched or simple.
Stigmas 3 or 1. Fruit a drupe, usually enclosed by the persistent, fleshy perianth.

A family of 4 genera and 12--17 species, but mainly in Central & South America.

1. Basella Linnaeus. 5/1. Twining herbs with much-branched stems to 9 m. Leaves
oblong, ovate to ovate-lanceolate, or circular, fleshy, stalked. Flowers bisexual in
usually simple spikes. Perianth urn-shaped, white, red or purple. Filaments erect and
straight in bud. Style branched, stigmas 3. *Tropical Africa & Asia.*

Species of **Anredera** Jussieu (filaments curved outwards in bud) and **Ullucus** Caldas
(filaments erect in bud, flowers in racemes, stigma 1) may occur cultivated out-of-
doors in very mild areas.

44. CARYOPHYLLACEAE

Herbs, more rarely small shrubs, with simple, entire, usually opposite leaves, sometimes with small stipules. Stems often weak and brittle with prominent nodes. Flowers
usually bisexual, radially symmetric, often in terminal dichasia. Sepals 4 or 5, free or
joined, sometimes in a long tube ending in calyx-teeth. Petals 4 or 5, free, sometimes
absent. Stamens usually 8 or 10. Ovary superior, 1-celled (at least above), of 2--5
fused carpels, with 1--many ovules on a basal or free-central placenta. Styles 2--5,

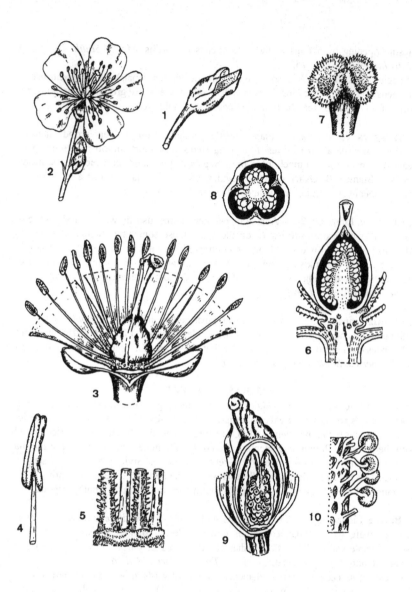

Figure 31. Portulacaceae. *Calandrinia grandiflora.* 1, Bud showing the characteristic two sepals. 2, Flower. 3, Flower with petals removed. 4, Upper portion of stamen. 5, Basal portions of filaments. 6, Longitudinal section of ovary. 7, Stigma. 8, Transverse section of ovary. 9, Longitudinal section of fruit showing persistent sepals. 10, Part of placenta with attached seeds.

152

free. Fruit usually a capsule opening by teeth or flaps, more rarely a berry or 1-seeded nutlet. Seeds usually kidney-shaped with a curved embryo.

A cosmopolitan family of about 90 genera and more than 2000 species, well represented in north temperate regions. See figures 32, p. 154 & 33, p. 156.

1a. Sepals free from each other	2
b. Sepals joined, sometimes into a long calyx-tube	9
2a. Stipules present; petals very small or absent	3
b. Stipules absent; petals usually well-developed	4
3a. Bracts surrounding flowers conspicuous, silvery	**18. Paronychia**
b. Bracts inconspicuous, greenish	**17. Herniaria**
4a. Flowers without petals; fruit a nutlet	**16. Scleranthus**
b. Flowers (with rare exceptions) with petals; fruit a capsule	5
5a. Petals 2-toothed or 2-lobed	**11. Cerastium**
b. Petals more or less entire	6
6a. Leaves linear-subulate, joined at base round stem; flower-buds spherical	
	15. Sagina
b. Leaves sometimes linear but not joined round stem; flower-buds more or less elongate	7
7a. Seed with an appendage; petals usually 4	**14. Moehringia**
b. Seed without an appendage; petals usually 5	8
8a. Capsule opening by twice as many teeth as there are styles	**12. Arenaria**
b. Capsule opening by as many teeth as there are styles	**13. Minuartia**
9a. Fruit a berry	**5. Cucubalus**
b. Fruit a capsule	10
10a. Styles 2	11
b. Styles 3--5	15
11a. Epicalyx present at base of calyx	12
b. Epicalyx absent	13
12a. Calyx with membranous bands of tissue between the veins	
	10. Petrorhagia
b. Calyx without membranous bands	**9. Dianthus**
13a. Petals with coronal scales	**7. Saponaria**
b. Petals without coronal scales	14
14a. Calyx unwinged, with membranous bands of tissue between the veins	
	6. Gypsophila
b. Calyx winged, without membranous bands	**8. Vaccaria**
15a. Calyx-teeth linear, more than 3 cm; coronal scales absent	**4. Agrostemma**
b. Calyx-teeth much less than 3 cm; coronal scales usually present	16
16a. Seeds with a tuft of hairs; styles 5	**3. Petrocoptis**
b. Seeds without a tuft of hairs; styles 3 or 5	17
17a. Styles 3 or rarely 5; capsule opening by twice as many teeth as there are styles	**1. Silene**
b. Styles always 5; capsule opening by 5 teeth	**2. Lychnis**

1. Silene Linnaeus. 500/25. Herbs, or sometimes small shrubs with woody stocks. Leaves opposite, linear to ovate or obovate, entire. Flowers usually conspicuous, solitary or in few- to many-flowered inflorescences. Calyx tubular, sometimes very

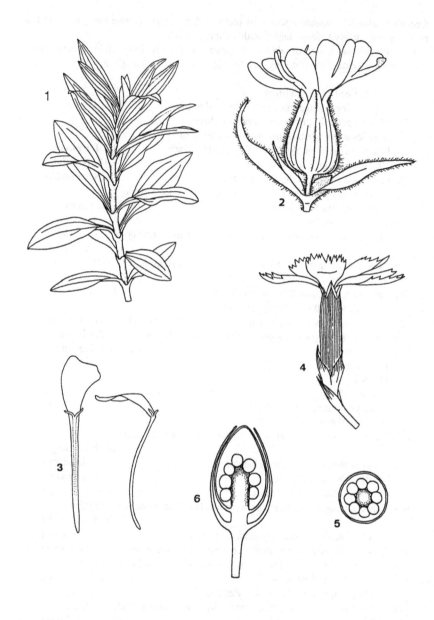

Figure 32. Caryophyllaceae. 1, Shoot of *Saponaria officinalis*. 2, Female flower of *Silene dioica*. 3, Petal of *Saponaria officinalis*, front and side views. 4, Flower of *Dianthus deltoides*. 5, Transverse section of ovary of *Silene dioica*. 6, Longitduinal section of calyx, carpophore and capsule of *Silene dioica*.

persistent and strongly inflated in fruit, with usually 10--30 (rarely 5) veins and 5 short teeth. Petals free, usually with a narrow claw and wide, spreading blade. Coronal scales often present at junction of claw and blade. Stamens 10. Styles 3 (rarely 5). Fruit a capsule with variably-developed basal septa, dehiscing apically with teeth twice the number of styles. A stalk of variable length (anthophore or carpophore) is often present between calyx and capsule. Seeds numerous, usually 1--2 mm. *Northern hemisphere, mainly in the Mediterranean area.* Figure 32, p. 154.

A wide definition of the genus is adopted here, including *Melandrium* Röhling and *Heliosperma* (Reichenbach) Reichenbach.

2. Lychnis Linnaeus. 15/6. Like *Silene*, but styles always 5 and ripe capsule opening apically with 5 teeth only. *North temperate areas.*

3. Petrocoptis Braun. 7/2. Like *Silene* but capsule opening with 5 teeth, and seeds with a tuft of hairs at the hilum. All are perennial, with 5 styles. *Pyrenees, mountains of north Spain.*

4. Agrostemma Linnaeus. 3/1. Annual herbs with erect stems and narrow leaves. Flowers large, solitary or in few-flowered dichasia. Calyx with long, linear teeth. Petals with long claw, slightly notched blade and no coronal scales. Styles 5. Capsule opening with 5 teeth, not stalked. *South Europe, western Asia; introduced into North America.*

5. Cucubalus Linnaeus. 1/1. Like *Silene* but fruit a berry. *Eurasia, North Africa.*

6. Gypsophila Linnaeus. 100 or more/8. Annual or perennial with somewhat woody stock. Flowers usually numerous and small in large spreading panicles; but some species with relatively large or solitary flowers. Calyx joined, 5-toothed, with 1 vein to each tooth and a band of membranous tissue between each vein. Petals 5, entire or notched, usually with no clear differentiation into blade and claw; coronal scales absent. Stamens 10. Styles 2. Fruit a capsule opening with 4 teeth. Ovary with or without a short carpophore. *Southeast Europe, western Asia, North Africa.*

7. Saponaria Linnaeus. 20/7. Perennial (rarely annual) herbs, sometimes with rather woody stocks; indistinguishable from *Silene* or *Lychnis* except by the possession of 2 styles (not 3 or 5) and the opening of the capsule by 4 (not 6 or 10) teeth. *South Europe, southwest Asia.* Figure 32, p. 154.

8. Vaccaria Medicus. 3/1. Like *Saponaria*, but calyx-tube winged and petals without coronal scales. *Mediterranean area.*

9. Dianthus Linnaeus. 300/45 and many hybrids. Herbs, rarely small shrubs. Leaves opposite, usually linear, entire, often glaucous. Flowers conspicuous, solitary or in few- to many-flowered inflorescences, sometimes capitate with distinct involucral bracts surrounding the base of the head. Calyx usually cylindric, enclosed in 1--3 pairs of epicalyx-bracts. Petals free, with narrow claw and wide spreading blade, often somewhat cut or toothed but not bifid. Petals often with a tuft of hairs ('bearded') in mouth of flower. Stamens 10. Styles 2. Fruit a capsule opening apically with 4 teeth. A short carpophore often present between calyx and capsule. Seeds numerous, concave on 1 side. *Europe, Asia, one species in arctic North America.*

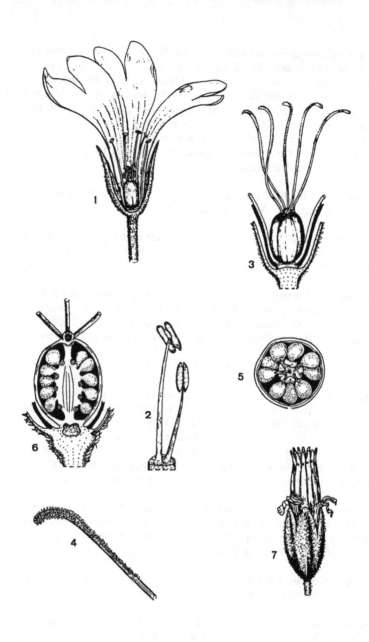

Figure 33. Caryophyllaceae. *Cerastium tomentosum*. 1, Flower with two sepals and two petals removed. 2, Two unequal stamens. 3, Flower with upper parts of sepals, petals and stamens removed to show the ovary. 4, Upper part of style showing stigmatic area. 5, Transverse section of ovary. 6, Longitudinal section of ovary. 7, Capsule surrounded by the persistent calyx.

10. Petrorhagia (de Candolle) Link. 29/1. Like *Dianthus* but with a membranous band of tissue between the veins of the calyx. *Europe, western Asia, North Africa.*

11. Cerastium Linnaeus. 100/few. Perennial or annual herbs of varied habit (the cultivated species tufted or mat-forming), sometimes woody at base, with stalkless leaves. Flowers in cymes or solitary. Sepals usually 5, free; petals usually 5, white, bifid or more or less deeply indented; stamens usually 10. Styles usually 5, on the same radii as the sepals. Fruit a many-seeded capsule, usually more or less cylindric and somewhat curved, opening by 10 short teeth. *North temperate & arctic areas.*

12. Arenaria Linnaeus. 160/few. Low-growing herbs, sometimes with woody stocks. Leaves very variable in shape, from linear-subulate to circular. Flowers usually in few-flowered cymes, sometimes solitary. Petals more or less entire, usually 5 and white; stamens 10; styles usually 3. Capsule opening with twice as many teeth as styles. *Temperate & arctic areas of the northern hemisphere.*

13. Minuartia Linnaeus. 100/2. Differs from *Arenaria* chiefly in the capsule, which opens with as many teeth as styles. Most species have linear leaves. *Temperate & arctic areas of the northern hemisphere.*

14. Moehringia Linnaeus. 20/1. Differs from *Arenaria* only in the presence of an appendage (strophiole) on the seed. Most species have 4 petals (a rare condition in *Arenaria*). *Temperate & arctic areas of the northern hemisphere.*

15. Sagina Linnaeus. 20/1. Small herbs with the general appearance of *Arenaria* and *Minuartia*, but always with linear-subulate leaves slightly joined at the base round the slender stem. Flowers small, solitary or in few-flowered cymes, spherical in bud. Sepals 4 or 5; petals 4 or 5, entire, white (sometimes very small or absent). Styles 4 or 5; fruit a more or less spherical capsule splitting to the base into 4 or 5 flaps. *Temperate northern hemisphere.*

16. Scleranthus Linnaeus. 10/1. Low-growing annual to perennial herbs with opposite, subulate or linear leaves slightly joined at base. Flowers small, solitary, paired or in branched cymes. Sepals 4 or 5, situated on the rim of a perigynous zone, persistent in fruit; petals 0; stamens 1--10; styles 2. Fruit a usually a 1-seeded nutlet enclosed in the perigynous zone. *Temperate areas of both hemispheres.*

17. Herniaria Linnaeus. 15/1. Low-growing annual to perennial herbs, sometimes with woody stocks. Leaves lanceolate to ovate or obovate, opposite near base but sometimes alternate above, with small, greenish stipules. Flowers small, greenish, in dense axillary cymes; sepals, petals and stamens perigynous. Sepals 5; petals 5, shorter than sepals; style 1, more or less stalkless, with bifid stigma. Fruit a nutlet inside the persistent sepals and perigynous zone, with a single black shining seed. *Europe, Asia, Africa.*
    Sometimes placed in the family *Illecebraceae.*

18. Paronychia Miller. 50/1--2. Like *Herniaria*, but with silvery stipules, conspicuous, often silvery bracts more or less concealing the flowers, and usually 2 styles free from each other or partly joined. The fruit, though 1-seeded, is partly dehiscent. *Cosmopolitan.*

45. CHENOPODIACEAE

Annual or perennial herbs or shrubs, rarely becoming tree-like, frequently succulent and with bladder-like hairs which give the plants a mealy appearance. Leaves alternate, rarely opposite, simple, lacking stipules. Flowers with bracteoles, bisexual or unisexual, radially symmetric. Perianth with 1--5 segments, sometimes absent in female flowers. Stamens 1--5, opposite the perianth-segments; anthers 2-celled, opening lengthwise. Ovary superior, or rarely partly inferior, 1-celled; stigmas 1--5; ovule solitary, basal. Fruit an achene with 2 horizontally or vertically flattened seeds. See figure 34, p. 160.

A cosmopolitan family of about 100 genera and 1400 species. Most are found in arid salty places. Relatively few have any ornamental value though several have been consumed as vegetables for many centuries.

1a. Stems segmented, with opposite branches; leaves apparently absent or
    rudimentary   2
  b. Stems not segmented; leaves present   3
2a. Stamens 3--5   **10. Anabasis**
  b. Stamens 1 or 2   **9. Salicornia**
3a. Flowers mostly unisexual, female flowers usually without perianth, but with
    2 bracteoles which become enlarged in fruit   4
  b. Flowers mostly bisexual, or if mostly unisexual, then female flowers with a
    perianth of 3 or more segments   5
4a. Bracteoles free in their upper half or more   **4. Atriplex**
  b. Bracteoles united almost to the apex   **3. Spinacia**
5a. Each perianth-segment (or at least some) bearing a spine on the back
    in fruit   **6. Bassia**
  b. Perianth-segments without spines in fruit, sometimes winged   6
6a. Ovary partly inferior, the lower part united with the thickened perianth in
    fruit   **1. Beta**
  b. Ovary completely superior, not united with perianth in fruit   7
7a. Plant smelling of camphor   **5. Camphorosma**
  b. Plant not smelling of camphor   8
8a. Leaves semi-cylindric in section, glaucous   **8. Suaeda**
  b. Leaves flat, not usually glaucous   9
9a. Perianth downy, segments winged in fruit   **7. Kochia**
  b. Perianth not downy, segments not winged in fruit   **2. Chenopodium**

1. Beta Linnaeus. 6/1. Annual, biennial or perennial herbs. Leaves alternate, entire, the basal long-stalked. Flowers bisexual, 1--few in axillary clusters in long spikes. Bracts 2, small. Perianth-segments 5, usually green, entire or laciniate; segments thickening in fruit, Stamens 5. Stigmas 2 or 3. Ovary half-inferior, united with the base of the thickened perianth in fruit. Fruits often held together by the swollen perianth and receptacle. Seed horizontal, lens- or kidney-shaped, glossy. *Mediterranean area.*

2. Chenopodium Linnaeus. 150/few. Annual to perennial herbs, mealy or with mostly glandular hairs. Stems leafy. Leaves mostly alternate, simple or pinnatifid. Flowers

158

bisexual (rarely female), greenish, in spikes or panicles of cyme-like clusters. Bracteoles absent. Perianth-segments 3--5, papery or red and fleshy in fruit. Stamens 1--5. Stigmas 2 or 3. Ovary superior; fruit an achene. *Cosmopolitan.*

3. Spinacia Linnaeus. 4/1. Erect, annual or biennial, dioecious herbs. Leaves alternate, stalked. Flowers unisexual; male flowers with parts in 4s or 5s, in dense spike-like inflorescences; female flowers axillary, without perianths, but with 2 (more rarely 4) persistent bracteoles which become enlarged, united almost to the apex and hardened in fruit. Stamens 4 or 5. Stigmas 4 or 5. Seeds vertical. *East Mediterranean area to Central Asia.* Figure 34, p. 160.

4. Atriplex Linnaeus. 200/few. Annual or perennial herbs, hairless to mealy or scurfy. Male and female flowers on the same or separate plants. Leaves alternate or opposite. Flowers unisexual, solitary or clustered, axillary or in terminal spikes or panicles. Male flowers with 3--5 perianth-lobes, stamens 3--5. Female flowers usually without perianth, but with 2 large persistent bracteoles, rarely some with 1--5 small scales or 3--5 perianth-lobes. Seed erect or inverted, rarely horizontal. *Temperate & subtropical areas.*

5. Camphorosma Linnaeus. 11/1. Annual or perennial herbs or small shrubs, smelling strongly of camphor (ours). Stems and leaves softly hairy. Leaves alternate, linear or awl-shaped. Flowers bisexual or female, solitary or in crowded ovoid cymes. Perianth-segments 4 or 5, 2 larger than the rest. Stamens 4 or 5, stigmas 2 or 3. Seeds vertically flattened. *Mediterranean Europe to Central Asia.*

6. Bassia Allioni. 90/2. Annual or perennial downy herbs. Leaves alternate, oblong to linear, flat or cylindric, entire and fleshy. Flowers bisexual or female, solitary or in cymes arranged in a panicle. Perianth-segments 5, becoming enlarged in fruit, stamens 5, stigmas 2 or 3. Fruiting perianth papery, smooth or with up to 5 horizontal spines developing on the backs of the segments. *Mediterranean area to Central Asia, Australia.*

7. Kochia Roth. 90/2. Similar to *Bassia* but flowers neither hidden in hairs nor spiny, sometimes with a short flat wing (rarely a tubercle) on the back of the fruiting perianth-segments. *Europe, temperate Asia, North & South Africa, Australia.*

8. Suaeda Scopoli. 110/1. Annual or perennial herbs or small shrubs. Stems naked or sometimes with floury scales. Leaves alternate, fleshy, semi-cylindric or flat. Flowers bisexual or female, solitary or in cymes of 2--5. Bracteoles 2, very small; perianth-segments 5, fleshy, sometimes tuberculate; stamens 5; stigmas 2--5. Fruit with thin papery pericarp, seeds horizontally or vertically flattened; embryo in a flat spiral. Two seed types are often found in the same species: the earlier are usually black, shiny and smooth, the later are larger, brown or green, dull and net-patterned. *Cosmopolitan in coastal situations or in salt deserts.*

9. Salicornia Linnaeus. 35/1. Succulent annual herbs with jointed stems. Leaves opposite, scale-like, inconspicuous, their bases joined and clasping the stem to form a segment. Flowers bisexual in groups of 3, sunken in the bracts of the segments, forming terminal spikes. Perianth with 3 or 4 lobes. Stamens 1 or 2, projecting. Ovary

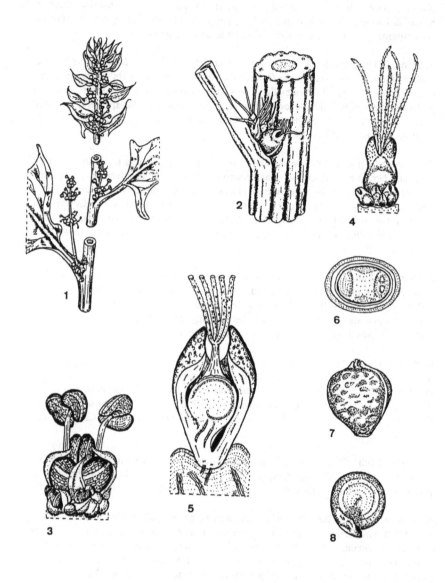

Figure 34. Chenopodiaceae. *Spinacia oleracea.* 1, Portion of stem with male inflorescences. 2, Portion of stem with female flowers. 3, Male flower with flower-buds below. 4, Female flower with buds below. 5, Longitudinal section of female flower. 6, Transverse section of female flower. 7, Fruit. 8, Seed.

1-celled with 2--4 style branches. Fruit bladder-like, surrounded by the perianth which hardens with age; seed coat thin, membranous. *Saline habitats in temperate & tropical areas.*

10. Anabasis Linnaeus. 30/1. Perennial herbs or shrubs, succulent, with jointed stems. Leaves opposite, joined and clasping the stem, inconspicuous, scale-like. Flowers bisexual or female, solitary or several in the upper leaf-axils. Female flowers with 2 bracteoles. Perianth-segments 5, each usually developing a transverse wing on the back in fruit. Stamens 5, alternating with 5 staminodes. Stigmas 2. Seeds vertical. *Mediterranean area to Central Asia.*

46. AMARANTHACEAE

Annual or perennial herbs or shrubs. Leaves alternate or opposite, usually entire, without stipules, often coloured in cultivated species. Flowers small and inconspicuous, in complex inflorescences made up of very condensed cymes or racemes which are normally aggregated into conspicuous heads or spikes, more rarely in panicles, each flower usually subtended by a bract and 2 bracteoles. Perianth of usually 5 segments (rarely fewer), these free or united below, usually scarious. Stamens as many as perianth-segments, their filaments usually united, at least below, often with staminodes in between them. Ovary superior, 1-celled, usually with a single ovule, more rarely with several; styles 1--3. Fruit a capsule opening by a lid or indehiscent, rarely a berry.

A family of about 60 genera and 900 species, mostly from the tropics and subtropics. Cultivated species have either conspicuous bracts or inflorescences (the flowers being usually small and insignificant) or brightly coloured leaves.

1a. Filaments united below; ovary containing several ovules; flowers bisexual
1. Celosia

  b. Filaments free; ovary containing a single ovule; flowers unisexual, both sexes on the same plant **2. Amaranthus**

1. Celosia Linnaeus. 50/1. Annual or perennial herbs, sometimes becoming woody at the base, occasionally ascending. Leaves alternate, simple, entire or variously lobed. Flowers bisexual, in compound spikes, with bracts and bracteoles. Perianth-segments 5, free, more or less equal. Stamens 5, filaments joined below, the free parts triangular above or swollen with teeth projecting on each side of the anther, anthers 4-celled. Ovary with few to many ovules, style long or occasionally almost absent; stigmas 2 or 3. Capsule with many black shiny seeds, seeds often strongly compressed with net-like, grooved or warty ornamentation. *Warmer areas of America & Africa.*

2. Amaranthus Linnaeus. 50/5. Large monoecious, annual herbs. Leaves alternate, stalked. Flowers borne in very condensed cymes which are aggregated into spherical or cylindric spikes, themselves aggregated into compound series of spikes; spikes mostly axillary but in several species aggregated near the stem-apex with subtending leaves reduced or absent. Flowers unisexual with a bract and 2 bracteoles which resemble the perianth-segments. Perianth-segments mostly 3--5 (rarely 1 or 2), scarious. Stamens 1--5, free. Ovary 1-celled with 3 styles. Fruit a 1-seeded capsule opening by a lid or rarely indehiscent. *Warmer areas of the world.*

161

Species of a few other genera may be grown in areas with very mild climates, or as half-hardy annuals: **Bosea** Linnaeus is an evergreen shrub with the fruit a berry; **Gomphrena** Linnaeus contains annuals and perennials with flowers in terminal heads with conspicuous, coloured bracts; and **Iresine** Browne rarely flowers but has leaves which are red to crimson on at least the midrib and stalk, the rest greenish or yellow.

47. CACTACEAE

Highly specialised perennials of diverse habit; roots fibrous or tuberous; stems terete, spherical, tubercled, ribbed, winged or flattened, often segmented, mostly leafless and variously spiny; spines, new growth and flowers always arising from cushion-like areoles. Flowers solitary or rarely clustered, appearing stalkless, nearly always bisexual, usually radially symmetric, receptacle enclosing and more or less produced beyond, the zone around the ovary ('pericarpel') and between ovary and perianth ('tube'), naked or invested with bract-like scales and areoles, areolar trichomes (felted hairs), hairs and/or spines. Perianth-segments numerous in a graded series; stamens often very numerous: anthers 2-celled, splitting longitudinally; ovary usually inferior, 1-celled, placentation parietal, ovules numerous; style single, stigma lobed, variously papillate. Fruit juicy or dry, naked, scaly, hairy, bristly or spiny, indehiscent or variously dehiscent, seeds numerous, sometimes with a caruncle or aril, coat variously patterned; embryo curved, usually strongly so, or nearly straight; cotyledons reduced or vestigial, rarely leaf-like.

An exclusively American family, apart from one species of *Rhipsalis* in Africa, Madagascar, the Mascarenes and Sri Lanka, comprising at least 1500 species. Amongst the most distinctive of flowering plants, the desert species have long attracted general curiosity, and the smaller-growing, at least, a worldwide clientele of specialist growers and collectors. Only a few species of a single genus are hardy in our area.

1. Opuntia Miller. 200/1--2. Small to large shrubs or trees with fleshy, cylindric, club-shaped, almost spherical, flattened or very rarely ribbed, segmented branches; areoles often raised on more or less prominent tubercles, with glochids (hairs barbed at the apex) and usually 1--many spines, sometimes sheathed and barbed, occasionally lacking. Leaves terete or subulate, usually small, falling early. Flowers solitary, stalkless, lateral or almost terminal, rarely terminal; pericarpel with leaves, areoles, glochids and often spines; perianth disc-shaped or spreading, rarely erect, without tube; stamens numerous, sometimes touch-sensitive; ovary inferior. Fruit fleshy or dry, with a depression at the top; seeds encased in a bony aril. *Central & North America.*

About 200 species, 1 or 2 grown out-of-doors in northern Europe.

48. MAGNOLIACEAE

Evergreen or deciduous trees or shrubs, the buds enclosed by stipules and the branchlets with conspicuous ring-like stipule-scars at the nodes. Leaves alternate, sometimes crowded into false whorls, with free or attached stipules, the blades simple, sometimes lobed. Flowers solitary, terminal or axillary, bisexual; perianth-segments 6--9, or numerous. free, all more or less similar, or of 2 forms; stamens numerous, inwardly or laterally dehiscent. The numerous carpels spirally arranged on the long floral axis, each with 2--6 ovules. The fruits are cone-like heads of persistent follicles or deciduous samaras. See figure 35, p.164.

About 12 genera and 200 species distributed in temperate and tropical east & south-east Asia and in the western hemisphere in eastern North America.

la. Fruits cone-like heads of closely overlapping 2-seeded samaras, the samaras falling at maturity leaving the persistent, spindle-shaped floral axis; plants deciduous, the leaves variously lobed **3. Liriodendron**
  b. Fruits cone-like heads of follicles, the follicles persistent on the floral axis; plants evergreen or deciduous, the leaves simple, occasionally cordate or auriculate at base or notched at apex 2
2a. Carpels with 2 ovules, the follicles with 2 seeds (sometimes 1 or 1 through abortion); plants evergreen or deciduous **1. Magnolia**
  b. Carpels with 4 or more ovules, the follicles with 4 or more seeds; plants evergreen **2. Manglietia**

1. Magnolia Linnaeus. 80/30. Evergreen or deciduous trees or shrubs; branchlets with septate or continuous pith, the buds enclosed by stipule-scales. Leaves alternate, sometimes in false whorls, stalked, with free or attached stipules which fall early, the blades thickly leathery to membranous, entire. Flowers fragrant, appearing before or with the leaves, terminal, solitary. Perianth of 6--9 or numerous (to 33) segments, white or pink to purple, occasionally greenish or pure yellow, in whorls of 3 (occasionally more), all more or less similar or sometimes those of the outer whorl reduced in size and sepal-like. Stamens numerous, spirally arranged, the linear pollen-sacs inwardly or laterally dehiscent. Carpels many, each with 2 ovules, spirally arranged on a short to elongated floral axis. Fruits cone-like heads of free, longitudinally dehiscent follicles, the 1 or 2 seeds suspended on thin threads, with orange, red or pink aril-like seed-coats. *Temperate & tropical Asia & America.* Figure 35, p. 164.

One species of **Michelia** Linnaeus (evergreen shrubs with axillary flowers and stalked ovaries) may be grown in especially mild areas.

2. Manglietia Blume. 25/1. Evergreen trees or large shrubs, very similar to *Magnolia.* Flowers terminal, carpels with 4 or more ovules. *Tropical & subtropical Asia.*

3. Liriodendron Linnaeus. 2/2 and a hybrid. Deciduous trees with the bark becoming fissured into longitudinal plates; young branchlets with septate pith; buds enclosed by fused stipules. Leaves alternate, long-stalked, the large, free stipules early deciduous; blades with truncate or widely notched apices and 1 or 2 lateral lobes on each side towards the base. Flowers odourless, appearing with the leaves, terminal. Perianth of 9 segments, the outer 3 reflexed and sepal-like, the inner 6 petal-like, in 2 whorls, simulating a tulip or cup-shaped corolla. Stamens numerous. Carpels many, each with 2 ovules, densely overlapping and spirally arranged. Fruits in a spindle-shaped, cone-like head of free, tightly overlapping, 2-seeded samaras which fall at maturity, the axis persistent. *China, Indochina, eastern North America.*

49. WINTERACEAE
Evergreen trees or shrubs, monoecious or some dioecious. Leaves aromatic, simple, entire, alternate, without stipules. Flowers bisexual or unisexual, in axillary or terminal compound inflorescences, floral parts indefinite in number; sepals 2--6; petals 2--many; stamens numerous, anthers opening inwards. Carpels 1--many, free or united, with 1--

163

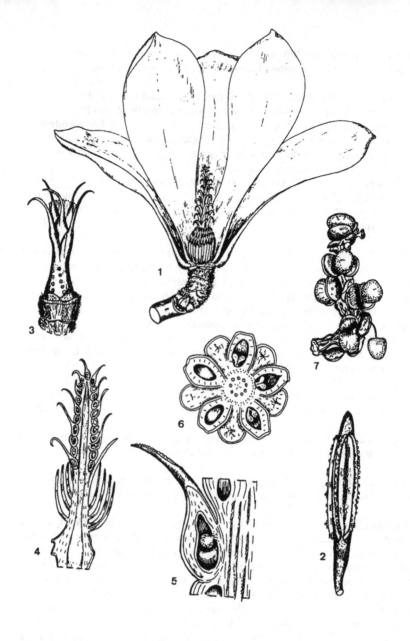

Figure 35. Magnoliaceae. *Magnolia* × *soulangeana*. 1, Longitudinal section of flower. 2, Stamen. 3, Lower part of receptacle with several free carpels. 4, Longitudinal section of receptacle. 5, Longitudinal section of a single carpel. 6, Transverse section of receptacle. 7, Fruit showing open follicles and seeds on dangling funicles.

many ovules; style absent. Fruit a capsule, follicle or berry. Embryo minute, endosperm absent.

A small family containing about 6 genera and perhaps 80 species, confined almost entirely to the southern hemisphere, but there is considerable disagreement about generic limits and species numbers. In Europe only 2 genera are represented.

1a. Leaves entirely green, or with red stalks; inflorescences terminal; calyx
   deciduous; stamens with terete filaments   **1. Drimys**
 b. Leaves coloured, not entirely green; inflorescences axillary; calyx not
   deciduous; stamens with flattened filaments   **2. Pseudowintera**

1. Drimys Forster & Forster. 30/2. Trees or shrubs, sometimes dioecious. Leaves hairless. Flowers bisexual or unisexual, terminal, solitary or in umbels. Calyx enclosing the bud, splitting into 2 or 3 lobes, deciduous. Petals few to numerous. Stamens with terete filaments. Fruit a berry. *Southern hemisphere.*

2. Pseudowintera Dandy. 3/1. Trees or shrubs. Leaves hairless, with obvious glands. Flowers in fascicles or solitary, bisexual. Calyx cup-shaped, not deciduous, not covering the petals in bud. Petals 5 or 6. Stamens to 15, filaments flattened. Fruit a berry with 2--6 seeds. *New Zealand.*
Formerly included in *Drimys*.

50. ANNONACEAE
Trees, shrubs and climbers, containing oil passages. Wood aromatic. Leaves alternate, arranged in 2 ranks, entire, without stipules, with pinnately arranged veins, aromatic. Flowers bisexual, usually solitary but sometimes in compound inflorescences, often fragrant. Sepals and petals often indistinguishable from each other, in whorls of 3. Stamens many, rarely few, usually spirally arranged. Carpels many, usually separate. ovules 1--many, basal or parietal, anatropous. Fruit a berry or, more commonly, an aggregate of berries. Seeds large with hard shiny testas, often with arils.

A family containing 120 genera and about 2100 species, found mainly in the tropics of the Old World but with some representatives in the New World. Only 1 genus is hardy in our area.

1. Asimina Adanson. 8/1. Evergreen or deciduous shrubs or small trees. Leaves alternate, simple, entire. Flowers usually axillary, solitary or in small clusters on pendulous stalks. Sepals 3, deciduous; petals 6; stamens numerous on short filaments. Carpels 3--15, separate, 1-celled. Fruit a berry, seeds many. *Eastern North America.*

51. SCHISANDRACEAE
Monoecious or dioecious woody climbers. Leaves alternate, simple, without stipules. Flowers unisexual, axillary, radially symmetric. Perianth made up of a single whorl of 5--20 free segments, usually all more or less of the same size, sometimes a few (outer) smaller. Stamens 4--80; the filaments united at least at the base, sometimes for the greater part of their length, or completely united into a more or less spherical mass on which the anthers are borne (occasionally on very short stalks). Carpels numerous, free, each with 2 or 3 ovules. Fruit a group of fleshy 'berries' borne in a head or on a long, spike-like receptacle.

A family of 2 genera, mainly from south-east Asia but with a single species in south-east USA.

   1a. Fruits forming a compact head   **2. Kadsura**
     b. Fruits borne spike-like along a long receptacle   **1. Schisandra**

1. Schisandra Michaux. 25/6. Deciduous or evergreen, monoecious or dioecious. Leaves entire or finely and distantly toothed. Perianth cup-shaped, segments 5--20. Stamens 4--60, filaments variously united. Fruit a group of 'berries' borne spike-like on a long receptacle. *Eastern Asia, southeast USA*.

2. Kadsura Jussieu. 20/1. Evergreen, monoecious, woody climbers. Leaves entire or slightly toothed. Flowers unisexual. Perianth of 7--24 segments which are overlapping in several series, the inner larger than the outer. Stamens 20--80, completely covering the surface of a fleshy column formed from the filaments. Carpels forming a head. Fruit a compact head of numerous 'berries'. *East & southeast Asia*.

52. ILLICIACEAE

Hairless shrubs or small trees with aromatic or fragrant bark, leaves more or less evergreen, alternate or sometimes clustered in false whorls, without stipules. Flowers solitary or in clusters of 2 or 3 in the leaf-axils, bisexual, radially symmetric. Perianth of numerous petal-like segments which often differ among themselves in size, white, yellow or red. Stamens 5--20, sometimes very fleshy and incurved. Carpels 5--many in a single whorl, free, each with a short style and a single, more or less basal ovule. Fruit a collection of follicles in a single whorl, each containing a single seed.
   A family of a single genus.

1. Illicium Linnaeus. 40/6. Description as for the family. *Southeast Asia, southeast USA*.

53. MONIMIACEAE

Trees or shrubs with aromatic bark, wood and leaves. Leaves usually evergreen, opposite, without stipules. Flowers solitary or in cymes or racemes, usually unisexual, radially symmetric. Perianth of 4--many segments, often in 2 or more series, but not clearly differentiated into calyx and corolla. Stamens generally numerous, usually each with a pair of glands or cup-like appendage at the base of the filament, anthers opening by slits or from the base upwards by flaps. Perianth and stamens borne on the rim of a perigynous zone. Ovary of usually several free carpels, each containing a single ovule. Fruit a group of achenes or drupes borne within the persistent perigynous zone.
   A family of 34 genera and about 450 species, mainly from the tropics and subtropics (extending south to Chile and New Zealand); only 1 genus is generally cultivated in northern Europe.

1. Atherosperma Labillardière. 2/1. Dioecious trees. Leaves evergreen, hairless, thick, often toothed, at least in the upper half. Flowers solitary, stalked, each subtended by an involucre of 2 bracts which are edge-to-edge in bud. Perianth-segments 8--10, more or less equal, overlapping in 2 series. Stamens 10--18, each with 2 long appendages at the base of the filament; anthers opening by flaps. Staminodes numerous in female flowers. Carpels many, each tapering into a long, hairy style. Fruit made up of the more or less spherical perigynous zone, containing many achenes. *Australia*.

A single species of **Laurelia** Jussieu, with flowers without involucres, in axillary cymes, may occasionally be grown in very mild areas.

54. CALYCANTHACEAE

Deciduous (ours) or evergreen shrubs with aromatic bark and wood. Leaves opposite, entire, without stipules. Flowers solitary, axillary on short shoots or terminal, subtended by bracts, fragrant, sometimes borne before the leaves. Perianth of numerous free segments. Fertile stamens 5--30, some stamens sterile. Carpels many, free, borne within a cup-shaped perigynous zone on whose rim the perianth-segments and stamens are borne. Ovules 1 or 2 per carpel, placentation marginal. Fruit a group of achenes borne within the persistent, enlarged perigynous zone.

A family of 2 genera from North America, Australia, Japan and China, both grown for their scented flowers.

    1a. Flowers borne before the leaves; perianth-segments mostly yellow

                                               2. Chimonanthus

     b. Flowers borne with the leaves; perianth-segments reddish brown to

       greenish purple   **1. Calycanthus**

1. Calycanthus Linnaeus. 5/2. Deciduous shrubs. Axillary buds hidden by the expanded bases of the leaf-stalks or exposed. Flowers fragrant, borne with the leaves, terminal on the branches or on short lateral branches, appearing axillary. Perianth-segments numerous, reddish brown, brownish or greenish purple. Fertile stamens 10--30, brown or yellowish brown. Fruiting perigynous zone bell-shaped or contracted at the apex. *North America, Australia*.

2. Chimonanthus Lindley. 4/1. Deciduous shrubs (ours). Axillary buds exposed. Flowers subtended by numerous bracts, borne before the leaves (ours) on leafless branches, almost stalkless. Perianth-segments numerous, the outer yellow, the inner yellow or yellow marked with purplish brown. Fertile stamens 5 or 6. Fruiting perigynous zone narrowed towards the mouth. *China, Japan*.

55. LAURACEAE

Trees or shrubs with hard, often foetid wood. Leaves alternate or almost opposite, leathery, often gland-dotted; veins pinnate or 3--5. Flowers bisexual or unisexual, usually small and inconspicuous and borne in umbels, cymes or panicles. Bracts often falling early or totally absent. Perianth of 4 or 6 segments in 2 series, sometimes enlarging in fruit. Stamens or staminodes usually twice as many as perianth-segments, in 4 or more whorls; anthers opening by flaps. Ovary stalkless, of 1 carpel, 1-celled. Style terminal, simple, stigma entire or irregularly lobed. Ovule solitary, pendulous, anatropous. Fruit a fleshy berry, a drupe or dry and indehiscent, sometimes partly covered by the persistent perianth. Seed pendulous, without endosperm. See figure 36, p. 168.

A large, homogeneous family of about 47 genera and about 2000 species, distributed in the tropics and subtropics.

    1a. Stamens 6 or 12; perianth-segments 4   2

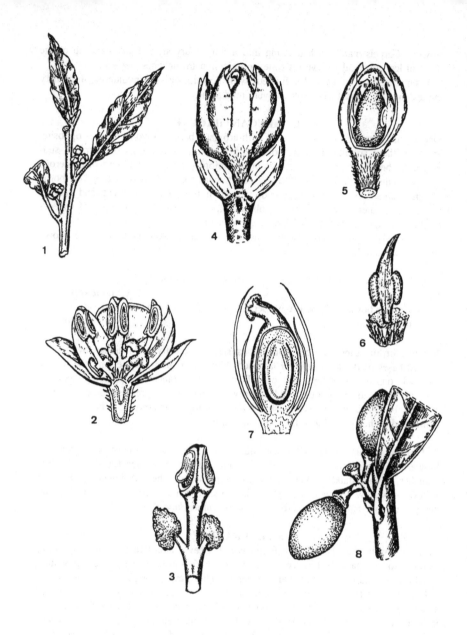

Figure 36. Lauraceae. *Laurus nobilis.* 1, Apex of shoot with male inflorescences. 2, Longitudinal section of male flower. 3, A single stamen. 4, Female flower. 5, Female flower with one perianth-segment removed to show staminodes. 6, A single staminode. 7, Longitudinal section of female flower. 8, Shoot with two developing fruits.

b. Stamens 9; perianth-segments 6   3

2a. Stamens 6   **5. Neolitsea**

   b. Stamens 12   **7. Laurus**

3a. Perianth persistent and hardening in fruit, enclosing the base of the fruit

                             2. Phoebe

   b. Perianth deciduous, or if persistent then not hardening in fruit   4

4a. Leaves evergreen, always entire   5

   b. Leaves deciduous, sometimes lobed   6

5a. Stigma peltate   **4. Umbellularia**

   b. Stigma not peltate   **3. Persea**

6a. Tree to 15 m or more; fruit a drupe   **1. Sassafras**

   b. Shrubs or small trees rarely exceeding 9 m; fruit a berry   **6. Lindera**

1. Sassafras Trew. 3/1. Deciduous aromatic trees to 15 m or more (to 30 m in the wild). Young branches smooth and green, bark deeply furrowed. Leaves alternate, entire or with 1--3 lobes, with prominent mid-vein and 2 lateral veins giving a 3-veined appearance. Flowers unisexual (when males and females usually on separate plants) or bisexual, borne in small racemes and appearing before the leaves. Perianth of 6 sepal-like segments. Male flowers with 9 stamens, female flowers with 6 staminodes. Style slender. Fruit-stalk thick, fleshy; fruit an ovoid drupe. *North America, China.*

2. Phoebe Nees. 70/1. Trees or shrubs. Leaf-buds small with few scales. Leaves alternate, pinnately veined. Flowers bisexual, in panicles or corymbs. Perianth of 6 persistent. sepal-like segments, hardening in fruit. Stamens in 3 series, anthers 4-celled. Fruit immersed in the hardened calyx. *Eastern Asia, tropical America & West Indies.*

3. Persea Miller. 150/2. Evergreen trees or shrubs. Leaves alternate, entire, pinnately veined. Flowers bisexual or unisexual, in axillary or terminal panicles. Perianth-segments 6. Stamens 9 in male flowers, anthers 4-celled. Ovary stalkless, style slender, terminating in a small, flattened stigma. Fruit a spherical or oblong berry. *Eastern Asia, tropical & subtropical America.*

4. Umbellularia Nuttall. 1/1. Strongly aromatic evergreen tree reaching 15--20 m. Leaves alternate, 5--12 cm, narrowly oblong-elliptic, leathery and glossy, entire and tapered at each end. Flowers bisexual in shortly stalked, many-flowered umbels to 2 cm across. Perianth with short tube and 6 equal lobes. Stamens 9, the 3 innermost each with 2 basal glands. Fruit a pear-shaped berry to 2.5 cm, green at first, becoming purple when ripe. *Western North America.*

5. Neolitsea (Bentham) Merrill. 80/1. Evergreen, dioecious trees. Leaves alternate, stalked, entire, normally 3-veined, rarely pinnately veined, softly hairy when young. Flowers unisexual, perianth-segments 4, soon falling. Stamens 6 (rarely 8) in male flowers, the inner 2 each with 2 basal glands, the outer 4 without. Fruit a berry. *East & southeast Asia.*

6. Lindera Thunberg. 100/7. Small trees or shrubs, mostly deciduous. Leaves pinnately 3- or 5-veined, alternate. Flowers unisexual, borne in false umbels and surrounded by 4 conspicuous persistent bracts. Perianth-segments 6. Male flowers with 9 (rarely 12) stamens. Female flowers with 9 staminodes; stigma peltate, conspicuous. Fruit fleshy or dry, splitting at maturity, containing a single stone and seated in a more or less developed cup or disc, sometimes surrounded by the persistent perianth. *Easern Asia from Himalaya to China, Japan & Malaysia, North America.*

7. Laurus Linnaeus. 2/2. Aromatic evergreen trees or shrubs. Leaves alternate, simple, pinnately veined. Flowers in axillary umbels, bisexual or unisexual. Perianth-segments 4. Male flowers with 12 (or more) stamens, anthers opening by 2 flaps. Fruit a berry. *Mediterranean area, Atlantic Islands.* Figure 36, p. 168.

56. TETRACENTRACEAE

Small to medium-sized trees with slender ascending primary branches, each marked with the closely packed ring-scars of fallen leaves. Buds long, slightly curved, pointed. Leaves stalked, alternate, ovate or heart-shaped, cordate at base, margins saw-toothed. Inflorescence pendulous, with 80--120 flowers, borne on a short spur. Flowers bisexual, very small, yellowish, parts in 4s, petals absent. Ovary superior. Fruit a 4-celled capsule.

A family closely allied to the *Trochodendraceae*; both have primitive characteristics in their wood anatomy (lack of vessels). Only a single genus is known, containing a single species.

1. Tetracentron Oliver. 1/1. Deciduous trees 17--30 m in the wild, rarely exceeding 15 m in cultivation. Buds 9--12 mm, light orange-brown. Leaves with stipules, stalks 1--3 cm, blades 4.5--7 cm, with 5 prominent veins radiating from a central point at the leaf-base, colouring a rich bluish red in autumn. Sepals 4, overlapping, petals absent. Stamens 4. Stigmas 4, arising from the base of the ovary. Carpels 4, united, each with up to 6 ovules. Fruit a capsule with a spur-like appendage on each cell. Seeds oily, linear-oblong. *Nepal, northeast India, Burma, southwest China.*

57. TROCHODENDRACEAE

Evergreen trees or shrubs with stalked, saw-toothed, almost whorled leaves. Flowers bisexual, regular or slightly asymmetric, in terminal, raceme-like clusters. Perianth absent. Stamens and carpels numerous; carpels free, in a ring. Fruit a ring of coalesced, many-seeded follicles. Seeds with oily endosperm.

A family of a single genus and a single species.

1. Trochodendron Siebold & Zuccarini. 1/1. Medium-sized trees or shrubs to about 10 m in cultivation, 20--25 m in the wild. Terminal bud to 2.1 cm × 8 mm, with papery scales. Leaves glossy above, leathery, broadly ovate to elliptic, 5--12 × 3--7 cm. Inflorescence 5--13 cm long, with 10--20 or more flowers. Stamens 40--70, 3.5--7 mm, spreading or reflexed, falling early. Carpels 6--11, the whole ovary obovoid, 2--2.5 mm, ovules 16--24 in each carpel. Style 0.5--2 mm. Fruits 7--10 mm in diameter, with 7--12 seeds per follicle. *Korea, Japan, Ryukyu Islands, Taiwan.*

58. EUPTELEACEAE

Small, erect to spreading, branched trees or shrubs. Leaves deciduous, in (false) whorls at the ends of short shoots, long-stalked, without stipules, irregularly and unequally toothed. Flowers bisexual, radially symmetric, borne in the leaf-axils. Perianth absent or forming a minute cup. Stamens 8--18, ovary of 8--18 free,

long-stalked carpels, each containing 1--3 ovules with marginal placentation; stigmatic area conspicuous. Fruit a group of stalked, 1--3-seeded samaras.
A family of a single genus.

1. Euptelea Siebold & Zuccarini. 2/2. Description as for the family. *Eastern Asia.*

59. CERCIDIPHYLLACEAE
Large deciduous, dioecious trees with pendulous branches and spirally twisted trunks. Leaves stalked, opposite on long shoots and alternate on short shoots. Flowers unisexual, the male almost stalkless, the female stalked. Male with 4 free perianth-segments and 15--20 stamens. Female with 4 free perianth-segments and 4--6 free carpels. Fruit a cluster of dehiscent follicles with woody endocarp. Seeds compressed, winged and 4-sided.
The family contains a single genus.

1. Cercidiphyllum Siebold & Zuccarini. 1/1. Tree 20--30 m with trunk solitary or branched from the base into 3--5 smaller boles. Leaves 5--10 cm, broadly ovate, cordate at base, margins scolloped, stalks 2--3 cm. Male flowers 1.8--2.2 cm, borne in the leaf-axils, either solitary or in bundles. Female flowers 5--8 mm, with the 4 sepals green and fringed; carpels 4--6, styles thread-like. Fruits pod-like, 1.2--2 cm, in clusters of 2--4 on stalks to 5 mm. *China, Japan.*

60. RANUNCULACEAE
Herbs or rarely woody and/or climbing plants. Leaves mostly alternate or all basal, rarely opposite, usually without stipules, often deeply divided into numerous segments. Flowers solitary and terminal or in racemes or panicles. Perianth of a single, usually petal-like whorl or of 2 whorls, calyx and corolla both present. Sepals or perianth-segments 3--many, free. Petals absent or 3 or more, sometimes flat and petal-like, sometimes small, variously shaped, almost always with a nectar-secreting area on the surface, at the base or at the apex. Stamens usually numerous, more rarely as few as 5; anthers opening by longitudinal slits. Carpels usually numerous (rarely 1-- few), usually free. Ovules 1--many per carpel, placentation marginal. Fruit usually group of follicles or achenes, more rarely a single follicle or berry-like. See figures 37, p. 178 & 38, p. 180.
A family of about 50 genera and 2000 species, mostly from temperate areas, a few from mountains in the tropics.

1a. Leaves opposite or rarely absent; plants usually woody and/or climbing
26. Clematis
  b. Leaves alternate or all basal (sometimes with a whorl of bracts below the flowers which can be mistaken for leaves); plants usually herbaceous and usually not climbing   2
2a. Plant a shrub with yellow wood; flowers small, brown-purple in drooping racemes; stamens 5--10   **3. Xanthorhiza**
  b. Combination of characters not as above   3
3a. Fruit of 1 or more follicles   4
  b. Fruit a group of achenes, or berry-like   22
4a. Flowers bilaterally symmetric   5
  b. Flowers radially symmetric   7

171

5a. Ovary of a single carpel; follicle 1   **20. Consolida**
   b. Ovary of 2 or more carpels; follicles 2 or more   6
6a. Upper perianth-segment hooded or helmet-like   **18 . Aconitum**
   b. Upper perianth-segment spurred   **19. Delphinium**
7a. Perianth of a single whorl of segments   8
   b. Perianth of 2 whorls of segments   10
8a. Perianth-segments 4, mauve or rarely white   **2. Glaucidium**
   b. Perianth-segments 5 or more, yellow or white   9
9a. Leaves 2- or 3-pinnate   **9. Isopyrum**
   b. Leaves simple   **15. Caltha**
10a. Petals as large as or larger than the sepals   11
   b. Petals smaller than the sepals (if as long or longer, then narrower)   12
11a. Each petal with a backwardly-pointing spur (except in a few spurless
     cultivars)   **11. Aquilegia**
   b. Petals not spurred but slightly pouched at base   **10. Semiaquilegia**
12a. Flowering stem leafless but with a whorl of bracts just below the solitary,
     terminal flower   13
   b. Flowering stems without such a whorl of bracts   14
13a. Annual; flowers blue; leaves very finely divided   **17. Nigella**
   b. Perennial; flowers white or yellow; leaves not finely divided   **8. Eranthis**
14a. Outer perianth-segments persisting around the group of follicles
                                      7. Helleborus
   b. Outer perianth-segments deciduous in fruit   15
15a. Carpels (and follicles) united, at least towards the base   **17. Nigella**
   b. Carpels (and follicles) completely free   16
16a. Sepals 3   **14. Anemonopsis**
   b. Sepals 4 or more   17
17a. Flowers very numerous, in long racemes or panicles   **5. Cimicifuga**
   b. Flowers 1--4, solitary or in short racemes   18
18a. Leaves evergreen, all basal; carpels and follicles stalked   **4. Coptis**
   b. Leaves deciduous; carpels and follicles not stalked   19
19a. Sepals mauve   **12. Paraquilegia**
   b. Sepals yellow, orange or white   20
20a. Annual herb; sepals white, 4--5 mm   **13. Leptopyrum**
   b. Perennial herb; sepals larger, yellow or orange or more rarely white   21
21a. Leaves 2- or 3-pinnate   **9. Isopyrum**
   b. Leaves palmately divided   **16. Trollius**
22a. Fruit berry-like   23
   b. Fruit dry, not berry-like   24
23a. Leaves 3, simple, palmately lobed, 1 basal, 2 on the stem; flower solitary,
     terminal   **1. Hydrastis**
   b. Leaves several, pinnate; flowers in dense racemes   **6. Actaea**
24a. Perianth of a single whorl   25
   b. Perianth of 2 whorls   28
25a. Flowers not subtended by a whorl of bracts   **21. Thalictrum**
   b. Flowers subtended by a whorl of 3 bracts   26
26a. Styles elongating and becoming feathery in fruit; nectar-secreting
     staminodes present   **24. Pulsatilla**

b. Styles neither elongating nor becoming feathery in fruit; nectar-secreting staminodes absent   27

27a. Style absent, stigma broad and depressed, borne directly on top of the carpel; achenes strongly 8--10-ribbed   **22. Anemonella**

b. Style present, stigma borne on its inner side; achenes not strongly ribbed   **23. Anemone**

28a. Sepals with backwardly-directed spurs adpressed to the flower-stalk   **28. Myosurus**

b. Sepals without backwardly-directed spurs   29

29a. Leaves all basal, main lobes 3--5, broad; sepals (or sepal-like involucral bracts) 3, green   **25. Hepatica**

b. Combination of characters not as above   30

30a. Petals without nectaries   **29. Adonis**

b. Petals with nectaries   31

31a. Petals white or pinkish, each with an orange spot at the base; leaves pinnately divided   **30. Callianthemum**

b. Petals yellow, white, pink or red, without orange spots, if petals white or pink, then leaves not pinnately divided   **27. Ranunculus**

1. Hydrastis Linnaeus. 2/1. Perennial rhizomatous herbs with usually a single basal leaf and a simple, hairy stem bearing a solitary flower and 2 stem-leaves. Sepals 3, soon falling. Petals absent. Stamens numerous. Carpels 12 or more, forming a head of scarlet, usually 1-seeded berries in fruit. *Northeast Asia, northeast North America.*

Sometimes treated as the sole genus of the family *Hydrastidaceae*.

2. Glaucidium Siebold & Zuccarini. 1/1. Herbaceous perennial with short, stout rhizomes. Basal leaves several, thin, kidney-shaped or rounded, palmately lobed with deeply toothed lobes, the base cordate. Flowering stems usually solitary, erect, with 2 alternate, stalked leaves in the upper half. Flowers 5--10 cm wide, solitary, shortly stalked, borne in the axils of stalkless, leaf-like bracts. Sepals 4, petal-like, rounded or obovate, mauve; petals absent. Stamens numerous. Carpels 2, ovoid or oblong, united at the base and spreading widely, stigmas almost capitate. Fruit of 2 oblong, compressed follicles united at the base. Seeds numerous, flat, obovate, winged. *Japan.*

Sometimes treated as the sole genus of the family *Glaucidiaceae*.

3. Xanthorhiza Marshall. 1/1. Deciduous shrubs with creeping rootstock and yellow wood. Stems 50--100 cm, erect, with a few branches from the upper parts. Leaves clustered at stem-apex, pinnate with 3--5 leaflets, each to 10 × 8 cm, stalkless, deeply lobed and irregularly toothed; stalk to 18 cm. Flowers numerous, 6--8 mm across, unisexual or bisexual, brown-purple, in simple or branched, drooping racemes to 15 cm, borne at the tops of the stems. Sepals 5, *c.* 3 × 2 mm, petal-like, ovate, spreading; petals 5, bilobed, very small. Stamens 5--10. Carpels 5--10. Fruit a single-seeded follicle. *Eastern USA.*

4. Coptis Salisbury. 10/4. Low-growing perennials with creeping yellow rhizomes. Leaves persistent, long-stalked, palmate or bipinnate, or divided into 3 leaflets. Flowering stems erect with 1--4 bisexual or unisexual flowers. Sepals 5--7, petal-like, often falling early. Petals 5--7, small, each with a narrow claw, either club-shaped or hooded with a nectary at the apex, or linear and with a nectary near the base. Stamens 10--25. Carpels 5--10, stalked. Fruit an umbel-like cluster of several stalked, oblong follicles, each with 4--8 seeds. *Temperate North America, Asia.*

5. Cimicifuga Linnaeus. 10/8. Large perennial herbs. Leaves all basal, or basal and on the stem, the lower long-stalked; blades 1--4 times divided into 3 segments, the ultimate segments toothed or lobed. Flowers radially symmetric in long racemes or spikes (sometimes branched below), each flower usually with a bract and 2 bracteoles (these sometimes inconspicuous or absent). Sepals 4 or 5 (rarely fewer), falling early, petal-like, usually whitish, sometimes brownish purple in bud. Petals to 8 (sometimes absent), usually bilobed or bifid, each bearing a nectary (the petals sometimes appear transitional to stamens). Stamens numerous. Carpels 1--8, free, shortly stalked or stalkless. Fruit a single follicle or a group of follicles; follicles stalked or not. Seeds smooth or scaly. *North temperate areas.*

6. Actaea Linnaeus. 8/8. Perennial herbs with short rhizomes. Leaves from the base and on the stem, alternate, bi- or tri-pinnate; stem-leaves similar to but smaller than the basal. Inflorescence a dense raceme, usually terminal, with small, persistent bracts, lengthening in fruit. Flowers many, small, white. Sepals 3--5, petal-like, deciduous. Petals 4--10 (rarely 0), spathulate. Stamens numerous, longer than petals. Carpel 1, with a 2-lobed stigma. Fruit a many-seeded berry. *North temperate areas.*

7. Helleborus Linnaeus. 15/15 and some hybrids. Rhizomatous perennials, herbaceous or with rather woody overwintering stems. Leaves basal or on the stem, pedate, palmate or with 3 leaflets. Flowers solitary or in cymes, with 5 large, persistent perianth-segments (sepals) and an inner whorl of 5--15 small, tubular or funnel-shaped nectaries (petals). Stamens numerous. Carpels usually 3--8, free or partly united. Fruit of 3--5 follicles, each containing several seeds. *Europe, southwest Asia.*

8. Eranthis Salisbury. 8/2. Low perennial herbs with tuberous rootstocks. Leaves mostly basal, palmately or pinnately divided, stalked; stem-leaves similar but stalkless, arranged in a whorl, forming an involucre just below the solitary, terminal flower. Sepals 5--8, yellow or white. Petals modified into 2-lipped, tubular nectaries. Stamens many. Carpels few to many, fruit a group of follicles. *Eurasia.*

9. Isopyrum Linnaeus. 20/2. Perennial herbs with rhizomatous or tuberous rootstocks. Stems slender, branching, hairless. Leaves bi- or tripinnate, the divisions all 3-parted, ultimate leaflets 2- or 3-lobed. Flowers axillary and terminal, white or pinkish. Sepals 5, petal-like, deciduous. Petals absent or reduced to small nectaries. Stamens 10--40, shorter than sepals. Carpels 2--6, occasionally more, styles pointed. Fruit a group of oblong to ovoid follicles, each containing 2 or more seeds. *Northern hemisphere.*

10. Semiaquilegia Makino. 6/1. Tufted herbaceous perennials with long-stalked basal leaves and leafy stems. Leaves pinnate or bipinnate with 3 main divisions, the leaflets usually long-stalked, the segments lobed. Flowering stems bearing several flowers in a loose panicle. Sepals 5, petal-like, somewhat spreading. Petals 5, erect, concave or pouched at the base. Stamens numerous with several small, membranous, lanceolate staminodes. Carpels 5--10, long-styled. Fruit of 3--5 erect, many-seeded follicles. *Eastern Asia.*

11. Aquilegia Linnaeus. 70/35. Perennial herbs. Basal leaves tufted, usually twice or more divided into 3 segments, on long stalks with expanded bases. Stem-leaves usually present but less divided than basal leaves. Stems often branched, with leaf-like bracts and few to many, terminal, erect or pendent flowers. Sepals 5, petal-like, usually spreading and smaller than petals. Petals 5 each with a flat, broad blade and a backwardly-projecting, hollow, nectar-producing spur (absent in a few cultivars). Stamens many, the innermost reduced to membranous staminodes. Carpels usually 5, free, stalkless. Fruit a group of erect, many-seeded follicles. *North temperate areas.*

12. Paraquilegia Drummond & Hutchinson. 6/1. Densely tufted perennials with the remains of the leaf-stalks persisting and crowded at the top of the rootstock. Leaves long-stalked, pinnate or bipinnate (each into 3 divisions), with long-stalked primary leaflets and deeply dissected ultimate segments. Flowering stems several, 1-flowered, with 2 small bracts above the middle. Sepals 5, petal-like. Petals 5, small, oblong or rounded, notched, slightly concave at the base. Stamens numerous. Fruit of 5--7, usually erect follicles with shining, slightly keeled seeds. *Central Asia to west China.*

13. Leptopyrum Reichenbach. 1/1. Annual, tufted herbs with slender stems, erect or spreading, 15--20 cm. Basal and stem-leaves divided into numerous fine segments. Stem-leaves often whorled. Flowers solitary at the tips of the shoots, to 1 cm across. Sepals 4 or 5, ovate, white, to 5 mm. Petals 4 or 5, yellow, minute. Stamens numerous. Carpels 12--20, narrow, pointed; follicles with 1--several seeds. *East Asia.*

14. Anemonopsis Siebold & Zuccarini. 1/1. Herbaceous perennial, 60--100 cm. Leaves hairless, mostly basal, 3 times pinnate with ovate, acute or acuminate, sharply toothed or lobed segments; stalk and axis blackish, blade light green, paler beneath; stalk expanded at the base and clasping the stem. Stem erect, almost black, hairless or sparsely woolly near the nodes, bearing smaller leaves above. Inflorescence a loose panicle or raceme, the branches with small leaf-like or undivided, oblong bracts at the base. Flowers *c.* 3 cm across, pendent. Sepals 3, oblong, *c.* 1.5 cm × 7 mm, pale mauve. Petals 8--10, almost square, *c.* 5 mm, deep violet-purple, white at the base. Stamens numerous, the filaments flattened below the anthers, greenish. Carpels 2--4, styles slender with small terminal stigmas. Fruit of 2--4 many-seeded follicles. *Japan.*

15. Caltha Linnaeus. 20/3. Low, fleshy, perennial herbs generally growing in damp places. Leaves alternate, stalked, entire or toothed. Flowers axillary or terminal, stalked. Sepals 5--9 or more, petal-like, yellow, white or rarely pink. Petals absent. Stamens numerous. Ovary of 4--many carpels. Fruit of 4--many follicles. *North & south temperate areas.*

16. Trollius Linnaeus. 20/8. Herbaceous perennials. Roots thickened, fibrous. Stems erect. Leaves basal and borne on the stem, palmately lobed or divided, the segments usually further divided or toothed. Flowers teminal, usually solitary, large. Outer perianth-segments (sepals) petal-like, 5--20 or more, yellow, white or purplish. Inner perianth-segments (petals) 5--15, longer or shorter than the outer, each with a nectary pit at the base. Stamens numerous. Carpels 5--20 or more. Fruit a group of follicles. *Eurasia.*

17. Nigella Linnaeus. 22/3. Annual, simple or branching herbs to 80 cm. Leaves alternate, 1--3 times pinnatisect, the ultimate divisions often thread-like. Flowers terminal and axillary. Sepals 5, often petal-like, blue, yellow or white, falling in fruit. Petals 5--10, much smaller than sepals, nectar-producing, 2-lipped, the outer lip bifid. Stamens numerous. Carpels usually 5, united, at least at the base, the ripened follicles forming a capsule with long, persistent styles. Seeds numerous. *Mediterranean area, north to Germany, east to Iran.*

18. Aconitum Linnaeus. 300/12. Perennial herbs, rarely climbing. Roots usually tuberous, sometimes long and clustered. Stems erect, or scrambling or climbing, often very leafy. Leaves ovate to almost circular in outline, base often cordate, the blade deeply palmately lobed or divided into 3--7 lobes or separate leaflets, the lobes or leaflets themselves toothed or shallowly to deeply lobed. Inflorescence a raceme or panicle; flower-stalks long or short, stiff or flexuous, each bearing a bracteole. Flowers showy, bilaterally symmetric. Sepals 5, petal-like (the conspicuous parts of the flower), the uppermost forming a large, erect, hemispheric to cylindric hood and known as the helmet. Petals 2--10, small, hidden within the sepals, the 2 uppermost with long, nectar-secreting spurs which project into the helmet. Stamens numerous. Carpels 3--5, free. Fruit a group of 3--9 follicles. Seeds angled or winged, sometimes with transverse plates or folds. *North temperate areas.*

19. Delphinium Linnaeus. 100/26 and many hybrids. Annual, biennial or perennial herbs with woody, fibrous or tuberous roots. Leaves palmately lobed, the lobes usually themselves toothed or lobed, sometimes the leaves made up of very narrow segments. Inflorescence a raceme or panicle, each flower stalked and subtended by a bract and 2 bracteoles. Flowers bilaterally symmetric. Sepals 5, the uppermost with a backwardly-pointing spur. Petals 4, sometimes paler than the sepals, sometimes darker to black or dark grey, the 2 uppermost with nectar-secreting projections which extend into the sepal-spur. Stamens numerous. Ovary of 3--5 free carpels. Fruit a group of 3--5 follicles. *North temperate areas, extending south of the equator in east Africa.* Figure 38, p. 180.

20. Consolida (de Candolle) S.F. Gray. 50/3. Annual herbs. Leaves divided into numerous, thread-like segments, alternate. Flowers in racemes, bilaterally symmetric, each subtended by a bract and 2 bracteoles. Sepals 5, petal-like, the uppermost spurred. Corolla made up of perhaps 4 united petals, the whole corolla 3--5-lobed or almost entire, with a nectar-producing spur projecting backwards into the spur of the upper sepal. Stamens numerous. Carpel 1. Fruit a single follicle. *Mediterranean area to Central Asia.*

21. Thalictrum Linnaeus. 130/20. Perennial herbs. Leaves 2--4-pinnate, sometimes divisions only 3, usually alternate, with stipules; leaflets lobed or toothed. Flowers small, usually bisexual in axillary or terminal racemes, panicles or corymbs. Sepals 4 or 5, petal-like, spreading, usually deciduous if small and inconspicuous. Petals absent. Stamens numerous, often coloured and showy. Carpels few. Fruiting achenes stalked or stalkless, ribbed, angled or winged. *North temperate areas.*

22. Anemonella Spach. 1/1. Perennial, hairless herbs to 25 cm with tuberous roots. Leaves basal, 2 or 3 times pinnate, each leaflet divided into 3 parts, 1--2 cm, ovate to round, 3-lobed at the tip. Umbels of 2--5 or more flowers, with an involucre of 2--3 stalkless leaves each with 3 stalked leaflets. Sepals 5--10, petal-like, ovate to elliptic, spreading, white to pink; petals absent. Stamens numerous. Stigma depressed, borne directly on the carpel. Achenes 4--15, 8--10-ribbed. *Eastern USA.*

23. Anemone Linnaeus. 120/35. Perennial herbs. Basal leaves stalked, usually lobed, dissected or compound. Stem-leaves shortly stalked or stalkless, often forming an involucre below the inflorescence, sometimes remote from it. Flowers radially symmetric, bisexual, solitary or in cymes. Perianth-segments petal-like, 5--20 or more, white, yellow, blue, purple or red. Stamens numerous. Carpels free, numerous, each with a single pendulous ovule and a short, persistent style. Fruit an achene, hairless to densely woolly. *Cosmopolitan, mostly from the northern hemisphere.*

24. Pulsatilla Miller. 30/7. Tufted, perennial herbs with woody, fibrous rootstocks. Leaves mostly basal, pinnately or palmately divided. Stem-leaves (involucral leaves) 3, united at the base, occasionally shortly stalked. Flowers solitary, perianth spreading or narrowly campanulate to openly campanulate, erect to pendent. Perianth-segments 6 (occasionally 5, 7 or 8), frequently silky on the outer surface. Stamens numerous, surrounded by whorls of nectar-secreting staminodes. Carpels numerous in a head, free; styles elongating and feathery in fruit. Ovule 1. *North temperate areas.*

25. Hepatica Miller. 6/6. Small perennial herbs with rhizomes. Leaves all basal, persistent, long-stalked, 3--5-lobed, the lobes broad, entire or further divided or toothed. Flowers solitary on long stalks arising from ground-level. Involucral bracts sepals-like, 3, green. Petals several, white, purplish, bluish or rose-coloured. Stamens numerous. Ovary of several to many carpels. Fruit a group of achenes. *Europe, Central & east Asia, eastern North America.*

26. Clematis Linnaeus. 200/50 and many hybrids. Woody climbers, low shrubs or perennial herbs. Leaves opposite or sometimes alternate, simple or compound, when divided into 3s or pinnate or bipinnate. Inflorescences 1--many-flowered cymes borne at the apices of young shoots or axillary on young or old shoots. Flowers with 4--8 (rarely many) free, petal-like perianth-segments. Staminodes sometimes present between perianth and stamens. Stamens numerous. Ovary of many free carpels. Fruit a group of achenes sometimes with long, feathery persistent styles. *Cosmopolitan, but centred in eastern Asia.*

27. Ranunculus Linnaeus. 400/35. Annual or perennial herbs. Leaves alternate, lobed, divided or toothed, sometimes simple and entire. Flowers bisexual, solitary or in cymose panicles, radially symmetric. Sepals 3--7, usually 5. Petals 0--16 (or more), usually 5, yellow, white, orange, red or purple, each with a nectar-producing scale towards the base. Stamens numerous. Carpels numerous, the style persistent as a beak in fruit. Fruit a head of achenes. *Chiefly in temperate areas.* Figure 37, p. 178.

28. Myosurus Linnaeus. 7/1. Small annual herbs with linear leaves in a basal rosette. Flowers small, solitary, on slender, upright stalks. Sepals 5 or more, each with a basal spur. Petals 5--7 or absent, tubular, secreting nectar. Stamens usually 5--10. Carpels numerous, ripening to achenes borne on a greatly elongate receptacle. *Northern hemisphere, Chile, New Zealand.*

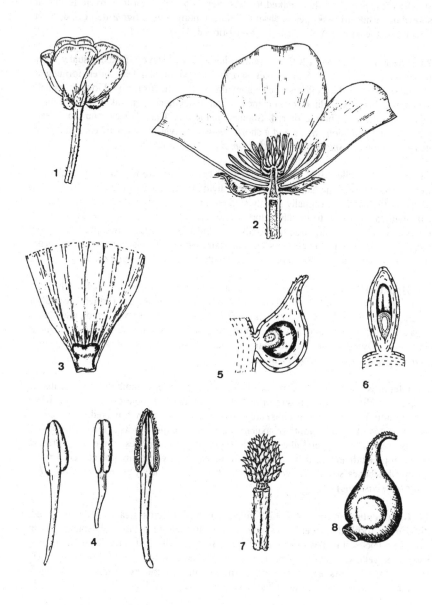

Figure 37. Ranunculaceae. *Ranunculus repens.* 1, Flower, slightly from below. 2, Longitudinal section of flower. 3, Base of petal to show nectary. 4, Stamen from various aspects. 5, Longitudinal section of carpel. 6, Transverse section of carpel. 7, Head of achenes. 8, A single achene.

178

29. Adonis Linnaeus. 20/10. Annual or perennial herbs. Leaves divided 1--3 times with narrow, more or less linear segments. Flowers solitary, radially symmetric. Sepals 5--8, often coloured and somewhat petal-like. Petals 5--20, red, yellow or whitish, nectaries absent. Stamens numerous, anthers blackish or yellow. Carpels borne on a long or roundish head. Fruit a loose or crowded head of achenes. Achenes beaked, not laterally compressed, often with a tooth below the beak. *Temperate Eurasia.*

30. Callianthemum Meyer. 10/4. Low-growing perennials with short rhizomes. Basal leaves long-stalked, pinnate with a terminal leaflet, the lobes pinnatifid. Flowering stems 1--3, leafless or bearing a few shortly stalked leaves. Flowers solitary or 2 or 3 in a short raceme. Sepals 5, broadly ovate or rounded, greenish, sometimes falling early. Petals 5--20, white or pink, each with an orange nectary at the base. Stamens numerous. Carpels numerous with very short styles. Fruit a head of achenes. *Mountains of central Europe & central & eastern Asia.*

61. BERBERIDACEAE

Shrubs or perennial herbs with rhizomes or tubers. Leaves alternate or more rarely opposite, simple or compound, usually divided into 3s, pinnately or palmately veined, Flowers bisexual, radially symmetric, solitary or in racemes, corymbs, panicles, spikes or clusters. Sepals and petals together in 2 or 3 whorls or rarely absent, each with 6 or 4 members, the outermost ('outer sepals') small, soon falling, the middle ('inner sepals') often petal-like, the innermost ('petals') often reduced and nectar-producing. Stamens as many as the petals and on the same radii as them, or rarely twice as many; anthers opening by longitudinal slits or by flaps up-rolling from the base and hinged at the top. Ovary superior, 1-celled; ovules basal and then 1, or lateral (obliquely basal) when 1--many; style long, short or absent; stigma small to large. Fruit a capsule or berry or an achene. See figure 39, p. 182.

The family is accepted here in its traditional sense, thus including *Nandinaceae, Diphylleiaceae, Leonticaceae* and *Podophyllaceae*. There are 14 genera and about 700 species, mostly from the northern hemisphere.

1a. Woody plants 60 cm or more high   2
  b. Herbaceous plants not more than 1 m high   5
2a. Leaves pinnately much divided, with more than 65, entire, spineless
     leaflets; flowers white; anthers opening by longitudinal slits   **1. Nandina**
  b. Leaves undivided or pinnate with 3--37 toothed or spiny leaflets;
     flowers yellow; anthers opening by 2 flaps up-rolling from base   3
3a. Leaves pinnate with 3--37 leaflets   **2. Mahonia**
  b. Leaves undivided or sometimes also with 3 leaflets   4
4a. Leaves always undivided; spines present on shoots   **4. Berberis**
  b. Leaves on same plant variable, undivided or with 3 leaflets; spines
     absent on shoots   **3. × Mahoberberis**
5a. Flowers stalkless in a many-flowered, slender spike; sepals and petals
     absent; anthers as broad as long   **9. Achlys**
  b. Flowers stalked, solitary or in racemes, corymbs or panicles; sepals and
     petals present; anthers longer than broad   6
6a. Flowers solitary on a basal, leafless stem; leaves all basal, with 2 stalkless,
     opposite leaflets or a single lobed blade   **8. Jeffersonia**

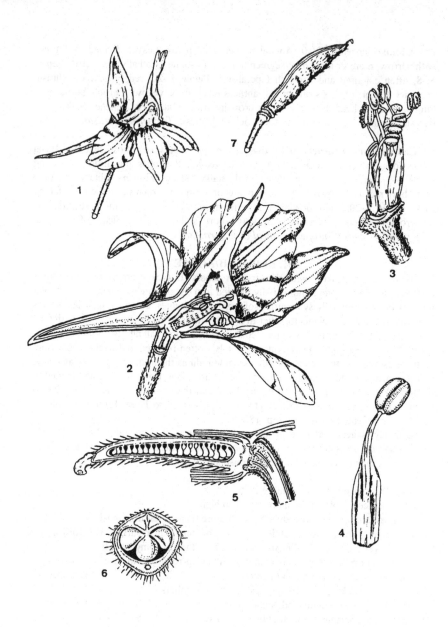

Figure 38. Ranunculaceae. *Delphinium ambiguum.* 1, Flower. 2, Longitudinal section of a flower. 3, Some of the stamens surrounding the ovary. 4, A single stamen. 5, Longitudinal section of the ovary. 6, Transverse section of the ovary. 7, Fruit.

b. Flowers several or many or, if solitary, then at apex of stem between 2 leaves   7

7a. Leaves undivided but sometimes lobed almost to stalk, peltate, major veins radiating palmately from top of stalk   8

b. Leaves divided into leaflets, not peltate, pinnately veined   9

8a. Stem-leaves 2, opposite, with the solitary flower arising between them or leaf 1 and flowers several, clustered, drooping; anthers opening by longitudinal slits; fruit a large red or yellow berry   **15. Podophyllum**

b. Stem-leaves 2, well-separated; flowers in erect, many-flowered cymes; anthers opening by up-rolling flaps; fruit a small blue berry   **14. Diphylleia**

9a. Leaflets stalkless; rootstock a tuber   10

b. Leaflets stalked; rootstock a rhizome   12

10a. Leaves all basal, simply pinnate, with 5--8 pairs of leaflets; inner sepals small; petals flat, conspicuous   **13. Bongardia**

b. Leaves basal and on stem, divided into 3 leaflets; inner sepals large, petal-like; petals small, nectar-producing   11

11a. Stems with several much-divided leaves, bearing terminal and axillary racemes; fruit large, inflated, membranous   **11. Leontice**

b. Stem with 1 leaf, not branched, bearing a terminal raceme only; fruit small, opening before ripening of seeds and exposing them   **12. Gymnospermium**

12a. Leaflets deeply 3--5-lobed   13

b. Leaflets entire or only slightly 3-lobed   14

13a. Flowers greenish, yellowish or purplish, in an erect, terminal raceme   **10. Caulophyllum**

b. Flowers lavender-violet on long, drooping stalks between 2 leaves   **5. Ranzania**

14a. Petals and stamens 4; petals various but not narrowly oblong   **6. Epimedium**

b. Petals and stamens 6; petals narrowly oblong with a flat or hooded nectar-producing tip   **7. Vancouveria**

1. Nandina Thunberg. 1/1. Shrub. Leaves alternate with numerous entire, spineless, pinnately-veined leaflets which are pinnately arranged or in 3s. Flowers very numerous in terminal panicles. Sepals numerous. Petals 6. Stamens 6, anthers almost stalkless, opening by longitudinal slits. Ovary with 1 lateral ovule; style short, stigma minute. Fruit a berry. *China, Japan.*

2. Mahonia Nuttall. *c.* 100/14 and some hybrids. Evergreen, hairless shrubs, occasionally tree-like. Stems spineless, with alternate leaves. Leaves with stipules, pinnate with a terminal leaflet often larger than the rest, sometimes only 3-lobed, leaflets with more or less wavy, spiny margins. Flowers bisexual, in clustered, erect, ascending or drooping panicles or racemes, borne in the leaf-axils but usually near the apices of lateral branches or terminal stems, Perianth-segments 15 in 2--3 series, the outer 3 often smaller, often greenish or purplish and sepal-like, the remainder concave, with a pair of glandular spots at the base, yellow in all but one of the species cultivated. Stamens 6, sensitive to touch, anthers opening by flaps. Ovary 1-celled, superior. Stigma circular, depressed. Fruit a bloomed berry, bluish black, purple or more rarely

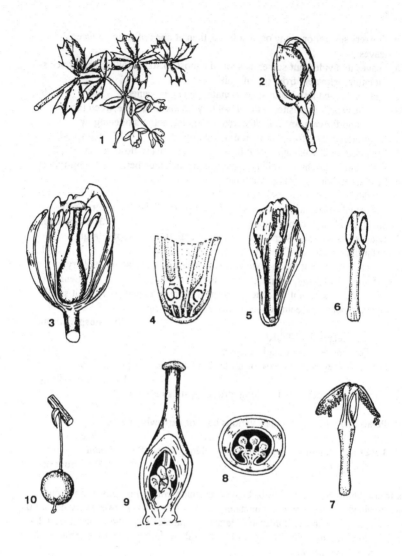

Figure 39. Berberidaceae. *Berberis darwinii.* 1, Shoot with pendent, racemose inflorescence. 2, Flower-bud, showing bracteoles. 3, Flower with some perianth-segments removed. 4, Base of perianth-segment, showing nectaries. 5, Stamen with adjacent petal. 6, Young stamen. 7, Dehisced stamen, showing valves. 8, Transverse section of the ovary. 9, Longitudinal section of the ovary. 10, Fruit.

182

red (in one species pale yellow). Seeds ellipsoid to pear-shaped with a hard coat. *Himalaya east to Japan & Indonesia, North & Central America.*

3. × **Mahoberberis** Schneider. 1/1. A genus of hybrids between various species of *Mahonia* and *Berberis*, known only in gardens. They are distinguished from *Mahonia* by having leaves simple and with 3 leaflets on the same plant, and from *Berberis* by the same character and by the lack of spines on the shoots. *Garden origin.*

4. Berberis Linnaeus. 600/100 and many hybrids. Spiny prostrate to erect shrubs; stems much-branched or arching, to 5 m; spines with 1--3 (rarely 7) prongs, occasionally leaf-like. Leaves persistent and leathery or deciduous and usually flexible, simple, entire or with spine-like teeth. Flowers borne in the axils of the leaves, solitary or in clusters, umbels, racemes or panicles. Flowers greenish to bright yellow or orange, 3--25 mm across. Sepals petal-like, in 2 (rarely 1 or 3) rows of 3, petals in 2 (rarely 1) rows of 3; stamens 6. Berries spherical to ellipsoid, fleshy, red or black, with or without a style; ovules 1--16. *Temperate areas of both Old & New Worlds.* Figure 39, p. 182.

5. Ranzania Ito. 1/1. Herbaceous perennials with slender rhizomes. Stems with 2 opposite leaves. Leaves divided into 3 shortly stalked, deeply lobed, palmately veined leaflets. Flowers 1--6 in a cluster between the 2 leaves. Outer sepals 3, small. Inner sepals 6, much larger than outer. Petals smaller than inner sepals, flat, each with 2 nectar-producing glands at the base. Stamens 6, anthers opening from the base by 2 up-rolling flaps. Ovary with numerous lateral ovules and with a large, stalkless stigma. Fruit a berry. *Japan.*

6. Epimedium Linnaeus. 25/18 and some hybrids. Herbaceous perennials with rhizomes. Stems leafless or with 1--6 leaves. Leaves with 2--60 leaflets, arranged in 3s or pinnately. Flowers in a raceme or panicle. Outer sepals 4, soon falling. Inner sepals 4, petal-like. Petals 4, flat or extended outwards into a nectar-producing pouch or spur, shorter than or longer than sepals. Stamens 4, anthers opening from the base by 2 up-rolling flaps. Ovary with numerous lateral ovules; style slender, with a slightly swollen stigma. Fruit a capsule opening by 2 flaps. Seeds with arils. *South Europe, North Africa, southwest Asia, Himalaya, China, Japan.*

7. Vancouveria Morren & Decaisne. 3/3. Herbaceous perennials with long slender rhizomes. Stems leafless. Leaves divided into 3s, with numerous stalked leaflets lacking marginal spines. Flowers pendulous in a raceme or panicle. Outer sepals 6--9, soon falling. Inner sepals 6, petal-like, reflexed. Petals 6, smaller than the sepals, reflexed, with a hooded or flat tip, nectar-producing. Stamens 6; anthers opening by 2 up-rolling flaps. Ovary with lateral ovules; style slender with slightly swollen stigma. Fruit a capsule opening by 2 flaps. Seeds with arils. *Western USA.*

8. Jeffersonia Barton. 2/2. Herbaceous, low-growing perennials with short rhizomes. Stems leafless, 1-flowered. Leaves basal, long-stalked; blade entire, lobed or divided into 2 stalkless, palmately veined leaflets. Sepals 4, soon falling. Petals 8, flat. Stamens usually 8; anthers opening from the base by 2 up-rolling flaps. Ovary with many lateral ovules; style short with small, 2-lobed stigma. Capsule opening either by an almost horizontal, incomplete slit below the top (thus appearing to have a lid), or by an oblique, downward slit from the top. Seeds numerous, each with a small, lacerate aril. *Eastern North America, China.*

9. Achlys de Candolle. 3/1. Herbaceous perennials with rhizomes. Stems leafless. Leaves basal with 3 stalkless, fan-shaped, palmately veined leaflets. Flowers numerous in a long, slender spike. Sepals and petals absent. Stamens 6--12, usually 9, with long filaments; anthers opening by 2 up-rolling valves. Ovary with 1 basal ovule; stigma stalkless, broad. Fruit a small achene. *Western North America, Japan.*

10. Caulophyllum Michaux. 2/1. Herbaceous perennials with short rhizomes. Stem with 1 leaf. Leaves compound, leaflets in 3s, stalkless but having the 3 primary divisions long-stalked with numerous, 2--5-lobed, pinnately veined leaflets. Flowers greenish, numerous in racemes or panicles. Outer sepals 3 or 4; inner sepals 6, petal-like. Petals 6, much shorter than the inner sepals, thick, fan-shaped, hooded, nectar-producing. Stamens 6, anthers opening by 2 up-rolling flaps. Ovary with short style; stigma gland-like, minute. Fruit almost non-existent because the 2 developing seeds burst the ovary-wall early and ripen naked, blue and berry-like. *Eastern North America, Japan.*

11. Leontice Linnaeus. 3/1. Perennial herbs with tubers. Leaves basal and on the stem, long-stalked, divided into 3s and/or pinnately, with numerous shortly stalked or stalkless, entire leaflets. Flowers numerous in axillary and terminal racemes, often forming a panicle. Sepals 6, flat, petal-like. Petals small, nectar-producing. Stamens 6, anthers opening by 2 up-rolling flaps. Ovary with 2--4 basal ovules; style short, stigma truncate. Fruit much inflated, bladder-like, net-veined, breaking open irregularly at the tip when dry. *Southeast Europe, North Africa, southwest Asia.*

12. Gymnospermium Spach. 6/2. Herbaceous perennials with tubers. Leaves basal and 1--3 on the stems, palmate with 4--7 narrow, entire leaflets. Flowers in a short, unbranched raceme. Sepals 6, petal-like. Petals much shorter than sepals, semi-cylindric, nectar-producing. Stamens 6, anthers opening from the base by 2 up-rolling flaps. Ovary with 2--4 basal ovules; style long, stigma minute. Fruit short, opening early by short lobes, the seeds ripening in an exposed state. *Balkan Peninsula to China.*

Formerly included in *Leontice.*

13. Bongardia Meyer. 1/1. Herbaceous perennials with tubers. Leaves all basal, pinnately divided, with paired or whorled, stalkless, lobed leaflets. Flowers numerous, long-stalked, in a loose, narrow panicle with ascending branches. Sepals 6, concave. Petals 6, flat, conspicuous, each with a small nectar-bearing pore at the base. Stamens 6, anthers opening from the base by 2 up-rolling valves. Ovary with 5 or 6 basal ovules; stigma almost stalkless, folded or lobed. Fruit an ellipsoid, slightly inflated, papery capsule opening by short, acute flaps at the top. *Southwest Asia.*

14. Diphylleia Michaux. 3/1. Herbaceous perennials with short rhizomes. Stem with 2 well-separated leaves. Basal leaves with centrally peltate blades, stem-leaves with marginally peltate blades, the lower long-stalked, the upper stalkless; blades circular to kidney-shaped, lobed and toothed, palmately veined. Flowers in an umbel-like cyme. Outer sepals 6, soon falling. Petals 6, flat. Stamens 6, anthers opening by 2 up-rolling valves. Ovary with few lateral ovules; stigma almost stalkless. Fruit a spherical berry with 2--4 seeds. *Eastern North America, Japan.*

15. Podophyllum Linnaeus. 9/4. Herbaceous perennials with rhizomes. Stem with 1 or 2 leaves. Leaves broad, peltate, deeply lobed to almost entire but then angled. Flower solitary between the 2 leaves or flowers several in a cluster. Sepals 6, soon falling. Petals 6 or 9, flat. Stamens 12--18, anthers opening by longitudinal slits. Ovary with numerous lateral ovules; stigma stalkless or almost so. Fruit a large berry with the seeds surrounded pulp. *Eastern North America, Himalaya, China.*

62. LARDIZABALACEAE

Woody plants, monoecious or dioecious, usually twining. Leaves alternate, without stipules, palmate or trifoliolate, rarely pinnate. Flowers radially symmetric, bisexual or unisexual, usually in racemes, occasionally solitary. Sepals usually 6, occasionally 3, petal-like. Petals 6 or absent. Nectaries often present. Stamens 6, united or not; anthers opening towards the outside of the flower. Carpels 3--9, superior, free, with many ovules in vertical rows. Fruit dehiscent or not.

A family of 7--9 genera of which 6 are in cultivation. The unisexual flowers generally have rudimentary organs of the other sex.

 1a. Erect, non-twining shrub; leaves pinnate; sepals greenish yellow
                                                           1. Decaisnea
   b. Climber with twining stems; leaves not pinnate; sepals white to purplish   2
 2a. Sepals 3; male and female flowers borne in the same inflorescence; female flowers with 5--10 carpels   **6. Akebia**
   b. Sepals 6; male and female flowers borne in separate inflorescences or, if in the same inflorescence, then sepals whitish; female flowers with 3, rarely 6 carpels   3
 3a. Stamens united   4
   b. Stamens free   5
 4a. Leaves palmate with 3--7 narrow leaflets; male and female flowers borne on separate plants; sepals white tinged with violet   **2. Stauntonia**
   b. Leaves once, twice or 3 times divided into 3 segments; male and female flowers borne on the same plant; sepals purplish brown   **3. Lardizabala**
 5a. Plant evergreen; male and female flowers borne on the same plant
                                                           4. Holboellia
   b. Plant deciduous; male and female flowers borne on separate plants
                                                   5. Sinofranchetia

1. Decaisnea Hooker & Thomson. 2/1. Deciduous shrubs. Flowers bisexual or functionally male, borne on the same plant, with 6 sepals in 2 whorls of 3, petals absent. Stamens united, sometimes shortly so. Fruit a many-seeded, cylindric, blue berry. *Himalaya, China.*

2. Stauntonia de Candolle. 16/1. Evergreen climbers with twining stems. Flowers unisexual, the male and female borne on separate plants, with 6 fleshy sepals and no petals. Stamens united. Fruit an ellipsoid berry. *Eastern Asia.*

3. Lardizabala Ruiz & Pavón. 2/1. Evergreen climbers with slender, twining tems. Flowers functionally unisexual, male and female borne on the same plant, each with 6 fleshy sepals in 2 whorls of 3, and reduced petals. Fertile stamens united. Fruit a many-seeded pulpy berry developed from 1 carpel. *Chile.*

4. Holboellia Wallich. 5/2. Evergreen, twining shrubs with long-stalked leaves composed of palmately arranged leaflets. Male and female flowers are produced in few-flowered racemes or corymbs on the same plant. Flowers with 6 rather fleshy sepals, petals reduced to nectaries. Stamens 6, free. Carpels 3. Fruit an indehiscent, fleshy pod containing many black seeds. *Himalaya, mainland southeast Asia.*

5. Sinofranchetia (Diels) Henry. 1/1. Deciduous climber with twining stems to 15 m. Leaves with 3 leaflets 6--14 cm, ovate, entire. Flowers unisexual, white, to 8 mm across, each with 6 sepals and 6 nectaries, in drooping racemes, male and female flowers produced on separate plants. Male flowers with 6 free stamens. Female flowers with 3 carpels. Fruit a blue-purple berry containing many black seeds; the berries are borne alternately along a stalk 20 cm or more. *China.*

6. Akebia Decaisne. 4/3. Deciduous or evergreen climbers with twining stems. Flowers unisexual, male and female borne in the same pendent raceme, the males smaller, numerous, the female few and larger and produced at the base of the raceme. Sepals 3 (occasionally 4), petals absent. Male flowers with 6--8 free stamens. Female flowers with 5--10 carpels. Fruit a fleshy follicle containing numerous seeds. *Eastern Asia.*

63. MENISPERMACEAE

Climbers, often woody, or erect shrubs or small trees, usually dioecious. Leaves alternate, without stipules, usually with long stalks, simple, entire, toothed or lobed, deciduous or evergreen. Flowers unisexual, in racemes or panicles, each usually with a bract and 2 bracteoles. Sepals 3--many, often in 2 whorls. Petals 0, 3 or 6. Stamens 3, 6 or more numerous, free or united; anthers opening by longitudinal or transverse slits. Staminodes often present in the female flowers. Carpels 1--6, free, each containing a single, marginally attached ovule, and each borne on a short stalk which elongates in fruit. Fruit a collection of stalked drupes (occasionally only 1 from each flower developing). Seeds weakly to conspicuously horseshoe-shaped.

A family of 67 genera and about 450 species, mostly tropical in origin.

1a. Stamens 6; plant an evergreen shrub or a climber	**1. Cocculus**
b. Stamens 9 or more; plants always climbers	2
2a. Anthers opening by longitudinal slits	**3. Menispermum**
b. Anthers opening by transverse slits	**2. Sinomenium**

1. Cocculus de Candolle. 7/3. Climbers, shrubs or small trees. Leaves very variable, often hairy beneath. Panicles axillary. Sepals 6, often hairy outside. Petals 6, small, often notched or bifid at the apex. Stamens 6, free, anthers opening by transverse slits. Carpels 3 or 6. Fruit of 1--6 drupes. Seeds horseshoe-shaped. *Mainly from tropical east Asia & Africa, 1 species from temperate North America.*

186

2. Sinomenium Diels. 1/1. Woody climber. Leaves variable, often angled or lobed. Panicles axillary. Sepals 6, hairy outside. Petals 6. Stamens 9 (rarely 12), each anther opening by a transverse slit at the top. Carpels 3. Drupes compressed. Seeds horseshoe-shaped. *Eastern Asia.*

3. Menispermum Linnaeus. 2/2. Woody climbers or perennials with woody stocks giving rise to annual, climbing stems. Leaves variable, sometimes slightly peltate. Flowers in axillary panicles. Sepals 4--10, rather irregularly arranged. Petals 6--9, also irregularly arranged. Stamens 12--18 or more, anthers opening by longitudinal slits. Carpels 2--4. Drupes compressed. Seeds horseshoe-shaped. *Eastern North America, eastern Asia.*

64. NYMPHAEACEAE

Annual or perennial aquatic herbs with short and erect or long and creeping rhizomes rooting at the nodes; sometimes compact, corm-like rhizomes also present. Mature leaves spirally arranged on the rhizome, submerged, floating or above water, long-stalked, peltate or with a deep sinus, simple (submerged leaves rarely finely divided), broadly ovate, kidney-shaped or circular, leathery; veins usually prominent beneath. Juvenile submerged leaves sometimes whorled or opposite, narrow and membranous. Flowers solitary, axillary on long stalks, bisexual, radially symmetric, mostly floating or above water; with usually many, spirally arranged parts. Sepals 4--6, sometimes petal-like, free or slightly joined to ovary at base. Petals 3--5 or numerous, often merging with staminodes and stamens. Stamens often with appendages. Ovary of 3 or more, free to partly or fully fused carpels, superior, to half or fully inferior. Carpels rarely individually sunk in pits in the persistent, corky receptacle. Styles absent; stigmas borne directly on top of the free carpels, or in the form of rays on top of ovary. Ovules 1--many, apical, marginal or scattered. Fruits many-seeded, berry-like, rarely nut-like, often ripening under water.

A cosmopolitan family of about 9 genera and over 100 species, found throughout temperate and tropical regions in still and slow-moving water.

   1a. Veins of leaf pinnate; ovary superior   **2. Nuphar**
     b. Veins of leaf radiating from top of stalk; ovary half-inferior   **1. Nymphaea**

1. Nymphaea Linnaeus. 50/20 and many hybrids. Perennial water-plants with sub-merged, rooted corms or creeping rhizomes, Leaves alternate, floating or rarely above the water, broadly ovate to circular, deeply cleft at base into 2 lobes, one on either side of the long stalk, somewhat leathery, entire, wavy or toothed; juvenile leaves narrower and thinner, submerged. Flowers bisexual, solitary, showy, fragrant, day- or night-opening, long-stalked, floating or held above water. Sepals 4 (rarely 5), green or sometimes petal-like. Petals numerous, often brightly coloured, ovate or obovate to elliptic, narrower towards centre of flower and grading into stamens. Stamens very numerous, sometimes with coloured appendages. Ovary superior, of 3--35 fused or partially fused carpels, with stigmatic rays on top corresponding to number of cells in ovary; ovules numerous. Fruit many-seeded, ripening under water by spiral contraction of the flower-stalk after flowering. *Cosmopolitan.*

2. Nuphar Smith. 25/4. Aquatic perennial herbs with stout, creeping rhizomes. Leaf-blades narrowly to broadly ovate or circular, with a deep basal sinus, not peltate, long-stalked; floating leaves leathery, submerged leaves membranous. Flowers more or less spherical, yellow and green, held above water. Sepals 4--6, broadly ovate to circular, the outer green, the inner yellow tinged with red or green. Petals numerous, yellow, much smaller than sepals, linear to narrowly oblong or spathulate, with nectaries on outer surface. Stamens numerous with broad filaments, borne on receptacle below ovary. Ovary superior, of 5--20 joined carpels, each with many ovules; stigmas in the form of rays on top of the ovary, styles absent. Fruit ripening above water, flask-shaped, berry-like. *North temperate areas.*

65. CERATOPHYLLACEAE

Aquatic herbs. Leaves whorled, without stipules. Flowers unisexual, usually solitary. Perianth with 1 whorl of united segments. Ovary superior. Fruit a nut.

A family containing 1 cosmopolitan genus only.

1. Ceratophyllum Linnaeus. 2/1. Rootless, submerged, aquatic herbs, perennating by buds. Stems branched with only 1 branch at a node. Leaves stalkless, in whorls of 6--12 (usually 8--10), thread-like, rather rigid and brittle, each leaf forked 1--4 times, the ultimate segments with 2 rows of tiny teeth and 2 apical bristles. Flowers unisexual, male and female borne on the same plant, usually solitary in a leaf-axil. Perianth with 8--13 linear lobes which are united at the base. Stamens 10--20; filaments absent or short; anther-connectives prolonged at the top into 2 spines. Ovary 1-celled. Fruit a nut with a terminal spine and with or without 2 basal spines.

About 30 species have been named, but recent study suggests that there are only 2 distinct species.

66. SAURURACEAE

Perennial, often aromatic herbs usually with short, thick rhizomes and creeping stolons bearing erect or ascending stems with conspicuous nodes. Leaves alternate or all basal, simple, with pinnate or palmate veins; stipules sheathing, partly fused to the leaf-stalks. Flowers radially symmetric, bisexual, in loose or dense terminal or leaf-opposed racemes or spikes, often with a whorl of petal-like bracts at the base (when the whole inflorescence resembles an *Anemone* flower). Smaller bracts may be present below each flower. Sepals and petals lacking. Stamens 3 (rarely 4), 6 or 8, filament-bases often fused to carpels. Carpels 3--5, superior or inferior and sunk in the inflorescence-axis, free to partly or completely fused; styles free. Free carpels each with 2--4 ovules and forming individual, dehiscent, capsule-like fruits. Fused carpels with 6--10 parietal ovules per cell, forming a thick, occasionally slightly fleshy capsule, opening at the apex.

A small family of moisture-loving and aquatic plants from Asia and North America, containing 5 genera and about 7 species.

1a. Flowering heads without petal-like bracts at the base		**3. Saururus**
b. Flowering heads with petal-like bracts at the base		2
2a. Leaves mostly basal, with pinnate veins		**1. Anemopsis**
b. Leaves alternate on stem, with palmate veins		**2. Houttuynia**

1. Anemopsis Hooker. 1/1. Leaves mostly basal; stalks 4--18 cm. hairy; blades slightly shorter, elliptic to oblong, obtuse to rounded, truncate to cordate at base, gland-dotted, slightly hairy on margins, with *c.* 6 pairs of pinnate side-veins. Stems to 50 cm, erect, woolly, round in section, with a stalkless, clasping leaf in the upper half and 1--3 smaller leaves in its axil. Flowers in a dense, conical head 1--4 cm long, with *c*. 6 unequal, persistent, whitish, spreading, petal-like bracts at its base; bracts 1--3 cm, some spotted with red, becoming brown and bent downwards in fruit. Each flower, except the lowest, with a small, white, spathulate bract at its base. Stamens 6, free, anthers 2-celled. Styles 3 (rarely 4), ascending, blunt, narrowly conical. Ovary sunk in inflorescence-axis, 1-celled, with 15--25 ovules. *Western North America.*

2. Houttuynia Thunberg. 1/1. To 60 cm, foetid. Leaves alternate, stalks 1--5 cm, often red; stipules 1.5--2.5 cm, blades bluish green, gland-dotted, ovate, acute to slightly acuminate, cordate at base, 3.5--9 × 3--8 cm, with *c*. 5 hairy, palmate veins, margins often red. Flowers in dense cylindric, terminal heads 1--3 cm long, with 4--6 whitish, oblong to obovate spreading bracts at the base; bracts 1--3 cm × 7--15 mm, withering but persistent. Inflorescence-stalk hairless, 2--3 cm. Stamens 3 (rarely 4), filaments fused to base of ovary. Ovary 1-celled, of 3 partly fused carpels, bearing 3 (rarely 4) hairy, curled styles. Fruit a capsule, splitting above, at the base of the styles. Seeds numerous, small. *East Himalaya, Taiwan, Japan, Indonesia (Java).*

3. Saururus Linnaeus. 2/2. Leaves alternate; blades with palmate veins converging towards the tip. Stipules indistinct. Flowers in moderately dense, slender racemes, without petal-like bracts at the base. Stamens 6--8; filaments free from carpels. Carpels 3--5, fused at base, with 2--4 ovules per carpel; styles free, curved. Fruit spherical, slightly fleshy, splitting into indehiscent, 1-seeded units. *Asia, North America.*

67. ARISTOLOCHIACEAE

Perennial herbs and climbers, occasionally with woody stems. Leaves simple, stalked, alternate. Flowers solitary or in axillary clusters, often foetid, bisexual. Perianth radially or bilaterally symmetric, more or less petal-like, 3-lobed at apex or with a single, unilateral lobe. Stamens 6 or 12, in 1 or 2 whorls, filaments free or united with stylar column. Styles 6, free or united to form a column with 6-lobed stigma. Ovary inferior, 6-celled, placentation axile, ovules numerous in each cell. Fruit a capsule. See figure 40, p. 192.

A family of 7 genera and about 600 species, mostly tropical and subtropical, a few from temperate areas.

   1a. Small herbs with rhizomes; flowers terminal; stamens 12   **1. Asarum**
     b. Erect herbs or climbers; flowers axillary; stamens 6   **2. Aristolochia**

1. Asarum Linnaeus. 75/7. Herbs with creeping rhizomes rooting at nodes. Flowers solitary, terminal, resin-scented. Perianth radially symmetric, 3-lobed, persistent in fruit. Stamens 12 in 2 whorls, filaments free (rarely fused). Styles 6, free or united into a column. Fruit nearly spherical, opening irregularly. Seeds boat-shaped. *North temperate areas.*

2. Aristolochia Linnaeus. 200/13. Perennial herbs or woody climbers. Leaves deciduous or evergreen, often heart-shaped or hastate at base. Flowers often unpleasantly and strongly scented, axillary, solitary, clustered or in racemes; Inflorescence-stalk usually with prominent bracts at base or below ovary. Perianth

189

bilaterally symmetric, tubular, deciduous, constricted at mouth. Stamens usually 6. Fruit a capsule. *Cosmopolitan.* Figure 40, p. 192.

68. DILLENIACEAE

Trees, shrubs, woody climbers or perennial herbs, rarely woody at base. Leaves spirally arranged (rarely opposite), simple (rarely pinnatifid or 3-lobed), entire or toothed; stipules absent or united to leaf-stalk. Flowers solitary or in variously cyme- or raceme-like inflorescences, usually bisexual. Sepals 3--20, usually 5, free, spiral, overlapping, persistent. Petals 2--5, free, overlapping, often crumpled in bud, white or yellow to orange, occasionally pink, falling, or sometimes absent. Stamens 3--numerous, free or in 3--15 bundles with filaments free or partly united, all fertile or some staminodial; anthers opening by slits or, rarely, pores. Ovary superior, of 1--20 carpels in 1 or 2 rows, free or sometimes partly united with each other or with receptacular apex, with 1--many ovules in each carpel. Fruit dry and splitting or sometimes not splitting and enclosed in a more or less fleshy calyx; seeds with or without a more or less fringed or lobed aril.

A family of 10--12 genera and about 350 species occurring throughout the tropics and in temperate Australia, but with a single genus in continental Africa. Only 1 genus has representatives marginally hardy in northern Europe.

1. Hibbertia Andrews. 122/1. Trees or shrubs, often straggling or twining, sometimes heather-like, evergreen, with leaf-scars encircling less than half the stem. Leaves simple, obscurely pinnately veined, margin entire to wavy or toothed; stalk unwinged or absent. Flowers solitary or in 1-sided, raceme-like cymes, terminal and often apparently axillary, bisexual. Sepals 5, not enlarging in fruit. Petals 5 (rarely 3 or 4), yellow or rarely white or pink. Stamens 3--many, free or in 2--5 bundles each with 2--4 stamens, sometimes with staminodes or all confined to one side of the flower; anthers opening by slits or pores. Ovary of 1--5 (rarely to 10) carpels, free or very slightly coherent basally; ovules 1--15 per carpel. Fruit a group of follicles. Seeds 1--4 per follicle, with arils. *Madagascar, New Guinea, Fiji, New Caledonia & Australia.*

69. PAEONIACEAE

Perennial herbs with fleshy roots, or soft-wooded shrubs. Leaves alternate, compound, at least the lower ones biternately or bipinnately divided into pinnately veined leaflets, segments and lobes. Flowers bisexual, radially symmetric, large, solitary and terminal or few from leaf-axils. Floral bracts 1--12, leafy, immediately below calyx; sepals 5, free, persistent, overlapping, unequal. Petals 5--9, or more (in garden variants). Stamens numerous; anthers opening by longitudinal slits. Carpels 1--8, free, on a fleshy disc which sometimes extends upwards into fleshy lobes or a sheath; ovules numerous, lateral, in 2 rows. Fruit of 1--5 diverging follicles, each with several seeds, opening by a slit on the upper side. Seeds large, the fertile ones black, the infertile red.

A family of 1 genus which has often been included in *Ranunculaceae*.

1. Paeonia Linnaeus. 30/30 and many hybrids. Description as for the family. *Mediterranean area, northern Asia, western North America.*

70. EUCRYPHIACEAE

Evergreen trees and shrubs (some usually deciduous in cultivation). Leaves opposite, simple or pinnate, entire or toothed, stipules falling early. Flowers bisexual, large, fragrant, solitary in the leaf-axils. Flower-stalks with bracts, usually at the base; flowering-stem not evident. Sepals 4, forming a cap over the petals which falls as the flower opens. Petals usually 4, white, rarely pink. Stamens numerous with pink anthers. Ovary superior. Fruit a woody capsule with numerous winged seeds.

A family of only 1 genus from Australia and South America.

1. Eucryphia Cavanilles. 5/5 and some hybrids. Description as for family. *South-east Australia, Chile.*

71. ACTINIDIACEAE

Trees, shrubs or rampant vines, dioecious or with female and bisexual flowers on the same plant. Leaves alternate, simple, evergreen or deciduous. Inflorescences axillary. Flowers usually with 4 or 5 sepals and petals, numerous stamens and a superior ovary with many ovules. Fruit a berry or capsule.

A family of 3 genera and about 300 species, mainly from tropical and subtropical Asia, with a few members in Australia and America.

1a. Stamens 10; ovary with 5 cells; style solitary **2. Clematoclethra**
  b. Stamens numerous; ovary with numerous cells; styles numerous
                                      1. Actinidia

1. Actinidia Lindley. 50/5. Deciduous, usually dioecious vines, climbing and clinging by spirally twisting stems. Leaves usually toothed. Flowers unisexual or very infrequently bisexual, solitary or in cymes. Sepals 5. Petals 5. Stamens numerous. Ovary with numerous cells; styles numerous. Fruit a juicy berry. *Eastern Asia.*

2. Clematoclethra (Franchet) Maximowicz. 10/1. Deciduous shrubs, with twining stems resembling *Actinidia*. Flowers in cymes. Sepals 5, not deciduous. Petals 5. Stamens 10. Ovary with 5 cells surmounted by a single style; stigma not lobed. Fruit a small berry. *China.*

Species of **Saurauia** Willdenow, which are upright trees or shrubs, with few unisexual flowers in cymes, may be grown in favoured areas.

72. THEACEAE

Trees and shrubs, occasionally lianas. Leaves evergreen or occasionally deciduous, stalked, simple, entire or toothed; spirally arranged, usually 2-ranked, without stipules. Flowers usually large, bisexual and regular, axillary and solitary, but sometimes small, unisexual, terminal, in racemes or panicles; occasionally fragrant. Bracteoles 2--12, sometimes intergrading with the sepals, deciduous or persistent. Sepals 4 or 5, basally united, overlapping. Petals 4--many, sometimes basally united, overlapping. Stamens usually numerous, free in 1--5 series, in 5 bundles on a fleshy disc, irregularly or basally united, attached to the base of the corolla; anthers versatile or basifixed. Ovary usually superior, rarely partly inferior, carpels 3--5 rarely 2--10, basally united. Styles free or partially united. Fruit a loculicidal capsule, sometimes woody, or a berry. Seeds 1--many, sometimes winged. See figure 41, p. 194.

A family of about 28 genera and 520 species; mainly tropical, a few warm temperate.

1a. Deciduous tree; capsule dehiscing at apex and base **1. Franklinia**

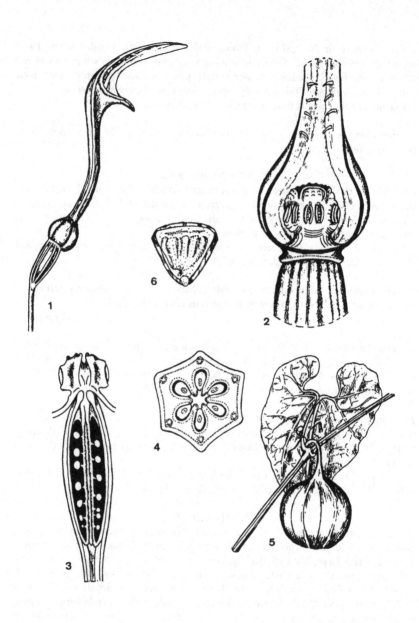

Figure 40. Aristolochiaceae. *Aristolochia clematitis.* 1, A single flower. 2, Longitudinal section of the lower part of the flower, showing downwardly-pointing hairs and stamens attached to the stylar column. 3, Longitudinal section of ovary. 4, Transverse section of ovary. 5, Fruit. 6, Seed.

b. Evergreen or rarely deciduous trees or shrubs; capsule usually dehiscing at apex only; if plant deciduous and capsule splitting down ribs, then winter buds silvery-hairy and seeds 2--4 per cell   2

2a. Deciduous or rarely evergreen plants; capsule without persistent central axis   **2. Stewartia**

  b. Plants evergreen; capsule with persistent central axis   3

3a. Seeds unwinged   **3. Camellia**

  b. Seeds winged   **4. Gordonia**

1. Franklinia Marshall. 1/1. Deciduous shrubs or small trees, erect with short trunk; branches stout; bark thin, smooth, grey to reddish brown. Leaves shortly stalked, blade obovate-oblong, blunt or acute, membranous, bright glossy green above, veins impressed, downy beneath; margin finely toothed. Flowers bisexual, axillary, solitary, almost stalkless, fragrant. Bracteoles 2, beneath calyx, minute, deciduous. Sepals 5, unequal, almost circular, white silky-hairy, persistent. Petals 5, snow-white, shortly united, obovate, margin wavy, membranous. Stamens yellow, numerous, free; anthers versatile. Ovary almost spherical, ridged, hairy, 5-celled. Styles 5, united; stigma disc-shaped. Fruit a woody capsule, almost spherical, with persistent central axis. Seeds 6--8 per cell, flat, angular, unwinged. *Southeast USA.*

The one species is probably extinct in the wild.

2. Stewartia Linnaeus. 20/7. Trees or shrubs, deciduous, rarely evergreen. Leaves with stalks winged, sometimes enclosing the winter leaf-buds; blades more or less hairless above, hairy beneath at least along the veins; margin toothed. Winter leaf-buds laterally compressed with 1--several overlapping scales, rarely naked. Flowers bisexual, axillary, solitary, rarely 2 or 3 together, stalked. Bracteoles 2 (rarely 1) beneath the calyx, usually leaf-like, almost opposite and persistent. Sepals 5, rarely 6, basally joined, usually persistent in fruit. Petals 5, rarely 6--8, snow- or cream-white, occasionally stained red outside, basally united in a short tube, margin wavy, outer surface usually finely silky-hairy. Stamens numerous, filaments basally joined, anthers versatile. Ovary superior, hairless or hairy, 5-, rarely 4- or 6-celled; styles 5 (occasionally 4 or 6), united for most of their length, rarely free, usually persistent in fruit. Fruit a woody capsule, almost spherical to ovoid, strongly angled and ribbed, often beaked. Seeds 2--4 per cell. *Korea, Japan, Thailand, Laos, Cambodia, Vietnam and south-west China; also in southeast USA.*

3. Camellia Linnaeus. 200/few but several hybrids with many cultivars. Evergreen trees and shrubs, erect, occasionally weeping. Leaves usually leathery, usually toothed, glossy deep green above, paler beneath. Flowers more or less terminal at the ends of new shoots, arising between the scales of axillary vegetative buds, solitary or clustered, stalked or almost stalkless, bisexual. Bracteoles 2--8; sepals 5--6, sometimes intergrading (perules). Petals 5--14, or in cultivars sometimes more, usually basally united, white, pink or red, rarely yellow. Stamens usually numerous, of unequal length, rarely hairy, the outer filaments partially, often irregularly united, often forming a cup; anthers versatile. Ovary superior, often hairy. Fruit a 1--5-celled capsule, leathery or woody, splitting from the apex, with a persistent axis. Seeds irregularly shaped, generally rounded, unwinged. *India, Nepal, Bhutan, Burma, Thailand, Laos, Cambodia, Vietnam, China, Korea, Japan, Indonesia & Philippines.* Figure 41, p. 194.

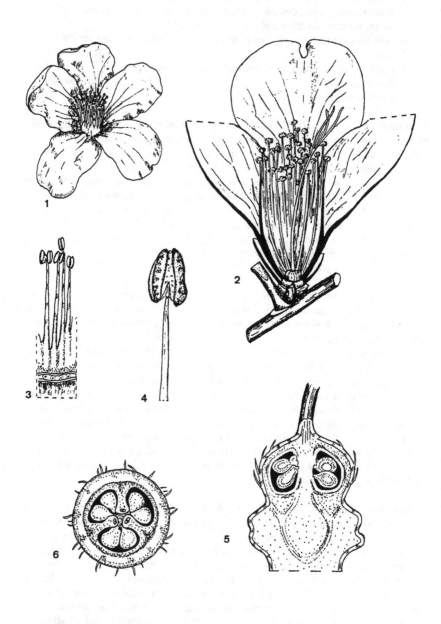

Figure 41. Theaceae. *Camellia* × *williamsii*. 1, Flower. 2, Oblique longitudinal section of flower. 3, Portion of cup formed by the united bases of the filaments. 4, Stamen. 5, Longitudinal section of ovary. 6, Transverse section of ovary.

4. Gordonia Ellis. 70/2. Evergreen trees and shrubs. Leaves usually shortly stalked, blade leathery, hairless, margin usually toothed. Winter buds naked, silky-hairy. Flowers usually at branch-tips, usually solitary, bisexual, stalked. Bracteoles 2--12, beneath calyx, sometimes intergrading with sepals, deciduous. Sepals 4 or 5, unequal, persistent. Petals 5--7, shortly united, outer petals somewhat transitional to sepals. Stamens numerous, the outer whorl at least on cup-shaped 5-lobed disc at base of corolla; filaments sometimes irregularly united; anthers versatile. Ovary almost spherical; styles 5, united; stigma lobed. Fruit a woody capsule, ovoid or cylindric, 3--6-celled, splitting from apex to base, central axis persistent. Seeds flat, apically winged. *Southeast Asia, southeast USA.*

Species of **Schima** Blume, which have the stamens free in 3--5 series and an almost spherical capsule, are occasionally found in gardens in very sheltered areas. One species of **Visnea** Linnaeus filius, which is a small shrub with basifixed anthers, may also occasionally be found.

73. GUTTIFERAE (*Clusiaceae*)

Trees, shrubs or herbs, rarely climbers, mostly containing a yellow to red or clear resinous latex, sometimes with simple or stellate hairs. Leaves usually opposite, rarely whorled or alternate, simple, entire or rarely with margin gland-fringed, nearly always without stipules, usually with glands and/or resin canals that are often translucent. Flowers bisexual or unisexual, radially symmetric, terminal or axillary, solitary or in cymes or false racemes. Sepals 2--6 or more, free or more or less united, overlapping, rarely not distinct from petals. Petals 3--6 or more, free, overlapping, usually in the same direction. Stamens basically in 2 whorls of 3--5 bundles each of 3--many stamens, the outer whorl usually sterile or absent, the inner (opposite petals) free or partly or wholly united, sometimes the ring of the bundles wholly merged, the stamens then appearing free; ovary superior, 1--12-celled, with 1--many ovules on axile, parietal, basal or rarely apical placentas. Styles 2--12, free or more or less united or, apparently, single. Fruit a capsule opening between cells (rarely also within cells), berry or drupe. Seeds without endosperm, sometimes winged or with arils or with a fleshy outer coat, the embryo sometimes with reduced or without cotyledons. See figure 42, p. 198.

A family of about 48 genera and more than 1000 species, almost confined to tropical regions apart from the large genus **Hypericum** (sometimes included in the separate family *Hypericaceae*), which is almost cosmopolitan.

1. Hypericum Linnaeus. 400/115. Small trees, shrubs or herbs, evergreen or deciduous, with translucent, amber, and often black or occasionally red glands (and/or canals), hairless or sometimes with simple hairs. Stems usually 2--6-lined when young, eventually terete. Leaves opposite or in whorls of 3--5, with pale and/or dark glandular dots, streaks or lines. Flowers bisexual, in cymes. Sepals 4 or 5, free or partly united, unequal or equal, entire, persistent or rarely falling. Petals 4 or 5, yellow, the parts visible in bud often tinged red, very rarely wholly carmine-red or white, each usually asymmetric and with a lateral projection, persistent or falling. Stamens usually numerous, in 4--5 equal bundles opposite petals or in 2 or rarely 1 bundle, united and larger, or all bundles merged; filaments in each bundle rarely united more than halfway and then bundles alternating with small bilobed scales; anthers small, oblong to elliptic, with an amber or dark gland on connective. Ovary 3--5-celled, or 1-celled with 2--5 placentae bearing 2--many ovules; styles 2--5, long, free or sometimes united; stigmas small. Fruit a 2--5-celled capsule splitting between the cells, rarely fleshy. Seeds small, usually curved, cylindric, sometimes keeled or laterally winged. *Cosmopolitan.* Figure 42, p. 198.

74. SARRACENIACEAE
Insectivorous perennial herbs with rhizomes. Leaves modified into flask-like structures (commonly called pitchers) capable of holding fluid and secreting enzymes, and in which insects are trapped; mouth of the pitcher sometimes covered by a hood. Flowers solitary, or few in a raceme, on erect stems, nodding, bisexual; sepals 4 or 5, sometimes petal-like or coloured, persistent, petals 5 and free, or absent. Stamens numerous, free. Ovary with 3--5 cells, style simple, sometimes expanded and umbrella-like. Fruit a capsule; seeds numerous.

An American family of 3 genera which inhabit swamps and bogs and are carnivorous.

1. Sarracenia Linnaeus. 8/few and some hybrids. Pitchers erect or decumbent, with a conspicuous hood which arches over the mouth; phyllodes produced by some species (these are not flask-like, but have a flat blade). Flowers solitary on erect stems. Sepals 5, usually green, persistent in fruit. Petals 5, longer than sepals, deciduous, distinctively folded (resembling an italic *N*) to fit over rim of the stigmatic disc; apical portion broadest; stamens numerous. Ovary 5-celled; style with a terminal, expanded, 5-lobed umbrella-shaped disc; stigmas on apices of lobes. *North America.*

75. DROSERACEAE
Herbs or low shrubs. Leaves alternate, or rarely whorled, usually in basal rosettes, usually simple, insectivorous with sticky glandular hairs and sometimes with traps. Flowers regular, bisexual, often in racemes, sometimes solitary. Sepals 4 or 5, fused at the base. Petals 4 or 5, free. Stamens 4--20, free. Ovary superior. Fruit a capsule.

A family of 4 genera found throughout the world but concentrated in Australia and New Zealand.

1. Drosera Linnaeus. 150/6. Carnivorous perennial, terrestrial herbs, rarely annual, sometimes overwintering as a resting bud or underground tuber. Roots usually fibrous, rarely stout and bootlace-like. Stems very short, sometimes erect and twining. Leaves at first in a basal rosette; some species then producing leafy stems (most tuberous species not producing basal leaves when re-emerging from dormancy). Leaf-shape often changing greatly from basal rosette to aerial stem, and very variable between species, always with numerous stalked sticky glandular hairs on upper surface. Stipules large and chaffy, often joined to basal part of leaf-stalk. Flowers in parts of 4 or 5, usually lasting a single day. Sepals united at the base, persistent in fruit. Petals free, obovate, white, purple or pink, rarely red or yellow, partly deliquescing then adhering together, forming a cap over the developing fruit. Stamens in a single whorl, anthers sometimes divided by the expanded connective. Ovary with 3 or 5 styles, varying greatly in branching and shape between species. Fruit a dehiscent capsule. Seeds dust-like, shape and sculpture varying greatly between species. *Cosmopolitan, but concentrated in southern Africa & south-western Australia.*

76. PAPAVERACEAE

Annual or perennial herbs, more rarely herbaceous or woody climbers or shrubs; sap milky or coloured with latex, or clear. Leaves usually alternate, rarely opposite or all basal, usually divided, often in complex ways; veins pinnate or palmate. Flowers solitary or in cymes, racemes, umbels or panicles, varying from small to large, radially symmetric (with 2 obvious planes of symmetry) or bilaterally symmetric. Sepals 2 or 3, usually falling early, sometimes united. Petals 4 or 6, rarely many or absent, usually borne in 2 distinct whorls, either of simple form or variously lobed, free or variably united. Stamens 8--many or 4, when 4 either simple or united in 2 bundles, each bundle consisting of a whole stamen united to 2 half-stamens. Sepals, petals and stamens rarely borne on a perigynous zone. Ovary superior with 2 or more carpels which are usually united; style present or absent, stigmas 2 or more; ovules usually numerous, parietal, more rarely 1 and almost basal. Fruit a capsule or an indehiscent nutlet, or breaking into 1-seeded segments. Seeds with oily endosperm, each often with an appendage (aril, caruncle, arilloid, elaiosome). See figure 43, p. 202.

A family of 41 genera from north temperate regions, South America, South Africa and Australia, of which a considerable number is grown. The family as recognised here in its broadest sense, is divided into 7 subfamilies (in a narrower sense it is divided into 2 or 3 families). Of these, 4 (*Papaveroideae, Platystemonoideae, Chelidonioideae* and *Eschscholtzioideae*, genera **1--19**) make up the *Papaveraceae* in the strict sense: plants with milky or coloured latex, radially symmetric flowers with no nectar, many stamens and the carpels 2--many. A fifth subfamily (*Fumarioideae*, often recognised as a separate family, *Fumariaceae*, genera **22--27**) consists of a group of genera looking superficially very unlike those of the strict *Papaveraceae*, having watery sap, flowers usually bilaterally symmetric, nectar usually secreted by the bases of the stamens (often prolonged into a nectar-secreting spur or peg), 4 stamens which are often united as described above, and the carpels always 2. These 2 groups are, however, linked by the final two subfamilies (*Hypecoideae*, sometimes recognised as a separate family, *Hypecoaceae*, and *Pteridophylloideae*, consisting only of the 2 genera *Hypecoum* and *Pteridophyllum*, genera **20 & 21**).

1a. Carpels 3 or more, usually united, occasionally more or less free   2
   b. Carpels always 2, always united   9
2a. Leaves opposite or whorled   3
   b. Leaves alternate or all basal   4
3a. Carpels several, becoming free from each other as they ripen
                 7. Platystemon
   b. Carpels 2--4, remaining united   **8. Stylophorum**
4a. Softly woody shrub with large white flowers; ovary covered with stiff, hard bristles   **1. Romneya**
   b. Herbs; combination of other characters not as above   5
5a. Stigmas forming radii on a flat, domed or conical disc borne on top of the ovary   **3. Papaver**
   b. Stigmas various, not as above   6
6a. Plants glaucous, with prickles on the stems and leaves or leaf-margins; sepals each with an almost apical, hollow horn   **2. Argemone**
   b. Plants not usually glaucous, not prickly but often bristly; sepals without horns   7

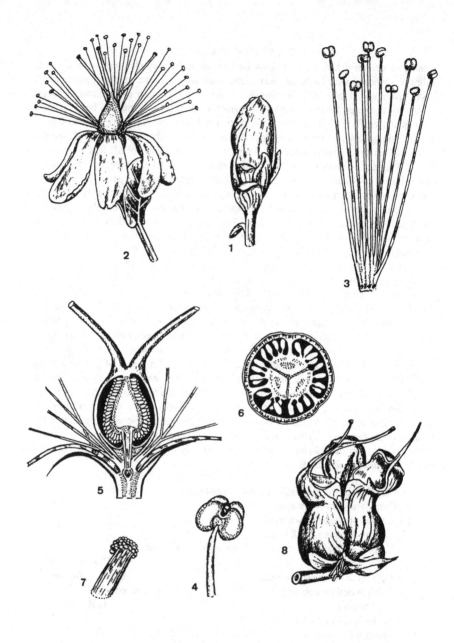

Figure 42. Guttiferae/Hypericaceae. *Hypericum perforatum.* 1, Flower-bud. 2, Flower. 3, A bundle of stamens. 4, Anther of a single stamen. 5, Longitudinal section of flower. 6, Transverse section of ovary. 7, Apex of style. 8, Fruit.

7a. Capsule more than 5 mm broad, spherical or cylindric, opening by a series of pores just below the apex 8
  b. Capsule less than 5 mm broad, linear-cylindric, opening by the walls splitting from the placental framework from the apex for some distance downwards **6. Roemeria**
8a. Style thread-like, stigma head-like; petals orange with a purplish blotch near the base, the claw below the blotch greenish; annual herb **5. Stylomecon**
  b. Style and stigmas not as above; petals various, but not coloured as above; perennial or monocarpic herbs **4. Meconopsis**
9a. Stamens more than 4, free; nectar absent 10
  b. Stamens 4, free or in 2 bundles; nectar usually secreted at the bases of the filaments 20
10a. Sepals, petals and stamens perigynous 11
  b. Sepals, petals and stamens hypogynous 12
11a. Sepals united, falling as a candle-snuffer-shaped unit; cotyledons bifid **17. Eschscholzia**
  b. Sepals free, falling individually; cotyledons not bifid **18. Hunnemannia**
12a. Petals absent 13
  b. Petals present 14
13a. Leaves palmately veined and lobed **15. Macleaya**
  b. Leaves pinnately veined, pinnately lobed or simple and entire **16. Bocconia**
14a. Leaves palmately veined and lobed 15
  b. Leaves pinnately veined and lobed 16
15a. Petals 8--16 or more; flowers solitary **14. Sanguinaria**
  b. Petals 4; flowers in cymes **13. Eomecon**
16a. Shrub; leaves evergreen **19. Dendromecon**
  b. Herb; leaves deciduous 17
17a. Leaves pinnate into distinct leaflets; seeds each with an appendage (aril) 18
  b. Leaves pinnatisect to lobed, but without distinct leaflets; seeds not appendaged 19
18a. Petals 2--2.5 cm; flowers solitary, without bracteoles **10. Hylomecon**
  b. Petals 7--15 mm; flowers in umbels, each with 2 bracteoles **9. Chelidonium**
19a. Seeds embedded in a spongy false septum **11. Glaucium**
  b. Seeds not embedded in a false septum **12. Dicranostigma**
20a. Stamens 4, free; no petals spurred 21
  b. Stamens 4 in 2 bundles of a whole stamen united to 2 half-stamens; upper petal (at least) usually spurred 22
21a. All 4 petals more or less similar; leaves all basal, pinnate into oblong leaflets; fruit a capsule **20. Pteridophyllum**
  b. The 2 outer petals entire to weakly 3-lobed, the 2 inner petals distinctly divided into 3 segments of which the central is very different from the laterals; leaves not as above; fruit breaking into 1-seeded segments (ours) **21. Hypecoum**
22a. Flowers with 2 distinct planes of symmetry; outer petals swollen or with spurs at the base 23

b. Flowers with only a single plane of symmetry; upper petal distinctly spurred, the lower sometimes swollen at the base but not spurred 24

23a. Outer petals rounded or slightly and equally swollen at the base but not spurred; perianth persistent on the fruit **23. Adlumia**

b. Outer petals both distinctly and equally spurred; perianth not persistent **22. Dicentra**

24a. Fruit a capsule containing 2 or more seeds **24. Corydalis**

b. Fruit a nutlet containing 1 or rarely 2 seeds 25

25a. Stigma crested or with 2 papillae **27. Fumaria**

b. Stigma not as above 26

26a. Inner petals without yellow lateral wings; nutlet not ribbed **26. Rupicapnos**

b. Inner petals with yellow lateral wings; nutlet ribbed **25. Sarcocapnos**

1. Romneya Harvey. 2/2 and the hybrid between them. Large, rather woody, glaucous perennials, with creeping, underground rootstocks. Leaves alternate, pinnatifid to pinnatisect. Flowers solitary, terminal, large and showy. Sepals 3. Petals 6, white. Stamens numerous, filaments thread-like, yellow. Ovary of 7--12 united carpels, placentation parietal, ovules borne on inwardly directed plates which almost meet at the centre. Stigmas stalkless on top of the ovary. Fruit a capsule opening at the top. *Western USA.*

2. Argemone Linnaeus. 30/4. Annual to short-lived perennial herbs. Latex usually yellowish. Stems glaucous, often prickly. Leaves alternate, glaucous, lobed and with prickle-tipped teeth, surfaces prickly or not, often pale or bluish over the veins, the uppermost leaves usually stem-clasping. Sepals usually 3, often prickly, each with an almost apical horn which is hollow at the base and prickle-like or flattened towards the apex. Petals usually 6, orange, yellow, white or rarely mauve. Stamens numerous, filaments mostly thread-like. Ovary of 3--5 carpels, 1-celled, placentas scarcely projecting inwards. Style short or absent, stigmas usually velvety, purplish. Capsule usually prickly, opening at the apex. Seeds numerous. *North & South America.*

3. Papaver Linnaeus. 70/30. Perennial, biennial or annual herbs, often bristly or hairy. Latex usually whitish, sometimes orange or reddish. Leaves alternate or all basal, usually divided and toothed, sometimes stem-clasping at the base. Flowers in raceme- or spike-like inflorescences or solitary, often large and showy, with or without bracts. Sepals usually 2, rarely 3, buds usually pendent until just before the flower opens, sepals falling as the flower opens. Petals usually 4, rarely 6, crumpled in bud, red, pink, orange, yellow or white, often blotched at the base with a contrasting colour. Stamens numerous, filaments thread-like or club-shaped. Ovary of 3 or more united carpels, the ovules borne on placentas which project into the ovary as plates, almost meeting in the centre. Styles borne on a flat, convex or pyramidal disc on the top of the ovary, as velvety rays. Fruit a capsule opening by pores just beneath the stigmatic disc. Seeds numerous, without arils. *North temperate areas, in the southern hemisphere in Australia and South Africa.* Figure 43, p. 202.

4. Meconopsis Viguier. 50/20 and some hybrids. Biennial or perennial herbs, some flowering only once and then dying (monocarpic), others flowering for several years before dying (polycarpic). Leaves simple or variously divided, often forming a dense basal rosette. Flowers stalked, borne singly or in a branched, often compound inflorescence, which opens from above downwards. Sepals 2, rarely 3 or 4, falling early. Petals 4--10. Stamens many. Ovary more or less spherical, ovoid or obovoid to cylindric, usually with a distinct style bearing a number of stigmatic lobes (style rarely expanded to form a disc on top of the ovary, or absent). Fruit a capsule (opening above by interplacental valves). *South-central temperate Asia, 1 species in western Europe.*

5. Stylomecon Taylor. 1/1. Annual herb with yellow sap. Leaves alternate, pinnatifid to pinnatisect, segments entire or themselves further divided. Flowers in the upper leaf-axils on long stalks. Sepals 2, free, falling individually. Petals 4. Stamens numerous. Ovary 1-celled with 4 or more placentas. Style slender, short, with a head-like stigma. Capsule narrowly top-shaped, opening by pore-like flaps towards the apex. Seeds kidney-shaped. *Western USA (California).*

Originally included in *Meconopsis*.

6. Roemeria Medikus. Few/1--2. Slender annual herbs with yellowish latex. Leaves alternate, 1--3 times pinnatisect. Flowers solitary. Sepals 2; petals 4. Stamens numerous, filaments club-shaped or thread-like. Ovary of 3 or 4 united carpels; placentation parietal. Fruit a linear-cylindric capsule opening towards the apex, bristly all over or at least in part, or with 3 or 4 long bristles at the apex which overtop the stigmas. *Mediterranean area, Middle East.*

7. Platystemon Bentham. 1/1. Small, hairy, annual herbs. Leaves opposite or whorled, entire, mostly borne in the lower parts of the stems, more or less stalkless, linear-oblong to lanceolate. Flowers solitary on long stalks. Sepals 3, rather slow to fall as the flower opens. Petals 6, white, yellowish or cream, 8--20 mm. Stamens numerous. Ovary of 6--25 carpels, each with many parietal ovules, at first united, ultimately separating and splitting transversely into 1-seeded segments. *Western USA, adjacent Mexico.*

8. Stylophorum Nuttall. 3/1. Perennial herbs with yellow-orange or reddish latex. Leaves mostly basal and long-stalked, generally with 2 opposite leaves on the stem between which is the inflorescence; blades pinnatisect (or the lowermost lobes distinct and separate) into 5--7 lobes which are themselves irregularly bluntly lobed or toothed, all green and sparsely hairy with curled hairs. Flowers few in an umbel-like cyme borne between the upper leaves; flower-stalks long, with small bracteoles at the base. Sepals 2, free, falling early. Petals 4, yellow. Stamens yellow, numerous. Ovary of 2--4 united carpels; style short, stigmas 2--4. Fruit a capsule covered with bristles. *East Asia, eastern North America.*

9. Chelidonium Linnaeus. 1/1. Perennial, somewhat glaucous, rhizomatous herbs with yellow, irritant latex. Leaves deciduous, alternate, long-stalked, pinnate with oblong-ovate or obovate leaflets which are lobed or scalloped. Flowers long-stalked in leaf-opposed umbels; flower-stalks with small bracteoles at the base. Buds pear-shaped. Sepals 2, falling quickly as the flower opens, somewhat yellowish. Petals 4, obovate, yellow, 7--15 mm. Stamens numerous, yellow. Ovary of 2 united carpels; style short; stigma 2-lobed. Fruit a linear-cylindric, wrinkled capsule to 6 cm. Seeds

201

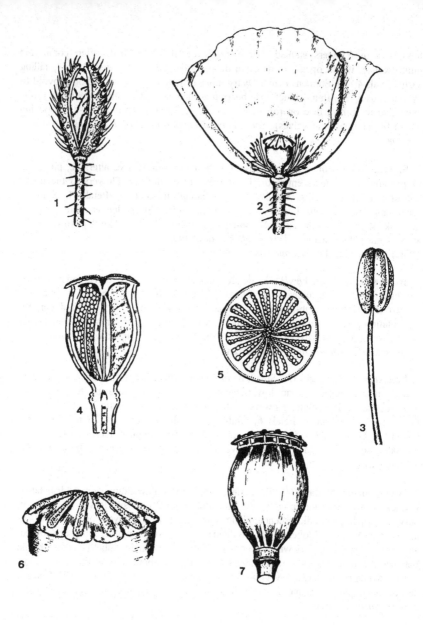

Figure 43. Papaveraceae. *Papaver rhoeas.* 1, Expanding flower-bud. 2, Longitudinal section of flower. 3, Stamen. 4, Longitudinal section of ovary. 5, Transverse section of ovary. 6, Stigmatic disc. 7, Fruit.

numerous, shiny, each with a small, whitish aril. *Eurasia; introduced into North America.*

10. Hylomecon Maximowicz. 1/1. Perennial herbs with short, oblique rhizomes and yellowish latex. Leaves soft, bright green, mostly basal and long-stalked, 1 or 2 alternate on the flowering stems, rather shortly stalked; blades pinnate with usually 5 oblong-obovate or obovate, sharply and irregularly toothed leaflets. Flower solitary, terminal, stalked, without bracteoles. Sepals 2, falling as the flower opens. Buds ovoid, pointed. Petals 4, yellow, elliptic to almost circular, 2--2.5 cm. Stamens numerous, yellow. Ovary of 2 united carpels, styles short, stigmas 2. Fruit a linear-cylindric capsule. *Eastern Asia.*

11. Glaucium Miller. 20/4. Annual, biennial or short-lived perennial herbs. Leaves often glaucous and somewhat fleshy, pinnatifid, pinnatisect, lobed or almost entire, alternate. Flowers solitary or in cymes. Sepals 2, falling as the flower opens. Petals 4, yellow, orange, red or mauve. Stamens numerous. Ovary of 2 united carpels containing many parietal ovules; style absent or very short, stigma mitre-like, conspicuous. Capsule linear-cylindric, usually more than 10 cm, often curved, sometimes contorted, opening from the base upwards or from the apex downwards. Seeds embedded in a spongy false septum. *Eurasia.*

12. Dicranostigma Hooker & Thomson. 3/2. Like *Glaucium* but always perennial, fruits not longer than 10 cm, the seeds not embedded in a spongy false septum. *Himalaya, China.*

13. Eomecon Hance. 1/1. Herbaceous perennial to 50 cm. Leaves all basal, long-stalked; blades ovate, main veins palmate, margins rather wavy or scalloped with blunt scallops, base deeply cordate. Inflorescence a loose cyme with several flowers; flower-stalks with small bracteoles at the base. Buds almost spherical but tapering to an acuminate apex. Sepals 2, completely united, falling as a whole as the flower opens. Petals 4, to 1.5 cm, white, obovate. Stamens numerous, yellow. Ovary of 2 united carpels; style present, lengthening in fruit, stigma bilobed. Fruit an ellipsoid capsule tapering into the long, persistent style. *China.*

14. Sanguinaria Linnaeus. 1/1. Herbaceous perennials with creeping rhizomes; latex red. Leaf 1 to each flowering shoot, basal, long-stalked and with small scale-leaves at the base, blade to 30 cm across, circular or somewhat broader than long, main veins palmate, palmately lobed to about half the radius or less, especially towards the apex, the lobes entire or rather bluntly toothed, base cordate, all greyish green. Flowers solitary, borne on a scape with a scale-leaf at the base, appearing just before the leaves. Sepals 2, free, soon falling. Petals 8--16, to 3 cm, obovate, white or white flushed with pink. Stamens many. Ovary of 2 united carpels; style present, stigma small. Fruit an ellipsoid capsule, tapering at both ends. *Eastern North America.*

15. Macleaya R. Brown. 2/2 and the hybrid between them. Tall, rhizomatous perennial herbs with stems woody below but dying back every year; latex orange. Leaves alternate, basal and on the stems, the lower long-stalked, all with the main veins palmate, blade oblong to circular, cordate at the base, rather deeply lobed, the lobes themselves lobed or bluntly toothed, all glaucous, hairy on the lower surface, at least when young. Flowers in clusters in large, upright, plume-like panicles. Buds narrowly obovoid, whitish or purplish. Sepals 2, falling as the flowers open. Petals absent. Stamens 8--30, with long or short filaments, anthers mucronate. Ovary of 2 united carpels; style short, stigma bilobed. Ovules few and parietal or 1 and basal. Fruit a small, flattened capsule containing 1--few seeds. Seeds each with a small, lateral aril. *Japan, China.*

16. Bocconia Linnaeus. 9/1. Shrubs, trees or climbers with persistent woody shoots, latex yellowish. Leaves alternate, pinnately veined, pinnately lobed or toothed, stalked. Flowers in erect, much-branched, rather tight terminal panicles. Buds ovoid to spherical. Sepals 2, falling as the flowers open, ovate or almost circular. Petals absent. Stamens 8--24, with filaments shorter than anthers. Ovary of 2 united carpels, long-tapered to a stalk below, containing a single, basally attached ovule; style present; stigma bilobed. Fruit flattened, elliptic, pointed at both ends, containing a single seed. Seed with its base enclosed in a cup-like aril. *Central & South America.*

Resembling *Macleaya* in being wind-pollinated, the flowers without petals.

17. Eschscholzia Chamisso. 10/2. Annual herbs (sometimes occasionally perennial in the wild) with clear sap. Leaves alternate, usually hairless, divided into 3 leaflets which are themselves further divided into linear-oblong segments, green or glaucous. Flowers numerous, cream, yellow, reddish or orange; sepals, petals and stamens borne on a cup- or funnel-shaped perigynous zone. Sepals 2, united, falling as a candle-snuffer-like unit. Petals large, spreading, sometimes crisped at the margins, usually 4 (rarely more). Stamens 16--many. Ovary cylindric, 1-celled, with 2 placentas, 10-veined. Styles 2, each divided into 2--3 thread-like segments. Capsule linear-cylindric, opening from the base upwards. *Western North America.*

18. Hunnemannia Sweet. 1/1. Very like *Eschscholzia*, but perennial, upright and very glaucous; sepals 2, free, falling individually. *Mexico.*

19. Dendromecon Bentham. 2/1. Evergreen shrub with clear sap. Leaves alternate, leathery, upper surface with a network of raised veins, yellowish or greyish green, entire or very finely toothed (a lens is necessary to see the teeth). Flowers solitary, terminal on short branches. Sepals 2, free, falling individually as the flower opens. Petals 4 (rarely 6), yellow. Stamens numerous, with short filaments. Ovary 1-celled with 2 placentas, styles 2. Capsule cylindric, opening from the base upwards. Seeds finely pitted, appendaged. *USA (California), adjacent Mexico.*

20. Pteridophyllum Zuccarini. 1/1. Perennial herb with rhizomes. Leaves all basal, dark green above, paler beneath, to 15 cm, oblong-obovate to narrowly elliptic, pinnately divided into numerous oblong leaflets, blunt and faintly toothed at the tip, truncate at the base, each with a tooth on the upper margin near the base; all with sparse adpressed bristles above and beneath. Scape to 30 cm, bearing a long raceme containing many flowers, the flower-stalks often arising in 2s or 3s at the lower nodes; bracts very small. Sepals 2, very small and very quickly deciduous. Petals 4, to 8 mm, white, spreading, all more or less similar or one pair narrower than the other pair. Stamens 4. Ovary of 2 carpels, 1-celled but with a style and 2 short stigmas and 2--4 ovules. Fruit a capsule opening by the sides falling from the central placental framework; seeds usually 2. *Japan.*

21. Hypecoum Linnaeus. 10/few. Annual herbs, often glaucous. Leaves in a basal rosette, deeply 2--4 times pinnate into narrow segments. Flowers in cymes, bracts similar to the leaves. Sepals 2, soon falling. Petals 4 in 2 opposite pairs, the outer pair entire to 3-lobed, broadly diamond-shaped, the inner pair each deeply 3-fid, the lateral lobes linear, spreading, the central lobe stalked, the blade folded, those of the pair enclosing the anthers and stigmas. Stamens 4, filaments winged towards the base. Ovary cylindric, of 2 carpels, 1-celled, containing several ovules, and with 2 styles. Fruit linear-cylindric, breaking into 1-seeded joints. *Mediterranean area to China.*

22. Dicentra Bernhardi. 19/10. Hairless perennial herbs with rhizomes, tubers or bulblets, sometimes climbing by means of tendrils which replace some leaflets. Leaves all basal or borne on the stem, usually 1--3 times pinnate or 2--3 times divided into 3, the ultimate segments often lobed or toothed. Flowers solitary or in axillary or terminal racemes or panicles, usually pendent, subtended by a bract and, often, 2 bracteoles. Flowers compressed, with 2 distinct planes of symmetry. Sepals 2, usually deciduous in flower. Petals 4, in 2 pairs, all variably united; outer pair large, enclosing the inner except at the apex, pouched or swollen at the base with free tips which are usually spreading or reflexed; inner pair clawed, crested towards the apex. Stamens 4, in 2 bundles of a central stamen attached to 2 half-stamens; filaments united to varying extents. Nectar secreted by the bases of the filaments or by a spur projecting backwards into the pouched part of the outer petal. Ovary of 2 carpels, 1-celled, with several ovules. Fruit a capsule, with persistent style. Seeds usually with an oily appendage. *North America, eastern Asia.*
The bizarrely-shaped flowers are easily recognisable.

23. Adlumia de Candolle. 2/1. Annual or biennial herbs, scrambling and climbing among other vegetation; stems brittle. Leaves 2- or 3-pinnate, the ultimate segments ovate or elliptic, entire or variously lobed and toothed; leaflet-stalks tendril-like, coiling around other vegetation. Flowers numerous, pendent, each subtended by a small bract, in axillary panicles whose stalks are united at the base for a short distance with those of the subtending leaves. Flowers with 2 distinct planes of symmetry. Sepals 2, toothed. Petals 4, united for most of their length to form a long urn-shaped corolla, the outer pair enclosing the inner pair except for their tips, swollen but not spurred at the base. Stamens 4 in 2 bundles of a whole stamen united to 2 half-stamens, each bundle united to the corolla for most of its length. Ovary with few ovules; stigma 4-lobed. Fruit a few-seeded capsule surrounded by the detached but persistent corolla. Seeds black, each with an appendage. *Eastern North America, eastern Asia.*

24. Corydalis Ventenat. 300/25. Annual to perennial, usually hairless herbs, rarely climbing; rootstock a taproot, a cluster of fleshy roots, a rhizome or a tuber. Stems leafy or not. Leaves basal, alternate or opposite, usually deeply divided (pinnate or successively divided into 3s) 2--4 times, usually thin and delicate, the terminal leaflet rarely modified into a branched tendril. Flowers in racemes (rarely aggregated into panicles), bracts entire or lobed towards the apex. Sepals 2, persistent in flower, usually somewhat peltate and toothed. Petals 4 in 2 pairs, variously united, the outer pair more or less enclosing the inner pair except at the tips; upper petal with a long or short spur, lower petal sometimes swollen at the base. Stamens in 2 groups of a whole stamen united to 2 half-stamens, the upper group with a nectary at the base which projects backwards into the petal-spur. Ovary of 2 carpels, with 2--many ovules. Fruit

205

a capsule containing 2--several seeds, opening by the sides separating from the persistent placental framework; style usually persistent. Seeds usually with a whitish or brownish appendage (caruncle, arilloid). *North temperate areas.*
A broad, traditional concept of the genus is used here.

25. Sarcocapnos de Candolle. 3/2. Cushion- or clump-forming annual to perennial herbs, leaves and stems fleshy and brittle, often glaucous. Leaves long-stalked, undivided or 1--3 times divided into 3. Flowers in corymb-like racemes with small bracts. Sepals 2, small. Petals 4, white or pink, the upper petal with a backwardly-projecting spur and the lower with a long claw and notched blade; lateral petals with yellow lateral wings, blotched internally with blackish red at the apex. Stamens 4 in 2 bundles (upper and lower) of 1 stamen plus 2 half-stamens, the upper with a nectary projecting back into the petal-spur. Ovary with 2 ovules; stigma expanded with receptive tissue along the margins. Fruit a 1- or 2-seeded nutlet with an apical appendage and ribbed faces, the fruit-stalks lengthening and bent downwards or inwards. *Spain, North Africa.*

26. Rupicapnos Pomel. 7/1. Annual or perennial (often short-lived) herbs forming clumps or loose cushions, stems and leaves very brittle. Leaves long-stalked, pinnate, segments deeply lobed or toothed. Flowers in flat-topped, shortly stalked racemes, each flower subtended by a small bract. Sepals 2, often lobed or toothed. Petals 4, white or pink, the upper with a small, purple-marked apex (ours) and a short, blunt, downwardly curved spur, the lower oblong-linear, greenish at the acute apex, the laterals winged, with purple apices. Stamens in 2 bundles (upper and lower) of a central stamen united to the anther with 2 half-stamens, the upper with a nectary projecting back into the petal-spur. Ovary with 1 ovule; stigma a flattened dome with lobed margins. Fruit a nutlet, the fruit-stalks lengthening and growing downwards or inwards, burying the fruit. *South Spain, North Africa.*

27. Fumaria Linnaeus. 55/1. Annual or perennial herbs with long, erect, spreading or scrambling stems. Leaves 2--4 times pinnately divided; leaf-segments flat or channelled, oblong-lanceolate to broadly ovate. Flowers in racemes subtended by bracts, or without bracts; sepals 2; corolla bilaterally symmetric, consisting of 1 spurred upper petal, 1 lower petal and 2 inner petals; ovary with 2 ovules, stigma crested or with 2 papillae. Fruit an indehiscent capsule with 2 apical pits, seeds 1--13 (in the species covered here), light brown or black. *Europe to Central Asia & the Himalaya, south to the east African highlands.*

77. CAPPARIDACEAE
Deciduous trees, shrubs or annuals. Leaves alternate, simple, with 3 leaflets or palmate. Stipules present or absent. Flowers bisexual, somewhat bilaterally symmetric, solitary or in terminal racemes. Sepals 4, free or united in the lower part. Petals 4, usually stalked. Stamens *c.* 6 or numerous. Ovary superior, stalked, the stalk lengthening in fruit. Fruit 1-celled, of 2 united carpels. Seeds few to many.
A mainly tropical family of 45 genera and about 700 species. Few species are cultivated.

1a. Leaves simple	**1. Capparis**
b. Leaves compound	**2. Cleome**

1. Capparis Linnaeus. 250/1. Shrubs with trailing or ascending branched stems to 2 m. Leaves ovate to almost circular, entire, rounded or cordate at base, obtuse to acuminate, mucronate, leathery or somewhat succulent, with or without hairs. Stipules rigid or bristle-like, spiny, often recurved. Flowers solitary in leaf-axils; stalks stout, longer than leaves. Sepals 4, broadly ovate, the outer pair longer, concave. Petals broadly obovate, obtuse, white, often tinged pink. Stamens numerous, longer than petals, purplish, especially towards the anthers; anthers purple. Stigma stalkless. Ovary stalked, the stalk lengthening in fruit. Fruit ovoid or ellipsoid, ribbed, leathery. Seeds kidney-shaped, smooth, brown. *Tropics; a few in north temperate areas.*

2. Cleome Linnaeus. 150/3. Erect annual or shrubby perennials, often with glandular hairs. Leaves compound. Stipules absent but leaf-stalks sometimes with a solitary spine at base. Flowers in terminal, bracteate racemes. Sepals 4; petals 4, entire; stamens usually 6; ovary stalked, the stalk lengthening considerably in fruit. Fruit a 2-celled, many-seeded capsule. Seeds kidney-shaped, smooth, brown. *Mainly tropics, a few in North temperate areas.*

One species of **Polanisia** Rafinesque (*eastern North America*), with 8--many stamens and notched petals, is occasionally found in specialist collections

78. CRUCIFERAE (*Brassicaceae*)

Annual to perennial herbs, rarely small shrubs. Leaves alternate, rarely opposite, without stipules. Flowers usually bisexual, sepals, petals and stamens hypogynous. Sepals 4, free, in two opposite pairs. Petals 4, free, usually clawed, alternating with the sepals. Stamens usually 6, rarely 4, 2 or absent, outer pair with short filaments, two inner pairs with long filaments; filaments sometimes winged or with tooth-like appendage. Nectaries variable and variously arranged around base of stamens and ovary. Ovary of 2 fused carpels, usually with 2 parietal placentas and 2-celled by the development of a membranous false septum formed from outgrowths of the placentas. Stigma capitate to 2-lobed. Fruit usually a dehiscent capsule opening by 2 valves from below; sometimes 1 or more seeds develop in an indehiscent beak at the base of the style. Fruit called a siliqua when at least 3 times as long as wide or a silicula if less than 3 times as long as wide; sometimes indehiscent or a lomentum breaking transversely into 1-seeded portions. Seeds often mucilaginous when wet. See figures 44, p. 210, 45, p. 212 & 46, p. 218.

A family containing about 3000 species in 390 genera, cosmopolitan but chiefly in north temperate regions.

Identification of genera requires knowledge of fruit characters, petal-colour and hairs type. A hand lens (× 10 or more) is necessary to see the hairs, and both leaves and stem should be examined: not only may hairs be commoner on leaves than stems, but different types may also occur on the different organs. Petal colour often changes on drying.

Key to groups

1a. Fruit less than 3 times as long as broad	*Group 1*
b. Fruit at least 3 times as long as broad	*Group 2*

Group 1

1a. Fruit with 2 segments, the upper flat, leaf-like, or with a conspicuous transverse appendage below the tip   2

b. Fruit without a terminal flat, leaf-like segment nor a conspicuous transverse appendage below the tip   3

2a. Small perennial shrubs; hairs unbranched   **51. Vella**

b. Annuals, though becoming woody; hairs stellate   **13. Anastatica**

3a. Fruit inflated, bladder-like   4

b. Fruit not inflated and bladder-like   10

4a. Petals white, pink or purple; hairs medifixed   **29. Lobularia**

b. Petals yellow; hairs stellate or absent   5

5a. Petals short-clawed   **36. Cochlearia**

b. Petals entire, distinctly clawed   6

6a. Leaves hairless   **38. Coluteocarpus**

b. Leaves with stellate hairs   7

7a. Fruits less than 8 mm   8

b. Fruits at least 10 mm   9

8a. Fruits 5--20 mm, usually 2-lobed or at least with a prominent apical sinus 2--4 mm deep   **35. Physaria**

b. Fruits less than 7 mm, never 2-lobed, usually without an apical sinus, or the sinus not more than 1 mm deep   **34. Lesquerella**

9a. Petals 1.4--1.5 mm; seeds 2 in each cell   **24. Degenia**

b. Petals 1.5--2 cm; seeds 4--8 in each cell   **23. Alyssoides**

10a. Plant hairless, or with hairs only on leaf-veins beneath   11

b. Plant hairy   30

11a. Petals yellow or creamy yellow   12

b. Petals white, pink, purple or mauve   15

12a. Fruits 2-lobed   **44. Biscutella**

b. Fruits not 2-lobed   13

13a. Fruits flattened and winged   **42. Aethionema**

b. Fruits flattened or not, wings absent   14

14a. Fruits dehiscent, seeds few to many   **32. Draba**

b. Fruits indehiscent, 1-seeded   **52. Crambe**

15a. At least some leaves pinnatifid, or pinnate   16

b. All leaves entire or with shallow lobes   21

16a. Plant annual; sepals pouched at base   **56. Heliophila**

b. Plant perennial; sepals not pouched at base   17

17a. Fruits strongly compressed at right-angles to septum   **39. Pritzelago**

b. Fruits not compressed, appearing ovoid to spherical   18

18a. Herbaceous perennials, robust, to 1.5 m   19

b. Annuals, perennial herbs or small shrubs to 70 cm, usually much smaller   20

19a. Styles short; seeds in 2 rows in each cell, though rarely produced   **16. Armoracia**

b. Styles absent; fruits 1-seeded   **52. Crambe**

20a. Style distinct; stigma capitate   **43. Iberis**

b. Style inconspicuous   **46. Lepidium**

21a. Straggling climber to 5 m   **56. Heliophila**

b. Plant not a climber   22

22a. Fruits compressed   23

b. Fruits not compressed   28

23a. Some sepals pouched at the base   24
   b. Sepals not pouched   25
24a. Outer sepals pouched   **42. Aethionema**
   b. Inner sepals pouched   **32. Draba**
25a. Leaves smelling of garlic when bruised   **41. Pachyphragma**
   b. Leaves not smelling of garlic when bruised   26
26a. Stamen-filaments expanded at the base   **32. Draba**
   b. Stamen-filaments not expanded, with or without appendages   27
27a. Stamen-filaments with a tooth-like appendage at the base; seeds 1 or 2
    in each cell   **28. Bornmuellera**
   b. Stamen filaments without an appendage; seeds 2--8, rarely only 1 in
    each cell   **40. Thlaspi**
28a. Style absent; seeds in 1 row in each cell   **52. Crambe**
   b. Style short; seeds in 2 rows in each cell   29
29a. Basal leaves 30--50 cm, ovate or ovate oblong   **16. Armoracia**
   b. Basal leaves variable, but usually 5--20 cm   **36. Cochlearia**
30a. Hairs simple   31
   b. Some hairs forked, branched or stellate   44
31a. Fruits compressed, with the septum across the widest diameter   32
   b. Fruits not compressed with the septum across the narrowest diameter   37
32a. Leaves palmately 3--5-lobed   **33. Petrocallis**
   b. Leaves simple to deeply toothed, or lyrate to pinnate   33
33a. Petals 1--2.5 cm   **21. Lunaria**
   b. Petals 1--8 mm, rarely to 1.4 cm   34
34a. Petals usually clawed, often with basal appendages   **56. Heliophila**
   b. Petals not clawed, without basal appendages   35
35a. Seeds with a membranous border   **37. Kernera**
   b. Seeds not bordered   36
36a. Basal leaves simple, entire or toothed   **32. Draba**
   b. Basal leaves long-stalked, lyrate, lobes toothed   **46. Lepidium**
37a. Fruits deeply notched, 2-lobed   38
   b. Fruits not deeply notched or 2-lobed   39
38a. Style long   **44. Biscutella**
   b. Style absent, stigma stalkless   **45. Megacarpaea**
39a. Outer 2 petals much larger than inner   **43. Iberis**
   b. Petals equal   40
40a. Leaves lobed or pinnate; seeds 1 in each cell   41
   b. Leaves simple; seeds 2--numerous, rarely only 1 in each cell   42
41a. Petals more than 3 mm, with a short claw   **52. Crambe**
   b. Petals to 3 mm, without a claw   **46. Lepidium**
42a. Fruit not strongly compressed or winged   **36. Cochlearia**
   b. Fruit strongly compressed and winged   43
43a. Leaves fleshy; seeds numerous in each cell   **47. Notothlaspi**
   b. Leaves not fleshy; seeds 2--8, rarely only 1 in each cell   **40. Thlaspi**
44a. Fruit almost spherical to obovoid or ellipsoid, not compressed   45
   b. Fruits compressed   46
45a. Hairs stellate   **34. Lesquerella**
   b. Hairs medifixed   **29. Lobularia**

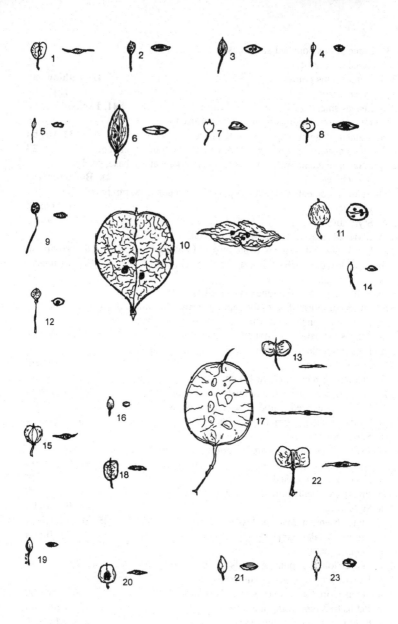

Figure 44. Cruciferae: siliculas. 1, *Aethionema*. 2, *Bornmuellera*. 3, *Cochlearia*. 4, *Kernera*. 5, *Armoracia*. 6 & 10, *Coluteocarpus* (different variants). 7, *Alyssum*. 8, *Aurinia*. 9, *Ionopsidium*. 11, *Alyssoides*. 12, *Crambe*. 13, *Biscutella*. 14, *Lobularia*. 15, *Thlaspi*. 16, *Petrocallis*. 17, *Lunaria*. 18, *Iberis*. 19, *Schivereckia*. 20, *Nothothlaspi*. 21, *Draba*. 22, *Pachyphragma*. 23, *Berteroa*.

46a. Fruits with septum across the narrowest diameter   47
  b. Fruits with septum across the widest diameter   49
47a. Fruits deeply notched, 2-lobed   **35. Physaria**
  b. Fruits not notched, or 2-lobed   48
48a. Leaves simple   **20. Aubrieta**
  b. Leaves pinnatisect   **39. Pritzelago**
49a. Petals yellow   50
  b. Petals white, pink or purple   53
50a. Inner sepals strongly pouched   **25. Fibigia**
  b. Inner sepals not or only very slightly pouched   51
51a. Lower leaves 4--10 cm, rarely only 2 cm, with persistent and swollen
    bases   **27. Aurinia**
  b. Lower leaves 0.2--1.5 cm, rarely to 2 cm, never with persistent and
    swollen bases   52
52a. Dwarf perennials with a distinct scape   **32. Draba**
  b. Annuals to perennials, without a scape   **26. Alyssum**
53a. Stamen-filaments not pouched, toothed or winged at base   54
  b. Stamen-filaments pouched or winged at base   55
54a. Seed 1 in each cell of the fruit   **29. Lobularia**
  b. Seeds in 2 rows in each cell   **32. Draba**
55a. At least some hairs stellate   56
  b. At least some hairs forked or simple   57
56a. Sepals erect to spreading, not pouched at base; petals often notched
    26. Alyssum
  b. Sepals spreading, slightly pouched at base; petals never notched
    31. Schivereckia
57a. Leaf-stalks of basal leaves with persistent and swollen bases   **27. Aurinia**
  b. Leaf-stalks of basal leaves never with persistent or swollen bases   58
58a. Petals deeply bifid   **30. Berteroa**
  b. Petals entire or slightly notched   59
59a. Plant herbaceous, mat-forming   **20. Aubrieta**
  b. Plant caespitose, or shrubby at base   60
60a. Seeds 1 or 2 in each cell   **28. Bornmuellera**
  b. Seeds 4--7 in each cell   **31. Schivereckia**

Group 2
  1a. Plants stemless; fruits stalkless, lengthening and curving down to bury the
    fruits in the soil   2
  b. Plants with obvious stem; fruits not lengthening and curving down to bury
    the fruits in the soil   3
  2a. Leaves pinnatisect with 1--4 pairs of lobes   **55. Raffenaldia**
  b. Leaves pinnatisect with 10 or more pairs of lobes   **53. Morisia**
  3a. Fruits pendent   4
  b. Fruits erect or spreading   5
  4a. Petals yellow   **8. Isatis**
  b. Petals lilac   **22. Ricotia**
  5a. Hairless or with unbranched hairs only   6
  b. At least some hairs forked, branched or stellate   24

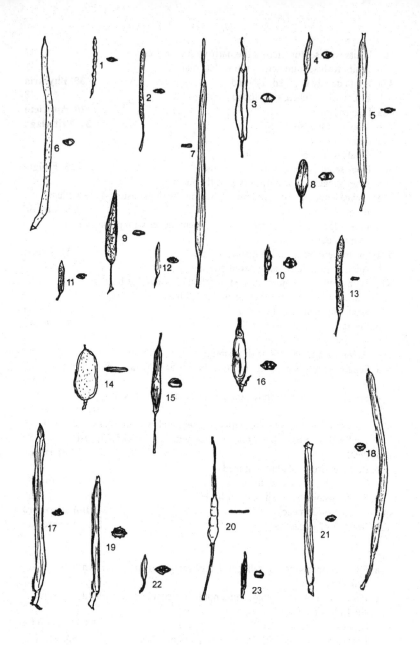

Figure 45. Cruciferae: siliquas. 1, *Heliophila*. 2, *Arabis*. 3, *Brassica*. 4, *Aubrieta*. 5, *Erysimum*. 6, *Malcolmia*. 7, *Orychophragmus*. 8, *Isatis*. 9, *Phoenicaulis*. 10, *Vella*. 11, *Nasturtium*. 12, *Hugueninia*. 13, *Schizopetalon*. 14, *Fibigia*. 15, *Cardamine*. 16, *Eruca*. 17, *Moricandia*. 18, *Hesperis*. 19, *Alliaria*. 20, *Raphanus*. 21, *Matthiola*. 22, *Dielsiocharis*. 23, *Barbarea*.

212

6a. Valves of fruit dehiscing explosively from the base   **18. Cardamine**
  b. Fruits not dehiscing explosively   7
7a. Plant with stout rhizomes 1--3 cm thick   **6. Wasabia**
  b. Rhizomes absent   8
8a. Fruit beaked with an upper seedless or seeded part and a lower (some-
    times very short) segment   9
  b. Fruit not beaked, not divided into 2 parts, though sometimes con-
    stricted between the seeds   12
9a. Fruit breaking transversely into 2 or more segments, not dehiscing by
    valves   **54. Raphanus**
  b. Fruit not breaking transversely into segments, dehiscing by valves   10
10a. Stigma capitate, inconspicuously 2-lobed   11
  b. Stigma with decurrent lobes joining the fruit, prominently 2-lobed
    50. Eruca
11a. Plant aquatic; stigma small   **17. Nasturtium**
  b. Plant terrestrial; stigma large   **49. Brassica**
12a. Petals yellow   13
  b. Petals white, pink, lilac, violet or blue   16
13a. Stem leaves auriculate and clasping the stem   **14. Barbarea**
  b. Stem leaves not auriculate or clasping the stem   14
14a. Stigma with decurrent lobes   **10. Hesperis**
  b. Stigma more or less capitate; style without wings   15
15a. Lower leaves pinnate   **2. Sisymbrium**
  b. Leaves all simple   **32. Draba**
16a. Basal leaves simple   17
  b. Basal leaves with 3 leaflets, pinnatifid, pinnate or bipinnate   23
17a. Petals white   18
  b. Petals coloured   20
18a. Plant smelling of garlic when crushed   **5. Alliaria**
  b. Plant not smelling of garlic when crushed   19
19a. Fruits less than 2 cm   **32. Draba**
  b. Fruits 2.5 cm or more   **56. Heliophila**
20a. Petals to 1.3 cm   21
  b. Petals 1.4 cm or more   22
21a. Fruits less than 2 cm   **32. Draba**
  b. Fruits 2.5 cm or more   **56. Heliophila**
22a. Leaves very glaucous and somewhat fleshy   **48. Moricandia**
  b. Leaves not glaucous or fleshy   **10. Hesperis**
23a. Perennial semi-aquatic herbs, often rooting at lower submerged nodes
    17. Nasturtium
  b. Not aquatic, not rooting at nodes   **56. Heliophila**
24a. Stigma deeply 2-lobed, the lobes sometimes erect and joined to form a
    beak on the fruit   25
  b. Stigma capitate, notched or slightly 2-lobed   27
25a. Stigma-lobes with a dorsal swelling or horn   **12. Matthiola**
  b. Stigma-lobes without a dorsal swelling or horn   26
26a. Style short, stigma-lobes free   **10. Hesperis**
  b. Style absent, stigma-lobes erect and joined   **11. Malcolmia**

27a. Leaves 2-pinnatisect   **4. Hueguenina**
   b. Leaves entire to 1-pinnatisect   28
28a. Petals yellow   29
   b. Petals white, pink, purple, orange, red, cream or violet   30
29a. Tufted perennials, rarely annuals, often forming tight cushions, seeds in
    2 rows   **32. Draba**
   b. Erect, herbaceous annual to perennial herbs, sometimes slightly woody;
     seeds in 1, rarely 2 rows   **9. Erysimum**
30a. Plant covered with medifixed hairs only   **9. Erysimum**
   b. Hairs various, but not all medifixed   31
31a. At least some leaves pinnatisect or pinnately divided   32
   b. Leaves simple, entire, or with a few teeth   33
32a. Petals pinnately lobed   **1. Schizopetalon**
   b. Petals notched, but not pinnately lobed   **3. Murbeckiella**
33a. Seeds in 1 row in each cell   34
   b. Seeds in 2 rows in each cell   35
34a. Fruit constricted between the seeds   **7. Braya**
   b. Fruit not constricted between the seeds   **19. Arabis**
35a. Inner sepals distinctly pouched   36
   b. Inner sepals not pouched   37
36a. Stem-leaves auriculate, clasping the stem, rarely all leaves basal **19. Arabis**
   b. Stem-leaves not auriculate, or clasping the stem, leaves never all basal
                         20. Aubrieta
37a. Stem-leaves auriculate, clasping the stem   **19. Arabis**
   b. Stem-leaves not auriculate, or clasping the stem   38
38a. Stigma 2-lobed   **7. Braya**
   b. Stigma capitate   39
39a. Fruits less than 2 cm   **32. Draba**
   b. Fruits 2--8 cm   **15. Phoenicaulis**

1. Schizopetalon Sims. 5/1. Annual herbs, downy with stellate hairs. Stems erect. Leaves alternate, simple to pinnately divided; margin wavy, toothed. Flowers in terminal racemes. Sepals erect, green; petals clawed, pinnately lobed, white to purple. Fruit a siliqua. *Chile*. Figure 45, p. 212.

2. Sisymbrium Linnaeus. 80/few. Annual to perennial herbs; hairless or with simple hairs. Leaves entire to pinnate. Sepals not pouched at base. Petals yellow, rarely white, entire. Fruit a siliqua; valves usually 3-veined; stigma more or less 2-lobed. Seeds usually less than 2.5 mm. *Eurasia, New World.*

3. Murbeckiella Rothmaler. 4/1. Perennial, hairless herbs with stellate, or long, unbranched hairs. Leaves entire to pinnatisect. Sepals unequal, the inner more or less pouched at base. Petals white, notched. Fruit a siliqua; valves with a distinct central vein; style very short; stigma slightly 2-lobed. Seed not more than 1.5 mm, often winged at apex. *Southwest Europe, North Africa, Turkey, Caucasus.*

4. Hugueninia Reichenbach. 1/1. Stout hairless to densely grey-hairy herbaceous perennial, 30--70 cm. Leaves 2-pinnatisect; lower to 30 cm, long-stalked; leaf-segments in 8--10 pairs, broadly linear to lanceolate, toothed to pinnatifid, ultimate lobes 4--10 mm wide. Flowers in crowded terminal racemes, lengthening in fruit. Petals *c.* 4 mm, yellow, exceeding sepals. Fruit a siliqua, 6--15 × 1.2--1.5 mm, erect to patent; valves strongly 1-veined; style very short; stigma slightly 2-lobed. Seeds *c.* 2 mm, unwinged, slimy when wet. *Alps, Pyrenees, mountains of Spain.* Figure 45, p. 212.

5. Alliaria Scopoli. 5/1. Annual or biennial herbs smelling of garlic, with simple hairs. Basal leaves simple. Sepals not pouched at base; petals with a short claw. Fruit a siliqua, usually 4-angled, unbeaked, lobes 3-veined, dehiscent; septum thin; style distinct; stigma shallowly 2-lobed; seeds large, *c.* 3 mm, winged, not mucilaginous. *Europe, North Africa, Asia east to Japan.* Figure 45, p. 212.

6. Wasabia Matsumura. 4/1. Perennial herbs with rhizomes; stems leafy, simple, decumbent or ascending. Basal leaves more or less circular, cordate with scalloped margins and long stalks with a broad sheathing base; stem-leaves much smaller, with short stalks and fewer, deeper teeth or lobes. Flowers small, white in simple racemes, at least the lowest with bracts. Petals obovate to oblong, each with a short claw. Fruit cylindric on a slender spreading or recurved stalk; stigma simple; seeds up to 8 in one row. *Eastern Asia.*

7. Braya von Sternberg & Hoppe. 20/1. Perennial, tufted herbs with both branched and unbranched hairs. Leaves undivided. Flowers small. Sepals not pouched at base; petals white or purplish, truncate. Fruit a siliqua or a silicula; valves 1-veined; style short; stigma slightly 2-lobed. Seeds in 1 or 2 rows in each cell, *c.* 1 mm. *Northern hemisphere, scattered.*

8. Isatis Linnaeus. 40/3. Annuals, biennials or perennials; hairless or with simple hairs. Stems erect, branched above, leafy. Sepals not pouch-like. Petals with a short claw, yellow. Fruit an indehiscent siliqua, flattened, winged, 1-seeded. *Mediterranean area, Middle East, central Asia.* Figure 45, p. 212.

9. Erysimum Linnaeus. 80/10 and some hybrids. Annuals, biennials or perennials, sometimes small shrubs, with forked or branched hairs. Leaves narrow, entire or toothed. Sepals erect, the inner usually pouched at base. Petals usually yellow. Fruit a siliqua. Seeds in 1, rarely 2, rows in each cell. *Eurasia, North America.* Figure 45, p. 212.
   This genus includes those plants formerly included in *Cheiranthus* Linnaeus.

10. Hesperis Linnaeus. 30/4. Biennial or perennial herbs with erect leafy stems; hairless or with simple, branched or glandular hairs. Leaves ovate to spathulate, entire to pinnatifid. Flowers usually with bracts, in racemes or panicles, sometimes strongly fragrant. Sepals erect, the inner pair with a pouch at the base. Petals with a long claw, white to violet, yellowish, greenish or brown. Fruit a slender siliqua, valves 3-nerved, style short; stigma with 2 decurrent lobes; seeds in 1 row in each cell. *West Europe to China.* Figure 45, p. 212.

11. Malcolmia R. Brown. 30/4. Annuals, biennials or perennials, hairs 2--4-branched. Leaves simple. Sepals erect, the 2 inner usually pouch-like. Petals each with a long claw, pink, purple or violet. Fruit a long, slender siliqua; valves 3-veined; stigma deeply 2-lobed. Seeds in 1 row in each cell. *Mediterranean area to Central Asia.* Figure 45, p. 212.

12. Matthiola R. Brown. 40/6. Annuals, biennials or perennials, sometimes almost shrubs, with branched or stellate hairs. Leaves oblong to linear, entire, sinuous or pinnatifid. Flowers conspicuous, fragrant. Sepals erect, the 2 inner pouch-like. Petals with a long claw. Fruit a long siliqua, usually hairy, the valves usually 1-veined; stigma-lobes persistent, often horn-like. Seeds many in each cell, broadly winged. *Mediterranean area.* Figure 45, p. 212.

13. Anastatica Linnaeus. 1/1. Much-branched annual becoming woody, 3--15 cm. Stems densely covered in stellate hairs. Leaves oblong to diamond-shaped, simple, toothed, 1--4 cm. Flowers numerous, almost stalkless, in short dense stalkless clusters in the angles of the branches. Petals white. Ovary stalkless, green, covered in stellate hairs, 4-seeded; style shorter than ovary, stigma large, bilobed. Fruit a hard silicula, ovoid to almost spherical, beaked; lobes strongly convex with a transverse appendage below the tip. *Mediterranean area.*

14. Barbarea R. Brown. 12/2. Biennial or perennial, erect branching herbs, hairless or with unbranched hairs. Basal leaves stalked, usually lyrate, rarely entire, forming a rosette. Upper leaves stalkless. Flowers in a dense raceme, much elongated in fruit. Petals small, yellow with a short claw. Filaments without appendages. Fruit a siliqua, 4-angled, round or flattened; style distinct, stigma small; seeds in 1 row in each cell. *North temperate areas.* Figure 45, p. 212.

15. Phoenicaulis Nuttall. 1/1. Perennial tufted herbs. Stems to 20 cm, covered in the remains of old leaf-stalk bases. Leaves in dense basal rosettes, 3--10 cm, grey with stellate hairs, with an entire margin. Inflorescence a showy raceme of small flowers. Sepals *c.* 4 mm. Petals long-clawed, 8--10 mm, pinkish, purple or rarely white. Fruit a siliqua, 2--8 cm × 2--6 mm, compressed, oblong-lanceolate, often more or less sickle-shaped, hairless. *Western North America.* Figure 45, p. 212.

16. Armoracia Gaertner et al. 3/1. Hairless perennial herbs with erect leafy branching stems. Rootstock stout with fleshy, cylindric roots. Leaves simple to pinnatifid; basal leaves very large with long stalks; stem-leaves much smaller, with short stalks or stalkless. Sepals without a basal pouch. Petals small, white with a short claw. Filaments broadened but without appendages. Fruit a spherical to ellipsoid silicula; seeds in 2 rows in each cell; style short; stigma large, scarcely lobed. *Balkans, southwest and northeast Asia.*

17. Nasturtium R. Brown. 2/2. Perennial, aquatic, almost completely hairless herbs. Stem often rooting at lower, submerged nodes. Leaves mostly pinnate. Petals white, longer than sepals. Median nectaries absent. Fruit a siliqua; midrib of valves somewhat inconspicuous; style short, beak seedless. Seeds brown, with net-veined patterns on surface, arranged in 1 or 2 rows in each cell. *Eurasia.* Figure 45, p. 212.

18. Cardamine Linnaeus. 200/11. Herbaceous perennials, usually with rhizomes and erect, leafy stems; hairless or with simple hairs; evergreen or deciduous. Leaves simple,

216

palmate or pinnate. Flowers in a compact terminal raceme or leafy panicle, lengthening in fruit. Petals white, pink or purple, rarely pale yellow. Fruit a linear or lanceolate flattened siliqua, the valves dehiscing elastically from the base to eject the seed. *Cosmopolitan but mostly in north temperate areas.* Figure 45, p. 212.

19. Arabis Linnaeus. 120/17 and some hybrids. Annuals, biennials or perennials; hairs forked, branched or stellate, rarely the plant hairless. Leaves simple. Inner sepals often pouched at base. Petals usually white or pink. Fruit a slender siliqua; valves with or without a median vein. Seeds in 1 row, rarely 2, in each cell, usually winged. *Europe, Southwest Asia, Mediterranean area.* Figures 45, p. 212 & 46, p. 218.

20. Aubrieta Adanson. 15/5. Perennials, usually forming mats; hairs stellate, forked or simple; stems short, weakly ascending. Leaves simple, entire or with a few teeth. Inner sepals pouch-like. Petals with a long-claw, pink, purple or violet. Filaments somewhat winged, the outer ones toothed. Fruit a siliqua or silicula. Seeds in 2 rows in each cell. *Sicily to Iran.* Figure 45, p. 212.

Most cultivated plants are variants or hybrids of *A. deltoidea* (Linnaeus) de Candolle.

21. Lunaria Linnaeus. 3/2. Erect, branching annual to perennial herbs. Leaves ovate or ovate-triangular, toothed. Flowers violet to purple, sometimes white, in terminal racemes. Petals long-clawed, sepals erect. Fruit with a persistent round, flat, papery septum. *Central & south-east Europe, naturalised elsewhere.* Figure 44, p. 210.

22. Ricotia Linnaeus. 9/2. Annuals or perennials, branched, grey, usually hairless. Leaves with 3 leaflets or pinnatifid. Sepals pouch-like. Petals lilac. Fruit a compressed siliqua. Seeds in 1 or 2 rows in each cell. *East Mediterranean area, southwest Asia.*

23. Alyssoides Miller. 2/1. Perennials, woody at the base; hairs branched to stellate. Leaves crowded, mostly in non-flowering rosettes, entire. Sepals erect to somewhat spreading, the inner pouch-like at the base. Petals with long claws. Fruit a strongly inflated silicula, the valves without a conspicuous vein; style long. Seeds 4--8 in each cell, winged. *Southeast Europe, Turkey.* Figure 44, p. 210.

24. Degenia Hayek. 1/1. Perennial herbs to 10 cm. Leaves linear-lanceolate in rosettes, some rosettes non-flowering. Flowers 1--1.2 cm. Petals yellow, 1.4--1.5 cm, long- clawed. Style long. Fruit a silicula, 1--1.4 cm × 7--8 mm, ovate, inflated, densely covered with stellate hairs. Seeds 2 in each cell, broadly winged, 5 mm across. *Northwest former Yugoslavia.*

25. Fibigia Medikus. 14/1. Erect perennial herbs and small shrubs. Leaves alternate, linear to spathulate. Plant covered with stellate hairs (occasionally a few simple). Sepals erect. Petals yellow, shortly clawed. Fruit a strongly compressed silicula. Seeds 2--8 in each cell, winged. *Mediterranean area to Afghanistan.* Figure 45, p. 212.

26. Alyssum Linnaeus. 150/20. Annuals, biennials or herbaceous to shrubby perennials; hairs simple, branched or stellate, sometimes scaly. Basal and stem leaves usually similar. Flower-buds ovoid or ellipsoid. Sepals erect to spreading, not pouch-like. Petals yellow, white or rarely purple, entire to notched or bifid. Filaments

217

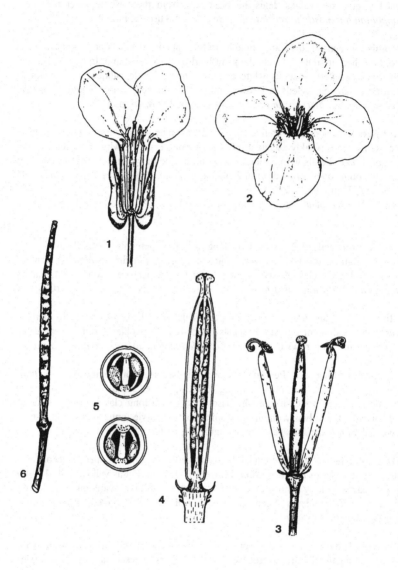

Figure 46. Cruciferae. *Arabis caucasica.* 1, Longitudinal section of flower. 2, Whole flower from above. 3, Ovary and two long stamens. 4, Longitudinal section of ovary. 5, Transverse section of ovary (two slightly differing views). 6, Fruit (siliqua).

218

of long stamens usually winged, those of short stamens usually with appendages. Fruit a silicula; valves with short, inconspicuous mid-veins; style often short. Seeds 1 or 2, rarely up to 6, in each cell, usually distinctly winged. *Europe, Mediterranean area, Asia.* Figure 44, p. 210.

The genus *Ptilotrichum* Meyer is here included within *Alyssum*, although some other species formerly included in *Alyssum* in horticultural texts have been removed to *Alyssoides, Aurinia, Lobularia* and *Schivereckia.*

27. Aurinia Desvaux. 12/5. Perennial herbs, rather woody at the base; stems somewhat woody, spreading or ascending, branched, forming a mat. Hairs simple and branched. Basal leaves forming loose rosettes, distinctly larger than the stem-leaves; stalks with persistent swollen bases. Flowers in corymb-like clusters; buds spherical. Sepals not pouched at base. Petals usually yellow. Fruit a flattened, almost circular silicula; style short. Seeds 2--8 in each cell, usually winged. *South Europe, southwest Asia.* Figure 44, p. 210.

28. Bornmuellera Haussknecht. 6/1. Small shrubby perennials; hairs forked, rarely 4--6-fid or the plant hairless. Sepals erect to spreading, not pouch-like at base. Petals entire, white. Filaments with a tooth-like appendage at base. Fruit a flattened silicula; style short. Seeds 1 or 2 in each cell. *Balkan Peninsula, Turkey.* Figure 44, p. 210.

29. Lobularia Desvaux. 5/1. Annuals or sometimes short-lived perennials; rather hairy with medifixed hairs, greyish. Stems ascending, much-branched near the base. Leaves linear-lanceolate, acute. Flowers fragrant, in dense, rounded clusters lengthening in fruit. Sepals spreading, not pouched at the base. Petals entire. Filaments of stamens not winged and without appendages. Fruit an obovate to almost circular silicula; valves slightly inflated, with more or less distinct mid-vein; style distinct; stigma capitate. Seed 1 in each cell. *Mediterranean area, Macaronesia.* Figure 44, p. 210.

30. Berteroa de Candolle. 8/2. Annuals to perennials; hairs simple, forked or stellate. Leaves more or less entire. Sepals not pouch-like. Petals deeply 2-lobed, white. Outer filaments toothed at base. Fruit a silicula; valves without conspicuous veins. Seeds 2--6 in each cell. *Europe, Mediterranean area, southwest Asia.* Figure 44, p. 210.

31. Schivereckia de Candolle. 2/2. Perennials with minute stellate and forked hairs. Leaves silvery-green; basal leaves in rosettes. Sepals 2--2.5 mm, spreading, the inner slightly pouched. Filaments toothed, the 2 inner winged. Fruit a convex, oblong to ovoid silicula; stigma capitate. Seeds *c.* 1 mm, 4--7 in each cell. *East Europe, west Turkey.* Figure 44, p. 210.

32. Draba Linnaeus. 300/30. Perennial herbs, occasionally annual, usually tuft- or cushion-forming, sometimes straggling. Rootstock woody. Leaves simple, frequently hairy, entire or toothed, often forming basal rosettes. Flowering stems naked or leafy, erect or ascending. Inflorescence a terminal raceme, often corymb-like at first, lengthening in fruit. Flowers usually small; sepals 4, erect or spreading, sometimes slightly pouched; petals yellow or white, longer than sepals, tip rounded, sometimes notched; stamens 6, filaments simple, somewhat expanded at base; ovary with few to many ovules. Fruit a broadly septate, dehiscent silicula or occasionally a siliqua. Seeds in 2 rows, not winged. *North temperate areas, subarctic areas, mountains of South America.* Figure 44, p. 210.

33. Petrocallis R. Brown. 2/1. Tufted perennial herbs resembling many species of *Draba* or *Saxifraga* in habit. Leaves in compact rosettes, palmately 3--5-lobed with simple unbranched hairs. Flowers few on leafless stalks. Fruit a silicula, obovate to elliptic, hairless with 1 seed, rarely 2, in each cell. *South Europe, north Iraq.* Figure 44, p. 210.

34. Lesquerella Watson. 40/2. Low annual to perennial herbs, densely covered with stellate hairs. Basal rosette present; leaves simple to deeply lobed. Inflorescence a raceme, lengthening in fruit. Flowers small, numerous. Fruit a silicula, often inflated, almost spherical to obovoid or ellipsoid; valves without veins, hairless or with stellate hairs. *North America, temperate South America.*

35. Physaria (Nuttall) Gray. 14/7. Perennial, tufted, herbs, silvery with stellate hairs. Stems simple, arising laterally from a somewhat elongated stock. Basal leaves numerous, stalked, oblanceolate to obovate or almost round, entire, toothed or divided into segments. Stem-leaves few, entire to toothed. Flowers in racemes; petals yellow, rarely purplish, spathulate. Fruit a silicula, often broader than long, more or less heart-shaped and notched at apex, often inflated. Seeds 2--6 in each segment, brown, wingless. *Western North America.*

36. Cochlearia Linnaeus. 25/1. Annual, biennial or perennial herbs. Leaves variable, usually simple, stalked stem-leaves often arrow-shaped with auricles. Flowers small in racemes; sepals erect to spreading, petals short-clawed, white, tinged yellow or purple. Fruit a silicula, inflated with convex valves. Seeds numerous. *North temperate areas.* Figure 44, p. 210.

37. Kernera Medikus. 1/1. Perennial, sometimes biennial; stems 10--30 cm, usually branched. Basal leaves in a rosette, stalked, lanceolate to obovate or spathulate, entire to deeply toothed, usually hairy; stem-leaves lanceolate or ovate, often clasping and arrow-shaped at base. Racemes many-flowered; bracts few. Sepals erect to spreading. Petals 2--4 mm, with a short claw, white. Filaments curved. Fruit a swollen siliqua, 2--4.5 mm, ovoid to almost spherical, stalked or not; valves convex, rigid, smooth with a more or less prominent mid-vein in lower part. *Mountains of central & southern Europe.* Figure 44, p. 210.

38. Coluteocarpus Boissier. 1/1. Tufted perennial herbs, 4--20 cm. Rosette leaves simple, oblong-lanceolate, hairless; margins with 3--5 sharp teeth; stem leaves stalkless, entire. Flowers numerous in flat-topped cymes; petals golden-yellow, 6--11 × 2.5--4 mm, clawed. Inflorescence lengthening after flowering; fruit-stalks to 9 mm. Fruit an inflated papery silicula, 1.4--3.3 × 1.6--2.1 cm; style persistent; 1--4 mm. Seeds 2--10. *East Mediterranean area.* Figure 44, p. 210.

39. Pritzelago Kuntze. 1/1. Small, branched perennial herbs to 15 cm; hairless or small, with both branched and unbranched hairs. Basal leaves in a rosette, pinnatisect with 3--9 divisions which are 1--5 mm × 1--3 mm. Sepals small, to 2 mm; petals 3--5

220

mm, white, clawed. Fruit a silicula, elliptic to lanceolate, with a narrow septum; seeds 2--4. *Mountains of central & southern Europe.*

The genus was formerly known as *Hutchinsia* R. Brown.

40. Thlaspi Linnaeus. 60/8. Annuals, biennials or perennials, more or less hairless. Leaves simple; basal leaves usually in a rosette, more or less spathulate; stem-leaves stalkless, usually clasping. Flowers in bractless racemes which lengthen in fruit. Sepals not pouch-like. Petals white, lilac or purple, with a short claw. Filaments without appendages. Fruit a compressed silicula, usually notched at apex; valves winged or keeled. Seeds 2--8, rarely only 1 in each cell. *Mostly Eurasia.* Figure 44, p. 210.

41. Pachyphragma Busch. 1/1. Rhizomatous perennial herb, smelling of garlic when bruised. Basal leaves with stem to 25 cm, the base dilated, persistent; blade 1.5--1.8 cm, ovate to almost circular, hairless except for short hairs on veins beneath. Flowering stems 15--40 cm, erect or almost so; stem-leaves few. Flowers slightly scented, in domed terminal clusters which lengthen in fruit. Petals 8--9 mm, white with fine green veins. Fruit a flattened silicula 8--16 mm, broadly winged, deeply notched at apex, with thick septum (fruit rarely produced in cultivation). *Caucasus.* Figure 44, p. 210.

42. Aethionema R. Brown. 70/12 and some hybrids. Usually woody-based perennials, rarely annuals or biennials. Leaves small, stalkless or nearly so, simple, entire, often leathery or somewhat fleshy. Sepals erect, the 2 outer pouch-like at base. Petals entire, white or pink, rarely yellow. Filaments of the 4 inner stamens winged, the wing sometimes toothed above. Silicula flattened and winged, either with a single indehiscent, 1--2 seeded cell, or with 2 cells, each with 1--4 seeds. *Mediterranean area, western Asia.* Figure 44, p. 210.

43. Iberis Linnaeus. 40/10. Annuals, perennials or almost shrubs. Leaves alternate, linear or obovate, entire or pinnately cut, hairless or with unbranched hairs. Inflorescence a corymb or raceme; petals white, pink or purple, the 2 outer much larger than the 2 inner. Fruit a flat silicula, rounded at base, entire or notched at apex. Seeds solitary in each cell, often winged. *South Europe, western Asia.* Figure 44, p. 210.

44. Biscutella Linnaeus. 40/2. Annual to perennial herbs or small shrubs. Leaves entire to pinnatifid. Petals yellow and usually clawed. Fruit a flat, indehiscent silicula. Seeds unwinged. *South & central Europe, North Africa.* Figure 44, p. 210.

45. Megacarpaea de Candolle. 7/3. Herbaceous perennials with unbranched hairs. Leaves pinnatisect. Sepals not pouched. Petals white, cream or violet. Stamens more than 6. Fruit a silicula, with a narrow septum, deeply 2-lobed; lobes flat, broadly winged; seed solitary. *Europe to central Asia, Himalaya and China.*

46. Lepidium Linnaeus. 140/2. Annuals to perennials, hairless or with small, simple hairs. Sepals not pouch-like. Petals usually small and inconspicuous, white, sometimes absent. Fruit a silicula, strongly compressed, keeled, often winged; style short or stigma stalkless. Seeds usually 1 in each cell. *Cosmopolitan.*

47. Notothlaspi J.D. Hooker. 2/1. Compact perennials or biennials with long taproots. Leaves spathulate, fleshy. Flowers crowded. Sepals erect. Fruit a reversed heart-shaped to obovate siliqua, strongly compressed, winged. Seeds *c*. 1 mm, many in each cell. *New Zealand.* Figure 44, p. 210.

48. Moricandia de Candolle. 7/2. Hairless annuals to perennials with simple, greyish fleshy leaves. Inner sepals pouch-like at base. Petals violet-purple. Fruit a linear, compressed siliqua; style short. Seeds many. *Mediterranean area, southwest Asia.* Figure 45, p. 212.

49. Brassica Linnaeus. 35/5. Usually perennials, hairless or with simple hairs. Leaves entire to pinnatifid. Sepals erect or spreading. Petals yellow or white, with a claw. Fruit a siliqua with a distinct beak; valves convex, prominently 1-veined. Seeds in 1 row (rarely 2 rows) in each cell, spherical. *Western Europe to China.* Figure 45, p. 212.

50. Eruca Miller. 5/1. Annuals, usually hairy. Leaves pinnatifid. Sepals erect, the inner somewhat pouched. Petals with a long claw. Fruit a siliqua, the valves 1-veined; beak long, flat. Seeds in 2 rows in each cell. *Mediterranean area.* Figure 45, p. 212.

51. Vella Linnaeus. 4/4. Small, much-branched shrubs, sometimes spiny; leaves entire; hairs unbranched. Flowers shortly stalked in loose, spike-like racemes. Sepals erect. Petals long-clawed, yellow, sometimes violet-veined. Filaments of inner stamens joined in pairs. Fruit transversely articulated, lower segment dehiscent by two lobes, ellipsoid; lobes convex, 3-veined; the upper segment sterile, strongly compressed in the form of a leafy, narrowly oblong, acute, 5-veined, hairless beak. *Spain, Morocco, Algeria.* Figure 45, p. 212.

52. Crambe Linnaeus. 25/5. Robust perennials, with large rootstock, hairless or with simple hairs. Stems erect, branched. Sepals erect to spreading. Petals with a short claw, white to cream. Filaments of the inner stamens usually with a tooth-like appendage. Fruit a 2-segmented silicula; lower segment short, sterile; upper segment ovoid to almost spherical, indehiscent, with 1 seed; stigma stalkless. *Mostly western & central Asia.* Figure 44, p. 210.

53. Morisia Gay. 1/1. A stemless perennial with a rosette of deeply cut, oblong-lanceolate, downy leaves, Flowers large, petals 9--12 mm, produced singly in axils of leaves. Flower-stalks erect when flowering, 5--25 mm, later arching and lengthening to 6 cm, causing the fruit to lie on the ground or be buried. Fruit in 2 segments, lower almost spherical, 2-celled dehiscing by 2 lobes with 3--5 (rarely to 12) seeds; upper ovoid, conical, indehiscent with 1 or 2 seeds and smaller than lower segment. *Corsica, Sardinia.*

54. Raphanus Linnaeus. 6/1. Bristly annuals, biennials or perennials. Leaves mostly lyrate-pinnatifid. Sepals erect, the 2 inner slightly pouched. Petals with a long claw, prominently veined. Fruits with 2 segments, the lower short, seedless, the upper usually constricted between the seeds; beak narrow, seedless. *Eurasia, Mediterranean area.* Figure 45, p. 212.

55. Raffenaldia Godron. 2/2. Stemless perennial herbs with all leaves in a basal rosette. Leaves pinnatisect with 1--4 pairs of lobes, the terminal largest, ovate to oblong, stalked. Flowers long-stalked, solitary amongst the leaves. Fruit-stalks becoming strongly reflexed. Sepals pouched at base. Fruit an indehiscent siliqua, squarish or compressed in cross-section, linear-oblong, dehiscing into 3--10 single-seeded segments, usually with thickenings between them. *North Africa*.

56. Heliophila Linnaeus. 72/6. Annual to perennial herbs or shrubs, to 3 m or more. Hairless or with simple hairs. Stems erect, decumbent or climbing. Leaves entire, lobed, pinnate or bipinnate. Stipules sometimes present, minute. Flowers usually in a raceme, sometimes in the leaf angles, or a few together on short side branches, or between stem nodes. Sepals sometimes pouched at the base. Petals usually clawed, lanceolate to round, often with appendages at base, white, blue-purple or pink. Fruit a siliqua or a silicula, rounded to linear, sometimes contracted between the seeds, round or compressed in cross-section, dehiscent, hairy or not. Seeds generally in 1 row, few to many, compressed, sometimes winged. *South Africa, Namibia*. Figure 45, p. 212.

79. RESEDACEAE

Annual to perennial herbs. Leaves alternate, simple or pinnately divided. Flowers in bracteate racemes, bisexual, bilaterally symmetric. Sepals 4--7; petals 2--8, fringed, shortly stalked. Stamens 3--40, filaments free or united at base, not covered by the petals in bud; anthers 2-celled. Ovary of 2--6 carpels. Fruit an open capsule. Seeds kidney-shaped. See figure 47, p. 226.

A family of 6 genera and some 75 species from the northern hemisphere, especially the Old World.

1. Reseda Linnaeus. 55/4. Description as family. *Northern hemisphere*. Figure 47, p. 226.

80. PLATANACEAE

Deciduous, bisexual trees to 50 m, with bark peeling away in thin flakes leaving a smooth, pale surface. Leaves alternate, simple, more or less palmately lobed and veined (rarely unlobed and pinnately veined) with the base of the leaf-stalk covering the axillary bud. Stipules usually prominent, often leaf-like, united at the base into a short tube around the stem. Young growth, especially the underside of leaves, often covered with a dense felt of stellate and simple hairs, sometimes persistent. Inflorescence of up to 12 dense, spherical, stalkless or stalked heads of flowers in hanging strings, each inflorescence either male or female. Sepals 3--8, free; petals 3--8, spathulate; both sometimes considered to be bracts rather than a true perianth. Male flowers with 3--4 almost stalkless anthers and occasionally 3--4 pistillodes; female flowers with 6--9 very short-stalked free ovaries, each with a long style and hooked stigma; 3--4 staminodes sometimes present. Achenes with often persistent styles, forming characteristic prickly balls; individual achenes are subtended by a tuft of bristly hairs derived from the perianth or scales.

A family of a single genus occurring throughout the north temperate zone.

1. Platanus Linnaeus. 8/4 and some hybrids. Description as for family. *North & Central America, southeast Europe to Iran, 'Indo-China'*.

81. HAMAMELIDACEAE

Shrubs or trees, often with stellate hairs, bisexual or unisexual. Leaves usually alternate, simple or palmate; stipules usually present. Flowers in spikes, clusters or pairs, bisexual or unisexual; petals usually 4 or 5, sometimes absent (rarely 2 or 3 or 6--10); sepals 4--5, rarely absent; anthers 4 or 5 (rarely to 25); carpels 2, inferior to superior, usually joined, with distinct styles; ovules 1--many, often aborting to 1; fruit a woody capsule, splitting along or between the radial walls. See figure 48, p. 228.

A family of 28 genera and 90 species, widespread but mainly in the subtropics and tropics of eastern Asia.

1a. Leaves deeply palmately divided	**7. Liquidambar**
b. Leaves simple	2
2a. Leaves at least partly evergreen	3
b. Leaves quite deciduous	5
3a. Flowers in racemes	**3. Distylium**
b. Flowers in heads	4
4a. Leaves evergreen, entire; bracts 5 mm	**12. Sycopsis**
b. Leaves semi-deciduous, shallowly toothed; bracts 10 mm or more	
	11. × Sycoparrotia
5a. Bark flaking	**8. Parrotia**
b. Bark not flaking	6
6a. All flowers bisexual	7
b. Some flowers unisexual	11
7a. Flower-clusters with conspicuous white bracts at base	**9. Parrotiopsis**
b. Flower-clusters or racemes without conspicuous bracts	8
8a. Leaves palmately veined, base cordate	**2. Disanthus**
b. Leaves pinnately veined, base not usually cordate	9
9a. Flowers in erect white spikes; petals lacking	**5. Fothergilla**
b. Flowers in clusters or racemes; petals present	10
10a. Flowers in hanging yellow racemes; petals ovate	**1. Corylopsis**
b. Flowers in clusters; petals strap-shaped	**6. Hamamelis**
11a. Racemes hanging, unisexual	**10. Sinowilsonia**
b. Racemes erect, bisexual in spring, male in autumn	**4. Fortunearia**

1. Corylopsis Siebold & Zuccarini. 7/5. Deciduous shrubs. Leaves elliptic-ovate to circular, green to glaucous, sharply and finely toothed, veins impressed. Flowers in hanging racemes, bell- to funnel-shaped, yellow, scented. *Eastern Asia*.

2. Disanthus Maximowicz. 1/1. Deciduous shrub to 8 m. Leaves alternate, palmately veined, long-stalked, circular to ovate, to 10 cm across, entire, hairless, turning deep red and orange in autumn. Flowers axillary in short-stalked pairs, dark purple, 1.5 cm across, faintly scented, bisexual; petals, sepals, stamens and staminodes 5, calyx softly hairy with recurved lobes. Capsule 2-celled with several seeds in each cell. *China, Japan*.

3. Distylium Siebold & Zuccarini. 12/1. Evergreen trees or shrubs. Leaves ovate to lanceolate, entire or remotely toothed above the middle. Flowers in racemes. *Eastern Asia, Central America*.

4. Fortunearia Rehder & Wilson. 1/1. Large deciduous shrub to 8 m. Leaves obovate, acuminate, toothed. Flowers greenish, in erect racemes; male flowers *c.* 2.5 cm formed in autumn, bisexual flowers *c.* 5 cm appearing with the leaves; petals 5, strap-shaped. *China.*

5. Fothergilla Linnaeus. 2/2. Deciduous shrubs with alternate, coarsely-toothed leaves. Flowers cream or white, without petals, in terminal spikes or heads like bottle-brushes, slightly before or with the leaves; perigynous zone bell-shaped, with 4--7 lobes; stamens 15--25, white, thick near the apex; ovary 2-celled; fruit a 2-seeded capsule. *Southeast North America.*

6. Hamamelis Linnaeus. 4/4 and some hybrids. Deciduous shrubs or small trees with stellate hairs. Leaves alternate, pinnately veined, short-stalked, unequal at base, ovate to obovate; margin irregularly toothed, sometimes wavy towards the tip. Flowers bisexual, fragrant, in many, very condensed inflorescences in clusters along the branches. Inflorescence usually with 3 flowers; bract 1; bracteoles 2, more or less fused. Flowers short-stalked, parts in 4s; calyx persistent, cup-shaped and hairy outside; petals strap-shaped, wrinkled and crisped, 5--25 × 1--2 mm, rolled in bud. Fruit a woody capsule splitting with great power, hurling away the black, shiny seeds. *Eastern USA, eastern Asia.* Figure 48, p. 228.

7. Liquidambar Linnaeus. 4/3. Deciduous unisexual trees with balsamic resin; twigs often with corky ridges. Leaves alternate, long-stalked, palmately lobed. Flowers without petals, numerous, in spherical heads, male heads in terminal catkins, female heads solitary. Fruiting heads spherical, woody, with conspicuous persistent styles; capsule splitting along the radial wall. Seed flattened, narrowly winged. *Eastern N America, Turkey, China.*
Sometimes placed in the separate family *Altingiaceae.*

8. Parrotia Meyer. 1/1. Deciduous tree or shrub to 12 m, with flaking bark; branches often hanging. Leaves 6--10 × 4--8 cm, oblong-elliptic to narrowly ovate, wavy and shallowly toothed, glossy green turning orange, yellow, crimson and purple in autumn, short-stalked, with large, early-falling stipules. Flowers bisexual, *c.* 1.2 cm across, in dense heads surrounded by large bracts; petals absent; anthers 5--7, red, conspicuous. *Southwest Caspian area.*

9. Parrotiopsis (Niedenzu) Schneider. 1/1. Deciduous tree to 6 m, or more usually an Alder-like shrub. Leaves 3--5 cm, circular, slightly truncate, toothed, turning yellow in autumn. Flowers bisexual, in clusters, *c.* 1 cm across, surrounded by 4--6 white bracts; petals absent; stamens 15--24. *Western Himalaya.*

10. Sinowilsonia Hemsley. 1/1. Deciduous bisexual tree to 8 m with stellate hairs. Leaves alternate, elliptic to broadly ovate, to 15 cm, entire. Flowers unisexual, without petals; racemes hanging, male to 6 cm, female to 3 cm. Flower-tube urn-shaped; styles projecting. *China.*

11. × Sycoparrotia Endress & Anliker. 1/1. Semi-deciduous shrub to 4 m. Similar to *Parrotia* but smaller; stems not flaking. Leaves oblong-elliptic, shallowly

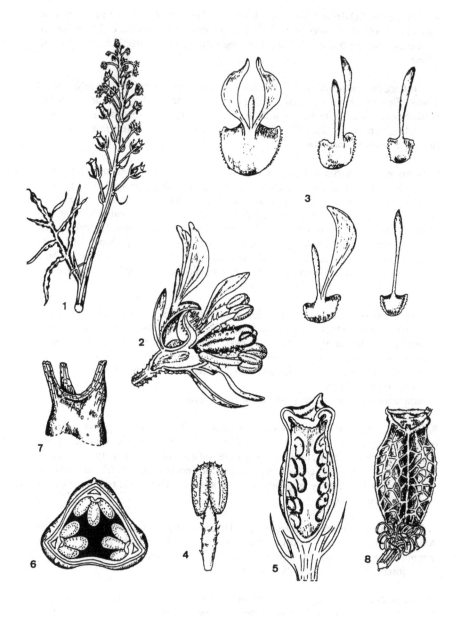

Figure 47. Resedaceae. *Reseda lutea*. 1, Inflorescence. 2, Flower from the side with some petals removed. 3, Five differently-shaped petals from the same flower. 4, Stamen. 5, Longitudinal section of ovary. 6, Transverse section of ovary. 7, Apex of ovary. 8, Fruit.

226

spiny-toothed, 5--8 × 2--4 cm, often persistent. Flowers as those of *Parrotia*. *Garden Origin.*

A hybrid between *Parrotia persica* and *Sycopsis sinensis.*

12. Sycopsis Oliver. 7/1. Evergreen shrubs or trees. Leaves entire, with pinnate veins, short-stalked. Flowers without petals, male or bisexual, in short racemes or heads, surrounded by softly hairy bracts; male flowers with 8--10 stamens and minute sepals; female flowers with an urn-shaped, 5-lobed tube. Fruit with 2 shining brown seeds. *NE India to China, Malaysia & New Guinea.*

82. CRASSULACEAE

Succulent shrubs and herbs, sometimes small trees. Leaves alternate, opposite or whorled, simple, usually entire; stipules absent. Flowers in cymes, spikes, racemes or panicles, or occasionally solitary, usually with 5 parts, sometimes with 3, 4 or more parts. Sepals almost free, persistent; petals usually as many; stamens usually in 2 whorls, twice as many or the same number as petals, rarely basally united; ovary superior, carpels as many as petals or sepals, usually free, each with a nectar-gland at base, placentation almost marginal. Fruit usually a group of follicles, rarely a capsule. See figure 49, p. 230.

A family of about 30 genera with an almost cosmopolitan distribution.

1a. Stamens as many as petals	2
b. Stamens twice as many as petals	3
2a. Leaves opposite; flowers borne in axils of most leaves	**1. Crassula**
b. Leaves opposite, alternate or whorled; flowers in terminal cymes or corymbs or in axils of upper leaves only	**2. Sedum**
3a. Leaves opposite	**1. Crassula**
b. Leaves alternate, in rosettes	4
4a. Flower parts 4 or 5 or, if 6 or 7, then leaves not in a rosette	5
b. Flower parts in 6s (rarely 5s) or more; leaves in rosettes (some also scattered in annual and biennial plants)	12
5a. Inflorescence lateral; leaves commonly in a rosette	**7. Rosularia**
b. Inflorescence terminal; leaves usually not forming a rosette	6
6a. Petals free or nearly so	7
b. Petals more or less united	9
7a. Inflorescence equilateral, with branches in all planes	**2. Sedum**
b. Inflorescence a 1-sided cyme with flowers all in one plane	8
8a. Perennial stem-base covered in appressed brown scale-leaves	**3. Rhodiola**
b. No scale-leaves at bases of stems	**2. Sedum**
9a. Plants annual	**4. Pistorinia**
b. Plants perennial	10
10a. Leaves opposite	**6. Chiastophyllum**
b. Leaves alternate	11
11a. Leaves persistent, abruptly replaced by bracts	**7. Rosularia**
b. Leaves deciduous, gradually replaced by bracts	**5. Umbilicus**
12a. Flower parts 6 or 7; flower bell-shaped	**9. Jovibarba**
b. Flower parts 8--16; flower star-shaped	**8. Sempervivum**

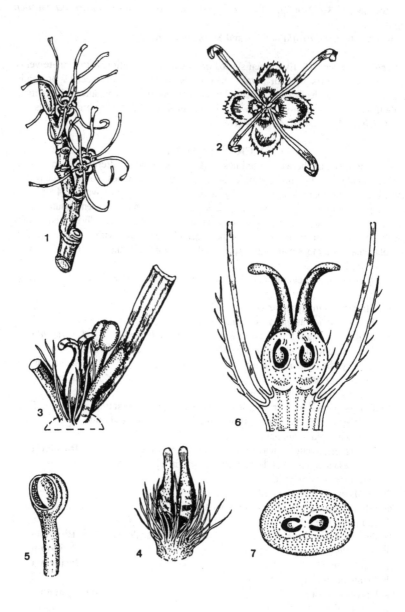

Figure 48. Hamamelidaceae. *Hamamelis mollis.* 1, Leafless shoot-apex with flower-clusters. 2, A single flower from above. 3, Ovary and parts of two stamens. 4, Young ovary. 5, Anther showing dehiscence by flaps. 6, Longitudinal section of ovary. 7, Transverse section of ovary.

228

1. Crassula Linnaeus. 150/6. Succulent annual or perennial herbs, shrublets or shrubs with cartilaginous or soft-wooded branches; rarely perennating in tubers. Leaves opposite, more or less united, membranous to thickly fleshy, persistent or deciduous, rarely regenerating from base. Inflorescence with 1--several dichasia, very rarely a single terminal flower; stem with leaf-like bracts. Flowers spreading or erect. Calyx of 4--7 or 12 sepals; corolla of 4--7 or 12 petals, sometimes with dorsal projection fused at the base into a short tube, or open and star-shaped; stamens 4, 5 or 12 in one whorl; anthers included or exserted, nectary scales 4, 5 or 12, free; carpels 4, 5 or 12 free and constricted into styles; stigma terminal. Seeds ellipsoid, smooth or covered with tubercles. *Europe, America, Australasia, Africa.*

2. Sedum Linnaeus. 280/112. Plants hairless, downy or glandular-hairy, fleshy, erect or decumbent, sometimes tufted or moss-like. Leaves very variable, opposite, alternate or whorled, entire or rarely toothed. Inflorescence usually terminal, rarely lateral, mostly cyme-like or flowers solitary, white, yellow or rose, rarely red or blue, bisexual, floral parts usually in 5s, rarely in 4s, 6s, 7s, 8s or 9s; petals usually free, sometimes fused for almost a third, stamens usually twice as many as petals, rarely equal in number to petals. *Northern hemisphere.* Figure 49, p. 230.

3. Rhodiola Linnaeus. 50/14. Perennial herbs usually with a stout, very short, fleshy, perennial stem, sometimes branched above, often partly exposed, covered in brown triangular scales; branches simple, annual, leafy with terminal flower heads developing laterally below each terminal bud on the perennial stem. Sometimes perennial stem lengthening, slender, much-branched forming tufts or shrublets, branches crowned with leaves distinct from those on the annual stems and scales present or absent. Flowers solitary or few to many in spherical, corymb-like, raceme-like or paniculate heads, bisexual or functionally unisexual, the plants then usually dioecious. Floral parts usually in 4s or 5s. Stamens twice as long as petals, absent in female flowers. Carpels opposite the petals in female flowers, reduced and opposite the sepals in male flowers. *Europe, eastern Asia, North America.*
This genus is sometimes included in *Sedum.*

4. Pistorinia de Candolle. 2/2. Erect annuals with alternate leaves. Stems hairless below, glandular-hairy above. Leaves linear, succulent, stalkless and soon falling. Flowers shortly stalked, numerous in dense corymbs. Sepals 5, petals 5, united into a funnel-shaped corolla, tube long and narrow, lobes spreading or erect. Stamens 10, 5 long, 5 short. Styles slender. Carpels 5. Fruit a many-seeded follicle. *Iberian Peninsula, North Africa.*

5. Umbilicus de Candolle. 18/4. Perennial, succulent herbs with new shoots arising annually from scaly rhizomes or tubers. Leaves alternate, hairless; stalked at base, those above progressively stalkless. Flowering stems terminal, usually solitary, but occasionally branched above into several racemes. Sepals 5, free, about half as long as corolla. Petals 5, united into a bell-shaped corolla; lobes more or less erect. Stamens 10, rarely 5; styles short or absent. Fruit a slender follicle; seeds numerous. *Europe, west Asia, North Africa & Atlantic Islands.*

6. Chiastophyllum Berger. 1/1. Fleshy, hairless, evergreen perennial with prostrate or ascending stems often rooting from the lower nodes. Leaves opposite, broadly elliptic

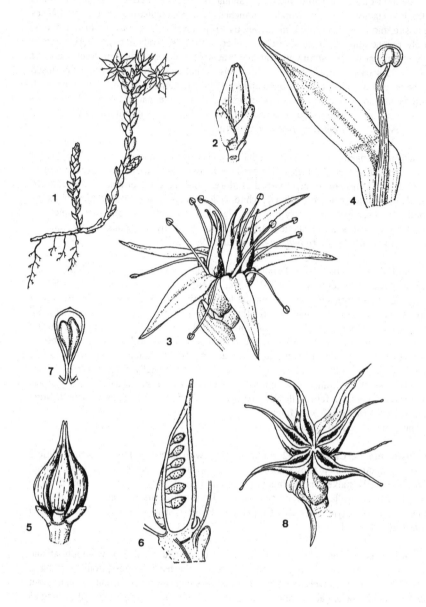

Figure 49. Crassulaceae. *Sedum acre.* 1, Flowering shoot. 2, Flower-bud. 3, Flower. 4, Inner stamen attached to a petal. 5, Ovary. 6, Longitudinal section of a single carpel. 7, Transverse section of a single carpel. 8, Opened fruit.

or rounded, scalloped, short-stalked. Inflorescence a loose panicle of arching spike-like branches with small linear bracts and almost stalkless yellow flowers. Calyx lobes 5, narrowly oblong; petals 5, elliptic, erect, rather fleshy. Stamens 10, filaments attached at base of corolla lobes. Carpels 5, each with a small scale at the base. Fruiting-head consisting of 5 erect oblong follicles. *Caucasus.*

7. Rosularia (de Candolle) Stapf. 25/13. Herbaceous, evergreen, rosette-forming perennials with or without rootstock and/or long taproot; rosettes single or highly tufted and mat-forming; leaves succulent-juicy, broadly spathulate to oblong, hairless or glandular-hairy; margin entire, glandular-hairy or minutely toothed, some with waxy bloom; inflorescence lateral or terminal, upright or decumbent, often drooping in bud, paniculate with sometimes curled lateral branches; perianth-segments 5--9, united for one-tenth to three-quarters of their length, tips more or less reflexed; stamens twice as many as petals, carpels free or basally united, upright. Seeds oblong-ellipsoid, 0.5--1.3 mm, dark brownish to ochre, longitudinally striped. *Turkey to Himalaya.*

Some species of the genus **Aeonium** Webb and Berthelot may be hardy in the mildest parts of northern Europe; they are perennials with terminal leaf-rosettes, the petals are free or only shortly united and the nectar-glands are more or less 4-sided.

8. Sempervivum Linnaeus. 50/50 and many hybrids. Succulent perennial herbs with pointed basal leaves in a rosette. Rosettes 1--15 cm across, monocarpic, plant increases by producing young offsets from the leaf axils, usually on runners (stolons). Flower-stems terminal, erect, usually with a covering of leaves. Flowers in a terminal cyme, with 8--16 parts in a star shape, twice as many stamens as petals, equal number of carpels to petals. Petals yellowish, pink, purple or red. *Europe, Morocco, southwest Asia.*

9. Jovibarba Opiz. 6/6. Like *Sempervivum* but petals 6 or 7 in number, pale yellow, keeled outside, fringed with glandular hairs forming a bell-shaped flower. *Europe.*

83. SAXIFRAGACEAE

Low to medium-sized herbs, usually perennial, and shrubs. Leaves usually simple and alternate, without stipules, with or without stalks. Flowers with regular or irregular symmetry, in cymes, racemes, or rarely solitary. Sepals 4 or 5, occasionally more showy than petals. Petals 4 or 5, rarely absent. Stamens usually twice as many as petals, occasionally fewer or more. Ovary superior or inferior, usually with 2 carpels, styles as many as carpels. Fruit usually a capsule, rarely a berry. See figures 50, p. 238 & 51, p. 242.

In the present wide interpretation the family contains about 80 genera and is almost cosmopolitan. The majority of species, however, occur in eastern Asia, the Himalayas and in North America. Currently, the family is often broken down into smaller, segregate families and some genera are removed to other established families. Thus, genus 1 is often placed in the *Penthoraceae*, genera 2--23, 25 & 26 are retained in the *Saxifragaceae* in the restricted sense, genus 27 is placed in the *Grossulariaceae*, genus 28 in the *Francoaceae*, genus 29 in the *Parnassiaceae*, genera 30--36 & 38--42 in the *Hydrangeaceae*, genera 37, 43, 44 & 46 in the *Escalloniaceae*, and genus 45 is sometimes moved to the existing family *Cunoniaceae* (see p. 244) or placed in a family of its own, *Baueraceae*.

1a. Plant herbaceous   2
  b. Plant a woody shrub, tree or climber   41
2a. Flowering stems with at least one leaf   3
  b. Flowering stems lacking leaves   31
3a. Carpels 4 or 5   4
  b. Carpels 2 or 3   5
4a. Staminodes present, carpels 4   **29. Parnassia**
  b. Staminodes absent, carpels 5   **1. Penthorum**
5a. Carpels 3   6
  b. Carpels 2   7
6a. Stamens 10, rhizome bearing bulb-like tubers   **14. Lithophragma**
  b. Stamens 15, rhizome not as above   **33. Kirengeshoma**
7a. Leaves peltate   8
  b. Leaves not as above   9
8a. Leaves palmately lobed; stamens 10   **25. Peltoboykinia**
  b. Leaves almost circular, not lobed; stamens 6--8   **4. Astilboides**
9a. Flowers in bracted dichotomous cymes on erect shoots, bracts leaf-like, greenish to bright yellow   **26. Chrysosplenium**
  b. Flowers not as above   10
10a. Basal leaves stalkless or with very short winged stalks   11
  b. Basal leaves stalked, often long-stalked   13
11a. Basal leaves with very short winged stalks; stigmas usually lacking styles   **5. Leptarrhena**
  b. Basal leaves stalkless; stigmas usually with styles   12
12a. Leaves lanceolate with a long gradually diminishing point at both ends   **1. Penthorum**
  b. Leaves not as above   **8. Saxifraga**
13a. Flowering stems with scale-like leaves at base and 2--4 large opposite leaves near the top   **24. Deinanthe**
  b. Flowering-stems not as above   14
14a. Flowers with a variable number of stamens present within a single inflorescence   **17. × Heucherella**
  b. Flowers with a constant number of stamens present within a single inflorescence   15
15a. Stamens 10, rarely 8 or 5, if 5, leaves 2- or 3-pinnate and petals absent   16
  b. Stamens 3 or 5, if 5, leaves not 2- or 3-pinnate and petals present   24
16a. Stamens 5   17
  b. Stamens 10, rarely 8   18
17a. Leaflets leathery, wrinkled   **3. Rodgersia**
  b. Leaflets not as above   **2. Astilbe**
18a. Fruit a few-seeded capsule with 2 unequal parts   **15. Tiarella**
  b. Fruit a many-seeded capsule with 2 equal parts   19
19a. Petals deeply divided, often comb-like   20
  b. Petals entire or rarely shallowly toothed   22
20a. Calyx usually 1--3 mm; styles less than 1 mm   **19. Mitella**
  b. Calyx 4--8 mm; styles over 1 mm   21
21a. Stamens 10   **20. Tellima**
  b. Stamens 5   **13. Elmera**

22a. Leaves usually 2 or 3 times divided into 3, in a single species simple and
　　　ovate, coarsely double-toothed and often somewhat lobed　　**2. Astilbe**
　　b. Leaves and inflorescence not as above　　23
23a. Styles partially fused　　**10. Telesonix**
　　b. Styles free above the ovule-bearing portion of the ovary　　**8. Saxifraga**
24a. Stamens 3　　**21. Tolmiea**
　　b. Stamens 5　　25
25a. Petals deeply divided　　26
　　b. Petals not as above　　27
26a. Calyx usually 1--3 mm; styles less than 1 mm　　**19. Mitella**
　　b. Calyx 6--8 mm; styles more than 1 mm　　**13. Elmera**
27a. Rootstock bearing bulbils　　**18. Bolandra**
　　b. Rootstock lacking bulbils　　28
28a. Plant with slender horizontal stolons　　**9. Sullivantia**
　　b. Plant lacking stolons　　29
29a. Leaves deeply palmate; petals absent　　**3. Rodgersia**
　　b. Leaves kidney-shaped or shallowly palmate; petals usually present　　30
30a. Ovary with 1 cell　　**16. Heuchera**
　　b. Ovary with 2 cells　　**12. Boykinia**
31a. Carpels 4, staminodes present　　**28. Francoa**
　　b. Carpels 2, staminodes absent　　32
32a. Leaves all peltate　　**7. Darmera**
　　b. Leaves not peltate　　33
33a. Capsule with 2 unequal cells　　**15. Tiarella**
　　b. Capsule with 2 equal cells　　34
34a. Petals deeply divided, often comb-like　　**19. Mitella**
　　b. Petals not deeply divided, or sometimes absent　　35
35a. Stamens 5 or 6　　36
　　b. Stamens 10, rarely 8　　38
36a. Petals fused into a short tube at base or absent　　**16. Heuchera**
　　b. Petals joined only at extreme base　　37
37a. Flowers on stems to 25 cm　　**22. Bensoniella**
　　b. Flowers almost stemless　　**23. Mukdenia**
38a. Leaves commonly 2--4 times divided into 3, in a single species simple
　　and then ovate, coarsely double-toothed and often somewhat lobed
　　　　　　　　　　　　　　　2. Astilbe
　　b. Leaves and inflorescence not as above　　39
39a. Petals absent; flowers inconspicuous　　**11. Tanakaea**
　　b. Petals present; flowers often conspicuous　　40
40a. Carpels almost free from each other and from the perigynous zone
　　　　　　　　　　　　　　　6. Bergenia
　　b. Carpels fused for most of their length　　**8. Saxifraga**
41a. Leaves alternate　　42
　　b. Leaves opposite or whorled　　51
42a. Ovary superior　　43
　　b. Ovary inferior or half-inferior　　47
43a. Ovary with 1 or 2 cells; flowers in unbranched racemes　　44
　　b. Ovary with 5 cells; flowers in umbels or branched panicles　　45

44a. Leaves with large, rounded teeth; flower parts in 6s or 9s; ovary 1-celled
46. Anopterus
  b. Leaves spiny or finely toothed; flower parts in 5s; ovary 2-celled **37. Itea**
45a. Ovary 1-celled; fruit a berry **27. Ribes**
  b. Ovary with 2--5 cells; fruit a capsule 46
46a. Evergreen or deciduous shrub; leaf-stalks to 1 cm **44. Escallonia**
  b. Evergreen tree, if a shrub then leaf-stalks *c*. 2 cm **43. Carpodetus**
47a. Deciduous or evergreen climber 48
  b. Deciduous or evergreen shrub 51
48a. Outer flowers of cluster sterile, or if all flowers fertile, then corymbs at
    first enclosed by 4 papery bracts 49
  b. All flowers similar, fertile 50
49a. Sterile flowers reduced to a single bract 2.5 cm or more long
35. Schizophragma
  b. Sterile flowers not as above or absent and then corymbs at first enclosed
    by 4 papery bracts **34. Hydrangea**
50a. Petals 7--10; stamens 20--30; ovary with 7--10 cells **36. Decumaria**
  b. Petals 4 or 5; stamens 8--10; ovary with 4 or 5 cells **42. Pileostegia**
51a. Stamens numerous 52
  b. Stamens 10 or fewer 55
52a. Evergreen shrub; ovary superior or mostly so 53
  b. Deciduous shrub; ovary inferior 54
53a. Leaves divided; flowers solitary, axillary; styles 2 **45. Bauera**
  b. Leaves simple; flowers 3--7 per cluster; style 1 **38. Carpenteria**
54a. Outer flowers of cluster sterile, with petal-like calyx; styles 2
41. Platycrater
  b. All flowers similar, fertile **32. Philadelphus**
55a. Outer flowers of cluster larger, sterile **34. Hydrangea**
  b. All flowers similar, fertile 56
56a. Sepals and petals 4; stamens 8 **40. Fendlera**
  b. Sepals and petals almost always 5; stamens 10 57
57a. Ovary more or less superior; leaves with thick down beneath **30. Jamesia**
  b. Ovary inferior; leaves without thick down beneath 58
58a. Deciduous shrub, rarely evergreen, usually with stellate hairs; filaments
    winged **31. Deutzia**
  b. Evergreen shrub without stellate hairs; filaments not winged **39. Dichroa**

1. Penthorum Linnaeus. 1/1. Erect perennials with stolons. Leaves alternate, lanceolate, tapering at both ends, stalkless, with minute teeth. Flowers in terminal, spirally coiled clusters, yellow-green; sepals 5; petals 5, or absent; stamens 10, carpels 5, united in the lower half; capsules flattened, 5-beaked, upper portion deciduous. *East & southeast Asia, North America.*

2. Astilbe D. Don. 14/8 and many hybrids. Herbaceous perennials with stout rhizomes, sometimes stoloniferous, generally with conspicuous brown hair-like scales. Leaves alternate, both basal and on the stem, divided into 3 or pinnately compound or rarely simple; stipules scarious. Leaflets lanceolate to broadly ovate, strongly toothed, teeth forwardly pointing. Inflorescence a terminal panicle, the branches usually spike-like. Flowers small, white, pink or purplish, sometimes unisexual. Calyx-tube short,

5-lobed; petals 5, linear, narrowly spathulate or rarely absent; stamens 5 or 10; carpels 2, more or less joined at the base, superior; capsule 2-celled, the cells dehiscing inwardly. *Himalaya, northeast Asia, North America.*

3. Rodgersia Gray. 6/4. Perennial, rhizomatous herbs. Flowering stems with reduced leaves, most of the foliage arising directly from the rhizome. Leaves long-stalked, pinnate, apparently pinnate or palmate, leaflets toothed, generally hairy or downy beneath. Inflorescence a large, corymb-like, flat-topped cyme with many flowers. Flowers shortly stalked, each with a bract. Sepals 5, greenish, whitish or red, united below. Petals absent. Stamens 5, spreading. Ovary of 2 carpels which are united in their lower part. Fruit a capsule. Seeds numerous (often not ripened in cultivation). *Eastern Asia.*

4. Astilboides Engler. 1/1. Perennial, brown-hairy rhizomatous herbs to 1.5 m. Leaves mostly basal, long-stalked, blade peltate, almost circular, to 90 cm across, lobed and irregularly toothed. Flowers to 8 mm across, creamy white, in spike-like, terminal cymes to 5 cm; stem-leaves very small. Sepals 4 or 5, triangular, blunt or notched. Petals 4 or 5, oblong or oblong-lanceolate, somewhat unequal. Stamens 6--8, about as long as the petals. Ovary usually 2-celled (rarely 4-celled), each cell with *c.* 8 ovules; styles 2 (rarely 4), long, with small, capitate stigmas. Capsule usually 2-celled, many-seeded. Seeds pointed at each end. *Eastern Asia.*

5. Leptarrhena R. Brown. 1/1. Herbs with horizontal, spreading rootstocks. Leaves in basal rosettes, oblong to obovate, narrowed to the short, winged stalk, toothed, leathery, dark glossy green above, paler beneath. Inflorescence-stalk with 1--3 small leaves clasping the stem. Flowers bisexual in dense, terminal inflorescences. Sepals 5, minute. Petals 5, spathulate, persistent, white or tinged with pink. Stamens 10, longer than petals. Stigmas borne on top of the carpels, more or less without styles; carpels 2. Fruit a group of 2 follicles, bright red. *North America.*

6. Bergenia Moench. 8/6 and some hybrids. Perennial herbs, with stout prostrate rooting stems terminating in leaf-rosettes. Leaves 6--25 × 5--17 cm, usually evergreen; stalk stout, sheathing at base; blade simple, shallowly toothed or nearly entire, leathery. Flowers numerous, in panicles supported on stout scapes, differing from those of *Saxifraga* in having the sepals, petals and stamens perigynous and the carpels nearly free from each other and from the perigynous zone. *Temperate Asia.*

7. Darmera Post & Kuntze. 1/1. Perennial herb to 65 cm, rhizomes stout, tips clothed by broad stipular leaf-sheaths. Leaves basal, almost circular, peltate, 5--40 cm across, with 7--15 shallow lobes, each lobe cut and sharply toothed, leaf-stalks 10--150 cm. Flowering stems 1--100 cm, with long hairs and shorter glandular hairs; flowers borne in many-flowered corymbs, calyx 5-lobed, 2.5--3.5 mm; petals white or pale pink, oblong-elliptic to obovate, 4.5--7 mm. Follicles purplish, 6--10 mm, joined to calyx for 1--2 mm. *Western North America.*

The genus was formerly known as *Peltiphyllum* Engler.

8. Saxifraga Linnaeus. 440/170 and many hybrids. Perennial, biennial or annual herbs, erect or forming mats or cushions. Leaves alternate (rarely opposite). Flowers solitary or in branched, usually cyme-like inflorescences. Sepals 5. Petals 5. Stamens usually

10, rarely 8, inserted at the junction of the ovary wall and floral-tube. Ovary usually of 2 carpels, with separate styles. *North temperate & arctic regions, with a few species as far south as Thailand, Ethiopia & the Andes south to Tierra del Fuego; the richest regions are the mountains of Europe, western North America, the Himalayan-Tibetan region and eastern Asia.* Figure 50, p. 238.

9. Sullivantia Torrey & Gray. 6/1. Herbaceous perennials with slender horizontal stolons. Leaves basal and on the stem, alternate, kidney-shaped or rounded, toothed and lobed. Flowers bisexual, radially symmetric in branched inflorescences. Perigynous zone bell-shaped. Sepals 5, erect. Petals 5, clawed. Stamens 5. Carpels 2, ovary 2-celled, stigmas without obvious styles. *Central & western USA.*

10. Telesonix Rafinesque. 1/1. Low-growing, herbaceous perennials, with short rhizomes, glandular-hairy. Basal leaves kidney-shaped, shallowly lobed and scalloped; stem-leaves smaller. Flowers bisexual, radially symmetric, *c.* 2 cm across, in clusters of 5--25 on reddish stems. Calyx 5-lobed. Petals 5, to 5 mm, ovate to rounded, reddish purple. Stamens 10. Styles 2, carpels 2. Fruit a many-seeded capsule. *Central USA.*

11. Tanakaea Franchet & Savatier. 2/2. Dioecious perennials with creeping rhizomes and slender stolons. Leaves all basal, evergreen, ovate-oblong, acute, base cordate to rounded, with forwardly pointing teeth, leathery, stalked. Scapes erect, with many small creamy-white flowers in a narrow or broad panicle, with small bracts. Sepals 5, lanceolate; petals absent; stamens 10, longer than sepals. Carpels 2, joined for most of their length, styles free. *Northeast Asia.*

12. Boykinia Nuttall. 8/3. Herbaceous perennials with short rootstocks, glandular-hairy. Leaves mostly basal (stem-leaves small), kidney-shaped, lobed and toothed; stipules present. Flowers bisexual in loose panicles. Perigynous zone bell-shaped to obconical. Sepals 5. Petals 5, often deciduous. Stamens 5. Carpels 2; styles 2; ovary 2-celled. Fruit a many-seeded capsule. *North America, eastern Asia.*

13. Elmera Rydberg. 1/1. Low-growing herbaceous perennials with slender, horizontal rhizomes. Basal leaves *c.* 2 × 3--5 cm, kidney-shaped, lobes rounded, toothed, somewhat hairy; stalk to 7 cm, glandular-hairy with large, membranous stipules; leaves on flowering stems 1--4, alternate, smaller. Flowers bisexual, 10--30 in a raceme to 25 cm. Sepals 5, triangular, erect, greenish yellow, *c.* 4 mm. Petals 5, erect, deeply 3--5-fid, yellowish white, 4--6 mm. Stamens 5, shorter than the sepals. Styles 2, thick, carpels 2. Fruit a many-seeded capsule. *Northwest North America.*

14. Lithophragma (Nuttall) Torrey & Gray. 9/1. Herbaceous perennial with rhizomes bearing bulb-like tubers. Leaves mostly basal, kidney-shaped or rounded, deeply or shallowly lobed, with long stalks. Flowers bisexual, slightly bilaterally symmetric, in few-flowered racemes. Calyx 5-lobed. Petals 5, clawed, often deeply divided. Stamens 10. Carpels and styles 3, ovary 1-celled. Fruit a many-seeded capsule. *Western North America.*

15. Tiarella Linnaeus. 7/6. Herbaceous perennial with rhizomes. Leaves mostly basal, simple, lobed or made up of 3 leaflets; stipules small; stalks long. Inflorescence a raceme or panicle. Flowers bisexual, small, white, radially symmetric. Sepals 5,

236

coloured. Petals 5, clawed. Stamens 10, protruding, sometimes of unequal size. Styles 2. Fruit a few-seeded capsule with 2 unequal flaps. *North America, eastern Asia.*

16. Heuchera Linnaeus. 55/10. Herbaceous perennials with branched, semi-woody, scaly rootstocks. Leaves mostly basal, simple, palmately lobed, usually with cordate bases and long stalks. Inflorescences paniculate, sometimes with leafy bracts. Flowers bisexual, usually radially symmetric, occasionally bilaterally symmetric where the perigynous zone is longer on one side. Sepals 5. Petals 5 or rarely absent. Stamens 5. Carpels 2, ovary 1-celled, styles 2. Fruit a many-seeded capsule. *North America.*

17. × Heucherella Wehrhahn. 2/2. Herbaceous perennials. Leaves mainly basal, ovate or rounded, long-stalked. Flowers small, numerous, in panicles. Sepals 5. Petals 5. Stamens 5--10, often 7 or 8. Carpels 2, slightly unequal. *Garden Origin.*

Hybrids between species of *Heuchera* and *Tellima*, which can be distinguished from the parents by the variable number of stamens within one inflorescence (5--10 but usually 7 or 8).

18. Bolandra Gray. 2/2. Herbaceous perennials with short rootstocks bearing bulbils. Leaves alternate with large stipules; lower leaves kidney-shaped, palmately veined and, long-stalked, upper leaves smaller, stalkless and with leaf-like stipules. Flowers bisexual, few, in loose panicles with conspicuous leafy bracts. Sepals 5, spreading, linear-lanceolate. Petals 5, linear, erect. Stamens 5, shorter than the petals. Ovary 2-celled; carpels 2. Fruit a capsule containing many seeds. *Western North America.*

19. Mitella Linnaeus. 20/8. Low-growing herbaceous perennials. Leaves mostly basal, simple, lobed, with cordate bases and long stalks. Inflorescences simple, often 1-sided, occasionally leafy racemes. Flowers numerous, small. Perigynous zone bell-shaped; sepals 5, small. Petals 5, usually deeply cut, green or white. Stamens 5 or 10, alternate or opposite to the petals. Styles 2, short, carpels 2, ovary 1-celled. Capsule containing numerous glossy, black seeds. *North America, eastern Asia.*

20. Tellima R. Brown. 1/1. Herbaceous perennials with thick, short rootstocks. Basal leaves to 10 cm across, kidney-shaped, stiffly hairy, 5--7-lobed, coarsely toothed, light green; stalks 15--25 cm, with long hairs; stem-leaves 2 or 3, smaller, almost stalkless. Flowers bisexual, radially symmetric, 15--30, more or less drooping in a raceme to 30 cm, stalk to 75 cm; flower-stalks to 4 mm. Perigynous zone 8--10 mm, inflated, glandular-hairy. Sepals 5, 3 mm, ovate, pale green. Petals 5, reflexed, 4--6 mm, lanceolate, fringed into linear segments, greenish white turning pinkish red. Stamens 10, filaments very short. Styles 2, divided to halfway, carpels 2. Fruit a many-seeded capsule. *Western North America.*

21. Tolmiea Torrey & Gray. 1/1. Herbaceous perennials with creeping rhizomes. Basal leaves 5--12 × 4--10 cm, heart-shaped, palmately veined, apex acute, shallowly lobed and toothed, hairy; stalks 5--20 cm, hairy; stem-leaves smaller with shorter stalks. Many leaves produce plantlets at the junction of blade and stalk. Inflorescence a narrow raceme to 60 cm, with usually 20--50 bisexual, shortly stalked, bilaterally symmetric flowers. Perigynous zone cylindric to funnel-shaped. Sepals 5, 3--4 mm, 2 smaller, ovate, greenish purple. Petals 4, *c.* 6 mm, thread-like, purplish brown.

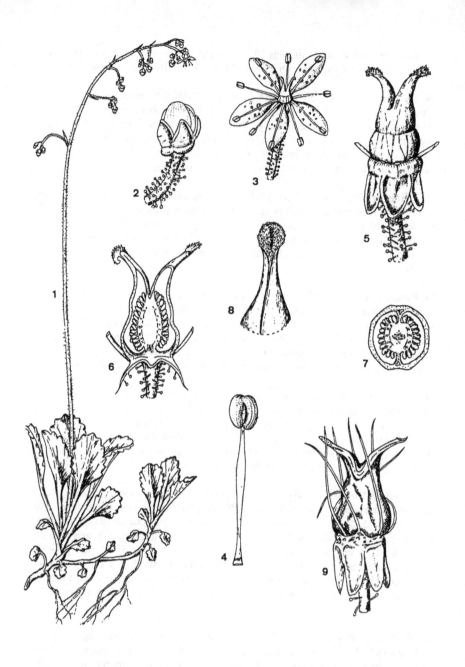

Figure 50. Saxifragaceae. *Saxifraga* × *urbicum*. 1, Rosette, stolon and flowering shoot. 2, Flower-bud. 3, Flower. 4, Stamen. 5, Flower with petals and stamens removed to show ovary. 6, Longitudinal section of ovary. 7, Transverse section of ovary. 8, Stigma. 9, Fruit.

Stamens usually 3, shorter than petals, unequal. Styles 2. Fruit a narrow, many-seeded capsule. *Western North America.*

22. Bensoniella Morton. 1/1. Herbaceous perennials with slender, scaly, branching rhizomes. Leaves all basal, 4--8 cm wide, shallowly 5--7-lobed, scalloped, base cordate, slightly hairy beneath; stalk to more than 7 cm. Flowers bisexual in rather narrow, dense racemes on stems to 25 cm. Perigynous zone saucer-shaped, creamy white. Sepals 5, *c.* 2 mm, creamy white. Petals 5, 2--3 mm, linear, white. Stamens 5, anthers pink. Styles 2. Capsules with many seeds. *Northwest USA.*
This genus was formerly known as *Bensonia* Abrams & Bacigalupi.

23. Mukdenia Koidzumi. 2/1. Herbaceous perennials with short, scaly rhizomes. Leaves 1 or 2, palmately lobed, with forwardly pointing teeth, hairless and slightly fleshy. Scape leafless. Inflorescence a compact bractless panicle of small, almost stalkless white flowers. Sepals 5 or 6, oblong, white; petals 5 or 6. Stamens 5 or 6, about equalling the petals; carpels 2, free to about the middle. Fruit a capsule. *China, Korea.*
The genus was formerly known as *Aceriphyllum* Engler.

24. Deinanthe Maximowicz. 2/2. Herbaceous perennials with woody, creeping rhizomes. Stems erect, with scale-like leaves at the base and 2--4 large leaves near the top. Leaves opposite, rather large, broadly ovate or elliptic, stalked, with forwardly pointing teeth, coarsely hairy. Inflorescence a hairless corymb, often with a few small sterile flowers consisting of 3 or 4 sepals. Fertile flowers nodding, with an inferior, conical ovary and 5 rounded sepals. Petals 5 or more, rounded, deciduous; stamens numerous, style 1, rather prominent with 5 small lobes. Ovary 5-celled, seeds numerous, with a short tail at each end. *Northeast Asia.*

25. Peltoboykinia (Engelmann) Hara. 2/1. Large herbaceous perennials with short, thick rhizomes. Basal leaves few, 10--25 cm, rounded, peltate on long stalks, palmate with 7--13 shallowly toothed lobes, almost hairless, glossy green; stem-leaves 2--3, smaller, almost stalkless. Flowers bisexual, radially symmetric, in terminal cymes on shoots to 60 cm. Calyx with 5 erect lobes, each 4--5 mm. Petals 5, 1--1.2 cm × 5 mm, oblanceolate, toothed at the tip, erect, pale yellow. Stamens 10, anthers dark brown. Styles 2, *c.* 3 mm. Capsules 1--1.3 cm. *Japan.*

26. Chrysosplenium Linnaeus. 60/2. Herbaceous perennials with underground rhizomes or rooting prostrate stems. Leaves stalked, alternate or opposite, slightly fleshy and often hairy. Erect stems with leaves often clustered towards the tips, the lower leaves much smaller. Flowers in bracted dichotomous cymes on erect shoots, the bracts leaf-like, sometimes coloured. Sepals 4 or 5, petals absent. Stamens 8 or 10, surrounded at the base by a fleshy, 8- or 10-lobed disc. Styles 2, free, carpels 2, united, more or less inferior, opening along the inner edge. Seeds numerous, black. *Europe, North Africa, northeast Asia, North America, temperate South America.*

27. Ribes Linnaeus. 150/70. Deciduous or occasionally evergreen, low to medium shrubs; branches with or without bristles and/or thorns. Leaves alternate, simple, usually lobed and toothed or scalloped, stalked. Flowers usually bisexual, sometimes unisexual and plants dioecious, usually 5-, rarely 4-parted, in few- to many-flowered racemes. Ovary inferior, styles 2. Fruit a berry, crowned by remains of calyx. *Temperate northern hemisphere, temperate South America.* Figure 51, p. 242.

28. Francoa Cavanilles. 1/1. More or less evergreen herb. Leaves in a basal rosette, lyrate, often with winged leaf-stalk, softly hairy, to 30 × 10 cm. Flowers white or pink, in dense terminal racemes on stems to 90 cm, occasionally branched. Sepals 4, acute, *c.* 5 mm. Petals 4, oblong, white or pink, with or without darker central markings, *c.* 10 × 3 mm. Stamens 8. Staminodes simple. Ovary superior, cylindric. Fruit a cylindric capsule. Seeds numerous, winged. *Chile.*

29. Parnassia Linnaeus. 50/4. Herbaceous perennials with short rootstocks and alternate, mostly basal leaves. Flowering stems bearing 1--6 leaves, flowers solitary. Sepals and petals 5, petals white. Stamens 5, alternating with the petals. Staminodes 5, yellowish, nectar-bearing, often fringed, opposite the petals. Ovary superior, ovoid, usually of 4 united carpels, style 1, short, stigmas 4. Fruit an ovoid or obovoid capsule, dehiscing along the middle of each cell. *North temperate areas.*

30. Jamesia Torrey & Gray. 1/1. Deciduous shrubs to 1 m or sometimes more, with peeling, brown, papery bark. Branches with solid pith, downy when young. Leaves 2--7 × 1.5--5 cm, opposite, simple, ovate, acute, coarsely toothed, wrinkled, dull green above, with thick, grey-white down beneath; stalk downy, 2--15 mm. Leaves on flowering shoots similar but smaller. Flowers *c.* 1.5 cm across, bisexual, slightly fragrant, in erect, terminal, many-flowered panicles to 5 cm. Sepals 5, *c.* 3 × 2 mm, ovate to lanceolate, downy. Petals 5, 8--10 × 5--6 mm, oblong to obovate, white or pinkish. Stamens 10. Styles 3--5, united at base. Ovary more or less superior. Fruit a many-seeded capsule to 4 mm. *North America.*

31. Deutzia Thunberg. 60/20 and many hybrids. Deciduous or rarely evergreen shrubs with pith-filled branches and peeling bark. Hairs often stellate with a varying number of rays; sometimes a long ray is borne erect, at right-angles to the others. Leaves opposite, usually shortly stalked, sometimes toothed, without stipules. Flowers in racemes, cymes, panicles or corymbs, or solitary on terminal or axillary shoots. Calyx teeth 5. Petals 5, edge-to-edge or overlapping in bud. Stamens 10, in 2 series of 5, those of the inner series smaller; filaments mostly broadly winged, with 2 teeth at the top on either side of the anther; anthers sometimes shortly stalked above the filament. Ovary 3- or 4-celled, inferior. Styles 3 or 4, thickened at apex. Fruit a capsule. *Himalaya to Japan and the Philippines.*

32. Philadelphus Linnaeus. 40/32 and some hybrids. Shrubs with mainly peeling bark; leaves usually deciduous, opposite, simple. Flowers in racemes, panicles or cymes, or solitary, often strongly scented. Sepals 4. Petals 4. Stamens numerous. Ovary inferior, surmounted by a nectar-secreting disc; carpels 4, united; styles 4, partially or wholly united. Fruit a many-seeded capsule. Seeds usually with tails. *South Europe, Caucasus, eastern Asia, Himalaya, North America.*

33. Kirengeshoma Yatabe. 2/1. Erect herbaceous perennial, with short and thick rhizomes. Leaves opposite, the lower long-stalked, the upper stalkless, all palmately lobed and coarsely sinuous-toothed. Flowers in terminal and axillary cymes, somewhat nodding on long stalks. Sepals 5, small, petals 5, narrowly ovate, rather thick, pale

240

yellow, not opening widely. Stamens 15, styles usually 3; ovary inferior, 3-celled. Capsule ovoid, 3-celled, seeds flat, with an irregular wing. *Northeast Asia.*

34. Hydrangea Linnaeus. 100/17 and some hybrids. Deciduous or evergreen shrubs, small trees or climbers. Leaves usually rounded-ovate and toothed, opposite or in whorls of 3. Bark often flaking when mature. Fertile flowers bisexual (rarely unisexual), radially symmetric, in panicles or corymbs. Sepals 4 or 5, small, inconspicuous. Petals 4 or 5, white, blue or pink. Stamens 8 or 10 (rarely more). Ovary inferior, 2--5-celled, containing many ovules. Fruit a 2--5-celled, many-seeded capsule. Many species also bear larger, sterile flowers borne at the outside of the corymb-like inflorescences. *Himalaya to Japan and the Philippines, North & South America.*

35. Schizophragma Siebold & Zuccarini. 2/2. Deciduous climbing shrubs climbing by aerial roots. Leaves opposite, on long stalks. Flowers white in large terminal cymes, the central flowers small and bisexual, outer ones sterile, reduced to a single large bract on a slender stalk. Fertile flowers with 4 or 5 sepals and petals. Stamens 10. Ovary inferior, united with the calyx tube, 10-celled. Stigma lobed. Fruit a ribbed capsule, dehiscing between the ribs. *Eastern Asia.*

36. Decumaria Linnaeus. 2/2. Deciduous or evergreen climbing shrubs clinging by aerial roots. Young shoots and buds downy. Leaves opposite, ovate, with or without shallow teeth. Flowers in terminal corymbs or panicles. Petals 7--10, oblong, white. Calyx-teeth 7--10, alternating with petals. Stamens 20--30. Ovary inferior with 7--10 cells. Stigma capitate. Fruit a ribbed capsule dehiscing between the ribs. *South-east USA, central China.*

37. Itea Linnaeus. 10/3. Deciduous or evergreen trees or shrubs. Branches with chambered pith. Leaves alternate, entire or toothed. Flowers small, bisexual, radially symmetric, in terminal or axillary racemes or spikes. Sepals 5, persistent. Petals 5, linear. Stamens 5. Styles 2, united; ovary superior, 2-celled. Fruit a many-seeded capsule. *North America, eastern Asia.*

38. Carpenteria Torrey. 1/1. Evergreen bushy shrubs to 6 m, stems angled, branches pithy. Leaves opposite, to 11 × 2.5 cm, simple, lanceolate, entire, acute, base narrowed, hairless and bright green above, glaucous with short white hairs beneath; stalk to 5 mm. Flowers 5--7 cm across, bisexual, fragrant, in terminal clusters of 3--7. Sepals usually 5, to 10 × 6 mm, ovate, downy. Petals 5, to 2.5 × 2.5 cm, circular, white. Stamens numerous with conspicuous yellow anthers. Style 1, 5--7-lobed. Ovary superior. Fruit a conical capsule containing numerous seeds. *Western USA.*

39. Dichroa Loureiro. 13/1. Evergreen shrubs. Leaves opposite, simple, usually coarsely toothed. Flowers in terminal panicles, white, blue or pink. Calyx lobes 5, rarely 4 or 6. Petals 5, sometimes 6, usually more brightly coloured on inner surfaces. Stamens 10. Ovary inferior, 4-celled. Styles 4. Fruit a small berry. Seeds numerous. *Southeast Asia to Indonesia.*

40. Fendlera Engelmann & Gray. 3/2. Deciduous bushy shrubs with ribbed branches. Leaves opposite, ovate to lanceolate, entire, more or less stalkless, 1--3-veined.

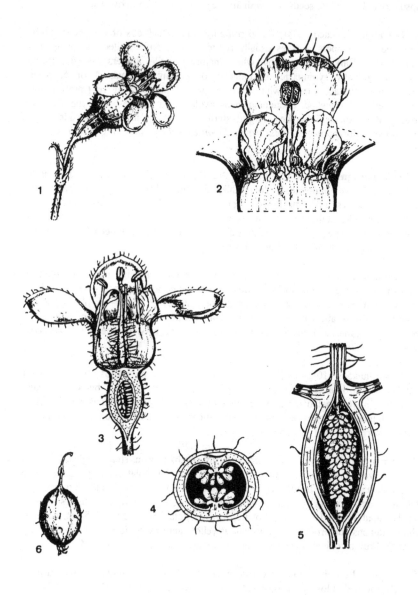

Figure 51. Saxifragaceae/Grossulariaceae. *Ribes uva-crispa.* 1, Single flower. 2, Part of the upper portion of the perigynous zone from inside, showing attachment of sepal, petals and stamen. 3, Longitudinal section of flower. 4, Transverse section of ovary. 5, Longitudinal section of ovary. 6, Fruit.

242

Flowers bisexual, 1--3, stalked, on short, lateral branches. Sepals 4, small. Petals 4, clawed, ovate with toothed margin. Stamens 8. Styles 4, completely free. Ovary half-inferior. Fruit a many-seeded capsule opening by 4 flaps, surrounded by the persistent calyx. *Southwest USA, Mexico.*

41. Platycrater Siebold & Zuccarini. 1/1. Deciduous shrub, sometimes prostrate with papery bark. Leaves opposite, 10--15 × 3--7 cm, oblong to broadly lanceolate, sparsely hairy above, hairy beneath, with forwardly pointing teeth, with slender points; stalk 5--30 cm. Flowers with long slender stalks; of 2 types; sterile ornamental, at the top of cyme branches, 1--3 cm across; calyx large, shallowly 3- or 4-lobed; fertile, less showy flowers lower down, 2--3 cm across; calyx smaller, deeply 4-lobed to near the base, sepals ovate-lanceolate; petals 4, white, very thick; stamens numerous; styles 2, persistent; ovary inferior; capsules cone-shaped. *Japan.*

42. Pileostegia Hooker & Thomson. 3/1. Evergreen climbing or prostrate shrubs, climbing by aerial roots. Leaves opposite, simple, usually entire, leathery. Flowers small, white, in terminal panicles. Calyx cup-shaped with 4 or 5 short lobes. Petals 4 or 5, coherent, falling quickly. Stamens 8--10, prominent. Ovary inferior, 4- or 5-celled. Fruit a spherical ribbed capsule, dehiscing between the ribs. *Eastern Asia.*

43. Carpodetus Forster & Forster. 10/1. Shrubs or small trees. Leaves alternate, without stipules. Flowers small, in few-flowered panicles. Calyx-tube united with ovary, bearing 5 or 6 deciduous lobes. Petals and stamens 5 or 6, alternating, at the margin of an epigynous disc. Ovary half inferior. Stigma capitate. Fruit a spherical, indehiscent capsule with 3--5 cells. Seeds numerous. *New Guinea, New Zealand.*

44. Escallonia Mutis. 39/11 and some hybrids. Evergreen or rarely deciduous shrubs, occasionally small trees; branches, leaves and inflorescences hairy, hairless or with stalked glands. Leaves alternate, usually somewhat leathery, toothed. Flowers in short racemes, panicles or solitary in axils of upper leaves. Calyx-tube short, lobes 5, erect or spreading; petals 5, obovate or circular and spreading or with an erect long narrow claw and spreading circular apical portion, the latter type of flower appearing tubular. Stamens 5, alternating with and about as long as petals. Ovary inferior, 2- or 3-celled; style 1, stigma capitate or rarely bifid; disc conical or cushion-shaped and surrounding base of style, or flat. Fruit a capsule with numerous seeds. *South America.*

45. Bauera Andrews. 3/1. Low evergreen shrubs, leaves opposite, with 3 leaflets, stalkless, appearing whorled. Flowers regular, solitary in upper leaf-axils, long-stalked or nearly stalkless. Sepals 4--10; petals 4--10, pink or white. Stamens numerous, on a conspicuous nectar-secreting disc. Ovary almost superior, 2-celled, styles 2, recurved. Fruit a 2-celled capsule. *Eastern Australia.*

46. Anopterus Labillardière. 2/2. Evergreen shrubs or small trees. Leaves alternate, simple, leathery, obovate to oblanceolate, tapered to both ends and with large rounded teeth, each gland-tipped. Flowers in upright terminal racemes, cup-shaped, white or pink-tinged. Petals 6--9 concave, obovate; calyx small, with 6--9 toothed, triangular lobes. Stamens as many as petals, with flattened tapering filaments. Ovary superior, 1-celled with numerous ovules. Fruit a cylindric capsule with 2 recurving valves. Seeds winged. *Australia.*

84. CUNONIACEAE

Evergreen trees or shrubs, rarely climbers. Leaves leathery, often glandular, opposite, rarely whorled, occasionally simple but more often compound, with 3 leaflets or pinnate. Stipules united in pairs, often conspicuous. Flowers unisexual or bisexual, in solitary or branched racemes or compact heads. Petals 4 or 5, free or united, sometimes absent, sepals larger than petals, 3--6, free or united. Stamens numerous, sometimes 4 or 5 (often alternating with petals), or 8--10. Ovary superior, with 2--5 free or united carpels; ovules in two ranks. Styles 2. Fruit a capsule, rarely a drupe or nut. Seed small, winged or hairy.

A southern hemisphere family of about 24 genera and 340 species, mostly from Australasia and the Pacific. Only 1 genus is hardy in northern Europe.

1. Weinmannia Linnaeus. 190/2. Trees or shrubs. Leaves opposite, simple, with 3 leaflets or pinnate with wings between the leaflets; stipules falling early. Flowers small, clustered in terminal or axillary racemes; sepals and petals 4 or 5; stamens 8 or 10; ovary superior, 2-celled; styles 2. Fruit a dry capsule; seeds often hairy, rarely slightly winged. *South America, Pacific Islands, New Zealand, Malaysia, Madagascar.*

85. PITTOSPORACEAE

Evergreen trees, shrubs or vines. Leaves alternate or whorled, usually entire. Inflorescences axillary or terminal; flowers bisexual (rarely unisexual), solitary or in clusters. Sepals 5; petals 5, longer than sepals; stamens 5, alternating with petals, all hypogynous. Ovary with 2--5 carpels, ovules numerous; style simple. Fruit a capsule or berry.

A family of about 9 genera and 300 species, concentrated in the southern hemisphere and Pacific region. Three genera contain hardy species.

1a. Robust shrubs; anthers ovate; fruit dehiscent   **2. Pittosporum**
  b. Twining plants with weak stems, woody at base; anthers linear or ovate;
     fruit an indehiscent berry   2
2a. Flowers blue, anthers linear, fused around style   **3. Sollya**
  b. Flowers not blue, anthers ovate, free   **1. Billardiera**

1. Billardiera Smith. 30/2. Vines or shrubs. Leaves entire. Flower solitary, or a few in clusters, pendent, bell-shaped, axillary or terminal. Sepals 5, petals 5, anthers 5, free. Fruit a berry. *Australia.*

2. Pittosporum Gaertner. 200/2. Trees or shrubs, usually evergreen. Leaves usually simple, often clustered towards tips of shoots. Flowers terminal or axillary, usually solitary. Sepals free or occasionally fused at base. *Pacific area, eastern Asia.*

3. Sollya Lindley. 2/1. Subshrubs with thin twining stems, usually hairless. Leaves stalkless, entire. Flowers in pendent clusters, or solitary, on slender stalks, blue. Anthers longer than filaments, joined to form cone around style; pollen shed inwards. Ovary with 2 cells. Fruit a juicy berry; seeds embedded in pulp. *Western Australia.*

86. ROSACEAE

Annual or perennial herbs, shrubs and trees. Branches often thorny. Leaves deciduous or evergreen, alternate or borne in apparently whorled clusters; entire, palmately compound or pinnately divided; stipules usually present, and attached to leaf-stalk, sometimes falling early. Flowers mostly bisexual, radially symmetric, sepals, petals and stamens (rarely perianth and stamens) perigynous to epigynous, solitary or in cymes. Sepals 5, rarely 3--10, often occurring as lobes on the perigynous zone; petals 5, rarely absent or 3--10, variously coloured, but mostly white, pink, red or yellow; stamens usually numerous, occasionally 1 or 5, free and attached to the perigynous zone; ovary superior or inferior, carpels 1--many, free or united; ovules 1--several; placentation basal, axile or parietal; styles as many as carpels. Fruit various, sometimes an achene, follicle, hip (a persistent, cup-shaped to tubular perigynous zone containing achenes), pome (a fruit made up of tissues from the ovary surrounded by those of the fleshy receptacle) or drupe. See figures 52--57, pp. 248, 254, 256, 262, 264 & 268.

A family of about 3200 species in 115 genera, widely distributed throughout the world, but most abundant in temperate regions. A few species of the genus *Quillaja* Molina (figure 52, p. 248) may occasionally be grown in very favourable areas; they are evergreen shrubs with toothed leaves, small, soon deciduous stipules, and flowers with 10 stamens.

1a. Leaves simple or shallowly lobed; if deeply lobed then at least part of the
    blade visible on either side of midrib   2
  b. Leaves pinnate or palmate (divided to mibrib), leaflets stalked or stalkless
                                       53
2a. Leaves evergreen, needle-like; solitary or clustered   **27. Adenostoma**
  b. Leaves deciduous, or, if evergreen, then not needle-like   3
3a. Deciduous perennial herbs, or plants with woody stock and annual leafy
    and flowering shoots   4
  b. Evergreen or deciduous trees and shrubs   5
4a. Leaves palmately divided, sepals yellowish green; petals absent
                             38. Alchemilla
  b. Leaves simple or 3--5 lobed; sepals reddish green; petals white **19. Rubus**
5a. Dwarf or prostrate shrublets; non-flowering stems not exceeding 10 cm 6
  b. Shrubs or trees; if herbaceous, leafy shoots at least 20 cm   9
6a. Leaves deeply 3--5-lobed in upper part, resembling those of a mossy
    saxifrage   **12. Luetkea**
  b. Leaves entire or margins scalloped or toothed   7
7a. Dense, cushion-forming tufted shrublet; leaves 2.5--4 mm; flowers almost
    stalkless, pink to purple   **11. Kelseya**
  b. Loose shrublets, stems not tufted; leaves 5 mm or more; flowers borne on
    long stalks, creamy white, white or rarely yellow   8
8a. Leaves stalkless, light green above and below; flowers borne in spike-like
    racemes   **10. Petrophytum**
  b. Leaves stalked; margins scalloped or toothed, dark green above, white-
    felted beneath; flowers solitary on long stalks   **31. Dryas**
9a. Leaves evergreen   10
  b. Leaves deciduous   24
10a. Leaves often deeply lobed to more than halfway towards midrib   11
  b. Leaves entire or shallowly toothed   13
11a. Leaves 6--15 mm, deeply divided into 3--9 very narrow lobes
                        30. Cowania
  b. Leaves 2 cm or more, lobes broad   12

12a. Flowers 2--5, in umbels; stamens 30--50   **41. Docynia**
  b. Flowers many, in corymbs or panicles, very rarely solitary; stamens
    5--25   **57. Crataegus**
13a. Ovary superior   14
  b. Ovary inferior   16
14a. Petals absent; style very long with conspicuous plumose hairs; fruit a
    1-seeded achene   **28. Cercocarpus**
  b. Petals present; other characters not as above   15
15a. Fruit a hip   **20. Rosa**
  b. Fruit a drupe   **58. Prunus**
16a. Leaf-margin entire   17
  b. Leaf-margin toothed   21
17a. Flowers borne in erect racemes or panicles; flower-stalks fleshy
      49. Rhaphiolepis
  b. Flowers solitary or in cymes at the end of lateral spurs   18
18a. Sepals with 2 bracteoles at base; fruit a dry capsule surrounded by
    calyx-lobes   **54. Dichotomanthes**
  b. Sepals without bracteoles; fruit fleshy, crowned by persistent calyx-lobes
      19
19a. Stamens 10   **52. Heteromeles**
  b. Stamens *c.* 20   20
20a. Branches conspicuously thorny   **55. Pyracantha**
  b. Branches thornless   **53. Cotoneaster**
21a. Fruits 1--4 cm, usually pear-shaped, yellow, fragrant   **46. Eriobotrya**
  b. Fruits 1 cm or less, orange, red, dark red or bluish black, rarely yellow   22
22a. Fruits bluish black   **49. Rhaphiolepis**
  b. Fruits orange, red, dark red or yellow   23
23a. Branches conspicuously thorny; fruits usually compressed laterally
      55. Pyracantha
  b. Branches usually thornless; fruits rounded or egg-shaped   **51. Photinia**
24a. Ovary or ovaries superior   25
  b. Ovaries inferior   40
25a. Leaves opposite; sepals and petals 4   **16. Rhodotypos**
  b. Leaves alternate; sepals and petals 5 or more   26
26a. Stipules absent   27
  b. Stipules present, but sometimes falling early   28
27a. Leaves lobed and toothed, widest below the middle; fruit of 5 hairy
    achenes enclosed in a persistent calyx   **14. Holodiscus**
  b. Leaves entire, widest above the middle; fruit of five, 2-seeded follicles
    joined at the base   **9. Sibiraea**
28a. Petals absent   **18. Neviusia**
  b. Petals present   29
29a. Petals yellow   30
  b. Petals white, pink, purple or red   32
30a. Branches thornless; fruit a group of 5--8 achenes surrounded by a
    persistent calyx   **17. Kerria**
  b. Branches thorny or with prickles; fruit a drupe or a hip   31

31a. Pith of young branches channelled, bark flaking; fruit a 1-seeded drupe

                                          59. Prinsepia

  b. Pith of young branches entire, bark smooth; fruit a hip   **20. Rosa**

32a. Fruit a 1-seeded drupe   33

  b. Fruit not as above   35

33a. Sepals 10, petals reddish brown   **60. Maddenia**

  b. Sepals 5, petals white, pink or purplish   34

34a. Pith of young branches channelled, each flower-stalk with 2 bracts

                                          61. Oemleria

  b. Pith of young branches entire, flower-stalks without bracts **58. Prunus**

35a. Leaves entire or lobed, or if divided then not more than halfway towards
    midrib   36

  b. Leaves divided almost to midrib   39

36a. Leaf-margins entire or toothed (rarely lobed)   37

  b. Leaf-margins lobed and toothed   38

37a. Leaf-margins entire; flowers at least 2 cm across; fruit a woody capsule

                                          1. Exochorda

  b. Leaf-margins toothed or lobed, or if entire then flowers not more than
    8 mm across; fruit a group of dehiscent follicles   **8. Spiraea**

38a. Flowers in rounded, umbel-like racemes; fruit bladder-like, opening by 2
    slits   **5. Physocarpus**

  b. Flowers in long conical racemes or panicles; fruit not bladder-like

                                          6. Neillia

39a. Leaves hairless or with few hairs, widest below the middle, ovate, doubly
    toothed and lobed; flowers in panicles; fruit an irregularly shaped follicle

                                       7. Stephanandra

  b. Leaves densely hairy especially beneath, widest above the middle, obovate,
    deeply divided into 3--7 lobes; flowers solitary; fruit an achene

                                    29. Purshia

40a. Leaves palmately or pinnately lobed   41

  b. Leaves entire but often toothed   44

41a. Fruit with 1--5 bony nutlets   **57. Crataegus**

  b. Fruit with 1--5 cells, each containing 1 or more seeds   42

42a. Styles free   **42. Pyrus**

  b. Styles joined at the base   43

43a. Leaves palmately lobed   **44. Eriolobus**

  b. Leaves never palmately lobed   **43. Malus**

44a. Stamens not more than 25   45

  b. Stamens numerous   47

45a. Stipules persistent, kidney-shaped   **40. Chaenomeles**

  b. Stipules deciduous, triangular   46

46a. Petals not more than 1 cm; fruit not more than 1 cm, red to black

                                          50. Aronia

  b. Petals 1.5--2 cm; fruit 2.5 cm across, green at first becoming brown

                                       56. Mespilus

47a. Stipules persisent, conspicuous, large and kidney-shaped **40. Chaenomeles**

  b. Stipules deciduous, usually inconspicuous   48

48a. Petals 2--3 cm   **39. Cydonia**

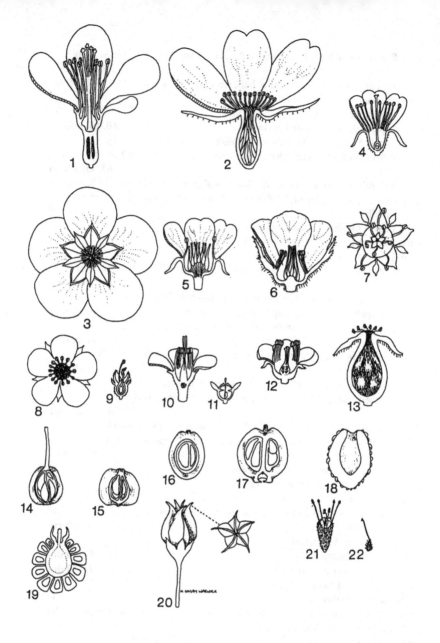

Figure 52. Rosaceae. Longitudinal sections of flowers: 1, *Chaenomeles*. 2, *Rosa*. 4, *Prunus*. 5, *Geum*. 6, *Rubus*. 9, *Acaena*. 10, *Rhaphiolepis*. 11, *Alchemilla*. 12, *Sorbaria*. Longitudinal sections of fruits: 13, *Rosa*. 14, *Malus*. 15, *Cotoneaster*. 16, *Prunus*. 17, *Sorbus*. 18, *Fragaria*. 19, *Rubus*. Flower from below: 3, *Potentilla*. Flowers from above: 7, *Quillaja*. 8, *Fragaria*. Whole fruits: 20, *Physocarpus*. 21, *Acaena*. Single achene: *Geum*.

248

   b. Petals 6--18 mm   49

49a. Fruit with 1--5 bony nutlets   **53. Cotoneaster**

   b. Fruit with 1--5 cells, each containing 1 or more seeds   50

50a. Flowers in racemes   **48. Amelanchier**

   b. Flowers in umbels, corymbs, panicles or solitary   51

51a. Leaves narrowly oblong, stalkless or very shortly stalked, clustered at the
     ends of short shoots   **47. Peraphyllum**

   b. Leaves not as above   52

52a. Calyx lobes deciduous   **45. Sorbus**

   b. Calyx lobes persistent on fruit   **51. Photinia**

53a. Deciduous perennial herbs or plants with woody stock and annual
     flowering shoots   54

   b. Evergreen or deciduous trees or shrubs   69

54a. Leaves with 3 leaflets   55

   b. Leaves pinnately or palmately divided with more than 3 leaflets   61

55a. Petals sometimes absent   **36. Sibbaldia**

   b. Petals present, usually 5--7   56

56a. Plants with spreading stolons or rhizomes   57

   b. Plants without stolons or rhizomes   58

57a. Flowers white or pink; fruiting receptacle juicy   **37. Fragaria**

   b. Flowers yellow or creamy-yellow; fruiting receptacle dry **33. Waldsteinia**

58a. Styles lengthening in fruit, often with feathery hairs, or jointed   **32. Geum**

   b. Styles not lengthening in fruit   59

59a. Calyx bell-shaped with 5 teeth   **2. Gillenia**

   b. Calyx 5-lobed, with 5 epicalyx segments between the lobes   60

60a. Stamens 4 or 5   **36. Sibbaldia**

   b. Stamens 10--30   **35. Potentilla**

61a. Petals absent   62

   b. Petals present   64

62a. Leaves palmately divided or lobed; sepals 4 with 4 alternating epicalyx-
     segments   **38. Alchemilla**

   b. Leaves pinnately divided; sepals 3--6 or if 4, epicalyx absent   63

63a. Stipules forming a sheath around leaf-stalk, with a leafy lobed apex on
     each side; receptacle usually with barbed spines   **25. Acaena**

   b. Stipules leaf-like, crescent-shaped, joined to leaf-stalk; receptacle without
     barbed spines   **22. Sanguisorba**

64a. Epicalyx present   65

   b. Epicalyx absent   67

65a. Leaves palmate or if pinnate then all leaflets similar in size   **35. Potentilla**

   b. Leaves pinnate, with terminal leaflet much larger than lateral leaflets   66

66a. Styles persistent as awns on the fruits   **32. Geum**

   b. Styles deciduous   **34. Coluria**

67a. Flowers yellow   **21. Agrimonia**

   b. Flowers cream, white, pink or purplish   68

68a. Flowers unisexual in spike-like racemes; fruit a group of follicles
                                   13. Aruncus

   b. Flowers bisexual in large panicles; fruit a group of achenes
                                 15. Filipendula

69a. Leaves evergreen   70
    b. Leaves deciduous   75
70a. Petals absent   71
    b. Petals present   72
71a. Fruit a white berry   **24. Margyricarpus**
    b. Fruit a group of achenes; receptacle with barbed spines   **25. Acaena**
72a. Epicalyx-segments present between calyx lobes   **35. Potentilla**
    b. Epicalyx-segments absent   73
73a. Stamens 1--20   **12. Luetkea**
    b. Stamens numerous   74
74a. Fruit a hip   **20. Rosa**
    b. Fruit a 1-seeded achene terminated by a feathery style   **30. Cowania**
75a. Ovary inferior; fruit a pome   **45. Sorbus**
    b. Ovary or ovaries superior; fruit not a pome   76
76a. Petals absent   77
    b. Petals present   79
77a. Stamens 2--7   **25. Acaena**
    b. Stamens 25 or more   78
78a. Leaves with 5--7 leaflets   **26. Polylepis**
    b. Leaves with 7--25 leaflets   **23. Sarcopoterium**
79a. Leaves 2-pinnate   80
    b. Leaves pinnate or with 3 leaflets   81
80a. Leaves fern-like, primary leaflets not more than 6 mm; fruit of 5 follicles   **4. Chamaebatiaria**
    b. Leaves not as above, primary leaflets 4--20 cm; fruit a berry of fleshy drupelets   **19. Rubus**
81a. Fruit a hip   **20. Rosa**
    b. Fruit not as above   82
82a. Fruit a berry of fleshy drupelets   **19. Rubus**
    b. Fruit not as above   83
83a. Leaves with 15--25 leaflets; fruit a capsule   **3. Sorbaria**
    b. Leaves usually with 3--9 leaflets; fruit a head of achenes   **35. Potentilla**

1. Exochorda Lindley. 4/3 and some hybrids. Deciduous shrubs. Leaves alternate, simple, stalked. Flowers frequently unisexual, in terminal racemes on previous year's shoots. Sepals 5. Petals 5, white. Perigynous zone broadly top-shaped with a large disc. Stamens 15--30, inserted on the margin of the disc. Styles 5, fused towards base. Ovary superior. Fruit with 5 wings or 5-angled, at maturity separating into 5 hard capsules; 1 or 2 seeds in each capsule, flattened, winged. *Eastern Asia.*

2. Gillenia Moench. 2/2. Perennial herbs with erect branches to 1.2 m. Leaves with 3 lanceolate leaflets, margins irregularly and sharply toothed. Stipules paired, small, inconspicuous or larger and leaf-like. Flowers bisexual, in long-stalked terminal panicles; petals 5, oblong-ovate, white or pinkish. Calyx bell-shaped with 5 teeth; stamens 10--20; styles 5; carpels with 2--4 ovules. Fruit of 5 leathery follicles, each with 1--4 seeds, testa leathery, endosperm striped. *Eastern North America.*

3. **Sorbaria** (de Candolle) Braun. 10/6. Deciduous shrubs, frequently with stellate hairs. Leaves large, pinnate with up to 25 toothed leaflets, alternate, with stipules. Inflorescence a large, terminal panicle. Flowers with sepals, petals and stamens perigynous, the perigynous zone cup-shaped, lined by a nectar-secreting disc. Sepals 5. Petals 5. Stamens 20--50. Ovary of 5 (rarely 4) carpels which are united towards the base, each with several ovules. Fruit apparently a capsule formed from the more or less united carpels, the follicles opening lengthwise along their outer margins. Seeds long, pale brown. *Mostly from Asia.* Figure 52, p. 248.

4. **Chamaebatiaria** (Brewer & Watson) Maximowicz. 1/1. Aromatic, deciduous, upright shrub to 1.5 m, with dense stellate hairs and stalked glands, at least on the young growth. Leaves alternate (sometimes appearing almost whorled on condensed lateral shoots), to 3 cm, ovate-lanceolate, bipinnate with very numerous small leaflets; stipules small, linear-lanceolate. Flowers in dense terminal panicles, to 1 cm across. Sepals 5, densely hairy, often greyish or whitish. Petals 5, circular or almost so, spreading, white. Stamens numerous, borne on the margins of the entire nectar-secreting disc. Carpels 5, hairy, united only at their extreme bases. Fruit a group of 5 follicles, each few-seeded, united only at their extreme bases. *Western USA.*

5. **Physocarpus** Maximowicz. 10/7. Deciduous shrubs with bark peeling in thin strips; hairs, when present, mostly stellate. Leaves alternate, with small, often toothed stipules, stalked, usually with 3--5 palmately arranged main veins (other venation pinnate), margins toothed or scalloped and often 3- or 5-lobed; buds in the leaf-axils several, one above the other. Flowers in corymb- or umbel-like racemes; bracts usually small and falling early. Sepals, petals and stamens perigynous, perigynous zone cup-shaped. Sepals 5, generally hairy on both surfaces. Petals 5, usually almost circular, sometimes irregularly notched or toothed. Stamens 20--40, anthers usually reddish. Ovary superior of 2--5 (rarely 1) carpels, united at least at the base, each containing 2--5 ovules. Fruit a group of usually bladder-like follicles opening along both sutures. Seeds usually 2 per follicle, yellowish and shining, each with an appendage (caruncle). *Eastern Asia, North America.* Figure 52, p. 248.

6. **Neillia** D. Don. 10/5. Arching deciduous shrubs to 3 m with brown, shredding bark. Leaves with variably deciduous stipules, ovate to lanceolate, toothed, variably divided into 1--5 (usually 3) lobes, usually hairy, at least on the veins beneath. Axillary buds multiple and one above the other on the vegetative shoots, sometimes also on flowering shoots. Inflorescence a terminal panicle or racemes borne on short, leafy, axillary shoots of the previous year. Flowers with bracts, shortly stalked. Sepals, petals and stamens perigynous, the perigynous zone bell-shaped or tubular, persisting and enlarging in fruit and often becoming bristly-glandular. Calyx-lobes 5, acute, downy within. Petals 5, white, pink or red. Stamens 15--30 (rarely as few as 10) in 2 or 3 whorls, inflexed in bud. Ovary superior. Carpels free, usually 1, rarely 2--5, each with 2--10 ovules. Fruit a follicle or group of follicles borne within the persistent, enlarged perigynous zone. *Himalaya, China, eastern Asia.*

7. **Stephanandra** Siebold & Zuccarini. 4/2. Arching shrubs to 2 m or more. Leaves deciduous, ovate to narrowly ovate, lobed and doubly toothed, with stipules and multiple buds, one above the other in their axils. Flowers in a panicle borne terminally on long or short shoots; sepals, petals and stamens perigynous. Perigynous zone

251

hemispherical, lined with a nectar-secreting disc. Sepals 5. Petals 5, white. Stamens 10--20. Ovary superior, hairy, of a single carpel which contains a single ovule. Fruit an irregularly shaped follicle. Seed 1, pale to dark brown. *Japan, China.*

8. Spiraea Linnaeus. 100/60 and many hybrids. Deciduous shrubs, buds small with 2--8 exposed scales. Leaves alternate, simple, lobed, toothed or occasionally entire; usually with short stalks, stipules usually absent. Flowers bisexual or unisexual in umbel-like racemes, corymbs or panicles; perigynous zone cup-shaped, lined with a nectar-secreting disc; sepals 5, petals 5, stamens 15--60, inserted between the disc and the sepals. Ovary superior, carpels 5, styles 5. Fruit a dehiscent follicle, seeds 2--10, oblong. *Temperate Europe, Asia, North America.* Figure 53, p. 254.

9. Sibiraea Maximowicz. 1/1. Deciduous, prostrate to erect shrubs to 1 m; young shoots dark reddish, hairless. Leaves alternate (sometimes appearing whorled on condensed lateral shoots), linear-oblong to narrowly obovate, shortly stalked, obtuse but slightly mucronate, 3--10 cm × 6--20 mm, slightly hairy beneath and on the margins. Stipules absent. Flowers in racemes clustered at the tips of the shoots, to 6 mm across, mostly functionally unisexual (plants effectively dioecious). Sepals 5, to 1 mm, triangular, erect. Petals white to greenish, 2--2.5 mm, almost circular but shortly clawed. Stamens numerous, borne on the margins of a cup-shaped, indented perigynous zone lined with a nectar-secreting disc. Ovary superior. Carpels 5, joined only at their extreme bases. Fruit a group of five 2-seeded follicles. *Eastern Asia, with disjunct occurrences in southern Europe.*

10. Petrophytum (Torrey & Gray) Rydberg. 3/3. Small, low, hummock-forming shrubs with dense mats of evergreen basal leaves. Stems bearing reduced, bract-like leaves. Inflorescence a very dense, spike-like raceme. Flowers with bracts, sepals, petals and stamens perigynous, perigynous zone top-shaped to hemispherical, lined with a nectar-secreting disc. Sepals 5. Petals 5, creamy white. Stamens 20--40. Ovary superior of usually 5 (more rarely 3 or 7) free carpels, each containing 2--4 ovules, style long. Fruit a group of few-seeded follicles. *North America.*

11. Kelseya (Watson) Rydberg. 1/1. Evergreen, cushion-forming, tufted shrublets, 5--8 × 7--8 cm. Leaves entire, 2.5--4 mm, leathery, densely overlapping, covered with fine silky hairs. Flowers solitary, pinkish to purplish, fading to brown, nearly stalkless; sepals 5; petals oblong-elliptic, 2--3 mm; stamens 10, longer than petals, reddish purple; styles 5. Fruit a follicle, *c.* 3 mm, splitting completely at 2 seams. *Western North America.*

12. Luetkea Bongard. 1/1. Low, evergreen shrubs with creeping woody stems and erect, herbaceous, sparsely hairy flowering shoots, to 20 cm. Leaves to 2 cm, mostly towards the bases of the flowering stems, alternate, stalked, usually divided into 3 narrow segments which are themselves deeply 3-lobed, the ultimate segments more or less linear; leaves on the flowering stems smaller and less divided. Flowers in terminal racemes, 5--8 mm across; stalks short. Sepals 5, triangular, acute. Petals white, *c.* 4 mm, exceeding the sepals. Stamens 20, somewhat united towards their bases, attached to the margins of a 10-lobed, fleshy, nectar-secreting disc. Ovary superior, of 5 free carpels. Fruit a group of 5 follicles, each containing several seeds. *Western North America.*

13. Aruncus Linnaeus. Few/3. Large herbs with rather thick, woody rhizomes, dioecious or almost so. Leaves alternate, arising from swollen nodes, divided into 3 (rarely 4 or 5) leaflets which themselves are similarly divided or more commonly pinnate, the ultimate segments irregularly doubly toothed. Flowers unisexual, in spike-like racemes which form a panicle (rarely a simple raceme). Sepals 5, small. Petals white or cream, slightly exceeding the sepals. Stamens numerous, small and vestigial in female flowers, attached at the margins of a cup-shaped, entire nectar-secreting disc. Carpels 3 (rarely more), free, absent in male flowers. Fruit a group of follicles. *Northern hemisphere.*

14. Holodiscus Maximowicz. 8/3. Shrubs or small trees to 8 m. Leaves alternate, deciduous, simple, toothed and/or lobed, usually whitish or greyish hairy at least beneath. Stipules absent. Inflorescence a large terminal panicle with many flowers, lateral branches spreading or drooping; axis, branches and calyces densely hairy. Flowers small, shortly stalked, each subtended by a bract. Sepals 5, triangular. Petals white, almost circular, shortly clawed, spreading. Stamens 20, borne on the margins of an entire, cup-shaped disc. Ovary superior, of 5 free carpels, usually hairy, each containing 2 ovules. Fruit a group of 5 indehiscent, 1-seeded achenes. *Western North America.*

15. Filipendula Miller. 10/7. Perennial herbs (often large) with tuberous roots or with rhizomes. Basal and lower stem-leaves pinnate with 3--5 or more major leaflets, often with smaller leaflets or lobes between them; terminal leaflet usually larger than the laterals, 3--5-lobed, the laterals lobed or not, all toothed. Inflorescence a large panicle with many flowers, branches arching erect or spreading. Flower-buds more or less spherical; flowers without bracts. Sepals 5 or 6, small, erect in flower, persistent and reflexed in fruit. Petals 5 or 6, white, cream, pink or purplish red, spreading. Stamens 20--40, attached at the margin of a small, cup-shaped perigynous zone lined with a nectar-secreting disc. Carpels 5--10, free, sometimes compressed, 2-seeded, usually attached by their bases but in 1 species attached on their inner faces, with part of the carpel projecting below the point of attachment. Fruit a group of achenes, erect and separate, or (in the species with laterally-attached carpels) those of each flower spirally coiling together. *North temperate areas.*

16. Rhodotypos Siebold & Zuccarini. 1/1. Deciduous shrub to 2 m or more; branches greyish, hairless. Leaves opposite, shortly stalked, 4--10 cm, ovate to ovate-oblong, apex acuminate, base rounded, irregularly doubly toothed, with long whitish hairs on both surfaces. Stipules linear or thread-like. Flowers terminating short, leafy, lateral shoots, 2.5--4 cm across. Sepals 4, toothed, alternating with 4 additional small lobes. Petals 4, white, almost circular. Stamens numerous. Ovary superior. Carpels 4, free. Fruit a group of up to 4 shining, black, almost dry drupes surrounded by the persistent and enlarged calyx. *China, Japan, Korea.*

17. Kerria de Candolle. 1/1. Deciduous shrub to 3 m; branches green, hairless. Leaves simple, alternate, stalks short, 2.5--11 cm, lanceolate to ovate, apex acuminate, base cordate to truncate, doubly toothed, hairless or with adpressed bristles above, sparsely white-hairy beneath. Stipules narrow, falling early. Flowers solitary, terminating short, leafy lateral shoots, 2--5 cm across. Sepals 5. Petals 5 (many in a widely grown cultivar), yellow or orange-yellow. Stamens numerous. Ovary superior; carpels 5--8,

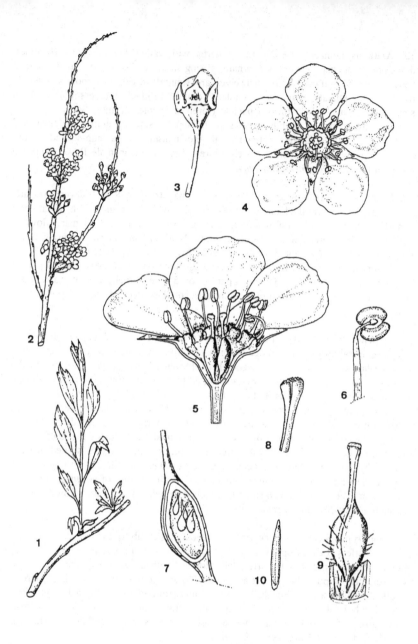

Figure 53. Rosaceae. *Spiraea* × *arguta*. 1, Leafy shoot. 2, Flowering shoot. 3, Flower-bud. 4, Flower from above. 5, Longitudinal section of flower. 6, Stamen. 7, Longitudinal section of a single carpel. 8, Styles and stigmas of two carpels. 9, A single carpel after fertilisation. 10, Seed.

free. Fruit a group of up to 8 achenes surrounded by the persistent calyx. *Probably China.*

18. Neviusia Gray. 1/1. Deciduous shrub, 1--2 m. Stems brownish, hairless. Leaves alternate, shortly stalked, ovate to ovate-oblong, 3--7 cm, acute or acuminate at the apex, rounded to slightly cordate at the base, irregularly doubly toothed, with adpressed, white bristles on both surfaces. Stipules very small, thread-like. Flowers in cymes or solitary, terminating short, leafy, lateral shoots, to 1.5 cm across. Sepals 5, spreading, toothed. Petals absent. Stamens numerous, yellowish white, exceeding the sepals. Carpels 2--4, free. Fruit a group of 2--4 achenes surrounded by the persistent calyx. *Southeast USA.*

19. Rubus Linnaeus. Many, some apomictic/over 50. Erect, scrambling, trailing or prostrate shrubs or low shrubs, shoots rarely herbaceous and dying down in winter, often with prickles on the stems, leaf- and leaflet-stalks and inflorescences. Leaves deciduous or evergreen, entire, lobed or divided into 3--many leaflets which are usually toothed and may themselves be lobed; stipules always present, often conspicuous, sometimes falling early. Flowers in clusters, racemes or panicles, sometimes solitary, usually bisexual. Sepals 4 or 5 or rarely more, spreading, erect or reflexed. Petals 4 or 5 or rarely more, very variable in size, white, pink, red or purple, spreading or erect and adpressed to the stamens. Stamens numerous; anthers sometimes hairy. Ovary superior; carpels 5--many, free, borne on a usually cylindric or conical receptacle. Fruit a 'berry' composed of 5--many variably coherent fleshy drupelets, which may separate from the receptacle as a (hollow) unit, or may fall from the plant by abscission of the flower-stalk while still attached to the receptacle. *More or less cosmopolitan.* Figure 52, p. 248.

20. Rosa Linnaeus. 150/many, and many hybrids. Deciduous, or sometimes evergreen, shrubs with erect, arching or scrambling, occasionally trailing stems, usually armed with prickles and/or bristles. Stipules usually present, persistent, usually joined to leaf-stalk for most of their length. Leaves alternate, usually odd-pinnate, rarely with 3 leaflets or simple; leaflets toothed. Perigynous zone spherical to urn-shaped. Flowers solitary or in corymbs, usually borne at the end of short branches, single to double. Sepals 5 (rarely 4), entire or the 2 outer and half the third one with lateral lobes, and the inner 2 and half of the third one entire, the tips acute to attenuate or broadened and leafy. Petals 5 (rarely 4) in single flowers, usually obovate, tip often notched, white, cream, pink, red, purplish, orange or yellow. Stamens (in single flowers) 30--200, in several whorls. Ovaries many, each with 1 ovule, free, carried on the base and/or sides of the enclosing perigynous zone. Styles free or united into a column, protruding from mouth of perigynous zone or not. Fruit (hip or hep) containing many achenes, usually red or orange, sometimes blackish or green, enclosed by the fleshy perigynous zone. Achenes 1-seeded, bony, often hairy. *Northern hemisphere.* Figures 52, p. 248, 54, p. 256.

Problems of classification and naming are rife in this genus, many species of which have been cultivated and hybridised for centuries. Species roses are, in fact, only commonly seen in botanic gardens or other specialist collections; most 'rose gardens' and general gardens contain only cultivars.

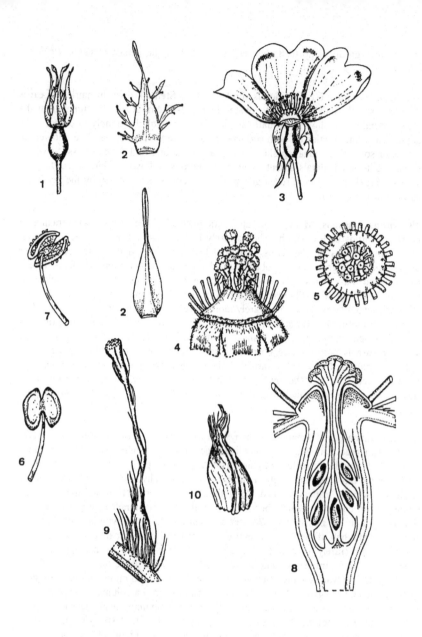

Figure 54. Rosaceae. *Rosa canina.* 1, Flower-bud. 2, Two sepals of different form. 3, Longitudinal section of flower. 4, Top of perigynous zone showing the bases of the filaments and the crowded but free styles, from the side. 5, As 4, but from above. 6, Undehisced stamen. 7, Dehisced stamen. 8, Longitudinal section of the perigynous zone showing the free carpels. 9, A single carpel. 10, A single achene.

256

21. Agrimonia Linnaeus. Few/3. Erect perennial herbs with rhizomes. Stems with short glandular hairs borne almost on the surface and longer, spreading or deflexed, simple hairs. Leaves sometimes forming a rosette at the base of the stem, pinnate with 7--13 large leaflets alternating in opposite pairs with much smaller leaflets; all deeply toothed. Stipules large, stem-clasping, deeply toothed. Flowers many in a raceme, each subtended by a bract and 2 bracteoles (all of which may be 3-lobed or entire), shortly stalked. Perigynous zone cylindric, bell-shaped or top-shaped, its upper part with numerous spreading or deflexed, hard, hooked bristles outside and below the calyx. Sepals 5, spreading in flower, persistent and more or less erect in fruit. Petals 5, yellow, golden yellow or orange-yellow, rarely whitish. Stamens 10--20, borne on the edge of a yellowish disc which roofs the perigynous zone. Carpels 2 (rarely more), free, their styles projecting through a small central hole in the roof of the perigynous zone. Fruit of 1 or 2 (rarely more) achenes enclosed in the hardened, often grooved, perigynous zone surmounted by the bristles and the persistent calyx. *Mainly north temperate areas, but also central Africa & South Africa.*

22. Sanguisorba Linnaeus. 10/4. Herbs, often large, with thick, woody stocks. Leaves pinnate, often with numerous, shortly stalked, toothed leaflets. Stipules leaf-like, often crescent-shaped. Flowers in dense heads (spikes) which are spherical to narrowly cylindric, flowers female or bisexual, opening from the top of the head downwards or the bottom upwards; each flower with a bract and 2 bracteoles which are often hairy. Perianth-segments 4, greenish, whitish or reddish, spreading. Stamens 4--30, filaments thread-like or dilated and flattened above, often projecting well beyond the perianth-segments. Carpels 1--3, free, borne within the rounded or 4-sided perigynous zone, the stigma a mop-like group of papillae. Fruit an achene or a group of achenes surrounded by the persistent perigynous zone, which is often winged or ornamented, surmounted by the remains of the perianth. *North temperate areas.*

Some species were formerly placed in *Poterium* Linnaeus.

23. Sarcopoterium Spach. 1/1. Mound-forming shrub to 75 cm, the ends of the branches persisting as spines, the outer bark often silvery, peeling in strips to reveal the brown under-bark. Leaves pinnate with 9--25 very small, oblong to ovate, entire to 3--5-toothed leaflets, 1--8 mm, downy above, whitish hairy beneath, margins turned under. Flowers in leaf-opposed or terminal spikes, unisexual, spikes often with male flowers above, female below. Perianth-segments 4, greenish and often white-margined, c. 2 mm. Stamens numerous. Carpels 2, free, enclosed in the perigynous zone. Fruit berry-like, the perigynous zone becoming fleshy, 3--5 mm across, red to yellowish brown. *Mediterranean area.*

24. Margyricarpus Ruiz & Pavón. 1/1. Creeping or prostrate, evergreen dwarf shrub; branches ascending towards apex. Stems straw-coloured, internodes covered for most of their length by clasping, papery stipules. Margins of stipules with white silky hairs. Leaves 1.4--1.8 cm × 8--12 mm, broadly ovate, pinnate with 3--5 pairs of glossy, linear lobes; margins strongly curved downwards and inwards, apex of lobes shortly pointed. Flowers solitary, stalkless, borne in the leaf-axils, but not on current year's growth. Sepals 5, persistent; corolla absent; stamens 1--3; carpel 1. Fruit 5--6 × 7--9 mm, an almost spherical, soft white berry, often tinged pale pink. Seeds ovate, c. 4 mm, reddish brown. *Andes.*

257

25. Acaena Linnaeus. 100/15. Perennial evergreen herbs or undershrubs. Stems usually prostrate, often rooting at nodes. Leaves alternate, pinnate with a terminal leaflet; stipules forming a sheath, often with an entire or divided leafy lobe on either side; leaflets usually toothed, asymmetric at the base. Plants bisexual or female and bisexual. Flowers borne in stalked spikes or dense ovoid or spherical heads. Cleistogamous flowers sometimes present in axils at base of flowering stem. Perigynous zone hollow, almost closed at the mouth, usually bearing barbed spines. Sepals 3--6, usually 4. Petals absent. Stamens 2--7, with dorsifixed anthers. Carpels 1 or 2, occasionally to 5, concealed in the perigynous zone except for the curved feathery stigmas. Fruit an achene, dispersed within the persistent perigynous zone. *South temperate areas, USA (California, Hawaii).* Figure 52, p. 248.

26. Polylepis Ruiz & Pavón. 15/1. Small trees or shrubs; branches twisted, covered with scars of fallen leaves. Leaves alternate, with 3 leaflets or pinnate; leaflets in few pairs, leathery; leaf-stalks broadly membranous, sheathing at base. Racemes slender, flowers mainly pendent, bracts present, sepals 3--5, persistent, forming a calyx with a constricted throat, usually with 3 or 4 wings; petals absent; stamens numerous, inserted in calyx throat. Fruit a leathery achene enclosed in a hardened, angular, spiny or winged perigynous zone. *Andes.*

27. Adenostoma Hooker & Arnott. 2/2. Evergreen, resinous shrubs. Leaves needle-like, borne in bundles or singly and alternately; stipules present or absent. Flowers small, bisexual, numerous, in erect or spreading terminal panicles. Sepals 5, translucent, erect. Petals 5, white, longer than the sepals. Stamens 10--15 borne at the mouth of the perigynous zone above some nectar-secreting glands. Perigynous zone cylindric, somewhat tapering to the base, 10-grooved. Carpel 1, containing a single ovule and with an oblique or lateral style. Fruit an achene surrounded by the persistent perigynous zone which hardens and becomes contracted towards the apex, the whole ellipsoid, black and hard. *USA (California) & adjacent Mexico.*

28. Cercocarpus Kunth. 6/2. Shrubs or small trees. Leaves more or less evergreen, alternate or clustered on short shoots, simple, entire or toothed, shortly stalked. Stipules rather small. Flowers solitary or in groups of 3 (rarely these groups somewhat aggregated), axillary or terminal on short lateral shoots. Perigynous zone hairy, narrow and tubular, broadening abruptly above into a 5-lobed perianth which ultimately opens very widely, the lobes and upper part of the widened tube reflexed. Petals absent. Stamens 10--25 (in our species), inserted in 2 or more series on the upper (reflexed) part of the expanded perianth; anthers hairy or hairless. Ovary superior, carpel 1, borne in the scarcely swollen base of the perigynous zone; ovule solitary; style terminal. Fruit an achene surmounted by the long, persistent, plumed-hairy style, surrounded at the base by the base of the perigynous zone which becomes dry and brownish. *Southwest USA, adjacent Mexico.*

29. Purshia de Candolle. 2/1. Shrubs to 3 m; shoots hairy when young. Leaves deciduous, alternate, mostly borne in clusters on short lateral shoots, obovate, long-tapered to the base, rounded to the apex which is divided into 3 oblong, blunt lobes, slightly hairy above, densely white-hairy beneath, margins rolled under. Stipules small, narrowly triangular. Flowers bisexual, solitary at the ends of the short lateral shoots, stalkless or almost so. Perigynous zone more or less funnel-shaped, white-hairy and glandular. Sepals 5, oblong, blunt, reflexed in flower, yellowish and hairless inside. Petals creamy white, spoon-shaped. Stamens *c*. 25. Ovary superior. Carpels 1 (rarely 2), style short. Achene(s) ovoid, hairy, *c*. 1 cm, projecting from the perigynous zone beyond the persistent sepals, tipped by the short, persistent style. *Western North America.*

Two species, 1 occasionally cultivated.

30. Cowania D. Don. 5/2. Evergreen shrubs or small trees. Leaves alternate, lobed or pinnatifid, leathery, dotted with glands on upper surface, densely felted with white hairs beneath, margins rolled downwards; stipules joined to the leaf-stalk. Flowers white, pale yellow or rich rose, solitary, very shortly stalked, terminal, on short leafy twigs; sepals 5, overlapping; petals 5, obovate, spreading; stamens numerous, inserted at the mouth of the calyx in 2 rows; styles 1--12, stalkless, covered with long hairs, lengthening in fruit; ovary superior. Fruit a 1-seeded achene, terminating with a long feathery style. *Southwestern USA to central Mexico.*

31. Dryas Linnaeus. 1--20/few. Low shrubs with creeping, branched, woody stems. Leaves evergreen, shortly stalked, toothed or scalloped, margins usually rolled under to some extent, usually dark green and conspicuously wrinkled above, densely white-hairy beneath with cobwebby hairs, often also with dark purplish or blackish stalked glands and/or long, bristle-like processes which themselves bear white hairs, sometimes also with stalkless, resin-secreting glands. Stipules lanceolate. Flowers solitary on long stalks which rise above the leaves, sometimes functionally unisexual. Perigynous zone short, cup-like; sepals 7--10, often densely covered with dark, stalked glands as well as cobwebby hairs. Petals 7--20, often 8, yellowish or white, forwardly directed or spreading. Stamens numerous, filaments sometimes hairy, deciduous. Ovary superior, carpels numerous. Fruit a head of achenes each with a long, persistent, plumed (silky-hairy) style, surrounded at the base by the persistent perigynous zone and sepals; the styles often twist together spirally while the fruits are ripening. *North temperate areas.*

A genus whose classification is very confused; many authors recognise only a single species (*D. octopetala* Linnaeus) which may be divided into subspecies, varieties, etc., whereas others recognise up to 20 or more species.

32. Geum Linnaeus. 40/12 and many hybrids. Perennial herbs with 2 types of leaves. Basal leaves usually unequally pinnate with the terminal leaflet often distinctly larger than the laterals. Stem-leaves much smaller. Flowers solitary or in corymbs on simple or branched stems, bisexual, yellow, orange, white or red. Sepals 5 with 5 smaller lobes between. Petals 5 in a saucer- or bell-shaped arrangement. Stamens many. Carpels many on a conical or cylindric receptacle. Fruit a group of achenes with long persistent styles; often feathery or jointed. *Temperate & cold areas.* Figure 52, p. 248.

33. Waldsteinia Willdenow. 3/3. Rhizomatous perennial herbs, often with stolons. Leaves mostly basal, with 3--7 lobes. Flowers terminal on slender branched stems. Bracteoles 5; sepals 5, narrowly lanceolate, apex acute; petals 5, rounded, yellow; stamens numerous; style lengthening, soon deciduous. Fruit of 2--6 achenes. *North temperate areas.*

34. Coluria R. Brown. 5/1. Herbs with rhizomes. Stems erect. Leaves mostly basal, pinnate with 7 or more large leaflets with much smaller leaflets in between, terminal leaflet largest, often lobed, all toothed; stem-leaves few, reduced; stipules lanceolate. Flowers solitary or few in racemes. Perigynous zone cylindric-bell-shaped to funnel-shaped, conspicuously 10-veined. Sepals 5, alternating with 5 smaller, persistent epicalyx segments. Petals 5--7, almost circular, very shortly clawed. Stamens numerous, filaments hardened and persistent in fruit. Carpels rather few. Styles long, hairy towards the base and constricted there, deciduous. Fruit a group of achenes in the persistent perigynous zone surmounted by the persistent filaments. *Eastern Asia.*

35. Potentilla Linnaeus. 500/45. Perennial, rarely annual or biennial herbs or small shrubs. Leaves pinnate, palmate, with narrow leaflets or divided into 3 leaflets. Flowers solitary or in cymes, usually 5-parted; epicalyx present. Perigynous zone more or less flat, receptacle dry or spongy. Petals yellow, white, pink, red, orange or bicoloured. Stamens 10--30. Carpels usually 10--80, rarely as few as 4. Fruit a head of achenes. Style usually deciduous. *Mainly in north temperate & Arctic areas.* Figure 52, p. 248.

36. Sibbaldia Linnaeus. 8/1. Tufted perennial herbs, often woody at base. Leaves deciduous, alternate, with 3 leaflets, hairy; stipules papery, attached at the base of leaf-stalks. Flowers bisexual, in cymes; sepals 5, epicalyx present; petals 5, sometimes absent; stamens 5, rarely 4 or 10; carpels 5--12. Fruit a cluster of 5--10 achenes, styles deciduous in fruit. *Northern hemisphere.*

37. Fragaria Linnaeus. 15/c. 10. Perennial herbs with stolons. Leaves all basal, divided into 3 leaflets. Flowers usually in cymes (rarely solitary) on axillary scapes. Perianth parts usually in 5s, additional petals sometimes present. Epicalyx present. Petals usually white, rarely partly or entirely pink. Stamens and carpels numerous, but one or other reduced or poorly formed in functionally unisexual flowers. Receptacle becoming fleshy and usually brightly coloured in fruit, either bearing achenes on its surface or sunk in pits. *North temperate & subtropical areas, South America.* Figure 52, p. 248.

One species of **Duchesnea** Smith may occasionally be grown in very mild areas; it is like *Fragaria*, but has yellow petals and a dry fruit.

38. Alchemilla Linnaeus. 300/25. Deciduous perennials with a more or less developed woody stock and annual flowering shoots. Basal leaves palmately 5--9- (rarely to 11-) lobed, sometimes compound with separate leaflets. Stem-leaves with fewer lobes and relatively large toothed or lobed stipules. Inflorescence cymose, much-branched, with numerous small flowers on short flower-stalks. Perianth greenish, of 4 outer epicalyx segments and 4 alternating sepals. Petals absent. Stamens 4, often with poorly developed anthers. Carpel 1, with basal style and head-like stigma, more or less protruding from a cup-like perigynous zone surrounded by a nectar-secreting disc. Fruit an achene enclosed in the dry perigynous zone. *Europe, Asia, rare in North America; also on mountains in Africa and the Andes.* Figures 52 p. 248 & 55, p. 262.

Many of the species are apomictic.

39. Cydonia Miller. 1/1. Deciduous shrubs or trees, without thorns, to 8 m; young shoots with grey-white felt. Leaves entire, ovate to elliptic, to 10 × 5 cm, densely felted beneath; stipules falling early, hairy, glandular. Flowers solitary, white to pink. Sepals 5, persistent, hairy outside, toothed. Petals 5. Stamens 15--25. Carpels 5, walls cartilaginous in fruit; ovules numerous. Styles 5, free. Fruit a pome, closed at top with persistent sepals, pear-shaped, aromatic, golden-yellow, covered with felt of short hair. *Western Asia; naturalised in southern Europe.*

Hybrids between this and species of *Pyrus* are grown as × **Pyronia** Veitch; they are intermediate between the parents. The equivalent graft-hybrid is also grown, as + **Pyrocydonia** Daniel.

40. Chaenomeles Lindley. 3/3 and several hybrids. Deciduous shrubs or trees, sometimes with thorns. Leaves simple, toothed; stipules persistent, large, kidney-shaped. Flowers in axillary clusters. Sepals 5, hairless, deciduous, erect, entire. Petals 5 (numerous in some double-flowered cultivars). Stamens more than 20. Carpels 5, ovules numerous; styles 5, free. Fruit a pome, closed at top, stalkless. *Himalaya to China & Japan.* Figure 52, p. 248.

41. Docynia Decaisne. 5/2. Evergreen or semi-evergreen trees or shrubs. Leaves ovate-elliptic to lanceolate, entire, lobed, or margins with forwardly pointing teeth; stipules present. Flowers 2--5, borne in stalkless umbels; calyx densely felted, lobes lanceolate; petals 5, broadly ovate, narrowed towards the base; stamens 30--50; ovary inferior, 5-celled with 3--10 ovules, styles 5, joined, with long hairs at base. Fruit fleshy, egg-shaped to pear-shaped, calyx persistent. *China, Vietnam, Malaysia, Indonesia.*

42. Pyrus Linnaeus. 25/20. Deciduous trees, sometimes with thorny branches. Buds with overlapping scales. Leaves alternate, inrolled in bud, stalked, entire or, very rarely, lobed; margins often with forwardly pointing teeth. Flowers in umbel-like racemes, appearing with or before the leaves. Petals 5, white, or rarely pinkish, clawed, rounded to broadly oblong; stamens 18--30, anthers usually reddish; styles 2--5, free, closely constricted at base by the nectariferous disc; ovules 2 per cell. Fruit a spherical or pear-shaped pome; flesh with grit cells; seeds black or brownish black. *Europe, North Africa, to eastern Asia.*

43. Malus Miller. 30/25 and many hybrids. Deciduous trees or shrubs, occasionally semi-evergreen; lateral shoots often ending in thorns. Leaves alternate, folded or rolled in bud, toothed or lobed, green to reddish purple. Flowers bisexual, stalked, borne in umbel-like clusters; calyx of 5 persistent lobes; petals 5, semicircular to broadly ovate, white, rose or crimson; stamens 15--50, anthers rounded. Ovary inferior, 3--5-celled, styles 2--5, always united at base, basal part with long shaggy hairs; fruit fleshy, with or without grit cells, almost spherical, calyx mostly persistent; seeds 1 or 2 per cell, brownish or black. *Temperate parts of northern hemisphere.* Figure 52, p. 248 & 56, p. 264.

44. Eriolobus (de Candolle) Roemer. 6/1. Deciduous trees. Leaves alternate, palmately lobed with long stalks; stipules falling early. Flowers bisexual in terminal stalkless or shortly stalked umbels. Calyx of 5 reflexed lobes, united at the base; petals 5, white; stamens *c.* 20; ovary inferior, 5-celled; styles 5, united at the base. Fruit spherical to pear-shaped with grit cells, not sunken at calyx; seeds 1 or 2, not compressed. *Southeast Europe, southwest Asia.*

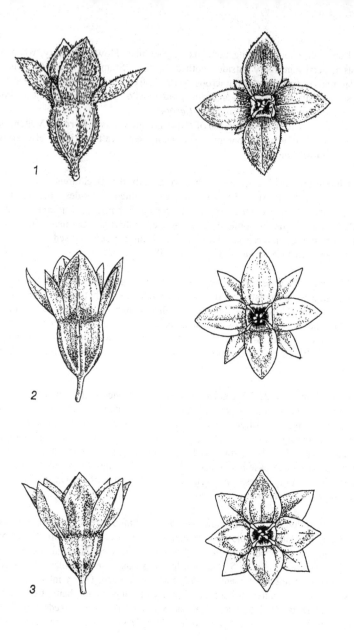

Figure 55. Rosaceae. *Alchemilla*. Flowers of three distinct species from the side (L) and from above (R).

262

45. Sorbus Linnaeus. 100/50. Deciduous small trees to rhizomatous shrubs. Buds ovoid-conic, reddish to black, with 2 or 3 bud-scales visible. Leaves alternate, simple and toothed or pinnate; stipules deciduous. Flowers usually bisexual in flat or pyramidal clusters, white or more rarely pink; sepals and petals 5; stamens 15--20; carpels 2--5, more or less united, each with 2 ovules; ovaries inferior or semi-inferior; styles free or united at the base. Fruit usually a small pome, spherical or ovoid-spherical, orange-red to crimson to white; cells 2--5, with cartilaginous walls, each with 1 or 2 seeds. *Northern hemisphere.* Figure 52, p. 248.

Many species are apomictic. Hybrids between species of *Sorbus* and *Aronia* are sometimes grown as × **Sorbaronia** Schneider; they are intermediate between the parents. Hybrids between *Sorbus* and *Amelanchier* are known as × **Amelosorbus** Rehder, and those between *Sorbus* and *Pyrus* as × **Sorbopyrus** Schneider.

46. Eriobotrya Lindley. 20/4. Evergreen tree or large shrub. Leaves spirally arranged, stalked or almost stalkless, simple, leathery, with prominent veins extending to leaf-margin; stipules present. Inflorescence a terminal panicle, to 15 cm. Calyx-tube 5-lobed, fused to receptacle. Petals 5, usually white; stamens *c.* 20. Ovary inferior, with 2--5 chambers, each containing 2 ovules; styles 2--5, fused at base. Fruit a fleshy pome containing 1 or 2 large seeds; calyx-lobes usually persistent, except 2. *Himalaya to Japan & southeast Asia.*

47. Peraphyllum Torrey & Gray. 1/1. Deciduous shrubs to 2 m, stems widely divergent, much-branched, bark grey. Leaves 2.5--4 cm × 4--10 mm, narrowly lanceolate to obovate, mucronate; margin smooth or with fine forwardly pointing teeth. Stipules falling early. Inflorescence an erect corymb of 2 or 3 flowers. Flowers white, bisexual, to 2 cm across, subtended by 2 small bracts. Calyx persistent, to 1 cm, covered with silky hairs, fused to bell-shaped receptacle. Petals 8--10 mm, circular. Stamens 20, as long as petals, arranged around rose-coloured dish; anthers yellow. Ovary 2--4-celled; stigma red, capitate; styles 2 or 3, not exceeding stamens. Fruit a yellow, fleshy, spherical drupe to 2 cm, containing 1 seed. *Western North America.*

48. Amelanchier Medikus. 30/12 and some hybrids. Deciduous trees or shrubs. Leaves alternate, simple, entire or with forwardly pointing teeth. Stipules deciduous. Flowers in terminal racemes, rarely solitary. Perigynous/epigynous zone bell-shaped; calyx-lobes 5, narrow, reflexed, persistent. Petals 5, white, rarely pink. Stamens 10--20, inserted on the throat of the epigynous zone; filaments awl-shaped. Styles 2--5. Ovary inferior, apex usually hairy, cells becoming double the number of styles. Fruit a small berry-like pome with 1--5 cells each containing 1 or more seeds. *North & central Europe, Japan, China, Korea, North America.*

49. Rhaphiolepis Lindley. 5/2. Small evergreen trees or shrubs. Leaves leathery, simple, stalked; young leaves densely covered in felty hairs. Inflorescence a terminal panicle, felted, flowers subtended by bracts. Floral parts in 5s. Stamens 15--20. Ovary inferior, 2-celled; styles 2. Fruit a dry ovoid or spherical berry with a distinct scar at apex, with 1 or 2 large seeds. *East & southeast Asia.* Figure 52, p. 248.

50. Aronia Medikus. 3/3. Deciduous shrubs; branches with closely adpressed, slender, pointed buds. Leaves alternate, stalked, simple with toothed margins and blackish glands on midrib above. Stipules small, falling early. Flowers white or pale pink in

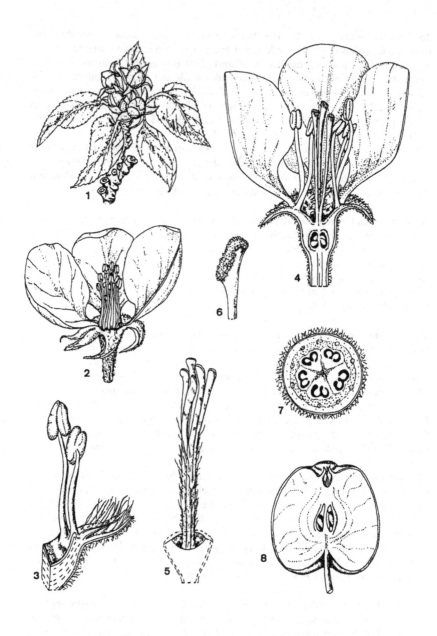

Figure 56. Rosaceae. *Malus* × *domestica.* 1, Short shoot with umbel of four flower-buds. 2, Flower from the side, two petals removed. 3, Two stamens attached to the rim of the short epigynous zone. 4, Longitudinal section of flower. 5, The five styles arising from the top of the inferior ovary. 6, Stigma. 7, Transverse section of ovary. 8, Longitudinal section of fruit (pome).

small corymbs. Calyx-lobes 5, joined at base. Petals 5, spreading; stamens numerous, anthers purplish pink; ovary inferior, 5-celled, with woolly hairs towards the apex; styles 5, joined at their bases, carpels partly free. Fruit apple-like, with persistent remains of calyx. *North America.*

51. Photinia Lindley. 40/14 and some hybrids. Evergreen or deciduous trees and shrubs. Leaves alternate, simple, usually with fine forwardly pointing teeth or entire, with short stalks, stipules sometimes almost leaf-like, free. Flowers normally white, woolly in bud, borne in terminal or axillary umbel-like panicles; petals 5 with a distinct claw, hairless or hairy at the base, sepals 5, persisting in fruit; stamens 15--25. Ovary semi-inferior, 2--5-celled, styles 2--5. Fruit more or less fleshy, seeds 1 or 2 to each cell. *Himalaya to Japan & Indonesia (Sumatra).*

52. Heteromeles Roemer. 1/1. Evergreen shrubs, 2--10 m; bark grey, branchlets slightly hairy. Leaves 5--10 cm, elliptic or lanceolate to oblong, leathery, sharply toothed, pale beneath; leaf-stalks 1--2 cm, downy. Flowers many, borne in flattish, cormyb-like panicles; sepals triangular, 1--1.5 mm; petals white, c. 3 mm; stamens 10; styles 2 or 3, separate. Ovary inferior. Fruit bright red, 5--6 mm, persistent throughout the winter. *North America.*

53. Cotoneaster Medikus. 200/100. Evergreen and deciduous shrubs and small trees. Leaves alternate, simple, entire. Flowers in cymes, small clusters, or solitary; petals white, pink or red; stamens 10--20; ovary inferior, carpels 1--5, usually with free styles. Fruit (pomes) somewhat succulent, containing 1--5 one-seeded, hard-walled nutlets. *Eurasia, North Africa, with a concentration of species in the Himalaya & western China.* Figure 52, p. 248.
Many of the species are apomictic.

54. Dichotomanthes Kurz. 1/1. Evergreen trees or shrubs, young branches covered with dense white woolly hairs. Leaves alternate, entire, ovate, pointed, tapered towards the base, 3--10 × 1.5--3 cm, dark green and hairless above, glossy with pale silky hairs beneath; stipules very small, thread-like, falling early. Flowers borne in terminal corymbs c. 5 cm across; sepals woolly outside, with 2 bracteoles at the base; petals 2--3 mm, white; stamens 15--20; ovary inferior. Fruit a dry, oblong capsule 6 mm, almost entirely surrounded by the calyx which becomes fleshy with maturity. *Western China.*

55. Pyracantha Roemer. 6/6. Evergreen shrubs; branches usually thorny; buds small, softly hairy. Leaves alternate, shortly stalked, margins entire, shallowly scalloped or with 5 forwardly pointing teeth; stipules minute, falling early. Flowers bisexual in compound corymbs. Calyx of 5 short triangular teeth; petals white, rounded, often with a notched apex and sometimes narrowed into a short basal stalk; stamens c. 20, anthers yellow. Ovary inferior, carpels 5, free on central axis, joined for about half their length to calyx-tube; styles 5. Fruit apple-like with persistent calyx, red, orange or yellowish, often slightly compressed laterally. Pyrenes 5. *Southeast Europe, Himalaya, China.*

56. Mespilus Linnaeus. 1/1. Shrub or small tree, 2--5 m; branches occasionally spiny, young shoots woolly. Leaves alternate, deciduous, entire; blade 6--12 × 3--6 cm, oblong-lanceolate; margin toothed, glandular; stalks 2--5 mm. Stipules deciduous. Flowers solitary, terminal on short shoots. Calyx softly hairy, with 2 bracteoles at the base, lobes 5, up to 4 times as long as tube. Petals 5, rounded, 1.5--2 cm, white; stamens 30--40, anthers almost joined. Carpel solitary at the base of the calyx tube, 1-celled; style lateral; stigma capitate; ovules 2. Fruit dry, often 1-seeded, 2--5 cm across, slightly projecting from the fleshy calyx which becomes enlarged in fruit, green at first, becoming brown. *Southeast Europe, southwest Asia.*

57. Crataegus Linnaeus. 1000/50 and some hybrids. Deciduous, or rarely evergreen trees or shrubs, often with spiny stems and branches. Leaves alternate with entire or lobed margins, stipules present. Flowers bisexual, solitary or in corymbs. Sepals 5, petals 5, stamens 5--25, ovary inferior, carpels solitary or 2--5 joined at base, free at apex. Fruit a drupe with 1--5 hard nutlets, each containing a single seed. *North temperate areas.*

Many species are apomictic. Hybrids between species of *Crataegus* and *Mespilus* are grown as × **Crataemespilus** Camus. A graft hybrid between species of the two genera is also grown as + **Crataegomespilus** Simon-Louis.

58. Prunus Linnaeus. 200/90. Deciduous or evergreen trees or shrubs of very variable habit, sometimes spiny. Leaves alternate, usually toothed, generally with 1 or more conspicuous glands (extra-floral nectaries) on the stalk; stipules present, sometimes falling early. Flowers solitary or borne in racemes, umbels, corymbs or clusters, with distinct campanulate or cylindric perigynous zones. Sepals 5, often toothed, sometimes reddish. Petals 5 (more in variants with semi-double or double flowers), white, cream, or pink to red, usually spreading. Stamens numerous. Ovary of a single carpel borne in the base of the perigynous zone; ovules 2. Fruit a drupe, generally fleshy, mostly 1-seeded, sometimes large. Stone pitted, ridged or smooth, sometimes keeled. *North temperate areas, a few in the tropics in the Andes and southeast Asia.* Figures 52, p. 248 & 57, p. 268.

59. Prinsepia Royle. 4/3. Deciduous, arching shrubs with axillary spines and flaking bark; branches with chambered pith; buds small, naked and/or covered with a few small hairy scales. Leaves alternate, often clustered, simple, membranous to leathery, entire or toothed. Stipules, if present, small, persistent. Flowers stalked, in axillary bracteate racemes of 1--8 (rarely to 13) in the axils of previous year's shoots. Flowers with perigynous zone tubular or cup-shaped; calyx 5-lobed, lobes equal or unequal, broad, short. Petals 5, distinct, equal, almost circular, clawed, white or yellow, spreading. Stamens 10 or many. Ovary superior. Fruit an oblique 1-seeded drupe, red or purple, juicy and edible. *Himalaya, China, Taiwan.*

60. Maddenia Hooker & Thomson. 4/1. Deciduous trees or shrubs. Leaves alternate, with glandular, forwardly pointing teeth; stipules large, glandular toothed. Flowers unisexual, on short stalks, borne in short terminal dense racemes; sepals and stamens inserted on the mouth of the perigynous zone; sepals 10, small, some long and petal-like; stamens 25--40, inserted in 2 more or less distinct rows. Male flowers with a solitary carpel, female flowers with 2. Ovary superior. Fruit a single-seeded drupe; stone ovoid, 3-keeled on one side. *Himalaya, China.*

61. Oemleria Reichenbach. 1/1. Deciduous, monoecious shrubs to 3 m, forming dense thickets. Bark purple-brown with prominent lenticels. Leaves 2.5--9 cm × 8--30 mm, obovate-lanceolate, dark green above, paler beneath. Inflorescence a raceme with up to 13 flowers, each subtended by 2 bracteoles. Flowers to 1 cm across, sepals, petals and stamens perigynous. Petals upright, oval. Flowers unisexual: males with numerous stamens erect-incurved, to 4 mm borne on the margin of the perigynous zone; females with shorter, sterile stamens and an superior ovary of 5 carpels. Fruit a thinly fleshy, bitter drupe. *Western North America.*

87. LEGUMINOSAE (*Fabaceae*)

Trees, shrubs, herbaceous or woody climbers (sometimes with tendrils) or herbs, often with nodules containing nitrogen-fixing bacteria on the roots. Leaves usually alternate, divided into 3 leaflets, pinnate (sometimes bipinnate or tripinnate) or palmate, rarely completely absent on the mature plant; stipules usually present. Flowers in inflorescences of various kinds, usually racemose or congested into heads, sometimes solitary, radially symmetric or more commonly bilaterally symmetric. Perigynous zone present or absent. Calyx usually of 5 united sepals, 5-toothed or -lobed at the apex, sometimes 2-lipped, sepals rarely free (or almost so). Petals usually 5 (sometimes up to 4 suppressed), often diversified among themselves, free at the base or variously united above the base. Stamens usually 10, sometimes numerous, more rarely fewer, the filaments free or more commonly variously united; anthers opening by longitudinal slits or more rarely by terminal pores. Carpel 1, with 1--many ovules. Fruit a legume containing 1--many seeds, sometimes modified so as to be indehiscent, or breaking transversely into 1-seeded segments.

A very large, diverse and cosmopolitan family (about 600 genera and 13,500 species) especially well developed in the tropics. It contains many important crop plants (peas, beans of various kinds, lentils, etc.), many tropical species are important timber trees and many species are grown as green manures because of their ability (through the bacteria present in the root nodules) to fix atmospheric nitrogen, thus improving the fertility of the soil. Many species are also grown as ornamentals. The family (often known alternatively as **Fabaceae**) is divided into 3 subfamilies, which are often treated as individual families. Though these differ conspicuously among themselves, they are held together by a number of characteristics, of which the possession of a legume as the fruit is the most important. The family is treated in the broad sense here, but, because of its diversity, the 3 subfamilies are briefly described and are keyed separately below.

Key to Subfamilies
    1a. Corolla radially symmetric, petals edge-to-edge in bud; leaves usually
        bipinnate or modified into phyllodes; seeds with a U-shaped lateral
        line   **Mimosoideae**
      b. Corolla bilaterally symmetric (sometimes weakly so), petals variously
        overlapping in bud; leaves various; seeds usually without a lateral line,
        rarely with a closed lateral line   2
    2a. Uppermost petal overlapped by the laterals (interior); seed usually with a
        straight radicle   **Caesalpinioideae** (p. 269)
      b. Uppermost petal overlapping the laterals (exterior); seed usually with a
        curved radicle   **Papilionoideae** (p. 270)

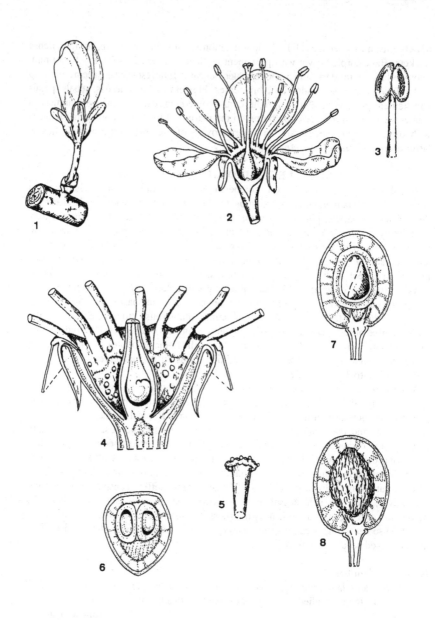

Figure 57. Rosaceae. *Prunus spinosa.* 1, Opening flower-bud. 2, Flower from the side, two petals removed. 3, Stamen. 4, Longitudinal section of flower, upper parts of petals, stamens and style removed. 5, Stigma. 6, Transverse section of ovary. 7, Longitudinal section of fruit. 8, Fruit, part of the flesh cut away to reveal the stone.

Mimosoideae (*Mimosaceae*). Species of only 2 genera are cultivated out-of-doors in northern Europe, and then only in very favoured locations. See figure 58, p. 272.

1a. Stamens free or almost so   **1. Acacia**
  b. Stamens united into a tube below   **2. Albizia**

1. Acacia Miller. 1200/1. Trees and shrubs, evergreen, Leaves compound, bipinnate and persistent in mature plants, or bipinnate leaves entirely absent after the seedling stage and replaced by phyllodes. Inflorescence fluffy, spherical or cylindric, solitary or in clusters, axillary; flowers yellow, bisexual. Sepals 3--5, minute. Petals 3--5, fused into bell-shaped corolla. Stamens numerous, much longer than corolla, thus conspicuous. Ovary stalkless; ovules numerous. Legumes usually dehiscent, with 2 valves; seeds usually with a fleshy aril. *Mainly Australia (especially Western Australia), but also in the tropical areas of Africa, Asia & America.* Figure 58, p. 272.

2. Albizia Durazzini. 150/2. Trees or shrubs without spines, rarely climbing. Leaves bipinnate, leaflets each with 1--many pairs of variably-sized ultimate segments; stipules usually small but sometimes large and leaf-like. Flowers in spherical heads or spikes or spike-like racemes, stalked, axillary and solitary or clustered. Flowers bisexual or occasionally some male only; 1 or 2 flowers in each head often larger and different in form from the others, apparently male. Calyx with 5 teeth or lobes. Corolla funnel-shaped or bell-shaped with 5 lobes. Stamens 19--50, fertile, their filaments united below into a tube which may project from the corolla. Ovary stalkless or shortly stalked, with many ovules; style thread-like, stigma minute. Pods oblong, straight, flat, usually dehiscent, without cross-walls, papery or leathery but not thickened or fleshy. Seeds ovate or circular, compressed. *Tropics & subtropics.*

Caesalpinioideae (*Caesalpiniaceae*). Species of 3 genera are cultivated out-of-doors in northern Europe; most of the species of the subfamily are tropical. See figure 59, p. 274.

1a. Leaves simple, consisting of a single leaflet   **5. Cercis**
  b. Leaves pinnate or bipinnate or consisting of 2 completely distinct leaflets   2
2a. Some leaves pinnate, others bipinnate; plants often spiny   **3. Gleditsia**
  b. All leaves bipinnate; plants not spiny   **4. Gymnocladus**

3. Gleditsia Linnaeus. 14/3. Deciduous trees, with trunks and branches usually armed with simple or branched spines. Leaves alternate, pinnate or bipinnate, with numerous, slightly scalloped leaflets, the terminal often absent. Flowers small, radially symmetric, greenish, unisexual and bisexual on the same plant, in racemes or more rarely panicles. Sepals 3--5. Petals 3--5, similar to the sepals. Stamens 6--10. Style short with a large terminal stigma. Fruit a flattened pod, indehiscent or very late-dehiscent. Seeds 1--many, flattened, ovate to almost circular. *Mainly Asia, some in America.*

4. Gymnocladus Lamarck. 4/1. Trees, without spines. Leaves deciduous, bipinnate, stipules present or absent, stipels present or absent. Inflorescence a terminal or axillary raceme or panicle. Flowers radially symmetric, functionally unisexual, the males with a rudimentary ovary, the females with abortive or sterile anthers; both sexes on the same

269

plant. Calyx tubular, 5-toothed with equal teeth, the calyx not completely covering the corolla in bud. Petals 4 or usually 5, similar to the sepals, hairy, cream or marked with purple outside. Stamens 10 in 2 whorls, those of one whorl longer than those of the other, filaments free. Ovary stalkless, with 4 or more ovules. Pod oblong, often curved, woody, pulpy inside, containing 1 or more seeds, opening along the upper suture. *Eastern North America, eastern China.*

5. Cercis Linnaeus. 6/5. Shrubs or small trees, flowering before the leaves expand. Leaves simple, palmately veined, ovate, circular or kidney-shaped; stipules deciduous. Flowers usually in umbel-like clusters (strictly short racemes but with very short axes) borne on short shoots on the previous year's growth, more rarely in pendent racemes. Calyx obliquely and widely bell-shaped, shallowly 5-toothed. Corolla usually rose-pink to purple, rarely white; petals clawed, keel longer than standard. Stamens 10, filaments free, curved downwards and included between the keel petals. Ovary several-seeded. Pods oblong, flat, with a narrow wing along the lower suture, several-seeded, eventually dehiscent. *Mediterranean area, eastern Asia, North America.* Figure 59, p. 274.

Papilionoideae (*Papilionaceae*). The largest subfamily, with many genera cultivated. The flowers have a characteristic pea-flower shape which is instantly recognisable. The calyx consists of 5 united sepals, and may be 5-toothed or obliquely 2-lipped. The uppermost petal usually has its limb reflexed backwards and is known as the standard; the two lateral petals are known as the wings, and the 2 lower petals are generally joined by their lower edges to form a keel within which the stamens and style and stigma are held. See figures 60 p. 286 & 61, p. 288.

1a. Filaments of all 10 stamens completely free for at least 90% of their length (sometimes all slightly fused at the extreme base)   2
  b. Filaments of all or most (usually 9) of the stamens united for much of their length, often the uppermost completely free at the base (sometimes united to the others higher up), rarely the stamens variably united in groups, or rarely the stamens 9 only, filaments all united   8
2a. Most leaves pinnately divided, with more than 3 leaflets   3
  b. Most leaves simple, or with 3 leaflets, or leaves with blades absent   5
3a. Base of the leaf-stalk swollen, enclosing the axillary bud   **7. Cladrastis**
  b. Base of the leaf-stalk not swollen as above, axillary bud not enclosed   4
4a. Pods flattened, linear-oblong, dehiscent; standard reflexed   **6. Maackia**
  b. Pods terete, constricted between the seeds, often winged, not dehiscent but breaking into 1-seeded segments; standard reflexed or forwardly-directed   **8. Sophora**
5a. Plant herbaceous, shoots dying back to the root in winter   6
  b. Plant woody, aerial shoots persistent through the winter   7
6a. Plants hairless or almost so, often glaucous; corolla violet-blue, yellow or white; pod inflated   **58. Baptisia**
  b. Plants usually conspicuously hairy, not glaucous; corolla deep purple or yellow; pod flat   **57. Thermopsis**
7a. Standard well-developed, about as long as wings; corolla more than 2.5 cm   **56. Piptanthus**
  b. Standard much shorter than wings; corolla 1.8--2.5 cm   **55. Anagyris**

8a. Stamens 10 with all their filaments united for a great proportion of their length, certainly all united in the lower part, forming a tube or sheath   9

b. Stamens 10 with the filament of the uppermost free at the base and for some distance above it, sometimes united with those of the other 9 well above the free base, or filaments variably united in groups or stamens 9, filaments all united   34

9a. Most leaves with more than 3 leaflets, mostly pinnate, occasionally palmate   10

b. All leaves simple or of 3 leaflets only   16

10a. Leaves palmate   **59. Lupinus**

b. Leaves pinnate   11

11a. Corolla reduced to the standard only, which envelops the stamens and style   **22. Amorpha**

b. Corolla with wings and keel as well as standard   12

12a. Leaves with conspicuous, brown, stalkless glands beneath; flowers *c.* 3 cm across   **23. Amicia**

b. Leaves without conspicuous, brown stalkless glands beneath; flowers to 2.5 cm, often much smaller   13

13a. Stipules and stipels persistent; corolla blue to purple   **17. Hardenbergia**

b. Stipules present but usually falling early, or absent, stipels absent; corolla variously coloured, not usually blue to purple   14

14a. Stems and leaves sticky, glandular-hairy; stipules united to the leaf-stalk   **49. Ononis**

b. Stems and leaves not glandular and sticky; stipules absent or free from the leaf-stalk   15

15a. Calyx-teeth about as long as tube; stipules persistent; corolla pale violet to white   **32. Galega**

b. Calyx-teeth shorter than the tube; stipules absent; corolla yellow or red, rarely creamy white   **38. Anthyllis**

16a. Mature leaves consisting of a single leaflet or mature plants without leaves   17

b. Mature leaves all or mostly with 3 leaflets   24

17a. Plants spiny, either the ends of the branches hardened and spiny, or leaves replaced by spiny phyllodes   18

b. Plants not at all spiny   20

18a. Corolla blue   **63. Erinacea**

b. Corolla yellow   19

19a. Leaves absent from mature plants, replaced by spine-tipped phyllodes   **70. Ulex**

b. Leaves present on mature plants, spine-tipped phyllodes absent   **68. Genista**

20a. Calyx divided to the base on the upper side   21

b. Calyx not divided to the base on the upper side   22

21a. Young stems broadly winged   **66. Chamaespartium**

b. Young stems not broadly winged   **65. Spartium**

22a. Seeds appendaged; upper lip of calyx with 2 very short teeth   **62. Cytisus**

b. Seeds not appendaged; upper lip of calyx deeply toothed or notched   23

23a. Pod dehiscent, not inflated   **68. Genista**

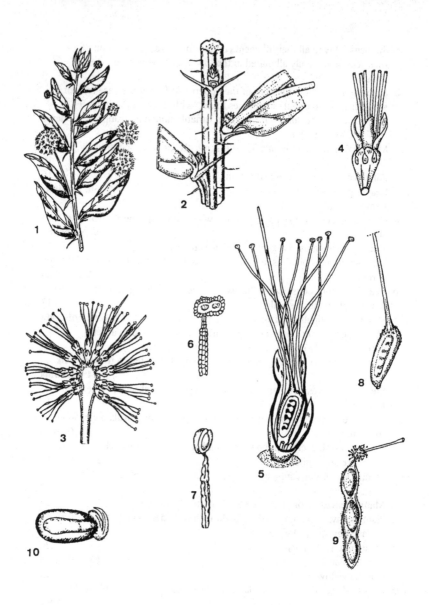

Figure 58. Leguminosae/Mimosaceae. *Acacia paradoxa.* 1, Shoot showing spines, cladodes and flower-heads. 2, Shoot enlarged, showing spines (modified stipules) and axillary buds. 3, Section of flower-head. 4, Single flower. 5, Longitudinal section of flower. 6, Undehisced stamen. 7, Dehisced stamen. 8, Ovary (single carpel). 9, Fruit. 10, Seed.

b. Pod indehiscent, inflated	**67. Retama**
24a. Leaflets toothed	**49. Ononis**
b. Leaflets not toothed	25
25a. Plant spiny	26
b. Plant not spiny	27
26a. Corolla blue	**63. Erinacea**
b. Corolla yellow	**68. Genista**

27a. Leaves dotted with dark glands, often with a resinous or bituminous smell
      28

  b. Leaves not dotted with dark glands, not smelling as above   29

28a. Foetid biennials to perennial herbs; leaves with 3 leaflets **20. Bituminaria**

  b. Scented herbs or shrubs; leaves pinnate with a terminal leaflet
      21. Psoralea

29a. Pod covered with glandular warts	**69. Adenocarpus**
b. Pod not covered with glandular warts	30
30a. Small tree with flowers in long, hanging racemes	**60. Laburnum**
b. Shrubs (rarely small trees) with flowers in erect inflorescences	31
31a. Upper lip of calyx with 2 very short teeth; seeds appendaged	**62. Cytisus**
b. Upper lip of calyx deeply 2-toothed; seeds not appendaged	32
32a. Tall shrubs; leaf-stalk 1.5--5 cm; pod 3.5--5 cm	**61. Petteria**
b. Small shrubs; leaf-stalk to 1.5 cm; pod less than 3.5 cm	33
33a. Leaves and branches mostly alternate; calyx not inflated	**68. Genista**

  b. Leaves and branches mostly opposite; calyx somewhat inflated
      64. Echinospartium

34a. Leaves of 3 leaflets, or of 5 leaflets and palmate, rarely simple or absent
    (sometimes replaced by a tendril)   35

b. Leaves pinnate with more than 3 leaflets	54
35a. Leaves simple or absent from mature plants	36
b. Leaves of 3 or 5 leaflets, present on mature plants	41
36a. Leaves simple	37
b. Leaves absent from mature plants	39
37a. Spineless annual; flowers yellow, several, in heads	**42. Coronilla**

  b. Spiny perennial or small shrub; flowers yellow, pink or red, solitary or
    in pairs   38

38a. Corolla yellow	**38. Anthyllis**
b. Corolla pink or red	**31. Alhagi**

39a. Leaves replaced by a tendril subtended by 2 enlarged leaf-like stipules
      45. Lathyrus

  b. Tendrils absent; stipules minute or absent   40

40a. Flower-stalk woolly; raceme with *c.* 20 flowers; pods indehiscent
      34. Chordospartium

  b. Flower-stalk hairless or finely downy; racemes usually with fewer flowers,
    if with *c.* 20, then pods dehiscent   **35. Carmichaelia**

41a. Leaflets with the veins running to the margin, which is usually toothed 42

  b. Leaflets with veins joining and looping near the margin, which is not
    toothed   46

42a. Standard and keel deep blue, wings pink; stipules free from the leaf-stalk
      50. Parochetus

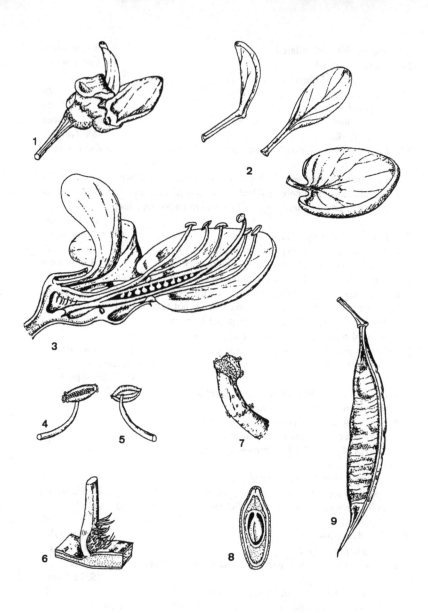

Figure 59. Leguminosae/Caesalpiniaceae. *Cercis siliquastrum.* 1, Flower from the side. 2, Three petals (from L to R, standard, wing and keel). 3, Longitudinal section of flower. 4, Stamen from front. 5. Stamen from back. 6, Filament-base attached to perigynous zone. 7, Stigma. 8, Transverse section of ovary (single carpel). 9, Fruit.

274

b. Flowers not of the above colours; stipules joined to leaf-stalk to some extent   43

43a. At least 5 of the filaments dilated below the anthers; pod usually 1- or 2-seeded, usually enclosed in the persistent corolla and calyx **54. Trifolium**

b. Filaments not dilated below the anthers; pods usually several-seeded, not enclosed as above, corolla usually falling   44

44a. Pods coiled or sickle-shaped, often spiny   **53. Medicago**

b. Pods not as above   45

45a. Pods nutlet-like, with 1--few seeds; plants smelling of new-mown hay (coumarin)   **51. Melilotus**

b. Pods straight or curved, long, many-seeded; plants smelling variously, but not of new-mown hay   **52. Trigonella**

46a. Standard with a projecting spur on the back   **18. Centrosema**

b. Standard without a spur on the back   47

47a. Stipels absent   48

b. Stipels present (rarely soon falling or gland-like)   51

48a. Shrub; flowers in axillary racemes congested into panicles at the tips of the branches   49

b. Herb or small, spiny shrub; flowers solitary or in pairs   50

49a. Keel somewhat curved, acute; bracts each subtending 1 flower; bracteoles quickly deciduous   **13. Campylotropis**

b. Keel straight, blunt; bracts each subtending 2 flowers; bracteoles persistent   **14. Lespedeza**

50a. Small, spiny shrub   **38. Anthyllis**

b. Herb, without spines   **39. Tetragonolobus**

51a. Style coiled through 2 or 3 coils   **19. Phaseolus**

b. Style straight or curved but not coiled as above   52

52a. Bracteoles present   **12. Desmodium**

b. Bracteoles absent   53

53a. Corolla blue-purple or whitish   **17. Hardenbergia**

b. Corolla reddish or yellow and black   **16. Kennedia**

54a. Most leaves terminating in a short or minute point, a spine or a tendril, without a terminal leaflet   55

b. Most leaves with a terminal leaflet   63

55a. At least some leaves terminating in tendrils   56

b. Tendrils absent   60

56a. Leaflets with parallel veins and/or stems winged   **45. Lathyrus**

b. Leaflets with pinnate veins; stems not winged   57

57a. Calyx-teeth all equal, all at least twice as long as the tube   **46. Lens**

b. Calyx-teeth unequal, at least 2 of them less than twice as long as the tube   58

58a. Calyx-teeth more or less leaf-like; stipules to 1 cm   **47. Pisum**

b. Calyx-teeth not at all leaf-like; stipules much smaller   59

59a. Style hairy all round, or on the lower side, or entirely hairless   **44. Vicia**

b. Style hairy on the upper side only   **45. Lathyrus**

60a. Leaves ending in a short or minute soft point which does not harden and persist   **27. Caragana**

b. Leaves ending in a spine which hardens and persists after the leaflets have fallen 61

61a. Creeping or mat-forming shrubs or herbs with shoots not clearly differentiated into long and short shoots **29. Astragalus**

b. Mostly upright or spreading shrubs with shoots clearly differentiated into short (flowering and leafing) shoots and long, extension shoots 62

62a. Flowers solitary or clustered on the short shoots on jointed stalks or rarely in a few-flowered umbel; corolla usually yellow, more rarely orange or pink **27. Caragana**

b. Flowers 2--5 in a raceme; corolla purplish pink to magenta **26. Halimodendron**

63a. Stipels present and persistent, though sometimes small 64

b. Stipels entirely absent 66

64a. Large woody climber with numerous flowers in long, pendent racemes **9. Wisteria**

b. Trees or shrubs; inflorescences various, not usually as above 65

65a. Hairs mostly medifixed; stipules not spine-like; shrubs **11. Indigofera**

b. Hairs basifixed or absent; stipules usually persisting as spines; trees or shrubs **10. Robinia**

66a. Veins of leaflets running out to the toothed margin **48. Cicer**

b. Veins of leaflets looping and joining within the entire margin 67

67a. Stipules absent or represented by minute glandular points, the lowermost pair of leaflets sometimes mimicking stipules 68

b. Stipules present, developed, not represented by minute glandular points 71

68a. Leaves more or less stalked, with a variable number of leaflets (from 1 to 27), but generally some with more than 5 leaflets 69

b. Leaves almost stalkless, all with 5 leaflets, the lowermost pair sometimes mimicking stipules 70

69a. At least some flowers in terminal or apparently terminal inflorescences **38. Anthyllis**

b. All flowers axillary or in axillary inflorescences **28. Calophaca**

70a. Keel beaked **40. Lotus**

b. Keel not beaked **41. Dorycnium**

71a. Pods inflated, translucent, papery **25. Colutea**

b. Pods not as above 72

72a. Pod indehiscent, breaking transversely into 1-seeded segments 73

b. Pod dehiscent or indehiscent, not breaking transversely into 1-seeded segments, sometimes the pod itself 1-seeded 75

73a. Flowers in racemes, the flower-stalks arising serially from the axis **36. Hedysarum**

b. Flowers in umbel-like clusters or heads, all the flower-stalks arising from more or less the same point 74

74a. Segments of the pod crescent-shaped, horseshoe-shaped or rectangular with an arched sinus **43. Hippocrepis**

b. Segments of the pod linear to oblong, straight or slightly curved **42. Coronilla**

75a. Twiners; roots bearing strings of tubers; keel coiled upwards within the standard **15. Apios**

b. Combination of characters not as above   76

76a. Shrubs; anthers consisting of a single sac; pod small, opening by the falling of the sides from a persistent framework to which the seeds are attached
                                                    35. Carmichaelia

b. Shrubs or herbs; anthers consisting of 2 sacs; pod not as above   77

77a. Standard reflexed, acuminate, keel and wings deflexed, narrow like a parrot's beak, all scarlet or rarely pink or white   **24. Clianthus**

b. Flowers not as above   78

78a. Leaves dotted with orange-brown glands   **33. Glycyrrhiza**

b. Leaves not dotted with orange-brown glands   79

79a. Pod indehiscent, more or less circular in outline, margins usually toothed, the sides often with toothed veins or pitted   **37. Onobrychis**

b. Pod not as above   80

80a. Keel toothed on the upper side; leaflets oblique at the base or, if narrow, then curved   **30. Oxytropis**

b. Keel toothed on the lower side or not toothed; leaflets symmetric at the base   **29. Astragalus**

6. Maackia Ruprecht & Maximowicz. 6/2. Deciduous trees. Leaves pinnate with up to 17 opposite leaflets and a terminal leaflet; axillary buds exposed. Flowers numerous, rather small, in terminal racemes or panicles; bracts at the bases of the flower-stalks. Calyx cylindric to bell-shaped, obliquely 5-toothed. Corolla whitish, blade of the standard reflexed upright. Stamens 10, their filaments united at the extreme base only. Pods compressed, linear-oblong, dehiscent, containing 1--5 seeds. *Eastern Asia.*

7. Cladrastis Rafinesque. 6/3. Trees or shrubs. Leaves pinnate with 7--13 alternate, shortly stalked leaflets including a terminal leaflet, the base of the leaf-stalk swollen and enclosing the axillary bud; stipels sometimes present. Flowers usually many, in erect or pendent panicles; bracts at the bases of the flower-stalks. Calyx usually bell-shaped to cylindric with 5 short, equal teeth or lobes. Corolla white or pinkish, sometimes marked with yellow. Stamens 10, filaments almost completely free. Pod membranous, flattened, sometimes winged, often somewhat constricted between the few seeds, indehiscent. *Eastern Asia (most), North America (1 species).*

8. Sophora Linnaeus. 50/7. Trees or shrubs. Leaves pinnate with up to 43 leaflets (including a terminal leaflet) which are generally opposite; stipules small; axillary buds small, exposed. Flowers bisexual, in terminal or axillary racemes or panicles, pea-flower-like (with blade of standard reflexed upright) or with all the petals forwardly directed; bracts borne at the bases of the flower-stalks. Calyx cup-shaped, truncate, usually obliquely so, with 5 small teeth. Stamens 10, filaments slightly united at the base. Pod stalked, terete, constricted between the seeds, sometimes longitudinally winged, often not dehiscing but breaking into 1-seeded segments. *Widespread.*

9. Wisteria Nuttall. 6/5. Large, deciduous woody climbers. Leaves pinnate with up to 19 opposite lateral leaflets and a terminal leaflet; stipules falling early, narrow stipels present; leaflet-stalks somewhat swollen. Flowers numerous in often large, pendent racemes. Calyx bell-shaped or somewhat cylindric, obliquely 5-lobed, the 2 upper lobes often united for most of their length. Corolla blue, purple, lilac, pink or white;

standard large, usually with 2 humps at the base of the blade. Stamens 10, 9 of them with their filaments united, that of the uppermost free. Ovary with several ovules. Pod large, compressed, with several seeds. *North America, eastern Asia.*

10. Robinia Linnaeus. 4/4 and many hybrids. Trees or shrubs. Leaves pinnate with opposite or somewhat alternate leaflets and a terminal leaflet; stipules often persistent as paired spines; stipels present, small, needle-like. Flowers in erect or pendent racemes; bracts present, soon falling. Calyx bell-shaped, unequally 5-toothed. Corolla white to pink or pale purple; standard broad, reflexed upwards, keel-petals united to each other at the base. Stamens 10, 9 of them with united filaments, that of the uppermost free or partly so. Pod flattened, bristly or not, rarely winged along the upper suture, containing several seeds. *East & southeast USA.*

11. Indigofera Linnaeus. 700/15. Shrubs (ours). Hairs usually medifixed, mostly adpressed. Leaves with a terminal leaflet and 7--27 opposite lateral leaflets (ours); stipules small to large, persistent or soon falling; stipels present, small. Racemes axillary, few- to many-flowered, usually borne on the current year's growth, more rarely borne on older wood; bracts usually small. Calyx bell-shaped, deeply 5-toothed with the lowermost tooth usually the longest, more rarely truncate and scarcely toothed. Corolla usually pink to purplish, rarely white or yellowish. Standard generally obovate, reflexed upwards, variously hairy outside. Wings borne alongside the keel or forming a horizontal plate above it, variously hairy towards the base. Keel with a projection or spur on each side, usually hairy, at least along the suture. Stamens 10, 9 of them with filaments united, the uppermost free (ours); anthers appendaged at their tips. Ovary several-seeded. Pod variable, usually linear-cylindric, scarcely compressed, more rarely compressed or constricted between the seeds, always with thin partitions between the individual seeds. *Mainly tropics & subtropics.*

12. Desmodium Desvaux. 300/2. Herbs with woody bases, or shrubs. Leaves usually with 3 leaflets, occasionally reduced to the terminal leaflets only; stipules present, attached by their bases, often early-deciduous, stipels present, small. Flowers in axillary and/or terminal racemes or panicles with more than 1 flower to each major bract; bracteoles present. Calyx funnel-shaped, deeply lobed. Standard large, reflexed upwards. Stamens 10, filaments of 9 of them united, that of the upper stamen free or free at the base but united to the others above the base. Pod generally segmented and breaking into 1-seeded segments, more rarely opening along the lower suture. Seeds with or without arils. *Tropics & subtropics.*

This genus has been split into several genera by recent authors.

13. Campylotropis Bunge. 65/1. Deciduous shrubs. Leaves made up of 3 leaflets, the terminal leaflet usually conspicuously stalked; stipules present, stipels absent. Flowers in axillary racemes crowded into panicles at the tips of the branches; bracts each subtending a single flower, bracteoles soon deciduous. Calyx bell-shaped, 5-toothed, the 2 upper teeth united for most of their length; flower-stalk jointed just beneath the calyx. Corolla purple; standard reflexed upright; keel curved, sickle-shaped, acute. Stamens 10, 9 of them with united filaments, that of the uppermost free. Pod short, 1-seeded, ovoid, ellipsoid or almost spherical, compressed, not opening. *Asia.*

14. Lespedeza Michaux. 40/10. Annual or perennial herbs or shrubs. Leaves borne in spirals or in 2 distinct ranks; leaflets 3, more or less equal; terminal leaflet stalked or not; stipules persistent, stipels absent; winter buds with spiral scales or scales in 2 ranks. Flowers axillary or in axillary racemes, each bract subtending 2 flowers; in several species corolla-less, cleistogamous flowers are produced as well as normal, opening flowers; bracteoles 2, closely overlapping the base of the calyx, persistent. Calyx funnel- or bell-shaped, 5-toothed, the upper teeth often united for much of their length. Standard clawed or not; keel straight, blunt. Stamens 10, filaments of 9 united, that of the uppermost usually free. Pod 1-seeded, not dehiscent. *Temperate & subtropical North America, Asia & Australia.*

A few species of the genus **Erythrina** Linnaeus may be grown in warm situations; they are small spiny shrubs with flowers (which sometimes drip nectar) in pairs.

15. Apios Medikus. 10/1. Twining perennial herbs, some (ours) with tuberous roots. Leaves alternate, pinnate with several leaflets, including a terminal leaflet; stipules subulate, persistent; stipels absent. Flowers in short racemes, sometimes leaf-opposed. Calyx cup-like, scarcely toothed. Corolla with a broad standard, the keel much longer but coiled upwards within the standard, blunt and notched at the apex. Stamens 10, 9 with their filaments united, that of the uppermost free. Ovary with coiled style. Pod flattened, several-seeded, the sides spiralling after dehiscence. *America, Asia.*

16. Kennedia Ventenat. 11/few. Woody scramblers or prostrate, scrambling herbs. Leaves with 3 (rarely more) leaflets; stipules sometimes falling early; stipels usually present, linear. Flowers in loose or dense axillary racemes (rarely solitary); bracteoles absent. Calyx tubular, deeply 5-toothed. Corolla reddish or black and yellow. Stamens with 9 of the filaments united, that of the uppermost free. Pods flattened, containing several seeds with spongy partitions between them. Seeds appendaged. *Australia.*

The name is often misspelled 'Kennedya'.

17. Hardenbergia Bentham. 2/2. Shrubs or woody twiners. Leaves of 3--5 leaflets, when the terminal leaflet stalked and the laterals opposite or whorled, or reduced to a single leaflet; stipules and stipels persistent (even when leaf reduced to a single leaflet). Flowers in pairs in axillary racemes. Bracteoles absent. Calyx cylindric to bell-shaped, 5-toothed, the upper teeth united for most of their length. Corolla blue or purple, rarely white. Stamens 10 with all their filaments united into a tube or the uppermost at least partially free. Pod flattened or not, with or without spongy partitions between the seeds. *Australia.*

18. Centrosema (de Candolle) Bentham. 45/1. Herbaceous climber. Leaves with 3 leaflets, stipules and stipels persistent. Flowers 1--5, in axillary racemes with swollen nodes. Bracteoles 2, conspicuous. Flowers borne upside-down (standard below). Calyx 5-toothed. Standard large, with a small projecting spur towards the base. Stamens 10, filaments of 9 united, that of the uppermost free, anthers alternately longer and shorter. Stigma hairy. Pods long-beaked, with thickened sutures. *Mostly tropical America, a few species from temperate North America.*

19. Phaseolus Linnaeus. 60/few. Annual to perennial herbs or climbers; hairs hooked. Leaves with 3 leaflets, stipules and stipels persistent. Flowers in racemes, with bracteoles. Calyx broadly bell-shaped, 5-toothed. Corolla often red, with broad

279

standard with auricles at the base; wings joined to the keel. Stamens 10, 9 with their filaments united, the uppermost free, at least at the base. Style coiled through a number of revolutions, stigma terete and bearded longitudinally. Pods narrow and more or less cylindric. *New World.*

Many (together with those of the allied genus **Vigna** Savi, some species having names in both genera) are widely cultivated as vegetables (beans of various kinds). Few of the species are cultivated purely as ornamentals.

20. Bituminaria Fabricius. 3/1. Foetid biennial to perennial herb, somewhat woody at the base, shortly and stiffly hairy; stems 30--120 cm, erect but often rather straggling, branched below. Stipules small, narrow. Leaves with 3 leaflets on long stalks, dotted with glands; leaflets 1--6 cm, narrowly lanceolate to ovate or almost circular, entire. Flowers in dense heads on axillary stalks longer than leaves, subtended by paired bracts. Calyx with 5 unequal teeth. Corolla 1.5--2 cm, bluish-violet, sometimes pink or white. Stamens with filaments united for much of their length. Fruit indehiscent, ovoid, flattened, 1-seeded, hairy, with long, curved beak; seed 5--6 mm. *Mediterranea area.*

21. Psoralea Linnaeus. Few/1. Scented herbs or shrubs. Leaves alternate, pinnate with a terminal leaflet, rarely simple, with translucent dots. Flowers solitary, in heads, racemes, spikes or sometimes clustered. Calyx lobes 5, nearly equal; corolla-wings about as long as keel; standard ovate or rounded; stamens 10, all united or 1 free. Fruit a short, 1-seeded, indehiscent legume. *Scattered in temperate & subtropical areas.*

22. Amorpha Linnaeus. 15/3. Shrubs. Leaves pinnate with up to 45 leaflets, including a terminal leaflet, leaflets usually with brownish glands on the lower surface; stipules soon falling, stipels thread-like. Flowers very numerous in dense, spike-like racemes which may be clustered to form an apparent panicle, flower-stalks short. Calyx funnel-shaped, 5-toothed, sometimes with glands near the top. Corolla reduced to the standard (wings and keel absent), which is purplish, clawed and envelops the stamens and style. Stamens 10, their filaments all united into a tube below; anthers yellow, brown or purplish. Pod 1-seeded, indehiscent, glandular. *North America, Mexico.*

23. Amicia Humboldt et al. 7/1. Softly wooded shrubs. Stems upright. Leaves pinnate, with 4 or 5 leaflets including a terminal leaflet, all with orange-brown glands on the lower surface; stipules large, almost circular, united at the base, enclosing the young growth, soon falling; stipels represented by dense tufts of hair. Flowers axillary, solitary or in few-flowered racemes; bracts similar to stipules but smaller. Calyx with a large, almost circular (stipule-like) upper tooth, the lateral teeth very small, the lower teeth narrow. Corolla large, yellow, keel blunt, standard glandular outside. Stamens 10, their filaments all united, the tube so formed split above. Pod several-seeded, indehiscent, breaking into 1-seeded segments. *Central & South America.*

24. Clianthus Lindley. 2/2. Herbs or soft-wooded shrubs. Leaves pinnate with numerous lateral leaflets and a terminal leaflet; stipules large, persistent, stipels absent. Flowers very showy in erect or pendent axillary racemes, each with a bract and 2 bracteoles. Calyx bell- or cup-shaped, 5-toothed. Corolla scarlet or scarlet and black, pink or white (a cultivar). Standard reflexed, narrow, acuminate, wings deflexed, shorter than the keel; keel large, acuminate, curved like a parrot's beak. Stamens 10, filaments of 9 of them united, that of the uppermost free. Ovary shortly stalked; style bearded. Pod firm, turgid and beaked, containing several seeds. *Australia, New Zealand.*

25. Colutea Linnaeus. 26/4 and a few hybrids. Shrubs or rarely small trees. Leaves pinnate, with 2--6 pairs of leaflets and a terminal leaflet, often somewhat glaucous; stipules persistent, stipels absent. Flowers rather few in axillary racemes, with bracts. Calyx usually broadly bell-shaped, with 5 short teeth. Corolla yellow to orange-red, the claws of the petals usually projecting from the calyx. Standard upright, with 2 small swellings above the claw; wings shorter than to longer than the keel, each with a distinct auricle on the upper margin and sometimes a spur on the lower; keel rounded or beaked at the apex. Stamens 10, 9 with their filaments united, that of the uppermost free. Ovary stalked. Pod inflated, papery, and translucent, indehiscent or splitting towards the apex, stalked, the stalk often projecting from the persistent calyx; seeds several. *Eurasia, east Africa (Ethiopia).*

26. Halimodendron de Candolle. 1/1. Round-headed shrub with pale brown, striped bark, with weakly distinguished long and short shoots. Leaves on the extension shoots persisting as stout spines to 6 cm, each with 2 much smaller stipular spines at the base. Foliage leaves with usually 4 leaflets, pinnately arranged, the axis tipped with a short spine; leaflets glaucous, oblong-obovate, tapered to the base, rounded to the apex, to 4 cm, hairy or hairless; stipules of the short-shoot leaves membranous. Racemes terminating the (axillary) short shoots, with 2--5 flowers each with a bract and 2 bracteoles. Calyx 5--7 mm, bell-shaped, finely hairy, truncate, with 5 small teeth. Corolla 1.5--2 cm, pale purplish pink or magenta, standard erect with its sides reflexed. Stamens 10, filaments of 9 of them united, that of the uppermost free. Ovary shortly stalked, with several ovules. Pod stalked, inflated, obovoid or obovoid-oblong, leathery, beaked, yellowish brown, containing several seeds. *East Europe (Russia) to central & eastern Asia.*

27. Caragana Lamarck. 65/14. Shrubs or small trees, often with widely spreading or pendent, little-branched branches. Shoots generally of 2 kinds: long, extension shoots on which the leaves are borne alternately, their axes often persisting as spines, their stipules also often spiny and persistent; and short shoots borne in the axils of the extension-shoot leaves, very long-lived, terminated by usually membranous stipules and bearing clusters of leaves and the flowers. Leaves with 4 or more leaflets, mostly pinnate but some with the leaflets borne in a close cluster (palmate), the terminal leaflet replaced with a spine. Stipules present, joined at the base, divergent at the apex, membranous at first, persistent, some with thickened midribs which persist as spines as the membranous tissue decays. Flowers solitary or in clusters, each with its own jointed stalk arising from the short shoot, rarely in few-flowered, stalked umbels; bracteoles sometimes persisting at the joint. Calyx bell-shaped or cylindric, truncate or oblique at the mouth, 5-toothed. Corolla usually yellow, more rarely orange or pink, the petals with long claws projecting from the calyx; standard erect, its margins reflexed, wings usually each with a backwardly-projecting auricle. Stamens 10, filaments of 9 united, that of the uppermost free. Ovary containing several ovules. Pod several-seeded, hairless or hairy inside. *Eastern Europe to Central Asia & China.*

28. Calophaca de Candolle. 10/1. Deciduous shrubs or low perennial herbs with alternate. Leaves pinnate with 3--13 pairs of leaflets and a terminal leaflet. Flowers solitary or in racemes; calyx tubular with 5 slender teeth; corolla violet or yellow; stamens 10, with 9 united and 1 free. Fruit a cylindric legume. *Asia.*

29. Astragalus Linnaeus. 2000/13. Herbs or small shrubs. Leaves pinnate, either with a terminal leaflet or the axis continuing and persistent as a hardened spine; leaflets symmetric at base; stipules persistent, often conspicuous; stipels absent. Hairs, when present, basifixed or medifixed, grey, white or dark brown to black, colours often mixed. Flowers in racemes or spikes, axillary or arising directly from the basal rosette on a scape or scape absent. Bracts usually present. Calyx bell-shaped to tubular, 5-toothed, sometimes very deeply so. Corolla with wings and keel usually shorter than the variably-shaped standard. Keel sometimes toothed on the lower side. Stamens 10, the filaments of 9 united, that of the uppermost free. Pod variously shaped, usually several-seeded, divided along its length by a septum developed from an infolding of the lower suture. *Northern hemisphere, scattered in the southern hemisphere.*

30. Oxytropis de Candolle. 100/10. Herbaceous perennials or subshrubs. Leaves pinnate; leaflets entire. Flowers in axillary spikes or racemes, similar to those of *Astragalus* but the keel with an acute beak at the apex. *Eurasia, North America.*

31. Alhagi Gagnepain. 3/3. Bushy, somewhat woody perennial herbs with rigid spines on the lower branches and with the upper, short, lateral flowering branches terminating in spines. Leaves simple, stalkless; stipules minute. Flowers borne singly or in pairs in the axils of minute bracts; flower-stalks very short or absent. Calyx bell-shaped, shallowly 5-toothed. Corolla pink to red, standard and keel longer than the wings, the keel obtuse. Stamens 10, 9 of them with their filaments united, that of the uppermost free. Ovary containing several ovules. Pod indehiscent, constricted between the 1--5 seeds. *Europe & North Africa to Central Asia.*

32. Galega Linnaeus. 6/2. Erect herbaceous perennials with a bushy habit. Leaves irregularly pinnate; leaflets obtuse or acute. Stipules conspicuous. Flowers many, in stalked racemes longer than leaves. Stamens 10, with filaments united. Fruit cylindric, slender, constricted between seeds, beaked; seeds many. *Europe, Asia, east Africa.*

33. Glycyrrhiza Linnaeus. 20/2. Glandular, rhizomatous perennial herbs. Leaves pinnate, with a terminal leaflet, the leaflets dotted with orange-brown glands; stipules lanceolate, minute; stipels absent. Flowers numerous in axillary racemes or spikes which are loose or dense and head-like. Calyx 5-toothed, 2-lipped, the tube narrowly funnel-shaped, the upper teeth short, the lower as long as or longer than the tube. Corolla white, yellow, mauve, violet or bluish; keel obtuse or acute. Stamens 10, 9 of them with their filaments united, that of the uppermost free. Ovary with few ovules. Pod compressed, dehiscent with the sides contorting, seeds 1--several. *Eurasia, North America, temperate South America.*

34. Chordospartium Cheeseman. 1/1. Shrub to 8 m, often with a trunk to 3 cm across in the wild. Branches drooping, leafless except for small, adpressed, triangular, brownish scales at the nodes, terete, grooved. Flowers usually numerous in cylindric racemes, borne singly or in groups of 2--5 at the nodes; inflorescence-stalk hairy, flower-stalks woolly. Flowers to 1 cm, pale lavender or whitish, the standard with purple veins. Calyx 3--4 mm, cup-shaped, hairy, with 5 minute teeth. Stamens 10, 9 of them with their filaments united, that of the uppermost free. Ovary hairy, containing 1--5 ovules; styles long, incurved, hairy on the upper side towards the apex. Pods hairy, indehiscent, swollen, *c.* 5 mm, usually 1-seeded. *New Zealand.*

35. Carmichaelia R. Brown. 39/11. Shrubs or small trees; habit diverse, ranging from small, compact, patch-forming plants to larger, erect shrubs or trees. Leaves pinnate with 3--7 often notched leaflets, usually absent from mature plants, their axils marked by small notches in the flattened or terete branchlets. Flowers solitary or in racemes borne in the notches, sometimes more than 1 raceme from each notch; flower-stalks hairy or finely downy. Calyx bell-shaped, 5-toothed. Petals distinctly clawed. Stamens 10, the filaments of 9 united, that of the uppermost free; anthers each with only a single pollen-sac. Pod dry with thickened margins which may project into the cell, hard, indehiscent or dehiscent only at the apex or by one or both sides separating from the persistent, thickened margins. Seeds 1--few. *New Zealand, Lord Howe Island.*

36. Hedysarum Linnaeus. 100/8. Annual or perennial herbs or low shrubs. Stems several, grooved, often covered with adpressed hairs. Leaves alternate, pinnate with a terminal leaflet, the leaflets almost stalkless; stipules united or free, papery. Flowers erect or nodding, borne in stalked axillary racemes; calyx with 5 teeth, corolla with 5 petals, standard longer than keel, pink, purple, yellow or white, ovary with 4--8 ovules. Fruit a stalked segmented pod, with or without wings; each segment containing 1 seed, segments flattened or rounded; seeds 2--6. *North temperate areas.*

37. Onobrychis Miller. 100/7. Herbaceous or subshrubby perennials, or annuals. Leaves irregularly pinnate; leaflets entire. Flowers in elongate, stalked axillary racemes. Calyx bell-shaped, with 5 equal linear teeth. Stamens united except for the uppermost which is free. Fruit indehiscent, flattened, almost circular, usually spiny on veins and margins; seed solitary. *Eurasia, North Africa.*

38. Anthyllis Linnaeus. 20/5. Annual or perennial herbs or low shrubs. Leaves usually pinnate with a terminal leaflet, rarely reduced to 3 leaflets or to a solitary (terminal) leaflet. Stipules small, falling early. Flowers usually in dense heads, each subtended by 2 usually large bracts, more rarely in clusters or solitary in the axils of single bracts. Calyx tubular, bell-shaped or unequally swollen, sometimes constricted towards the mouth, with equal or unequal teeth. Corolla usually yellow, more rarely cream or red. Stamens with the filaments all united or the filament of the uppermost stamen free for up to half its length. Pod shortly stalked or stalkless, often indehiscent, usually enclosed in the persistent, papery calyx, containing 1--many seeds. *Most of Europe, North Africa, south-west Asia, extending to east Africa (Ethiopia).*

39. Tetragonolobus Scopoli. Few/2. Plants similar to *Lotus* (see below) but leaves with 3 leaflets, genuine stipules better-developed and leaf-like. Flowers solitary or 2 together. Calyx-teeth equal. Corolla yellow or dark red. Pod almost square in section, the angles winged. *South Europe, Mediterranean area.*
A small genus often included in *Lotus.*

40. Lotus Linnaeus. 150/5. Annual or perennial herbs, sometimes woody below. Leaves pinnate with 5 leaflets (ours) including a terminal leaflet, axis often short, the 2 lowermost leaflets often resembling stipules. Genuine stipules very small, glandular.

Flowers in heads or solitary. Calyx bell-shaped to almost tubular, with 5 equal or unequal teeth. Corolla scarlet, crimson to dark purplish brown or yellow, the keel with a conspicuous beak. Stamens 10, filaments of 9 united, that of the uppermost free. Pod cylindric, sometimes broadly so, straight or curved. Seeds numerous. *Temperate areas.*

41. Dorycnium Miller. 8/2. Perennial herbs or small shrubs. Leaves almost stalkless, pinnate with 5 leaflets, the lower pair often appearing superficially like stipules; true stipules minute. Flowers in axillary heads. Calyx bell-shaped with 5 equal or unequal teeth. Corolla whitish, often with red or pink spots or lines; keel obtuse, dark red. Stamens 10, 9 with the filaments united, that of the uppermost free. Pod oblong, ovoid or narrowly cylindric. Seeds 1--many. *Mediterranea area.*

42. Coronilla Linnaeus. 25/7. Annual or perennial herbs or low shrubs. Leaves pinnate with a terminal leaflet, rarely simple or with only 3 leaflets; stipules variable, free or united to each other. Flowers in heads in the leaf-axils. Calyx bell-shaped, more or less 2-lipped. Petals yellow or pinkish, keel acute. Stamens 10, filaments of 9 united, that of the uppermost free. Pod breaking up into 1-seeded linear to oblong segments, which are not constricted between them, round in section or ridged or angled. *Canary Islands, Europe, western & central Asia.*

43. Hippocrepis Linnaeus. 15/2. Herbaceous or shrubby perennials (cultivated species). Leaves irregularly pinnate; leaflets entire. Stipules small, linear to lanceolate. Flowers in umbels on long axillary stalks. Calyx elongately bell-shaped with 5 teeth. Corolla yellow. Stamens with filaments united except for that of the uppermost, which is free. Fruit flattened, indehiscent, of a single segment or segmented and breaking up when ripe, the segments with deep sinuses. *Europe, Mediterranean area.*

44. Vicia Linnaeus. 150/6. Perennial and annual herbs; stems angular in cross-section but never winged. Leaves usually with many pairs of leaflets, the axis ending in a tendril or short point; leaflets folded in bud. Flowers blue, purple, yellow, orange or whitish. Wing petals attached to keel. Style cylindric or flattened, hairy all round or with a tuft or hairs on the outer face. Pod diamond-shaped or flattened-cylindric. Germination hypogeal. *Temperate northern hemisphere extending into tropical Africa & South America.* Figure 60, p. 286.

45. Lathyrus Linnaeus. 150/15. Herbaceous perennials and annuals with unbranched climbing or sprawling stems which are often winged. Leaves with 2 or more leaflets, the axis ending in a tendril or a short point; leaflets rolled up in bud. Racemes axillary, 1--many flowered. Wing petals attached to keel. Style flattened, with a brush of hairs on the inner face, often twisted through 90°. Germination hypogeal. *Temperate northern hemisphere extending into tropical Africa and South America.* Figure 61, p. 288.

46. Lens Miller. 5/1. Low, hairy annuals, with stems angular but not winged. Leaves with a few pairs of small leaflets and tendrils; leaflets folded in bud. Racemes few-flowered; flowers small, pale and inconspicuous; calyx teeth equal, as long as the corolla, wing petals adhering to the keel; style flattened, hairy on the inner surface. Pod stalked, oblong, 1--3-seeded; seeds lens-shaped. Germination hypogeal. *Mediterranean area, south-west Asia, tropical Africa.*

47. Pisum Linnaeus. 2/1. Hairless annuals with stems circular in cross-section. Leaves with large leafy stipules, 1--4 pairs of leaflets and a strong, branched tendril; leaflets folded in bud. Racemes few-flowered, flowers showy. Calyx-teeth almost equal, wing petals united with the keel; style with retroflexed margins, hairy on the inner face. Pod cylindric. Germination hypogeal. *Mediterranean area, southwest Asia.*

48. Cicer Linnaeus. 40/1. Perennial and annual herbs, often spiny, conspicuously glandular-hairy. Leaves ending in a leaflet, spine or tendril; leaflets 3--many, toothed, the veins terminating in the teeth. Racemes 1--few-flowered. Wing petals free from the keel. Style hairless. Fruit inflated, seeds beaked. Germination hypogeal. *Mediterranean area to Himalaya and central Asia.*

49. Ononis Linnaeus. 75/6. Annual or perennial herbs or dwarf shrubs, usually sticky, glandular-hairy. Leaves usually with 3 leaflets, rarely reduced to a single leaflet or pinnate with a terminal leaflet; leaflets usually toothed. Stipules united to the leaf-stalk. Flowers in spikes, racemes or panicles. Calyx bell-shaped or tubular. Corolla yellow, pink or purple, rarely almost white; keel more or less beaked. Stamens 10, their filaments all united. Pod oblong or ovate. Seeds 1--many. *Mediterranean area, east to India and Mongolia, west to the Canary Islands.*

50. Parochetus D. Don. 1/1. Creeping herbaceous perennial. Leaves with 3 palmate, heart-shaped, finely toothed leaflets, veins running out to margins. Stipules free from the leaf-stalk. Flowers blue or pale purple, 1--4, stalked, in umbels, those in the lower leaf-axils very small and not opening, with pods ripening on or below the soil surface. Calyx bell-shaped, deeply cleft; lobes 5, acute, almost equal. Petals free, obovate, short-clawed; standard deep blue (rarely purple) wings pinkish, long-clawed, oblong-obovate; keel shorter than wings, turned inwards. Stamens 10, filaments united except for that of the uppermost, which is free. Style bent; stigma small. Pod linear, acutely beaked; seeds 8--20. *Asia, tropical east Africa, southern Africa.*

51. Melilotus Miller. 20/3. Annual or biennial herbs, smelling of newly mown hay. Leaves with 3 pinnate leaflets; leaflets usually linear to elliptic-oblong, short-stalked, veins running out to the margin; stipules lanceolate to awl-shaped, fused to the leaf-stalk to some extent. Flowers small, yellow, white or white tipped with blue, in erect, axillary, spike-like racemes. Calyx shortly bell-shaped; lobes 5, almost equal, awl-shaped to lanceolate, acute to acuminate, shorter than the tube. Petals falling early; standard obovate-oblong, narrow at base, nearly stalkless; wings oblong, auricled at the base. Keel blunt, clawed, obtuse. Stamens 10, filaments of 9 united, that of the uppermost free. Style thread-like; stigma small. Pod nutlet-like, straight, beaked, spherical or obovoid. Seeds 1--few. *Europe, southwest Asia.*

52. Trigonella Linnaeus. 80/3. Erect annual herbs, often strongly scented. Leaves with 3 toothed leaflets, veins running out to the leaflet-margins; stipules joined to the leaf-stalk. Flowers yellow, blue or white in heads, umbels or short dense racemes in leaf-axils, rarely solitary. Calyx 5-toothed, teeth ovate, acuminate. Petals free, standard obovate or oblong; wings oblong, auricled; keel oblong, shorter than wings, obtuse. Stamens free from petals, filaments of 9 united, that of the uppermost free. Fruit variable, oblong or oblong-linear, compressed or terete, with 1--many seeds. *Eurasia, southern Africa, Australia.*

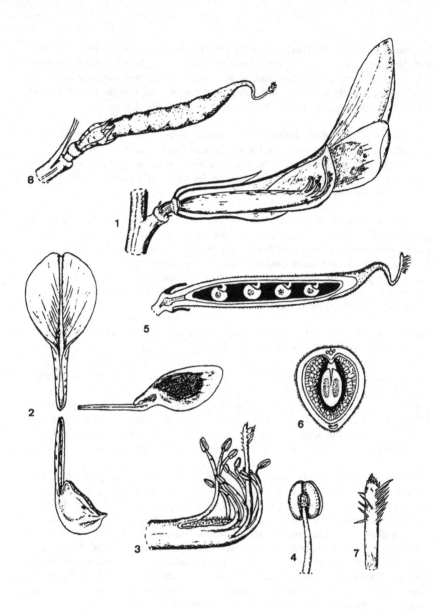

Figure 60. Leguminosae/Papilionaceae. *Vicia faba*. 1, Longitudinal section of flower. 2, Petals from above (clockwise: standard, wing, keel). 3, The ten stamens. 4, Upper part of a single stamen. 5, Longitudinal section of ovary. 6, Transverse section of ovary. 7, Stigma. 8, Fruit.

53. Medicago Linnaeus. 50/3. Annual or perennial herbs or shrubs. Leaves with 3 pinnate leaflets; leaflets obovate, margins toothed, veins running out to margins. Stipules fused to leaf-stalk to some extent. Flowers small, yellow or violet, rarely variegated, in axillary spikes or small heads. Calyx bell-shaped; lobes 5, almost equal. Petals free from staminal tube, standard obovate or oblong, narrowed at the base; wings oblong, auricled, clawed, longer than the obtuse keel. Stamens 10, filaments of 9 united, that of the uppermost free. Style awl-shaped, smooth; stigma almost capitate, oblique. Pod spirally coiled, or sickle-shaped, often covered with spines, with 1--several seeds. *Eurasia, Mediterranean area, Africa.*

54. Trifolium Linnaeus. 300/12. Annual, biennial or perennial herbs, occasionally somewhat woody. Leaves usually with 3 leaflets, rarely palmately divided with 5--8 leaflets, the leaflets usually toothed, the veins running out to the margins; stipules large, persistent, joined to the leaf-stalk to some extent. Flowers in heads or short spikes, rarely solitary. Calyx tubular, 5-toothed, teeth equal or unequal. Petals persistent or deciduous, attached to each other and to the staminal tube. Stamens 10, 9 with their filaments united, that of the uppermost free, with all or 5 of the filaments swollen towards the apex. Pod enclosed in the persistent calyx or shortly protruding, indehiscent or dehiscent by the inner suture or by a hardened lid; seeds 1--4, rarely to 10. *Mainly from north temperate areas, a few from the southern hemisphere.*

55. Anagyris Linnaeus. 1/1. Deciduous shrub 1--3 m, foetid and poisonous. Twigs green. Leaves with 3 leaflets 3--8 × 1--3 cm, elliptic, more or less obtuse, hairy beneath. Stipules minute, lanceolate. Flowers up to 20 in short axillary racemes on previous year's wood. Calyx bell-shaped, with 5 triangular teeth. Corolla 1.8--2.5 cm, yellow, the standard about half as long as the other petals, usually with a blackish spot; petals of keel free. Stamens free. Fruit 10--18 cm, shortly stalked, pendent, flat, constricted between seeds, hairless; seeds few, large. *Mediterranean area, west Asia.*

56. Piptanthus Sweet. 2/2. Shrubs or small trees to 4 m, branches with wide pith. Leaves stalked, of 3 leaflets; stipules conspicuous, fused for two-thirds or more of their length. Flowers in a loose or dense, stalked terminal raceme, 3 arising from each bract. Calyx bell-shaped, 5-toothed, the upper 2 teeth fused for most of their length. Corolla yellow, standard reflexed, as long as the wings, with a conspicuous claw, blade notched at the apex. Stamens 10, their filaments free. Ovary hairless or hairy, with 3--10 ovules. Pod oblong, flattened, leathery, with up to 10 seeds. *Himalaya, southwest China.*

57. Thermopsis R. Brown. 20/8. Tall, usually hairy, perennial herbs with woody rhizomes. Leaves with 3 leaflets which vary from linear to elliptic, oblanceolate or obovate; stipules conspicuous, persistent, often leaf-like. Flowers in terminal or axillary, compact or loose racemes. Calyx 5-toothed, sometimes 2-lipped with the upper lip truncate or notched. Corolla deep purple or yellow. Stamens 10, filaments free. Ovary more or less stalkless, containing numerous ovules. Pod flat, straight or recurved, many-seeded. *North America, Asia.*

58. Baptisia Ventenat. 50/4. Perennial herbs with woody rhizomes, often glaucous, stems somewhat woody towards the base when old. Leaves with 3 leaflets (ours), shortly stalked; stipules minute, thread-like and deciduous or large, leaf-like and

Figure 61. Leguminosae/Papilionaceae. *Lathyrus*: leaves and stipules (1, *L. vernus*; 2, *L. japonicus*; 3, *L. grandiflorus*; 4, *L. latifolius*; 5, *L. sativus*, 6, *L. nervosus*).

persistent. Flowers in racemes. Calyx bell-shaped, 2-lipped, the upper lip of 2 almost completely fused teeth, the lower of 3 distinct teeth. Corolla white, cream, yellow or blue; standard reflexed. Stamens 10, their filaments free. Ovary with numerous ovules. Pod inflated, variably shaped, many-seeded. *Eastern & central USA.*

59. Lupinus Linnaeus. 200/18 and some hybrids. Annual, biennial or perennial herbs or woody low shrubs. Leaves palmate, stipules attached to the base of the stalk. Flowers in racemes or spikes, in whorls or alternate. Calyx 2-lipped, each lip either deeply cleft to about the mid-point or each with 2 or 3 tiny teeth (almost entire) or entire, bracteoles attached to the calyx. Stamens 10, filaments of all joined together in a tube. Fruit a flat hairy pod. *Widely distributed, absent from South Africa & Australia.*

60. Laburnum Fabricius. 2/2 and their hybrid. Small trees. Leaves made up of 3 leaflets. Flowers in simple, axillary or apparently terminal, leafless racemes, pendent while flowering. Calyx bell-shaped, slightly 2-lipped, lips undivided or shortly toothed. Corolla yellow. Stamens 10, all filaments united into a tube. Pod dehiscent, flattened, slightly constricted between the seeds. Seeds numerous, compressed. *Central & southeast Europe.*

A graft hybrid between species of *Laburnum* and *Cytisus* (see below), is known as + **Laburnocytisus** Schneider; it can be recognised by its Laburnum-like appearance, but with racemes of pink flowers as well as the more normal yellow ones.

61. Petteria Presl. 1/1. Non-spiny shrubs. Leaves made up of 3 leaflets. Flowers in terminal, erect, leafless racemes. Calyx bell-shaped to tubular, 2-lipped; upper lip divided to about two-thirds, lower 3-toothed. Corolla yellow. Pod linear-oblong, dehiscent, straight or slightly curved, somewhat inflated. Seeds without appendages. *Balkan Peninsula.*

62. Cytisus Linnaeus. 100/40 and some hybrids. Shrubs or small trees without spines, from 10 cm to more than 6 m. Leaves with 1 or 3 leaflets, alternate, sometimes crowded, often soon falling, the branches and branchlets performing most of the photosynthesis. Flowers axillary, forming leafy or leafless, terminal or lateral racemes. Calyx 2-lipped, upper lip with 2 usually short teeth (rarely deeply cleft). Corolla white, yellow, purple or dark-brown; keel more or less sickle-shaped. Stamens 10, filaments all united into a tube. Stigma curved upwards, or rarely rolled up. Pods linear or oblong, seeds numerous, with appendages (strophioles). *Europe, North Africa.*

63. Erinacea Adanson. 1/1. Spiny shrubs with opposite or alternate branches. Leaves simple or sometimes made up of 3 leaflets; shortly stalked. Flowers 1--3, in axillary or more or less terminal clusters. Calyx inflated, bell-shaped, 2-lipped; upper lip with 2 teeth; lower with 3 teeth one-third as long as the tube. Corolla blue-violet. Pod narrow-oblong, dehiscent. Seeds without appendages. *South Europe, North Africa.*

64. Echinospartum (Spach) Rothmaler. 5/1. Small shrubs with opposite branches. Leaves made up of 3 leaflets, shortly stalked or stalkless. Calyx somewhat inflated, bell-shaped, 2-lipped, upper lip deeply bifid; lower with 3 prominent teeth; all teeth as long as or longer than the tube; corolla yellow. Pod dehiscent. Seeds without appendages. *Pyrenees, Iberian Peninsula.*

65. Spartium Linnaeus. 1/1. Deciduous, unarmed shrub 1--4 mm. Branches many, erect, rush-like, grooved, green, hairless. Leaves few, falling early, 1--3 cm, linear-oblong to narrowly elliptic or lanceolate, with appressed-silky hairs beneath. Stipules absent. Flowers in loose terminal racemes, fragrant. Calyx sheath-like, split above, 1-lipped, with 5 small teeth. Corolla 2--3 cm, bright yellow. Stamens united. Fruits 3--8 cm, linear-oblong, flat, silky-hairy when young. Seeds many. *Mediterranean area; naturalised elsewhere.*

66. Chamaespartium Adanson. 4/1. Dwarf shrubs without spines, the young stems distinctly winged and flattened. Leaves simple or absent. Flowers in dense, terminal racemes. Calyx tubular, 2-lipped; upper lip deeply bifid, lower with 3 distinct teeth; corolla yellow, the standard broadly ovate, equalling the wings and keel. Pod dehiscent. Seeds with or without appendages. *Central Europe, Mediterranean area.*

67. Retama Rafinesque. 4/1. Shrubs without spines. Leaves simple; falling early. Flowers in racemes. Calyx urn-shaped, bell-shaped or obconical, 2-lipped. Corolla white to yellow. Pod ovoid to spherical, indehiscent or finally incompletely dehiscent along ventral suture. Seeds without appendages. *Canary Islands, Mediterranean area, western Asia.*

68. Genista Linnaeus. 100/21. Spiny or non-spiny shrubs with alternate or opposite branching. Leaves mostly deciduous, sometimes very early so, shortly stalked, simple or made up of 3 leaflets, alternate or opposite. Flowers bisexual, alternate or opposite in racemes or axillary clusters or heads. Calyx tubular, usually with prominent upper and lower lips, the upper lip bifid, the lower with 3 distinct teeth. Corolla yellow. Standard broadly ovate or triangular, acute, hairless or downy, as long as or shorter than the keel. Keel narrowly oblong, hairless or downy. Wings as long as standard, hairless. Stamens 10, their filaments all united into a tube. Pod either narrowly oblong and compressed, or sickle- or diamond-shaped and more or less inflated, several-seeded, more rarely ovoid-acuminate and 1--2-seeded, hairless or downy. Seeds without appendages. *Canary Islands, Madeira, Europe, North Africa, south-west Asia; introduced elsewhere.*

69. Adenocarpus de Candolle. 15/6. Unarmed shrubs with alternate branches. Leaves divided into 3 leaflets, the leaves sometimes on short shoots. Flowers in terminal racemes or clusters. Calyx tubular, 2-lipped, sometimes with glandular tubercles, the upper and lower lips prominent, the upper deeply bifid, the lower with 3 distinct teeth. Corolla orange-yellow. Stamens 10 with their filaments united into a tube. Pod oblong, dehiscent, covered with glandular warts. Seeds numerous, oblong, each with a small appendage along one of the shorter sides. *Canary Islands, Spain, north & tropical Africa.*

70. Ulex Linnaeus. 20/3. Spiny, dense shrub. Leaves usually alternate, to 1.5 cm, trifoliolate, linear, sharply pointed when young, reduced to green spines or scales when mature; stipules absent. Flowers fragrant, 1.5--2.5 cm, solitary or few in small axillary clusters or racemes, produced in the leaf axils of the previous year's growth, shortly stalked, golden-yellow; bracteoles 2, small, below flower. Calyx persistent, 2-lipped, lower lip with 3 small teeth, upper lip with 2 small teeth. Corolla persistent, standard ovate, wing and keel obtuse. Stamens 10, all united, alternating in 2 lengths; style slightly curved. Fruit a dehiscent, small, hairy, broadly-ovate to linear-oblong, explosive pod containing 1--6 seeds. *Europe, North Africa.*

88. LIMNANTHACEAE

Low annual herbs. Leaves alternate, pinnately divided, without stipules. Flowers bisexual, radially symmetric, solitary on axillary stalks. Sepals 3--6, more or less free; petals 3--6, free, usually clawed. Stamens twice as many as the petals, in 2 whorls, the outer ones alternating with the petals and often with a basal nectar-gland. Ovary superior, of 3 or 5 free carpels, united by a single style arising in the centre; stigmas 3 or 5. Ovules solitary in each carpel. Fruit of indehiscent, 1-seeded nutlets.

A family of 2 genera from temperate North America; only one is cultivated.

1. Limnanthes R. Brown. 7/1. Sepals usually 5, sometimes 4 or 6; petals usually 5, sometimes 4 or 6, with a U-shaped band of hairs on the claw. Ovary of 5 carpels; stigmas 5, capitate. *Western North America.*

89. OXALIDACEAE

Annual or perennial herbs, sometimes shrubs or trees. Leaves alternate, usually palmate or pinnate, rarely simple. Flowers bisexual, radially symmetric. Sepals and petals 5, sometimes more or less united at base; stamens in 2 whorls of 5, filaments fused towards base, anthers opening by slits. Ovary superior, 5-celled, with 2--many ovules per cell, placentation axile; styles 5, free. Fruit a dry dehiscent capsule or fleshy berry. See figure 62, p. 292.

A family of 3 genera and about 900 species, widely spread throughout the world, mainly tropical or subtropical.

1. Oxalis Linnaeus. 800/10. Annual or perennial, stemmed or stemless herbs and shrubs, often with tubers or bulbs; very rarely aquatic. Leaves palmate; leaflets 3--20 or more (rarely 1), often folding down at night. Flowers with differing style and stamen lengths, axillary, often in cymes or umbel-like cymes, sometimes solitary; petals white, pink, red or yellow; filaments fused into a tube. Fruit a capsule; seed enclosed in a fleshy aril which springs the seed from the capsule at maturity. *Mainly Southern Africa, South America; naturalised worldwide.* Figure 62, p. 292.

90. GERANIACEAE

Herbs or shrubs, sometimes succulent. Leaves with stipules, alternate or opposite, simple, pinnate or palmate, usually toothed. Flowers in umbels, sometimes composed of only 2 or 3 flowers and then referred to as cymules, occasionally solitary. Flowers radially or bilaterally symmetric. Sepals 5. Petals 5 or sometimes fewer. Nectaries usually 5, at bases of stamens or 1 at the apex of a spur. Stamens 10 or 15, shortly united at the base, some occasionally staminodial. Ovary superior, of 5 carpels, each containing 2 ovules and united to a central column; style simple, divided at apex into 5 linear lobes covered on the inwards-facing side, stigmatic surface with papillae and more or less recurved when receptive. Fruit with the central column greatly elongated. Mature carpels consisting of a carpel-body containing 1 seed (rarely 2) and a ribbon-like piece, the awn; they separate from the axis of the column on drying out. Splitting involves 3 stages: (1) the carpel-bodies gently separate from the column, (2) an interlude, the 'pre-explosive interval', (3) the peeling away of the awns from below upwards, which takes place with explosive force. Seeds either expelled from the

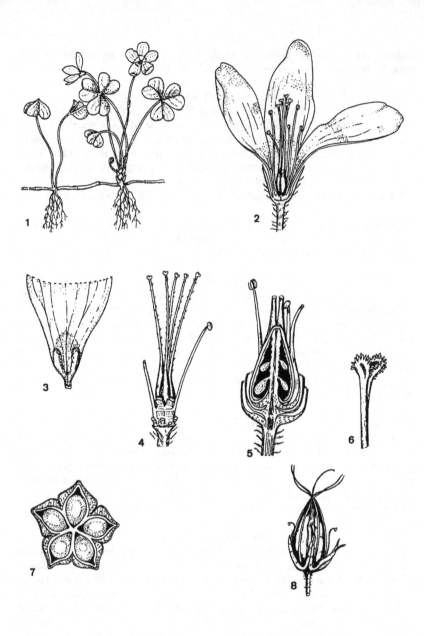

Figure 62. Oxalidaceae. *Oxalis acetosella*. 1, Non-flowering and flowering shoots. 2, Transverse section of flower. 3, Base of petal showing nectaries. 4, Flower from the side with sepals and petals removed. 5, Longitudinal section of ovary. 6, Stigma. 7, Transverse section of ovary. 8, Fruit.

carpels or dispersed within them. In the former case the tip of the awn sometimes remains attached to the tip of the column; in the latter the awn may or may not remain attached to the carpel-body. See figure 63, p. 294.

A family now usually restricted to 5 genera with about 700 species. They are plants of temperate climates in all continents.

1a. Sepal on the upper side of the flower with a slender spur produced at its
    base and attached to the flower-stalk   **3. Pelargonium**
  b. Sepals all the same, without spurs   2
2a. Stamens all with anthers   **1. Geranium**
  b. Alternate stamens without anthers   **2. Erodium**

1. Geranium Linnaeus. 300/80 and some hybrids. Plants herbaceous, sometimes woody at the base; annual, biennial or perennial. Vegetative parts, sepals and fruits usually hairy. Stem-leaves usually paired but the lowest and uppermost sometimes alternate. Leaves palmately divided or palmate, usually with the divisions lobed and the lobes toothed. Flowers usually in pairs on a Y-shaped structure (the cymule), sometimes solitary or in umbels, bisexual or apparently so, radially symmetric. Petals 5. Stamens 10, almost free, usually about as long as the sepals. Seeds expelled from the carpels or dispersed within them; in the latter case the awn may or may not remain attached to the carpel-body. Awn curved or coiled, not twisted more than half a turn. *Temperate areas, on mountains in the tropics.* Figure 63, p. 294.

2. Erodium L'Héritier. 100/30 and some hybrids. Annual, biennial or perennial herbs, usually evergreen. Leaves either simple and toothed, usually also lobed, or pinnate with variously dissected leaflets. Flowers in umbels or occasionally solitary on a jointed stalk, bisexual or unisexual (plants then dioecious). Petals 5, frequently distinctly unequal, the 2 upper being shorter and broader than the 3 lower and sometimes bearing a dark blotch. Stamens 10, united into a tube at the base, alternately fertile (with an anther) and sterile (staminodial, with no anther). Seeds dispersed within the carpels. Awn remaining attached to carpel-body and in our species divided into a lower corkscrew-like part and an upper tail-like part bent to one side. *Temperate areas, especially the Mediterranean area.*

3. Pelargonium L'Héritier. 200/40 and many hybrids. Shrubs, herbaceous perennials, sometimes woody at base, or annuals, some with tuberous roots. Stems erect or spreading, sometimes succulent. Leaves alternate, simple or pinnate or palmate, sometimes fleshy or aromatic. Flowers in umbels or pairs, bisexual, bilaterally symmetric, sometimes fragrant. Sepals 5, the upper with a slender spur at its base that is joined to the flower-stalk throughout its length and is swollen at its extremity where there is a nectar gland. Petals 5, sometimes 4 or 2, occasionally absent, clawed, upper 2 often larger than lower 3. Stamens 10, of which no more than 7 bear fertile anthers. Seeds dispersed within the carpels. Awn remaining attached to carpel-body, twisted, plumed. *Mainly South Africa, also tropical Africa, Australia & the Middle East.*

91. TROPAEOLACEAE

Herbs, mostly climbing. Leaves alternate, entire, palmately lobed or divided into leaflets, stalked, with or without stipules. Flowers usually solitary in leaf-axils, borne

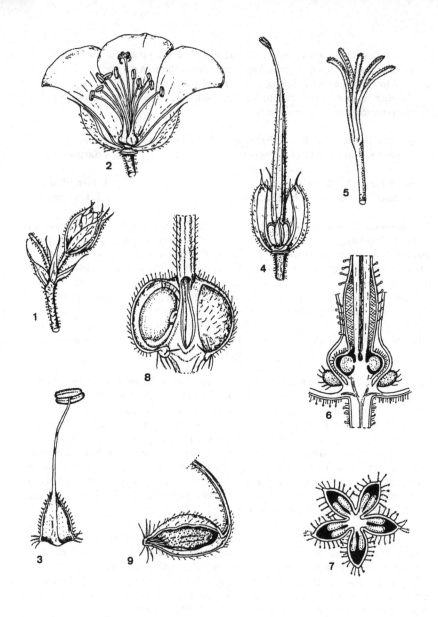

Figure 63. Geraniaceae. *Geranium pratense.* 1, Flower-bud. 2, Longitudinal section of flower. 3, A single stamen. 4, Flower form the side with petals and stamens removed. 5, Stigma. 6, Longitudinal section of ovary. 7, Transverse section of ovary. 8, Detail of ovary, one mericarp removed. 9, Ripened mericarp.

294

on long stalks. Calyx usually spurred. Stamens 8. Style 3-lobed. Ovary 3-celled. Fruit usually a schizocarp of 3 mericarps, or rarely 1-seeded with a broad wing.

There are 3 genera from Central & South America, of which only *Tropaeolum* is in cultivation.

1. Tropaeolum Linnaeus. 86/18. Annual or perennial, herbaceous, somewhat succulent climbers or sometimes procumbent; perennials with tubers. Leaves with twining stalks usually much longer than blade. Stipules usually very small and falling early, often only present in seedlings. Flowers usually borne on pendent stalks. Bracteoles usually absent. Calyx spurred; sepals 5, the 2 lower often larger than the 3 upper. Petals usually 5, sometimes 2, equal or the 2 upper different from the 3 lower. Fruit a schizocarp with 3 fleshy indehiscent carpels. *New World, from Mexico to temperate South America.*

92. ZYGOPHYLLACEAE

Small shrubs, annual or perennial herbs, usually succulent. Leaves opposite or alternate, usually deciduous, simple or pinnately divided. Stipules present, often spiny. Flowers bisexual, usually radially symmetric. Disc usually present. Sepals 4 or 5, free; petals 4 or 5, free; stamens 8--10 or 12--15, free; ovary superior, of 3--5 united cells; styles solitary. Fruit a capsule or schizocarp, more rarely a berry.

A family of some 250 species in 27 genera, largely from the tropics and warm temperate areas, particularly in arid and saline habitats.

1a. All leaves alternate	**1. Peganum**
b. Leaves mostly opposite	**2. Zygophyllum**

1. Peganum Linnaeus. 6/1. Branched, hairless, perennial herbs. Leaves alternate, irregularly divided, stipules narrowly pointed. Flowers solitary, leaf-opposed or lateral. Sepals 4 or 5, persistent; petals 4 or 5; disc a ring; stamens 12--15; ovary spherical. Fruit a capsule; seeds with endosperm. *Mediterranean area to Mongolia, southern North America.*

2. Zygophyllum Linnaeus. 80/2. Small shrubs or branched perennial herbs with stout rootstocks. Leaves opposite, simple or with 2 leaflets, succulent. Sepals and petals 4 or 5; stamens 8--10. Fruit an inflated 4- or 5-angled capsule; seeds with endosperm. *Mediterranean area to central Asia, South Africa & Australia.*

93. LINACEAE

Trees, woody climbers, shrubs or herbs. Leaves simple, alternate or opposite; stipules present or absent, sometimes gland-like. Flowers bisexual, usually radially symmetric. Sepals 4 or 5, free or partly united at base. Petals 4 or 5, spirally twisted in bud, short-lived, often clawed. Stamens 5, 10 or 15, staminodes often present. Ovary superior, with 2--5 cells; cells often subdivided; ovules 2 in each cell. Styles 1--5. Fruit a capsule, splitting at maturity into its component segments, or a drupe, nut or pair of 1-seeded mericarps. See figure 64, p. 296.

A cosmopolitan family of 15 genera with some 300 species. Only 1 genus has species hardy in northern Europe.

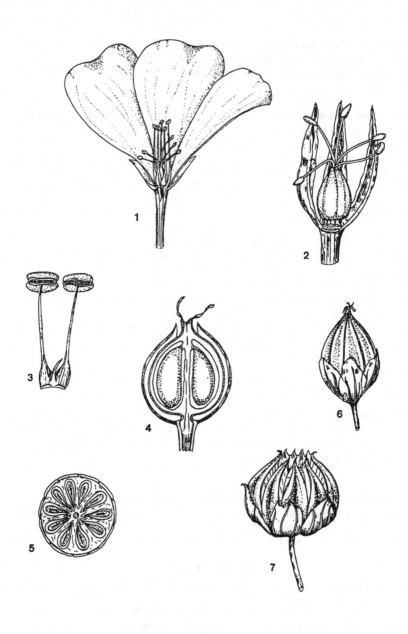

Figure 64. Linaceae. *Linum perenne.* 1, Flower with two sepals and two petals removed. 2, Flower with all petals and stamens removed. 3, Two stamens and one staminode. 4, Longitudinal section of ovary. 5, Transverse section of ovary. 6, Unopened fruit. 7, Opened fruit.

1. Linum Linnaeus. 200/14. Annual to perennial herbs or small shrubs. Leaves stalkless, often narrow with single or parallel veins, entire. Stipules sometimes present, modified to paired glands at base of leaves. Flowers short-lived, but new ones open daily, often heterostylous, parts in 5s. Petals free, rarely united at base, blue, yellow, red, pink or white. Petals alternate with 5 tooth-like staminodes. Filaments united at base. Ovary with 5 cells in each of which a cross-wall develops during maturation. Capsule usually spherical to almost spherical with a beak dehiscing into 10 flaps. Seeds brown or black. *Temperate areas.* Figure 64, p. 296.

94. EUPHORBIACEAE

Trees, shrubs or herbs (sometimes succulent), often with milky, irritant sap, sometimes spiny. Leaves alternate or opposite (rarely replaced by flattened cladodes), stipules present (sometimes very small) or absent. Flowers in inflorescences of various kinds, the inflorescence in one group (Tribe *Euphorbieae*) mimicking a flower and known as a cyathium (see below). Flowers usually unisexual, often very small. Calyx present or absent. Corolla usually absent. Stamens 1--many. Ovary superior, generally 3-celled (more rarely 2- or 4-celled), with 1 or 2 ovules in each cell. Styles as many as the cells, often divided. Fruit generally a capsule or schizocarp, splitting into units which separate from a persistent central column. Seeds usually appendaged. See figures 65, p. 300 & 66, p. 302.

A very large family of over 200 genera and 8000 species, occurring in most parts of the world, though mainly concentrated in the tropics. As the flowers are generally small, most are cultivated as foliage plants or as succulents.

The inflorescences of plants of the Tribe *Euphorbieae* (the only genus included here is *Euphorbia*) are remarkable structures which can superficially be mistaken for individual flowers. Each ultimate inflorescence (cyathium) consists of a cup-like structure which has an irregular or 5-toothed margin, and bears around the top a number of coloured, nectar-secreting glands (these may be united and not distinguishable into separate glands). In *Euphorbia*, there are 1--5 yellow or orange-red glands which spread in a remarkable petal-like formation. In the centre of the cup is a single, stalked female flower, consisting of a naked ovary which, when mature, projects from the cup, leaning to one side. Attached to the inside of the cup are several male flowers, each consisting of a solitary, stalked stamen, often subtended by a small, feather-like bract (bracteole); the stamen is jointed to the stalk, and this joint is clearly visible at a magnification of times 15 (in some tropical, non-cultivated genera related to *Euphorbia*, a small perianth is borne at the joint). In fruit, the cyathium remains, with the ripe schizocarp projecting from it. In all these genera the cyathia (the primary inflorescences) may be grouped into more complex inflorescences, which are often umbel-like, though there is usually a solitary cyathium at the centre of each pseudo-umbel.

1a. Flowers borne in cyathia   **5. Euphorbia**
  b. Flowers borne in inflorescences of various kinds, but not in cyathia   2
2a. Flowers (at least the male) with a perianth consisting of sepals and petals
                                                2. Andrachne
  b. Flowers with a perianth of a single whorl, petals absent   3
3a. Leaves palmately veined and lobed; stamens much-branched   **4. Ricinus**
  b. Leaves not palmately veined and lobed; stamens not as above   4
4a. Leaves mostly opposite   **3. Mercurialis**

b. Leaves alternate   **1. Securinega**

1. Securinega Commerson. 25/1. Deciduous shrubs or small trees, often spiny. Normally bisexual or unisexual. Leaves often small, alternate, simple, entire, short-stalked, with stipules. Flowers small, greenish, without petals, axillary. Male flowers in clusters; sepals 5 or 6, greenish white; stamens 5 or 6. Female flowers sometimes solitary, with 5 or 6 sepals; ovary 3-celled with 3 free or shortly joined and bifid styles. Fruit a capsule containing 3--6 seeds. *Temperate and subtropical parts of south Europe, Africa, Asia and South America.*

2. Andrachne Linnaeus. 20/2. Low, deciduous shrubs or perennial herbs. Leaves alternate, 2-ranked, usually simple and entire, with or without stipules. Flowers bisexual or incompletely unisexual, on long stalks. Male flowers in axillary clusters; sepals 5, rarely 6, larger than the petals, free or shortly joined; petals 5, rarely 6, free or joined, with disc-glands; stamens shorter than and equal in number to the sepals, free or joined around a rudimentary ovary. Female flowers solitary; petals minute or absent; sepals 5, rarely 6, free or joined; styles 3, more or less deeply bifid, free or shortly joined. Ovary 3-celled. Fruit more or less spherical, 6-seeded, separating into 3 carpels, each opening into 2 parts. *South Europe, Africa, Asia, America.*

3. Mercurialis Linnaeus. 8/2. Annual or perennial herbs with watery latex. Leaves opposite, with small stipules. Flowers usually unisexual; petals absent and sepals 3 in both sexes. Male flowers in long axillary spikes; stamens 8--15. Female flowers axillary, solitary, in stalkless clusters, or in spikes; 2 or 3 sterile filaments sometimes present; styles 2; ovary of 2 cells each with 1 seed. *Europe, Mediterranean area, temperate Asia.*

4. Ricinus Linnaeus. 1/1. Annual herbs, shrubs or small, short-lived trees, 1--7 m. Stems green or reddish, often glaucous, becoming hollow. Branching occurs at the nodes immediately below an inflorescence. Leaves alternate, spherical, peltate, 10--75 cm across, palmately 5--11-lobed, the lobes up to half the length of the leaf. Stipules united, sheathing, 1--3 cm. Leaf-stalks 8--50 cm, with 2 nectaries at the base, 2 at the junction with the leaf-blade and 1 or more towards the base on the upper side. Leaves and stalks green or reddish, often glaucous; blade lobed, sharply toothed and acuminate. Flowers in narrow terminal panicles 10--40 cm. Male and female flowers separate within the panicle, male below, female above. Male flowers in 3--16-flowered cymes on stalks 5--15 mm; sepals 3--5, ovate, 5--7 mm; petals absent; stamens numerous, 5--10 mm, filaments much-branched with many anthers. Female flowers in 1--7-flowered cymes on stalks 4--5 mm; sepals 3--5, united, soon falling; petals absent; ovary superior, 3-celled; ovules 1 per cell; stigmas 3, style short. Ovary covered with fleshy spines, enlarging in fruit, finally rigid. Fruit a 3-lobed capsule, 1.5--2.5 cm wide, green or reddish, becoming brown and woody. Capsule splitting into 3, often explosively. Seeds 5--15 mm, ovoid, pale brown to black, strongly mottled, with a yellowish white caruncle. *Origin uncertain; now widely naturalised in the tropics & subtropics.* Figure 65, p. 300.

5. Euphorbia Linnaeus. 1600/40. Trees, shrubs, perennial, biennial or annual herbs, often succulent and spiny or cactus-like. Copious milky, irritant latex present in all parts. Leaves on the stem usually alternate (often opposite or whorled in the

298

inflorescence), more rarely opposite, variable in shape, occasionally lobed, often small and very short-lived in succulent species; stipules present or absent, when present often glandular or represented by spines. Ultimate inflorescences flower-like (cyathia), consisting of a cup formed from 5 fused bracts and containing a central, stalked female flower surrounded by several male flowers; bracteoles often present between the male flowers; margins of the cup 5-lobed, and bearing 1--5 (rarely more) nectaries which may be flat or pouched and entire or horned, in some cases with petal-like extensions giving the cyathium the appearance of a flower with a corolla. Cyathia sometimes unisexual, with males and females usually on separate plants. Each male flower consists of a single stamen whose filament is jointed to a short stalk. The female flower consists of a 3-celled naked ovary with 1 axile ovule in each cell; styles 3, united at the base, often further divided above. The cyathia may be solitary or borne in more complex inflorescences (see below). Fruit a hard, brittle or spongy capsule opening along the septa, borne on the usually elongate female flower-stalk. Seeds usually with appendages (caruncles). *Cosmopolitan.* Figure 66, p. 302.

95. DAPHNIPHYLLACEAE
Evergreen trees or shrubs. Buds with several overlapping outer scales. Leaves alternate or almost whorled, entire, often with a glaucous bloom beneath, stalked. Plants dioecious, flowers without petals, borne in axillary racemes. Male flowers with 3--8 sepals; stamens 6--12, anthers with 2 cells, opening by lateral slits. Female flowers with 3--8 sepals; staminodes small or absent; ovary with 2 cells, ovules 4, styles 2, short recurved. Fruit a 1-seeded drupe.
A family of one genus with about 24 species native to eastern Asia and Malaysia.

1. Daphniphyllum Blume. 24/2. Description as for family. *Temperate and tropical eastern Asia.*

96. RUTACEAE
Evergreen or deciduous trees or shrubs, rarely herbaceous. Leaves opposite or alternate, simple or compound, usually with stipules and glands, often aromatic. Flowers in cymes, or in terminal or axillary umbels, heads, spikes, panicles, racemes, occasionally solitary; mostly radially symmetric and bisexual, occasionally unisexual. Sepals 3--5, often fused, overlapping in bud. Petals 3--5, overlapping or edge-to-edge, rarely absent. Stamens much more numerous than petals, borne at the base of a thick disc or rarely on the disc-margin. Styles distinct or united, as many as carpels. Ovary superior or half-inferior; carpels 2--5, usually each with 2 ovules. Fruit a drupe, capsule, follicle, samara or berry; berries often leathery skinned, breaking into segments. Seeds oblong or kidney-shaped. See figure 67, p. 304.
A family of 150 genera consisting of about 1600 species found mainly in Africa, Australia and America. Some important ornamental plants belong to this family, but it is chiefly of note because of citrus fruits (oranges, grapefruit, etc.).

    1a. Herbaceous perennials, shrubs or trees, without spines; fruit a capsule, berry or samara   2
      b. Trees or shrubs, with thorns or spines on shoots; fruit usually fleshy or juicy   12
    2a. Herbaceous perennials or small shrubs; leaves 1--3-pinnate; flowers in panicles, cymes or racemes   3

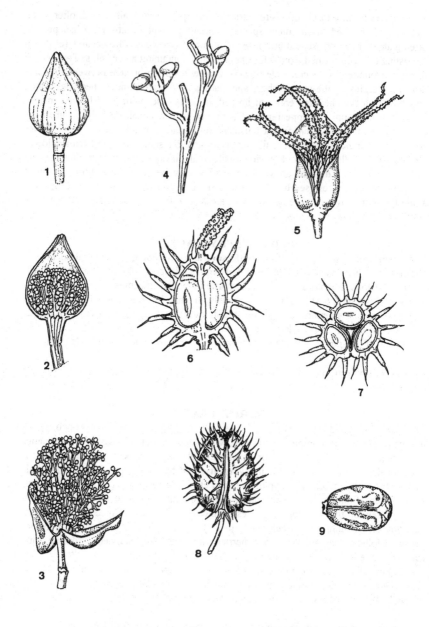

Figure 65. Euphorbiaceae. *Ricinus communis.* 1, Bud of male flower. 2, Longitudinal section of unopened male flower. 3, Opened male flower. 4, Part of branched stamen. 5, Female flower. 6, Longitudinal section of ovary. 7, Transverse section of ovary. 8, Fruit. 9, Seed.

300

b. Shrubs or trees; leaves simple, sometimes divided; flowers often solitary 5
3a. Leaves pinnate; petals yellow   **1. Ruta**
 b. Leaves 1--3-pinnate; petals white or purplish   4
4a. Leaves 2- or 3-pinnate; flowers less than 1 cm across; sepals and petals 4
                                                      2. Boenninghausenia
 b. Leaves pinnate; flowers at least 2.5 cm across; sepals and petals 5
                                                            3. Dictamnus
5a. Leaves simple   6
 b. Leaves compound   8
6a. Leaves opposite; sepals fused into a cup; petals 4, fused to form a
    campanulate corolla; stamens 8   **5. Correa**
 b. Leaves alternate, rarely opposite; sepals 4 or 5, free; petals 4 or 5, free
    or fused only at base   7
7a. Deciduous shrub; flowers unisexual; male flowers in axillary racemes,
    stamens 4; female flowers solitary   **6. Orixa**
 b. Evergreen shrub; flowers bisexual   **12. Skimmia**
8a. Stamens as many as petals or absent   9
 b. Stamens twice as many as petals   11
9a. Leaves usually with 3 leaflets; fruit a winged samara   **11. Ptelea**
 b. Leaves with more than 3 leaflets; fruit a berry or capsule, not winged   10
10a. Buds in leaf-axils exposed; fruit of 4 or 5 dehiscent carpels   **7. Tetradium**
 b. Buds in leaf-axils concealed by swollen leaf-stalk base; fruit an
    indehiscent drupe with 5 seeds   **10. Phellodendron**
11a. Leaves with 3 leaflets; petals white with velvety hairs   **4. Acradenia**
 b. Leaves simple, with 3 leaflets or compound; petals without hairs, white,
    pink, brown or yellow-green   **8. Choisya**
12a. Flowers unisexual or bisexual; petals 3--5, stamens 3--5   **9. Zanthoxylum**
 b. Flowers always bisexual; 4 or 5, stamens 20--60   **13. Poncirus**

1. Ruta Linnaeus. 5/1. Aromatic perennial herbs; shoots becoming woody towards base. Leaves spotted with glands, alternate, pinnately divided, segments linear to obovate. Flowers in a cyme. Sepals 4. Petals 4 (except central flower of each cyme, which has 5), dull, dark yellow, hooded, margins toothed or hairy (rarely entire). Stamens 8--10 with hairless filaments. Styles fused. Fruit a 4- or 5-celled capsule. *Europe, west Asia, Macaronesia.*

A closely related genus is **Haplophyllum** Jussieu, which may be distinguished from *Ruta* by its roughly hairy filaments, less divided leaves (simple or with 3 leaflets), and 5 petals in all flowers.

2. Boenninghausenia Meissner. 1/1. Herbaceous perennials or woody, low shrubs to 50 cm, hairless, stems and foliage glaucous; shoots terete. Leaves alternate, 2 or 3 times divided into 3. Leaflets 1--2.5 cm × 7--20 mm, entire, obovate to elliptic, very glaucous (rarely white) beneath, with glandular spots; stalks 5--10 cm. Flowers in panicles, white, bisexual. Sepals 4, oblong, fused at base, to 1 mm, persistent in fruit. Petals 4, oblong, to 4 mm. Stamens 6--8, longer than petals. Disc urn-shaped. Ovary with 4 cells each with 6--8 ovules. Style short, deciduous. Fruits of 4 follicles, to 3 mm; seeds black, tubercled, kidney-shaped. *India to Japan.*

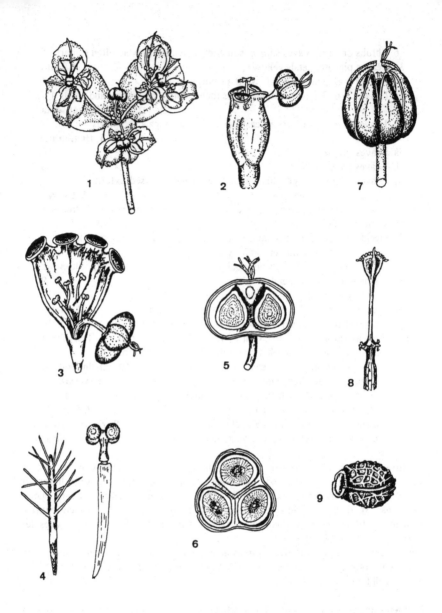

Figure 66. Euphorbiaceae. *Euphorbia cyparissias*. 1, Part of compound inflorescence. 2, Single cyathium showing stalked female flower. 3, Section of cyathium showing glands, male flowers and stalked, reflexed, female flower. 4, Male flower, bract (L) and stamen (R). 5, Longitudinal section of female flower. 6, Transverse section of ovary. 7, Fruit dehiscing. 8, Central column of capsule after the walls and seeds have fallen. 9, Seed.

3. Dictamnus Linnaeus. 1/1. Deciduous, perennial herb to 1 m, with woody rootstock. Leaves to 7 cm, alternate, pinnate, gland-dotted. Leaflets 9--11, ovate, 2.5--7.5 cm with fine, forwardly pointing teeth, dark green, lemon-scented. Flowers to 2.5 cm across, bilaterally symmetric, white to purple, in terminal racemes; stalks with bracts. Sepals 5, lanceolate, minute. Petals 5, pointed, narrow, the lowest bent downwards. Stamens 10, curved upwards. Fruit a 5-lobed capsule. *South Europe to Siberia and north China.*

4. Acradenia Kippist. 2/1. Trees or shrubs with evergreen, opposite leaves; leaflets 3. Flowers borne in panicles, bisexual. Sepals 5 or 6, fused towards base, persistent in fruit. Petals 5 or 6, deciduous. Stamens 10--12, those alternating with the petals equalling the petals, those opposite the petals shorter; filaments tapering; anthers versatile, mucronate. Ovary with 5 cells, each with 2 ovules, fused at base, and joined on inner side at the middle by the style, each with prominent gland on upper side. Fruit composed of 1--5, one-seeded follicles, fused at base. *Australia.*

Two species, but only 1 is in general cultivation.

5. Correa Andrews. 11/1. Shrubs (rarely trees) with stellate hairs on shoots, foliage and flowers. Leaves simple, opposite, usually entire. Flowers solitary or clustered in terminal cymes, hanging or erect. Calyx cup-shaped with an entire or lobed margin. Corolla of 4 fused petals, sometimes splitting to the base. Stamens 8, inserted at base of 8-lobed disc; filaments linear, those opposite the petals usually shortest and with broadened bases, hairless; anthers projecting or included. Ovary hairy, with 4 cells each with 2 ovules. Style slender, almost equalling stamens. Stigma with 4 minute lobes. *Australia.*

6. Orixa Thunberg. 1/1. Shrubs to 3 m, dioecious, deciduous; branches without spines, with grey felt when young. Leaves simple, alternate, dark green (turning pale yellow in autumn), aromatic, obovate to elliptic, 5--12 × 3--7 cm, entire, hairless except when young. Flowers unisexual; males in short axillary racemes, c. 3 cm, green; females solitary; flower-stalks 1--2 cm. Sepals 4, fused at base. Petals 4, spreading, c. 0.3 mm. Male flowers with 4-lobed disc and 4 stamens. Females flowers with deeply 4-lobed ovary; single style; 4-lobed stigma. Fruit of 4 carpels; 1 seed per carpel, explosively expelled when ripe. *China, Japan, Korea.*

7. Tetradium Loureiro. Few/2. Deciduous trees; young shoots with pith, and prominent lenticels and axillary buds. Leaves compound, pinnate. Inflorescence a terminal corymb. Flowers unisexual; sepals 4 or 5; petals 4 or 5. Stamens 4, 5 or absent. Fruit with 4 or 5 follicles, each with 1 or 2 smooth, shiny, black seeds. *E Asia, from Nepal to China & Korea.*

Often formerly included in the genus *Euodia* Forster & Forster, which is otherwise not cultivated in northwest Europe.

8. Choisya Kunth. 5/1. Shrubs with aromatic, opposite, palmately divided leaves; leaflets 3--15. Flowers in terminal or axillary panicles, or solitary. Sepals usually 5, deciduous. Petals usually 5, white, hairless. Stamens usually 10, inner whorl (opposite petals) shorter than outer whorl. Ovary hairy, with 5 carpels fused at base; carpels with hairless apical horn joined by the centrally attached 5-furrowed style; stigmas 5,

303

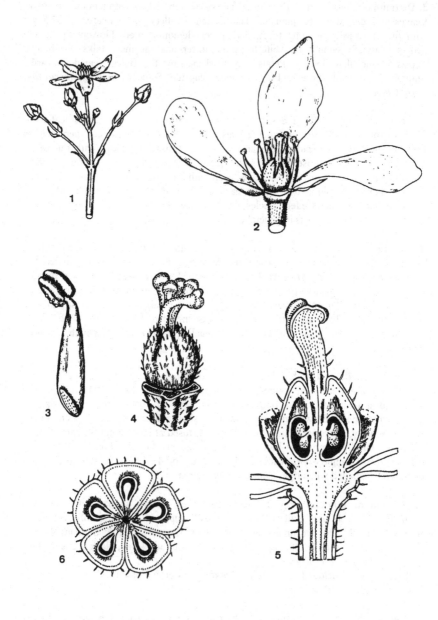

Figure 67. Rutaceae. *Choisya ternata.* 1, Part of inflorescence. 2, Flower with two petals and several stamens removed. 3, Stamen. 4, Ovary. 5, Longitudinal section of ovary. 6, Transverse section of ovary.

head-like. Fruit of 2 carpels each with 1 or 2 seeds. *Southwest USA, Mexico*. Figure 67, p. 304.

9. Zanthoxylum Linnaeus. 200/7. Deciduous or semi-evergreen trees or shrubs, more or less spiny, bark aromatic. Leaves alternate, pinnate, dotted with glands. Flowers unisexual or bisexual, usually in cymes or panicles, yellow-green. Sepals 3--5. Petals absent or up to 5. Stamens 3--5. Fruit a capsule or follicle. *America, Africa, Asia, Australia.*

10. Phellodendron Ruprecht. 10/5. Tall, deciduous trees; bark often thick and corky. Leaves to 40 cm, opposite, pinnate with a terminal leaflet; base of stalks swollen, hiding the bud completely. Flowers unisexual, inconspicuous, yellow-green, in terminal panicles. Petals 5--8; sepals 5--8, small. Male flowers with 5 or 6 stamens, alternating with and twice as long as sepals; ovary rudimentary. Female flowers with 5 or 6 staminodes; ovary 5-celled with short stout style and 5-lobed stigma. Fruit *c.* 1 cm, an orange-like, indehiscent drupe with 5 seeds and a tough black skin. *Northeast Asia.*

11. Ptelea Linnaeus. ?60/few. Aromatic, spineless, deciduous shrubs and trees. Leaves alternate, with 3--5 stalkless leaflets. Flowers green or yellow-white, mostly unisexual, in terminal corymbs. Sepals 4 or 5, minute. Petals 4 or 5. Stamens 4 or 5; filaments hairy on inner surface. Fruit a 2-seeded samara with a distinct, thin, flat, broadly encircling wing. *North America.*

12. Skimmia Thunberg. 4/4. Unisexual or bisexual evergreen shrubs to trees, more or less aromatic and almost hairless. Basal half of shoots with oblong to lanceolate, acute bracts which fall early; lenticels inconspicuous. Young stems green, brownish or reddish becoming creamy grey to yellow. Leaves clustered near tips of each season's growth, alternate, simple, entire or slightly scalloped near tip, dotted with translucent glands, persisting for 1 or more years. Flowers many, small, in short terminal compound panicles, white or yellow, sometimes tinged with pink, normally unisexual, dioecious; female and bisexual inflorescences often with only 1--5 flowers; flower-parts usually in 4s or 5s. Carpels 2--5, fused. Style short or about equal to the ovary. Stigma irregular with 2--5 lobes. Fruit fleshy, red or black, more or less spherical, containing 1--4 (rarely 5) seeds. *Himalaya, China, Japan.*

13. Poncirus Rafinesque. 1/1. Deciduous shrubs to 7 m. Branches stiff, flattened, smooth, green, with green spines 2.5--5 cm. Leaves compound, often borne on old wood; stalks slightly winged; leaflets 3--5, elliptic to obovate, to 4 cm, gland-dotted. Flowers solitary or in pairs, axillary, borne before leaves on previous year's shoots, fragrant. Sepals 5. Petals 4 or 5, white, concave, obovate, narrowed towards base. Stamens 20--60, free, pink, irregular. Fruit with 6--8 cells, spherical, 3--5 cm across, yellow, densely downy, very fragrant, pulp scant, acidic; seeds numerous. *North China, Korea.*

Occasionally, species of **Citrus** Linnaeus may survive out-of-doors in the mildest areas of northern Europe. In this genus, with species of which *Poncirus* hybridises, the leaves are simple but with a winged stalk.

97. CNEORACEAE

Small shrubs, sometimes with medifixed hairs (not ours). Leaves alternate, simple, usually rather leathery, without stipules. Flowers few, in cymes, bisexual, radially symmetric, without a disc. Sepals 3 or 4, free, small. Petals 3 or 4, free, rather upright or spreading. Stamens 3 or 4, anthers opening by slits. Ovary superior, shortly stalked, 3- or 4-celled; ovules axile, 1 or 2 per cell. Fruit a schizocarp breaking into 3 or 4 mericarps.

A family of 2 genera from Cuba, the Canary Islands and the west Mediterranean area. Only 1 is cultivated.

1. Cneorum Linnaeus. 2/1. Evergreen shrubs to 1 m, usually rather upright. Leaves entire, leathery, greyish green, margins somewhat rolled under. Flowers small; petals yellow. Fruit usually of 3 mericarps which fall from a central axis; mericarps fleshy at first outside, very hard within. *West Mediterranean area, Cuba*.

98. SIMAROUBACEAE

Trees or shrubs, with alternate compound leaves, dioecious, monoecious or with bisexual flowers. Flowers small, radially symmetric, usually borne in axillary or terminal racemes, panicles, or cyme-like spikes. Sepals 3--7, free or united at base; petals similar and sometimes absent. A ring or disc occurs between the petals and stamens. Stamens free, usually twice the number of petals or sepals, filaments often with a basal appendage. Ovary superior, carpels 2--5, fused or free at base. Styles 2--5. Fruit a drupe, samara or capsule, or splitting into 4. Endosperm scarce or absent.

A family of 20 genera and about 120 species. Most are found in the tropics and subtropics and include several species of economic importance.

1a. Leaves with 3--7 leaflets; petals 4--5 cm, red   **3. Quassia**
  b. Leaves with 7--35 leaflets; petals 3--30 mm, white, greenish white,
     cream or greenish yellow   2
2a. Leaves with one or more pairs of glands at the base; fruit a winged
     samara, *c*. 3.5 cm   **1. Ailanthus**
  b. Leaves without glands; fruit a drupe or separating into 4 parts, 3--25 mm
     2. Picrasma

1. Ailanthus Desfontaines. 10/1. Deciduous, dioecious tall trees, often with fissured grey bark. Leaves alternate, foetid, pinnate with a terminal leaflet, the basal leaflets with conspicuous glands near the base. Stipules minute, falling early. Flowers usually unisexual, in terminal panicles. Calyx of 5 sepals (rarely 6). Petals 5 (rarely 6), not overlapping; disc hemispherical. Stamens lacking appendages, 2--5 in bisexual flowers, 10 in male flowers. Ovary 5-celled; stigmas 5; ovule solitary, pendent. Fruits of 1--5 samaras, containing a single seed. Seed without endosperm. *Temperate & tropical Asia & Australia*.

2. Picrasma Blume. 6/1. Trees or shrubs with very bitter constituents. Leaves alternate, pinnate with a terminal leaflet. Flowers inconspicuous, in axillary panicles. Calyx of 4 or 5 teeth, often enlarging in fruit; disc thickened. Stamens 4 or 5, hairy. Ovary with 3--5 cells, free. Styles distinct at apex and base, but joined in middle. Ovules erect, solitary. Fruit of 1--3 fleshy or leathery drupes. Seeds erect, almost transparent. *East Asia, Pacific Islands*.

3. Quassia Linnaeus. 40/1. Trees or shrubs, bisexual or monoecious. Leaves simple or compound, alternate, stalked, usually with pitted glands on lower surface; stipules absent. Inflorescences axillary or terminal, clustered, racemes, panicles or umbels. Flowers with 4--6 floral parts. Calyx lobed; petals overlapping or twisted, longer than sepals. Stamens twice as many as petals, with appendages near the base. Disc cylindric or almost spherical. Carpels free or apparently joined basally; styles fused, stigmas stellate or head-like; ovule apical. Fruit a drupe, or sometimes woody. *Tropics and subtropics.*

99. MELIACEAE

Trees, shrubs or woody-based perennials. Leaves usually pinnate, more rarely with 1--3 leaflets, when pinnate with or without a terminal leaflet or bud which continues the development of the leaf. Flowers in axillary panicles, usually radially symmetric, bisexual or unisexual. Sepals small, 4--6, often united at base. Petals usually 4--6, free. Stamens and ovary borne above the insertion of the petals on a column in some genera. Stamens 8--12, filaments usually united into a tube, anthers opening inwards. Nectary present within the bases of the stamens. Ovary superior with 3--20 cells, each with 2--many ovules with axile placentation; style single, stigma head-like. Fruit a capsule or indehiscent and fleshy.

A family of 52 genera and about 550 species, largely restricted to the tropics, where many are important timber trees. Few are hardy in Europe and only a very small number is generally grown.

1a. Flowers sweetly scented; leaflets toothed or entire, with oblique or
obtuse bases **1. Toona**
 b. Flowers smelling of rotten onions; leaflets entire, rounded or obtuse at
base **2. Cedrela**

1. Toona (Endlicher) Roemer. 6/2. Deciduous or semi-deciduous, small to medium monoecious trees, ill-smelling when bruised. Branches with prominent lenticels and large leaf-scars. Leaves pinnate with more than 5 pairs of oblique or obtuse, sometimes toothed, stalked leaflets. Inflorescences hanging with long branches and up to several hundred white, sweetly scented flowers, which appear bisexual but are unisexual. Calyx saucer-shaped, sometimes deeply 5-lobed. Petals 5, small. Stamens 5, filaments free. Ovary 5-celled, each with 6--10 ovules. Ovary and stamens borne on a short stalk. Fruit ellipsoid; seeds winged at one or both ends. *India to north Australia.*

2. Cedrela Browne. 8/1. Very similar to *Toona*, but flowers ill-scented (of rotting onions); calyx cup-shaped, often divided to the base; petals 3--7 mm; seeds winged at one end only. *Tropical & subtropical America.*

100. POLYGALACEAE

Herbs, shrubs, woody climbers or small trees. Leaves alternate, sometimes in bundles, rarely opposite, simple. Flowers bisexual, bilaterally symmetric, usually solitary, axillary or terminal, sometimes arranged in a spike or raceme, rarely a panicle. Sepals 5, free, overlapping, the inner larger and sometimes wing-like, the upper 2 sometimes joined. Petals 3--5, the upper 2 free or joined to the lower one, the 2 lateral free, often absent or vestigial. Stamens 8, rarely 4 or 5; filaments usually united into a slit tube, rarely free; anthers usually opening by pores or flaps. Ovary superior, usually 2-celled, rarely 3--5. Style simple, often dilated and 2-lobed at the apex. Stigma frequently lateral. Fruit a capsule, samara or drupe. Seeds pendent, often silky, usually with a caruncle. See figure 68, p. 310.

A cosmopolitan family of 18 genera and about 950 species.

1a. Herb or shrub; fruit a capsule	**1. Polygala**
b. Tree; fruit a winged samara	**2. Securidaca**

1. Polygala Linnaeus. 500/9. Annual or perennial herbs or shrubs, sometimes spiny. Leaves alternate, more rarely opposite or whorled, ovate, lanceolate or linear, sometimes rudimentary. Flowers in a terminal or lateral raceme, spike or head. Sepals 5, falling early or persistent, unequal, the inner 2 petal-like, sometimes shortly clawed, all free, or the inner 2 united, stalkless. Petals 3 (rarely 5), the upper 2 basally joined to the staminal tube; the lower boat-shaped, clawed, entire or bearing a crest of 2 lobes, both entire or divided into a number of appendages; lateral petals minute, simple, bifid or deeply 2-lobed, sometimes absent. Disc sometimes present. Stamens 8 (rarely 4, 5 or 9), sometimes only 6 fertile with 2 staminodes; filaments linear, joined into a split tube, sometimes with marginal hairs; anthers opening at apex. Fruit a membranous capsule, compressed, elliptic or obovate, often notched. *Cosmopolitan.* Figure 68, p. 310.

2. Securidaca Linnaeus. 80/1. Shrubs, small trees, or climbers, sometimes spiny. Leaves alternate, entire. Flowers in terminal and axillary racemes or panicles. Sepals 5, unequal, inner 2 larger, petal-like. Petals 3, free, central petal usually clawed, or sometimes 2 extra scale-like petals present. Stamens 8, filaments linear, joined into a split tube. Style borne obliquely on ovary, flattened above; stigma terminal. Fruit a samara; seeds spherical. *Asia, America, 1 species in Africa.*

101. CORIARIACEAE

Shrubs often dying back to a woody base, branches angular, the lower opposite or 3 in a whorl. Leaves opposite, entire, with 3 or more veins; blades ovate-cordate to lanceolate; stipules absent. Flowers bisexual or unisexual, axillary when borne on previous year's growth or terminal on current year's growth, solitary or in racemes. Sepals 5, persistent, overlapping, ovate-triangular. Petals 5, shorter than sepals, 3-angled, keeled inside, eventually fleshy. Stamens 10, free or fused to the petal-keels. Ovary superior with 5--10 free carpels, each containing a single ovule; styles long, conspicuous. Fruit comprising carpels which have become embraced by the very succulent pesistent petals, seeds flattened.

A family of a single genus with a wide distribution from central and western South America, Mediterranean, Himalayas and eastern Asia to New Zealand.

1. Coriaria Linnaeus. 5/5. Description as for the family. *Widespread.*

102. ANACARDIACEAE

Trees or shrubs with resinous bark. Leaves alternate, rarely opposite, simple or compound. Flowers unisexual or bisexual, usually radially symmetric. Calyx with 3--5 lobes. Petals 3--5, rarely absent. Disc usually in the form of a ring; stamens as many or twice as many as petals, rarely more, inserted at base of disc, filaments separate. Ovary with 1--5 cells and 1 ovule in each cell, placentation axile; styles 1--5. Fruit usually a drupe, rarely dehiscent. See figure 69, p. 312.

A family of about 73 genera mostly tropical or subtropical but with some genera from the Mediterranean area and temperate North America. The family provides a number of ornamental and fruit trees.

1a. Petals absent	**3. Pistacia**
b. Petals present	2
2a. Stamens 8--20	3
b. Stamens 1--6	6
3a. Stamens 12--20	**7. Sclerocarya**
b. Stamens 8--10	4
4a. Petals not overlapping	**6. Spondias**
b. Petals overlapping	5
5a. Ovary 1-celled	**4. Schinus**
b. Ovary with 4 or 5 cells	**8. Harpephyllum**
6a. Style or stigma more or less at apex of ovary	**1. Rhus**
b. Style or stigma on the side of the fruit	7
7a. Ovary sunk into a cup-shaped or tubular hollowed receptacle; fruiting panicles not feathery	**5. Semecarpus**
b. Ovary not sunk as above; fruiting panicles feathery	**2. Cotinus**

1. Rhus Linnaeus. 150/14. Deciduous or evergreen shrubs, climbers or trees, dioecious or with both unisexual and bisexual flowers. Leaves alternate, odd-pinnate, with 3 leaflets or simple. Inflorescence a panicle; flowers small. Calyx 5-lobed. Petals 5, longer than calyx-lobes. Stamens 5, reduced to staminodes in female flowers. Stigmas 3, at apex of ovary. Fruit a drupe, often with thin red flesh. *Temperate & subtropical North America, southern Africa, east Asia & north Australia.*

2. Cotinus Adanson. 3/2. Deciduous trees or shrubs with yellow wood. Leaf-blades simple, entire or slightly toothed, stipules absent. Inflorescences large, loose terminal panicles often with feathery divisions. Flowers small, rarely bisexual, often yellow. Sepals 5, overlapping. Petals 5, twice as long as sepals. Stamens 5. Ovary 1-celled with 3 lateral styles. *South Europe to China and southeast USA.* Figure 69, p. 312.

3. Pistacia Linnaeus. 11/4. Dioecious trees or shrubs. Leaves alternate, pinnate with or without a terminal leaflet, rarely of 3 leaflets or simple, membranous or leathery. Flowers minute, unisexual, in axillary, loose or dense panicles or clustered racemes. Bracts leaf-like in texture or membranous. Bracteoles often membranous, unequal in size and form. Petals absent. Stamens 3--8, anthers ovate or oblong; filaments very short, inserted on the disc. Ovary spherical or ovoid, 1-celled with a short style and 3 stigmas. Fruit a 1-seeded drupe, obovoid to spherical or rarely transversely ovoid or oblong. *Mediterranean area, Asia to Malaysia.*

4. Schinus Linnaeus. 28/4. Evergreen, dioecious, resinous trees or shrubs. Leaves alternate, simple or compound; leaflets stalkless. Flowers small in axillary or terminal panicles, with bracts. Sepals 5. Petals 5, overlapping, yellow or green. Stamens 10, inserted on a disc. Ovary stalkless, 1-celled, styles 3. Fruit a drupe. *South America; naturalised in North America, Canary Islands & China.*

309

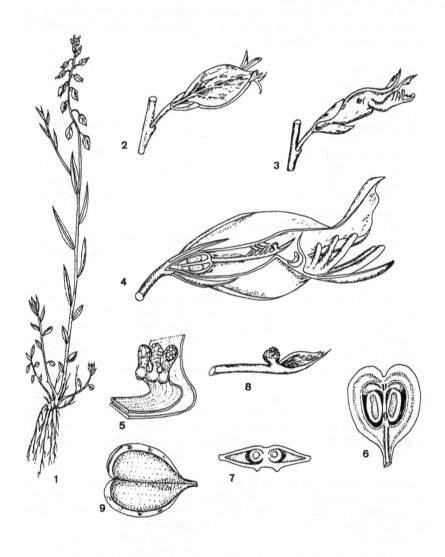

Figure 68. Polygalaceae. *Polygala vulgaris.* 1, Part of plant with inflorescence. 2, Single flower. 3, Flower with the two petaloid sepals removed. 4, Longitudinal section of flower. 5, Upper parts of three stamens. 6, Longitudinal section of ovary. 7. Transverse section of ovary. 8, Upper part of style. 9, Fruit.

310

5. Semecarpus Linnaeus. 38/1. Trees. Leaves alternate, simple, entire, leathery; stipules absent. Flowers small, unisexual and bisexual, in terminal (rarely axillary) panicles. Sepals 3 or 5, deciduous. Petals 3 or 5, ovate or oblong-ovate, overlapping. Stamens 5 or 6; filaments thread-like, longer than petals in male flowers, short in bisexual flowers; anthers usually oblong. Ovary sunk into a cup-shaped or tubular hollow receptacle, absent or vestigial in male flowers, superior in bisexual flowers, 1-celled; styles 3, divergent; stigmas head-like or bilobed. Fruit a drupe, oblong or nearly spherical, oblique, on a fleshy receptacle. Seed pendent. *Asia, Australia.*

6. Spondias Linnaeus. 12/2. Trees. Leaves usually clustered near the ends of the branchlets, alternate, odd-pinnate. Flowers small, short stalked in racemes or panicles. Sepals 4 or 5, deciduous. Petals 4 or 5, not overlapping. Stamens 8--10. Ovary stalkless, with 4 or 5 cells; styles 4 or 5. Fruit a fleshy drupe. *Southeast Asia, tropical America.*

7. Sclerocarya Hochstetter. 3/1. Trees or shrubs. Bark grey. Leaves alternate, pinnate with a terminal leaflet, stalked; leaflets circular to broadly oblong, toothed to entire. Inflorescence a panicle; sepals 4 or 5, oblong or round, overlapping; petals 4 or 5, oblong or obovate, obtuse, overlapping; stamens 12--20; ovary with 2 or 3 cells; styles 2 or 3. Fruit a drupe. *Tropical & southern Africa, Pacific Islands.*

8. Harpephyllum Bernhardi. 1/1. Evergreen dioecious tree, 6--15 m. Leaves to 30 cm, alternate, dark green, leathery and shiny, with 4--8 pairs of leaflets; terminal leaflet 5--10 × 1.3--2.5 cm, lanceolate or broader; lateral leaflets slender, slightly sickle-shaped, sharply pointed and narrowed to the base with the midrib to one side. Flowers whitish, borne on lateral racemes. Stamens 10. Ovary with 4 or 5 cells. Fruits *c.* 2.5 × 1.3 cm, red when ripe, long, plum-shaped. *Southern Africa.*

103. ACERACEAE

Monoecious or dioecious trees and shrubs. Buds with or without 4--many scales. Leaves opposite, stalked, simple, with 3--13 lobes, palmate or pinnate, with 3 leaflets, entire or toothed, evergreen or deciduous. Inflorescences terminal or lateral, racemose, corymbose, paniculate or sometimes umbellate. Flowers unisexual, male and female in the same inflorescences or in separate inflorescences. Sepals 4 or 5. Petals 4 or 5, rarely united or absent, disc outside, inside or around the insertion of the filaments, usually in a ring, rarely lobed or absent. Stamens 4--12, mostly 8. Ovary superior, 2-celled, with 2 axile ovules in each cell; stigmas 2. Fruit composed of 2 single-seeded samaras. Germination epigeal, rarely hypogeal. See figure 70, p. 314.

A family of 2 genera mainly from temperate and subtropical areas of the northern hemisphere.

1a. Buds with scales; leaves simple, lobed or pinnate with 1--5 pairs of
  leaflets; leaf-stalks, if leaves pinnate, not containing latex; wing of samara on
  one side of nutlet **1. Acer**
  b. Buds without scales; leaves pinnate with 4--9 pairs of leaflets; leaf-stalks
  containing latex; nutlet encircled by a broad wing **2. Dipteronia**

1. Acer Linnaeus. 120/65 and some hybrids. Trees or shrubs, monoecious, occasionally dioecious, deciduous, occasionally (semi-)evergreen. Bud scales 2--many-paired,

311

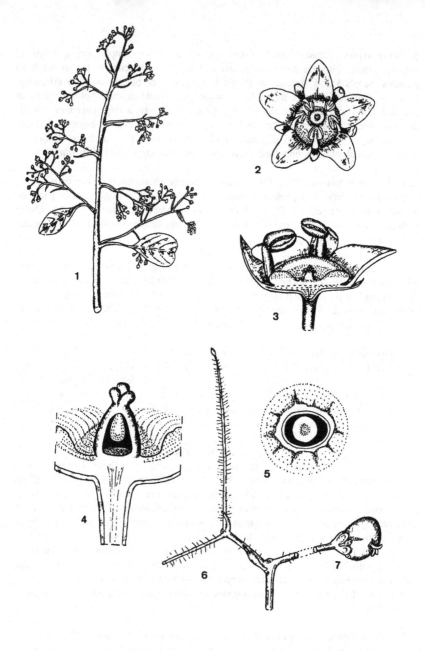

Figure 69. Anacardiaceae. *Cotinus coggygria.* 1, Part of inflorescence. 2, Flower from above. 3, Longitudinal section of flower. 4, Longitudinal section of ovary. 5, Transverse section of ovary and part of disc. 6, Details of plumose fruit-stalk. 7, Fruit.

edge-to-edge or overlapping. Leaves usually palmately lobed with 3--13 (usually 5) lobes, sometimes simple or with 3 leaflets, rarely palmate or pinnate, stalks sometimes containing latex. Inflorescences terminal and/or axillary, racemose, corymbose, paniculate and sometimes umbellate. Flowers unisexual. Sepals usually 5, less often 4. Petals usually 5, rarely 4, occasionally absent or united. Stamens mostly 8, less often 4 or 5, occasionally 10--12. Ovary superior, 2-celled, each cell with 2 ovules. Fruit mostly composed of 2, sometimes 3 or 1, single-seeded samaras. Germination epigeal, occasionally hypogeal. *Temperate northern hemisphere, subtropical southeast Asia.* Figure 70, p. 314.

2. Dipteronia Oliver. 2/1. Trees or shrubs, deciduous, monoecious. Buds naked. Leaves opposite, pinnate; stalks containing latex. Inflorescences terminal, large, paniculate. Flowers unisexual. Sepals and petals 5, white. Disc outside the filaments. Stamens 8; ovary downy. Fruits composed of 2 single-seeded samaras; nutlet flat, encircled by a broad samara. *Central & southwest China.*

104. SAPINDACEAE

Trees, shrubs or climbers, when climbing often with irregular stems with anomalous secondary growth. Leaves usually alternate and compound, stipules present in climbers, absent in upright species. Flowers unisexual, though often appearing bisexual (apparently both stamens and ovary present, both sexes on the same plant), radially or bilaterally symmetric, in cymes, racemes or panicles; the climbing species generally have tendrils in the inflorescence. Sepals 4 or 5, often unequal. Petals 4 or 5, or rarely absent, often each or some with an internal scale or patch of hairs. Stamens 5--8. Disc present, sometimes elaborate. Ovary usually 3-celled; ovules 1 or 2, rarely more per cell; stigmas usually 3. Fruit usually a large capsule, drupe or schizocarp. Seeds usually with arils.

A family of 145 genera and about 1300 species, mainly from the tropics. Only a few genera and species are grown.

1a. Flowers bilaterally symmetric, yellow   **1. Koelreuteria**
  b. Flowers radially symmetric, white   **2. Xanthoceras**

1. Koelreuteria Laxmann. 3/2. Deciduous trees with thick, fissured bark. Leaves alternate, without stipules, pinnate or bipinnate, with alternate or opposite ultimate segments. Flowers bilaterally symmetric, in large, loose terminal panicles, unisexual, but male and female in the same inflorescence. Calyx 5-lobed, 3 of the lobes longer than the other 2, margins glandular or hairy. Petals 4 or 5, clawed, the base reflexed, the base of the blade bearing lobed outgrowths. Stamens mostly 8, filaments hairy. Ovary 3-celled, ovules 2 per cell, attached to the outer walls of the ovary (placentation parietal). Fruit an inflated capsule, papery. Seeds without arils. *China & Taiwan to Fiji.*

2. Xanthoceras Bunge. 1/1. Deciduous tree or shrub to 8 m. Leaves stalkless, alternate, pinnate, with a terminal leaflet. Flowers appearing with the leaves, in racemes, those of the terminal raceme usually female, those of the lateral racemes usually male, though all appearing to have stamens and ovaries. Sepals 5. Petals 5, much larger than the sepals. Stamens 8, inserted on a disc which is 5-lobed, each lobe

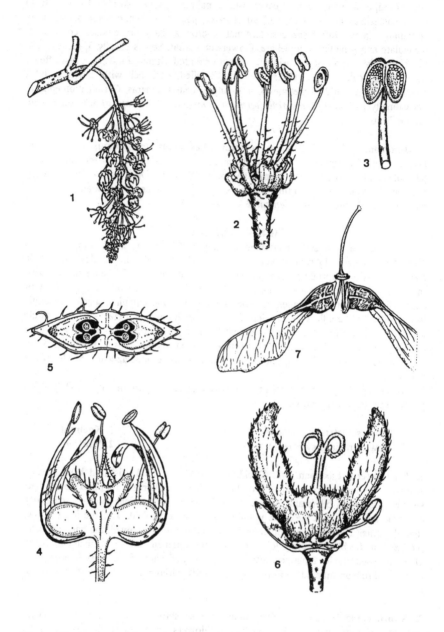

Figure 70. Aceraceae. *Acer pseudoplatanus.* 1, Inflorescence. 2, Single male flower. 3, Upper part of dehisced stamen. 4, Longitudinal section of a bisexual flower. 5, Transverse section of ovary. 6, Young fruit. 7, Mature fruit.

with an upright appendage half as long as the stamens. Ovary 3-celled, with a short, thick style. Capsule opening by 3 flaps, each cell with several seeds. *North China.*

105. HIPPOCASTANACEAE

Trees, with large distinctive winter buds enclosed by usually resinous scale-leaves. Leaves opposite, palmately divided, lacking stipules. Inflorescence a terminal panicle. Flowers with 4 or 5 sepals united at base and 4 or 5 large, showy, clawed petals. Stamens 5--8, free. Ovary superior, 3-celled, each cell with 2 ovules on the axile placenta. Style long; stigma simple. Fruit a leathery capsule, splitting into 3, usually with a single large seed. See figure 71, p. 316.

A small family comprising 2 genera, scattered in temperate parts of the world. Only 1 genus is cultivated.

1. Aesculus Linnaeus. 23/12 and some hybrids. Deciduous trees and shrubs with opposite, palmate leaves and prickly or smooth fruit. Leaves with 5--7 (rarely to 9) unequal leaflets borne on a long stalk. Flowers borne in large panicles occuring at the apex of branches formed during the current season's growth. Corolla somewhat irregular, white or brightly coloured, free. Ovules 6, producing 1 (or rarely 2) seeds by abortion. Fruit with a fleshy coat containing a shiny brown seed. *South Europe, temperate North America, Himalaya to China.* Figure 71, p. 316.

106. SABIACEAE

Trees or shrubs. Leaves evergreen or deciduous, simple or compound, alternate and without stipules. Flowers small, in panicles, bisexual, bilaterally symmetric. Sepals 3--5 or calyx with 3--5 lobes, or sepals apparently more than 5 (see below). Petals usually 5, free, usually 3 large and 2 smaller. Stamens 5, on the same radii as the petals, often only 2 of them fertile, these on the same radii as the 2 smaller petals. Disc usually present, lobed. Ovary superior, 2-celled, ovules 2 per cell, placentation axile; style short, stigma simple. Fruit a berry or drupe.

A family of 4 genera, of which only 1 is grown. They occur in southeast Asia and tropical America.

1. Meliosma Blume. 25/5. Trees or shrubs with open branching and stout twigs. Leaves simple or pinnate. Flowers small, white or cream, fragrant, in large, usually terminal panicles. Sepals 3--5 but sometimes apparently more (to 13), the additional 'sepals' being formed from small, sterile bracts, sometimes the lowermost a little distant from the rest. Petals usually 5, the 3 outer unequal, much larger than the 2 inner which are notched or bifid. Stamens 5, 2 fertile, with strap-shaped filaments broadening above into a wide cup containing the anther which has a broad connective. Fruit a drupe. *South & central America, eastern Asia.*

107. MELIANTHACEAE

Perennial herbs, shrubs or small trees. Leaves alternate, pinnately compound with leaflets in opposite pairs; stipules present between leaf and stem, often large. Inflorescence a terminal or axillary raceme. Flowers bilaterally symmetric, bisexual or unisexual. Calyx and corolla perigynous, stamens hypogynous. Sepals 5, unequal, overlapping. Petals 5, free, clawed, unequal. Stamens 4, 5 or 10, free or fused at base, inserted within disc lining the inside of calyx. Ovary superior, 1--5-celled; ovules

315

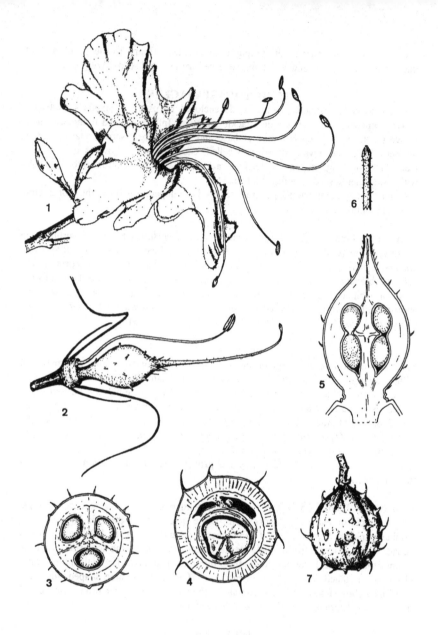

Figure 71. Hippocastanaceae. *Aesculus hippocastanum.* 1, Whole male flower from the side. 2, A bisexual flower cut longitudinally along the side of the ovary. 3, Transverse section of the young ovary. 4, Transverse section of the ovary after fertilisation, one ovule developing. 5, Longitudinal section of ovary. 6, Upper part of style, with stigma. 7, Ripe fruit.

1--several per cell, placentation axile. Style with 1, 4 or 5 stigmatic lobes or toothed or truncate. Fruit a papery or woody capsule.

A family of 3 genera and 16 species found in southern and tropical Africa. Only 1 genus is grown out-of-doors in northern Europe.

1. Melianthus Linnaeus. 5/2. Evergreen perennial herbs in cultivation, becoming woody, with unpleasant odour when bruised. Leaflets toothed, asymmetric, hairless or hairy. Flowers conspicuous, solitary or 2--4 together at nodes or in racemes, profusely nectariferous. Petals 4 or 5, upper ones partly fused into a hooded tube, the lower ones forming a short hairy spur. Stamens 4, protruding, free. Ovary 4-lobed. Style usually hairy at base. Fruit a papery capsule, 4-lobed or 4-winged. Seeds black, shining. *South Africa.*

108. BALSAMINACEAE

Annual or perennial herbs, rarely almost shrubby. Stems simple or branched, often thick and fleshy. Leaves alternate or whorled, usually stalked; blade simple with a scalloped or saw-toothed margin, sometimes with stalkless or stalked glands at the base or running onto the leaf-stalk; stipules absent, but sometimes nectariferous glands in their place. Flowers solitary or in clusters or borne in racemes (sometimes umbel-like), usually lateral but occasionally terminal by suppression of the stem-apex. Sepals 3 or 5, the lateral 1 or 2 pairs small and inconspicuous but the lower sepal larger, boat-shaped to deeply pouched, very variable in shape, with a short or long, swollen or thread-like spur. Petals 5, free or the lateral fused into 2 pairs in which the petals may be of similar or dissimilar size and shape; upper petal flat, concave or forming a hood or helmet, often crested. Stamens 5, closely united around the ovary and falling as a single unit, often before the stigmas are receptive. Ovary with a very short style; stigma generally 5-lobed. Fruit a berry or a 5-celled fleshy capsule which explodes elastically at maturity to expel the seeds forcibly from the plant. See figure 72, p. 318.

A family of 2 genera found primarily in tropical and subtropical areas of Africa, Madagascar, India, Sri Lanka and Asia as far southeast as New Guinea and the Solomon Islands, but with many species in temperate areas of the Himalaya, China and Japan, with a few species found in Europe and North America south as far as Mexico. Only 1 of the genera is in cultivation.

1. Impatiens Linnaeus. 900/20. Description as for family but lateral petals always united into 2 pairs and fruit a capsule. *Widespread.* Figure 72, p. 318.

109. CYRILLACEAE

Hairless trees or shrubs. Leaves spirally arranged, simple, entire. Stipules absent. Flowers borne in racemes, each with a bract and often 2 bracteoles, bisexual, regular. Sepals 5, rarely 7, joined at base, overlapping, persistent and often hardening in fruit. Petals 10--14, alternating with sepals, sometimes united at base; stamens 10--28, anthers with longitudinal slits or apical pores, nectary-disc present. Ovaries 2--5, placentation axile; ovules 1--3 per cell. Fruit indehiscent, a drupe, capsule or samara.

A family of 3 genera from warm America, with several ornamental species grown for their fragrant flowers and good autumn colour.

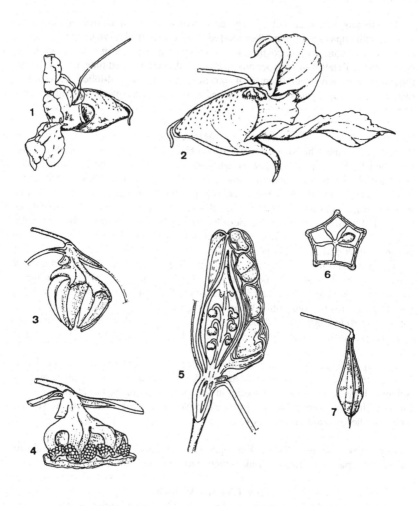

Figure 72. Balsaminaceae. *Impatiens glandulifera.* 1, Flower from the side. 2, Longitudinal section of flower. 3, Stamens before dehiscence. 4, Stamens after dehiscence. 5, Longitudinal section of ovary. 6, Transverse section of ovary (only a single ovule shown). 7, Fruit.

1a. Calyx persistent, becoming larger and hard in fruit; stamens 5   **1. Cyrilla**

  b. Calyx deciduous, not enlarging after flowering; stamens 10   **2. Cliftonia**

1. Cyrilla Linnaeus. 1/1. Evergreen or deciduous tree or shrub, 1--2 m (rarely to 6 m). Leaves 4--10 cm, oblanceolate to obovate, glossy deep green above, paler beneath. Flowers borne in racemes 8--15 cm on previous year's wood. Bracts acute, *c.* 2 mm; flower-stalks 2--4 mm. Calyx persistent, enlarging and becoming hard after flowering. Petals *c.* 5 mm. Stamens 5; style thick, short, stigma with 2 or 3 lobes. Fruit *c.* 2 × 3.5 mm, green at first, eventually pinkish red. Seeds 2 or 3. *South-east USA to northern South America.*

One very variable species.

2. Cliftonia Gaertner. 1/1. Evergreen shrub or small tree to 5 m. Leaves 4--5 cm, obtuse, wedge-shaped at base, dark green above, shortly stalked. Flowers fragrant, in racemes 3--6 cm; calyx deciduous; petals 5, white; stamens 10; style short, stigmas 3. Fruit *c.* 6 mm, wings 2--5, papery. *South USA.*

110. AQUIFOLIACEAE

Evergreen or deciduous trees and shrubs, rarely climbers. Bark smooth with hairy or hairless twigs. Stipules minute, falling early. Leaves alternate, rarely opposite, simple, entire, spiny or toothed. Flowers radially symmetric, unisexual, rarely bisexual, axillary, solitary or in clusters or cymes. Sepals usually 4--8, fused at the base. Petals usually 3--8, fused at the base or rarely free, overlapping. Stamens 4--8 sometimes attached to the petals. Ovary superior, with 2--9 cells; ovules 1 or 2 per cell, placentation axile. Fruit a berry, mostly containing 2--10, rarely to 20 pyrenes.

A family of 2 genera and over 400 species distributed in the tropical, subtropical and temperate regions of the world.

1a. Petals free; stamens distinct from petals; calyx-lobes deciduous or semi-
    persistent in fruit   **1. Nemopanthus**

  b. Petals fused at the base with attached stamens; calyx-lobes persistent in fruit
         2. Ilex

1. Nemopanthus Rafinesque. 1/1. Deciduous shrubs to 3.5 m, stoloniferous. Leaves 2.5--6 × 1--3 cm, alternate, oblong-ovate, entire or slightly toothed. Flowers unisexual. Calyx-lobes 4 or 5, deciduous or semi-persistent in fruit. Petals 4 or 5, free, greenish yellow. Stamens free. Fruits spherical, red, borne on long solitary stalks; pyrenes 3--5. *Eastern North America.*

2. Ilex Linnaeus. 400/53 and some hybrids. Description as for family but petals fused at the base and stamens attached to them. *Widespread.*

111. CELASTRACEAE

Small trees, shrubs or woody climbers without tendrils. Leaves alternate or opposite, simple, usually stalked; stipules usually stalked or absent. Flowers bisexual or unisexual, radially symmetric, usually insignificant, in axillary or, more rarely, terminal panicles or racemes, or solitary; outer parts in 4s or 5s, usually hypogynous. Petals free, usually small, mostly white to cream or greenish. Stamens alternating with petals, outside an annular disc. Ovary superior, occasionally immersed in disc, with 1--5 cells;

styles as many as cells, free or united, or absent; ovules usually 2 per cell, placentation axile. Fruit a capsule, berry, drupe or samara. See figure 73, p. 322.

A widespread family of 60--70 genera and about 1300 species. Species of 5 genera are cultivated out-of-doors in northern Europe.

1a.	Vigorous climbing shrubs	2
b.	Erect trees, or erect or weakly trailing shrubs	3
2a.	Fruit a more or less spherical capsule	**3. Celastrus**
b.	Fruit a papery 3-winged samara	**5. Tripterygium**
3a.	Fruit with 3--5 valves; seeds winged or with arils	**1. Euonymus**
b.	Fruit 2-valved; seeds (of cultivated species) with arils, not winged	4
4a.	Leaves alternate, petals usually 5	**2. Maytenus**
b.	Leaves opposite, petals usually 4	**4. Paxistima**

1. Euonymus Linnaeus. 160/30. Erect or climbing small trees or shrubs, deciduous or evergreen. Leaves opposite, stalked. Flowers in axillary cymes, outer parts in 4s or 5s. Fruit a 4- or 5-lobed capsule. Seeds with arils. *Temperate areas of the Old World.* Figure 73, p. 322.

2. Maytenus Molina. 200/3. Evergreen trees or shrubs; cultivated species unarmed, but some species (formerly separated as *Gymnosporia* (Wight & Arnott) Hooker) are spiny. Leaves alternate, leathery, bluntly toothed, almost stalkless. Flowers in few-flowered axillary clusters or solitary, small; outer parts usually in 5s. Fruit a 2-valved capsule in cultivated species (usually 3-valved in the genus as a whole). Seeds of cultivated species with 1 or 2 arils, not wholly immersed in aril. *Temperate & tropical areas.*

3. Celastrus Linnaeus. 30/6. Deciduous, usually unarmed, twining or climbing shrubs, tall-growing. Leaves alternate, stalked, elliptic to oblong or ovate to circular. Flowers terminal or axillary, solitary or in cymes or panicles, often unisexual, tiny and insignificant, outer parts in 5s. Fruit a 3-valved, more or less spherical capsule opening to reveal seeds immersed in bright red arils. *Tropical and warm temperate areas, mostly in the northern hemisphere.*

4. Paxistima Rafinesque. 2/2. Erect to spreading, low, evergreen shrubs from decumbent rooting stems, to 1 m; leaves opposite, leathery, with very short stalks or more or less stalkless, margins entire. Flowers solitary or in axillary cymes, their stalks being shorter than the subtending leaf, minute; outer parts in 4s. Fruit a 2-valved capsule, to 5 mm, white. Seeds 1 or 2, immersed in white membranous arils. *North America.*

They have often been grown under the variant names *Pachistima* Rafinesque or *Pachystima* Endlicher, both of which are incorrect.

5. Tripterygium J.D. Hooker. 3/2. Deciduous unarmed, climbing shrubs, reaching 10 m when suitably supported. Leaves alternate, stalked, margin toothed. Flowers in terminal, many-flowered panicles, *c.* 5 mm across, greenish white, outer parts in 5s. Fruit a papery 3-winged samara, *c.* 1.5 cm. *Eastern Asia.*

112. STAPHYLEACEAE

Trees or shrubs. Leaves opposite or alternate, compound, pinnate, with 3--7 leaflets or simple. Stipules falling early or absent. Flowers usually bisexual, in drooping panicles or racemes. Sepals and petals 5, overlapping. Stamens 5, alternating with petals, inserted on or below disc which lines the perigynous zone. Ovary superior, with 2--4 cells each with one or a few ovules; placentation axile. Styles 2--4. Fruit a capsule, drupe or berry.

A small family of 5 genera and 50--60 species found in north temperate areas, South America and Asia. Only 1 genus is cultivated.

1. Staphylea Linnaeus. 11/3 and 1 hybrid. Deciduous shrubs. Branches terete. Bark grey to black, mottled. Leaves opposite, with 3--7 leaflets, lateral leaflets almost stalkless. Flowers in terminal panicles; flower-stalks with stipule-like bracts. Petals pink or cream. Stamens about as long as petals. Ovary with 3 or 4 cells. Styles 3 or 4, fused. Fruit an inflated or flattened, indehiscent capsule. Seeds nearly spherical. *Eurasia, temperate America.*

113. BUXACEAE

Evergreen shrubs and trees, rarely herbs. Leaves opposite or alternate, simple. Stipules absent. Flowers usually unisexual, radially symmetric, in spikes, racemes or clusters. Perianth-segments 4--12, or absent. Stamens 4 or more, free. Ovary superior, cells 2--4 (usually 3); styles 3; ovules 1 or 2 per cell. Fruit a capsule or drupe. Seeds black and shiny. See figure 74, p. 324.

A family of about 5 genera and 60 species with an almost cosmopolitan distribution.

1a. Plant a prostrate shrub or herb; leaves toothed	**1. Pachysandra**
b. Plant an erect shrub or small tree; leaves entire	2
2a. Leaves generally alternate; fruit drupe-like	**2. Sarcococca**
b. Leaves opposite; fruit a capsule	**3. Buxus**

1. Pachysandra Michaux. 5/4. Perennial, prostrate, evergreen herbs or subshrubs. Leaves arranged spirally, usually crowded at the end of stems, toothed. Flowers unisexual, in erect or drooping, axillary or terminal spikes with the females below the males. Male flowers with 4 projecting stamens with thick filaments on the same radii as the perianth-segments. Female flowers with 4--6 sepals. Fruit a 2- or 3-celled capsule with persistent styles. *East Asia, North America.*

2. Sarcococca Lindley. 14/4. Shrubs. Leaves generally alternate except towards end of shoots, hairless and entire, stalked and fleshy. The flowers are borne mainly in dense axillary clusters with separate male and female flowers in the same or different clusters. Flowers small, creamy white or tinged pinkish, the male with 4 perianth-lobes and 4 stamens; female flowers with 4--6 perianth-lobes, styles 2 or 3. Fruit spherical, fleshy, red, purple or black, with 1--3 seeds. *West China to Malaysia.*

3. Buxus Linnaeus. 30/7. Evergreen small trees and shrubs with opposite, simple, fleshy, aromatic leaves. Flowers small and inconspicuous, produced in dense clusters in leaf-axils, with each female flower usually adjacent to a pair of male flowers. Male flowers with 4 perianth-segments and 4 stamens. Female flowers with 6

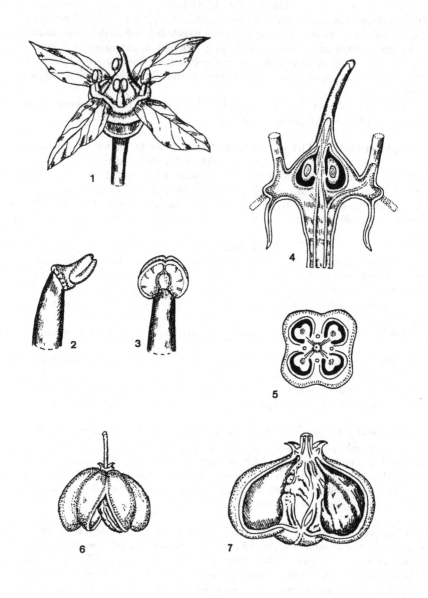

Figure 73. Celastraceae. *Euonymus europaeus.* 1, Flower obliquely from the side. 2, Side view of dehisced stamen. 3, Front view of dehisced stamen. 4, Longitudinal section of flower. 5, Transverse section of ovary. 6, Fruit just at dehiscence. 7, Longitudinal section of mature fruit.

perianth-segments and 3 styles. Fruits 3-celled, armed with 3 horns. *West Europe, Mediterranean, central Asia to central America & South Africa.* Figure 74, p. 324.

One species of **Simmondsia** Nuttall may occasionally be cultivated in warmer areas; it is like *Buxus*, but has male flowers with 5 or 6 perianth-segments and 10 or 12 stamens.

114. RHAMNACEAE

Mainly trees and shrubs, often spiny, sometimes climbers and rarely tender herbs. Leaves simple, spirally arranged or opposite, veins pinnate or palmate, stipules small or spiny. Flowers small and mainly unisexual, perianth and stamens usually perigynous. Sepals and petals 4 or 5. Stamens 4 or 5, borne on the same radii as the petals. Ovary 1--4-celled, superior or inferior, cells 1- or 2-seeded. Fruit a berry or dry capsule, rarely winged. See figure 75, p. 326.

A mainly tropical and subtropical family with 53 genera and 875 species. Several are important in horticulture,

1a. Branches 2- or 3-angled, leathery, mainly leafless except when young
                                                                        1. Colletia
  b. Branches terete, woody and leafy   2
2a. Branches with large, opposite, equal and straight spines; leaves *c.* 2 cm ×
    5 mm, stipules connected by a thin line   **2. Discaria**
  b. Branches with spines either not straight or not opposite; leaves generally
    larger, stipules not connected   3
3a. Flowers blue, rarely white or pink, bisexual, in panicles, racemes or umbels
                                                3. Ceanothus
  b. Flowers mainly greenish or yellowish, mainly in clusters or umbels   4
4a. Fruit-stalks conspicuously thick   **4. Hovenia**
  b. Fruit-stalks not conspicuously thick   5
5a. Fruits dry, disc-shaped with expanded wing   **5. Paliurus**
  b. Fruits often fleshy, not disc-shaped or winged   6
6a. Stipules not spine-like; fruit with 1 or 2 seeds if fleshy   **11. Berchemia**
  b. Stipules often spine-like; fruit with 3--several stones if fleshy   7
7a. Branches spiny; leaves strongly 3-veined from base; flowers in axillary
    racemes; petals 5   **6. Ziziphus**
  b. Not with the above combination of characters   8
8a. Leaves alternate or opposite, pinnately veined; flowers often unisexual,
    4-parted, in axillary cymes; fruit fleshy, styles 3 or 4   **7. Rhamnus**
  b. Combination of characters not as above   9
9a. Flowers bisexual, 5-parted, in axillary cymes, style 1   **8. Frangula**
  b. Flowers not as above   10
10a. Hairs stellate   **9. Pomaderris**
  b. Hairs not stellate   **10. Sageretia**

1. Colletia Jussieu. 5/4. Thorny shrubs, some reminiscent of gorse (*Ulex* species). Seedlings or young growth leafy, but plants leafless when mature, with branches often spine-tipped, at successive right-angles to adjoining branch. Leaves (when present) ovate, opposite. Flowers in clusters or solitary, at base of spines or on the spines. Perigynous zone tubular or campanulate. Petals absent. Stamens 4--6. Fruit a 3-celled capsule. *Temperate South America.*

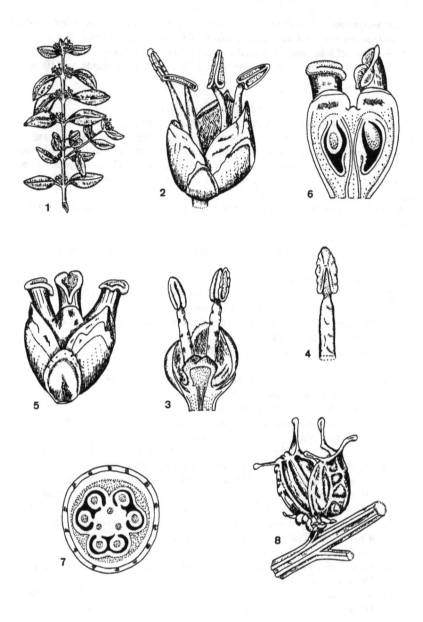

Figure 74. Buxaceae. *Buxus sempervirens.* 1, Young shoot with axillary inflorescences. 2, Male flower. 3, Longitudinal section of male flower. 4, Upper part of stamen. 5, Female flower. 6, Longitudinal section of ovary. 7, Transverse section of ovary. 8, Fruit.

2. Discaria W.J. Hooker. 15/1. Deciduous shrubs or small trees with long, slender, opposite thorns. Leaves opposite or clustered, small; stipules connected by a thin line. Flowers many, in axillary clusters, 4- or 5-parted; petals often absent. Fruit a dry, 3-celled capsule with a 3-lobed stigma. *Temperate South America, Australia, New Zealand.*

3. Ceanothus Linnaeus. 55/39 and some hybrids. Evergreen or deciduous shrubs, sometimes spiny. Leaves alternate or opposite, often fleshy, pinnate or 3-veined, margins toothed or entire; stipules present. Flowers small, 5-parted, in stalkless umbels, racemes or panicles. Fruits 3-celled, 3-styled, fleshy at first, later breaking into 3 dry nutlets. *Southwest North America to central America.* Figure 75, p. 326.

4. Hovenia Thunberg. 2/1. Deciduous shrubs or small trees to 20 m. Leaves alternate, stipulate, 3-veined, toothed or sometimes entire. Flowers very small, 5-parted, in large terminal or axillary cymes. Fruit a 3-celled, indehiscent berry, borne on a thickened stalk. *Eastern & southern Asia.*

5. Paliurus Miller. 8/2. Spiny, deciduous or evergreen trees or shrubs. Leaves alternate, 3-veined, entire or toothed. Inflorescence axillary. Flowers 5-parted, small. Styles 2 or 3. Fruits dry, disc-shaped with expanded wing. *South Europe to Japan.*

6. Ziziphus Miller. 8/1. Evergreen or deciduous trees and shrubs. Stipules often represented by thorns, one straight, the other recurved. Flowers 5-parted, bisexual, styles 2 or 3. Fruit a berry or rarely a drupe. *Tropical and warm temperate areas.*

7. Rhamnus Linnaeus. 160/14. Evergreen or deciduous, small trees or shrubs. Leaves usually alternate, rarely opposite. Flowers unisexual, males and females either on the same or separate trees, small, yellowish green or green to brown. Buds with scales. Petals, sepals and stamens similar in number, usually 4 or 5, sometimes petals absent. Styles 3 or 4. Fruit dark purple, a round or top-shaped drupe to 5 mm, 2--4-seeded. *Cosmopolitan.*

8. Frangula Miller. 50/6. Like *Rhamnus* but plants usually not spiny; winter buds not protected by scales; flowers 5-parted, bisexual, with 1 style. *North temperate areas.*

9. Pomaderris Labillardière. 45/1. Unarmed, evergreen shrubs or small trees, 2--5 m. Leaves alternate, oblong to needle-like, with dense stellate hairs; stipules small. Flowers in cymes or panicles. Calyx 5-lobed. Petals 5 or absent. Fruit a dry capsule with 3 segments. *Australia, New Zealand.*

10. Sageretia Brongniart. 35/1. Deciduous or evergreen shrubs, often climbing, twigs spiny. Leaves opposite or almost so, small and pinnately veined, entire or finely toothed; stipules small, deciduous. Flowers 5-parted, small, white, shortly stalked or stalkless, in terminal or axillary spikes. Petals hooded. Fruit a small, fleshy drupe with 2--3 seeds. *Southwest Asia to Taiwan, Somalia.*

11. Berchemia de Candolle. 12/4. Deciduous, unarmed climbers or shrubs. Leaves alternate, veins parallel. Flowers small, 5-parted. Fruit a berry with one stone. *Western North America, East Africa to eastern Asia.*

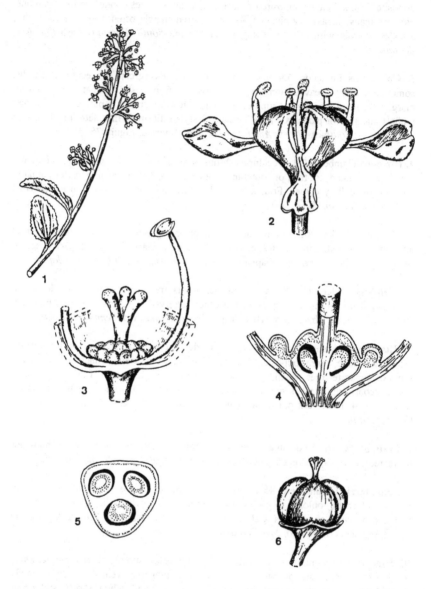

Figure 75. Rhamnaceae. *Ceanothus thyrsiflorus*. 1, Part of inflorescence. 2, Flowers from the side. 3, Longitudinal section of flower, perigynous zone, sepals and petals only partially shown. 4, Longitudinal section of ovary and disc. 5, Transverse section of ovary. 6, Young fruit.

326

115. VITACEAE

Climbers with tendrils opposite leaves, erect shrubs or small trees, rarely succulent. Leaves usually alternate, simple or compound; stipules present or absent. Flowers numerous, small, greenish, bi- or unisexual, 4-, 5- or 6-parted; calyx, corolla and stamens perigynous. Calyx entire or minutely toothed. Petals free, not overlapping, often united at tips and falling as a hood when the bud opens. Stamens on the same radii as the petals. Ovary superior, of 2 fused carpels; placentation axile. Nectariferous disc usually present below the ovary. Fruit a berry. See figure 76, p. 330.

A widespread family of 13 genera and about 800 species.

1a. Nectariferous disc absent or inconspicuous; tendrils usually with expanded
   tips **4. Parthenocissus**
  b. Nectariferous disc prominent in flower; tendrils without expanded tips 2
2a. Petals cohering into a cap, detaching at base and falling together **3. Vitis**
  b. Petals separating and spreading 3
3a. Petals usually 5 or 6 4
  b. Petals 4 5
4a. Calyx 4-lobed; disc 4-lobed **5. Ampelopsis**
  b. Calyx more or less entire; disc entire **6. Rhoicissus**
5a. Petals hooded at apex **1. Cissus**
  b. Petals not hooded at apex **2. Cayratia**

1. Cissus Linnaeus. 350/2. Erect or climbing herbs, shrubs or vines including several stem-succulents. Tendrils opposite leaves or absent. Leaves deciduous or persistent, simple or palmate with 3--7 leaflets, thin and herbaceous to succulent, margins variously toothed, rarely entire. Stipules present. Succulent species rarely flower in cultivation; climbing species produce insignificant white to green flowers in terminal or leaf-opposed compound cymes made up of umbels. Flowers unisexual or bisexual, parts in 4s. Calyx entire to 4-lobed; bud conical, not constricted in the middle. Petals hooded at apex, deflexed after the flower opens and falling early. Filaments short; disc ring-shaped, entire or lobed, more or less united to the ovary; style simple to subulate; stigma subulate or almost head-like. Fruit a berry, generally inedible; seed usually 1, oblong to ovoid or nearly spherical, often abruptly narrowed at one end. *Subtropics & tropics.*

2. Cayratia Jussieu. 45/1. Herbs, shrubs or vines. Tendrils opposite leaves, forked 1--3 times. Leaves alternate, palmately compound; leaflets 3--12, margins scalloped or with sharp, forwardly pointing teeth. Petals 4, free, spreading or reflexed, green; perigynous zone cup-shaped, fused to ovary-base. Ovary 2-celled; style cylindric, tapering. Fruit a berry, spherical to thickly disc-shaped or transversely ellipsoid. Seeds 2--4. *Asia.*

3. Vitis Linnaeus. 65/25. Deciduous, woody vines or vine-like shrubs, climbing by leaf-opposed tendrils, these often absent at every third node. Bark flaking in strips; pith brown, usually interrupted by nodal diaphragms. Leaves mostly simple, lobed or unlobed, toothed, densely hairy to hairless; stipules deciduous. Inflorescence a panicle. Flowers male on some plants, bisexual on others. Calyx minute. Petals 5, cohering into a cap, detaching at base and falling together, alternating with nectar-bearing glands;

style short. Fruit a pulpy berry, with 1--4 pear-shaped seeds. *Northern hemisphere.* Figure 76, p. 330.

4. Parthenocissus Planchon. 10/5. Woody vines, usually deciduous, trailing or ascending with tendrils; tendrils usually with expanded, adhesive discs. Leaves alternate, pinnate or partly 3-lobed. Flowers in compound cymes opposite leaves. Petals 5, sometimes 4, separate, short and thick. Disc absent or inconspicuous. Fruit a berry, dark blue or blue-black, 1--4-seeded. *North America, Himalaya, east Asia.*

5. Ampelopsis Michaux. 25/5. Deciduous shrubs or climbers, often with tendrils opposite leaves. Twigs with lenticels and white pith. Leaves alternate, simple or compound. Flowers in cymes, small, green-tinged, 4- or 6-parted. Calyx saucer-shaped, shallowly 4-lobed. Petals separate, spreading, usually 5 or 6. Disc 4-lobed, cup-like, surrounding base of pistil (reduced and sterile in male flowers). Fruit a berry. Seeds 2--4. *North America, Asia.*

6. Rhoicissus Planchon. 12/1. Shrubs or more or less woody vines. Tendrils usually present, opposite leaves. Leaves simple or with 3 or 5 leaflets, margins entire to toothed. Stipules usually present. Inflorescence opposite leaves. Calyx more or less entire. Petals 5 or 6, more or less thickened to succulent at tip. Disc entire; style simple, entire. Seeds 1--4. *Tropical & southern Africa.*

116. ELAEOCARPACEAE

Evergreen or deciduous trees or shrubs. Leaves opposite or alternate, occasionally whorled, simple, entire, sometimes with domatia; stipules persistent or not. Flowers in racemes, panicles, cymes or solitary, bisexual. Sepals 4 or 5, free or joined. Petals 4 or 5, sometimes absent, free or partly united, often fringed at the edge. Stamens usually numerous, arising from the disc; anthers dehiscing by 2 terminal pores or short slits. Ovary superior, cells 2--many; ovules 2--many in each cell, pendent. Style solitary, mostly lobed. Fruit a capsule or drupe.

1a. Flowers solitary or in 2s or 3s on inflorescence-stalks   **1. Crinodendron**
  b. Flowers usually in racemes, panicles, clusters or cymes   **2. Aristotelia**

1. Crinodendron Molina. 5/2. Evergreen large shrubs or small trees. Leaves opposite or alternate, simple, margins with forwardly pointing teeth, somewhat leathery. Flowers solitary or in 2s or 3s, borne on long pendent stalks arising from leaf-axils, cup- or urn-shaped. Petals 5, somewhat fleshy, usually with 3 apical teeth, white or crimson; stamens 15--20. Fruit a tough capsule. *Temperate South America.*

2. Aristotelia L'Héritier. 5/3. Small evergreen or deciduous trees or shrubs. Leaves opposite or alternate, simple or pinnately divided, entire or toothed. Flowers borne in terminal panicles or cymes or arising from leaf-axils, unisexual. Sepals 4 or 5, free. Petals 4 or 5, 3-lobed, apex toothed or lobed. Stamens free, downy. Ovary with 2--4 cells. Fruit a berry. *Eastern Australia, New Zealand, Peru and Chile.*

117. TILIACEAE

Herbs or, more usually, trees or shrubs, often with stellate hairs. Leaves alternate, simple, usually palmately veined, sometimes evergreen; stipules present, often deciduous. Flowers usually radially symmetric, bisexual, in cymes; calyx, corolla and stamens hypogynous. Calyx of 4 or 5 free segments or united into a tube at the base and 4- or 5-lobed. Corolla usually of 4 or 5 free petals, rarely absent, petals sometimes with nectaries on the claws. Stamens 10--many, free or with the filaments somewhat united at the base into a ring or bundles, sometimes the outermost sterile. Stamens and ovary sometimes borne on an elevated receptacle. Ovary superior, 4- or 5-celled, with 1--many ovules per cell, placentation axile; style 1, stigma usually with 4 or 5 lobes. Fruit a capsule, drupe or indehiscent. See figure 77, p. 332.

A family of 48 genera and 725 species, occurring in most parts of the world. Few of the genera are cultivated, the most important being *Tilia*, which is widely planted in gardens or as a street tree.

1a. Inflorescence-stalk united to a generally tongue-shaped bract for at least
    part of its length; generally large trees   **1. Tilia**
  b. Inflorescence and bract not as above; herbs, shrubs or small trees   2
2a. Annual herb   **4. Corchorus**
  b. Shrubs or small trees   3
3a. Evergreen shrub; fruit a bristly capsule   **3. Entelea**
  b. Deciduous shrub; fruit a drupe   **2. Grewia**

1. Tilia Linnaeus. 25/25. Deciduous trees with a strong tendency to sprout from the base. Leaves alternate, often cordate and asymmetric, with a slender stalk; stipules present, falling shortly after the leaves expand. Inflorescence-stalk fused to the upper surface of the narrowly elliptic or oblanceolate bract; flowers in a dichasial cyme, usually about 1 cm across, strongly scented. Sepals 5, boat-shaped, with hair-covered nectaries on their upper surfaces. Petals 5, strap-shaped, pale to bright yellow. Stamens numerous in 5 bundles; petal-like staminodes present or absent. Ovary spherical with 5 cells, each containing 2 ovules; style slender with a small, 5-lobed stigma. Fruits nut-like, usually indehiscent and containing 1--3 seeds, the whole shed attached to the bract. *North temperate areas.*

2. Grewia Linnaeus. 150/1. Shrubs or trees, rarely climbing (not ours). Leaves entire or toothed, deciduous, somewhat oblique at the base, with 3--9 veins; stipules persistent. Flowers in axillary or terminal cymes, sometimes few-flowered, or flowers rarely solitary. Sepals 5, free. Petals 5, each with a nectary on the claw. Stamens many, all fertile, filaments free. Ovary 5-celled with several ovules per cell. Fruit a drupe with 1--4 stones. *Tropical & subtropical Africa, Asia & Australasia.*

3. Entelea Brown. 1/1. Evergreen shrub or small tree, densely hairy. Leaves characteristically drooping, broadly ovate, acuminate, cordate at the base, sometimes obscurely 5--7-lobed, margin doubly scalloped or toothed; stipules persistent. Flowers to 2.5 cm across, few to many, in umbel-like cymes. Sepals 4 or 5, white. Petals 4 or 5, white, crumpled in bud. Stamens all fertile, free or slightly united at the base. Ovary 5--7-celled with numerous ovules per cell. Capsule more or less spherical, to 2 cm across, covered with bristles. *New Zealand.*

5. Corchorus Linnaeus. 40/1. Herbs (ours) or small shrubs. Leaves toothed but not lobed, the lowermost teeth often bristle-tipped; stipules persistent, long and bristle-like. Flowers 1--3, in stalked cymes which are axillary or leaf-opposed. Sepals 4

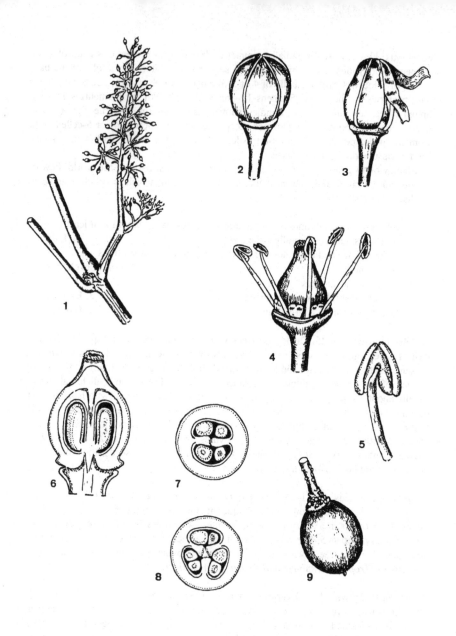

Figure 76. Vitaceae. *Vitis vinifera.* 1, Inflorescence. 2, Unopened flower-bud. 3, Flower-bud just opening. 4, Flower, the apically united petals having just fallen. 5, Stamen. 6, Longitudinal section of ovary. 7, Transverse section of two-celled ovary. 8, Same of three-celled ovary. 9, Mature fruit.

or 5. Petals 4 or 5, yellow, without nectaries on their claws. Stamens numerous, borne, with the ovary on an elevated receptacle. Ovary 3--5-celled, ovules many per cell; style short, stigma 3--5-lobed. Fruit a capsule. *Tropics & subtropics.*

118. MALVACEAE

Herbs, shrubs and trees, usually with stellate hairs. Leaves spirally arranged, simple to dissected, usually palmately veined; stipules usually present, often persistent. Flowers usually bisexual, solitary and axillary or in cymes. Epicalyx usually present. Calyx with 5 sepals, usually free, sometimes united at base, often with nectaries on the inner surfaces. Corolla radially symmetric, with 5 petals, usually free, often united near base to the staminal tube. Stamens 5--numerous, filaments united in a tube for most of their length; anthers opening by longitudinal slits; stamens at the top of the tube sometimes without anthers. Ovary superior, with 1--numerous fused carpels (but often 5); styles usually as many as carpels, usually united at base; cells as many as carpels; placentation axile; ovules 1--numerous per cell. Fruit a capsule or schizocarp, rarely a berry or samara. See figure 78, p. 336.

A family of 116 genera and 1550 species from all over the world but particularly the tropics.

1a. Epicalyx absent	2
b. Epicalyx present	9
2a. Plants rosette-forming	**9. Nototriche**
b Plants not rosette-forming	3
3a. Style branches terminating in head-like stigmas	4
b. Style branches thread-like, longitudinally stigmatic on inner surfaces	7
4a. Ovules 2 or more per cell; seeds usually 2 or more per mericarp; mericarps usually dehiscent and dropping the seeds at maturity	
	15. Abutilon
b. Ovules 1 per cell; seeds 1 per mericarp; mericarps releasing the seed by withering of the wall or by dehiscence at the apex	5
5a. Petals blue-violet to lavender or white	**13. Anoda**
b. Petals white, cream, ivory or yellowish	6
6a. Leaves simple; mericarps 5--15	**12. Hoheria**
b. Leaves palmately lobed; mericarps 8--10	**14. Sida**
7a. Flowers *c.* 5 mm across; petals yellow or white	**11. Plagianthus**
b. Flowers more than 5 mm across; petals white to deep red-purple	8
8a. Staminal column simple, the filaments all arising equally from near the apex of the column	**7. Callirhoe**
b. Staminal column apically divided into an inner and outer series of filaments	
	8. Sidalcea
9a. Fruit a capsule; carpels 3--5	10
b. Fruit schizocarpic; carpels 3--many	11
10a. Calyx not persistent in fruit; fruits more than twice as long as wide	
	20. Abelmoschus
b. Calyx persistent in fruit; fruits not more than twice as long as wide	
	19. Hibiscus
11a. Staminal column with 5 teeth at apex; mericarps 5; styles and stigmas 10	
	18. Malvaviscus

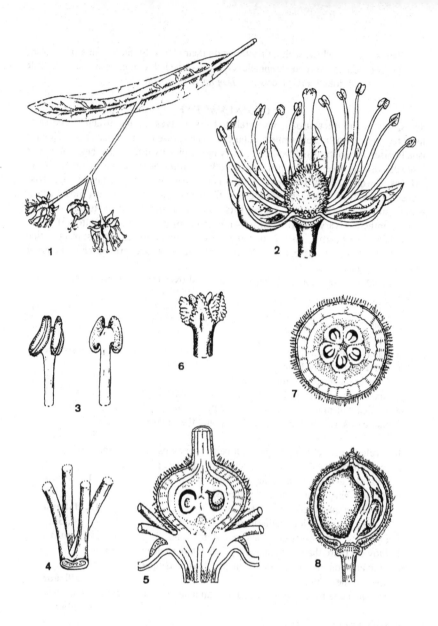

Figure 77. Tiliaceae. *Tilia platyphyllos.* 1, Inflorescence with its stalk attached to a bract. 2, Flower with three petals and some stamens removed. 3, Stamen from front and back. 4, Basal parts of fused filaments. 5, Longitudinal section of ovary. 6, Stigma. 7, Transverse section of ovary. 8, Longitudinal section of mature fruit.

332

b. Staminal column bearing anthers at apex; mericarps 3--60; styles and
stigmas as many as mericarps   12
12a. Epicalyx-segments 3--11   13
 b. Epicalyx-segments 1--3   15
13a. Mericarps in up to 5 superimposed whorls, forming a spherical head
    2. Kitaibela
 b. Mericarps in a single whorl, forming a circular disc   14
14a. Staminal tube cylindric; flowers not more than 3 cm across, at least some
of them on conspicuous stalks   **3. Althaea**
 b. Staminal tube 5-angled; flowers at least 3 cm across, more or less
stalkless in a spike-like inflorescence   **4. Alcea**
15a. Epicalyx-segments united at least at base in bud   **5. Lavatera**
 b. Epicalyx-segments free   16
16a. Mericarps horseshoe- or kidney-shaped   17
 b. Mericarps not as above   18
17a. Petals obcordate; carpels without beaks or appendages   **6. Malva**
 b. Petals usually truncate; carpels beaked and with transverse appendages
inside and underneath the beak   **7. Callirhoe**
18a. Mericarps arranged in a spherical head   **1. Malope**
 b. Mericarps in a single ring, separating at maturity from the central axis   19
19a. Fruits more than 1.2 cm   **17. Iliamna**
 b. Fruits to 1.2 cm   20
20a. Mericarp-edges with irregular projections or apical beaks, each mericarp
with 1--6 seeds; petals white or flushed pink to deep rose-magenta or
purple   **10. Anisodontea**
 b. Mericarp-edges not as above, each mericarp with 1--3 seeds; petals
yellow, orange, red, lavender, pink or white   **16. Sphaeralcea**

1. Malope Linnaeus. 4/2. Herbaceous hairy or hairless annuals or perennials. Leaves unlobed or palmately lobed. Flowers solitary in leaf-axils, stalks long. Epicalyx-segments 3, free, broader than sepals. Sepals 5. Petals 5. Staminal column bearing filaments at apex. Stigmas many, thread-like. Fruit hairless; mericarps numerous, arranged in a spherical head. *Europe, Mediterranean area.*

2. Kitaibela Willdenow. 1/1. Robust, upright, herbaceous perennials to 3 m; stems, leaf-stalks and inflorescence with rough white hairs. Leaves to 18 cm across, almost circular, with 5--7 toothed, triangular lobes, base truncate to cordate; stalk to 15 cm. Flowers 5--7 cm across in axillary clusters of 1--4; stalks 4--8 cm. Epicalyx-segments 6--9, joined at base, *c.* 2 × 1 cm, ovate, acuminate, hairy, exceeding sepals. Calyx 5-lobed, hairy. Petals 5, to 2.5 × 2 cm, broadly triangular, white. Stamens numerous, joined into a column which bears filaments at apex. Mericarps numerous in up to 5 superimposed whorls forming a spherical head, dark brown, hairy. *Former Yugoslavia.*

3. Althaea Linnaeus. 15/2. Annual or perennial herbs to 2 m. Leaves shallowly to deeply lobed or parted. Flowers rather small in racemes or panicles, usually on distinct stalks. Epicalyx-segments 6--9, joined at base. Petals 5, obovate, notched or entire, pale pink to lilac. Staminal column cylindric, hairy. Stigmas lateral, thread-like. Fruit a schizocarp; mericarps 1-seeded, indehiscent, 1-whorled. *Europe, Mediterranean area.*

4. Alcea Linnaeus. 60/2 and some hybrids. Tall biennials or short-lived, perennial herbs with densely felted hairs, sometimes becoming hairless or nearly so at maturity. Stems erect and sometimes branched. Leaves palmately lobed or unlobed. Flowers large, almost stalkless with leaves at base or in long, leafless, spike-like racemes. Epicalyx-segments 6 or 7, joined at base. Petals 5, pink, purple, yellow or white and sometimes notched. Staminal column 5-angled, hairless, with filaments at apex. Styles thread-like, lateral. Fruit a schizocarp; mericarps 18--40, indehiscent, 1-whorled, divided into 2 unequal cells, upper one empty, lower one 1-seeded. *Europe, Mediterranean area, Middle East.*

5. Lavatera Linnaeus. 25/7 and some hybrids. Shrubs, herbaceous perennials, biennials or annuals, usually with stellate hairs. Leaves lobed or angled, toothed, palmately veined, stalked, with stipules. Flowers solitary or clustered in leaf-axils often forming a tall inflorescence. Epicalyx-segments 3, united at least at base in bud. Sepals 5. Petals 5, usually clawed, white, pink or pinkish purple. Stamens numerous with filaments fused to form a column which bears filaments at the apex. Styles with thread-like stigmas; fruit of many 1-seeded, indehiscent mericarps arranged in a single whorl. *Mostly Mediterranean area, but also USA (California), Canary Islands, Himalaya, Australia.*

6. Malva Linnaeus. 30/6. Herbs and subshrubs, annual, biennial or perennial. Stems erect or decumbent, simple or branched. Leaves angled, lobed or dissected. Flowers axillary, solitary or clustered, small or showy. Epicalyx with 1--3 free segments. Sepals 5. Petals 5, white, pink or purple, notched or 2-lobed, obcordate. Staminal column bearing filaments at apex. Stigmas thread-like, linear. Fruit a disc-shaped schizocarp with numerous, 1-seeded, non-beaked, horseshoe- or kidney-shaped mericarps. *Europe, Asia, Africa, North America.* Figure 78, p. 336.

7. Callirhoe Nuttall. 8/4. Annual or perennial herbs to 1.2 m, erect or trailing in habit; perennial species with thick, carrot-like taproots. Leaves alternate, lobed, palmate or finely divided; stipules present. Flowers bisexual, *c.* 5 cm across, cup-shaped, solitary or few, in upper leaf-axils. Petals 5, truncate at the apex, pale pink to purple or red. Epicalyx 3-parted. Calyx 5-lobed in cultivated species. Staminal column bearing filaments at apex. Fruit a schizocarp; mericarps 10--25, more or less beaked, horseshoe- or kidney-shaped, 1-seeded. *North America.*

8. Sidalcea Gray. 25/4. Annual or perennial herbs. Stems simple or branched, hairless or hairy. Upper leaves rounded, palmately lobed or dissected, lower leaves more roughly toothed or shallowly lobed. Flowers normally showy, pink, purple or white, in terminal racemes or spikes, often female by abortion of the anthers. Epicalyx absent. Calyx 5-lobed. Petals 5, with margins toothed, or truncated. Stamens joined into groups in 2 series. Style branches thread-like. Fruit a schizocarp; mericarps 5--9, 1-seeded, without beaks and indehiscent. *North America.*

9. Nototriche Turczaninow. 70/1. Perennial, sometimes annual, plants with stellate hairs and thick, woody, branched, usually underground stems forming rosettes at the surface. Leaves usually palmately lobed to repeatedly divided; stipules present, joined to the persistent leaf-stalks to form a tube protecting the buds. Flowers on short stalks which are attached to the leaf-stalks. Epicalyx absent. Sepals 5. Petals 5, sometimes with a corolla-tube; stamens numerous. Styles 8--14; carpels beaked, dehiscent. *South America (Andes from Ecuador to Chile).*

10. Anisodontea Presl. 19/2. Evergreen, perennial herbs and shrubs. Leaves stalked, linear to elliptic, 3-, 5-, or 7-lobed, or with 3 leaflets, stipulate. Flowers bisexual, axillary, solitary or in small clusters. Epicalyx of 3 segments which may be free, joined at the base, or united with the calyx. Calyx usually longer than epicalyx, ovoid at the bud stage but becoming campanulate as the flower opens. Petals exceeding calyx, white to deep pink, each often with a dark basal spot. Staminal column bearing filaments at apex. Fruit a schizocarp; mericarps 5--26, in 1 whorl, laterally compressed, beaked or with irregular projections. *Southern Africa.*

11. Plagianthus Forster & Forster. 2/2. Evergreen or deciduous trees and shrubs. Leaves simple; stipules deciduous. Flowers unisexual or bisexual in terminal or axillary panicles, or solitary. Epicalyx absent. Calyx 5-lobed. Petals 5. Stamens 8--20. Styles 1--5. Fruit irregularly 1--5-celled. *New Zealand.*

12. Hoheria Cunningham. 5/5. Evergreen or deciduous shrubs or small trees with alternate, simple, usually toothed leaves, stellate-hairy. Seedlings and young plants shrubby with more slender branches, and juvenile foliage with leaves much more deeply toothed or even lobed and broader than in the adult state. Flowers solitary or in cymose clusters of up to 10, axillary or occasionally terminal, papery white. Epicalyx absent. Calyx campanulate, 5-lobed. Petals 5, notched at apex. Staminal column split above into 5 bundles of filaments. Style branches as many as carpels; stigmas head-like. Carpels 5--15, 1-seeded, winged or not, surrounding and, when seed is ripe, separating from a central axis. *New Zealand.*

13. Anoda Cavanilles. 10/1. Herbaceous annuals and perennials, some forming small shrubs. Leaves variable even on one plant, entire or palmately lobed, hastate, stalked. Flowers bisexual, solitary or paired in leaf-axils in our species, forming an upright spike. Epicalyx absent. Calyx of 5 sepals. Petals 5, equal to or longer than calyx. Staminal column tubular. Stigmas terminal, head-like. Fruit a disc-like schizocarp, mericarps 10--20, 1-seeded. *North & South America.*

14. Sida Linnaeus. 150/1. Annual and perennial herbs and small shrubs. In our species leaves 3-, 5-, or 7-lobed, or toothed, with stipules. Flowers bisexual, axillary, in loose cymes. Epicalyx usually absent. Calyx 5-lobed or toothed. Styles 5--10. Stigmas head-like. Fruit a schizocarp; mericarps equal to styles in number, 1-seeded, in one row, dehiscent at the tip, awned or beaked on the inner edge. *Tropical & warm temperate areas.*

15. Abutilon Miller. 150/6 and some hybrids. Annual or perennial herbs, shrubs or small trees, hairless or downy. Leaves alternate, long-stalked, ovate to heart-shaped or elliptic, entire or with 3, 5 or 7 palmate lobes, margins scalloped or toothed. Flowers axillary or terminal, solitary or grouped in racemes or panicles. Bracteoles and epicalyx absent. Calyx 5-lobed, lobes lanceolate, heart-shaped or ovate. Petals 5, yellow, orange, white, lavender, rose or purple, sometimes clearly veined. Stamens many, fused in a tubular column. Styles 5 or more; stigmas each with a spherical head. Fruit a

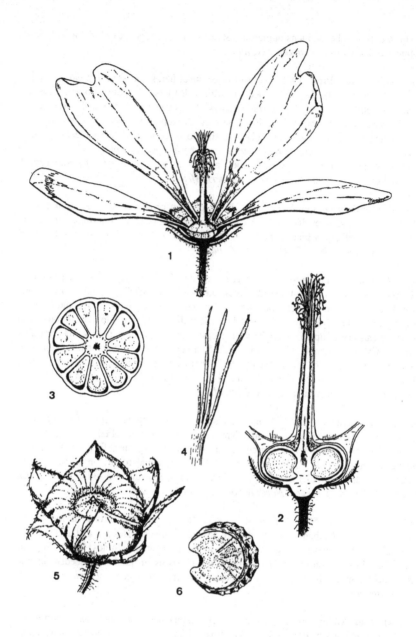

Figure 78. Malvaceae. *Malva sylvestris.* 1, Flower from the side with one petal removed. 2, Longitudinal section of ovary and staminal tube. 3, Transverse section of ovary. 4, Top of the style with separate stigmas. 5, Fruit. 6, A single mericarp from the side.

schizocarp; mericarps usually dehiscent and shedding the seeds at maturity; seeds 2--9 per mericarp, kidney-shaped, hairless or slightly downy. *Subtropics & tropics.*

16. Sphaeralcea St. Hilaire. 60/4. Annual to perennial herbs, woody at base with closely stellate down throughout. Leaves varying from simple and circular to deeply divided and linear-lanceolate, rarely 3--5-lobed with lobes toothed, these sometimes pointing forwards. Flowers in axillary clusters or in terminal, panicled racemes or occasionally corymbs. Epicalyx-segments usually 3 (rarely 1 or 2 or absent). Calyx 5-lobed. Petals 5, normally notched, white, yellow, orange, rose, purple or magenta. Staminal column bearing filaments at apex. Styles equal in number to the mericarps. Stigmas head-like. Fruit spherical or nearly so, stellate dorsally, wrinkled to netted and indehiscent on lower third to four-fifths, upper portion dehiscent and smooth; mericarps 5--20 in a single ring, with 1--3 kidney-shaped, smooth to bristly seeds per mericarp. *North & South America, South Africa.*

17. Iliamna Greene. 7/1. Perennial herbs and small shrubs. Leaves alternate, palmate. Flowers conspicuous, bisexual, solitary or in tall interrupted spikes. Epicalyx-segments 3, free. Petals 5, clawed. Staminal column roughly hairy. Styles 10--15. Fruit a schizocarp; mericarps 10--15, in 1 whorl, 2--4-seeded, dehiscent, roughly hairy with stellate hairs towards the apex. *North America.*

18. Malvaviscus Adanson. 3/1. Shrubs, sometimes vine-like. Leaves alternate, simple, entire or palmately lobed. Flowers solitary in upper axils or in few-flowered, terminal racemes or cymes. Epicalyx-segments 6--16. Corolla funnel-shaped, red with prominent basal auricles. Staminal column exceeding petals, with 5 teeth at apex. Styles 10. Stigmas head-like. Ovary 5-celled, each cell with 1 ovule. Fruit schizocarpic; mericarps 5, sticky or fleshy; seeds red, soon drying. *Central & South America.*

19. Hibiscus Linnaeus. 250/2. Annual or perennial herbs, shrubs or trees. Leaves stalked, palmate or entire, varying in shape from lanceolate to ovate, elliptic and heart-shaped; stipules present. Flowers bisexual, stalked, axillary, solitary or clustered, usually large and campanulate. Epicalyx-segments 4--15, sometimes united with the calyx. Calyx 5-lobed and often enlarging to enclose (at least partly) the capsule, persistent in fruit. Petals 5, from white to yellow, pink or red, each often with a prominent basal red to purple spot. Styles 5, often protruding. Stigmas head-like. Fruit a dehiscent capsule, ovoid to oblong, with 5 cells. *Warm temperate, subtropical & tropical areas.*

20. Abelmoschus Medikus. 15/3. Annual and perennial herbs, more or less hairy. Leaves large, palmate to compound or lobed, rarely entire, with stipules. Flowers bisexual, solitary in leaf-axils, or in terminal, racemose inflorescences. Epicalyx of 4--16 segments. Calyx splitting longitudinally, lobed; lobes spathulate, with 5 teeth at the tip. Calyx, petals and staminal column joined at the base, and falling together as a cap after flowering. Petals 5, usually yellow with a dark purple spot at the base; some varieties have pink or red petals with white basal markings. Fruit a 5-celled capsule, splitting longitudinally when ripe. *Old World tropics & subtropics.*

119. STERCULIACEAE

Trees, shrubs, herbs or sometimes vines. Leaves alternate, simple or palmate; stipules mostly deciduous. Flowers bisexual or unisexual, clustered or solitary, radially symmetric. Calyx 5-lobed, lobes not overlapping. Petals 5 or absent. Stamens 5 or more in 2 whorls, those opposite the sepals sterile (staminodes) or absent; anthers opening by longitudinal slits. Ovary superior, usually 5-celled, each cell with 2 or more ovules or axile placentas; styles 2--5, free or united. Fruit dry, usually dehiscent.

A family of 73 genera, mostly tropical in distribution, although a few are from more temperate regions.

1a. Fruit a follicle or samara	**1. Brachychiton**
b. Fruit a capsule or a drupe	2
2a. Petals conspicuous	**2. Reevesia**
b. Petals absent or scale-like	3
3a. Sepals with 3--5 prominent ribs	**4. Guichenotia**
b. Sepals not ribbed	**3. Fremontodendron**

1. Brachychiton Schott & Endlicher. 31/1. Mostly trees, often with swollen trunks, less often shrubs, evergreen or deciduous, monoecious. Leaves usually hairless, alternate, entire to deeply palmatifid, very variable on the same plant; stalks long. Flowers unisexual, in axillary panicle-like cymes, often borne on leafless stems. Calyx petal-like, 5-lobed, cream to red; petals absent. Fruit composed of 5 or fewer woody follicles, ovoid. Seeds few to several, with a loose, brittle, stellate-hairy outer coat. *Australia, Papua New Guinea.*

2. Reevesia Lindley. 4/2. Deciduous shrubs or small trees. Leaves simple, alternate, entire. Flowers bisexual, white, in dense, many-flowered terminal cymes or corymbs. Calyx with 3--6 unequal teeth. Petals 5, clawed. Receptacle extended, bearing 5 groups of 3 stalkless anthers alternating with 5 staminodes, and a stalked 5-celled ovary. Fruit a woody capsule. Seeds winged. *Himalaya to Taiwan.*

3. Fremontodendron Coville. 3/2. Evergreen shrubs or small trees. Leaves simple, alternate, unlobed or with up to 7 lobes; stipules falling early. Flowers solitary, radially symmetric, subtended by 2 or 3 small bracts. Petals absent. Calyx petal-like, campanulate, yellow to orange to copper red; lobes 5, longer than tube. Stamens 5; filaments united for about half their length. Fruit a capsule with stellate hairs. *Southwest USA, adjacent Mexico (Baja California).*

4. Guichenotia Gay. 6/2. Evergreen, stellate-hairy shrubs. Leaves opposite, entire, stalked, narrow, with margins curved downwards and inwards; stipules absent or similar to leaves but shorter and almost stalkless. Bracts and bracteoles present. Flowers in racemes. Calyx conspicuous, pink to purple, divided to or below the middle, with 3--5 raised ribs, membranous. Petals 5, small, scale-like. Stamens 5, united at the base or free. Ovary 5-celled. Fruit a 5-valved capsule. *Southern part of Western Australia.*

120. THYMELAEACEAE

Trees or shrubs, rarely herbs. Leaves alternate or opposite, simple, entire; stipules absent. Flowers radially symmetric, usually bisexual, in terminal or axillary heads, racemes, spikes or rarely solitary. Perianth and stamens perigynous; perigynous zone tubular, often coloured and petal-like. Sepals 4 or 5, rarely 6. Corolla small, segments scale-like or more usually absent. Stamens as many as or twice as many as sepals, or reduced to 2, usually inserted at mouth of tube. Ovary superior, 1- or rarely 2-celled; ovules 1 per cell. Fruit an indehiscent nut, berry or drupe, rarely a dehiscent capsule. See figure 79, p. 340.

A family of about 40 genera.

1a. Stamens up to 4	2
b. Stamens 8 or more	3
2a. Stamens 4	**8. Drapetes**
b. Stamens 2 (rarely 1)	**1. Pimelea**
3a. Leaves *c*. 1 mm wide	**7. Passerina**
b. Leaves over 2 mm wide	4
4a. Calyx obscurely 4-lobed	**4. Dirca**
b. Calyx conspicuously 4–6-lobed	5
5a. Plants evergreen; petals minute or absent	**9. Gnidia**
b. Plants evergreen or deciduous; petals absent	6
6a. Fruit a nutlet	**6. Stellera**
b. Fruit a fleshy or dry drupe	7
7a. Style very short or almost absent	**2. Daphne**
b. Style long and thin	8
8a. Fruit exposed when mature	**3. Ovidia**
b. Fruit surrounded by persistent base of calyx-tube	**5. Edgeworthia**

1. Pimelea Gaertner. 80/7. Small evergreen shrubs or subshrubs, branches tough and wiry with prominent leaf-scars. Leaves opposite (rarely alternate), stalkless and often 4-ranked, not toothed. Inflorescence a terminal head or short, spike-like raceme, rarely axillary, often subtended by 4 or more leaf-like or coloured bracts. Flowers small, often scented, short-stalked. Perigynous zone tubular, white, pink, purplish or yellow, hairy outside; sepals 4, spreading, petals absent, rarely small, scale-like, opposite the bases of the sepals. In several species, the perigynous zone parts along a line round the circumference, just above the ovary, the lower portion remaining to surround the developing fruit. Stamens 2, attached at mouth of perigynous zone; ovary 1-celled, with 1 ovule; style slender, arising to one side of the apex of the ovary; stigma small. Fruit a 1-seeded indehiscent capsule, rarely fleshy, often enclosed in the persistent base of the perigynous zone. *Australia, New Zealand, Timor & Lord Howe Island.*

2. Daphne Linnaeus. 70/35 and some hybrids. Dwarf to medium-sized shrubs, evergreen or deciduous, usually with tough, flexible branches. Leaves usually alternate, entire, more than 5 mm wide. Flowers usually bisexual, in terminal or axillary clusters (rarely in terminal or axillary panicles), often scented. Perigynous zone tubular at base, expanding into 4 sepals above. Petals absent. Stamens 8, in 2 rows of 4. Style very short or absent. Fruit usually a fleshy drupe, exposed when mature. *Mostly Europe and temperate to subtropical Asia, a few in North Africa.* Figure 79, p. 340.

3. Ovidia Meissner. 4/2. Deciduous shrubs similar to *Daphne* but plants dioecious with very flexible twigs; style long and thin. *Chile.*

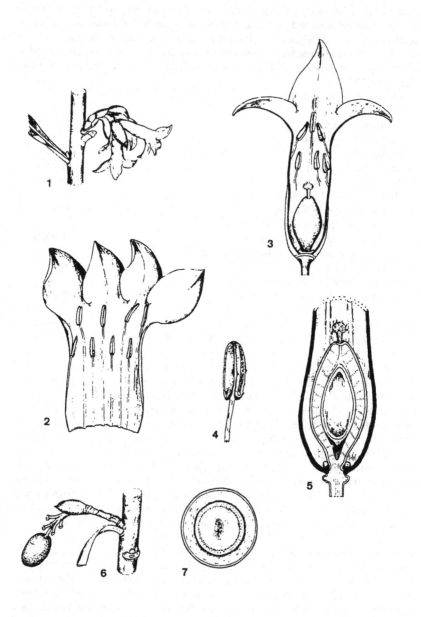

Figure 79. Thymelaeaceae. *Daphne laureola.* 1, Part of axillary inflorescence. 2, Perianth opened out to reveal stamens in two whorls of four. 3, Longitudinal section of flower. 4, Stamen. 5, Longitudinal section of ovary. 6, Fruit. 7, Transverse section of fruit.

4. Dirca Linnaeus. 2/1. Deciduous shrubs with pliable branches, soon hairless. Buds hidden in bases of leaf-stalks. Leaves almost stalkless, alternate, elliptic to obovate, entire. Flowers very small, in clusters of 2--4 at nodes on previous year's growth, appearing before the leaves. Bud-scales persistent, hairy. Perigynous zone pale yellow, narrowly funnel-shaped, slightly narrowed above the ovary, mouth sinuous or obscurely 4-lobed; petals absent. Stamens 8, borne at top of perigynous zone and protruding slightly. Ovary superior, 1-celled, ovule 1. Style slender, longer than stamens, stigma minute. Fruit an ovoid or rounded drupe, pale green or reddish when ripe. *North America.*

5. Edgeworthia Meissner. 2/2. Deciduous or evergreen shrubs with tough pliable shoots. Leaves alternate, lanceolate to ovate, narrow but more than 2 mm wide. Inflorescence a dense spherical head of many flowers, on a short stalk, arising from a leaf-axil near the end of the previous year's shoot, often clustered. Bracts few, ovate, hairy, soon falling. Perigynous zone hairy outside, with a cylindric tube and 4 spreading lobes, often scented. Petals absent. Stamens 8, in 2 whorls at different levels in the perigynous zone. Ovary 1-celled, hairy; style slender with a long, papillose stigma. Fruit a dry drupe, surrounded by the persistent base of the perigynous zone. *Eastern Asia.*

6. Stellera Linnaeus. 1/1. Herbaceous perennials 30--50 cm, bases woody, supporting numerous unbranched stems of about equal length. Leaves simple, narrowly elliptic to broadly ovate, alternate but arranged in a whorl below the inflorescence, more than 2 mm wide. Inflorescences *c.* 4 cm across, terminal, spherical, many-flowered. Flowers bisexual, white, red in bud or yellow. Perigynous zone tubular with 4--6 spreading lobes. Petals absent. Stamens 10, in 2 series attached at middle and at mouth of tube; filaments short. Styles short or absent. Fruit a nutlet. *Temperate Asia.*

7. Passerina Linnaeus. 18/2. Evergreen, heath-like shrubs. Leaves in pairs that alternately cross each other at right-angles, thus making 4 rows, simple, entire, *c.* 2 × 1 mm. Inflorescence a spike. Flowers 4-parted, conspicuously woolly. Perigynous zone tubular; sepals 4. Petals absent. Stamens 8, projecting. *South Africa.*

8. Drapetes Banks. 10/2. Low-growing, evergreen shrubs with small, overlapping, alternate leaves. Flowers stalkless, in heads at the ends of the branches, bisexual or unisexual. Perigynous zone white, hairy externally, funnel-shaped or campanulate, with 4 spreading lobes, each with 1 or 2 small scales at the base. Stamens 4, protruding, alternating with the sepals. Ovary 1-celled, with a single ovule. Stigma head-like; style slender. Fruit dry, indehiscent, or a small ovoid berry. *Australia, New Zealand, New Guinea, Borneo, southern South America.*

9. Gnidia Linnaeus. 140/2. Evergreen shrubs often with a heath-like habit. Leaves opposite or alternate, usually small, more than 2 mm wide. Flowers in terminal clusters, solitary or in leaf-axils, yellow, white, red or violet. Sepals 4 or 5, petal-like. Petals 4, 8 (ours) or 12, small or absent. Stamens 8--10. Style slender. Fruit small, dry, enclosed in persistent base of calyx *Tropical & southern Africa, Madagascar, Arabia, Sri Lanka.*

121. ELAEAGNACEAE

Shrubs or small trees, with more or less conspicuous, silvery or brown, stellate or peltate scales on shoots, foliage and flowers. Leaves deciduous or evergreen, alternate or opposite, entire; stipules absent. Flowers axillary, solitary, clustered, in racemes or spikes, bisexual or unisexual. Perigynous zone tubular; perianth-segments 2 or 4. Stamens 4 or 8, inserted on perianth. Ovary superior (sometimes appearing inferior because of a constriction in the perigynous zone above it) 1-celled, style 1, ovule solitary, basal. Fruit 1-seeded, dry but appearing berry-like because of the persistent fleshy remains of the perigynous zone.

A family of 3 genera and about 45 species, mostly native to temperate parts of Europe, Asia and North America, a few in tropical Asia and one extending to north-eastern Australia.

1a. Leaves opposite; stamens 8   **2. Shepherdia**
  b. Leaves alternate; stamens 4   **2**
2a. Flowers unisexual; perigynous zone very short, lobes 2   **1. Hippophae**
  b. Flowers mostly bisexual; perigynous zone long, lobes 4   **3. Elaeagnus**

1. Hippophae Linnaeus. 3/3. Deciduous suckering shrubs or small trees to 12 m, at least the young growth densely covered with silvery or brown peltate scales or stellate hairs. Twigs often rather stout, stiff and usually spine-tipped. Leaves short-stalked, alternate, linear or lanceolate to oblong. Flowers opening in spring, male and female on separate plants, usually in bracteate axillary racemes, often with leafy tips, on wood of the previous season. Flowers small, in the male the perigynous zone split to the base into 2 concave lobes enclosing 4 short-stalked anthers, in the female the perigynous zone tubular with 2 minute lobes, the cylindric style projecting, stigmatic on one side. Fruit berry-like, the perigynous zone becoming fleshy and surrounding a single hard seed. *Temperate Eurasia.*

2. Shepherdia Nuttall. 3/2. Deciduous or evergreen dioecious shrubs. Shoots scaly when young, sometimes spiny. Leaves opposite, entire, short-stalked, with dense brown or silvery scales, at least on lower surface. Flowers small, borne in short axillary spikes or racemes, on bare twigs of deciduous species. Female flowers with a tubular perigynous zone surrounding the ovary and closed by a fleshy disc at the base of 2--6 (usually 4) spreading ovate lobes; style 1, stigma simple. Male flowers without perigynous zone, lobes cupped, finally reflexed; stamens 8, alternating with lobes of the disc. Fruit ovoid, red, yellow or brownish and scaly. *North America.*

3. Elaeagnus Linnaeus. 40/8 and some hybrids and selections. Deciduous or evergreen shrubs or small trees, sometimes climbing, young branchlets with dense scales or stellate hairs, sometimes spiny. Leaves alternate, entire, short-stalked, closely covered, at least beneath, with peltate scales or stellate hairs. Flowers solitary or few together in axillary umbels, bisexual or unisexual. Perigynous zone tubular, more or less constricted above the ovary; lobes 4, ovate to triangular, outer surface densely covered with silvery scales or stellate hairs, inner surface usually hairless except for sparse stellate hairs on the lobes. Stamens 4, almost stalkless, inserted in throat of perigynous zone, alternating with the lobes. Nectariferous disc above constriction in base of calyx-tube, inconspicuous in most species. Ovary superior, ovule 1; style 1, with

342

oblique stigma, hairless or with stellate hairs towards base. Fruit drupe-like, the ellipsoid or narrowly obovoid achene or stone ribbed and embedded in the fleshy or mealy, swollen base of the perigynous zone. *Mostly temperate eastern Asia, also Europe & North America.*

122. FLACOURTIACEAE

Trees, shrubs or woody climbers, sometimes spiny. Leaves alternate, simple, pinnately veined in most genera, usually toothed, with stipules which are often quickly deciduous. Flowers bisexual or unisexual, radially symmetric, borne in racemes, panicles or axillary clusters. Calyx of 4 or 5 (rarely 3 or more than 6) sepals, free or somewhat united below. Petals absent or 2--15. Stamens usually numerous, anthers opening by slits. Ovary superior (in all cultivated species) or inferior. Carpels usually 3, usually united, with parietal placentation. Ovules numerous. Fruit a berry or capsule (in the cultivated species).

A large, mainly tropical family of 89 genera and 1250 species. The few genera in general cultivation are from temperate or subtropical areas and are not typical of the family as a whole.

1a. Climber; perianth of 9 or more red segments   **1. Berberidopsis**
  b. Shrubs or trees; perianth of 4 or 5 yellowish or greenish segments   2
2a. All inflorescences axillary   3
  b. Some inflorescences terminal   4
3a. Leaves of 2 types, mostly alternating larger and smaller; style 1   **3. Azara**
  b. Leaves all similar; styles 2 or more   **2. Xylosma**
4a. Leaf-stalk with 2 nectaries; styles 3--5, each undivided at the apex   **4. Idesia**
  b. Leaf-stalk without nectaries; styles 3 or 4, each with 2 or 3 lobes at apex   5
5a. Flowers unisexual; styles more or less spathulate, each 2-lobed
                                       5. Poliothyrsis
  b. Flowers bisexual; styles each 3-lobed   **6. Carrierea**

1. Berberidopsis J.D. Hooker. 1/1. Evergreen climber to 3 m or more; shoots sparsely hairy. Leaves oblong, stalked, tapering to the apex, truncate at the base, toothed in at least the lower half, hairless, prominently veined beneath. Flowers more or less spherical, pendent, borne in terminal racemes and singly in the axils of the upper leaves; flower-stalks red. Perianth of 9--15 red segments, of which the 3 outermost are small, the others larger, 8--15 mm. Stamens 7--10, with very short filaments. Ovary stalkless, tapering into the thick style which is weakly 3-lobed at the apex. Fruit a berry. *Chile.*

2. Xylosma Forster. 60/1. Small trees and shrubs, sometimes spiny. Leaves shortly stalked, toothed, often leathery, with stipules. Flowers small, usually unisexual, in axillary racemes or clusters. Sepals 4 or 5, usually hairy. Petals absent. Stamens numerous. A nectar-producing disc or distinct nectaries usually present. Ovary 1-celled with 2 or 3 (rarely 6) placentas. Styles 2 or 3 (rarely 6), free or united below. Fruit a few-seeded berry. *Mostly tropical America.*

3. Azara Ruiz & Pavón. 10/5. More or less evergreen shrubs to 5 m (rarely to 8 m). Leaves stalkless, of 2 kinds, borne in 2 rows, the larger alternate, the smaller more or less opposite to them; stipules very small, deciduous. Flowers in axillary racemes (sometimes umbel-like), spikes or clusters, small, vanilla-scented. Sepals 4 or 5, downy inside or at least around the margins. Petals absent. Stamens usually numerous, rarely 4 or 5, yellow, exceeding the sepals, the outer sometimes sterile. Rod-like nectaries often present, usually on the same radii as the sepals. Ovary hairless; style simple, stigma obscurely to distinctly 3-lobed. Fruit a spherical berry with persistent style. *South America.*

4. Idesia Maximowicz. 1/1. Deciduous tree to 10 m or more with variably hairy shoots. Leaves to 15 × 15 cm or more, ovate, cordate, tapering to an acuminate tip; stalks long, with 2 conspicuous nectaries. Lower surface of leaves whitish with numerous closely set papillae, the sides of the midrib and the vein-axils often hairy. Flowers unisexual, in large, terminal, pendent panicles. Sepals 5, 4--10 mm, yellowish, densely downy inside and outside. Petals absent. Stamens numerous; filaments downy; staminodes in female flowers very short. Styles 3--5, spathulate, free. Ovary 1-celled, placentas usually 5. Fruit a spherical, red, many-seeded berry, often black in less sunny areas. *South & south-central China.*

5. Poliothyrsis Oliver. 1/1. Tree or shrub to 15 m; young shoots hairy. Leaves to 15 × 10 cm, ovate, narrowly ovate or elliptic, tapering to an acuminate tip, rounded at the base, hairy beneath, not papillose; stalks without nectaries. Flowers mostly unisexual, yellowish, in fragrant, terminal panicles. Sepals 5, 3--6 mm, enlarging in fruit, densely hairy inside and outside. Petals absent. Stamens many, of several lengths; staminodes in female flowers very short. Styles 3, free, more or less spathulate, 2-lobed at the apex, reflexed along the 1-celled ovary. Fruit a spindle-shaped, densely downy capsule, 1.5--2.5 cm, opening by 3 teeth. Seeds numerous. *South & south-central China.*

6. Carrierea Franchet. 3/1. Tree to 20 m; shoots hairless. Leaves long-stalked. Flowers bisexual, in terminal racemes. Sepals 5. Petals absent. Stamens numerous. Ovary 1-celled with 3 or 4 placentas. Styles 3 or 4, broad, recurved, each 3-lobed towards the apex. Fruit a large, hard, very woody capsule containing many seeds which are usually somewhat winged. *Southeast Asia.*

123. VIOLACEAE
Herbs or shrubs with alternate, simple, stipulate leaves. Flowers radially or bilaterally symmetric, with a pair of bracteoles on the flower-stalk, usually bisexual. Sepals 5, free or nearly so. Petals 5, the lowest in bilaterally symmetric flowers with a spur. Stamens 5, anthers opening to the interior of the flower, connective-appendages and nectaries variably developed. Ovary superior, 1-celled, with 3 (rarely 2 or up to 5) placentas and a single style. Fruit a capsule or berry; seeds often with an oil-rich aril. Cleistogamous flowers, in which self-pollination takes place and from which abundant seed is set, occur in many *Viola* species. See figure 80, p. 346.

A cosmopolitan family with some 20 genera and nearly 1000 species. The largest genus, *Viola* (about 500 species) is mainly temperate, occurring in both hemispheres.

1a. Flowers more or less radially symmetric; fruit a berry; shrubs **1. Melicytus**
  b. Flowers markedly bilaterally symmetric; fruit a capsule; mostly herbs
2. Viola

1. **Melicytus** Forster & Forster (*Hymenanthera* Hooker). 14/4. Evergreen or semi-evergreen shrubs (rarely deciduous); sometimes dioecious. Leaves simple, alternate or clustered, with minute stipules. Flowers solitary or in clusters, small and inconspicuous, radially symmetric, usually unisexual. Sepals 5, joined; petals 5, small; stamens 5, with very short filaments and variably developed, flat, membranous connective-appendages, sometimes fused into a tube. Styles 2- or 3- or rarely 6-fid. Fruit a spherical, few-seeded berry. *Australia, New Zealand, Pacific Islands.*

2. **Viola** Linnaeus. 500/37 and many hybrids. Herbs, rarely dwarf shrubs. Leaves alternate, simple, often in a basal rosette; stipules always present, sometimes leaf-like. Flowers solitary, bilaterally symmetric, usually on a long stalk bearing a pair of bracteoles. Sepals 5, persistent, sometimes enlarging in fruit. Petals 5, free, lower spurred or pouched and usually with a wide blade. Stamens 5 with triangular connective-appendages protruding apically, the 2 lowermost stamens with nectar-secreting spurs running into the petal-spur. Style thickened above, stigma head-like, narrowed or beaked, sometimes bilobed. Ovary superior, 1-celled with 3 placentae and numerous ovules. Capsule opening by 3 valves. Seeds with an oil-body, dispersed by ants. *Widespread in north temperate areas, also in the Andes, Australia & New Zealand.* Figure 80, p. 346.

124. STACHYURACEAE

Shrubs or small trees, deciduous or evergreen. Branches often straggly. Leaves alternate, simple, toothed, stalked; leaf-blade membranous to leathery. Stipules small, falling early. Flowers small, radially symmetric, bisexual or unisexual, stalked or stalkless, in erect or pendent axillary racemes. Sepals, petals and stamens hypogynous. Sepals and petals 4, free, overlapping. Stamens 8, free, in 2 whorls. Ovary 4-celled, ovules numerous in each cell, placentation axile; style short, simple. Fruit a berry with many seeds.

A family of 1 genus from Eastern Asia.

1. **Stachyurus** Siebold & Zuccarini. 6/2. Description as for family. *Himalaya to China, Taiwan & south Japan.*

125. PASSIFLORACEAE

Trees, shrubs or lianas with axillary tendrils, often with unusual secondary growth. Leaves spirally arranged, entire or palmately lobed. Leaf-stalks usually with nectaries. Stipules deciduous. Flowers radially symmetric, usually bisexual, solitary or in cymes; perigynous zone flat to tubular, often with an elongated androgynophore (stalk bearing ovary and stamens). Calyx usually with 5 (rarely 3--8) free or sometimes basally united sepals. Petals usually 5. Corona of 1 or more units of filaments or scales. Stamens 4, 5 or numerous, when few usually alternating with petals. Ovary superior, 1-celled, carpels 2--5. Styles usually united only at base. Placentation parietal. Ovules usually numerous. Fruit a berry or capsule. Seeds with oily endosperm. See figure 81, p. 348.

A family of 18 genera and about 600 species from tropical and warm temperate areas, especially the Americas. Only one genus is commonly cultivated.

1. **Passiflora** Linnaeus. 350/1. Woody or sometimes herbaceous climbing plants with tendrils formed in leaf-axils. Leaves alternate, simple or lobed, sometimes toothed, often variable in shape and size even on one plant; stipules present. Nectar-producing

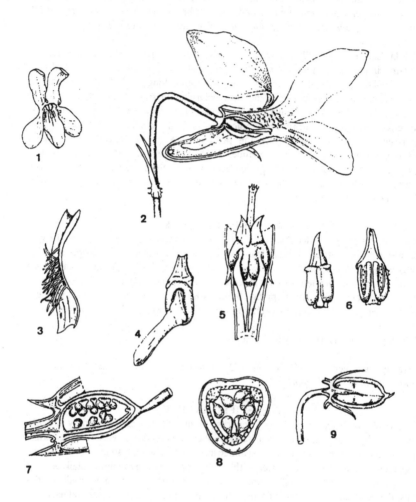

Figure 80. Violaceae. *Viola reichenbachiana.* 1, Flower from the front. 2, Longitudinal section of flower with two of the petals removed. 3, Part of the base of an upper petal showing hairs lining the stigmatic groove. 4, Staminal spur. 5, As for 4, a different view. 6, Anther of one of the upper stamens, front and back views 7, Longitudinal section of ovary. 8, Transverse section of ovary. 9, Fruit.

346

glands usually present on the leaf-stalks, bracts or undersides of leaves. Flowers usually stalked, sometimes unisexual, usually with bracts on flower-stalk often forming a conspicuous involucre. Sepals 5, sometimes 4, often keeled and with an awn, joined at base; perigynous zone bowl-shaped or long and cylindric, with variable membranes attached to inside. Petals 5, sometimes 4 or absent; corona usually of several rows of filaments joined to apex of perigynous zone. Stamens 5, sometimes 4, attached to a stalk raised above sepals, petals and corona. Ovary with 3 styles and head-like stigmas, borne on the androgynophore above the stamens. Fruit fleshy, sometimes with a hard skin, sometimes edible, containing numerous seeds surrounded by fleshy arils. *Mostly tropical America & Australasia*. Figure 81, p. 348.

126. CISTACEAE

Evergreen shrubs or herbs. Leaves usually opposite, simple, entire, often with simple glandular hairs and stellate hairs, stipules sometimes present. Flowers bisexual, radially symmetric, solitary, in cymes or in cyme-like racemes. Sepals 3--5, overlapping. Petals usually 5 (rarely 4--11), free, overlapping. Stamens many. Ovaries 3--10, though usually 5 carpels, 1-celled, or incompletely 3--5-celled (ours); ovules few to many, placentation parietal; styles simple, sometimes jointed, with 1--5 stigmas. Fruit a capsule, usually woody. See figure 82, p. 350.

A family of 7 genera and about 180 species of sun-loving plants native to temperate areas but especially prominent in the Mediterraean area.

1a. Ovary and capsule with 5--10 cells; flowers white to purplish red or pink
                                         1. Cistus & 3. × **Halimiocistus**
  b. Ovary and capsule with 3 cells; flowers white, pink or yellow   2
2a. Style short, straight or absent; shrubs; sepals 3 or 5
                                2. Halimium & 3. × **Halimiocistus**
  b. Style long, or if short or absent then plant a herb; sepals 5   3
3a. Annual or perennial herb, with a basal leaf-rosette (which may be withered at flowering); stigma stalkless or almost so   **6. Tuberaria**
  b. Dwarf shrubs or annuals, without basal leaf-rosettes; style present   4
4a. Leaves all opposite, oblong to linear; all stamens fertile **4. Helianthemum**
  b. Upper leaves usually alternate, more or less linear; outer stamens sterile
                                          5. Fumana

1. Cistus Linnaeus. 16/16 and several hybrids. Shrubs, many of which exude an aromatic, sticky resin. Flowers usually in cymes, rarely solitary. Sepals 3--5, petals 5, usually stained yellow or red at base. Ovary with 5--10 cells. Capsules splitting into 5 (rarely 6--12) segments. *Mediterranean area, Canary Islands.*

2. Halimium Spach. 9/6. Evergreen shrubs or subshrubs. Leaves opposite, without stipules. Sepals 3 or 5, the 2 outermost much smaller than the 3 inner. Petals yellow or white. Capsules 3-valved. *Mediterranean area.*

3. × **Halimiocistus** Janchen. 3/3. A genus of hybrids between *Cistus* and *Halimium* with characters which are intermediate between the parents.

4. Helianthemum Miller. 120/10. Herbs or small shrubs. Leaves opposite or stem-leaves alternate, entire. Flowers in simple or compound terminal racemes. Sepals 5,

347

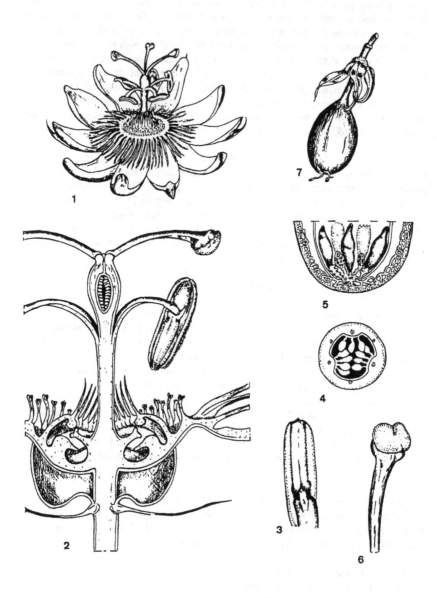

Figure 81. Passifloraceae. *Passiflora caerulea.* 1, Flower obliquely from above. 2, Longitudinal section of the central part of the flower. 3, Upper part of stamen. 4, Transverse section of ovary. 5, Transverse section of part of mature fruit. 6, Upper part of style and stigma. 7, Mature fruit.

inner 3 ovate, outer 2 smaller, usually linear or oblong, enlarging in fruit. Petals 5, sometimes very small or absent. Stamens numerous. Ovary with 2--12 ovules; style slender and curved or bent. Fruit a capsule, splitting into 3 segments; fruit-stalks often recurved. *Mediterranean area, North & South America*. Figure 82, p. 350.

5. Fumana (Dunal) Spach. 9/2. Low shrubs. Leaves usually alternate, linear, ovate-lanceolate to linear. Outer 2 sepals small, inner 3 large, scarious, prominently veined. Petals yellow. Outer stamens sterile, beaded. Style thread-like, more or less bent at base. Capsule splitting into 3 segments which are usually spreading after dehiscence. *Europe, north Africa, southwest Asia.*

6. Tuberaria (Dunal) Spach. 12/2. Annuals or perennials with a basal leaf-rosette. Flowering stems erect. Leaves 3-veined. Flowers in terminal cymes, yellow. Sepals 5, outer 2 usually smaller than inner 3. Stigma more or less stalkless. Fruit a capsule, splitting into 3 segments. *Central & southern Europe.*

127. TAMARICACEAE

Shrubs or small trees, often adapted to saline conditions and resistant to drought. Leaves small, often scale-like and with salt-excreting glands, alternate, without stipules. Flowers radially symmetric, bisexual, mostly small, in bracteate racemes, sometimes aggregated into panicles, or solitary. Sepals 4 or 5 (rarely 6), usually free, persistent. Petals 4 or 5 (rarely 6), alternating with sepals, sometimes persistent. Stamens 4--many, often inserted on or beneath a nectariferous disc surrounding the base of the ovary, or joined at the base into a short tube or into 5 bundles. Ovary superior, of 2--5 fused carpels, usually 1-celled with many ovules; placentation parietal or basal. Fruit a capsule. Seeds hairy or with a tuft of hairs at one end, sometimes stalked; embryo straight.

A family of 5 genera from Africa through Europe to central Asia and China, especially in the eastern Mediterranean area and central Asia.

1a. Flowers solitary; stamens numerous; styles 5   **3. Reaumuria**
  b. Flowers in racemes or panicles; stamens 4--10 (rarely to 15); styles 3
    or 4 or stigma stalkless   2
2a. Stamens 10, united at the base into a short tube; stigma stalkless
                2. Myricaria
  b. Stamens 4 or 5 (rarely to 15), free, often inserted on a fleshy nectariferous
    disc around the base of the ovary; styles 3 or 4   **1. Tamarix**

1. Tamarix Linnaeus. 54/10. Deciduous shrubs or small trees to 10 m, erect and bushy, hairless, papillose or rarely hairy. Shoots slender, flexible. Leaves stalkless, alternate, entire, small and scale-like, forward-pointing, giving the appearance of a conifer or heather; many species with salt-secreting glands. Flowers small, pink or white, in spike-like racemes, borne singly or in terminal panicles, either in spring on the previous season's wood or in summer on new growth, or both; each flower subtended by a single bract (rarely 2). Sepals and petals 4 or 5, petals sometimes persistent. Stamens variable in number: all species have 4 or 5, borne opposite the sepals, usually on the edge of a nectariferous disc surrounding the base of the ovary and some have an additional 1--10, borne opposite the petals. The nectariferous disc either merges smoothly with the filament bases or is lobed between them, the lobes often being

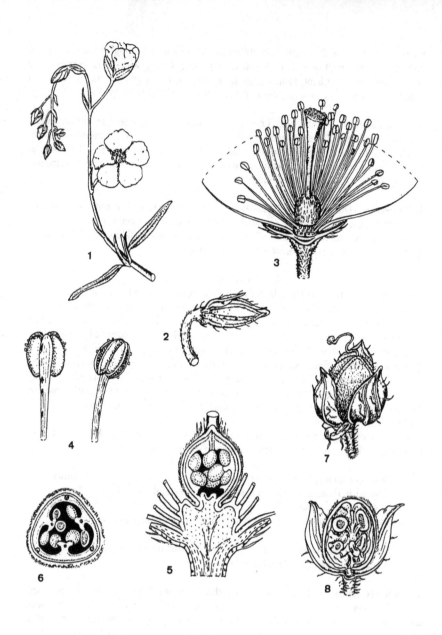

Figure 82. Cistaceae. *Helianthemum nummularium*. 1, Inflorescence. 2, Flower-bud. 3, Central part of flower with some sepals and petals removed. 4, Stamen from front and back. 5, Longitudinal section of ovary. 6, Transverse section of ovary. 7, Fruit just before dehiscence. 8, Fruit cut longitudinally.

350

indented. Ovary more or less conical, with 3 or 4 short styles. Fruit a capsule. Seeds numerous, hairless except for a tuft of hairs at one end, this usually longer than the seed. *Temperate Eurasia.*

2. Myricaria Desvaux. 10/2. Deciduous shrubs with slender, flexible shoots. Leaves alternate, small and scale-like, forward-pointing, giving the appearance of a conifer or heather. Flowers small, red, pink or white, each subtended by a bract, in narrow terminal or axillary racemes. Sepals and petals 5. Stamens 10, the filaments joined at the base to form a short tube. Ovary narrowly conical, with a terminal stalkless stigma, which is sometimes 3-lobed. Fruit a dehiscent capsule. Seeds numerous, each with a usually stalked tuft of hairs at one end, often several times longer than the seed. *Temperate Eurasia.*

3. Reaumuria Linnaeus. 20/1. Small deciduous shrubs. Leaves alternate, small and more or less spreading. Flowers solitary, terminal. Sepals and petals 5, the petals each with 2 scale-like appendages at the base. Stamens numerous, united at the base into 5 bundles opposite the petals. Ovary spherical to conical, with 5 thread-like terminal styles. Fruit a dehiscent capsule. Seeds numerous, hairy. *Mediterranean area to central Asia.*

128. FRANKENIACEAE

Herbs or shrubs. Leaves opposite, simple and often *Erica*-like, with inrolled margins and salt-excreting glands; stipules absent. Flowers axillary, solitary or in cymes, with 2 bracteoles, usually bisexual, radially symmetric. Calyx-lobes 4--7, united at the base, touching above. Petals 4--7, overlapping, with long claws. Stamens 4--7, rarely to 24, but usually in two rows of 3, anthers opening by longitudinal slits. Ovary superior, of 2--4 carpels, 1-celled, style 1; ovules 2--6 to many. Fruit a capsule, enclosed in the persistent calyx.

A family of 3 genera from temperate and subtropical, mostly saline habitats. Only one is cultivated.

1. Frankenia Linnaeus. 25/3. Annual or perennial herbs or small shrubs with wiry branches and usually inrolled, entire, hairy leaves. Flowers stalkless, solitary or in leafy cymes. Petals and sepals usually 5; calyx tubular, persistent. Stamens of the outer whorl shorter than those of the inner. Ovary stalkless, of 3 or 4 carpels. *Temperate & subtropical areas.*

129. ELATINACEAE

Annual or perennial, hairless or glandular-hairy herbs or occasionally shrub-like plants, of aquatic or semi-aquatic habitats. Stems upright when underwater, more or less prostrate on land. Leaves simple, opposite or whorled, entire or coarsely toothed with small, paired stipules. Flowers small, bisexual, radially symmetric, solitary in the leaf-axils or in cymes, usually inconspicuous. Sepals 2--5 (rarely 6), free or barely joined at base. Petals 2--5 (rarely 6), free, overlapping. Stamens as many as or twice the number of petals, in 1 or 2 whorls, the outer alternating with the petals; anthers opening by longitudinal slits. Ovary superior, with 2--5 cells, placentation axile, ovules numerous; styles very short, distinct, free; stigma head-like. Fruit a 3- or 4-celled capsule, spherical, sometimes depressed, seeds numerous, pitted, straight or curved.

A small family of 2 genera with an almost cosmopolitan distribution but mainly from temperate and subtropical areas; only *Elatine* is found in cultivation.

1. Elatine Linnaeus. 25/2. Hairless annuals or short-lived perennials. Leaves narrowly elliptic to almost circular, apex rounded, base tapered to slightly rounded, entire. Flowers with parts mainly in 2s or 4s, membranous, sepals very pale green or pinkish, inconspicuous, without visible midribs, petals pale greenish white or pale red. Stamens 2--8; styles 3--5. Fruit a capsule, membranous, spherical or depressed-spherical, seeds straight or curved, numerous. *Temperate areas.*

130. LOASACEAE

Usually herbs, sometimes shrubs or even small trees, occasionally with twining stems, usually covered with rough, bristly or often stinging hairs. Leaves alternate or opposite, simple or more usually toothed or lobed. Stipules absent. Flowers radially symmetric, bisexual, solitary or in cymes. Calyx-lobes usually 5 (more rarely 4--7), rolled or overlapping, persistent. Petals 5 or 10 (more rarely 4--7), flat or hooded, free or sometimes joined at base. Stamens usually numerous, free or with a basal tube or in bundles, anthers with longitudinal slits; petal or thread-like staminodes often present. Ovary inferior, with 1--3 cells and 3--5 parietal placentas. Nectary-scales present or absent. Fruit a capsule or indehiscent.

A family of 15 genera with about 250 species mostly in North & South America.

1a.	Stamens in bundles; nectary-scales present	2
b.	Stamens not in bundles; nectary-scales absent	4
2a.	Capsule not twisted	**3. Loasa**
b.	Capsule twisted	3
3a.	Nectary-scales each with a dorsal wing	**1. Blumenbachia**
b.	Nectary-scales lacking dorsal wings	**2. Caiophora**
4a.	Petals united at base; ovary with 5 placentas	**5. Eucnide**
b.	Petals free at base; ovary with 3 placentas, or if placentas 5, then petals 10	**4. Mentzelia**

1. Blumenbachia Schrader. 6/2. Annuals or biennials with bristly, stinging hairs. Stems trailing or climbing, usually 4-angled. Leaves opposite, lobed. Flowers solitary, axillary, bracteate, usually borne on long stalks. Petals 5, strongly concave. Stamens numerous, in bundles opposite the petals. Staminodes 10, thread-like, alternating with petals. Nectary-scales with dorsal wings. Fruit a spherical to cylindric, strongly twisted capsule, opening by 3--10 spirally twisted valves. *South America.*

2. Caiophora Presl. 70/1. Annuals or short-lived perennials. Plants usually with stinging hairs. Stems terete, often twining. Leaves opposite, pinnately lobed. Flowers white, yellow or red, solitary, axillary. Petals hooded. Nectary-scales without dorsal wings. Stamens numerous, in bundles opposite petals. Fruit a spirally twisted capsule opening by spiral valves. *South America.*

3. Loasa Adanson. 105/4. Herbs or subshrubs, usually with stinging hairs. Leaves opposite or alternate, simple to dissected. Flowers yellow, white or red, usually nodding, solitary or in short racemes or panicles. Petals 5, sac-like or boat-shaped, hooded, protecting stamens. Nectary-scales 5, conspicuous, each with a dorsal wing.

352

Stamens in bundles opposite the petals. Fruit a capsule opening by apical valves. *Mexico to South America.*

4. Mentzelia Linnaeus. 60/3. Annual, biennial or perennial herbs or shrubs with barbed (never stinging) hairs. Leaves mostly alternate, usually coarsely toothed or pinnatifid. Flowers white, yellow or orange, solitary or in short racemes or cymes. Petals free, ovate-elliptic. Stamens numerous, free, not in bundles, outer sometimes petal-like. Ovary with 3 (rarely to 5) placentas. Fruit a capsule, opening at the top. *North America, Mexico.*

Species still found in nurserymen's catalogues under the name *Bartonia* Sims.

5. Eucnide Zuccarini. 8/2. Biennial or perennial herbs, often becoming woody at the base. The genus differs from *Mentzelia* in having petals united at the base and ovaries with 5 placentas. *Southwest USA, Mexico.*

131. DATISCACEAE

Herbs, shrubs or trees. Leaves alternate, without stipules, simple or compound. Flowers usually in racemes or raceme-like inflorescences, radially symmetric, unisexual, plants usually dioecious. Perianth of 3--8 segments or more rarely calyx of 3--8 free or united segments, petals up to 8, free. Stamens 4--many. Ovary inferior, of 3 carpels, 1-celled, placentation parietal, ovules numerous. Fruit a capsule containing many seeds.

A small family of 4 genera and a small though uncertain number of species. Only *Datisca* is grown.

1. Datisca Linnaeus. 2/1. Tall dioecious herbs. Leaves pinnate with a terminal leaflet, leaflets deeply toothed. Flowers numerous in long, terminal and axillary racemes, without petals. Stamens 4--25. Styles 3, each deeply divided into 2. *Scattered.*

132. BEGONIACEAE

Herbs, shrubs or climbers, usually succulent. Roots rhizomatous, tuberous or fibrous. Leaves alternate, stalked, usually asymmetric. Flowers unisexual, regular or irregular. Male flowers with 2--4 petal-like perianth-segments, edge-to-edge in bud. Female flowers with 2--many petal-like, overlapping perianth-segments. Stamens usually many in several whorls; anthers 2-celled. Ovary inferior or partly inferior, with 1--3 wings, usually 3-celled. Styles 2--5. Placentation usually axile. Fruit a capsule or sometimes a berry. Seeds many, small. See figure 83, p. 356.

A family of 3 genera, of which only *Begonia* is grown out-of-doors in northern Europe.

1. Begonia Linnaeus. 1000/250 and many hybrids. Monoecious, succulent herbs, shrubs or climbers, the latter with aerial roots, rarely epiphytic. Roots fibrous, often arising from rhizomes or tubers, rarely from a bulb-like structure. Aerial stems often swollen and conspicuously jointed. Leaves usually alternate in 2 rows, asymmetric or rarely regular, membranous to leathery, often brightly multi-coloured; stipules present; small axillary bulbs rarely present. Inflorescences basal, axillary or terminal, cymose or rarely racemose, erect or pendent, bearing few or many separate male and female flowers. Male flowers often with 4 perianth-segments, in opposite pairs at right-angles, stamens few to numerous, female flowers often with 5 perianth-segments, ovary

353

inferior, carpels 2 or 3, fused. Fruit a tough, fleshy or rarely leathery capsule, usually winged; seeds without endosperm. *Mainly tropics & subtropics.* Figure 83, p. 356.

More than 10,000 hybrids and cultivars have been recorded (mostly not hardy in northwest Europe).

133. CUCURBITACEAE

Usually herbaceous, often fleshy, bristly-papillose monoecious or dioecious trailers or climbers of considerable size, rarely woody, characteristically with coiled tendrils, one at each node, these rarely absent or transformed into spines. Leaves alternate, simple and palmately veined or pedately compound, spirally arranged, often with extrafloral nectaries. Flowers unisexual, axillary, solitary or in inflorescences of various types, usually radially symmetric. Sepals normally 5 (rarely 3, 4 or 6), usually united. Corolla yellow or white, lobes normally 5 (rarely 3, 4 or 6), rarely free. Stamens 5, often reduced to 3, 2 double and 1 single. Anthers twisted or contorted. Ovary inferior, of united carpels, usually 1--3-celled, placentation parietal; stigmas as many as carpels, usually forked. Fruit normally a fleshy, indehiscent berry, sometimes of considerable size and then hard and thick-walled, rarely a capsule. Seeds 1 to usually numerous, large and often flattened, sometimes winged. See figure 84, p. 358.

A cosmopolitan family, mainly from tropical and warm temperate areas, with about 120 genera and 800 species, grown mainly as annuals for their fruits which are edible and/or decorative.

1a. Stamens 4 or 5	**2. Thladiantha**
b. Stamens 3 or fewer	2
2a. Petals free	**1. Telfairia**
b. Petals united at least at base	3
3a. Tendrils absent	**3. Ecballium**
b. Tendrils present	4
4a. Filaments more or less united into a column	5
b. Filaments free or united only at base, sometimes anthers united	6
5a. Fruit dehiscent	**7. Echinocystis**
b. Fruit indehiscent	**8. Sicyos**
6a. Fruit less than 2.5 cm, red	**10. Mukia**
b. Fruit more than 2.5 cm, usually not red	7
7a. Corolla 5-lobed to middle of tube or just beyond	**6. Cucurbita**
b. Corolla 5-lobed to base or almost to base	8
8a. Male flowers in racemes	**4. Bryonia**
b. Male flowers in clusters or solitary	9
9a. Connective not produced beyond the cells of the anthers	**5. Citrullus**
b. Connective produced beyond the cells of the anthers	**9. Cucumis**

1. Telfairia J.D. Hooker. 3/2. Dioecious perennial climbers to over 30 m, often woody. Stems hairless, or slightly downy; tendrils bifid. Leaves pedate, with 3--7 leaflets, terminal leaflet large. Male inflorescence racemose, bracteolate; calyx tubular, short, downy or hairless, toothed, sometimes with nectaries; petals 5, free; stamens 3. Female flowers solitary, larger than male, ovary ribbed; stigma 3-lobed. Fruit large, 10-ribbed; seeds ovate to almost circular, flattened. *Africa.*

354

2. Thladiantha Bunge. 23/1. Annual or perennial, dioecious, climbing herbs. Stems hairy; tendrils simple. Leaves ovate to heart-shaped, simple, toothed in cultivated species. Flowers axillary, solitary, stalked; calyx short, campanulate, lobes recurved, enclosed within by scale-like appendages; corolla campanulate, petals free; ovary oblong, hairy. Fruit oblong, indehiscent, fleshy. Seeds obovoid, flattened, small. *E Asia, Philippines, Africa.*

3. Ecballium Richard. 1/1. Low, spreading, monoecious perennial covered with rough grey hairs and lacking tendrils. Leaves fleshy, heart-shaped to triangular, toothed or shallowly lobed. Flowers yellow, *c*. 2.5 cm, slightly campanulate; male in axillary clusters, with 3 stamens; female solitary. Fruit green, 4--5 cm, like a small cucumber, explosive when ripe, ejecting seeds up to several metres. *Mediterranean area.*

4. Bryonia Linnaeus. 12/3. Dioecious, rarely monoecious, climbing, perennial herbs with annual growth from fleshy or tuberous rootstocks. Leaves rough-hairy, 5--7-lobed. Tendrils simple. Flowers greenish white to greenish yellow in axillary, racemose or corymbose panicles. Calyx-tube campanulate, 5-lobed. Corolla shallowly cup-shaped, deeply 5-fid. Stamens 3, female flowers with 3--5 staminodes which are often extremely small. Fruit a small, spherical, few-seeded berry. Seeds ovoid to oblong with scalloped margin. *Europe, adjacent Asia, north Africa.*

4. Bryonia Linnaeus. 12/3. Dioecious, rarely monoecious, climbing, perennial herbs with annual growth from fleshy or tuberous rootstocks. Leaves rough-hairy, 5--7-lobed. Tendrils simple. Flowers greenish white to greenish yellow in axillary, racemose or corymbose panicles. Calyx-tube campanulate, 5-lobed. Corolla shallowly cup-shaped, deeply 5-fid. Stamens 3, female flowers with 3--5 staminodes which are often extremely small. Fruit a small, spherical, few-seeded berry. Seeds ovoid to oblong with scalloped margin. *Europe, adjacent Asia, North Africa.*

5. Citrullus Ecklon & Zeyher. 3/2. Annual or perennial, monoecious or rarely dioecious trailers or climbers with simple or branched tendrils. Leaves simple, usually deeply pinnately lobed. Flowers solitary in the leaf-axils, yellow. Calyx-tube short. Corolla-lobes 5, united below the middle. Male flowers with 3 stamens, anthers twisted; female flowers with staminodes and a downy, almost spherical ovary. Fruit broadly cylindric to almost spherical, greenish or yellowish, firm-walled with fleshy interior, sometimes very large (to 1 m × 40 cm). Seeds numerous, ovate, compressed. *Tropical & southern Africa.*

6. Cucurbita Linnaeus. 27/6. Annual or perennial, climbing, trailing or bushy (due to shortening of nodes), usually monoecious herbs with fibrous or tuberous roots. Leaves entire to deeply palmately lobed, commonly hairy, downy or warty. Tendrils simple or forked. Flowers large, yellow, solitary in the leaf-axils. Corolla and calyx campanulate and 5-lobed. Stamens 3, long and converging into a single, conspicuous, often somewhat twisted body. Stigma 1 with 3 bilobed parts. Ovary constricted at the top, 3- or 5-celled. Fruit a berry, sometimes very large, the fibrous interior often breaking away from the hard, durable external wall. Seeds ovate to ovate-oblong, hairless when dried, white to tawny or black to brown, often with a raised margin. *Tropical & subtropical America.* Figure 84, p. 358.

355

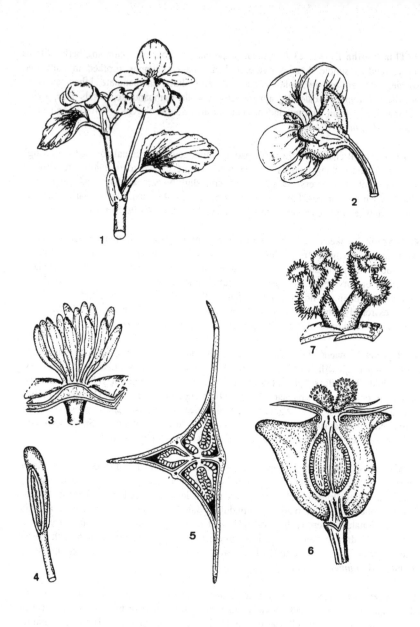

Figure 83. Begoniaceae. *Begonia semperflorens.* 1, Inflorescence. 2, Back view of female flower. 3, Central cluster of stamens of male flower. 4, A single stamen. 5, Transverse section of ovary. 6, Longitudinal section of ovary. 7, Stigmas.

7. Echinocystis Torrey & Gray. 1/1. Monoecious annual vine to 6 m, almost hairless, with parted tendrils. Leaves light green, 5--15 cm, thin, palmately 5-lobed, toothed, lobes triangular, acuminate-mucronate. Flowers 8--10 mm across, greenish white. Calyx narrowly 6-lobed. Corolla-lobes 6. Male flowers in long racemes; stamens 3, united into a column. Female flowers in the same axil, solitary or clustered, ovary 2-celled with 2 ovules in each cell, stigmas 2, large, converging. Fruit 3--5 cm, almost spherical, inflated, membranous, weakly prickly, at first watery and bursting at the top, becoming dry. Seeds 1--1.9 cm, obovate-oblong, flat, slightly 2-toothed at base. *North America.*

8. Sicyos Linnaeus. 15/1. Annual, monoecious herbs, climbing by tendrils. Leaves alternate, shallowly lobed or angular. Male flowers in axillary corymbs; stamens 3, filaments united into a column. Female flowers in axillary, rounded clusters. Calyx campanulate, 5-lobed. Corolla campanulate or with a spreading limb and short tube, 5-lobed almost to base. Ovary with a single cell and ovule. Stigmas 3. Fruit dry, spiny, indehiscent, 1-seeded. *North America, Asia, Australia.*

9. Cucumis Linnaeus. 30/6. Annual or perennial, mostly monoecious, trailing or climbing herbs with bristles or short stiff hairs on stems or leaves. Tendrils simple. Leaves highly variable, especially in size when cultivated, simple, but often angled, lobed or occasionally palmate. Flowers deep yellow, 1--2 cm across; male flowers normally clustered; female solitary. Calyx and corolla deeply 5-lobed. Stamens 3, free, with twisted anthers, with connective produced beyond anthers. Female flowers with staminodes. Disc basal, almost spherical in male flowers, ring-like at the base of the style in female flowers. Ovary ovoid to spherical, downy. Fruit a pepo, indehiscent, firm-walled, fleshy, smooth or spiky, almost spherical to cylindric, green to yellow. Seeds many, compressed, lens-shaped. *Old World tropics.*

10. Mukia Arnott. 4/1. Monoecious climbing herbs, somewhat woody at base. Stems bristly; tendrils simple in cultivated species. Leaves ovate, cordate at base, toothed. Calyx campanulate. Corolla tubular at base. Male flowers in axillary clusters, stamens 3, projecting. Female flowers mostly solitary, style club-shaped, inserted on a ring-like disc. Fruit small, ellipsoid to spherical, red at maturity. *Mostly Old World tropics.*

134. LYTHRACEAE

Herbs, shrubs or trees, with 4-angled or terete stems. Leaves simple, entire, opposite, rarely whorled or alternate; stipules vestigial or absent. Flowers bisexual, often heterostylous, solitary, in axillary clusters, terminal racemes or panicles, bilaterally or radially symmetric, sepals, petals and stamens perigynous. Perigynous zone campanulate or cylindric. Sepals 4--6, sometimes alternating with appendages. Petals 2--6, free, conspicuous or not, inserted on inside of perigynous zone. Stamens as many or twice as many as the petals, or numerous, inserted on inner surface of perigynous zone below the petals. Ovary superior, free, of 2--4 (rarely to 6) fused carpels, often subtended by a curved disc; ovules usually numerous, placentation axile. Fruit a membranous capsule enclosed by the persistent perigynous zone, dehiscing variously or indehiscent; seeds 3--many. Figure 85, p. 360.

A family of 26 genera and 580 species native to the tropics with a few temperate species, of which only a small number is cultivated.

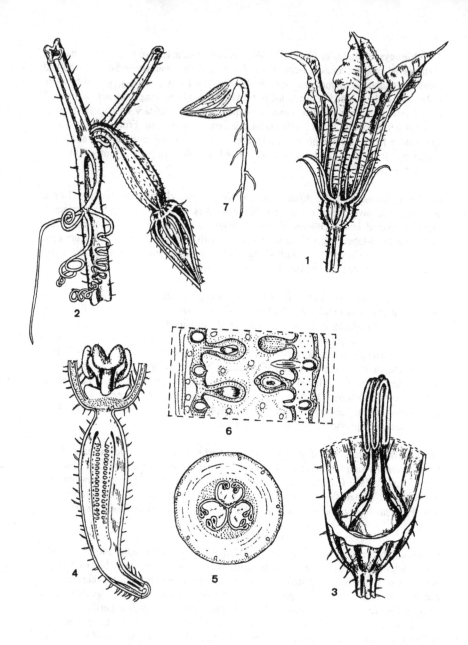

Figure 84. Cucurbitaceae. *Cucurbita pepo.* 1, Male flower from the side. 2, Part of stem with axillary female flower. 3, Lower part of a male flower, perianth cut away. 4, Longitudinal section of ovary. 5, Transverse section of ovary. 6, Longitudinal section of a placenta of a maturing fruit showing seeds at different stages embedded in pulp. 7, Seedling showing peg at base of the hypocotyl.

1a. Perigynous zone campanulate or obconical, about as long as wide 3
  b. Perigynous zone cylindric, about twice as long as wide 2
2a. Flowers radially symmetric; perigynous zone entire in fruit **1. Lythrum**
  b. Flowers bilaterally symmetric; perigynous zone and capsule splitting
    longitudinally in fruit **2. Cuphea**
3a. Flowers in terminal panicles 4
  b. Flowers solitary in leaf-axils or in dichasial cymes 5
4a. Stamens 8 **3. Lawsonia**
  b. Stamens numerous **4. Lagerstroemia**
5a. Perennial herb; inflorescence cymose **5. Decodon**
  b. Shrub; flowers solitary in leaf-axils 6
6a. Petals yellow or yellow-orange **6. Heimia**
  b. Petals pale lilac or rose-purple **2. Cuphea**

1. Lythrum Linnaeus. 38/3. Annual or perennial herbs or small shrubs, with 4-angled or winged, branched, often woody stems. Leaves opposite, alternate or whorled, ovate to linear, entire, stalkless. Flowers radially symmetric, either solitary, paired in leaf-axils, or numerous in clusters in terminal, leafy, false-spikes. Perigynous zone cylindric. Petals deciduous, rose-purple, pink or white. Stamens and styles of unequal length. Fruit a 2-valved capsule. *Old World, North America.* Figure 85, p. 360.

2. Cuphea Browne. 260/3. Annual or short-lived perennial herbs or small shrubs, to 2 m, branching or unbranched, hairless or with glandular hairs on stems and flowers. Leaves opposite or whorled, ovate to lanceolate or elliptic, entire or slightly toothed. Inflorescence a leafy raceme or panicle, terminal or axillary. Flowers bilaterally symmetric, 6-parted, 1--3 at a node. Perigynous zone cylindric. Sepals 6, the upper or lower often larger than the others. Petals 2, 6 or apparently absent, crisped, sometimes deciduous. Stamens mostly equal or alternately unequal. Style projecting, hairless or woolly; ovary subtended by a curved disc. Fruit a membranous capsule, enclosed by the perigynous zone, these splitting longitudinally together. *Central & South America.*

3. Lawsonia Linnaeus. 1/1. Evergreen shrub to 6 m, silvery. Leaves opposite with each pair of leaves at right-angles to those above and below it, elliptic or elliptic-lanceolate, entire, 1.3--5 cm. Inflorescence a pyramidal terminal panicle, to 40 cm; flowers 4-parted, to 5 mm broad, fragrant. Perigynous zone obconical, to 5 mm. Sepals one-quarter of the length of the perigynous zone. Petals kidney-shaped, crisped, white, rose, or cinnabar-red. Stamens 8, shorter than style. Fruit a spherical, indehiscent capsule. *Asia.*

4. Lagerstroemia Linnaeus. 55/1 (selections). Deciduous or evergreen shrubs or trees, to 24 m. Leaves opposite, in 2 rows, obovate to ovate-oblong, elliptic, rounded or oblong-lanceolate, to 19.5 × 8.5 cm, entire, nearly stalkless or stalks to 9 mm. Flowers in terminal panicles. Perigynous zone campanulate. Petals 6, purple or red to white, crisped, clawed. Stamens numerous, projecting. Style long, curved. Fruit a woody dehiscent capsule joined to calyx, with 3--6 valves; seeds winged. *South & eastern Asia, Australia.*

5. Decodon Gmelin. 1/1. Deciduous, perennial herb or small shrub, to 2.5 m; stems somewhat woody, arched, rooting at apex. Leaves lanceolate, 5--12 cm, opposite or

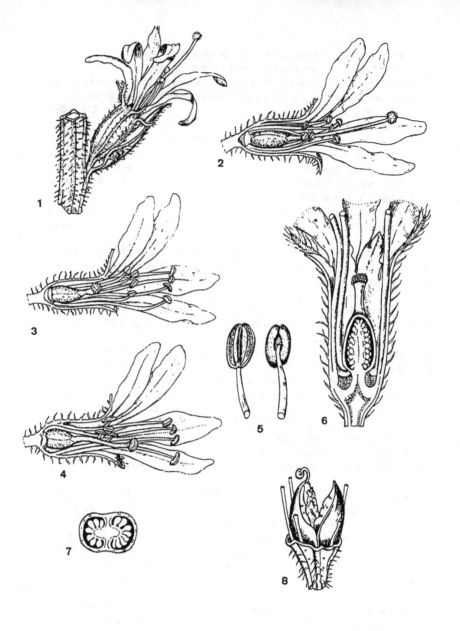

Figure 85. Lythraceae. *Lythrum salicaria.* 1, Flower from the side. 2, 3 & 4, Long-itudinal sections of the three forms of flower (note differing lengths of stamens and styles). 5, Stamen, front and back view. 6, Longitudinal section of lower part of flower. 7, Transverse section of ovary. 8, Fruit.

whorled, hairless above, somewhat downy beneath. Flowers in axillary cymes. Perigynous zone campanulate, bearing 5--7 sepals and as many horned appendages. Petals 4 or 5 (rarely to 7), to 1.3 cm, purple. Stamens 8--10, alternately long and short. Fruit a capsule. *Eeastern USA.*

6. Heimia Link. 3/2. Small, deciduous, hairless shrubs, branches 4-sided. Leaves opposite or alternate, entire. Flowers solitary in leaf-axils, yellow or yellow-orange in cultivated species. Perigynous zone campanulate, with horn-like appendages alternating with sepals. Petals 5--7. Stamens 10--18. Fruit a 4-celled capsule. *South USA to southern South America.*

135. TRAPACEAE

Annual, bottom-rooted, aquatic herbs with floating and submerged leaves. Stems long, mostly simple, flexible, submerged. Leaves opposite below, alternate above; when submerged stalkless, linear, entire, falling off early; when floating, diamond-shaped, in terminal rosettes, glossy above, more or less hairy beneath, with toothed margin; base tapered to truncate; stalk often spongy and partly inflated; stipules absent. Flowers bisexual, radially symmetric, inconspicuous, solitary, axillary, with short stalks. Sepals 4, free. Petals 4, white or lilac, falling early. Stamens 4. Ovary inferior, 2-celled, each cell with 1 axile ovule. Fruit *c.* 4 cm across, woody or bony nut, indehiscent, obconical with 2--4 large spines.

A family consisting of a single genus of about 30 species from temperate to tropical Old World (Europe, Asia, Africa).

1. Trapa Linnaeus. 30/1. Description as for family. *Eurasia, Africa.*

136. MYRTACEAE

Trees and shrubs with abundant, scattered secretory cavities containing aromatic oils. Leaves usually opposite, often leathery; stipules inconspicuous or absent. Flowers bisexual, radially symmetric. Calyx and corolla 3--6-parted (usually 4- or 5-parted), segments free or united, sometimes much reduced, splitting at flowering or falling as a cap. Stamens numerous, free or in bundles, usually bent inwards in bud. Ovary inferior or half-inferior; cells 2--5 (sometimes up to 16); placentation axile; ovules 2--many per cell. Fruit a berry, capsule or drupe. Seeds few to many, endosperm often absent. See figure 86, p. 366.

A family of 120 genera and 3850 species from tropical and subtropical areas, mainly America and Australia.

1a. Fruit a fleshy berry, rarely a drupe, indehiscent; ovary inferior; leaves always opposite	2
b. Fruit a dry capsule or nut-like, dehiscent; ovary inferior to half-inferior; leaves opposite or alternate	9
2a. Flowers borne in racemes, panicles or cymes	3
b. Flowers solitary or in a dichasial inflorescence	4
3a. Calyx 5-lobed	**1. Ammomyrtus**
b. Calyx 4-lobed	**2. Myrtus**
4a. Calyx and corolla 4-lobed	5
b. Calyx and corolla 5-lobed	8
5a. Calyx-lobes unequal	**3. Blepharocalyx**

361

b. Calyx-lobes equal   6
6a. Prostrate mat-forming shrub to 25 cm   **4. Myrteola**
  b. Shrub or small tree to 5 m or more   7
7a. Flowers solitary or in a dichasium; ovary 2--4-celled; fruit yellow-brown to orange-brown when dry   **5. Myrceugenia**
  b. Flowers always solitary; ovary 2- or 3-celled; fruit red, violet or black
           6. Lophomyrtus
8a. Ovary 2-celled   **7. Luma**
  b. Ovary 3--5-celled   **8. Ugni**
9a. Calyx or corolla or both fused to form a cap; leaves not heath-like
           9. Eucalyptus
  b. Calyx or corolla not fused into a cap; petals narrowed to base or leaves heath-like   10
10a. Embryo with cotyledons much smaller than hypocotyl; ovary with 1--3 cells   11
  b. Embryo with cotyledons equalling or longer than hypocotyl; ovary with 1--many cells   13
11a. Ovary 2- or 3-celled   **10. Baeckea**
  b. Ovary 1-celled   12
12a. Stamens numerous, in one or more series   **11. Calytrix**
  b. Stamens 10, alternating with 10 staminodes   **12. Darwinia**
13a. A distinct internode present below final groups of 1, 3 or 7 flowers; flower-stalks distinct; leaves usually relatively large   **13. Metrosideros**
  b. A distinct internode absent below final groups of 1 or 3 flowers; flower-stalks absent; leaves usually narrow or small (if broad then with strong parallel venation and with flowers aggregated in spikes or heads)   14
14a. Stamens not all united or aggregated into bundles on the same radii as the petals   15
  b. Stamens usually more or less united or at least aggregated in bundles on the same radii as the petals   16
15a. Stamens greatly exceeding petals   **14. Kunzea**
  b. Stamens more or less equalling petals   **15. Leptospermum**
16a. Anthers versatile   17
  b. Anthers erect, basifixed   18
17a. Stamens grouped (usually fused) into 5 bundles on the same radii as the petals   **16. Melaleuca**
  b. Stamens not in 5 bundles, usually free but occasionally fused into a short tube at base   **17. Callistemon**
18a. Ovules 1--4 per cell, peltate and laterally attached; leaves usually opposite
           18. Beaufortia
  b. Ovules 2 or more per cell, erect or ascending, linear or wedge-shaped; leaves alternate   **19. Calothamnus**

1. Ammomyrtus (Burret) Legrand & Kausel. 2/1. Evergreen shrubs or small trees with smooth or scaly bark. Leaves opposite, leathery, lanceolate or ovate, strongly aromatic. Flowers solitary or *c.* 6 to raceme-like clusters. Sepals 5. Petals 5. Ovary 2- or 3-celled. Fruit a berry with 4--6 hard, woody seeds. *Central & southern Chile, Andes of southern Argentina.*

The one cultivated species is often found under the name *Myrtus luma.*

2. Myrtus Linnaeus. 2/1. Evergreen shrubs. Leaves small, shining, dark green, opposite, entire with aromatic oil-glands, pinnately veined. Flowers solitary, in axils or in clusters, white or pink. Calyx 4- or 5-lobed, lobes separate, persistent. Petals 4, white, spreading. Stamens numerous, longer than petals. Ovary enclosed in epigynous zone. Style single. Fruit a berry, with persistent calyx-lobes. *Mediterranean area.*

3. Blepharocalyx Berg. 3/1. Evergreen shrubs or small trees. Bracteoles falling early. Inflorescence a dichasium with 3--15 flowers. Calyx 4-lobed, the lobes borne in 2 unequal pairs, margins feathery-hairy towards the apex, persistent. Stamens numerous. Fruit a globe-shaped berry with 1--9 seeds. *Caribbean to southern South America.*

4. Myrteola Berg. 3/1. Evergreen subshrubs or shrubs. Leaves opposite, 4-ranked, leathery. Bracts leaf-like, persistent. Flowers solitary, arising from the leaf-axils. Petals 4 or 5. Stamens numerous. Ovary 2 or 3-celled, the septa between the cells not complete. Fruit a berry, 5--8 mm across. *South America.*

5. Myrceugenia Berg. 40/1. Evergreen shrubs or small trees. Leaves opposite, not strongly aromatic. Bracts 4; bracteoles clasping, persistent until fruit matures. Flowers solitary or in a dichasium. Petals 4. Stamens numerous. Ovary 2--4-celled. Fruit a berry with 1--5 seeds. *Argentina, Brazil, Chile, Paraguay, Uruguay.*

6. Lophomyrtus Burret. 2/2. Shrubs or small trees. Leaves opposite, simple, leathery, glandular. Flowers solitary, axillary with glandular ovary. Sepals and petals in 4s. Calyx-tube with persistent lobes, not exceeding ovary. Stamens numerous; anthers dorsifixed. Ovary inferior, with 2 or 3 cells. Fruit a berry. Seeds kidney-shaped. *New Zealand.*

7. Luma Gray. 2/2. Evergreen shrubs or small trees. Leaves leathery, usually elliptic. Calyx 4-lobed, lobes ovate-triangular to ovate-circular; bracteoles narrow, deciduous. Flowers solitary or in a dichasium of 3. Stamens numerous. Ovary 2-celled; ovules 6--14 per cell. Fruit a fleshy berry with 1--16 seeds. *Chile, Argentina.*

8. Ugni Turczaninow. 5/1. Evergreen shrubs. Leaves opposite, leathery, elliptic to lanceolate. Bracteoles persistent. Flowers solitary, nodding, arising from the axil of a leaf or bract. Calyx usually with 5 lobes. Petals usually 5. Stamens numerous, not exceeding the corolla. Ovary usually 3-celled. Fruit a berry. *Tropical & warm temperate America.*

9. Eucalyptus L'Héritier. 700/few. Evergreen or sometimes deciduous woody shrubs or trees to 90 m. Bark smooth, fibrous, stringy, flaky or tessellated. Leaves usually of differing shapes in seedling, juvenile, intermediate and adult phases; adult leaves alternate, stalked, lanceolate or sickle-shaped, pendent, rarely erect, with distinct midrib. Flowers usually in axillary umbels, sometimes on leafless, second-year branchlets inside crown, or on leafless stalk in upper axils. Flowers 1, 3, 7 or more in each umbel, usually bisexual, calyx and corolla each or together united to form a bud-cap which is shed during the opening of the flower; epigynous zone sometimes present; stamens numerous, white, pink, red, or orange; ovary inferior or partly superior, with 2–7 cells; ovules numerous. Fruit a capsule, within an often woody epigynous zone, opening by flaps to release numerous small seeds; disc convex. *Australia.*

10. Baeckea Linnaeus. 70/1. Prostrate undershrubs to tall shrubs. Leaves small, opposite, entire or slightly lobed. Flowers small, white to pink, solitary, axillary on jointed inflorescence-stalks, each with 2 bracteoles at joint or in clusters on short inflorescence-stalks with a small bract at base of each flower-stalk. Petals 5, broadly obovate or circular, longer than calyx-lobes, spreading. Stamens free, in single row around lower part of epigynous zone or enclosed in it. Ovary with 2 or 3 cells. Style inserted in a tubular or shallow depression in the centre of the ovary. Fruit a small woody capsule, dehiscent. Seeds with embryo with cotyledons much smaller than hypocotyl. *South-east Asia, Australia, Pacific Islands.*

11. Calytrix Labillardière. 40/3. Dwarf shrubs to small trees. Leaves small, alternate, opposite or whorled. Inflorescences 1-flowered, in axils of leaves or bracts. Flower-stalks short. Calyx-lobes 5, each ending in a long rigid or hair-like awn. Petals 5, spreading, deciduous, white, pink, purple, blue, cream or yellow. Stamens numerous, in 1 or more series; filaments often bright yellow. Ovary 1-celled. Fruit a small nut, enclosed in lower part of epigynous zone. Seeds with embryo with cotyledons much smaller than hypocotyl. *Australia.*

12. Darwinia Rudge. 23/1. Dwarf to medium-sized shrubs. Leaves usually opposite, often keeled, 3-angled, oil-glands prominent, aromatic. Flowers small, arranged in terminal heads or pairs, erect or pendent. Bracteoles usually in pairs at base of calyx. Stamens 10, alternating with 10 staminodes; anthers spherical, opening by terminal pores. Style often projecting beyond floral parts, hairy just below stigma. Ovary 1-celled. Fruit a nut. Seeds with embryo, with cotyledons much smaller than hypocotyl. *Australia.*

13. Metrosideros Banks. 20/5. Erect or climbing trees or shrubs. Leaves opposite, leathery, hairy or hairless. Flowers white, red or crimson in terminal cymes. Flower-stalks distinct. Epigynous zone campanulate, obconical or urceolate, fused to base of ovary; calyx-lobes 5, overlapping. Petals 5, spreading. Stamens numerous, longer than petals; filaments thread-like; anthers versatile. Style thread-like. Capsule leathery, enclosed within or projecting beyond persistent epigynous zone, dehiscent. Seeds with embryo, with cotyledons equalling or longer than hypocotyl. *Malaya to New Zealand.*

14. Kunzea Reichenbach. 35/few. Evergreen shrubs or small trees. Leaves alternate or opposite. Flowers usually in head-like clusters, solitary or in 2s or 3s; bracts sometimes forming an involucre. Sepals and petals 5; petals free, white, yellow or pink to purple. Free part of epigynous zone exceeding ovary. Stamens numerous, in 1 or more series, greatly exceeding petals. Ovary inferior with placentation apical or axile. Fruit a capsule or indehiscent, with persistent calyx-lobes. Seeds with embryo, with cotyledons equalling or longer than hypocotyl. *Australia, New Zealand.*

15. Leptospermum Forster. 70/7. Evergreen shrubs or small trees. Bark smooth and flaking, fibrous or papery. Leaves small, alternate, entire, variable in shape, mostly

364

long-elliptic to diamond-shaped, thick, often aromatic, base tapered. Flowers solitary or in 2s or 3s, axillary or terminal, unisexual and bisexual with translucent bracts. Epigynous zone campanulate or obconic, fused to ovary below. Petals 5, spreading, deciduous, white, pink or red. Stamens numerous, free, usually shorter than petals. Ovules anatropous. Fruit a woody capsule with flaps opening at top. Seeds numerous, pendent, ovoid; embryo with cotyledons equalling or longer than hypocotyl. *Australia, New Zealand.*

16. Melaleuca Linnaeus. 220/3. Evergreen shrubs or trees. Leaves opposite, some with 3 or more pronounced longitudinal veins, stalked or stalkless, entire, flat, concave or semi-cylindric, leathery. Inflorescences terminal or axillary spikes, or head-like. Flowers bisexual or male, stalkless. Sepals 5. Petals 5, white, yellowish, pink, red or mauve, free, concave, with short claw. Stamens numerous, projecting, fused into bundles on the same radii as the petals, basal part of bundle flattened into a claw; anthers versatile. Fruit a capsule, woody, remaining on stem and sometimes embedded in it. Seeds with embryo with cotyledons equalling or longer than hypocotyl. *Australia, New Guinea, New Caledonia, Indonesia.*

17. Callistemon Brown. 30/2. Small trees or shrubs; bark fissured or papery. Leaves alternate, entire, stalkless or shortly stalked. Young foliage with silky hairs, reddish or silvery. Inflorescence a cylindric spike, whose axis continues to grow into a leafy shoot. Flowers usually subtended by non-persistent bracts. Calyx and corolla 5-lobed, inconspicuous, calyx fused to campanulate epigynous zone. Stamens much longer than petals, numerous, free or slightly fused at base; anthers versatile. Stigma head-like; ovary inferior, 3- or 4-celled, ovules numerous. Fruit a dry capsule persisting for many years. Seeds with embryo with cotyledons equalling or longer than hypocotyl. *Southern Australia.* Figure 86, p. 366.

18. Beaufortia Brown. 16/3. Shrubs. Leaves small. Inflorescence terminal, flowers many, small. Stamens prominent, in spherical heads or dense brush-like spikes, mauve to red or green. Anthers attached at base, opening by transverse flaps at top. Ovules 1 per cell. Capsules usually fused together around stems, retaining seeds for a number of years. Seeds with embryo with cotyledons equalling or longer than hypocotyl. *Southwestern Australia.*

19. Calothamnus Labillardière. 24/6. Woody shrubs, usually medium-sized to tall. Leaves usually crowded, alternate, oil-glands prominent. Flowers stalkless, in small clusters or dense spikes. Calyx-lobes sometimes persistent. Petals 4 or 5, usually deciduous. Staminal bundles 4 or 5, equal or unequal (sometimes upper 2 fused and lower ones reduced); anthers oblong or linear, cells parallel, opening inwards. Ovary inferior, 3-celled; ovules 2 or more per cell. Fruit a capsule. Seeds with embryo, with cotyledons equalling or longer than hypocotyl. *West Australia.*

137. PUNICACEAE

Deciduous, densely branched, somewhat spiny shrubs or small trees to 6 m. Leaves 2--8 cm, almost opposite, simple, oblong to obovate, entire, hairless, pale green, shiny above; stipules absent. Flowers bisexual, radially symmetric, solitary or in clusters of 2--4, stalkless, 2.5--4 cm across. Calyx, corolla and stamens epigynous. Calyx funnel-shaped, with 5--7 lobes, red. Petals 5--7, crinkled, orange-red or white.

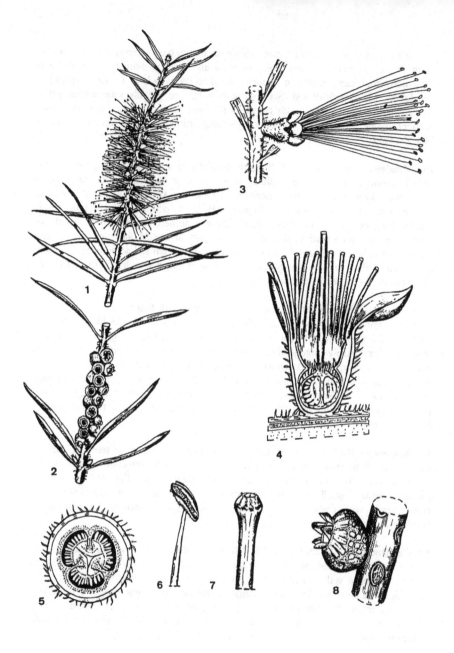

Figure 86. Myrtaceae. *Callistemon citrinus.* 1, Part of shoot showing stalkless flowers in the dense inflorescence. 2, The same in fruit. 3, Flower from the side. 4, Longitudinal section of base of flower. 5, Transverse section of ovary. 6, Stamen. 7, Stigma. 8, Fruit.

Stamens many. Ovary usually with 8--12 cells, placentation axile. Fruit 5--10 cm across, spherical, with a hard, leathery, yellowish brown to red rind, the apex crowned by the persistent calyx; seeds many, each embedded in sweet, juicy pulp.
A family of one genus originally from Asia.

1. Punica Linnaeus. 2/1. Description as for family. *West & south-west Asia; now naturalised throughout the Mediterranean area.*

138. ONAGRACEAE
Herbs and shrubs, rarely trees. Leaves alternate, opposite, whorled or basal, simple and entire to pinnatifid; stipules sometimes present. Flowers usually bisexual, radially symmetric, solitary or in spikes, racemes or panicles, usually with a long epigynous zone (floral tube in much of the literature). Sepals usually 4, rarely 2--7, free or united at base, deciduous or sometimes persistent. Petals generally 4, rarely 2--7 or absent, free, often clawed. Stamens twice as many as sepals or as many, rarely 1; anthers opening by slits. Ovary inferior, of 4 fused carpels and 4-celled, or half-inferior and 2-celled, ovules few to many per cell, placentation axile or parietal. Fruit a capsule, berry or nut. Seeds usually many, sometimes each with a tuft of hairs. See figure 87, p. 370.
A cosmopolitan family of 16 genera and about 650 species, with particular concentrations of diversity in temperate and warm America.

1a. Sepals persistent; epigynous zone absent; petals quickly deciduous
                                                  4. Ludwigia
  b. Sepals deciduous with corolla; epigynous zone generally conspicuous, rarely absent; petals not quickly deciduous   2
2a. Petals and sepals 2; fruit covered with hooked hairs   **3. Circaea**
  b. Petals and sepals 3 or 4; fruit not covered with hooked hairs   3
3a. Fruit a fleshy berry; flowers mostly red   **1. Fuchsia**
  b. Fruit dry; flowers mostly yellow, white, pink, rarely red   4
4a. Stamens 1 (with staminode) or 2; fruit a dry spherical or club-shaped capsule opening at apex   **2. Lopezia**
  b. Stamens 8 or 4; fruit mostly dry cylindric capsules, capsule, rarely 4-winged, sometimes indehiscent   5
5a. Fruit a short, indehiscent, nut-like capsule, often shortly stalked   **9. Gaura**
  b. Fruits dehiscent, mostly cylindric capsules, rarely shortly stalked   6
6a. Seeds with tuft of hairs attached to one end; petals pink or rose-purple to white, very rarely cream-yellow   7
  b. Seeds lacking tufts of hairs; petals mostly yellow or white, sometimes pink   8
7a. Leaves spirally arranged; epigynous zone absent   **11. Chamerion**
  b. Leaves opposite at least below inflorescence; epigynous zone present
                                                  10. Epilobium
8a. Stigma 4-lobed   9
  b. Stigma entire, head- or disc-like   10
9a. Anthers attached at base; petal-tip often lobed, or petal-base clawed
                                                5. Clarkia
  b. Anthers attached at middle, petal-tip notched or toothed, not lobed, petal-base not clawed   **6. Oenothera**

10a. Stigma head-like or hemispheric   **7. Camissonia**
    b. Stigma disc-like   **8. Calylophus**

1. Fuchsia Linnaeus. 110/20 and many hybrids. Erect to climbing shrubs, epiphytes, small trees or prostrate creepers. Leaves simple, alternate, opposite or whorled, with small, generally deciduous stipules. Flowers mostly bisexual, when unisexual plants usually dioecious; radially symmetric, stalked, axillary or in racemes or panicles. Epigynous zone cylindric or obconic, deciduous in fruit. Sepals 4. Petals 4 or absent. Stamens 8, unequal, those opposite sepals usually longer than those opposite petals, all erect, or those opposite petals reflexed within epigynous zone. Anthers oblong or kidney-shaped, 2-celled. Nectary at base of epigynous zone, lobed or not. Stigma usually projecting, head-like, club-shaped, or 4-lobed. Ovary 4-celled. Fruit a berry; seeds 8--400 per fruit. *Central & South America.* Figure 87, p. 370.

2. Lopezia Cavanilles. 22/1. Annual or perennial herbs, sometimes woody, often with tuberous roots; stems branched, hairy or hairless. Leaves stipulate, alternate or opposite. Flowers in terminal racemes, with more or less stalkless, leaf-like bracts; bilaterally symmetric, usually anthers ripening before stigmas. Epigynous zone absent or present in a few species. Sepals 4, more or less equal, lanceolate, green or red; petals 4, mostly unequal, the upper 2 usually free or basally fused with upper 3 sepals in some species, red, orange, pink, purple, or white, sometimes with glands or hairs on margins, or other appendages. Stamens 2, usually 1 fertile, the other sometimes enveloping the fertile stamen and releasing it explosively, shedding pollen; stigma more or less head-like, papillose. Fruit a club-shaped or spherical dry capsule. Seeds mostly papillose, rarely winged. *Mexico, central America.*

3. Circaea Linnaeus. 8/2 and their hybrid. Herbaceous perennials, from rhizomes, sometimes with stolons tipped by tubers. Leaves simple, opposite, with acute apices and mostly rounded bases, stalked, entire to toothed. Inflorescence a raceme or panicle, with bracts. Flowers bilaterally symmetric. Sepals 2, spreading or reflexed. Petals 2, erect, alternate with sepals. Stamens 2, opposite sepals, pollen yellow, grains shed singly. Ovary 1 or 2-celled, ovules 1 per cell. Fruit club-shaped, indehiscent, falling with stalk, covered with hooked hairs. *Northern hemisphere.*

4. Ludwigia Linnaeus. 82/10. Perennial and annual herbs or subshrubs (small trees), sometimes floating on water or creeping and rooting at nodes. Leaves alternate or opposite, simple; stipules present, generally deciduous. Inflorescence a bracteate spike or flowers solitary in leaf-axils. Flowers radially symmetric; epigynous zone absent. Sepals 4 or 5 (rarely to 7), persistent on fruit. Petals as many as sepals or absent, yellow or white; stamens as many as sepals, or twice as many, in 2 series. Ovary 4- or 5-celled; stigma club-shaped or spherical. Fruit a thick- or thin-walled cylindric to spherical capsule, irregularly dehiscing. *Cosmopolitan.*

5. Clarkia Pursh. 41/6. Annual herbs, with prostrate to erect stems. Leaves simple, pinnately veined, linear to elliptic or ovate, entire or finely toothed, hairless to sparsely downy, stalkless or with stalks as long as blades. Inflorescence a leafy spike or raceme, nodding in bud and later straightening, or erect. Flowers 4-parted; sepals often reddish, reflexed singly, in pairs, or united at tips. Corolla often bowl-shaped, petals oblanceolate to obovate, often lobed or clawed, lavender or pink to dark red,

368

sometimes yellow or white, often spotted, flecked or streaked with red, purple, or white. Stamens 8 in 2 whorls, or reduced to 1 series of 4. Ovary cylindric or spindle-shaped, often 4- or 8-grooved; stigma 4-lobed. Fruit generally a dry capsule, usually opening through the cells, rarely indehiscent and nut-like. Seeds many (rarely 1 or 2), angled, tan to dark brown. *Western North America, 1 species in southwestern South America.*

6. Oenothera Linnaeus. 120/30. Erect, ascending, or rarely decumbent annual to perennial herbs. Leaves in a basal rosette or alternate, entire, toothed or pinnatifid, stipules absent. Flowers showy, solitary in leaf-axils or in corymbose, racemose or spike-like inflorescences, radially symmetric, 4-parted, usually opening in the evening. Epigynous zone cylindric with a flared apex, deciduous. Petals white, yellow or rose to purple, becoming orange, pink or red. Stamens 8. Stigma usually with 4 linear lobes, the lobes, otherwise disc-like. Ovary inferior, 4-celled. Fruit a long capsule, terete or 4-angled, winged or not, straight or curved, dehiscent. Seeds many, arranged in rows or not. *New World; introduced elsewhere.*

7. Camissonia Link. 62/1. Annual or perennial herbs (rarely subshrubs), from taproot or lateral roots. Leaves basal, on the stems, or both, alternate, simple to bipinnate. Inflorescence with bracts; flowers in a spike, raceme or solitary, radially symmetric, usually 4-parted, usually opening near sunrise. Sepals reflexed. Petals yellow, white, lavender, often with darker basal spots, fading purplish. Stamens 8 in 2 unequal sets, or rarely one set of 4; stigma head-like or hemispheric. Fruit a capsule opening through the cells, straight to coiled, generally stalkless. Seeds in 1 or 2 rows in each of the 4 cells. *Western North America, adjacent Mexico, Peru, Chile & Argentina.*

8. Calylophus Spach. 6/1. Herbaceous or shrubby perennials, often from a woody stock. Stems almost prostrate or decumbent to erect. Leaves alternate, entire or with forwardly pointing teeth, stalkless or nearly so. Flowers from axils of upper leaves, radially symmetric, 4-parted, with long epigynous zone above ovary. Sepals greenish yellow, often with reddish spots or streaks; petals reflexed at flowering, yellow but often fading to reddish, orange or purplish. Stamens 8, more or less equal or in 2 unequal sets of 4 each; stigmas disc-like. Capsules stalkless, cylindric, with 2 rows of seeds in each of 4 cells. *Southern North America.*

9. Gaura Linnaeus. 21/1. Annual, biennial, or perennial herbs from woody stock, rhizome, or taproot. Leaves basal and on stems, basal usually pinnatifid, with winged stalks, stem-leaves alternate, stalkless, margin generally wavy-toothed, hairy at least on veins and margins. Inflorescence a terminal, bracteate spike. Flowers generally bilaterally symmetric, opening at dusk or dawn, 4- (rarely 3-) parted. Epigynous zone 6--15 mm, narrow. Sepals opening wide. Petals white or yellow, fading reddish, clawed. Stamens 8, more or less equal. Ovary with 4 cells (1 in fruit); stigma deeply 4-lobed, mostly projecting beyond anthers. Fruit an indehiscent, nut-like capsule, erect, mostly 4-angled or -winged, often shortly stalked. Seeds mostly 3 or 4, ovoid, angled, 2--3 mm, pale brown. *Temperate North America to Central America.*

10. Epilobium Linnaeus (*Boisduvalia* Spach, *Zauschneria* Presl). 164/15. Perennial herbs with rosettes, shoots, or stolons from rootstock, annuals from taproot, or rarely subshrubs. Leaves opposite at least below inflorescence, simple, often finely toothed to

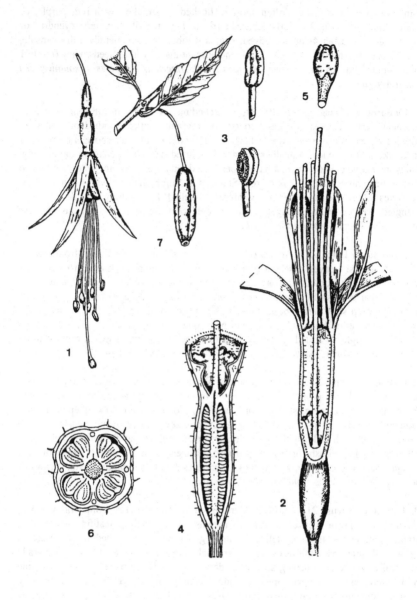

Figure 87. Onagraceae. *Fuchsia magellanica.* 1, Pendulous flower. 2, Longitudinal section of lower part of flower. 3, Stamen undehisced (above) and dehisced (below). 4, Longitudinal section of ovary. 5, Stigma. 6, Transverse section of ovary. 7, Fruit.

more or less entire. Inflorescence a bracteate spike or raceme, generally erect. Flower radially symmetric, rarely slightly bilaterally symmetric, 4-parted; epigynous zone usually short, rarely more than 3 cm. Sepals erect, green to reddish. Petals each with an apical notch, rose-purple to pink or white, rarely cream or red-orange. Stamens 8 in 2 unequal sets, anthers attached at middle. Ovary inferior, 4-celled; stigma entire, club-shaped or 4-lobed. Fruit a straight, cylindric or club-shaped, mostly many-seeded dry capsule. Seeds mostly with tufts of hairs (rarely without). *Almost cosmopolitan.*

As now understood, the genus includes *Boisduvalia* and *Zauschneria* as sections, but the Willow-herbs (Fireweeds) are now segregated as *Chamerion.*

11. Chamerion (Rafinesque) Rafinesque (*Chamaenerion* Séguier). 8/4. Perennial herbs, often clumped, basally woody, with few to many erect stems, rarely branched, hairy to more or less hairless. Leaves spirally arranged, simple, the basal ones small, scale-like and leathery; stipules absent. Inflorescence a more or less erect, simple raceme; bracteoles absent. Flowers bisexual, 4-parted, slightly bilaterally symmetric; epigynous zone absent. Petals entire. Stamens 8, more or less equal in single whorl, erect at outset of flowering, recurved later; pollen bluish to yellow. Style deflexed at outset of flowering, straightening later; stigma deeply 4-lobed, the lobes becoming gradually curled under. Fruit a long, slender, 4-celled capsule, opening through the cells. Seeds many, with terminal tuft of silky hairs. *Colder parts of Europe, Asia & North America.*

139. HALORAGACEAE

Aquatics or herbs of moist terrestrial habitats, annuals or perennials, sometimes with persistent woody base. Leaves spirally arranged, opposite and in pairs that alternately cross each other at right-angles, thus making 4 rows, alternate or in whorls, entire to lobed, finely pinnate when submerged; stipules absent or present. Flowers inconspicuous, bisexual or unisexual, radially symmetric, usually very small, solitary in leaf-axils, in axillary dichasia or in terminal spikes, racemes or panicles. Perianth and stamens epigynous. Calyx-lobes 2--4 or absent. Petals free, 2--4, concave, often hood-shaped, or absent. Stamens 8, in 2 series of 4 or 2--6, anthers basifixed. Ovary inferior, of 2--4 fused carpels, 1-, 2- or 4-celled with pendent ovules. Styles 1--4, free with feathery or papillose stigmas. Fruit usually a very small nut or drupe, indehiscent or splitting into 1-seeded nutlets.

A family of 9 genera and 120 species, cosmopolitan in distribution but chiefly in the southern hemisphere.

    1a. Aquatic herbs, submerged except for inflorescences; leaves without
        stipules, whorled, pinnate with thread-like segments; ovary 4-celled;
        fruit a nut or drupe separating into 1-seeded nutlets   **1. Myriophyllum**
      b. Terrestrial marsh herbs; leaves circular to ovate, stalked, with stipules;
        ovary 1-celled; fruit a leathery or fleshy indehiscent drupe   **2. Gunnera**

1. Myriophyllum Linnaeus. 45/2. Annual or perennial aquatic herbs, submerged except for sometimes emergent inflorescences, hairless, stems erect or curving upwards. Leaves in whorls of 3--6, opposite or alternate, pinnatisect, emergent leaves and bracts sometimes simple, toothed or entire. Flowers bisexual, unisexual or sometimes male and bisexual on the same plant (rarely males and females on separate plants), in terminal spikes, stalkless in axillary whorls or solitary in leaf-axils. Calyx

4-lobed in male flowers, minute in female flowers. Petals 2--4 in male flowers, minute or absent in female flowers. Stamens 8. Ovary 4-celled. Styles 2--4. Stigmas oblong, recurved. Fruit grooved, splitting into 2--4 nutlets. *Cosmopolitan.*

2. Gunnera Linnaeus. 50/5. Very large clump-forming or very short mat-forming, rhizomatous perennial herbs. Leaves mostly alternate, circular to ovate, often lobed, stalked, with stipules. Inflorescence a spike or panicle; upper flowers male, central flowers bisexual and lower flowers female, or occasionally plant dioecious; flowers minute, usually green or yellow. Calyx-lobes 2 or 3 or absent. Petals 1 or 2 or absent. Stamens 1 or 2. Ovary 1-celled with 1 pendent ovule. Fruit a leathery or fleshy drupe. *South America, South Africa, eastern Asia, Australia (Tasmania), New Zealand.*

140. HIPPURIDACEAE
Perennial aquatic herb with a creeping rhizome bearing erect leafy shoots. Leaves simple, entire, clustered in whorls, if submerged then linear, pale and flaccid, if emergent then broader, darker and rigid. Flowers inconspicuous, minute, solitary in the leaf-axils of emergent leafy shoots, which die after flowering, very variable, bisexual, unisexual or apparently sterile. Perianth much reduced or absent. Stamen 1. Ovary inferior, of 1 carpel; style long and slender, papillose. Fruit a smooth, ovoid, 1-seeded nutlet.

A family containing a single genus.

1. Hippuris Linnaeus. 1/1. Stem length dependent on water depth. Leaves 6--12 per whorl, the submerged to 10 cm, the emergent 5--40 mm. Flowers pinkish. *Widely distributed in the cooler parts of the northern hemisphere, Argentina, Antarctica.*

141. ALANGIACEAE
Deciduous or evergreen trees, shrubs or lianas, with latex, sometimes thorny. Leaves alternate, margins entire or lobed. Flowers in axillary cymes, whitish. Calyx small, campanulate, sepals 4--10, sometimes almost absent. Petals 4--10, linear, usually slightly reflexed, sometimes fused at base. Stamens 4--30. Ovary inferior, 2-celled and long; ovules solitary, pendent in each cell or one cell empty. Fruit a small rounded drupe. Seed solitary; endosperm copious.

A family of one genus mostly from the Old World tropics.

1. Alangium Lindley. 18/2. Description as for family. *Central Africa, China to Eastern Australia.*

142. NYSSACEAE
Deciduous trees or shrubs. Leaves simple, entire, alternate; stipules absent. Flowers unisexual or bisexual, radially symmetric. Male flowers in heads, racemes or umbels. Perianth and stamens epigynous. Sepals absent or 5, minute. Petals absent or 5--8, overlapping. Stamens 5--16. Female or bisexual flowers solitary or in heads of 2--12 flowers. Sepals 5. Petals overlapping, very small, fleshy, much larger than in male flowers. Ovary 1-celled, inferior; ovules 1 per cell, placentation axile. Style simple or divided. Fruit an ovoid or elliptic drupe or samara.

A family of 3 genera and 8 species from Eastern North America and China.

1. Nyssa Linnaeus. 10/2. Deciduous trees. Leaves stalked, crowded at ends of branches. Flowers minute, greenish white. Male flowers numerous, slender-stalked; sepals ovate or linear-oblong; petals 5, alternate; stamens 8--16 in 2 alternate whorls, protruding, disc cushion-shaped, entire or lobed. Female flowers solitary or in clusters; petals 5--8, often minute; style cylindric, tapering, with 2 adpressed (becoming divergent) branches; calyx tubular-campanulate or slightly urn-shaped; disc depressed in centre. Fruit a fleshy drupe with a 1-seeded stone. *North America, China.*

143. DAVIDIACEAE

Deciduous trees to 20 m. Branches ascending and spreading, lower branches horizontal in mature trees. Shoots of young plants dark red. Leaves 7.5--15 cm, ovate or roundish, acuminate, base cordate, upper surface with silky hairs, lower surface felted with thick grey down or hairless with tufts of hair in vein-axils, margins toothed; leaf-stalk 3.8--5 cm; young leaves strongly scented. Flowers in a rounded head *c.* 1.8 cm across, terminal, on drooping stalk *c.* 7.5 cm. Bracts 2 or 3, white or creamy white, boat-shaped, oblong, long-pointed; lower bract *c.* 20 × 10 cm; upper bract *c.* 10 × 5 cm, forming a canopy over flower-head. Central flower of head bisexual, the others male. Perianth absent. Stamens 1--7, epigynous in bisexual flowers, ovary inferior, with 6--10 cells with a single axile ovule in each cell. Fruit a solitary drupe *c.* 3.8 cm with a hard, ridged nut containing 3--5 seeds.

A family of a single genus.

1. Davidia Baillon. 1/1. Description as for family. *Western China.*

144. CORNACEAE

Trees and shrubs, rarely herbs. Leaves opposite or occasionally alternate, simple, sometimes evergreen; stipules absent. Inflorescences usually corymbs or umbels, sometimes surrounded by large showy bracts. Flowers small, bisexual or, if unisexual, those of each sex borne on separate plants, radially symmetric. Calyx-lobes 4 or 5. Petals 4 or 5, free, rarely absent. Stamens equal in number to and alternating with petals. Anthers 2-celled, opening lengthwise. Carpels 1--4, fused, forming a 1--4-celled inferior ovary, each cell with 1 pendent ovule; placentation axile or parietal. Style simple with lobed stigma. Fruit a drupe or berry, with 1--4 cells and 1 or 2 stones. See figure 88, p. 376.

A family of 12 or 13 genera and about 100 species distributed mainly in the north temperate area with some species in the tropics and subtropics. This family is sometimes split into several segregate families, with genus **1** being placed in *Griseliniaceae*, **2** in *Cornaceae* in the restricted sense, **3** in *Helwingiaceae*, **4** in *Aucubaceae*, and **5** in *Escalloniaceae* (the rest of which is here considered as being part of the *Saxifragaceae*).

1a. Petals overlapping in male flower-buds; leaves evergreen, shining
                                                      1. Griselinia
  b. Petals edge-to-edge in male flower-buds; leaves usually not as above   2
2a. Fruit a drupe   3
  b. Fruit a berry   4
3a. Ovary 2-celled; flowers in cymes, umbels, panicles or a compact head
    terminating the branches or in the leaf-axils   **2. Cornus**
  b. Ovary 3--4-celled; flowers in inconspicuous clusters on the upper

side of the midrib of the leaf   **3. Helwingia**
4a. Petals 4; flowers maroon to green   **4. Aucuba**
  b. Petals 5; flowers yellow   **5. Corokia**

1. Griselinia Forster. 8/4. Evergreen trees and shrubs, dioecious. Leaves alternate, asymmetric, leathery, sometimes sharply toothed, leaving prominent leaf-scars. Inflorescence an axillary panicle or raceme. Male flowers minute: calyx 5-toothed; petals 5 overlapping; stamens 5, alternate with petals; disc fleshy and 5-angled. Female flowers: calyx 5-toothed; petals absent or 5, not overlapping; styles 3; stamens rudimentary or absent. Fruit a 1- or 2-celled berry with 1 seed. *New Zealand, Chile, south-east Brazil.*

2. Cornus Linnaeus. 45/27. Perennial semi-woody herbs, shrubs or small trees, deciduous or evergreen. Leaves opposite or alternate, simple, entire with conspicuous veins and sometimes adpressed hairs. Inflorescence a terminal cyme, umbel, panicle or compact head; bracts 1--4. Flowers usually bisexual, white, green-white or yellow. Sepals 4. Petals 4. Fruit a drupe, 2-celled, 2-seeded, white, blue or in shades of red through to black. *Temperate northern hemisphere.* Figure 88, p. 376.

3. Helwingia Willdenow. 3/1. Deciduous shrubs, hairless, dioecious. Leaves simple, alternate, toothed, with free, thread-like stipules. Flowers in inconspicuous clusters on the midrib of upper side of leaves. Petals 3--5, mostly 4; stamens 3--5; disc in male flower flat, depressed-conical in female; ovary 3- or 4-celled; styles fused; stigmas 3--5. Fruit a drupe with 1--4 stones. *Himalaya, eastern Asia.*

Grown as a botanical curiosity in that the inconspicuous flowers and showy fruits are borne on the leaves.

4. Aucuba Thunberg. 3/2. Evergreen shrubs, dioecious. Leaves opposite, leathery, simple, stalked, toothed, without stipules. Flowers in terminal panicles or cymes. Sepals and petals 4; male with petals brownish, edge-to-edge, ovate or lanceolate, disc fleshy; female with style thick, short, stigma entire, ovary 1-celled, ovules pendent; placentation parietal. Fruit an ellipsoid, shiny berry, red, white or yellow, crowned with the calyx-teeth and style; stones oblong. *Himalaya, eastern Asia.*

5. Corokia Cunningham. 3/3 and a hybrid. Evergreen shrubs with alternate leaves. White or silvery T-shaped hairs present on stems, on leaf undersides, flower-stalks, backs of petals and calyx-lobes. Flowers yellow, in a terminal or axillary panicle, raceme or cluster, or solitary. Ovary top-shaped. Sepals, petals and stamens 5, the petals usually each with a small incised scale at the base. Ovary inferior, 1- or 2-celled; style slender, with a somewhat 2-lobed stigma. Fruit a spherical or oblong berry with persistent sepals. *New Zealand.*

145. GARRYACEAE

Evergreen shrubs or trees. Leaves simple, stalked, without stipules. Male and female flowers borne on separate plants; flowers inconspicuous, small, clustered in catkin-like, hanging inflorescences; inflorescences simple or branched, clustered at tips of leafy shoots. Floral bracts in male and female catkins in pairs, fused at least at base, to form cup-shaped or boat-shaped structures protecting the flowers, pairs of bracts alternating along axis of catkins. Male flowers concealed in axils of bracts; each flower on a short stalk, with 4 perianth-segments, joined at apex but free below and 4 stamens alternating with perianth-segments. Female flowers concealed in axils of bracts, stalks minute; flowers composed of a 2-chambered ovary with 2 persistent styles; ovules 2, apical. Fruit a berry, dry at maturity; cluster of berries black, dark blue or grey at maturity.

A family comprising a single genus restricted to North and Central America and the Caribbean Islands.

1. Garrya Lindley. 14/3 and some hybrids. Description as for family. *North & Central America, Caribbean Islands.*

146. ARALIACEAE

Trees, shrubs, climbers or perennial herbs, evergreen or deciduous, many more or less prickly. Adult leaves usually stalked, spirally arranged or alternate (rarely opposite), simple or pinnately or palmately lobed or compound, often crowded towards or in conspicuous rosettes at ends of branches; juvenile (and intermediate) leaves sometimes distinct from those of adults, with gradual to abrupt transitions. Inflorescences frequently large, simple or compound and then paniculate or umbellate. Flowers generally small, bisexual or unisexual, occurring singly or in small umbels or heads, stalked or not. Sepals very small or almost absent. Petals usually 5 or more (rarely fewer), usually spreading at opening of flower but sometimes wholly or partly fused and then falling as a cap. Stamens 5 or more, the anthers usually with 4 (rarely 8) cells; ovary wholly or largely inferior (rarely superior) with 2 (occasionally 1) to 15 (rarely up to 90) cells each with 1 axile ovule. Fruit usually fleshy and drupaceous (occasionally developing as a schizocarp with 2 mericarps), with 1 or more pyrenes; styles persistent, free or partially or wholly united into a more or less conspicuous stylar column or more or less entirely reduced. See figure 89, p. 380.

A family of 50 or so genera and at least 1400 species, mainly in the tropics (including tropical montane areas), subtropics and warm temperate areas of both northern and southern hemispheres, with the greatest generic diversity in Asia, Malesia, Australia and the Pacific Islands.

1a. Plants with herbaceous annual growth, dying back at end of season		2
b. Plants woody, with aerial buds		4
2a. Stems very prickly, almost woody	**1. Oplopanax**	
b. Stems unarmed, herbaceous or mostly so		3
3a. Leaves more than once compound; ovary 5-celled	**2. Aralia**	
b. Leaves once compound; ovary 2-celled	**3. Panax**	
4a. Plants with spines		5
b. Plants without spines		7
5a. Leaves bipinnate; petals overlapping in bud	**2. Aralia**	
b. Leaves palmately lobed or compound; petals edge-to-edge in bud		6
6a. Shrubs or small trees, often much-branched, or scrambling; leaves divided into 3 or more distinct leaflets; inflorescences relatively small	**4. Eleutherococcus**	
b. Small to medium-sized trees, not or little-branched when young; leaves palmately lobed or apparently so with a hub at the top of the common stalk and 5 or more distinct segments	**5. Kalopanax**	
7a. Climbers or plants sprawling on ground		8

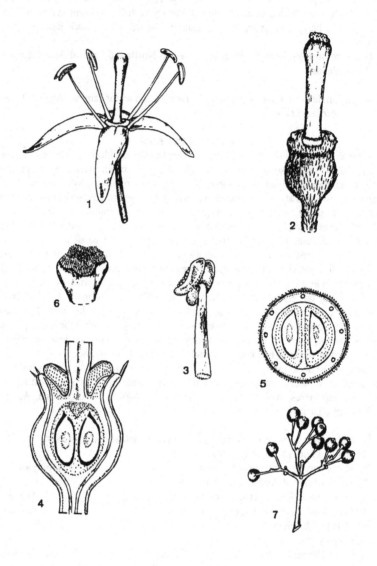

Figure 88. Cornaceae. *Cornus sanguinea.* 1, Flower. 2, Flower with petals and stamens removed. 3, Stamen. 4, Longitudinal section of ovary. 5, Transverse section of ovary. 6, Stigma. 7, Fruits.

b. Self-supporting trees or shrubs   9
8a. Leaves of 2 different kinds; adult leaves usually without lobes; fruit black, blue-black or orange at maturity   **6. Hedera**
  b. Leaves not of 2 different kinds; adult leaves with 5 lobes; fruit never produced   **7. × Fatshedera**
9a. Leaves divided into distinct leaflets   10
  b. Leaves simple or merely lobed   14
10a. Leaflets 3   **4. Eleutherococcus**
  b. Leaflets 4 or more   11
11a. Leaflets pinnatifid to palmatifid and segmented; flowers tightly aggregated into long, pincushion-like spikes on thick branches in large inflorescences; ovary 2-celled   **8. Cussonia**
  b. Leaflets entire or finely or coarsely toothed or occasionally somewhat pinnatifid or with large projections; flowers in small umbels or heads, rarely spike-like or racemose and then on slender inflorescence branches; ovary 2--75-celled   12
12a. Leaf-stalk without sheathing base   **9. Pseudopanax**
  b. Leaf-stalk with sheathing base   13
13a. Flower-stalks jointed below ovary; ovary 2-celled   **9. Pseudopanax**
  b. Flower-stalks not jointed below ovary; ovary 5--75-celled   **10. Schefflera**
14a. Leaves entire or with up to 3 lobes   15
  b. Leaves with 5 or more lobes or leaflets   19
15a. Leaves narrow to very narrow, sometimes toothed   16
  b. Leaves at least 5 cm wide   17
16a. Leaves very thin (juvenile stage only)   **11. Meryta**
  b. Leaves rather thick, stiff, more or less toothed   **9. Pseudopanax**
17a. Leaves with pinnate venation, large   **11. Meryta**
  b. Leaves with 3 veins   18
18a. Margins of leaves irregularly toothed; leaves entire to 3-lobed   **12. Metapanax**
  b. Margins of leaves entire   **13. Dendropanax**
19a. Leaves very large, hairy, soft in texture   **14. Tetrapanax**
  b. Leaves moderately large to large, hairless at least on the upper surface; texture firm   **15. Fatsia**

1. Oplopanax (Torrey & Gray) Miquel. 3/1. Deciduous shrubs with annual growth to 4 m; prickles to 1.3 cm, very sharp. Leaves bright green, simple, palmately veined, variously toothed and more or less lobed, the upper surface hairless, the undersurface usually slightly hairy. Inflorescences near branch ends, woolly-hairy, narrowly conic; flowers in small, umbel-like clusters, greenish white; stalks not jointed. Calyx-rim very short, obscurely 5-toothed. Petals 5, edge-to-edge in bud. Stamens 5. Ovary 2-celled, the styles mostly free. Fruit remaining inferior through maturity, drupe-like, red, round, somewhat compressed laterally. *Northeast Asia, western & central North America.*
Formerly widely known as *Echinopanax.*

2. Aralia Linnaeus. 55/10. Deciduous or evergreen rhizomatous herbs, subshrubs, shrubs or small to medium-sized trees (rarely climbers), often prickly or thorny. Aerial parts annually dying back or persistent and in a few species developing a substantial trunk. Leaves spirally arranged to alternate, simple to 3--5-partite to 3 times pinnately

377

compound. Leaflets thin, margins usually toothed. Inflorescences terminal, simple to compound, sometimes large and conspicuous. Flowers usually white or greenish, in umbels or heads; calyx obscurely 5-lobed; petals usually 5, overlapping in bud; stamens 5; ovary with 2--5 (rarely to 8) cells. Fruit drupe-like, usually black at maturity with 2--8 pyrenes. *North & South America, eastern & southern Asia.*

3. Panax Linnaeus. 5/2. Perennial herbs with thickened or tuberous roots, spreading by offsets. Stems annual, terminating in a solitary umbel (this sometimes accompanied by 1--6, stalked lateral umbels), unbranched, to 1 m, arising from a scaly stock, bearing a single whorl of generally 2--4 (sometimes as few as 1 or as many as 6) palmately compound leaves, each with 3--7 stalkless or stalked leaflets. Leaflets singly or doubly toothed, hairless or with scattered or abundant hairs along the veins. Umbels with bisexual and/or male flowers. Flower-stalks of bisexual flowers jointed at top. Calyx 5-toothed. Petals 5, slightly overlapping in bud. Stamens 5. Ovary 2- or 3-celled; styles in bisexual flowers free, in male ones united and non-functional. Fruit spherical or somewhat compressed, yellow or red, remaining inferior, topped by the persistent styles and with 2 or 3 pyrenes. *East, south & southeast Asia, eastern North America.*

4. Eleutherococcus Maximowicz. 30/11. Generally deciduous shrubs or small trees, sometimes more or less climbing or sprawling and often bristly or prickly; branches generally slender, sometimes cane-like. Leaves stalked, usually palmately compound (rarely simple), with 3--5 more or less toothed leaflets; bases more or less sheathing the twigs but without stipules. Inflorescences terminal, simple or compound, often (and sometimes only) on side or short shoots. Flowers in umbels or heads, solitary or in clusters with the central or terminal unit always larger; stalks not jointed at the base of the ovary; petals usually 5, free, edge-to-edge in bud, spreading at opening of flowers; stamens as many as the petals; ovary with 2--5 cells; styles free or variously united. Fruit remaining inferior, drupe-like, black or purplish black at maturity, ellipsoid or round, or more or less compressed laterally; styles persistent. *Eastern & central Asia, south to Vietnam & Philippines.*

5. Kalopanax Miquel. 1/1. Sparingly branched trees, branches and trunk prickly, particularly in young growth. Leaves to 35 cm across but usually less; stalk to 50 cm; blade dark green above, light green beneath, hairy when young, lobes triangular to ovate to ovate-elongate, the sinuses of variable depth. Inflorescence above the leaves, paniculate, umbel-like, to 30 cm or more across; umbels terminal at ends of primary rays or racemosely arranged along them in their upper portions, the latter *c*. 1.5 cm across on stalks to 3.5 cm. Flowers white; petals 5, free, edge-to-edge in bud, spreading at opening of flower; stamens as many as the petals; ovary with 2 cells. Fruit remaining inferior, bluish black at maturity, *c*. 4 mm across, the stylar column bifid at top, persistent. *Eastern Asia.*

6. Hedera Linnaeus. 22/12. Evergreen, often climbing shrubs; aerial stems often attaching by adventitious roots. Hairs on new growth either large and white or small and reddish. Juvenile sterile leaves alternate, lobed or heart-shaped, covered with small hairs. Adult leaves arranged spirally, mostly entire. Flowers radially symmetric, in spherical umbels or panicles. Calyx 5-lobed. Petals 5. Stamens 5, alternating with petals. Fruit a drupe, black or orange. *Europe, North Africa, Asia.* Figure 89, p. 380.

7. × **Fatshedera** Guillaumin. (*Fatsia* × *Hedera*). 1/1. Unarmed, low, evergreen shrubs with weak branches; young parts at first rusty-hairy. Leaves palmately lobed, entire or nearly so, stalked, the bases sheathing the twig, the free portion of the stipules 2-lobed. Inflorescences terminal, paniculate, once-compound, covered with rusty hairs when young. Calyx-rim obscure, slightly overgrown by the disc. Petals 5, edge-to-edge in bud, opening from the top and spreading before falling. Stamens 5, the anthers without pollen. Ovary 5-celled; disc broad; styles 5, free. Fruit not set. *Garden Origin.*

A bigeneric hybrid, which arose spontaneously in the nurseries of the Lizé brothers in Nantes, France, and was first exhibited in 1912.

8. Cussonia Thunberg. 20/2. Armed or unarmed, usually thick-stemmed evergreen or deciduous trees or shrubs, rarely rhizomatous, woody perennials. Leaves in conspicuous palm-like rosettes at the ends of stems or branches, long-stalked, variously lobed, palmatifid or palmately compound, the 5--12 segments or leaflets sometimes toothed, irregularly lobed or pinnatisect. Inflorescences terminal, once- or twice-compound. Flowers small, spicately or racemosely arranged along the primary or secondary branches, sometimes closely crowded in the manner of a cob, each possessing a single, more or less conspicuous bracteole. Calyx usually 5-toothed. Petals usually 5, edge-to-edge in bud. Stamens usually 5. Ovary usually 2-celled; styles united only at base or slightly above it, otherwise free. Fruit drupe-like; pyrenes 2. *Africa south of the Sahara.*

9. Pseudopanax Koch. 20/10. Unarmed monoecious or dioecious, hairless trees or shrubs, some with different kinds of foliage in different periods of growth. Leaves usually simple or palmately compound; stipular ligule absent or only slightly developed; blades entire or variously toothed; stems sometimes covered with reflexed 'armour-leaves'. Inflorescences terminal or, less commonly, lateral on short shoots, simple or once- or twice-compound. Flowers in umbels, clusters or racemes or mixtures of these; stalks jointed at the top; petals and stamens 4 or 5; ovary 2--5-celled. Fruit drupe-like, wholly inferior at maturity; stones 2--5; seeds with smooth endosperm. *Southern South America, New Zealand, Australia (Tasmania).*

10. Schefflera Forster & Forster. 900/several. Unarmed, mostly evergreen, trees, shrubs, or climbers; juvenile foliage sometimes differing markedly from that in adults. Leaves usually palmately compound, stalked, with the base generally forming a sheath around the twig and commonly developing 2 stipules fused into a ligule of varying length (to 10 cm); leaflets usually 3--20, usually stalked, developing in a single plane or occasionally clustered. Inflorescences terminal or becoming falsely lateral but never on short shoots, compound, basically paniculate or umbellate, any bracts if present not usually developing into leaves. Flowers in small umbels or heads or racemosely or spicately arranged; calyx-rim conspicuous to obscure; petals usually 5, rarely more or 4, edge-to-edge in bud or partly or wholly fused into a cap; stamens as many as the petals or more numerous (to 500); ovary with usually 2--30 cells, the styles free or variously united or entirely reduced. Fruit a drupe, elongate, spherical or more or less compressed, at maturity inferior or up to half or more superior; pyrenes smooth. *North & South America, south & south-east Asia, Pacific Islands.*

11. Meryta Forster & Forster. 30/2. Small to medium, evergreen, dioecious trees of a generally stout habit. Leaves spirally arranged, simple, entire or (rarely) lobed or

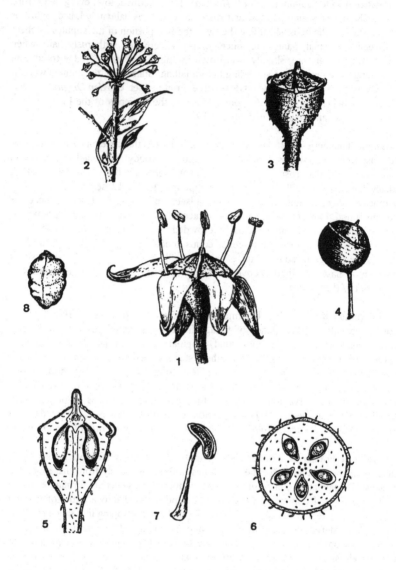

Figure 89. Araliaceae. *Hedera helix*. 1, Flower. 2, Inflorescence after the fall of petals. 3, Ovary after the fall of the other floral parts. 4, Fruit. 5, Longitudinal section of ovary. 6, Transverse section of ovary. 7, Stamen. 8, Seed.

pinnatifid, more or less tufted. Inflorescences terminal, paniculate. Flowers in heads or umbels, the stalks when present not jointed below the ovary. Fruit drupe-like, fleshy, remaining wholly or mostly inferior. *Pacific Islands from Micronesia through Vanuatu to New Zealand and southeastern Polynesia.*

12. Metapanax Frodin. 4/1. Evergreen shrubs or trees of bushy habit with relatively slender branches, hairless throughout. Leaves simple or palmately lobed or compound, without stipules; blades broadly and shallowly toothed to almost entire. Inflorescence terminal, paniculate, once- or twice-compound, conical to pyramidal in outline; main axis to 18 cm, exceeding the lowest primary branch; branches few to several. Flowers in small umbels at ends of main axis and branches as well as on stalks along the latter; flower-stalks jointed just below the ovary. Petals 5, free, edge-to-edge in bud. Stamens as many as the petals. Ovary with 2 cells, the disc slightly to strongly elevated. Fruit compressed, more or less transversely oblong or ellipsoid, thin-walled, expanding entirely or mostly below the disc; styles partially or wholly united into a column, persistent. *China, northern Vietnam.*

Formerly included in *Nothopanax* and, more recently, *Pseudopanax*.

13. Dendropanax Decaisne & Planchon. 80/1. Hairless, evergreen shrubs or small trees with slender branches. Leaves stalked, simple or rarely palmately 3--5-lobed, often 3--5-veined from the base. Inflorescences terminal, usually small, simple or compound, umbellate or paniculate and then mostly umbel-like or with raceme-like branches. Flowers in umbels or heads, their stalks not jointed below the ovary. Petals 5--8, free, edge-to-edge in bud, spreading. Stamens as many as the petals. Ovary with 5--8 cells, the styles free or partly or wholly united into a column. Fruit a many-seeded drupe, spherical or ovoid when fresh but often markedly grooved on drying. *Mexico & West Indies to central South America, eastern, southern and south-eastern Asia.*

14. Tetrapanax Koch. 1/1. Unarmed, unbranched or little-branched, evergreen shrubs or small trees, spreading by growth of successive, initially rhizomatous stems; young aerial parts covered with loose pale floss. Leaves large, palmately lobed, hairy, quite pale beneath. Inflorescence terminal, paniculate, pale-woolly with conspicuous bracts. Flowers in small, stalked umbels racemosely arranged along inflorescence branches. Petals 4 or 5, edge-to-edge in bud. Stamens as many as the petals. Ovary 2-celled, styles free. Fruit small, drupe-like. *Eastern Asia.*

15. Fatsia Decaisne & Planchon. 3/1. Unarmed, stoutly-branched, evergreen shrubs or small trees, the foliage crowded towards branch tips; branches few to many. Leaves leathery, palmately lobed, the lobes 7--11, more or less narrowed below. Inflorescences terminal, appearing above the leaves, usually twice-compound, remaining hairy or becoming hairless, the main axis long. Flowers sometimes unisexual, in umbels; stalks jointed below the ovary or not. Petals 5, free, edge-to-edge in bud, spreading. Stamens as many as petals. Ovary with 3--10 cells. Fruit a 3--10-seeded drupe, usually spherical and remaining mostly inferior, initially green but maturing blue-black to black, topped by the persistent free styles. *East & north-east Asia.*

147. UMBELLIFERAE (*Apiaceae*)
Herbaceous annuals, biennials or perennials, rarely shrubs. Leaves simple or divided, hairless or hairy, often bristly. Stipules usually absent. Flowers bisexual, or sometimes

381

only functionally male (plants monoecious or dioecious), usually clustered in simple umbels which are sometimes subtended by bracts and borne on stalks or rays. The rays may be solitary or, more usually, clustered together to form compound umbels, which may be subtended by bracts. Calyx-teeth 5. Petals 5, overlapping in bud. Stamens 5. Ovary 2-celled, with 1 apical ovule per cell. Styles 2, usually arising from a basal swelling (stylopodium); stigmas minute or head-like. Fruit a dry schizocarp, usually splitting from the sterile carpophore or stalk into 2 mericarps.

A family of about 420 genera with a mainly north temperate distribution, several of which are cultivated. The ripe fruits are the most important feature for identification (see figures 90, 91, 92 & 93, pp. 384, 386, 388 & 396). The number and position of ridges, the presence or absence of wings and the plane of compression, i.e. dorsal or lateral, are all important diagnostic features and are frequently used to determine generic limits. Many have a very characteristic aniseed odour and others smell strongly of celery. Most are very easy to grow, the majority being easily raised from seed.

1a. Leaves rigid or leathery, usually persistent; margins and apices spiny or
    bristle-tipped   2
 b. Leaves not rigid or leathery, usually deciduous; margins and apices
    without spines   5
2a. Stipules absent; flowers borne in conspicuous cone-like heads; never
    cushion-forming   **10. Eryngium**
 b. Stipules present, although sometimes minute; flowers borne in simple or
    compound umbels; sometimes cushion-forming   3
3a. Leaves pinnately divided, at least twice as long as wide; umbels
    compound   **40. Aciphylla**
 b. Leaves simple, though sometimes divided almost to the base or midrib,
    about as long as wide; umbels simple   4
4a. Leaves and fruit with stellate hairs; flowers greenish white   **6. Bolax**
 b. Leaves and fruit with simple hairs or hairless; flowers yellow or brownish
                                                                 5. Azorella
5a. Leaves simple, almost entire, though blade sometimes palmately divided
    or deeply 3--5-lobed, but not cut to midrib   6
 b. Leaves compound, blade often pinnately divided to midrib 1--5 times,
    or 1--6 times divided into 3   14
6a. Leaf-blades at least twice as long as wide, usually linear, oblong or
    lanceolate   **31. Bupleurum**
 b. Leaf-blades about as long as wide, almost circular, kidney-shaped or
    broadly ovate or triangular in outline   7
7a. Umbels simple, surrounded by conspicuous broadly lanceolate or obovate
    bracts   8
 b. Umbels simple or compound; bracts linear, inconspicuous or absent   10
8a. Plant densely covered with stellate hairs; leaves divided 2 or 3 times
    almost to midrib; petals absent   **4. Actinotus**
 b. Plant hairless or with scattered hairs; leaves palmately divided or lobed;
    petals present   9
9a. Petals yellow or greenish yellow; bracts 5--7, obtuse, toothed towards
    apex, green, veins inconspicuous   **8. Hacquetia**
 b. Petals white or pink or dark red; bracts 10 or more, acute, entire, pink or
    white, with conspicuous greenish veins   **9. Astrantia**

10a. Stem-leaves perfoliate, margins entire; flowers bright yellow
16. Smyrnium
  b. Stem-leaves, if present, not perfoliate, entire to deeply divided; flowers green, white, pinkish or blue   11
11a. Stems slender, creeping, rooting at the nodes; stipules present   12
  b. Stems erect, sometimes woody at base, not rooting at the nodes; stipules absent   13
12a. Petals edge-to-edge, white or green, sometimes tinged pink; fruit surface without net-veins   **1. Hydrocotyle**
  b. Petals overlapping, deep red or pink; fruit surface net-veined   **2. Centella**
13a. Calyx-teeth minute, mericarps flattened, semicircular, smooth or hairy, covered with papillae or tubercules, but lacking hooked bristles
3. Trachymene
  b. Calyx-teeth conspicuous, persistent, longer than the petals; mericarps not as above, covered with rigid, hooked bristles   **7. Sanicula**
14a. Plants covered with fine or coarse hairs, especially below   15
  b. Plants hairless or with few scattered hairs   27
15a. Low-growing, cushion-forming perennial; umbels mostly obscured by the leaves   **41. Anisotome**
  b. Plants erect, annual, biennial or perennial; umbels held clear of leaves   16
16a. Plant smelling strongly of aniseed when crushed   17
  b. Plant not smelling of aniseed when crushed   18
17a. Perennial herb; leaf-lobes ovate-lanceolate, flecked with white; fruit 1.5--2.5 cm, glossy black   **14. Myrrhis**
  b. Annual herb; leaf-lobes linear-lanceolate, not flecked with white; fruit 3--5 mm, brownish   **18. Pimpinella**
18a. Base of stem covered with the persistent remains of old leaf-stalks; at least some ultimate leaf-lobes linear or thread-like   19
  b. Base of stem without fibrous remains; ultimate leaf-lobes ovate to lanceolate, rarely linear   20
19a. Bracteoles absent   **18. Pimpinella**
  b. Bracteoles conspicuous, sometimes united   **22. Seseli**
20a. Flowers yellow; bracts and bracteoles falling early; leaves usually simply pinnate   **46. Pastinaca**
  b. Flowers white, reddish or pinkish; bracts and or bracteoles persistent; leaves 1--5-pinnate or 1 or 2 times divided into 3   21
21a. Bracts 2--5 cm, pinnately divided   22
  b. Bracts 1--18 mm, linear, or absent   23
22a. Fruit densely covered with soft hairs   **24. Athamanta**
  b. Fruit covered with hooked spines   **48. Daucus**
23a. At least some leaves 1 or 2 times divided into 3   24
  b. All leaves 2--5 times pinnately divided   25
24a. Fruit at least 4 times as long as wide, narrowly ellipsoid, covered with long silky hairs   **13. Osmorhiza**
  b. Fruit less than twice as long as wide, broadly ellipsoid, smooth or sparsely hairy   **47. Heracleum**
25a. Fruit densely covered with soft hairs   **24. Athamanta**
  b. Fruit sparsely hairy or hairless   26
26a. Flower-stalks without a ring of hairs at the apex   **11. Chaerophyllum**

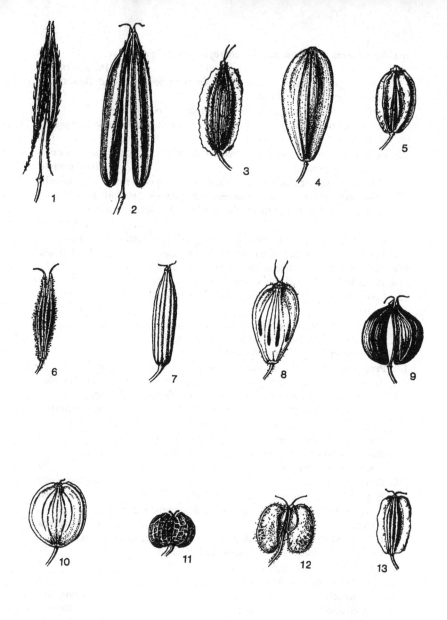

Figure 90. Umbelliferae: fruits. 1, *Osmorhiza*. 2, *Myrrhis*. 3, *Melanoselinum*. 4, *Ferula*. 5, *Bupleurum*. 6, *Athamanta*. 7, *Chaerophyllum*. 8, *Heracleum*. 9, *Smyrnium*. 10, *Pastinaca*. 11, *Centella*. 12, *Trachymene*. 13, *Angelica*.

           b. Flower-stalks with a ring of stout hairs at the apex   **12. Anthriscus**
27a. Leaves very succulent; plant woody at base   **21. Crithmum**
    b. Leaves not succulent; plant herbaceous   28
28a. All leaves with very fine linear or thread-like lobes   29
    b. At least some leaves with diamond-shaped, ovate or lanceolate lobes 40
29a. Flowers yellow   30
    b. Flowers white or tinged pink or purple   33
30a. Base of plant surrounded by conspicuous fibrous remains of old
    leaf-stalks   **45. Peucedanum**
    b. Base of plant not surrounded by remains of leaf-stalks   31
31a. Plant not smelling of aniseed when crushed; stems solid   **44. Ferula**
    b. Plant smelling strongly of aniseed when crushed, stems hollow or
       becoming so with age   32
32a. Fruit unwinged   **25. Foeniculum**
    b. Fruit with conspicuous papery wing   **26. Anethum**
33a. Base of plant surrounded by conspicuous fibrous remains of old leaf-
    stalks   34
    b. Base of plant smooth   36
34a. Stems solid; leaves bluish green, stiff   **22. Seseli**
    b. Stems hollow; leaves bright green, soft   35
35a. Basal leaves 3- or 4-pinnate with 3--10 pairs of lobes   **28. Meum**
    b. Basal leaves simply pinnate, with more than 20 pairs of palmatisect lobes,
       appearing whorled   **37. Carum**
36a. Bracts longer than the rays; fruit hairy   **32. Cuminum**
    b. Bracts shorter than the rays or absent; fruit smooth   37
37a. Bracts many, 1.5--5 cm, pinnatifid; fruit 1.5--2.5 mm   **36. Ammi**
    b. Bracts few, 10 mm or less, entire, sometimes absent; fruit 2.5--4 mm 38
38a. Annual or biennial herb, all parts with an aniseed-like smell (caraway);
    stems hollow   **37. Carum**
    b. Perenial herbs, without aniseed-like smell; stems solid, at least after
       flowering   39
39a. Plant arising from a tuber; stems hollow after flowering; fruit unwinged
                                                                            17. Conopodium
    b. Plant without tubers; stems solid; fruit with a broad lateral wing
                                                                            38. Selinum
40a. Base of plant surrounded by conspicuous remains of old leaf-stalks   41
    b. Base of plant smooth   44
41a. Base of plant surrounded by numerous scale-like remains of old leaf-
    stalks; leaf-lobes wedge-shaped to diamond-shaped, divided in upper
    part; flowers greenish yellow   **43. Levisticum**
    b. Base of plant surrounded by fibrous remains of old leaf-stalks; flowers
       white, pink or, if yellow, leaf-lobes linear, lanceolate or ovate   42
42a. Flowers white, greenish white, pink or red   **39. Ligusticum**
    b. Flowers yellowish or creamy yellow   43
43a. Rays sharply angled; fruit oblong-ovoid, 4--5 mm, narrowly winged
                                                                            27. Silaum
    b. Rays smooth; fruit ovoid-ellipsoid, 2--2.5 mm, ridged   **34. Petroselinum**

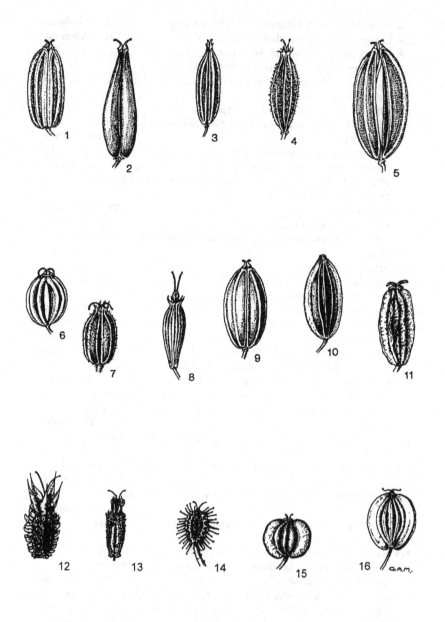

Figure 91. Umbelliferae: fruits. 1, *Foeniculum.* 2, *Anthriscus.* 3, *Cryptotaenia.* 4, *Cuminum.* 5, *Meum.* 6, *Sium.* 7, *Seseli.* 8, *Oenanthe.* 9, *Crithmum.* 10, *Anethum.* 11, *Aciphylla.* 12, *Eryngium.* 13, *Astrantia.* 14, *Daucus.* 15, *Hydrocotyle.* 16. *Peucedanum.*

386

44a. Stems bright light green, distinctly blotched with purple; fruit with wavy
ridges   **29. Conium**
  b. Stems greenish or if reddish purple then not blotched; fruit ridges not
normally wavy   45
45a. Bracts and bracteoles with a conspicuous white papery wing
    30. Pleurospermum
  b. Bracts and bracteoles absent or if present, without a papery wing   46
46a. Leaf-stalks, especially of upper stem-leaves, inflated or winged and often
sheathing stem   47
  b. Leaf-stalks not conspicuously inflated or winged   55
47a. Leaf-stalks winged or only slightly inflated; stems not more than 1 m,
usually less, slender   48
  b. Leaf-stalks greatly inflated, blade often reduced or absent; stems usually
1--3 m, stout   53
48a. Rhizomatous perennial; leaves divided into 3 ovate lobes; inflorescence
very loose with unequal rays   **35. Cryptotaenia**
  b. Biennial, or perennial either arising from a tuber or rooting at the nodes,
or sterile rosettes present at flowering time; leaves 1--3 times pinnately
divided; inflorescence not as above   49
49a. Biennials with a strong characteristic smell of celery or aniseed   50
  b. Perennials without a characteristic smell; arising from a tuber or rooting at
the nodes or sterile rosettes present at flowering time   51
50a. Plant, especially fruit, smelling of celery; stem solid; fruit *c*. 1.5 mm
    33. Apium
  b. Plant with a strong aniseed-like (caraway) smell; stems hollow; fruit
3--4 mm   **37. Carum**
51a. Plant creeping or ascending, rooting from the nodes; leaves usually
simply pinnate   **33. Apium**
  b. Plant erect, sometimes arising from a tuber, never rooting at the nodes;
leaves 1--3-pinnate   52
52a. Plant arising from a tuber; leaf-lobes linear-lanceolate   **17. Conopodium**
  b. Plant not arising from a tuber, sterile rosettes present at flowering time;
leaf-lobes ovate   **18. Pimpinella**
53a. Leaf-lobes lanceolate to broadly linear, margins smooth; fruit *c*. 1.2 cm
    44. Ferula
  b. Leaf-lobes ovate to triangular in outline, margins coarsely toothed; fruit
5--8 mm   54
54a. Flowers creamy yellow; bracteoles absent or few, very small; fruit ridged
but not winged, laterally compressed, black   **16. Smyrnium**
  b. Flowers white, pink, purple or greenish; bracteoles numerous; fruit
conspicuously winged, dorsally compressed, brownish   **42. Angelica**
55a Leaves 1--3 times divided into 3   56
  b. Leaves simply or 2--4 times pinnately divided   57
56a. Rhizomatous perennial with stems to 1 m; leaves sometimes variegated;
flowers white; fruits 3--4 mm, shallowly ridged   **19. Aegopodium**
  b. Biennial or monocarpic perennial; stems 1--3.6 m; flowers greenish
yellow; fruits 7--9 mm, winged   **45. Peucedanum**
57a. Rays 2--5; fruit almost spherical, not splitting at maturity **15. Coriandrum**
  b. Rays 5--40; fruit not spherical, splitting at maturity   58

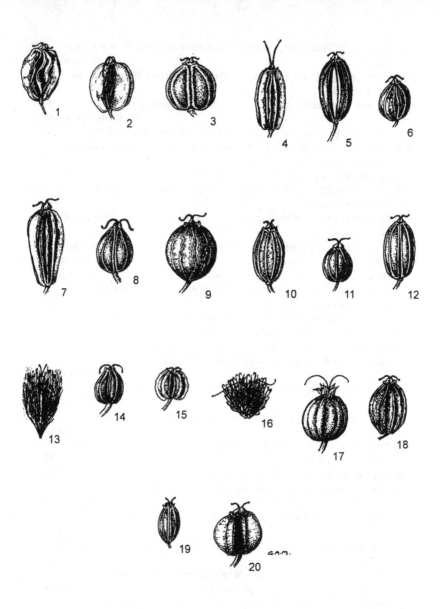

Figure 92. Umbelliferae: fruits. 1, *Ligusticum*. 2, *Selinum*. 3, *Conium*. 4, *Pleuro-spermum*. 5, *Conopodium*. 6, *Petroselinum*. 7, *Levisticum*. 8, *Aegopodium*, 9, *Coriandrum*. 10, *Azorella*. 11, *Pimpinella*. 12, *Carum*. 13, *Actinotus*. 14, *Ammi*. 15, *Apium*. 16, *Sanicula*. 17, *Hacquetia*. 18, *Silaum*. 19, *Anisotome*. 20, *Bolax*.

388

58a. Plants with tubers; fruit with persistent calyx-teeth   **23. Oenanthe**

  b. Plant without tubers, or if tubers present then fruit without persistent calyx-teeth   59

59a. Stems solid; bracts absent or 1 or 2, falling early   **38. Selinum**

  b. Stems hollow; bracts 2--10, persistent   60

60a. Stems green; stems leaves simply pinnate; fruit laterally compressed   **20. Sium**

  b. Stems often purplish; stem-leaves 2- or 3-pinnate; fruit strongly dorsally compressed   **45. Peucedanum**

1. Hydrocotyle Linnaeus. 100/5. Creeping perennial herbs, stems usually rooting at nodes. Stipules present. Leaves simple, blades more or less circular. Bracts present. Umbels simple. Flowers small, white or greenish. Calyx-teeth almost absent. Petals edge-to-edge. Fruit ellipsoid to rounded, strongly flattened; ridges prominent to obsolete, surface without net-veins. *Cosmopolitan but concentrated in the southern hemisphere*. Figure 91, p. 386.

2. Centella Linnaeus. 20/1. Creeping perennial herbs of wet habitats; stems usually rooting at nodes; stipules present. Leaves simple, circular to kidney-shaped with slender stalks. Flowers borne in compact simple umbels sheathed by small bracts. Calyx-teeth almost absent. Petals overlapping. Fruit strongly flattened, rounded, ridges prominent, net-veined. *South Africa, Chile, New Zealand.* Figure 90, p. 384.

3. Trachymene Rudge. 12/3. Annual, biennial or perennial erect herbs. Basal leaves simple, almost entire, lobed but not cut to the midrib. Stem-leaves sometimes entire, broadly triangular to ovate, stalks with sheathing bases. Bracts linear. Umbels simple, terminal, sometimes head-like. Calyx-teeth minute. Petals ovate, outer radiating, blue, white or pinkish. Fruit with flattened semi-circular mericarps which are smooth or hairy, covered with papillae or tubercles. *Australia, west Pacific area.* Figure 90, p. 384.

4. Actinotus Labillardière. 17/1. Annuals, perennials and shrubs. Leaves simple, toothed to deeply divided or lobed 2 or 3 times but not cut to midrib. Flowers many, in dense, head-like simple umbels, surrounded by large, petal-like bracts. Petals absent. *Australia, New Zealand.* Figure 92, p. 388.

5. Azorella Lamarck. 70/2. Evergreen perennial herbs, often woody at base, or low shrubs. Leaves entire, toothed or sharply divided into 3 leaflets, sometimes palmately cut, about as long as wide, with simple hairs or hairless; stalks sheathing stems at base. Stipules very small. Umbels mostly stalkless, simple. Bracts conspicuous, united at their bases. Calyx-teeth toothed. Petals entire, obtuse, overlapping, golden brown to deep yellow. Fruit terete or laterally compressed, with simple hairs or hairless. *Temperate South America, New Zealand, Antarctic Islands.* Figure 92, p. 388.

6. Bolax Jussieu. 3/1. Perennial herbs resembling *Azorella*, forming dense cushions. Leaves simple, sometimes bristle-tipped, with stellate hairs, crowded, overlapping; stalk with sheathing base. Stipules very small. Umbels simple. Flowers greenish white. Calyx of 5 petal-like teeth. Fruit compressed; mericarps ridged, stellate-hairy. *Temperate South America.* Figure 92, p. 388.

7. Sanicula Linnaeus. 40/1. Perennial herbs. Leaves palmately 3--7-parted. Inflorescence of a number of simple umbels in a cyme, often forming a false compound umbel. Bracts small. Calyx-teeth conspicuous, linear. Petals white or flushed pink. Fruit ovate to spherical, mericarps densely covered in rigid, hooked bristles. *More or less cosmoplitan.* Figure 92, p. 388.

8. Hacquetia de Candolle. 1/1. Herbaceous perennial, 10--25 cm, with creeping rootstock. Leaves palmately divided into 3, each lobe 2--4 cm, sometimes further divided, ovate, wedge-shaped at base, toothed. Basal leaves long-stalked. Inflorescence a simple, stalked umbel. Bracts 5--7, 1--2 cm, obtuse, toothed towards the apex, green, obovate, like the leaf-lobes. Calyx-teeth conspicuous. Petals yellow or greenish yellow. Fruit *c.* 4 mm, hairless, ovoid, slightly compressed laterally, ridges stout, prominent. *European Alps.* Figure 92, p. 388.

9. Astrantia Linnaeus. 10/5. Perennial herbs. Leaves palmately lobed or palmatisect. Inflorescence a simple umbel. Bracts usually more than 10, large, pink or white with conspicuous greenish veins, lanceolate to acute, entire. Calyx-teeth conspicuous, pinkish. Petals lanceolate, whitish, pink or dark red, apex inflexed. Fruit cylindric with prominent ridges, densely covered with scales. *Central & southern Europe, west Asia.* Figure 91, p. 386.

10. Eryngium Linnaeus. 230/25 and some hybrids. Perennial or annual herbs. Leaves spine-toothed, simple but often lobed. Flowering stems ascending and branching, or plants stemless or with stocky base. Leaves mostly basal, entire to 3-pinnatisect, linear-lanceolate to ovate, leathery or membranous, stalked or stalkless, variably spiny, often with conspicuously silvered veins. Bracts spiny, 3 or more. Flowers borne in spherical to cylindric cone-like heads, solitary or arranged in loosely candelabriform cymes or racemes. Bracteoles entire or with 3 or 4 teeth. Flowers stalkless. Calyx of 5 stiff teeth, often longer than petals. Petals to 4 mm, notched, white, blue, mauve or green. Fruit ovoid, scaly. *Europe, temperate & tropical America.* Figure 91, p. 386.

11. Chaerophyllum Linnaeus. 35/2. Biennial or perennial herbs usually covered with fine or coarse hairs; stems solid. Leaves 2- or 3-pinnate. Bracts few, linear or absent. Rays 12--25. Bracteoles few. Flower-stalks without a ring of hairs at apex; flowers radially symmetric. Calyx-teeth minute or absent. Petals white or pink. Fruits usually more than 3 times as long as wide, slightly compressed laterally, hairless, ridges low, rounded. *Europe, Asia.* Figure 90, p. 384.

12. Anthriscus Persoon. 12/2. Erect annual, biennial or perennial herbs, covered with fine or coarse hairs, especially below. Leaves usually 2--3-pinnate. Bracts linear, few or absent. Bracteoles entire, usually reflexed. Flower-stalks with a ring of stout hairs at the apex. Calyx-teeth minute or absent. Petals white. Fruit narrowly cylindric to ovoid, with ridged beak or non-seed-bearing part, usually hairless. *Europe, North Africa.* Figure 91, p. 386.

13. Osmorhiza Rafinesque. 10/1. Hairy perennial herbs with thick fleshy roots, stems erect. Leaves 2 or 3 times divided into 3, or 2-pinnate, lobes lanceolate to circular, with forwardly pointing teeth, or pinnatifid; stalks sheathing. Umbels compound. Bracts linear, 7--12 mm, few or absent. Bracteoles several, narrow or absent. Flowers

390

white (in ours); petals spathulate to obovate. Fruit ellipsoid, at least 4 times as long as wide, stiffly hairy or hairless; mericarps with thread-like ridges. *North & South America, Asia.* Figure 90, p. 384.

14. Myrrhis Miller. 1/1. Perennial herbs, with strong aniseed smell when crushed; stems to 1.8 m, hollow, softly hairy, erect. Leaves 2--4-pinnate, with conspicuous whitish flecks; lobes ovate-lanceolate, acute. Bracts absent. Bracteoles present. Calyx-teeth absent or minute. Petals white. Fruit 1.5--2.5 cm, more than 3 times as long as wide, glossy black when ripe, laterally compressed, hairless or with minute bristles, ridges acute, well-developed. *Europe.* Figure 90, p. 384.

15. Coriandrum Linnaeus. 2/1. Annual herbs, hairless. Stems solid. Lower leaves simply lobed, lobes ovate. Upper leaves 1--3-pinnate, lobes linear. Umbels compound, sometimes leaf-opposed. Petals white, pink or lilac. Fruit almost spherical, hard, not splitting at maturity. *South Europe, southwest Asia.* Figure 92, p. 388.

16. Smyrnium Linnaeus. 8/2. Erect, rather stout biennial herbs, hairless or nearly so. Lower leaves usually 1--3 times divided into 3. Upper leaves usually simple. Calyx-teeth absent. Petals lanceolate to reversed-heart-shaped, apex inflexed, creamy yellow or bright yellow. Fruit ovoid or almost spherical, ridges slender, the marginal ridges usually inconspicuous. *Europe, Mediterranean area.* Figure 90, p. 384.

17. Conopodium Koch. 20/1. Erect perennial herbs. Stems solid, becomimg hollow after flowering, solitary, underground parts pale and fragile, developing from a spherical tuber. Leaves 2- or 3-pinnate; upper leaf-stalks slightly winged and sheathed at base. Umbels compound. Calyx-teeth inconspicuous. Petals white (occasionally pink), obovate, notched. Fruit oblong-ovoid, compressed laterally, finely ridged, but not winged. *Europe, north Africa, southwest Asia.* Figure 92, p. 388.

18. Pimpinella Linnaeus. 150/3. Annual or perennial herbs, with hollow or solid stems and fibrous remains of old leaf-stalks at base. Basal leaves entire or 1--3-pinnate. Bracts absent. Flowers radially symmetric. Calyx-teeth usually minute. Petals white or pinkish (our species). Fruits ovoid-oblong to broadly ovoid, slightly laterally compressed, hairless, with low narrow ridges. *Eurasia, Africa.* Figure 92, p. 388.

19. Aegopodium Linnaeus. 7/1. Hairless perennial herbs with creeping rhizomes. Leaves 1 or 2 times divided into 3, lobes broadly ovate. Flowers mostly bisexual. Bracts and bracteoles absent or few. Calyx of 5 minute teeth. Petals white or yellow. Fruit ovoid, laterally compressed; mericarps with fine ridges. *Europe, west Asia.* Figure 92, p. 388.
The only species grown is considered a pernicious weed by many, but the variant with variegated foliage is less aggressive and is more widely grown.

20. Sium Linnaeus. 15/3. Hairless perennials of wet and marshy places. Stems hollow. Submerged leaves, if present, 2- or 3-pinnate, with linear lobes. Stem-leaves usually 1-pinnate, with broad leaflets. Calyx-teeth present, sometimes very small. Petals reversed-heart-shaped with inflexed apex, white. Fruit ovoid or ovoid-oblong, slightly compressed laterally; ridges slender or broad, sometimes with a narrow wing. *Northern hemisphere, Africa.* Figure 91, p. 386.

21. Crithmum Linnaeus. 1/1. Semi-succulent plants, stems solid, woody below. Leaves 2- or 3-pinnate, triangular in outline; lobes linear; stalks narrowly sheathing at base. Bracts lanceolate, membranous, deflexed at maturity. Umbels compound. Bracteoles present, similar to bracts. Calyx-teeth minute. Petals yellowish green. Fruit ovoid-oblong, not compressed, spongy; ridges thick and prominent. *Europe*. Figure 91, p. 386.

Difficult in cultivation because it requires somewhat saline conditions.

22. Seseli Linnaeus. 65/3. Annual, biennial or perennial herbs, hairy or hairless; stems solid with fibrous remains of old leaf-stalks at base. Leaves 1--4-pinnate, lobes usually linear or thread-like, more rarely ovate. Upper stem-leaves usually less divided, with short sheathing bases. Umbels compound. Bracts absent or up to 15. Bracteoles 5--12. Calyx absent or reduced to small teeth. Petals white, often tinged pink, broad. Fruit oblong to ovoid, slightly compressed; mericarps with 5 prominent ridges. *Eurasia*. Figure 91, p. 386.

23. Oenanthe Linnaeus. 35/4. Hairless perennial herbs, often rooting at lower nodes. Roots fibrous or tuberous. Leaves 1--4-pinnate or pinnatisect; lobes toothed to pinnately cut. Umbels compound. Bracts absent or small. Bracteoles numerous, small. Calyx-teeth small, acute, persistent. Petals white to flushed pink, outer sometimes enlarged. Fruit cylindric to oblong; ridges usually thickened and corky. *North temperate areas, mountains in tropical Africa*. Figure 91, p. 386.

24. Athamanta Linnaeus. 10/2. Perennial herbs, mostly downy. Leaves 2--5-pinnate, lobes ovate, narrowly lanceolate, rarely linear or thread-like. Umbels with bracts and bracteoles. Petals white to creamy yellow, notched or 2-lobed, sometimes hairy beneath or along the margins. Fruit oblong or shortly ovoid, scarcely compressed, hairy, narrowed to a short beak, ridges low. *Mediterranean area*. Figure 90, p. 384.

25. Foeniculum Miller. 1/1. Aromatic biennial or perennial to 2 m, smelling of aniseed. Stems finely grooved, hollow when mature, glaucous, hairless. Leaves 20--40 cm, triangular in outline, 3- or 4-pinnate, lobes to 5 cm, thread-like, very fine. Umbels with 10--40 rays. Bracts and bracteoles absent. Flowers yellow. Fruit scarcely compressed; mericarps ridged. *Europe, Mediterranean area*. Figure 91, p. 386.

26. Anethum Linnaeus. 2/2. Annual or biennial herbs, glaucous, hairless and all parts smelling stongly of aniseed; stems hollow. Leaves 3--4-pinnate, lobes linear to thread-like. Bracts and bracteoles usually absent. Calyx-teeth absent. Petals oblong with recurved apices, yellow. Fruit ovoid, dorsally compressed, mericarps with slender dorsal ridges and conspicuous winged lateral ridges. *South-west Asia; naturalised in Europe and North America*. Figure 91, p. 386.

27. Silaum Miller. 10/1. Perennial hairless herbs with fibrous remains of old leaf-stalks at base. Leaves 1--4-pinnate, narrowly sheathing at base. Calyx-teeth absent. Petals ovate, yellowish, apex short, with margins curved upwards and inwards. Fruit ovoid-oblong to nearly cylindric, slightly compressed; mericarps with prominent, slender ridges, the lateral forming narrow wings. *Temperate Eurasia*. Figure 92, p. 388.

28. Meum Miller. 1/1. Hairless, strongly aromatic perennial, with aniseed-like smell; stems hollow, 7--60 cm. Stock surrounded by coarse, fibrous remains of stalks. Leaves mostly basal, bright green, soft, 3- or 4-pinnate, with 3--10 pairs of crowded, thread-like lobes, 2--5 mm. Bracts absent or to 2, bristly. Rays 3--15. Bracteoles few, bristly, often small. Calyx-teeth absent. Petals white or purplish, ovate, apex more or less inflexed. Fruit 4--10 mm, ovoid-oblong, scarcely compressed. Ridges very prominent, stout. *Europe.* Figure 91, p. 386.

29. Conium Linnaeus. 3/1. Hairless biennials. Stems hollow, often spotted. Leaves 2--4-pinnate, lobes sharply toothed to pinnatifid. Umbels compound. Bracts and bracteoles few to several. Calyx absent. Petals with inflexed apices, white. Fruit ovate, slightly compressed; mericarps strongly ridged, ridges usually wavy. *Eurasia.* Figure 92, p. 388.
Very poisonous.

30. Pleurospermum Hoffmann. 15/1. Biennial or perennial herbs; stems solid, hairless. Leaves 2- or 3-pinnate. Inflorescence a compound umbel. Bracts few to many, with whitish papery wing. Bracteoles several, usually pale with a winged toothed margin. Calyx-teeth minute. Flowers white or pinkish, petals obovate. Fruit ovoid-oblong, somewhat compressed laterally, ridges prominent often narrowly winged. *Eastern Europe, Asia.* Figure 92, p. 388.

31. Bupleurum Linnaeus. 100/8. Annual or perennial herbs or small shrubs, hairless. Leaves simple, entire. Bracts present or absent. Umbels compound. Calyx-teeth usually absent. Petals entire, yellow or purple-tinged, obovate, apex inflexed. Fruit usually ovoid or oblong, slightly compressed; ridges usually conspicuous. *Europe, north Africa, Asia.* Figure 90, p. 384.

32. Cuminum Linnaeus. 1/1. Annual herb to 40 cm; stems solid, slender, hairless. Leaves alternately pinnately divided into thread-like divisions which are 2--5 mm. Umbels compound. Bracts 2--4, thread-like or 3-fid, usually longer than the rays. Rays 1--5. Bracteoles usually 3, very unequal. Flowers 3--5 in each ultimate umbel. Calyx-teeth conspicuous. Petals white or pinkish, apex long, inflexed. Fruit 4--5 mm, narrowly cylindric to ovoid, dorsally compressed, hairy or covered with short bristles. *Eastern & southern Mediterranean area; naturalised elsewhere.* Figure 91, p. 386.

33. Apium Linnaeus. 20/2. Biennial or perennial herbs. Leaves pinnate or twice divided into 3 towards the inflorescence; stalks slightly inflated or winged, sheathing at base. Umbels compound. Calyx-teeth minute or obsolete. Petals whitish, apex sometimes inflexed. Fruit ovoid or elliptic-oblong, laterally compressed; ridges stout. *North temperate areas.* Figure 92, p. 388.

34. Petroselinum Hill. 3/1. Biennial herbs with strong taproots. Stems solid, hairless, often with fibrous remains of old leaf-stalks at base. Leaves 1--3-pinnate. Bracts small. Bracteoles small. Fruit ovoid-ellipsoid, ridges slender, not wavy or winged. *Europe.* Figure 92, p. 388.

35. Cryptotaenia de Candolle. 4/1. Hairless perennials. At least basal leaves divided into 3. Flowers in irregularly branched, compound umbels. Bracts absent. Rays 3--10.

Bracteoles small or absent. Calyx-teeth inconspicuous. Petals incurved at apex, white. Fruit linear-oblong, slightly compressed, ridges slender, not wavy. *North temperate areas, on mountains in Africa.* Figure 91, p. 386.

36. Ammi Linnaeus. 10/2. Hairless annual or biennial herbs; stems solid. Leaves 1--3-pinnate or rarely 1--3 times divided into 3; stalk broadened into a sheathing base. Umbels compound. Bracts 1--1.5 cm, pinnatifid, leaf-like with linear lobes. Rays 3--7 cm. Bracteoles linear. Calyx-teeth minute to absent. Petals white. Fruit ovoid-oblong, slightly laterally compressed, ridges very slender. *Macaronesia, Mediterranean area, western Asia, Chile, Brazil.* Figure 92, p. 388.

37. Carum Linnaeus. 30/2. Annual, biennial or perennial herbs, usually hairless, often with fibrous remains of old leaf-stalks at the base. Leaves pinnate or compoundly divided into 3. Umbels compound. Petals white or pinkish. Fruit laterally compressed, ellipsoid. *Temperate & subtropical areas.* Figure 92, p. 388.

38. Selinum Linnaeus. 6/2. Perennial herbs, hairless, stems solid. Leaves 2- or 3-pinnate. Bracts absent or to 5, entire, falling early. Bracteoles numerous. Calyx absent or minute. Petals radially symmetric, white. Fruits dorsally compressed, about as long as wide, hairless; ridges with conspicuous wings. *Temperate Europe & Asia.* Figure 92, p. 388.

39. Ligusticum Linnaeus. 50/4. Perennial herbs with fibrous remains of old leaf-stalks at base; stems hollow. Leaves 2--5-pinnate or 1 or 2 times divided into 3. Umbels compound. Bracts and bracteoles absent or small. Calyx-teeth triangular, small. Fruit not compressed; ridges very prominent to narrowly winged. *North temperate areas.* Figure 92, p. 388.

40. Aciphylla Forster & Forster. 40/11. Erect hairless, spiny, evergreen, usually dioecious perennial herbs, often forming rosettes, clumps or tufts. Leaves at least twice as long as wide, simple or pinnate, with thick needle- or bayonet-like, often stiff leaflets, with a conspicous sheathing base bearing a pair of stipules which may be divided and leaf-like, often the same length as blade or leaflet and indistinguishable except by position, or small and narrow. Flowers borne in compound umbels on a stout flowering-spike. Bracts and bracteoles conspicuous, spiny. Fruit dorsally flattened; ridges narrowly winged. *New Zealand, Australia.* Figure 91, p. 386.

41. Anisotome J.D. Hooker. 13/1. Perennial herbs, covered with long fine hairs. Leaves mostly basal, 1--4-pinnate; leaflets ovate to lanceolate; leaf-stalks sheathing at base. Stem-leaves reduced or absent. Inflorescence robust, mostly obscured by the leaves, umbels large; rays few to numerous. Bracts leaf-like, linear to ovate, papery. Bracteoles few to several. Calyx-teeth unequal. Petals white, red or purple. Fruit oblong-ovoid, dorsally compressed; mericarps ridged, without wings. *New Zealand & subantarctic islands.* Figure 92, p. 388.

42. Angelica Linnaeus. 50/4. Stout biennial, monocarpic or perennial herbs, more or less hairless, usually with stout taproots. Leaves twice divided into 3 or with 2- or 3-pinnate lobes, broadly ovate, margins toothed and often lobed; leaf-stalk sheathing and inflated. Bracts few or absent. Rays numerous. Bracteoles numerous, linear. Calyx of minute teeth or absent. Flowers white, greenish white, pinkish or deep purple. Fruit rounded, dorsally flattened, brownish; mericarps with broad wings and 3 dorsal ridges. *Northern hemisphere.* Figure 90, p. 384.

43. Levisticum Hill. 1/1. Hairless perennial herb to 2 m, with scale-like remains of old leaf-stalks at base, celery-scented; stems hollow, finely grooved, purple-pink at base. Leaves to 70 cm, triangular to diamond-shaped, 2- or 3-pinnate, bronze when young, becoming dark green, lobes to 11 cm, diamond-shaped, base wedge-shaped, margins irregularly toothed and lobed above middle. Umbels with 12--20 rays. Bracts linear-lanceolate, deflexed, margins translucent. Bracteoles similar to bracts, united at base. Flowers bisexual, green-yellow. Fruit 5--7 mm. *East Mediterranean area.* Figure 92, p. 388.

44. Ferula Linnaeus. 170/2. Perennial herbs, arising from a thick rootstock. Lower leaves 2--4-pinnate or 3 or 4 times divided into 3, lobes linear or lanceolate, margins smooth. Upper leaves often reduced to a sheathing base. Umbels compound, central umbel fertile, lateral umbels often sterile. Bracts few or absent. Flowers regular; calyx-teeth minute or absent; petals yellow or whitish. Fruit elliptic or oblong-elliptic, dorsally compressed; mericarps with thread-like dorsal ridges and narrowly winged lateral ridges. *Mediterranean area to central Asia.* Figure 90, p. 384.

45. Peucedanum Linnaeus. 170/3. Biennial or perennial herbs. Stems solid or hollow, hairless. Leaves several times pinnate or twice divided into 3. Calyx-teeth absent or conspicuous. Petals white, yellow or rarely pink or purplish, broadly ovate, apex long, inflexed. Fruit strongly compressed dorsally. Lateral ridges winged, wings closely adpressed to one another; dorsal ridges usually prominent. *Eurasia, tropical & southern Africa.* Figure 91, p. 386.

46. Pastinaca Linnaeus. 14/1. Biennials or perennials. Stems solid or hollow, angled. Leaves simply pinnate (in ours), lobes simple to pinnatisect; stem-leaves reduced. Bracts absent or 1 or 2, short-lived. Umbels compound, with 3--30 rays. Bracteoles absent or 1 or 2. Flowers yellow. Fruit elliptic, strongly compressed dorsally; mericarps with unwinged dorsal ridges, and narrowly winged ridges. *Eurasia.* Figure 90, p. 384.

47. Heracleum Linnaeus. 60/few. Biennials or often monocarpic perennials. Stems hollow. Leaves pinnately divided and sometimes compoundly divided into 3 on the same plant; lobes broadly ovate. Bracts absent or several, entire. Bracteoles several. Calyx-teeth minute or conspicuous and persistent. Petals white, purplish or pink, sometimes radiating. Fruits strongly dorsally compressed, usually longer than wide, lateral ridges winged. *Europe, north Africa, Asia.* Figure 90, p. 384 & 93, p. 396.

48. Daucus Linnaeus. 25/1. Annual, biennial or perennial herbs; stems solid, hairy. Leaves 2- or 3-pinnate, lobes often finely dissected. Bracts and bracteoles several, distinctively pinnatifid into long, linear lobes. Umbels compound. Calyx-teeth triangular. Petals white or purplish. Fruit ovoid, slightly compressed, densely covered with hooked spines. *Cosmopolitan.* Figure 91, p. 386.

One species of **Melanoselinum** Hoffmann (figure 90, p. 384), with stout stems to 3 m, coarsely toothed bracts and winged fruits, may occasionally be grown.

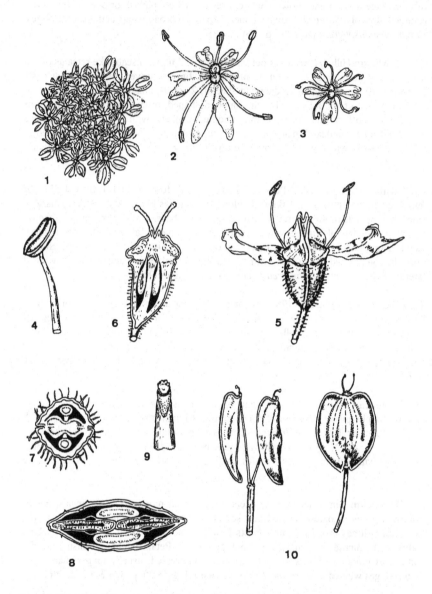

Figure 93. Umbelliferae. *Heracleum sphondylium*. 1, Inflorescence showing outer flowers each with large outer petal. 2, Outer flower with large outer petal. 3, Inner flower. 4, Stamen. 5, Flower with two petals removed. 6, Longitudinal section of ovary. 7, Transverse section of ovary. 8, Transverse section of young fruit. 9, Stigma. 10, Fruit (before splitting into mericarps to right, after splitting to left).

396

148. DIAPENSIACEAE

Evergreen shrublets or stemless perennial herbs. Leaves alternate or basal, simple, without stipules. Flowers solitary or in racemes, bisexual, radially symmetric, sepals petals and stamens hypogynous. Calyx 5-lobed. Corolla 5-lobed. Stamens 5, inserted on corolla, sometimes alternating with 5 staminodes. Ovary superior, 3-celled; ovules several per cell, placentation axile. Fruit a 3-valved capsule.

A family of 6 genera from the Arctic and north temperate areas.

1a. Flowers in spike-like racemes   **3. Galax**
  b. Flowers solitary or few in a loose inflorescence   **2**
2a. Leaves overlapping, narrow   **1. Diapensia**
  b. Leaves long-stalked, circular   **2. Shortia**

1. Diapensia Linnaeus. 4/1. Prostrate or mat-forming evergreen subshrubs. Leaves overlapping, mostly opposite, leathery, entire, narrow. Flowers solitary, stalks long or very short at first, lengthening in fruit, with 1 bract and usually 2 bracteoles. Calyx 5-lobed, persistent and surrounding the mature fruit. Corolla white, pink or pale yellow with a short tube and 5 rounded, somewhat spreading lobes. Stamens 5, alternating with calyx-lobes, filaments flattened, inserted near top of corolla-tube; staminodes 5 or absent. Style 1, about equalling corolla-tube. Fruit a 3-celled capsule. *Northern hemisphere, arctic & subarctic areas, also Sino-Himalaya.*

Difficult to cultivate.

2. Shortia Torrey & Gray. 6/3. Mat-forming perennials with long-stalked, evergreen, alternate, rounded or elliptic, more or less toothed leaves (the teeth are generally sharper and more prominent on immature leaves). Flowers solitary or in a short raceme on a scape with several bud-scales at the base and a few small bracts near the top. Calyx-lobes 5, free to near the base; corolla funnel-shaped with 5 more or less spreading, toothed or fringed lobes. Stamens 5, attached near the mouth of the corolla-tube, usually alternating with 5 short staminodes at the base. Capsule spherical, 3-valved, dehiscent, style persistent in fruit. Seeds numerous, small, oblong or ovoid. *Eastern Asia, eastern USA.*

3. Galax Sims. 1/1. Hairless perennial with a slender, branched rhizome bearing ovate, brown, membranous scales. Leaves long-stalked, leathery, rounded or broadly triangular-ovate, base deeply cordate, apex blunt, margins scalloped or bluntly toothed. Flowers very numerous, white, in a dense, narrow spike or raceme on a leafless scape; bracts ovate, minute. Calyx cut to near the base into 5 blunt, lanceolate lobes. Corolla deeply 5-lobed, lobes oblong, rounded. Filaments united into a tube attached to the corolla near its base and bearing 5 anthers alternating with 5 linear teeth. Anthers 1-celled. Ovary 3-celled, style very short, stigma head-like. Fruit a 3-celled capsule. *Southeastern USA.*

149. CLETHRACEAE

Small trees or shrubs, evergreen or deciduous. Leaves simple, spirally arranged, alternate, toothed, with stellate hairs. Flowers small, white or cream, fragrant, bisexual, petals and stamens hypogynous, in long terminal racemes or panicles. Sepals 5 or 6, united below with overlapping lobes. Petals 5, overlapping. Stamens 10--12, free or shortly attached to the base of the petals; anthers opening by pores. Ovaries of 3 fused carpels, superior, often with nectary at the base; ovules many per cell, placentation axile; style 3-fid. Fruit a capsule enclosed by a persistent calyx, dehiscent with often winged seeds.

A family of a single genus.

1. Clethra Linnaeus. 64/8. Description as for family. *North America, tropical America & southeast Asia.*

150. PYROLACEAE

Perennial, evergreen herbs and dwarf shrubs with creeping rhizomes, some slightly woody at base. Leaves simple, ovate to elliptic or tending to lanceolate, often alternate, usually basal, without stipules. Flowers radially symmetric, solitary or in racemes or cymes, bisexual, speals, petals and stamens hypogynous. Sepals and petals usually 5, sometimes 4. Stamens twice the number of petals, anthers opening by pores. Ovary superior. Fruit a many-seeded capsule.

A family of 4 genera and about 42 species, found in north temperate areas and temperate South America; sometimes included in the *Ericaceae*. All species have mycorrhizal associations and can be difficult to cultivate.

1a. Flowers solitary	**1. Moneses**
b. Flowers in racemes or umbels	2
2a. Leaves basal	**2. Pyrola**
b. Leaves arising from the stem, opposite or whorled	3
3a. Leaves lanceolate to oblanceolate with forwardly pointing marginal teeth	
	3. Chimaphila
b. Leaves ovate-elliptic, entire or with fine, rounded teeth	**4. Orthilia**

1. Moneses Salisbury. 1/1. Perennial, evergreen, hairless, stoloniferous herb, to 10 cm. Leaves to 2.5 cm, opposite or whorled, near base of stem, blades ovate to obovate, dark green, paler beneath, margins slightly scalloped; stalks to 8 mm. Flowers solitary, nodding, *c.* 1.5 cm across. Sepals 5. Petals 5, white or tinged pink. Stamens 10. Ovary nearly spherical; style long; stigma 5-lobed. *Europe, Asia, North America.*

2. Pyrola Linnaeus. 15/3. Perennial, evergreen, hairless, dwarf shrubs, with creeping rhizomes. Leaves in basal clusters, alternate, usually long-stalked, simple. Scapes erect, with bracts. Flowers nodding, in loose racemes. Sepals 5, persistent. Petals 5, concave, incurved, stalkless. Stamens 10, filaments tapering. Ovary 5-celled, disc absent. Style protruding, usually curved. Fruit a drooping spherical capsule, splitting into 5; seeds minute. *Northern hemisphere, extending south to Indonesia & tropical America.*

3. Chimaphila Pursh. 8/2. Perennial, evergreen shrubs or herbs, to 30 cm, with creeping rhizomes. Leaves lanceolate to oblanceolate, opposite or whorled, leathery, with teeth pointing forwards; stalks short. Flowers bisexual, arranged in loose terminal umbels, white or pink, nodding; flower-stalks long; styles short. *Eurasia, north & tropical America.*

4. Orthilia Rafinesque. 3/1. Perennial, evergreen, herbs, somewhat woody at base. Leaves ovate to elliptic, leathery, in a basal rosette. Raceme 1-sided, dense, drooping, with bracts. Petals greenish white; style long, straight, protruding from corolla. *North temperate areas.*

151. ERICACEAE

Trees, shrubs, lianas or sometimes almost herbaceous, sometimes epiphytic, mycorrhizal. Leaves simple, spirally arranged, opposite or whorled, usually evergreen; stipules absent. Flowers in racemes, sometimes solitary, usually bisexual, usually radially symmetric, less often bilaterally symmetric; sepals, petals and stamens hypogynous. Calyx with usually 5 (sometimes 3 or 4, or 6 or 7) lobes, persistent. Corolla-lobes usually 5 (sometimes 3 or 4, or 6 or 7), occasionally lobes free or almost so. Stamens in 2 whorls, usually twice as many as corolla-lobes, sometimes up to 28, rarely 4 or 5. Ovary superior or inferior, of 2--10 (usually 5), united carpels, many-celled (some 1-celled); placentation axile basally, parietal apically or all axile. Fruit a capsule, berry, drupe or rarely a nut. Seeds small, usually numerous, sometimes winged. See figure 94, p. 406.

A family of 103 genera and about 3350 species, nearly cosmopolitan, absent from deserts.

1a. Ovary inferior	2
b. Ovary superior or half-inferior	3
2a. Fruit a drupe	**37. Gaylussacia**
b. Fruit a berry crowned by the persistent calyx	**36. Vaccinium**
3a. Petals free	4
b. Petals united	9
4a. Flowers 6- or 7-parted; stamens 12--14	**1. Bejaria**
b. Flowers 3--5-parted; stamens 10 or fewer	5
5a. Petals usually 3; sepals 3--5; fruit 3-celled	**7. Tripetaleia**
b. Petals usually 4 or 5; sepals 4 or 5; fruit usually with more than 3 cells	6
6a. Deciduous shrubs; petals 4 or 5	7
b. Evergreen shrubs; petals 5	8
7a. Petals usually 4, white	**8. Elliottia**
b. Petals 5, pink with yellow tips	**6. Cladothamnus**
8a. Stamens 10; anthers dehiscing by longitudinal slits	**17. Leiophyllum**
b. Stamens 5 or 10; anthers each opening by 2 apical pores	**5. Ledum**
9a. Corolla persistent, enclosing the capsule	10
b. Corolla deciduous	12
10a. Calyx longer than and concealing corolla	**22. Calluna**
b. Calyx shorter than corolla	11
11a. Stamens inserted at base of prominent disc	**20. Erica**
b. Stamens inserted at base of corolla, disc minute	**21. Bruckenthalia**
12a. Leaves scale-like and overlapping	**26. Cassiope**
b. Leaves not scale-like	13
13a. Fruit a berry or a drupe enclosed in a succulent, enlarging calyx	14
b. Fruit a capsule; calyx not succulent	16
14a. Inflorescence glandular-hairy; anther-appendages bifid	**35. Gaultheria**
b. Inflorescence hairless or slightly hairy; anther-appendages simple	15
15a. Fruit a granular or warty berry	**23. Arbutus**
b. Fruit not as above	**24. Arctostaphylos**
16a. Plant prostrate or decumbent	17

b. Plant upright   24
17a. Leaves 2--10 cm   **9. Epigaea**
   b. Leaves to 2 cm   18
18a. Leaves linear   19
   b. Leaves lanceolate to circular   21
19a. Calyx and corolla 4-parted   **16. Bryanthus**
   b. Calyx and corolla 5-parted   20
20a. Corolla deeply 5-lobed; anthers opening by longitudinal slits
    18. Loiseleuria
   b. Corolla shallowly 5-lobed; anthers opening by apical pores
    11. Phyllodoce & 12. × Phyllothamnus
21a. Calyx fleshy, surrounding the drupe   **35. Gaultheria**
   b. Calyx not as above   22
22a. Stamens 5   **3. Tsusiophyllum**
   b. Stamens more than 5   23
23a. Stamens 6--8   **13. × Phylliopsis**
   b. Stamens 10   **14. Kalmiopsis**
24a. Corolla with 10 pouches which contain the anthers when in bud
    10. Kalmia
   b. Corolla without pouches   25
25a. Calyx fleshy, surrounding the drupe   **35. Gaultheria**
   b. Calyx not as above   26
26a. Corolla with fewer than 5 lobes   27
   b. Corolla of 5 or more lobes   28
27a. Evergreen shrubs to *c.* 65 cm   **19. Daboecia**
   b. Deciduous shrubs to *c.* 2 m   **4. Menziesia**
28a. Seeds 1--few   **25. Enkianthus**
   b. Seeds many   29
29a. Corolla bell- to funnel-shaped   30
   b. Corolla urceolate to cylindric   34
30a. Anthers awned   **33. Zenobia**
   b. Anthers not awned   31
31a. Capsule dehiscence through the cells   **29. Craibiodendron**
   b. Capsule dehiscence through the cross-walls   32
32a. Bracts leaf-like or leathery, persistent   **15. Rhodothamnus**
   b. Bracts scale-like, deciduous at flowering   33
33a. Stamens usually projecting from corolla; anthers opening by terminal
   pores   **2. Rhododendron**
   b. Stamens included; anthers opening by short slits   **4. Menziesia**
34a. Anthers opening by longitudinal slits   **28. Oxydendrum**
   b. Anthers each opening by a terminal pore or slit   35
35a. Leaves densely scaly beneath   **32. Chamaedaphne**
   b. Leaves not scaly beneath   36
36a. Capsule with thickened sutures separating from walls; filaments abruptly
   bent   **30. Lyonia**
   b. Capsule without thickened sutures separating from walls; filaments not
   abruptly bent   37
37a. Anthers with 2 reflexed awns   38

b. Anthers with 4 awns or awns absent   39
38a. Margins of leaves toothed or scalloped   **31. Pieris**
  b. Margins of leaves entire   **27. Andromeda**
39a. Seeds not winged   **4. Menziesia**
  b. Seeds winged   40
40a. Stamens with abruptly bent filaments or spurred anthers
                                              29. Craibiodendron
  b. Stamens with more or less straight filaments, anthers never spurred
                                              34. Leucothoe

1. Bejaria Linnaeus. 25/1. Evergreen shrubs, with branching stems. Leaves alternate, entire. Flowers in terminal racemes or corymbs. Calyx often glutinous; lobes 6 or 7. Corolla white, yellow or red; lobes 6--7, free, spreading, or ascending, somewhat unequal, overlapping. Stamens 12--14; filaments thread-like; anthers opening by terminal pores. Ovary superior, 6--7-celled. Capsule depressed, 6--7-lobed. Seeds not winged. *Mainly South & Central America, but extending north to Florida*

2. Rhododendron Linnaeus. 1000/300 and many hybrids. Trees, shrubs or low shrubs (some very small), usually with hairs and often with lepidote scales. Leaves alternate but usually crowded into false whorls at the ends of the branches, simple, usually entire (rarely finely scalloped), margins often rolled under, mostly evergreen. Bud-scales of vegetative buds usually deciduous, more rarely persistent. Flowers in lateral or terminal racemes (racemes sometimes 1-flowered), enclosed in bud in usually hairy bud-scales (bracts) which fall as the raceme expands; bracteoles 2, bract-like or thread-like, falling with the bud-scales. Calyx 5- or more-lobed, or very reduced and forming a small, sometimes wavy rim, often coloured and petal-like when large. Corolla tubular, campanulate, funnel-shaped or almost flat and disc-like, rarely with a parallel-sided tube and spreading limb, usually 5-lobed (sometimes 6--10-lobed), the lobes shorter than or longer than the tube, variously coloured, often with blotches or spots of contrasting colour on the upper part of the tube or the upper lobes, almost radially symmetric to strongly bilaterally symmetric. Nectar produced at the base of the ovary, either a sticky smear or copious and watery, when sometimes collecting in in darker-coloured pouches at the base of the corolla-tube. Stamens usually 10, but rarely 4, sometimes 5, or sometimes as many as 27, radially arranged or declinate; anthers opening by distinct terminal pores, pollen released as a sticky, thread-forming mass. Ovary with usually 5 cells (sometimes as many as 12); style straight, declinate or sharply deflexed just above the ovary, inserted in a shouldered depression on top of the ovary or the ovary tapering smoothly into it; stigma head-like, often large, composed of a sheath and as many receptive lobes as there are ovary-cells. Capsules many-seeded, woody or soft. Seeds sometimes winged or with long, tail-like appendages at one or both ends. *North temperate areas, especially the Himalayas and western China, extending south of the equator in Asia to north Australia, and numerous in Papua New Guinea.*

Cultivars are listed and briefly described in Fletcher, H.R., *The International Rhododendron Register*, Royal Horticultural Society, London (1958), and supplements to it issued annually by the Society (which is the international registration authority for the genus).

3. Tsusiophyllum Maximowicz. 1/1. Dwarf, densely branched, semi-evergreen shrub to 50 cm. Leaves simple, alternate, 1--2 cm, ovate to lanceolate or oblanceolate, adpressed-hairy above, sparsely so beneath. Flowers to 1.5 cm, campanulate, white, borne in dense clusters of 2--6. Calyx and corolla each 5-lobed. Stamens 5, dehiscing by longitudinal slits. Fruit a 3-celled capsule. *Japan.*

Often included within the genus *Rhododendron.*

4. Menziesia Smith. 7/5. Low, deciduous shrubs. Leaves alternate, entire. Flowers long-stalked, in terminal clusters. Calyx small, flat or saucer-shaped. Corolla campanulate to urceolate, 4- or 5-lobed. Stamens 5--10, sometimes projecting, appendages absent. Anthers opening by slits at the apices. Ovaries 4- (rarely 5-) celled. Fruit a leathery capsule. Seeds many, linear. *Eastern Asia, North America.*

5. Ledum Linnaeus. 10/4. Low, densely branched evergreen shrubs. Leaves alternate, short-stalked, entire, margins often inrolled, aromatic, densely felted-hairy or glandular beneath. Calyx 5-parted. Corolla 5-parted, white. Stamens 5 or 10. Anthers opening by 2 pores at apex. Ovaries 5-celled; styles filamentous; stigma 5-lobed. Fruit a capsule. Seeds numerous, winged. *Northern hemisphere.*

Now often placed in *Rhododendron.*

6. Cladothamnus Bongard. 1/1. Deciduous shrubs to 3 m. Leaves alternate, thin, pale green, slightly glaucous, 2--5 cm, elliptic-oblanceolate to oblanceolate, entire, almost stalkless. Flowers solitary, terminal. Calyx fused only at base, lobes 5, 7--10 mm, narrowly linear-lanceolate. Corolla 1--1.5 cm, lobes 5, oblong-elliptic. Stamens usually 10. Ovary superior, style long and recurved. Fruit 5- or rarely 6-valved; seeds numerous. *North America.*

7. Tripetaleia Siebold & Zuccarini. 2/2. Deciduous shrub. Leaves alternate, entire. Flowers in terminal panicles and racemes. Sepals 3--5. Corolla with 3 free petals (rarely 5) which are somewhat spirally twisted. Stamens 6; anthers dehiscing longitudinally. Fruit a capsule, 3-celled. *Japan.*

8. Elliottia Elliott. 1/1. Deciduous shrub 1--3 m, or rarely a small tree to 6 m, with erect branching stems. Leaves 5--12 × 2--4.5 cm, alternate, ovate or obovate, apex and base acuminate, entire, sparsely soft-hairy beneath. Flowers in terminal racemes or panicles, fragrant. Sepals 3--5, usually 4. Corolla white, petals 6--16 mm, free, hairy on margins, unequal, overlapping, reflexed. Stamens 4--10. filaments flattened, anthers arrow-shaped, opening longitudinally. Ovary superior, 4- or 5-celled. Capsule *c.* 1 cm across, 3--5 lobed. Seeds winged. *Southeast USA.*

9. Epigaea Linnaeus. 2/2. Low evergreen shrubs; stems creeping. Leaves alternate; blades leathery, entire. Flowers in axillary clusters. Bracts several, subtending calyx. Sepals 5, overlapping, persistent. Corolla salver-shaped, white or pink, usually with 5 lobes. Stamens 10, included; filaments fused to the base of the corolla tube. Anthers dehiscing through long slits, appendages absent. Disc 10-lobed. Ovary 5-celled, 5-lobed, hairy; ovules numerous in each cell. Stigma 5-lobed. Capsule depressed-spherical, 5-valved, sparsely hairy. *North America, Japan.*

10. Kalmia Linnaeus. 7/6. Evergreen or deciduous shrubs, rarely small trees; branches terete or 2-angled. Leaves alternate, opposite or in whorls of 3, leathery, margins entire. Flowers solitary or in terminal or axillary racemes and panicles, each flower subtended by a persistent bract. Calyx 5-lobed, green or reddish, hairless, hairy or glandular. Corolla bell- or funnel-shaped, white to rose or purple, petals united, with 5 shallow lobes. Stamens 10, anthers reddish brown, each held in a pouch at the base of corolla. Stigma capitate, with 5-lobes. Fruit a small spherical or more or less spherical 5-valved capsule; seeds numerous. *Eastern North America & Cuba.*

11. Phyllodoce Salisbury. 7/7. Dwarf evergreen shrubs; leaves alternate, linear, leathery, stalks short. Flowers pendent to almost erect, solitary, in terminal clusters, stalks glandular, with 2 bracts at the base. Calyx 5-lobed, divided to near the base. Corolla 5-lobed, urceolate or campanulate, usually shortly lobed; stamens 8--12. Anthers opening by apical pores, lacking appendages. Ovary superior. Fruit a capsule. *North temperate areas.*

12. × Phyllothamnus Schneider. 1/1. Evergreen shrublet, *c.* 30 cm; stems erect. Leaves alternate, crowded, dark green, *c.* 1 cm × 2.5 mm, linear, margin slightly toothed, apex acute. Flowers solitary, in clusters of 6--10 at stem apex, *c.* 1 cm across, campanulate, rose-pink; stalks slender, glandular. *Garden Origin.*

Phyllodoce empetriformis × *Rhodothamnus chamaecistus.*

13. × Phylliopsis Cullen & Lancaster. 1/1. Evergreen shrublet to 30 cm. Bark shiny brown. Leaves 1.5--2 cm × 6--8 mm, alternate, oblong-obovate, apex rounded, base wedge-shaped, shiny, dark green above, pale green with brownish yellow glands beneath. Flowers numerous in long racemes; bracts leaf-like. Sepals 5, margins hairy. Corolla *c.* 1 cm across, campanulate, red-purple, 5-lobed. Stamens 6--8, falling early, not projecting from the corolla. Ovary almost spherical. *Garden Origin.*

Phyllodoce brewerii × *Kalmiopsis leachiana.*

14. Kalmiopsis Rehder. 1/1. Low evergreen shrubs, 15--30 cm, branching from base, branches hairy. Leaves alternate, 7--18 mm, elliptic-obovate, gland-dotted, almost stalkless; bracts 5--6 mm, ovate. Flowers erect, in terminal racemes. Calyx lobes 5, 3--4 mm, ovate, reddish purple, margin with short hairs. Corolla rose-purple, campanulate, tube equalling calyx, lobes 5, *c.* 6 mm, ovate, obtuse, hairy on the back. Stamens 10. Capsule almost spherical. *Western North America.*

15. Rhodothamnus Reichenbach. 1/1. Evergreen dwarf shrubs to 40 cm. Leaves alternate, almost stalkless, margins with stiff, white hairs. Flowers in axils of foliage leaves; stalks long. Bracts persistent. Sepals 5, fused at base. Corolla 2--3 cm across, flattish, lobes 5, pale pink; tube short. Stamens 10; anthers without appendages. Style 1.5--2 cm. Fruit a capsule splitting through the cross-walls. *Eastern Alps.*

16. Bryanthus Gmelin. 1/1. Small evergreen shrubs, 3--8 cm, mat-like, twigs wiry. Leaves 2--3 mm, linear, sparsely and finely toothed, densely white-felted-hairy beneath. Flowers solitary, terminal. Sepals 4. Corolla 3--4 mm across, flattish, pink-red; petals 4. Stamens 8. Fruit a dry capsule. *North Japan, former USSR (Kamchatka & neighbouring islands).*

17. Leiophyllum Hedwig. 1/1. Evergreen shrubs, usually 5--30 cm. Leaves 6--12 mm, alternate or opposite, oblong, ovate or obovate, entire, hairless, glossy, leathery. Inflorescences terminal, umbel-like corymbs. Flowers *c*. 6 mm across. Calyx 5-parted. Petals 5, free, ovate, white to pale pink, spreading. Stamens 10; anthers dehiscing by longitudinal slits. Fruit a capsule, *c*. 3 mm, hairless. Seeds numerous. *Eastern USA.*

18. Loiseleuria Desvaux. 1/1. Low, evergreen shrubs, much branched. Leaves hairless, leathery, entire. Flowers in umbellate clusters at the tips of the branchlets. Bracts leaf-like. Calyx 5-parted, campanulate. Corolla 3--4 mm across, broadly campanulate, deeply 5-lobed, white, rarely pale rose; lobes overlapping in bud. Stamens 5; filaments thread-like, flattened near base; anthers dark purple, each opening by a longitudinal slit. Ovary spherical. Fruit a capsule, *c*. 2.5 mm. Seeds numerous. *Arctic & alpine areas of northern hemisphere.*

19. Daboecia D. Don. 2/2. Evergreen, low-growing shrubs, with sprawling, branched stems. Leaves alternate, on short stalks, not scale-like but with broad, entire blade, leathery, the margins slightly curved downwards and inwards. Flowers large, nodding, in elongate racemes. Sepals 4, not becoming fleshy. Corolla urn-shaped, with 4 short, recurved lobes, deciduous. Stamens 8; anthers opening by pores, without awns. Ovary superior. Fruit a dry capsule. *Azores, western Europe.*

20. Erica Linnaeus. 800/30. Evergreen shrubs or shrublets, usually erect, sometimes prostrate; leaves undivided, in whorls, usually linear, margins curved downwards and inwards concealing some or all of the lower surface, hairless or with hairs; hairs often with glandular tips. Flowers clustered in terminal umbels or racemes, or in panicles. Flower-stalks with 2 or 3 bracteoles. Sepals 4, not fused, usually very short, less than half the length of corolla, green or coloured. Corolla cylindric to spherical with 4 apical lobes, persisting in fruit. Stamens 8, filaments longer or shorter than corolla so anthers visible or concealed; anthers opening by pores, 2-celled, sometimes with a pair of very prominent basal awns. Ovary with 4 carpels, ovules numerous. Fruit a dry, sometimes woody capsule; seeds numerous. *Mainly South Africa, also east & north Africa, Mediterranean area, Europe.* Figure 94, p. 406.

The vast majority of the species are native in the Cape region of South Africa. The African species were popular plants at the beginning of the nineteenth century in western Europe, but they declined in popularity and now very few Cape heaths are cultivated generally in European gardens. Most of the native European species are widely cultivated, and some have spawned numerous cultivars. Hybrids, both accidental and deliberate crosses, are also frequent.

21. Bruckenthalia Reichenbach. 1/1. Evergreen, tufted, dwarf shrub to 20 cm; shoots erect, downy with glandular hairs, becoming hairless. Leaves in whorls of 4 or 5, scale-like, linear, spreading, to 6 mm, with fine, short hairs and some longer glandular hairs, and a long glandular bristle at tip, margins curved downwards and inwards almost concealing the white undersurface. Flowers in dense, erect, terminal racemes. Sepals and petals of similar texture and colour, bright pink (very rarely white); flower-stalk *c*. 3 mm. Calyx campanulate, sepals 4, fused for about half their length *c*. 1.5 mm; corolla to 3 mm, campanulate, not constricted at mouth, with 4 rounded lobes, not deciduous. Stamens 8, fused at base to corolla, anthers without awns. Style projecting. *South-east Europe, western Asia.*

22. Calluna Salisbury. 1/1. Evergreen shrub. Leaves minute, scale-like, opposite. Flowers small, in raceme-like inflorescence produced at tips of shoots. Sepals and petals of similar size, texture and colour; sepals 4, scarious, oblong-ovate, to 4 mm, pale purple; petals 4, not fused together, not deciduous, shorter than sepals, pale purple. Stamens 8; each anther with 2 horn-like awns, opening by pores; nectar produced by swellings at base of stamens. Fruit a capsule. *Europe, Morocco, western Asia.*

Selected clones are numerous and widely available.

23. Arbutus Linnaeus. 12/4. Evergreen or semi-deciduous trees or bushy shrubs, often with attractive bark; some species producing woody, swollen, underground lignotubers from which shoots readily sprout. Leaves leathery, margins entire or toothed. Flowers numerous, in erect or pendent clusters. Sepals 5, fused at base. Corolla usually urn-shaped, with 5 apical lobes. Stamens 10, each anther with 2 awns. Fruit a fleshy berry with numerous seeds, dry to juicy, edible and popularly considered to resemble a strawberry. *Europe, Mediterranean area, Canary Islands, western N America.*

24. Arctostaphylos Adanson. 50/13. Prostrate or erect shrubs and small trees with sinuous twigs. Leaves typically evergreen, alternate, stalked or stalkless with entire or occasionally toothed margins. Flowers small, pendent, borne in racemes or panicles. Sepals usually 5, occasionally 4, free. Corolla usually 5-lobed, bell- or urn-shaped. Stamens usually 10, opening by terminal pores. Ovary 2--10-chambered, surrounded at the base by a nectar disc. Style 1. Fruit a round, berry-like drupe containing 2--10 nutlets that are freely separable or strongly fused. *North & central America.*

25. Enkianthus Loureiro. 10/4. Deciduous or more or less deciduous shrubs. Bracts and bracteoles poorly developed. Corolla 5-parted, urn-shaped to campanulate. Stamens 10, strongly dimorphic, filaments swollen at base; anthers awned, each dehiscing by a longitudinal slit. Ovary 5-celled. Fruit a capsule. Seeds few to several, usually winged. *Himalaya to Japan.*

26. Cassiope D. Don. 12/7. Dwarf, evergreen, often sprawling shrubs. Stems horizontal-ascending. Leaves small, overlapping, arranged in 4 distinct ranks, often adpressed to and concealing the stems, grooved or convex on the back; stalks absent or very short. Flowers small, white to pink, solitary, axillary; stalks hairy or hairless, erect or pendent. Sepals usually 5 (occasionally 4), shorter than corolla, often persistent. Petals fused into a usually 5-, occasionally 4-lobed urn- or bell-shaped tube below. Stamens 10 (occasionally 8); anthers almost spherical, dehiscing through large pores and bearing 2 large projections. Ovary 4- or 5-celled. Fruit a spherical, many seeded, dry, capsule. *North temperate and Arctic areas.*

27. Andromeda Linnaeus. 2/2. Evergreen, dwarf shrubs, with creeping rhizomes, and erect, usually unbranched stems. Leaves alternate, on short stalks, not scale-like but with broad, entire blades, leathery, stiff and with margins curved downwards and inwards. Flowers in umbel-like clusters at tips of stems. Sepals 5, not becoming fleshy. Corolla urn-shaped, with 5 short lobes, deciduous. Stamens 10. Anthers with 2 horn-shaped, recurved awns, opening by pores. Ovary superior, with nectar-secreting swellings around base. Fruit a dry capsule. *Northern hemisphere.*

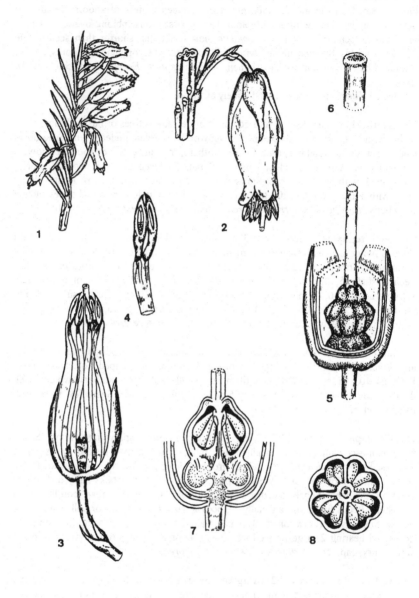

Figure 94. Ericaceae. *Erica carnea*. 1, Inflorescence. 2, Single flower. 3, Longitudinal section of flower (shown upright). 4, Upper part of stamen showing anther-pores. 5, Base of flower with calyx, corolla and stamens removed. 6, Stigma. 7, Longitudinal section of ovary. 8, Transverse section of ovary.

28. Oxydendrum de Candolle. 1/1. Deciduous tree or shrub; bark deeply furrowed. Leaves alternate, 8--20 cm, oblong-lanceolate, with fine forward-pointing teeth, veins slightly hairy beneath, bronze-green when young, light green above and grey-white beneath in summer, turning scarlet-red in autumn. Flowers in 1-sided terminal panicles. Calyx 5-parted to the base. Corolla small, tubular, white, 5-lobed. Stamens 10, each dehiscing by a longitudinal slit at apex. Fruit a capsule, 5-celled, dehiscing through the cells. Seeds many. *Eastern USA.*

29. Craibiodendron W.W. Smith. 5/1. Evergreen shrubs or trees; branches often with sparse, gland-tipped hairs. Leaves alternate, simple, leathery, often reddish when young, with gland-tipped hairs on both surfaces (hairs often falling early). Inflorescences axillary. Flowers pendent, stalk elongating in fruit. Calyx usually of 5 overlapping lobes. Corolla urn-shaped to cylindric or bell-shaped; lobes usually 5, overlapping. Stamens usually 10, in 2 whorls, inserted at base of corolla. Anthers dehiscing by terminal pores. Fruit a capsule. *China, eastern India, Burma, Laos, Thailand, Cambodia & Vietnam.*

30. Lyonia Nuttall. 35/4. Deciduous or evergreen shrubs or trees. Leaves alternate, entire or toothed. Flowers in axillary clusters, racemes or terminal panicles. Corolla urceolate or tubular-campanulate; lobes 5, short. Filaments abruptly bent. Ovary superior. Fruit a capsule with thickened sutures. *East & south-east Asia, eastern North America, West Indies.*

31. Pieris D. Don. 7/5. Evergreen trees, shrubs or woody vines. Branches terete, bark brown or grey, longitudinally striped. Leaves alternate or whorled, entire or toothed, sparsely covered in glandular or multicellular hairs. New growth prominently red or bronze. Inflorescence a terminal or axillary raceme or panicle, produced in year prior to flowering. Flowers stalked, often scented, subtended by 1 bract and 2 bracteoles. Calyx and corolla 5-lobed. Corolla urceolate, usually white. Stamens 10, in 2 whorls, filaments flattened or bent, with pair of spur-like appendages, base of stigma sometimes sunken into ovary. Capsule erect, more or less spherical, with 5 sutures; seeds numerous, with thin testa. *East Asia, Himalaya, eastern USA, West Indies.*

32. Chamaedaphne Moench. 1/1. Evergreen, much-branched shrubs, 17--50 cm. Leaves leathery, dull green above, rust-green and scaly beneath, 1--1.5 cm × 3--15 mm, oblong-oval to oblong-lanceolate, obtuse or almost acuminate, margin faintly toothed. Flowers pendent, in 1-sided leafy terminal clusters. Calyx-lobes 5, 1.5--3 mm, oblong-ovate, persistent in fruit. Corolla white, 5-lobed, 4.5--6.5 mm, oblong-ovoid. Stamens 10, filaments swollen at base, anthers opening by terminal pores. Capsule flat-spherical. *Northern hemisphere.*

33. Zenobia D. Don. 1/1. Shrubs with grooved branches. Leaves alternate; blade entire or irregularly toothed, net-veined. Flowers in axillary clusters. Sepals 5, united. Corolla campanulate, white or pink; lobes 5, rounded. Stamens 10, included; filaments expanded at base; anthers awned. Disc 10-lobed. Ovary superior, 5-celled. Fruit a capsule dehiscing through the cells. *Southeast USA.*

34. Leucothoe D. Don. 8/6. Evergreen or deciduous shrubs. Leaves more or less toothed. Inflorescence axillary; bracts and bracteoles present. Calyx-segments 5, free.

Corolla urceolate, 5-lobed. Stamens 10; filaments swollen towards base, more or less straight; anthers with 1 or 2 pairs of terminal awns, opening by a terminal pores. Ovary 5-celled. Fruit a capsule. Seeds winged. *North & South America, Himalaya, eastern Asia, Madagascar.*

35. Gaultheria Linnaeus. 135/45. Evergreen, creeping to erect shrubs; stems prostrate, arching or erect. Leaves leathery, entire to finely toothed, variable, apex occasionally mucronate. Flowers usually bisexual but occasionally unisexual; solitary or in racemes or panicles, usually in the axils of leaves. Corolla white, pink or red, 2--15 mm, lobes 5 (rarely 4). Stamens 10 (rarely 8 or 5), often with terminal awns. Fruit a capsule surrounded by an enlarged fleshy calyx, a capsule with the calyx unchanged or a berry. Some species smell of oil of wintergreen. *East Asia, Australia, New Zealand, South, Central and North America.*

Includes *Chiogenes* Torrey and *Pernettya* Gaudichaud.

36. Vaccinium Linnaeus. 450/35. Evergreen or deciduous shrubs, sometimes trees. Leaves alternate, entire or with fine, forwardly pointing teeth, shortly stalked. Flowers borne in terminal or axillary racemes, clusters or solitary. Calyx 4- or 5-lobed or unlobed. Corolla 4--5-lobed, urceolate, campanulate or cylindric, white, pink or red. Stamens 8--10. Fruit a many-seeded berry with a persistent calyx. *Northern hemisphere, from the Arctic Circle to the tropics.*

37. Gaylussacia Kunth. 48/4. Shrubs with erect or underground stems. Leaves alternate; blades usually entire, usually glandular. Flowers in axillary, drooping racemes. Sepals 5. Corolla campanulate to tubular-conical, grooved or 5-angled; lobes 5, erect or reflexed. Stamens 10, usually included in the corolla; filaments distinct, more or less winged. Disc ring-like or swollen. Ovary inferior, 10-celled. Fruit a drupe with a 10-celled stone; nutlets bony or horny. Seeds solitary, flattened. *New World.*

152. EMPETRACEAE

Evergreen, heath-like shrubs. Leaves small, alternate or almost arranged in whorls, simple, each with a swelling at the base; margins curved downwards and inwards with a deep furrow beneath; stipules absent. Flowers small, radially symmetric, perianth and stamens hypogynous, bisexual or unisexual. Perianth of 3--6 segments more or less in 2 series; outer segments 2 or 3, overlapping, inner segments 2 or 3, overlapping, or 3 or 4 in a single series. Stamens 2--4, alternate with inner perianth-segments when these are distinct from outer segments. Fruit a berry or drupe with 2--9 seeds.

A family of 3 genera and 3--9 species confined to northern temperate and arctic areas, and southern South America.

1a. Flowers solitary in leaf-axils; stamens 3; fruit a drupe with 6--9 nutlets
1. Empetrum
 b. Flowers in terminal inflorescences; stamens 3 or 4; fruit a berry with 3 seeds
2. Corema

1. Empetrum Linnaeus. 6/2. Dwarf shrubs, much branched. Flowers inconspicuous, greenish or purplish, almost stalkless, solitary, axillary or scattered; stamens 3. Fruit a drupe with 6--9 nutlets. *Widespread in temperate areas.*

2. Corema D. Don. 2/1. Diffusely branched, mostly dioecious, small shrubs. Leaves stiff, linear, mostly in whorls of 3, numerous and crowded. Flowers usually unisexual, in terminal clusters. Fruit a 3-seeded berry. *Azores, Canary Islands, southwest Europe, eastern USA.*

153. EPACRIDACEAE

Shrubs or small trees. Leaves alternate, rigid, simple, stalked or stalkless, without stipules. Flowers small, radially symmetric, bisexual, solitary or in spikes, racemes or panicles, subtended by bracts. Sepals, petals and stamens hypogynous, or stamens and petals perigynous. Sepals 4 or 5, free, persistent. Corolla tubular, with 4 or 5 lobes, overlapping or edge-to-edge in bud. Stamens borne on the corolla or not, alternate with corolla-lobes, with tufts of hair or glands between, anthers usually 1-celled. Ovary superior, often with a glandular disc at the base, with 1--10 cells, ovules 1--many per cell, placentation axile. Style simple. Fruit a capsule or drupe.

A family of 31 genera and 400 species largely confined to Australia with some species in New Zealand, Indo-Malaysia and South America.

1a. Leaves usually more than 2 cm with broad sheathing bases; stems
   with annular leaf-scars; corolla-lobes united in the form of a cap **1. Richea**
  b. Leaves usually less than 2 cm; stems without annular leaf-scars; corolla-
   lobes not united in form of a cap   2
2a. Style inserted in a deep depression at apex of ovary; fruit a capsule;
   corolla-lobes with indentations near base   **2. Epacris**
  b. Style tapering from the ovary; fruit indehiscent; corolla-lobes without
   indentations near base   3
3a. Flowers solitary or in racemes; fruit a drupe, not separating into pyrenes
                                   3. Cyathodes
  b. Flowers in spikes; fruit separating into 8--10 pyrenes   **4. Trochocarpa**

1. Richea R. Brown. 11/2. Evergreen shrubs or small, often unbranched and palm-like trees. Stems with annular leaf-scars. Leaves to 1 m or more, tapered, hard and leathery, clasping stem with broad sheathing bases, closely adpressed then spreading, upright or reflexed and mostly confined to ends of branches, apex hard, abrupt or tapered to a long, hard, sharp point, margins with minute, forwardly pointing teeth. Flowers in a terminal spike or panicle, pink, orange or white. Corolla-lobes not opening, remaining joined in the form of a conic, truncate or ovoid cap, splitting transversely near base and shedding. Stamens 4 or 5, persisting, arising from receptacle; anthers 2-lobed or entire, continuous with filament or free at the lower end. Style embedded in a depression in the top of the ovary; stigma small, head-like. Fruit a dehiscent 5-celled capsule. Seeds small, numerous. *Southeast Australia, Tasmania.*

2. Epacris Cavanilles. 35/1. Evergreen shrubs, erect, rigid. Leaves to 1.5 cm, crowded or overlapping, jointed at the base, lanceolate, sharp-pointed. Flowers solitary, axillary, white, red or pink. Bracts covering inflorescence-stalk and base of calyx. Corolla-tube much longer than calyx, cylindric or campanulate with 5 overlapping, spreading lobes with indentations near base. Stamens fully or partly included in corolla-tube; filaments inserted on corolla-tube; anthers attached above middle. Ovary 5-celled, with several ovules in each cell. Style inserted in a deep, apical depression in ovary. Fruit a capsule, seeds numerous. *Southeast Australia, New Zealand, New Caledonia.*

409

3. Cyathodes Labillardière. 15/3. Evergreen shrubs, prostrate to erect. Leaves small, scattered or overlapping, flat or convex, striped, often white beneath. Bracts few to numerous. Corolla-tube slightly longer than calyx; in all cultivated species the lobes are hairy. Sepals *c.* 2 mm, pointed. Stamens borne on corolla-tube. Fruit a crimson to pink drupe with 2--10 seeds. *Australia, New Zealand.*

4. Trochocarpa R. Brown. 12/1. Evergreen trees and shrubs. Leaves leathery, mostly alternate, usually stalked, flat or convex. Flowers stalkless, subtended by 1 bract and 2 bracteoles, in terminal or axillary spikes. Corolla-tube cylindric or campanulate, hairless or with reflexed hairs inside; lobes spreading, usually shorter than the tube. Stamens 5, inserted at top of corolla-tube and alternate with lobes. Ovary 10-celled, 1 ovule per cell. Fruit fleshy with a single stone which separates into 8--10 one-seeded nutlets or pyrenes. *North Borneo to Australia.*

154. MYRSINACEAE

Trees or shrubs, sometimes prostrate or climbing, sometimes dioecious, usually with translucent or coloured dots or stripes in the soft parts. Leaves without stipules, alternate and spirally arranged or rarely whorled, sometimes clustered on axillary short shoots, simple, entire or toothed. Flowers usually radially symmetric, small, usually with parts in 4s or 5s. Sepals hypogynous, petals and stamens perigynous. Sepals often and petals usually united at base. Stamens as many as and on the same radii as corolla-lobes and usually united with the corolla at base. Anthers opening towards the centre of flower by slits or apically by pores. Ovary superior or half-inferior, 1-celled with 1 style. Ovules few to many, usually embedded in the free-central placenta. Fruit a berry or drupe.

A family of 39 genera and about 1250 species, mostly in warm regions and the tropics, a few in temperate parts of the Old World. Only 1 is cultivated out-of-doors in northern Europe.

1. Myrsine Linnaeus. 155/3. Evergreen trees or shrubs, the latter sometimes trailing. Flowers usually in dense axillary clusters, without bracteoles, unisexual or bisexual, with parts in 4s in our species. Sepals and petals not twisted in bud. Petals free or united. Anthers opening by slits. Ovules in 1 row. Fruit a 1-seeded drupe, dry or fleshy. *Subtropics & tropics.*

Here defined in a broad sense (including *Rapanea* Aublet and *Suttonia* Richard).

155. PRIMULACEAE

Herbs, rarely dwarf shrubs. Leaves usually simple, often toothed or lobed, more rarely pinnate, alternate or all basal, rarely whorled; stipules absent. Flowers 5--7-parted, usually radially symmetric. Calyx hypogynous, corolla and stamens perigynous or rarely all whorls epigynous. Calyx of 5--7 united sepals, lobed at the apex. Corolla 5--7-lobed, rarely absent. Stamens as many as and on the same radii as the corolla-lobes, borne on the corolla-tube; staminodes sometimes present; anthers opening by slits. Ovary superior (rarely semi-inferior), 1-celled, with 1--many ovules; placentation basal or free central; style simple. Fruit a capsule. See figure 95, p. 414.

A family of about 22 genera, almost cosmopolitan in distribution but mainly from the northern hemisphere.

1a. Plants aquatic; leaves pinnate   **1. Hottonia**

b. Plants terrestrial; leaves simple   2
2a. Corolla-lobes sharply deflexed   3
  b. Corolla-lobes not sharply deflexed   4
3a. Rootstock not tuberous; stamens forming a ring around the protruding
      stigma   **8. Dodecatheon**
  b. Rootstock tuberous; stamens not as above   **9. Cyclamen**
4a. Flowers bilaterally symmetric and calyx spiny   **14. Coris**
  b. Flowers radially symmetric, rarely bilaterally symmetric; calyx not spiny   5
5a. Leaves basal, often in a rosette; flowers solitary or on a leafless scape   6
  b. Stems leafy, bearing flowers   10
6a. Corolla-lobes laciniate   **7. Soldanella**
  b. Corolla-lobes not laciniate   7
7a. Filaments with dilated bases, fused   **6. Cortusa**
  b. Filaments without dilated bases, free   8
8a. Corolla-throat constricted   **4. Androsace**
  b. Corolla-throat not or scarcely constricted   9
9a. Flowers 5-parted; bracts present   **2. Primula**
  b. Flowers 5--8-parted; bracts absent   **3. Omphalogramma**
10a. Flowers with corolla-like calyx and no corolla   **12. Glaux**
  b. Flowers with calyx and corolla   11
11a. Ovary partly inferior   **15. Samolus**
  b. Ovary superior   12
12a. Corolla shorter than calyx   **13. Anagallis**
  b. Corolla equalling or longer than calyx   13
13a. Flowers 7-parted   14
  b. Flowers 5- or 6-parted   15
14a. Leaves in a whorl at stem-apex, sometimes with a few alternate
      below; flowers white, solitary   **11. Trientalis**
  b. Leaves alternate or opposite, occasionally in whorls; flowers yellow,
      purple or white, solitary or in clusters   **10. Lysimachia**
15a. Capsule parting or splitting along a line around the circumference
      13. Anagallis
  b. Capsule opening by valves   16
16a. Flowers with stamens and styles of different relative lengths   **5. Dionysia**
  b. Flowers all with stamens and styles of the same relative length   17
17a. Corolla-throat constricted   **4. Androsace**
  b. Corolla-throat open   **10. Lysimachia**

1. Hottonia Linnaeus. 2/1. Aquatic perennial herbs. Leaves more or less whorled, pinnate. Flowers aerial. Corolla longer than calyx, 5-lobed. Stamens 5. Seeds numerous. *Europe, west Asia, western North America.*

Two species, 1 grown in clean, more or less neutral water.

2. Primula Linnaeus. 425/130 and many hybrids. Rhizomatous or stoloniferous perennial herbs (sometimes annual). Leaves in basal rosettes, very rarely on the stem, usually simple, revolute (margins curved downwards and inwards) or involute (margins curved upwards and inwards) in bud, with 1- or many-celled hairs or hairs absent. Scapes commonly present and often lengthening after opening of flower. Inflorescence terminal, whorled or umbellate with involucral bracts, or a simple raceme, sometimes

411

spike-like. Most species are heterostylous with all flowers on the plant having either a long style with the stigma sitting at the mouth of the tube like a pin and anthers deep in the corolla ('pin flower') or, a short style with the stigma buried deep in the corolla-tube and anthers attached near the mouth of the corolla ('thrum flower'). Calyx tubular or campanulate, persistent, 5-lobed. Corolla salver- or funnel-shaped to more or less campanulate; tube usually exceeding calyx, its mouth often with a thickened rim; lobes 5, entire, bifid or toothed to lacerate. Stamens 5. Style slender; stigma head-like. Capsule spherical to tubular-cylindric, contained within or exceeding calyx, dehiscing by splitting apically at maturity into 5 or 10 lobes or breaking irregularly, or rarely with a dehiscent cap. Seeds many, ovoid angular to almost spherical or compressed, with papillae. *Northern hemisphere, mainly the Himalaya & west China, a few in north-east Africa, Arabia & South America.* Figure 95, p. 414.

3. Omphalogramma (Franchet) Franchet. 12/3. Perennial herbs, overwintering as resting buds; rootstock often woody. Leaves usually arising from a sheath of scales, to 20 × 5 cm, more or less ovate, margin curved upwards and inwards, covered with stalkless glands; stalks to 30 cm. Flowers emerging from base before or at same time as leaves, solitary on stalks to 35 cm, covered with long jointed hairs, lacking bracts. Calyx- and corolla-lobes 5--8; calyx to 2 cm, divided to base into narrow segments; corolla to 6 cm, funnel-shaped, tube usually shorter than the lobes, dark blue to violet-rose. Stamens at rear of tube erect, those at front bent across tube. Capsule with upper one-quarter projecting from calyx, dehiscing from apex almost to base. Seeds flat-winged. *Asia, mainly Himalaya & western China.*

4. Androsace Linnaeus. 150/4. Annual, biennial or perennial, mat- or cushion-forming alpine herbs, often dwarf, usually with simple basal, somewhat fleshy leaves. Flowers with bracts, solitary or in umbels. Calyx campanulate with 5 united sepals. Corolla-lobes 5; tube usually shorter than the calyx-tube, throat constricted and closed by a ring of coloured scales. Filaments free, very short, attached to upper corolla-tube. Style usually shorter than corolla-tube. Capsule spherical, dehiscing almost to the base. Seeds usually numerous and minute. *West Europe to China, North America.*

Sometimes split into the smaller genera *Douglasia* Lindley, *Gregoria* Duby and *Vitaliana* Sesler.

5. Dionysia Fenzl. 44/15. Woody-based perennials or dwarf compact shrubs, forming loose to dense tufts, usually sweetly aromatic. Stems usually becoming thick and woody below in older plants, covered for part or all of their length in spreading or overlapping, persistent withered leaves or leaf-bases. Leaves simple, spirally arranged, evenly spaced or crowded into whorls, but stems always terminating in a loose or congested leaf-rosette; covered on each or either surface with jointed hairs and stalkless or stalked glands, rarely almost hairless; with or without meal. Inflorescence terminal, stalked, bearing a simple umbel or 2--7 whorls of flowers, or reduced to a single stalkless or shortly stalked flower. Bracts leaf-like to linear or lanceolate. Flowers usually heterostylous, radially symmetric. Calyx tubular to campanulate, 5-lobed. Corolla yellow, pink or violet; tube slender, at least 3 times length of calyx, swollen at insertion of stamens; limb 5-lobed, flat. Stamens 5. Fruit capsule dehiscing by 5 flaps when mature, usually about half as long as persistent calyx. Seeds few to numerous. *Southwest Asia to Afghanistan, Pakistan & Oman.*

6. Cortusa Linnaeus. 8/2. Perennial herbs; leaves basal, blade heart-shaped to circular, lobed, toothed; stalk long. Inflorescence an umbel, with bracts. Corolla much longer than calyx, funnel-shaped to campanulate, divided to half or more into 5 lobes. Filaments with swollen bases. Seeds numerous. *Central Europe to north Asia.*

7. Soldanella Linnaeus. 10/10 and hybrids. Perennial herbs to 30 cm; young parts often glandular-hairy. Leaves basal, 2--10 cm wide, circular to kidney-shaped, entire, leathery, evergreen; stalk long. Scapes 1--8-flowered; flowers pendent, 5-parted; corolla much longer than calyx, funnel-shaped to campanulate, lobes fringed, blue, violet or occasionally white. Fruit a capsule, seeds numerous. *Mountains of Europe.*

8. Dodecatheon Linnaeus. 13/10. Herbaceous perennials, to 60 cm, consisting of basal rosette of leaves on short underground rootstock. Bulbils sometimes produced during flowering season. Leaves entire to toothed, ovate to elliptic or spathulate, dying down towards the end of summer. Flowers characterised by the reflexed petals, borne singly or in terminal umbels. Corolla 4- or 5-lobed, magenta to lavender or white. Filaments short, free or united; stamens forming a ring around protruding stigma. Fruit an ovoid, 1-chambered capsule, opening by blunt or sharp-toothed flaps. *North America, east Asia (Siberia).*

9. Cyclamen Linnaeus. 19/18. Perennial herbs, with tuberous rootstock varying in size (1.5--24 cm across), shape and surface texture. Leaves simple, alternate, long-stalked, arising directly from points on the upper surface of the tuber or at the ends of underground, woody, simple or branched trunks called floral trunks. Leaf-blade heart-shaped to kidney-shaped or almost circular, entire or toothed, lobed or angled. Flowers solitary, nodding. Stalks long, erect or decumbent, covered with brownish, reddish or purplish glands which are densest towards the top, often coiling spirally in fruit. Calyx deeply 5-lobed, closely adpressed to corolla-tube, persistent in fruit. Corolla purple, pink or white, sometimes red, scarlet or mauve, often with a darker zone or blotching at the mouth; lobes 5 (occasionally 6), fused near base, twisted in bud, but reflexed at flowering time, each with or without an auricle at point of reflexion. Stamens 5, not usually projecting. Style solitary. Ovary spherical or almost so, superior, of 5 fused carpels. Fruit an almost spherical capsule, splitting irregularly into 5--7 recurved, triangular teeth. Seeds 8 or more, relatively large, spongy, covered with a sticky sugary coating. *Europe, North Africa (to Somalia), west Asia.*

10. Lysimachia Linnaeus. 180/15. Perennial or annual herbs, rarely dwarf shrubs, with creeping or erect stems. Leaves alternate or opposite, occasionally arranged in whorls, linear-lanceolate to ovate, margins entire. Flowers borne in leaf-axils, solitary or in clusters, or in terminal racemes or panicles. Corolla-lobes 5--7, dark-red or purple, yellow or white. Stamens 5--7, anthers projecting from corolla or not. Fruit a more or less spherical capsule with 5--7 chambers containing numerous wrinkled seeds. *Northern hemisphere, South Africa.*

Several of the species have been placed in the segregate genera, *Steironema* Rafinesque and *Naumburgia* Moench.

11. Trientalis Linnaeus. 1/1. Hairless perennials. Stems 5--30 cm, usually unbranched. Leaves usually in a whorl at stem-apex, simple. Flowers usually 7-parted, axillary,

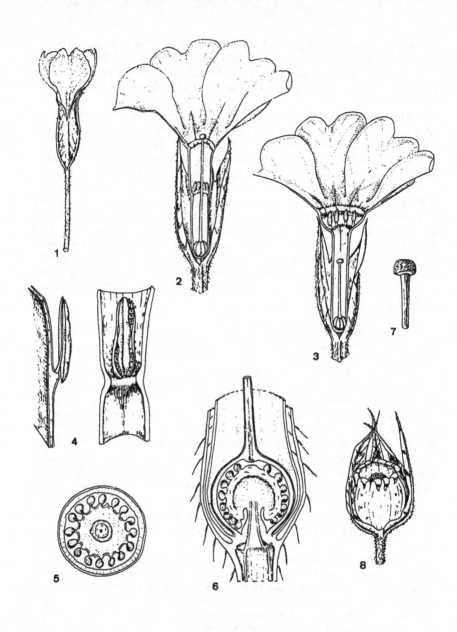

Figure 95. Primulaceae. *Primula vulgaris.* 1, Flower from the side, just opening. 2, Longitudinal section of pin-eyed flower (style exceeding stamens). 3, Longitudinal section of thrum-eyed flower (stamens exceeding style). 4, Attachment of stamen to corolla-tube. 5, Transverse section of ovary. 6, Longitudinal section of ovary. 7, Stigma. 8, Opened fruit.

414

solitary, lacking bracts. Calyx divided almost to base; corolla usually white. Capsule spherical. Seeds few. *Northern Europe.*

13. Glaux Linnaeus. 1/1. Hairless perennial herb, somewhat succulent, rhizomatous. Stems to 30 cm, creeping or erect, rooting at nodes. Leaves opposite, simple, 4--14 × 1.5--6 mm. Flowers solitary in leaf-axils; calyx corolla-like, 5-lobed, white to purple or pink; corolla absent. Seeds few. *North temperate area.*

13. Anagallis Linnaeus. 20/3. More or less annual or perennial herbs, stems erect or creeping, leafy, leaves opposite or alternate, rarely in whorls of 3, simple, stalkless or shortly stalked. Flowers 5-parted, solitary and axillary in all cultivated species; stamens usually free, filaments bearded. Fruit a spherical capsule; seeds numerous, very small. *Europe, Africa, South America; 1 species pantropical.*

14. Coris Linnaeus. 2/2. Perennials or biennials. Stems woody at base. Leaves simple, alternate. Flowers bilaterally symmetric. Calyx campanulate, with 2 rows of teeth. Stamens projecting. Fruit a spherical capsule. Seeds 4--6. *South Europe, Somalia.*

16. Samolus Linnaeus. 15/2. Hairless perennial herbs with fibrous roots. Leaves basal and alternate on stem, simple. Inflorescence a raceme, with bracts. Flowers 5-parted; calyx campanulate; corolla twice as long as calyx. Stamens alternating with clusters of up to 3 staminodes. Ovary at least partly inferior. Capsule opening by 5 teeth. *Cosmopolitan.*

156. PLUMBAGINACEAE

Herbs, shrublets and climbing shrubs, often bearing chalk-glands. Leaves simple, entire or lobed, usually forming a basal rosette, lacking stipules. Inflorescences simple or compound, bracteate, racemose or cymose; bracts usually sheathing, dry and membranous. Flowers bisexual, often heterostylous, often in more or less dense spikelets, each usually subtended by 1 or 2 sheathing bracteoles. Calyx hypogynous, corolla and stamens perigynous. Calyx forming a 5- or 10-ribbed tube which is usually translucent and sometimes petal-like, rarely of 5 almost free sepals. Corolla more or less united but lobes often nearly free, each lobe overlapping and overlapped by 1 other in bud. Glandular disc present, or 5 glands alternating with the sepals. Stamens 5, on the same radii as the corolla-lobes, basally adherent to petals or free. Ovary 1-celled with solitary basal ovule and 5 styles or styles fused with a 5-lobed stigma, in fruit persistent and indehiscent or splitting around the circumference, rarely opening by valves. See figure 96, p. 418.

A family of between 15 and 27 genera depending on circumscription, cosmopolitan but with two distinct subfamilies, *Staticoideae* and *Plumbaginoideae*; the former is centred in the Mediterranean area and southwest Asia, while the latter is widely distributed in tropical and temperate areas. Generally sun-loving and insect-pollinated, the family includes many plants which are found in the wild in salty soils.

1a. Corolla with slender tube equalling or exceeding calyx		2
b. Corolla divided almost to base; petals more or less free		3
2a. Calyx-ribs with prominent stalked and stalkless glands	**1. Plumbago**	
b. Calyx 5-ribbed, ribs not glandular	**2. Ceratostigma**	
3a. Flowers in a dense head borne on scapes	**6. Armeria**	

b. Flowers in loose, much-branched inflorescences, not as above   4
4a. Stem leaves spiny; plant forming dense cushions   **5. Acantholimon**
  b. Leaves all in a basal rosette or stem-leaves scale-like; plant not forming dense cushions   5
5a. Bracts entire   **3. Limonium**
  b. Bracts 3-lobed   **4. Goniolimon**

1. Plumbago Linnaeus. 15/1. Annual herbs, perennial subshrubs or shrubs, sometimes climbing. Leaves alternate, simple, sometimes clasping and with auricles at the base, often scurfy with limescale. Inflorescence a terminal, spike-like raceme, loose or compact, flowers usually with 2 or 3 bracts, occasionally paired. Calyx tubular, somewhat translucent, 5-ribbed, usually with prominent stalked glands. Corolla with a slender tube exceeding the calyx and 5 spreading lobes. Stamens 5, rarely 4, borne on the same radii as the corolla-lobes, the filaments often broad-based and shortly united. Ovary superior, spherical to oblong; style 1, with 5 stigma-lobes. Fruit a 1-seeded capsule enclosed by the calyx, splitting into 5 segments from the base. *Warm temperate & tropical areas.*

2. Ceratostigma Bunge. 8/5. Perennial herbs or shrubs. Stems wiry, often ribbed and angled at the nodes in zig-zag fashion. Leaves alternate, simple, margins hairy or spine-toothed, often bristly and with scurfy scales. Inflorescence a compact terminal or axillary head comprising spikelets of 1 or 2 flowers, each subtended by 3 bracts with hairy margins. Calyx tubular, somewhat translucent, 5-ribbed. Corolla with a slender tube and 5 spreading lobes, obovate, truncate or notched and mucronate. Stamens 5, free, sometimes inserted on the corolla-tube. Style 1, with 5 stigma-lobes, gland-dotted or, more commonly, with peg-like glandular protuberances. Ovary superior. Fruit a membranous 1-seeded capsule enclosed by the calyx, splitting into 5 segments from the base. Seeds with scattered stellate hairs. *Himalaya, China, east Africa.*

3. Limonium Miller (*Statice* Linnaeus). 300/15. Shrubs or perennial herbs, more rarely biennials or annuals. Leaves simple, lobed or entire, usually in basal rosettes (though leafy branches do occur in some species); leaves often withered by flowering time. Inflorescence an upright, spreading or arching complex panicle, made up of ultimate branches (spikes); flowers borne in 1--8-flowered spikelets on the spikes, each spikelet being subtended by 3 bracts of which the inner is generally smaller than the 2 outer; flowers stalkless. Calyx cylindric to funnel-shaped, limb usually papery and coloured, irregularly 5-lobed, persistent in fruit. Corolla with a short tube and 5 lobes, usually papery, persistent in fruit. Stamens 5, inserted in the base of the corolla-tube. Styles 5. Fruit dehiscing irregularly or around its equator. *Temperate areas.*

  A very complex genus from most temperate parts of the world, growing mainly in salt-marshes or salty flats. The classification of the species is very difficult, because of the occurrence of apomixis in many species, hybridisation in some areas and the existence of isolated populations, especially in coastal areas.

4. Goniolimon Boissier. 20/2. Perennial herbs with short, woody stems. Branches angled or winged. Leaves few, in basal rosettes, leathery or fleshy. Flowers in panicles or in spike-like corymbs. Spikes with mucronate bracts; spikelets 2--6-flowered. Calyx funnel-shaped, scarious, sometimes 5- or 10-lobed. Corolla-lobes fused at base.

Stamens 5, fused to petal bases, filaments dilated below. Stigma head-like. *Europe, north Africa to central Asia.*

5. Acantholimon Boissier. 120/9. Spiny, cushion-forming shrublets or herbs. Leaves alternate, simple, sometimes variable on the same plant, summer leaves linear, sharp, often 3-angled in cross-section, spring leaves (when different) shorter, broader, flatter. Inflorescence a simple or branched spike, spikelets arranged in 2 rows along axis. Spikelets with 3 bracts. Calyx funnel-shaped, 5-nerved, 5--10-lobed, papery, pleated. Petals 5, free; styles 5, stigmas head-like. Fruit indehiscent or splitting at angles. *East Mediterranean area to central Asia.*

6. Armeria (de Candolle) Willdenow. 80/8. Tufted perennial herbs or cushion-forming shrubs with branched aerial stems. Leaves simple, in basal rosettes or on densely leafy branches. Inflorescence a head, with an involucre of overlapping bracts and a tubular sheath of fused bracts enclosing the top of the scape. Spikelets cymose, flowers with secondary bracts. Calyx funnel-shaped, with a basal spur; tube 5--10-lobed; limb translucent, lobes usually awned. Corolla funnel-shaped, lobes 5, fused at base, pink or occasionally white. Stamens 5, free. Styles 5, stigmas feathery. *Europe, North Africa, Turkey, Pacific coasts of North America.* Figure 96, p. 418.

157. EBENACEAE

Evergreen or deciduous trees or shrubs with hard, dark-coloured wood; sap watery. Leaves simple, usually alternate, without stipules. Flowers solitary or in cymes in the upper leaf-axils, generally unisexual, plants usually dioecious (a few bisexual flowers may occur sporadically in some species). Calyx with 3--7 lobes. Corolla with 3--7 lobes, lobes spreading. Stamens 6--20, borne on the corolla-tube; staminodes present or absent in female flowers. Ovary superior, with 2--16 cells, each cell with 1 or 2 pendent ovules; ovary absent in male flowers. Fruit a berry, surrounded at the base by the persistent calyx.

A family of an uncertain number of genera (2--5) and about 500 species, mainly from the tropics. Only 1 genus is generally grown.

1. Diospyros Linnaeus. 400/4. Evergreen or deciduous trees or shrubs, usually dioecious. Leaves alternate, stalked, entire. Flowers unisexual (rarely a few bisexual). Male flowers usually 2--5 together, shortly stalked in axillary cymes, female flowers usually solitary. Calyx usually 4-lobed, often increasing in size in fruit. Corolla campanulate to urn-shaped, usually 4-lobed, lobes spreading. Stamens 6--20. Ovary usually 8-celled, with 1 or 2 ovules per cell. Fruit a juicy berry, often sweet when ripe. *Mainly tropics, a few from subtropical & temperate areas.*

158. STYRACACEAE

Shrubs and trees. Leaves alternate, simple, usually entire, without stipules. Flowers regular, usually bisexual, in panicles, racemes or clusters. Sepals hypogynous, petals and stamens perigynous or all whorls epigynous. Calyx 2--5-toothed or with 5--10 lobes, campanulate, cup- or top-shaped, persistent in fruit. Corolla tubular at base, often only shortly so, with 4--8 free lobes. Stamens 4--7, alternate with corolla-lobes or double the number of corolla-lobes, usually attached to corolla-tube or united to tube. Ovary superior or inferior, of 2--5-fused, 2--5-celled carpels, each cell with 1--many anatropous ovules on axile placentas. Style simple with head-like or lobed

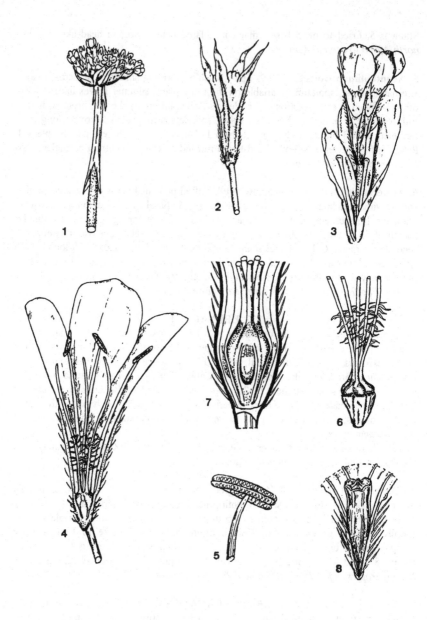

Figure 96. Plumbaginaceae. *Armeria maritima.* 1, Upper part of scape with inflorescence. 2, A single flower. 3, A single flower with adjacent bracts. 4, Longitudinal section of flower. 5, Upper part of stamen. 6, Ovary and styles. 7, Longitudinal section of ovary. 8, Fruit.

stigma. Fruit a dry or fleshy drupe or capsule. Seeds with copious endosperm. Embryo straight or slightly curved.

A family of 12 genera and 165 species distributed in East and southeast Asia, southeastern North America, Central and South America and with 1 species in the Mediterranean area.

1a.	Stamens alternating with corolla-lobes; ovary superior	**1. Styrax**
b.	Stamens double the number of corolla-lobes; ovary inferior	2
2a.	Fruit a drupe	3
b.	Fruit a capsule	4
3a.	Flowers 4-parted; corolla campanulate; stamens 8--16	**2. Halesia**
b.	Flowers 5-parted; petals distinct or barely fused at base; stamens 10	
		3. Pterostyrax
4a.	Fruit ribbed; stamens 10	**4. Rehderodendron**
b.	Fruit not ribbed; stamens 10--14	**5. Sinojackia**

1. Styrax Linnaeus. 120/10. Deciduous or evergreen shrubs or small trees. Leaves alternate, entire or toothed, stellate-downy. Flowers white, solitary or in pendent, terminal or axillary racemes or panicles or in clusters on short branchlets. Calyx 2-, 3- or 5-toothed, almost campanulate or cup-shaped, fused to base of ovary or free with 5--10 shallow lobes or truncate. Corolla with 5--10 deep lobes, radially symmetric. Stamens 10--16, at base of and alternate with corolla. Ovary superior, 3--5-celled; style with 3--5 lobes. Fruit a dry or fleshy spherical to oblong, dehiscent drupe with 1 or 2 large seeds. *Mainly tropical & warm temperate Asia & America; 1 species in the Mediterranean area.*

2. Halesia Linnaeus. 4/3. Deciduous small trees or shrubs. Branches with stellate hairs becoming hairless. Leaves alternate, toothed, thin, stalked. Flowers pendent in axillary clusters or racemes on year-old wood, white, rarely rose. Calyx cup-shaped, tube obconic with 4 minute teeth and 4 faint ribs. Corolla campanulate, with 4 lobes. Stamens 8--16 with shaggy-haired filaments. Style downy to hairless; ovary inferior, 2--4-celled. Fruit a dry oblong drupe, with 2 or 4 longitudinal wings; stone with 1--3 seeds. *China, eastern North America.*

3. Pterostyrax Siebold & Zuccarini. 4/2. Deciduous trees and shrubs. Branchlets stellate-downy, becoming hairless; winter buds with 2 outer scales. Leaves alternate, bristly-toothed, stalked, simple, membranous, sparsely downy. Flowers in 1-sided panicles, white, bisexual, radially symmetric. Calyx campanulate or obconical, with 4 or 5 teeth. Petals 5, distinct or barely fused at base. Stamens 10, protruding, fused below into a tube or almost free. Ovary inferior or almost so, 3--5-celled; style longer than stamens. Fruit a dry drupe, indehiscent, ribbed or winged, with 1 or 2 seeds. *China, Burma, Japan.*

4. Rehderodendron Hu. 10/1. Deciduous shrubs or trees. Winter buds with 2 or 3 outer scales. Leaves alternate, stalked, with fine, forwardly pointing teeth. Flowers white, in axillary leafless panicles or racemes. Calyx-lobes 5. Petals 5, fused at base. Stamens 10, unequal, filaments united at base and attached to corolla. Ovary 3- or 4-celled, inferior except for conical apex; style longer than stamens. Fruit oblong or ellipsoid, 5--10 ribbed, woody, indehiscent. Seeds 1--3, cylindric. *Indo-China, China.*

5. Sinojackia Hu. 2/2. Deciduous shrubs or small trees. Winter buds with 2 outer downy scales; branchlets stellate-downy. Leaves toothed, stipules absent, usually elliptic to elliptic-obovate, acuminate. Flowers in leafy racemes, white with yellow anthers; stalks stellate-downy. Calyx top-shaped, 5--7-lobed. Petals 5--7, united at base. Stamens 10--14, stellate-hairy; filaments attached to base of corolla. Ovary inferior, 3- or 4-celled; style longer than stamens. Fruit ovoid to cylindric-oblong, woody, usually 1-seeded, remains of calyx limb below conical apex. *China.*

159. SYMPLOCACEAE

Deciduous and evergreen trees and shrubs with simple, alternate leaves; stipules absent. Inflorescence a terminal or axillary spike, raceme or panicle, or flowers solitary. Flowers radially symmetric, usually bisexual, sepals, petals and stamens epigynous. Calyx-lobes 4 or 5, not overlapping. Corolla with 4--10 lobes. Stamens 4--numerous, free or often partly united in bundles, attached to corolla. Ovary more or less inferior, with 2--5 cells each with 2--4 ovules; placentation axile. Style 1, slender. Fruit a drupe or berry with persistent calyx-lobes.

A family of a single genus.

1. Symplocos Jacquin. 250/1. Description as for family. *Tropical & warm temperate areas of America, Asia & Australia.*

160. OLEACEAE

Trees, shrubs or woody climbers. Leaves opposite or rarely alternate, simple, with 3 leaflets or pinnate, without stipules. Inflorescences terminal or axillary, various. Flowers radially symmetric, bisexual or unisexual. Calyx small, usually 4-lobed, rarely absent. Corolla usually 4-lobed, rarely absent. Stamens 2, rarely 4, inserted on the corolla and generally borne within the corolla-tube. Ovary small, superior, 2-celled; ovules 1--many, usually 2 per cell. Style terminal; stigma 2-lobed or more or less head-like. Fruit a drupe, berry, capsule or samara. See figure 97, p. 424.

A family of about 28 genera and 900 species, almost cosmopolitan, but with its greatest representation in east and southeast Asia.

1a. Fruit a winged samara 2
  b. Fruit a capsule, drupe or berry 4
2a. Leaves pinnate, rarely simple; flowers often without corolla and/or calyx;
    fruit elliptic to linear-oblong with an apical wing **9. Fraxinus**
  b. Leaves simple; flowers with calyx and corolla; fruit more or less
    spherical with a wing all round 3
3a. Flowers white sometimes tinged pink; corolla-lobes *c.* 1 cm; stamens not
    protruding **11. Abeliophyllum**
  b. Flowers creamy yellow to whitish; corolla-lobes 2--3 mm; stamens
    protruding **10. Fontanesia**
4a. Fruit a capsule 5
  b. Fruit a drupe or berry 6
5a. Flowers yellow; corolla-lobes overlapping in bud **12. Forsythia**
  b. Flowers not yellow; corolla-lobes edge-to-edge in bud **8. Syringa**
6a. Flowers without corolla, usually functionally unisexual **6. Forestiera**
  b. Flowers with corolla, usually bisexual or functionally unisexual 7

7a. Fruit a 2-parted berry, sometimes single by abortion; corolla-tube 9--22 mm; leaves simple or compound, opposite or alternate **13. Jasminum**
 b. Fruit a drupe or simple berry; corolla-tube less than 12 mm, or apparently absent; leaves simple, opposite   8
8a. Corolla-lobes not overlapping in bud; flowers axillary or terminal   9
 b. Corolla-lobes overlapping in bud; flowers mostly axillary   11
9a. Corolla-lobes 1.5--4 cm   **5. Chionanthus**
 b. Corolla-lobes less than 6 mm   10
10a. Leaves with silvery scales beneath   **3. Olea**
 b. Leaves without silvery scales beneath   **7. Ligustrum**
11a. Drupe more or less spherical, 6--10 mm, endocarp thinnish; corolla greenish white   **2. Phillyrea**
 b. Drupe more or less ellipsoid-ovoid, 8 mm or more across, endocarp hard; corolla white or yellow to orange   12
12a. Flowers in clusters or cymes; bracts somewhat insignificant **1. Osmanthus**
 b. Flowers in racemes, in opposite pairs; bracts prominent, late in falling   **4. Picconia**

1. Osmanthus Loureiro. 35/10 and some hybrids. Evergreen trees or shrubs. Leaves simple, opposite, often leathery, margins entire or armed with sharp, spiny teeth. Flowers axillary, in clusters or paniculate cymes, small, bisexual or functionally unisexual, usually fragrant. Calyx irregularly 4-lobed. Corolla-tube usually more or less campanulate; corolla-lobes 4, white or yellow to orange, rounded triangular to strap-shaped, overlapping in bud. Stamens 2, filaments often short. Ovary usually flask- or bottle-shaped. Fruit a 1-seeded drupe, ellipsoid-ovoid, dark blue or purple when ripe, endocarp hard. *Asia (especially China), New Caledonia, eastern North America.*

2. Phillyrea Linnaeus. 2/2. Evergreen shrubs or small trees. Leaves opposite, simple, toothed or entire. Flowers axillary, clustered, in short racemes, bisexual. Calyx small, more or less 4-lobed. Corolla united, greenish white, tube somewhat cup-shaped, lobes 4, overlapping in bud. Stamens 2, anthers largish, protruding. Ovary bottle-shaped; stigma 2-lobed. Fruit a somewhat spherical, dry drupe, endocarp thinnish. *Mediterranean area.*

3. Olea Linnaeus. 20/2. Evergreen trees or shrubs. Leaves opposite, simple, leathery, with silvery scales beneath, margins usually entire. Flowers terminal or axillary, in racemes or panicles, bisexual or functionally unisexual. Calyx small, 4-lobed. Corolla-tube more or less campanulate, lobes 4, usually slightly longer than wide, edge-to-edge in bud. Stamens 2, borne in the corolla-tube. Ovary usually bottle-shaped. Fruit a drupe with hard endocarp, dark blue to black when ripe. *Warmer parts of the Old World, including the Mediterranean area.*

4. Picconia de Candolle. 2/1. Evergreen trees or shrubs to 15 m or more. Leaves opposite, simple, entire, leathery, elliptic to broadly elliptic, apex obtuse to usually acute, sometimes slightly acuminate, hairless. Flowers bisexual, in axillary or sometimes terminal, in tufted racemes with paired, opposite flowers; bracts concave, with woolly margins, slow to fall. Calyx 4-lobed. Corolla white, lobes 4, strap-shaped, 3--4 mm, scarcely joined at the base, irregularly notched, overlapping in early bud.

Stamens 2, anthers 2--3 mm. Ovary bottle-shaped, stigma shortly bilobed. Fruit a drupe, ellipsoid, 1.2--2 cm, blue-black when ripe, endocarp hard. *Macaronesia, Azores.*

5. Chionanthus Linnaeus. 100/2. Deciduous or evergreen shrubs or trees. Leaves opposite, simple. Flowers axillary or terminal on side-shoots, cymose-paniculate or racemose with flowers in opposite pairs, bisexual. Calyx more or less campanulate, 4-lobed. Corolla-lobes 4, not overlapping in bud, usually narrow, strap-shaped, divided almost to the base or united in pairs by the staminal filaments. Stamens 2 (in our species). Ovary flask- or bottle-shaped. Fruit a 1-seeded drupe. *Tropical America, Africa and Asia, but with 2 species native to temperate east Asia and temperate North America.*

6. Forestiera Poiret. 20/2. Deciduous or evergreen shrubs, usually dioecious. Leaves opposite, simple, entire or toothed. Flowers axillary, clustered or in short racemes from the previous year's branches, unisexual, without petals. Calyx minute or absent. Corolla absent. Male flowers usually stalkless, stamens 1--4, ovary rudimentary or absent. Female flowers with 1--4 abortive stamens or these absent, ovary with slender style, stigma simple or shortly 2-lobed. Fruit a drupe, usually 1-seeded, black to dark blue when ripe. *Southern USA, Central America, West Indies.*

7. Ligustrum Linnaeus. 40/17. Evergreen or deciduous shrubs or small trees. Leaves opposite, simple, entire. Flowers terminal on side-shoots, cymose-paniculate, bisexual, usually scented. Calyx campanulate, entire or with 4 shallow teeth. Corolla united, with 4 lobes not overlapping in bud, white or creamy white. Stamens 2, attached near the top of the corolla-tube. Ovary with 2 ovules per cell; style more or less elongate, stigma more or less 2-lobed or grooved. Fruit a 1- or 2-seeded drupe. *Mainly east & south-east Asia, also Australia; 1 species in Europe.*

8. Syringa Linnaeus. 20/several and many hybrids and selections. Deciduous shrubs or small trees. Leaves simple, lobed or pinnate, opposite, margins entire. Inflorescence terminal or lateral, paniculate-cymose; flowers bisexual, white, pink, or red to purple. Calyx small, more or less campanulate, 4-toothed, irregularly so or nearly truncate, persistent. Corolla with a cylindric or slightly funnel-shaped tube and 4 shortish, spreading lobes, edge-to-edge in bud (not overlapping). Stamens 2, borne within the corolla-tube and included or sometimes protruding. Style with 2-lobed stigma, within the corolla-tube; ovary 2-celled. Fruit an oblong capsule with 2 seeds in each cell. *Europe, Himalaya, east Asia to Japan.*

9. Fraxinus Linnaeus. 65/28. Trees or shrubs, mostly deciduous. Leaves opposite, rarely in whorls of 3, pinnate with a terminal leaflet, rarely simple. Inflorescence paniculate, terminal or lateral from buds of the previous year, appearing before or with the leaves. Flowers unisexual or bisexual, small. Calyx, if present, small, campanulate, united, entire or with 4 lobes. Corolla, if present, with 4, oblong or linear lobes. Stamens 2. Ovary 2-celled. Fruit a 1-seeded samara, elliptic to linear-oblong with an apical wing. *North temperate areas.*

10. Fontanesia Labillardière. 1/1. Deciduous shrubs or small trees, twiggy. Leaves simple, hairless, entire or with very fine forwardly pointing teeth. Inflorescences axillary and terminal, shortly paniculate, flowers bisexual. Calyx small, irregularly 4-lobed. Corolla creamy yellow to whitish, lobes 4, joined in pairs at the base, 2--3 mm. Stamens 2, inserted at the base of the corolla, protruding. Ovary 2-celled, stigmas 2-lobed. Fruit an ovoid samara, narrowly winged, 1- or 2-seeded. *Sicily, Turkey, Syria, China.*

11. Abeliophyllum Nakai. 1/1. Deciduous shrubs to 2 m. Young stems slightly 4-angled. Leaves opposite, simple, entire, ovate to broadly elliptic, 5--8.5 × 2--4 cm, base rounded to obtuse to tapering gradually, apex acuminate, with scattered hairs above and beneath; stalks 2--5 mm. Racemes axillary, 3- to 15-flowered; flowers bisexual, in opposite pairs. Calyx lobes 4, rounded, hairy on margins. Corolla white, sometimes tinged pink; lobes 4, strap-shaped, *c.* 1 cm × 2 mm, apex notched. Stamens not protruding. Fruit an almost spherical winged samara, 2--2.5 across, very rarely seen in cultivation. *Korea.*

12. Forsythia Vahl. 9/7 and some hybrids. Deciduous shrubs. Leaves simple or of 3 leaflets, entire or toothed. Inflorescences axillary; flowers appearing before the leaves, yellow, bisexual, heterostylous. Calyx small, deeply 4-lobed, ciliate, persistent. Corolla 4-lobed, lobes longer than the tube, overlapping in bud. Stamens 2, borne at the base of the corolla-tube, not or scarcely protruding. Ovary 2-celled, many ovules. Fruit a capsule, seeds slightly winged on one side. *One species from southeast Europe, the rest from eastern Asia.*

13. Jasminum Linnaeus. 200/13 & some hybrids. Shrubs, often climbing with support. Leaves evergreen or sometimes deciduous, simple, of 3 leaflets or pinnate, opposite or alternate. Inflorescences axillary or terminal, cymose or flowers solitary, bisexual. Calyx tubular, 5-lobed. Corolla white, red or yellow with narrow tube 9--22 mm and 5--9 (rarely more) usually spreading lobes. Stamens 2, included in the corolla-tube. Ovary 2-celled, each cell with 2 ovules. Fruit 2-parted (or 1 by abortion), 1-seeded fleshy berries, black when ripe. *Old World, mostly tropical.* Figure 97, p. 424.

161. LOGANIACEAE

Trees, shrubs, climbers and herbs. Leaves opposite, simple, entire to lobed, pinnately-nerved; stipules absent or present, often reduced to, or with lines linking leaf-stalk bases; leaf-stalks sometimes united into a membranous sheath (ochrea). Flowers usually bisexual, usually radially symmetric, solitary or in terminal cymes. Calyx hypogynous, corolla and stamens usually perigynous. Calyx with 5 overlapping lobes (ours). Corolla 5-lobed (ours). Stamens 5 (ours), attached to the base of the corolla-tube, usually alternating with corolla-lobes. Ovary superior, 2-celled; style 2-lobed. Fruit a capsule, berry or drupe; seeds 1--numerous, fleshy, often winged.

A family of about 30 genera and 600 species from the tropics, subtropics and temperate areas. Few species are cultivated. *Strychnos nux-vomica* Linnaeus from India and Burma is also used as a timber source and ornamental but its main use is as the commercial source of strychnine; many members of Loganiaceae are extremely poisonous, causing death by convulsions.

1a. Twining shrubs to 6 m   **1. Gelsemium**
  b. Perennial herbs 30--60 cm   **2. Spigelia**

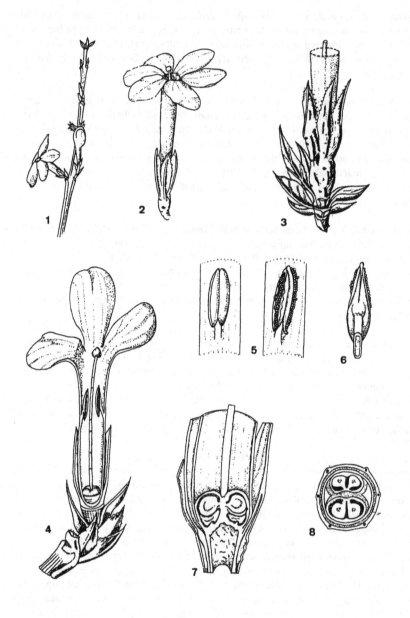

Figure 97. Oleaceae. *Jasminum nudiflorum.* 1, Part of inflorescence. 2, Single flower. 3, Base of flower showing bracts and calyx. 4, Longitudinal section of flower. 5, Stamen from front, undehisced (L) and dehisced (R). 6, Stamen from back. 7, Longitudinal section of ovary. 8, Transverse section of ovary.

424

1. Gelsemium Jussieu. 3/2. Perennial, evergreen, twining hairless shrub. Leaves entire; stipules small. Flowers axillary or terminal, in 1--many-flowered clusters. Calyx 5-lobed. Corolla 5-lobed, funnel-shaped, lobes short and overlapping. Fruit a capsule; seeds flattened, sometimes winged. *America, southeast Asia.*

2. Spigelia Linnaeus. 50/2. Hairless, hairy or stellate-hairy, annual or perennial herbs, rarely woody at base. Leaves opposite, stipules or transverse membranes connecting the leaf-bases; blade entire, often membranous, usually with 3 or 5 veins. Inflorescence of few- to many-flowered clusters. Calyx segments 5, narrow. Corolla-lobes 5, red, yellow or purple, tubular to salver-shaped. Fruit a 2-lobed capsule. *North & tropical America.*

162. DESFONTAINIACEAE

Small to medium-sized, evergreen shrubs. Leaves stalked, opposite, leathery, with spiny teeth; stipules absent. Flowers bisexual, solitary, terminal or axillary, drooping, radially symmetric. Calyx hypogynous, corolla and stamens perigynous. Calyx 5-lobed, united at base. Corolla shallowly 5-lobed, funnel-shaped. Stamens 5, inserted at base of corolla-lobes. Ovary superior with 3--5 cells; ovules numerous, with axile placentation. Style 1, stigma more or less head-like. Fruit a berry, surrounded by the persistent calyx.

A family closely allied to the *Loganiaceae* and sometimes included in it, containing a single genus from temperate South America.

1. Desfontainia Ruiz & Pavón. 1/1. Shrub 1--3 m, branched, twigs greyish, bark pale brownish, shiny. Leaves 2.5--4 × 1--2.3 cm, ovate, wedge-shaped at base, with 2--8 pairs of broadly triangular spiny teeth, deep green and glossy beneath, closely resembling those of holly; stalks 8--10 mm. Flowers to 4 cm on stalks 9--10 mm; bracts 6 mm, linear. Calyx-lobes 7--8 × 3--3.5 cm with marginal hairs. Corolla 4.5--6 cm with rounded lobes, tube scarlet-red, corolla-lobes yellow-orange. Stigma included or slightly longer than corolla. Fruit 1.2--1.5 cm × 9--11 mm, spherical, greenish purple eventually black; seeds 2--25 × 1--1.2 mm, smooth. *Temperate South America.*

163. GENTIANACEAE

Annual to perennial herbs, rarely small shrubs or twiners. Leaves usually opposite, sometimes in whorls of 3--8, occasionally alternate, simple (ours), without stipules, veins often parallel. Flowers solitary or in cymes or panicles, sometimes very condensed or head-like, bisexual, usually radially symmetric. Calyx hypogynous, corolla and stamens perigynous. Calyx 4--12-lobed, the fused part small and cup-like or larger and forming a tube, occasionally split down one side, lobes equal or unequal, their bases sometimes joined by a translucent membrane. Corolla 4--12-lobed, lobes longer or shorter than the tube, sometimes with accessory lobes (plicae -- see under *Gentiana*, below) between them. Stamens 4--12, attached to the corolla-tube or at the sinuses between the lobes; anthers sometimes united, usually opening by slits, occasionally by pores. Ovary 1-celled with parietal placentas, rarely the placentas so intrusive as to make the ovary appear 2-celled. Fruit usually a capsule, more rarely fleshy and berry-like. Seeds numerous, often winged. See figure 98, p. 428.

A cosmopolitan family of about 80 genera and 700 species.

1a. Corolla with 5 or 6 keeled and twisted lobes **7. Eustoma**

b. Corolla with 4--12 lobes which are not keeled and twisted   2
2a. Plants twining   3
  b. Plants upright, not twining   4
3a. Stamens equal in length, symmetrically arranged; fruit a dry capsule
    2. Crawfurdia
  b. Stamens unequal in length, asymmetrically arranged; fruit fleshy, berry-like
    3. Tripterospermum
4a. Calyx-lobes, corolla-lobes and stamens 6--12; glaucous annual herb with
    yellow flowers   **9. Blackstonia**
  b. Calyx- lobes, corolla-lobes and stamens usually 4 or 5, if more, other
    features not as above   5
5a. Calyx lobed almost to the base, the joined part scarcely tubular, without a
    translucent membrane between the bases of the lobes   6
  b. Calyx usually lobed at most to halfway, if further lobed then the united
    part clearly tubular and the bases of the lobes joined by a translucent
    membrane   7
6a. Each corolla-lobe bearing 1 or 2 usually fringed nectary glands near the
    base   **8. Swertia**
  b. Corolla-lobes entirely without nectary glands, nectar secreted from 5 pits
    within the corolla-tube   **6. Gentianella**
7a. Nectary glands entirely absent; corolla usually pink or pinkish purple
    1. Centaurium
  b. Nectary glands present, either in the corolla-tube or around the base of
    the ovary; corolla not pink   8
8a. Nectary glands in pits in the corolla-tube; calyx 4-lobed, the lobes in 2
    opposite pairs, those of 1 pair larger than those of the other
    5. Gentianopsis
  b. Nectary glands at the base of the ovary; calyx 4- or 5-lobed, various but
    never as above   **4. Gentiana**

1. Centaurium Hill. 12/6. Annual, biennial or perennial herbs with basal rosettes and narrow stem-leaves. Flowers in cymes which are often compressed or spike-like. Calyx with 5 (rarely 4) keeled, narrow, acute lobes. Corolla with a funnel-shaped tube and 5 spreading lobes, usually pink, more rarely white or yellow. Stamens 5, anthers slender, twisting spirally as the flower ages. Ovary with a thread-like, persistent style and 2 stigmas which fall early. *Europe, Mediterranean area.*

2. Crawfurdia Wallich. 16/2. Perennial twiners. Leaves opposite. Flowers terminal and axillary, few, or terminal and axillary in cymes. Calyx 5-lobed, the tube 10-veined. Corolla tubular-campanulate (ours), plicae present. Stamens 5, symmetric, equal. Nectaries 5, at the base of the ovary. Fruit a capsule. Seeds compressed, each with a disc-like wing. *Himalaya, China.*

3. Tripterospermum Blume. 17/1. Perennial twiners. Leaves opposite. Flowers terminal and axillary, few or in terminal and axillary cymes. Calyx 5-lobed, 5-veined. Corolla tubular, lobes 5, plicae present. Stamens 5, asymmetric, unequal in length, curved. Nectary forming a collar surrounding the base of the ovary. Seeds 3-angled, sometimes winged. *East & southeast Asia.*

4. Gentiana Linnaeus. 300/75 and some hybrids. Annual, biennial or perennial herbs, often with basal leaf-rosettes; stems erect to prostrate. Leaves usually opposite, more rarely in whorls of 3--8, stalked or not, margins sometimes finely toothed or papillose-warty (rough to the touch). Flowers solitary or in few- to many-flowered terminal and axillary cymes sometimes congested into head-like clusters. Calyx usually tubular with 4 or 5 (rarely to 8) lobes, sometimes split partially or completely down one side. Corolla campanulate to tubular, usually with 4 or 5 (rarely to 8) lobes which are generally spreading and with additional lobes (plicae) between the major lobes. Stamens usually 4 or 5, rarely to 8, attached to the corolla-tube. Nectary borne below the ovary. Fruit a capsule, stalked or not. Seeds numerous, variable in shape and surface ornamentation. *Almost cosmopolitan.* Figure 98, p. 428.

In the past, species here included in *Crawfurdia*, *Gentianella*, *Gentianopsis* and *Tripterospermum* have formed part of *Gentiana*, but are now generally excluded.

5. Gentianopsis Ma. 25/2. Like *Gentiana*, but plants always annual or biennial, buds large, somewhat flattened, 4-angled; calyx 4-lobed, the lobes in 2 pairs, those of one pair longer, those of the other shorter, without a translucent membrane between the bases of the lobes; corolla-lobes usually fringed; plicae absent; nectaries borne in pits on the corolla tube. *Mostly North America, but extending into Eurasia.*

6. Gentianella Moench. 200/6. Like *Gentiana*, but nectary glands borne in pits on the corolla-tube; corolla broadly campanulate to cup-shaped, without plicae; calyx without a continuous translucent membrane between the lobes. *Widely distributed in temperate parts of the world.*

7. Eustoma Salisbury. Few/1. Annual, biennial or perennial herbs with well-developed taproots, usually glaucous. Leaves opposite, stalkless, sometimes clasping the stem. Flowers long-stalked, solitary or in panicles. Calyx with 5 or 6 keeled lobes. Corolla funnel-shaped to bell-shaped, 5- or 6-lobed, the lobes keeled and twisted. Stamens 5 or 6. Style long, with 2 stigmatic lobes. Fruit a capsule. *South USA, central America.*

One species commonly grown as a cut flower, and often sold as 'Lisianthus'.

8. Swertia Linnaeus. 150/4. Annual or perennial herbs; stems sometimes winged or angled. Leaves usually crowded into a rosette at the base of the stems, stem-leaves few, usually opposite, sometimes alternate or whorled. Flowers (in our species) in panicles, with parts in 4s or 5s. Calyx deeply 4- or 5-lobed. Corolla usually widely cup-shaped, deeply 4- or 5-lobed, each lobe with 1 or 2 conspicuous, often fringed nectaries at the base. Stamens 4 or 5. Stigmas 2 on a short or long style. Fruit a capsule. *Almost cosmopolitan.*

9. Blackstonia Hudson. 4/1. Annual herbs. Leaves broad, opposite, sometimes joined in pairs. Calyx deeply divided into 6--12 lobes. Corolla with a short tube and 6 or more spreading lobes, yellow. Stamens as many as corolla lobes, anthers narrow, basifixed, untwisted or very slightly spirally twisted after flowering. Style thread-like with 2 stigmas which are themselves 2-lobed. Fruit a capsule. *Europe, Mediterranean area.*

164. MENYANTHACEAE

Aquatic or marsh perennials. Leaves usually alternate, sometimes opposite on flowering stems, simple or made up of 3 leaflets; stalks sheathing, stipules absent.

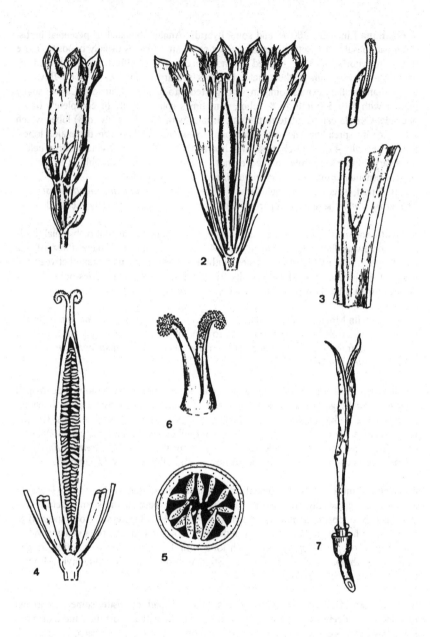

Figure 98. Gentianaceae. *Gentiana semptemfida*. 1, Single flower subtended by two bracteoles. 2, Flower with corolla-tube opened out. 3, Stamen. 4, Longitudinal section of ovary. 5, Transverse section of ovary. 6, Stigmas. 7, Opening fruit.

Flowers with parts in 5s, radially symmetric. Calyx hypogynous, deeply lobed. Corolla deeply lobed, lobes edge-to-edge in bud, often fringed. Stamens borne on corolla-tube. Ovary superior, of 2 united carpels, 1-celled, placentation parietal. Fruit a capsule.

A family of 5 genera and about 33 species, cosmopolitan, of which only a few are grown.

    1a. Leaves with 3 leaflets; corolla pink or white   **1. Menyanthes**
      b. Leaves simple; corolla yellow   **2. Nymphoides**

1. Menyanthes Linnaeus. 1/1. Emergent aquatics with creeping rhizomes. Stems to 50 cm. Leaves alternate, long-stalked, made up of 3 shortly stalked leaflets; leaflets obovate to diamond-shaped, hairless, to 10 cm. Flowers 10--20 in a raceme. Calyx-lobes ovate, recurved. Corolla to 1.5 cm across, pinkish outside, white inside, lobes fringed. *Europe, North America.*

2. Nymphoides Seguier. Several/1. Aquatic perennial herbs with creeping rhizomes. Leaves mostly alternate, sometimes opposite on flowering stems, simple, floating, long-stalked, ovate to almost circular, deeply cordate at the base. Flowers solitary or in axillary clusters of 2--5, the inflorescences sometimes bearing tuber-like adventitious roots. Calyx 5-lobed. Corolla white or yellow with a short tube and 5 spreading lobes which are entire or fringed. Stamens 5, inserted at the base of the corolla-lobes. Fruit a capsule which opens irregularly. *Tropical & temperate areas.*

165. APOCYNACEAE

Trees, shrubs or woody climbers, rarely perennial herbs or stems succulent, usually with milky latex. Leaves simple, entire, opposite or whorled, occasionally alternate, very rarely with stipules but often with scales at base. Flowers bisexual, radially symmetric, often large, showy and fragrant, in terminal cymes, sometimes on short side-branches and appearing axillary. Flower-parts in 5s. Calyx and corolla usually fused into a tube with 4 or 5 lobes. Calyx-tube often with basal glandular scales inside. Corolla-lobes usually spirally twisted in bud and characteristic in their direction of overlapping, spreading to form a salver-shaped flower; corona of 10 smaller lobes sometimes borne at base of corolla-lobes. Disc ring- or cup-shaped, lobed or of separate glands, sometimes absent. Stamens 4 or 5, borne in the corolla-tube, almost stalkless or with short filaments; anthers 2-celled (sometimes with sterile basal spur), forming a cone around the stigma and often fused to it. Carpels 2, fused or joined only by the style. Ovary 1- or 2-celled, each with 1--many ovules; style simple, or divided below, stigma (clavunculus) club-shaped, 2-lobed, about level with the anthers. Fruit usually a pair of cylindrical follicles, sometimes a dry or fleshy drupe or berry. Seeds various, often winged or with a tuft of hairs (coma). See figure 99, p. 432.

    A family of about 250 genera and 2000 species, widespread but largely tropical; most members are trees or woody climbers. The family is a rich source of drugs and poisons, e.g. strophanthine for the treatment of heart disease, *Catharanthus* alkaloids in leukaemia, and arrow-poisons from *Acokanthera*. The latex in some species has been used to make rubber.

    1a. Twining woody climbers   **5. Trachelospermum**
      b. Herbs, sometimes woody at base   2

2a. Plant with stolons; corolla violet or rarely blue, pink or white   **3. Vinca**
  b. Plant without stolons; corolla blue, pink or white   3
3a. Corolla campanulate or bowl-shaped, lobes twisted to right   **4. Apocynum**
  b. Corolla salver-shaped, lobes twisted to left   4
4a. Leaves hairless or more uniformly hairy   **1. Amsonia**
  b. Leaves with spreading hairs on margins and midrib   **2. Rhazya**

1. Amsonia Walter. 20/3. Herbaceous perennials. Leaves alternate or almost whorled. Inflorescence terminal or occasionally lateral, with several to many flowers. Calyx 5-lobed, lobes slightly overlapping, scales absent. Corolla with a tube and abruptly expanded regular lobes. Anthers not converging, without enlarged connective, not projecting. Fruit cylindric, tapering, constricted between the seeds. Seeds numerous. *North America, Japan.*

2. Rhazya Decaisne. 2/1. Erect perennial herbs, woody at base. Leaves alternate, deciduous. Flowers in terminal corymbs. Corolla with long slender tube and abruptly expanded lobes; throat hairy, scales absent; lobes overlapping to the left. Anthers surrounding stigma but not attached to it. Carpels surrounded by a disc. Seeds few, hairless. *South Europe to northwest India.*

3. Vinca Linnaeus. 7/5. Low-growing, evergreen, woody-based shrubs or herbaceous perennials, usually with trailing or arching vegetative shoots. Leaves opposite, entire. Flowers solitary in leaf-axils, on long stalks. Sepals narrow, more than a quarter the length of corolla-tube, without internal glands. Corolla large; tube funnel-shaped, gradually widened, without internal appendages but with a ring of hairs above the insertion of the stamens; mouth wide and open; lobes horizontally spreading or slightly ascending, as long as the tube, oblique, joined at base by a low ridge. Stamens attached halfway up the corolla-tube; filaments short, abruptly bent at base; anthers crowned with a hairy, flap-like appendage. Carpels 2, pressed together, each with 4--8 ovules. Follicles (rarely produced in cultivation) spreading, slender. Seeds hairless, cylindric, grooved on one side. *Portugal & north Africa through the Mediterranean area and central Europe to the Caucasus & west & central Asia.* Figure 99, p. 432.

4. Apocynum Linnaeus. 7/2. Herbaceous perennials. Leaves opposite, rarely whorled. Flowers few to many, in 1-sided compound cymes. Calyx 5-lobed, divided nearly to base, slightly overlapping. Corolla campanulate to urn-shaped or cylindric; tube with arrow-shaped appendages at base opposite lobes. Anthers forming a cone around and fused to the stigma, with enlarged, 2-lobed connectives. Follicles usually 2, separate or joined at tip, slender, cylindric and tapering. *Temperate America.*

5. Trachelospermum Lemaire. 20/3. Evergreen climbing shrubs or lianas with white latex; leaves opposite, often rather leathery. Flowers salver-shaped, white or red, fragrant, born in terminal or axillary corymbose cymes. Calyx deeply divided almost to base, glandular within. Corolla-tube cylindric, narrow at base, widening around the stamens then contracting slightly towards the mouth, lobes obliquely obovate-oblanceolate, tightly rolled spirally to the left but overlapping to the right. Disc deeply lobed. Stamens almost stalkless, inserted on the corolla-tube, anthers not projecting or just tips projecting, forming a cone around and fused to the stigma. Ovary of 2 free carpels, ovules many. Fruit a pair of long cylindric follicles, divergent or incurved. Seeds linear-oblong, flattened, not beaked, hairless with white apical tuft of hairs. *India to Korea & Japan.*

166. ASCLEPIADACEAE

Trees, shrubs, herbs, herbaceous or woody climbers or scramblers, stems sometimes twining, sometimes succulent, usually with milky sap. Plants with a succulent, swollen, perennial water-storage organ from which arise slender, usually annual, erect, twining or climbing stems are referred to as 'caudiciform'; plants with very thickened stems adapted for water-storage are termed 'pachycaul'. Leaves opposite, whorled, rarely spiral, simple but often reduced, stipules absent. Flowers in axillary or terminal cymes or umbels, usually bisexual, 5-parted, radially symmetric. Corolla usually with a short tube, lobes 5, sometimes with smaller lobes in between, each overlapped by and overlapping one other or edge-to-edge in bud. Corona single or double of 5 or 10 free or more or less joined segments, inserted at the base of the filaments. Stamens joined to the petals or united in a short sheath around and often joined to the stigma forming a column, the gynostegium. Pollen in pollinia or tetrads, becoming attached to pollen-transfer devices called translators (solidified secretions of the anthers, style-head, or both) with arms (retinacula) joined by a 2-parted gland (the corpusculum), the whole being termed a pollinarium and transferred as a whole by the pollinator. Ovary superior, of 2 carpels, free below, united at the stigma. Fruit of 2 distinct follicles. Seeds with a terminal head of long hairs. See figure 100, p. 434.

A very large and diverse family with 315 genera and about 2900 species, especially numerous in the tropics and warm regions (particularly Africa). They possess elaborate 'lock and key' insect-pollination mechanisms. The nectar-seeking insects remove the pollinia, guided by grooves in the gynostegium.

1a. Anthers not joined to stigma, opening by longitudinal slits; pollen granular,
   in tetrads   **1. Periploca**
  b. Anthers joined to stigma, opening apically; pollen in pollinia   2
2a. Stems upright, not climbing   **2. Asclepias**
  b. Stems twining or climbing   3
3a. Deciduous herbaceous perennial   **3. Vincetoxicum**
  b. Evergreen herbaceous perennial   **4. Oxypetalum**

1. Periploca Linnaeus. 11/1. Hairless shrubs with twining stems. Poisonous sap present. Leaves deciduous, opposite. Flowers in long-stalked corymbs, ill-smelling. Corolla-lobes purplish violet, downy above, each with a white papillose streak near base, greenish yellow and hairless beneath. Corona-segments violet, thread-like, shorter than the lobes, more or less hairless. Stamens free, filaments very short, free as far as the stigma. Fruits yellow, cylindric, slightly joined at the apex, poisonous. *Mediterranean area to east Africa.*

2. Asclepias Linnaeus. 90/10. Shrubs or annual or perennial herbs, mostly with milky sap. Leaves opposite, alternate or in whorls. Stipules absent. Flowers bisexual, borne in terminal or axillary umbels. Calyx 5-lobed, sometimes united into a tube at the base, the lobes acute, often glandular within. Corolla 5-parted or divided, the lobes often edge-to-edge in bud, reflexed at flowering. Crown usually with a column, the lobes 5, concave and hood-like, spreading or erect, each bearing within a slender or subulate, included or projecting horn. Filaments united into a tube; each anther tipped with an

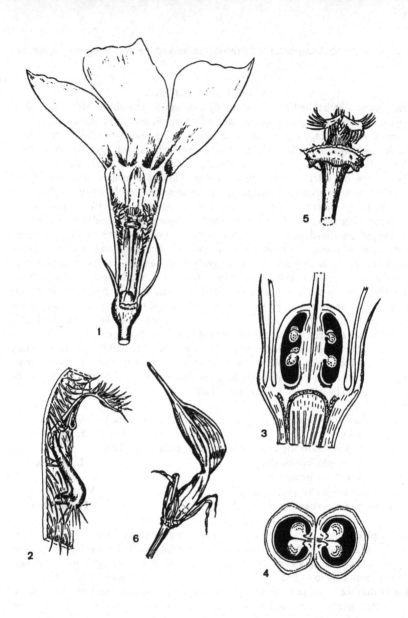

Figure 99. Apocynaceae. *Vinca minor.* 1, Flower with part of the calyx and corolla removed. 2, Stamen attached to corolla-tube, with adjacent hairs. 3, Longitudinal section of lower part of flower. 4, Transverse section of ovary. 5, Upper part of style. 6, A single, opened follicle.

432

inflexed scarious membrane, winged, wings broadest below the middle. Stigma almost flat, 5-lobed or -angled. Fruit a follicle, acuminate at apex. Seeds compressed, each with a tuft of silky hairs at one end. *Mostly from South Africa & the New World.* Figure 100, p. 434.

3. Vincetoxicum Wolf. 15/2. Rhizomatous herbs or small shrubs, often with twining shoots. Leaves opposite, shortly stalked, the lower often rounded, the upper lanceolate. Flowers in axillary few-flowered cymes. Corolla contorted in bud; lobes spreading when mature. Corona single, the segments not awned, united at least at the base by a more or less folded toothed membrane; lobes 5, fleshy, flat. Pollinia pendent. Fruits usually fusiform, smooth. *Europe, Asia.*

4. Oxypetalum R. Brown. 125/1. Perennial herbs and low shrubs with erect, decumbent or climbing stems, with white milky latex. Leaves opposite. Flowers in open, axillary and terminal cymes. Calyx 5-parted, lobes lanceolate, acute. Corolla deeply lobed, tube campanulate, very short, lobes lanceolate or strap-like. Corona-scales 5, joined to base of corolla and staminal tube; stamens 5, filaments united into a tube, surrounding the 2 styles. Styles united by a narrowly conical stigma. Fruits fusiform or ovoid, smooth or with stiff hairs. *Tropical & subtropical South America.*

A genus of about 125 species only 1 of which is widely cultivated, sometimes as an annual.

167. RUBIACEAE

Herbs, shrubs or trees, sometimes scrambling or climbing, deciduous or evergreen. Leaves opposite or more rarely in whorls of 3 or more, entire; stipules present, either borne on the stems between the leaf-bases (interpetiolar) or between the stem and the leaf-stalk (intrapetiolar; does not occur in cultivated genera), or sometimes of the same size and shape as the leaves so that, superficially, the leaves appear to be borne in whorls of 4 or 6 (only 2, the actual leaves, buds in their axils). Flowers usually bisexual, occasionally unisexual, solitary or in cymes, panicles or clusters which are sometimes very congested and head-like, often fragrant. Calyx, corolla and stamens epigynous. Calyx generally with 4 or more lobes, occasionally very reduced, sometimes united at the base above the top of the ovary. Corolla with a long or short tube (occasionally very short) and usually 4 or more spreading or erect lobes. Stamens as many as the corolla-lobes, inserted in the base of the corolla-tube or more usually at its throat. Ovary mostly or completely inferior, usually 2-celled, more rarely 1--5- or more -celled, ovules 1--many per cell. Fruit a capsule, berry or drupe, or of 2 dry mericarps (sometimes reduced to 1), rarely a syncarp of all the coalesced fleshy fruits in the inflorescence. Seeds winged or unwinged. See figure 101, p. 440

A large, cosmopolitan, though mainly tropical, family with some 631 genera and 10,700 species. For its size, it contributes very little to ornamental horticulture. The account below covers all that are in general cultivation in Europe. More genera and species are grown in North America, and more may also be found in specialist collections, especially those with a strong interest in medicinal plants.

In making use of the key to the genera it is necessary, at No. 1, to distinguish between whorls formed from leaves plus stipules which are similar in size and shape to the leaves, and whorls of true leaves. True leaves generally have their own stipules and have buds in their axils, whereas stipules mimicking leaves have neither.

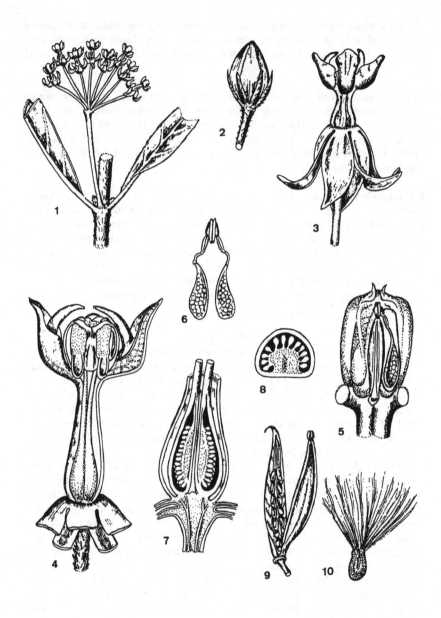

Figure 100. Asclepiadaceae. *Asclepias curassavica.* 1, Inflorescence. 2, Flower-bud. 3, Flower. 4, Longitudinal section of central portion of flower. 5, Detail of gyno-stemium, showing a pollinium. 6, A pair of pollinia connected by a translator. 7, Long-itudinal section of ovary. 8, Transverse section of a single carpel. 9, Open follicle. 10, Seed.

434

1a. Stipules similar to the leaves in size and shape, forming apparent whorls of 4 or 6 'leaves'; plants mostly herbs, sometimes woody below; fruit of 2 mericarps (rarely reduced to 1 fleshy mericarp); corolla-lobes edge-to-edge in bud   2
  b. Stipules various, but not similar to the leaves, which may be opposite or in whorls of 3 or more; plants usually woody, less often herbaceous; fruit various but not as above; corolla-lobes edge-to-edge or overlapping in various ways in bud   7
2a. Corolla-tube as long as the lobes or longer than them   3
  b. Corolla-tube shorter than the lobes   5
3a. Annual herb; calyx present, persistent and enlarged in fruit   **16. Sherardia**
  b. Perennial herbs; calyx minute or absent, not as above   4
4a. Corolla-lobes 5; style very shortly 2-lobed at the apex   **12. Phuopsis**
  b. Corolla-lobes 3 or 4; styles 2, united only at the base   **11. Asperula**
5a. Evergreen, scrambling plants with prickly stems; leaves and stipules distinctly stalked   **15. Rubia**
  b. Annual or perennial, deciduous herbs, stems rarely prickly; leaves and stipules not stalked   6
6a. Flowers in short cymes in the leaf-axils, shorter than the leaves which subtend them, the central flower of each cyme bisexual, the others male (or absent); leaves and stipules ovate   **14. Cruciata**
  b. Flowers in axillary or terminal panicles, all bisexual; leaves and stipules linear   **13. Galium**
7a. Ovary 2--several-celled, with a single ovule in each cell   8
  b Ovary 2--several-celled (rarely 1-celled), with several (usually many) ovules in each cell   13
8a. Petals edge-to-edge in bud   9
  b. Petals variously overlapping in bud   12
9a. Flowers mostly unisexual, inconspicuous, green   **9. Coprosma**
  b. Flowers bisexual, conspicuous, usually not green   10
10a. Upright shrubs or trees; bracteoles 2, united into a cup below each flower   **7. Leptodermis**
  b. Mat-forming or trailing herbs; bracteoles not united into a cup   11
11a. Stigmas 4; stamens inserted in the throat of the corolla-tube; flowers with ovaries joined in pairs   **6. Mitchella**
  b. Stigmas 2; stamens inserted in the base of the corolla-tube; ovaries not joined   **10. Nertera**
12a. Much-branched, mat-forming shrub   **8. Putoria**
  b. Erect trees or shrubs, not mat-forming   **5. Cephalanthus**
13a. Corolla-lobes variously overlapping in bud; corolla pink, greenish white or the tube red and the lobes white; fruit a capsule   **3. Luculia**
  b. Corolla-lobes edge-to-edge in bud; corolla not as above   14
14a. Stipules divided into several narrow, bristle-like segments   **2. Bouvardia**
  b. Stipules not divided into narrow, bristle-like segments   15
15a. Perennial herbs, not at all woody at the base   **1. Houstonia**
  b. Shrubs or trees   **4. Emmenopterys**

1. Houstonia Linnaeus. 50/1. Delicate perennial herbs. Basal leaves in a rosette, stem-leaves opposite; stipules entire, borne on the stems between the leaf-bases. Flowers numerous in cymes or solitary, axillary or terminal. Calyx with 4 persistent lobes. Corolla with a narrow tube and 4 spreading lobes (ours), which are edge-to-edge in bud. Stamens 4, inserted in the throat of the corolla-tube. Ovary 2-celled, only partially inferior, ovules numerous in each cell. Fruit a capsule. Seeds unwinged. *North America, Mexico.*

2. Bouvardia Salisbury. 31/2 and many hybrids. Perennial herbs or shrubs. Leaves usually in whorls of 3 or more (rarely opposite), stalked or not; stipules borne on the stems between the leaf-bases, united at the base, divided into bristle-like segments. Flowers in terminal cymes or solitary. Calyx 5-lobed, lobes usually erect and persistent. Corolla tubular to funnel-shaped, lobes usually 5, shorter than the tube, edge-to-edge in bud. Stamens 5 with long or short filaments (flowers heterostylous). Neither anthers nor style projecting from the corolla-tube. Ovary entirely or mostly inferior, 2-celled, with numerous ovules in each cell; style thread-like, divided at the apex into 2 short, linear, stigmatic lobes. Fruit a capsule. *Mexico, southern USA.*

3. Luculia Sweet. 5/2. Shrubs or small trees. Leaves opposite, stalked; stipules borne on the stem between the leaf-bases, falling early. Flowers numerous in terminal panicles, fragrant, white, pink or red. Calyx-lobes 5, not persistent, linear to oblong, unequal. Corolla-tube hairless, funnel-shaped to parallel-sided, lobes 5, overlapping in bud, spreading in flower, sometimes each lobe with 2 appendages at the base, not ciliate. Stamens 5, inserted in the throat of the corolla-tube. Ovary 2-celled, with numerous ovules in each cell. Fruit a woody capsule. *Himalaya, China.*

4. Emmenopterys Oliver. 2/1. Deciduous trees. Leaves opposite, stalked, entire, rather leathery; stipules falling early. Flowers numerous in flat-topped, terminal panicles. Calyx 5-lobed, lobes ovate, ciliate, sometimes 1 of the lobes in some flowers enlarged. Corolla white or yellow, funnel-shaped to campanulate, lobes 5, ovate, spreading, downy. Stamens 5, included in the corolla-tube. Ovary 2-celled with many ovules in each cell; style thread-like. Fruit a capsule containing many winged seeds. *Burma, Thailand, China.*

5. Cephalanthus Linnaeus. 10/1. Deciduous or evergreen shrubs or small trees. Leaves opposite or in whorls of 3 or 4; stipules borne on the stem between the leaf-bases. Flowers in dense, more or less spherical heads, axillary or terminal, all flowers opening together. Calyx with 4 or 5 teeth or lobes. Corolla tubular-funnel-shaped, sometimes hairy inside the throat, lobes 4 or 5, erect or spreading, overlapping in bud. Stamens 4 or 5, inserted in the throat of the corolla-tube. Ovary 2-celled, with a single ovule in each cell; style thread-like, long-projecting from the corolla, stigma head-like or club-shaped. Fruit indehiscent, 2-seeded. *Tropical America, Africa & Asia.*

6. Mitchella Linnaeus. 2/1. Evergreen trailing herbs, forming mats. Leaves opposite; stipules minute. Flowers in pairs, terminal on the branches, each pair very shortly stalked, the ovaries sometimes somewhat united in flower, coalesced in fruit. Calyx with 4 lobes, persistent on the fruits. Corolla funnel-shaped, with a slender tube and 4 spreading lobes which are edge-to-edge in bud, densely hairy inside. Stamens 4. Ovary

436

4-celled with 1 ovule in each cell; style thread-like, stigmas 4; ovaries of each pair of flowers united. Fruit a fleshy drupe formed from the coalescence of 2 ovaries, with up to 8 seeds. *North America, Japan, South Korea.*

7. Leptodermis Wallich. 30/2. Deciduous shrubs. Leaves opposite, entire; stipules acute, persistent. Flowers almost stalkless in terminal or axillary panicles; bracteoles 2, united into a tube. Calyx 5-lobed, lobes persistent. Corolla funnel-shaped to tubular, hairy inside; lobes 5, edge-to-edge in bud. Stamens 5, inserted in the throat of the corolla-tube. Ovary 5-celled with a single ovule in each cell; style thread-like, stigma 5-lobed. Fruit a capsule. *Himalaya, western & central China.*

8. Putoria Persoon. 3/1. Small shrubs. Leaves opposite, smelling unpleasantly when crushed, shortly stalked; stipules borne on the stem between the leaf-bases, more or less fused. Flowers solitary or few in clusters. Calyx 4- or 5-lobed, persistent. Corolla funnel-shaped with a long tube and 4 or 5 lobes. Stamens 4 or 5. Ovary 2-celled with a single ovule in each cell; style thread-like, stigma 2-lobed. Fruit a drupe containing 2 stones. *Mediterranean area.*

9. Coprosma Forster & Forster. 90/15 and some hybrids. Shrubs or small trees, usually dioecious. Branches sometimes developing spine-tips. Leaves opposite or sometimes clustered, evergreen; stipules borne on the stem between the leaf-bases, sometimes united into a tube at the base or for most of their length. Flowers solitary or in cymes, sometimes reduced to clusters, terminal or axillary, small and inconspicuous, greenish, each subtended by 2 partially fused bracts, unisexual, though occasional apparently bisexual flowers may occur in some species. Calyx with 4--6 lobes, the whole calyx often absent in male flowers. Corolla tubular, funnel-shaped or campanulate, 4- or 5-lobed, lobes edge-to-edge in bud. Stamens 4 or 5 (rarely 6), inserted near the base of the corolla-tube, usually projecting well beyond the corolla-lobes. Ovary usually 2-celled (sometimes 3- or 4-celled) with a single ovule in each cell; styles usually 2, projecting. Fruit a drupe with usually 2 (sometimes up to 4) stones. *New Zealand, extending to Indonesia, Australia & Hawaii.*

10. Nertera Banks & Solander. 15/3. Creeping perennial herbs rooting at the nodes. Leaves opposite; stipules small, borne on the stems between the leaf-bases. Flowers solitary, terminal or axillary. Calyx 4- or 5-lobed. Corolla funnel-shaped to campanulate, the lobes 4 or 5, edge-to-edge in bud. Stamens 4 or 5, filaments inserted in the base of the corolla-tube, long-projecting. Ovary 2-celled with a single ovule in each cell. Fruit a coloured drupe. *South America, Malaysia, Taiwan.*

11. Asperula Linnaeus. 150/15. Annual or perennial herbs, sometimes rather woody at the base. Leaves and stipules forming whorls of 4 or 6; dried leaves and stems often aromatic. Flowers bisexual in cymes or panicles, the cymes sometimes head-like or forming whorls in the leaf-axils, with bracts and often bracteoles. Calyx almost absent. Corolla 3- or 4-lobed, the tube usually as long as to longer than the lobes. Ovary 2-celled with a single ovule per cell, hairless or with hair-like processes. Styles 2, equal, stigmas club-shaped or spherical. Fruit a schizocarp with 2 hairless, papillose or hairy mericarps. *Mediterranean area, southwest Asia.*

12. Phuopsis (Grisebach) Hooker. 1/1. Mat-forming perennial, stems sprawling to erect, to 25 cm, 4-angled, winged and with downwardly pointing, hook-like hairs. Leaves and stipules forming whorls of 6, narrowly lanceolate to very narrowly elliptic, to 15 × 4 mm, margins translucent and with forwardly pointing, hook-like hairs. Flowers fragrant, in a terminal, head-like cyme. Corolla magenta, very narrowly funnel-shaped, tube to 1.5 cm, lobes 5, spreading, acute, to 2.5 mm. Ovary 2-celled with a single ovule per cell. Style 1, magenta, projecting from the corolla, the stigmatic area swollen, very narrowly ellipsoid, papillose, shortly 2-lobed at the apex. Fruit a schizocarp composed of 2 hairless, smooth mericarps. *Iran, Caucasus.*

13. Galium Linnaeus. 200/few. Annual or perennial herbs. Stems and leaf-margins often with reflexed, hook-like hairs. Leaves and stipules similar, forming whorls of 4 or 6. Flowers axillary or in axillary or terminal whorls, cymes or panicles, usually bisexual. Corolla with a very short tube and usually 4 spreading lobes which are longer than the tube. Stamens 4. Ovary 2-celled with a single ovule per cell. Fruit a mericarp splitting into 2 schizocarps which are often hairy or papillose. *Temperate areas.* Figure 101, p. 440.

14. Cruciata Miller. 6/1. Like *Galium* and *Asperula*, but flowers yellow, borne in short cymes in the axils of the upper leaves; central flower of each cyme bisexual, the lateral flowers male or absent; corolla-lobes longer than tube. *Europe, east Mediterranean area.*

15. Rubia Linnaeus. 40/2. Herbs, sometimes rather woody at the base, ours scrambling or climbing. Leaves and stipules forming whorls of 4 or 6 (ours), with reflexed, hook-like hairs along the margins and midrib beneath, more or less evergreen (ours). Flowers in axillary cymes or panicles. Corolla with a short tube and 5 widely spreading lobes. Ovary 2-celled with a single ovule per cell. Style deeply bifid with head-like stigmas. Fruit reduced to a single, fleshy, berry-like mericarp. *Temperate areas.*

16. Sherardia Linnaeus. 1/1. Small annual herbs; stems to 40 cm, branched at the base, hairy. Leaves and stipules forming whorls of 4--6, lanceolate, to 20 × 5 mm, with thickened, whitish margins. Flowers in dense head-like cymes, surrounded by fused bracts which form an involucre. Calyx of 4--6 small teeth which enlarge and harden in fruit. Corolla funnel-shaped with a long tube and 4 short spreading lobes, 4--5 mm in diameter, lilac, pink or white. Stamens 4. Ovary 2-celled with a single ovule per cell. Style thread-like, with 2 equal branches, stigmas capitate. Fruit a schizocarp splitting into 2 dry mericarps. *A weed in the northern hemisphere.*

168. POLEMONIACEAE

Annual or perennial herbs, shrubs, small trees or climbers. Leaves alternate or opposite, occasionally whorled or in a basal rosette, simple, pinnately or palmately lobed, sometimes pinnately compound, with or without stalks, without stipules. Flowers bisexual, axillary or terminal, in cymes, heads or rarely solitary, radially or sometimes bilaterally symmetric. Calyx with 5 (sometimes 4 or 6) lobes, these often connected by a translucent membrane which is split by the expanding capsule. Corolla of 5 (sometimes 4 or 6) petals, united, radially symmetric or almost 2-lipped. Stamens usually 5, attached to the corolla-tube, sometimes at different levels, alternate with the corolla-lobes. Ovary superior, often inserted on a basal disc; usually 3-celled, each cell with 1--many ovules, placentation axile, styles 1, stigma usually 3-lobed. Fruit a capsule, usually dehiscent; seeds often mucilaginous when wet. See figure 102, p. 442.

A family of about 20 genera and 300 species occurring in America, Europe and northern Asia.

1a. Calyx not ruptured by expanding fruit ... 2
 b. Calyx ruptured by expanding fruit ... 3
2a. Leaves pinnate ... **1. Polemonium**
 b. Leaves variously divided but not pinnate ... **2. Collomia**
3a. Leaves usually opposite, entire; stamens attached at different heights in
  corolla-tube ... **3. Phlox**
 b. Leaves usually incised, alternate or opposite; stamens attached at the same
  height in corolla-tube or at throat ... 4
4a. Leaves opposite, sometimes sharply pointed but not spine-tipped
  ... **4. Linanthus**
 b. Leaves usually alternate, but if opposite also spine-tipped ... 5
5a. Leaves spine-tipped ... **5. Leptodactylon**
 b. Leaves not spine-tipped ... **6. Gilia**

1. Polemonium Linnaeus. 20/10. Herbaceous perennials. Leaves basal and alternate on the flowering stems, pinnate with a terminal leaflet, leaflets ovate to lanceolate. Flowers in terminal and axillary racemes or panicles. Flowers tubular-campanulate or saucer-shaped. Calyx 5-lobed, persistent, enlarging after flowering to enclose capsule. Petals 5, joined at base into a tube. Stamens 5; anthers with white, yellow or orange pollen. Style divided into 3 stigmas at the tip. Ovary 3-celled. Seeds few to numerous. *Northern hemisphere*. Figure 102, p. 442.

2. Collomia Nuttall. 15/3. Annual or perennial herbs. Leaves alternate, obovate or ovate to linear or lanceolate, entire or lobed. Flowers congested at ends of branches in terminal or axillary clusters. Bracts usually leaf-like. Calyx 5-lobed, expanding but not splitting in fruit. Corolla trumpet- or funnel-shaped, 5-lobed. Stamens inserted at the same or different levels within the corolla-tube. Fruit a capsule. Seeds of annual species mucilaginous when wet. *North & South America*.

3. Phlox Linnaeus. 67/22 and many hybrids. Mostly perennial herbs or shrubs. Leaves mostly opposite, upper stem leaves and bracts sometimes alternate, hairy or hairless, midrib usually prominent, margins thin or thickened to a pale cartilaginous rim, entire or rarely toothed and hairy, occasionally lobed at base, often sheathing stem with united leaf-bases. Inflorescences mostly cymose, or flowers in panicles or rarely solitary. Flowers showy, 5-parted, purple to lilac, pink, violet, lavender, rarely light yellow or white. Calyx narrowly campanulate or tubular; lobes triangular to linear, usually hairy. Corolla-tube long, thin, often more deeply coloured than lobes; lobes radially symmetric, flattened, obovate, elliptic or circular, sometimes notched or bilobed at apex. Stamens usually included, unequal, joined to corolla-tube at petal-junctions. Style divided into 3 short branches at tip. Fruit an ovoid to oblong, 3-celled capsule, rupturing through calyx at maturity. Seeds 1 or 2, rarely 5 per cell, ellipsoid. *Western North America (Alaska to Mexico)*.

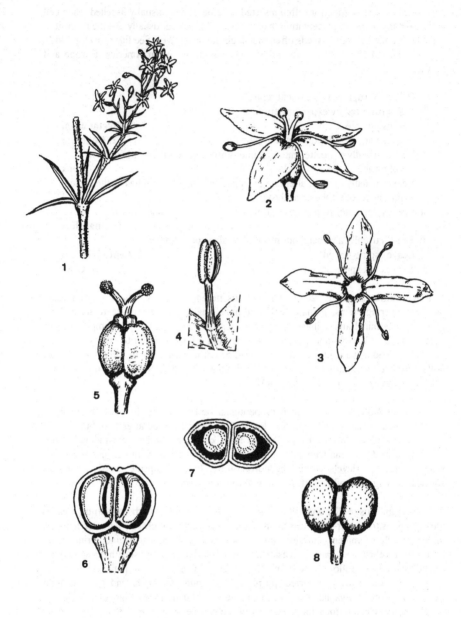

Figure 101. Rubiaceae. *Galium verum.* 1, Partial inflorescence. 2, A single flower. 3, Corolla and stamens from above. 4, Young stamen. 5, Ovary with bifid style. 6, Longitudinal section of ovary. 7. Transverse section of ovary. 8, Fruit.

4. Linanthus Bentham. 35/4. Annual or perennial herbs. Stems simple or branched. Leaves usually opposite, palmately divided into linear segments, rarely simple. Flowers solitary in leaf-axils, in loose cymes or more or less dense clusters, stalkless or on long stalks. Calyx 5-lobed, with membranous sinuses. Corolla campanulate, funnel-shaped or with tubular base and expanded limb, white, pink, purplish or yellow. Stamens inserted at top of corolla-tube, equal or unequal, sometimes protruding. Stigma 3- or 4-lobed. Capsule opening only at top, seeds few. *Western North America, Chile*.

5. Leptodactylon Hooker & Arnott. 10/2. Erect or decumbent perennials or sub-shrubs. Leaves small, alternate or opposite, often in clusters at nodes, deeply palmately or pinnately lobed into narrow spine-tipped segments. Flowers in dense cymes or solitary, usually stalkless. Calyx usually of 5 linear lobes with broad translucent membrane, ruptured by expanding fruit. Corolla usually of 5 spreading lobes, funnel-shaped with narrow tube. Stamens 5, inserted at the same level, included within corolla-tube; style included. Fruit a dehiscent capsule. *Western North America*.

6. Gilia Ruiz & Pavón. 25/few. Annual (rarely perennial) herbs, often hairy and glandular. Leaves alternate, often in a basal rosette, much smaller on stems, mostly pinnately divided, rarely entire. Flowers either solitary in leaf-axils or in panicles or dense, more or less spherical heads. Calyx usually equally 5-lobed, often with membranous sinuses, splitting in fruit. Corolla tubular to funnel-shaped, often with a cupped or flat, 5-lobed limb, blue, purple, pink, white or yellow. Stamens inserted in throat of corolla, sometimes unequal, usually protruding. Style thread-like, often protruding. Capsule with few to many seeds per cell, splitting in upper half only. *Southern USA to southern South America*.

169. CONVOLVULACEAE

Annual and perennial herbs and shrubs, often with milky sap. Stems typically climbing by twining. Leaves alternate, entire or divided, usually lacking stipules. Flowers radially symmetric, occasionally bilaterally symmetrical, borne in terminal or axillary inflorescences or sometimes solitary. Calyx hypogynous, corolla and stamens perigynous. Sepals 5, typically free, sometimes persistent and enlarging in fruit. Corolla usually radially symmetric, campanulate or funnel-shaped or tubular with an abruptly expanded limb, 5-lobed, sometimes deeply so, the lobes often folded and each overlapped by and overlapping 1 other. Stamens 5, borne on the corolla-tube, alternating with corolla-lobes. Ovary superior, 1--4-celled with 1 or 2 ovules per cell. Style terminal, occasionally divided to base; stigma usually distinctly 1- or 2-lobed. Fruit a capsule or indehiscent. See figure 103, p. 444.

A widespread family of around 55 genera and 1500 species. Most of the species cultivated in Europe are climbers and are grown for their large, trumpet-like flowers; few are totally hardy. Some of the species may become invasive weeds, notably *Convolvulus arvensis*. Care should be take to differentiate between bracts (which subtend inflorescences) and bracteoles (which subtend flowers).

1a. Leaves less than 2 cm; stems spreading or erect; flowers small, less
 than 2 cm, corolla somewhat obscured by sepals; style divided to the
 base or almost so **1. Evolvulus**

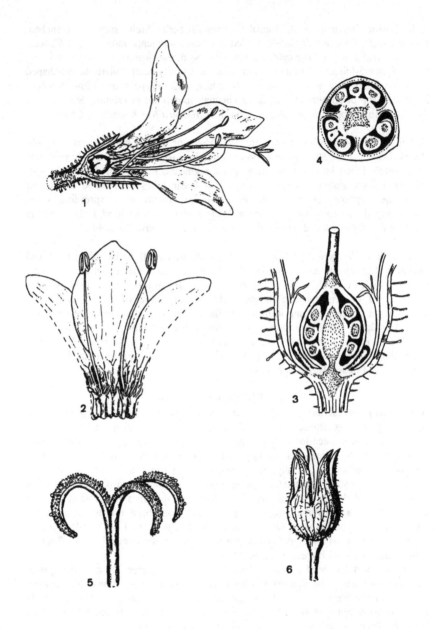

Figure 102. Polemoniaceae. *Polemonium caeruleum.* 1, Longitudinal section of flower. 2, Part of corolla opened out to show attachment of two stamens. 3, Longitudinal section of base of flower. 4, Transverse section of ovary. 5, Top of style showing three stigmas. 6, Capsule after opening.

b. Leaves usually more than 2 cm; stems usually climbing; flowers large and conspicuous, corolla not obscured by sepals; style bearing a simple or lobed stigma, not divided   2
2a. Stigma with 1--3 head-like or globular lobes   **4. Ipomoea**
b. Stigma with 2 fine or narrowly club-shaped lobes   3
3a. Flowers not subtended by broad bracteoles, solitary or in clusters; annual or perennial herbs or shrubs   **2. Convolvulus**
b. Flowers subtended by a pair of broad bracteoles, solitary; rhizomatous perennial herbs   **3. Calystegia**

1. Evolvulus Linnaeus. 100/1. Annual and perennial woody-based herbs. Stems typically erect, sometimes spreading and rooting at the nodes. Leaves typically small, short-stalked or stalkless. Inflorescence resembling a dichasium but sometimes reduced to a single flower or forming a spike-like structure. Flowers small. Sepals generally persistent and not, or only slightly, enlarged in fruit. Corolla funnel- to salver-shaped, shallowly or deeply lobed, blue, violet or white in colour, to 1.2 cm across. Stamens typically not exceeding the corolla tube. Styles 2, free or united at the very base. Stigmas long and terete in cross-section. Fruit a dehiscent, 4-celled capsule containing 1--4 dark seeds. *Warmer parts of the New World, Old World tropics.*

2. Convolvulus Linnaeus. 250/9. Annual and perennial erect, trailing or climbing herbs and shrubs to 1 m that sometimes produce latex. Stems erect, ascending or climbing. Leaves stalked or not, simple, usually entire, sagittate, hastate or tapering to the base. Inflorescences cymose, racemose or reduced to a single flower. Bracteoles present or absent, when present not large and broad. Sepals typically unequal, free. Corolla funnel-shaped (or rarely campanulate, but not in our species), often with 5 dark (sometimes downy) stripes that mark the mid-lines of the fused petals. Stamens never protruding. Pollen not spiny. Ovary 2-celled. Stigma with 2 fine or narrowly club-shaped lobes. Capsule spherical to ovoid, 1- or 2-celled, typically 4-seeded. *Cosmopolitan.* Figure 103, p. 444.

3. Calystegia R. Brown. 25/5. Rhizomatous perennial herbs, to 5 m. Stems climbing or prostrate, exuding white latex when cut. Leaves stalked, more or less sagittate, with the sinus between the basal lobes varying in shape. Flowers axillary, solitary or in small clusters (not in our species), each subtended by a pair of large, broad bracteoles. Corolla large, tubular or funnel-shaped, hairless or ciliate, white, pink or yellow, 5-lobed. Ovary with a single cell. Stigma with two elongated lobes. Fruit a spherical capsule. Seeds hairless. *Temperate areas.*

4. Ipomoea Linnaeus. 500/16 and some hybrids. Annual and perennial, climbing herbs and shrubs; forming tubers or woody at the base. Climbers can reach 30 m. Some species produce latex. Stems usually climbing or less often prostrate to erect. Leaves usually stalked, entire or palmately lobed, occasionally compound, variable on the same plant. Flowers 5-lobed, borne singly or in axillary panicles or cymes of few to many. Sepals often unequal, herbaceous or papery, sometimes enlarging in fruit. Corolla funnel-shaped or campanulate, sometimes with a long, slender tube that abruptly expands into the limb, white, red, purple or yellow. Stamens protruding or included within the corolla, hairy at the base. Stigma 2- or 3-lobed. Fruit a 4--6-lobed, dehiscent capsule. Seeds few, hairy or hairless. *Tropics & subtropics.*

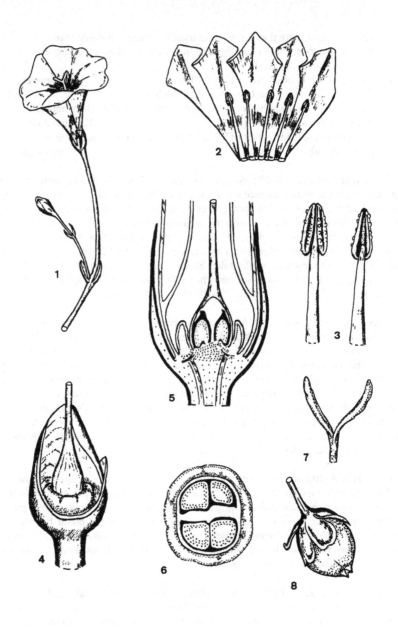

Figure 103. Convolvulaceae. *Convolvulus arvensis.* 1, Part of inflorescence. 2, Corolla opened out. 3, Stamens, undehisced (L) and dehisced (R). 4, Lower part of corolla removed to show ovary and disc. 5, Longitudinal section of lower part of flower. 6, Transverse section of ovary. 7, Top of style with two stigmas. 8, Fruit.

444

170. HYDROPHYLLACEAE

Herbs or, more rarely, shrubs, sometimes with spines or with rough, irritant or stinging hairs. Leaves usually alternate, more rarely opposite, simple, entire, toothed or lobed, sometimes deeply so. Flowers in cymes, these sometimes very loose and open, sometimes coiled and dense. Calyx hypogynous, corolla and stamens perigynous. Calyx 5-lobed, sometimes almost to the base. Corolla radially symmetric, 5-lobed, bell-, funnel- or saucer-shaped, often hairy inside the tube. Stamens 5, equal or unequal, filaments often hairy below, borne on the corolla-tube. Ovary 1-celled with 2 parietal placentas with few to many ovules, the placentas often intrusive, the ovary appearing superficially to be 2-celled, or, more rarely, properly 2-celled with numerous ovules on axile placentas. Fruit a capsule with few to many seeds.

A small family of 22 genera and 275 species, almost cosmopolitan but centred in western North America.

1a. Plant with leaves all basal, flowering stems leafless; leaves simple, entire
                                                   6. Hesperochiron
  b. Leaves not all basal, some borne on stems, often pinnately or palmately lobed   2
2a. Style undivided, or divided just below the 2 stigmas   3
  b. Style divided for at least the upper quarter of its length, stigmas 2   4
3a. Ovary clearly 1-celled, placentas not intrusive; stamens projecting from corolla-tube   **1. Hydrophyllum**
  b. Ovary apparently 2-celled by intrusion of the placentas; stamens included in the corolla-tube   **7. Romanzoffia**
4a. Ovary clearly 1-celled, placentas not intrusive   5
  b. Ovary apparently 2-celled by intrusion of the placentas   6
5a. Ovary and capsule bristly hairy   **3. Pholistoma**
  b. Ovary and capsule hairless or hairy but not bristly   **2. Nemophila**
6a. Flowers pendent; corolla white, yellow or pink, persistent, becoming papery   **5. Emmenanthe**
  b. Flowers erect or spreading; corolla generally blue or purplish, usually not persistent and papery   **4. Phacelia**

1. Hydrophyllum Linnaeus. 7/5. Rhizomatous perennial herbs with fleshy, fibrous or tuber-like roots. Leaves long-stalked, pinnately or palmately lobed almost to the midrib, the lobes themselves less deeply lobed. Flowers in coiled cymes, sometimes congested. Calyx 5-lobed, sometimes very deeply so, with or without appendages between the calyx-lobes. Corolla bell-shaped, lobed to halfway or more, each lobe bearing 2 scales internally. Stamens 5, borne at the base of the corolla-tube, projecting well beyond the corolla-lobes, filaments usually hairy. Ovary 1-celled, placentas 2, parietal, ovules usually 2 per placenta; style projecting beyond the corolla-lobes; stigmas 2. Capsule spherical. *North America.*

2. Nemophila Nuttall. 17/2. Annual herbs, often hairy. Leaves opposite (ours), pinnately lobed or toothed. Flowers solitary in the leaf-axils, stalks long, elongating in fruit. Calyx deeply 5-lobed, with small appendages in the sinuses between the lobes. Corolla bell- to saucer-shaped, exceeding the calyx, white to bright blue, the venation usually blue, sometimes the lobes dark-tipped. Stamens 5, not exceeding the corolla-lobes. Ovary 1-celled, not bristly-hairy; style 1, divided into 2 in the upper half. Capsule spherical to ovoid. *Western North America.*

3. Pholistoma Lilja. 3/1. Annual herbs. Stems sprawling, much-branched, angular, the angles generally bristly or with hooked, prickle-like bristles. Lower leaves opposite, upper alternate, all broadened and stem-clasping at the base, pinnately lobed, bristly. Flowers solitary, axillary or few together in cymes. Calyx deeply 5-lobed, the lobes bristly, appendages present in the sinuses between the lobes (ours). Corolla bowl-shaped, with 5 hairy lobes. Ovary 1-celled; style 2-lobed in the upper half. Fruit a spherical capsule. *Western North America.*

4. Phacelia Jussieu. 100/10. Annual or perennial herbs. Leaves alternate (occasionally the lowermost opposite or almost so), simple, lobed or compound. Flowers in open or congested cymes which are sometimes strongly coiled. Calyx divided almost to the base into 5 narrow lobes. Corolla saucer-, bell- or funnel-shaped, lobed to halfway or more, rarely the tube slightly constricted above so that the corolla is somewhat urn-shaped, usually blue, violet, purple or white; 10 scales are usually present in the base of the corolla-tube. Stamens 5, usually projecting from the corolla; filaments often hairy, sometimes with teeth at the base. Ovary 1-celled but apparently 2-celled by intrusion of the placentas, ovules 2--many; style divided above into 2 stigmatic lobes. Fruit a capsule with 2--many seeds. *North America, mostly in the west.*

5. Emmenanthe Bentham. 1/1. Glandular-sticky, aromatic annual herb; stems simple or branched. Leaves in a basal rosette, few on the stems, alternate and stem-clasping at the base, 1--12 × 1--3 cm, toothed to deeply pinnately lobed. Flowers axillary, pendent on recurved stalks which elongate in fruit. Calyx 5-lobed, lobes 4--11 × 1--4 mm, glandular. Corolla bell-shaped, 6--15 mm, white, cream, yellow or pink, persistent and becoming papery, enclosing the fruit. Ovary 1-celled but appearing 2-celled by intrusion of the placentas; stigma 2-lobed in the upper third. Fruit a small, spherical, hairy capsule. *Western North America.*

6. Hesperochiron Watson. 2/2. Perennial, stemless herbs, flowers borne on scapes. Leaves in a basal rosette, long-stalked, entire, usually ciliate. Flowers solitary. Calyx 5-lobed almost to the base, lobes somewhat unequal. Corolla bell-, funnel- or saucer-shaped, 5-lobed to the middle or further, the tube hairy inside, the lobes hairy or hairless. Stamens 5, somewhat unequal, bases of filaments broadened. Ovary hairy, 1-celled; style 1, divided above into 2 stigmas. Fruit a hairy capsule. *Western USA.*

7. Romanzoffia Chamisso. 6/4. Herbs, leaves mostly basal, few on the stems, with rhizomes bearing bulbils or with a bulb-like structure at the base formed from expanded leaf-bases. Leaves long-stalked, kidney-shaped to almost circular, palmately lobed. Flowers solitary or few in cymes. Calyx 5-lobed almost to the base. Corolla bell-shaped to funnel-shaped, 5-lobed to halfway, white or pale purple, often yellowish at the base. Stamens 5, equal, included in the corolla-tube. Ovary apparently 2-celled by intrusion of the placentas; style 1, bearing 2 stigmas. Fruit a capsule. *Western North America.*

446

171. BORAGINACEAE

Usually herbs, more rarely evergreen or deciduous small shrubs or small trees; hair-covering often conspicuous and bristly; shoots terete. Leaves alternate, without stipules. Inflorescences usually of coiled cymes (often 1-sided and appearing racemose), more rarely flowers solitary; bracts present or absent. Flowers bisexual. Calyx usually 5-lobed, lobes sometimes extending to the base or almost so, often persistent and enlarged in fruit. Corolla usually radially symmetric, 5-lobed, variously shaped, often with a very short tube and spreading, flat limb; appendages (scales, folds or hair-tufts) often present in the corolla-throat. Stamens 5, attached to the corolla-tube, often 1 or more projecting from the corolla. Ovary superior, of 2 carpels which are usually each divided by a false septum, to produce a 4-lobed ovary; style usually arising from the depression between the 4 lobes (gynobasic), more rarely apical; stigma 2-lobed or capitate. Fruit a group of 1--4 separate nutlets, variously ornamented, attached to the flat to conical receptacle, rarely a fleshy drupe with 2--4 seeds. See figures 104, p. 448, 105, p. 450 & 106, p. 456.

A family of about 150 genera, mainly found in north temperate areas, but extending into South America, South Africa and Australasia; a few are restricted to the tropics.

1a. Tree to 12 m or more; fruit a fleshy drupe with 2--4 seeds   **1. Ehretia**
  b. Herbs or small shrubs, fruit usually a group of 1--4 nutlets   2
2a. Shrubs with woody stems, leaves evergreen   3
  b. Herbs, sometimes woody at the base only, leaves evergreen or deciduous
                                                                     5
3a. Corolla bilaterally symmetric   **29. Echium**
  b. Corolla radially symmetric   4
4a. Calyx funnel-shaped, deeply divided but not to the base; nutlets usually
      solitary   **13. Lithodora**
  b. Calyx tubular, divided to the base or almost so; nutlets usually 4
                                                            14. Lobostemon
5a. Leaves hairless or almost so, fleshy or glaucous   6
  b. Leaves with a more or less dense covering of hairs and/or bristles   9
6a. Corolla with a long thin tube and abruptly spreading limb; anthers unequal,
      1 longer than the other 4   **4. Caccinia**
  b. Corolla with short tube and flat, spreading limb, or tubular; anthers all
      equal   7
7a. Corolla tubular, yellow or red; nutlets fused into 2 separate pairs
                                                       24. Cerinthe
  b. Corolla with tube 1--2 mm and flat, spreading limb; nutlets 4, separate   8
8a. Leaf-blades of lower leaves to 20 cm, fleshy, green; nutlets with sinuous
      margins   **3. Myosotidium**
  b. Leaf-blades of lower leaves not more than 10 cm, not fleshy, glaucous;
      nutlets each with a thickened or incurved wing   **6. Omphalodes**
9a. At least 1 of the stamens projecting well beyond the mouth of the corolla
                                                               10
  b. All stamens enclosed in the corolla-tube or included in the throat   17
10a. Throat of corolla smooth, lacking scales or hairs   11
  b. Throat of corolla with scales or hairs   13
11a. Corolla funnel-shaped, bilaterally symmetric with an oblique, open limb
                                                            29. Echium

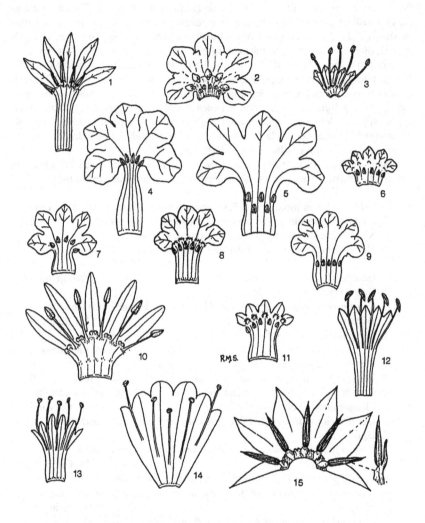

Figure 104. Boraginaceae: corollas (opened out) and stamens. 1, *Caccinia*. 2, *Omphalodes*. 3, *Solenanthus*. 4, *Arnebia*. 5, *Arnebia* (another species). 6, *Lithospermum*. 7, *Lithodora*. 8, *Nonea*. 9, *Alkanna*. 10, *Trachystemon*. 11, *Anchusa*. 12, *Moltkia*. 13, *Echium*. 14, *Echium* (another species). 15, *Borago*.

b. Corolla tubular, radially symmetric, a distinct limb lacking, lobes very
   short   12
12a. Corolla long-tubular, pendent; anthers arrow-shaped, united at their bases
   or to the middles   **23. Onosma**
b. Corolla short-tubular, erect; anthers not or only slightly arrow-shaped,
   free from each other   **28. Moltkia**
13a. Corolla-lobes linear, their margins recurved; scales shaggy or papillose, in
   2 series of 5, one in the corolla-throat, the other lower in the tube
                                                              20. Trachystemon
b. Corolla-lobes not as above; scales neither hairy nor papillose, not as above
                                                                             14
14a. Annual; anthers pressed together to form a conical beak; nutlets ridged
                                                                   31. Borago
b. Perennial; anthers not as above; nutlets winged or spiny   15
15a. Calyx woolly, divided for part of its length; nutlets flat, each with a broad,
   membranous wing   **7. Rindera**
b. Calyx not woolly, divided almost to the base; nutlets densely clothed with
   spines tipped with anchor-like barbs   16
16a. Corolla campanulate or funnel-shaped, 1--1.5 cm   **5. Lindelofia**
b. Corolla almost cylindric, lobes indisinct, at most 7 mm   **8. Solenanthus**
17a. Nutlets smooth, not fleshy   18
b. Nutlets variously ornamented with papillae, spines, ridges, etc., or if more
   or less smooth, then fleshy   20
18a. Nutlets less than 2.5 mm, dark-coloured, each usually with a distinct rim
                                                                30. Myosotis
b. Nutlets more than 2.5 mm, ivory-white or yellowish, each without a
   distinct rim   19
19a. Throat of corolla with downy or glandular appendages
                                                           12. Lithospermum
b. Throat of corolla with 5 longitudinal bands of hairs   **15. Buglossoides**
20a. Plants with rhizomes; each nutlet with a thickened, collar-like basal ring
                                                              21. Brunnera
b. Plants without rhizomes; nutlets without basal collars   21
21a. Corolla-throat without scales, folds or hairs   22
b. Corolla-throat with distinct appendages (scales, folds or tufts of hair)   24
22a. Flowers pendent; corolla tubular, lobes minute   **23. Onosma**
b. Flowers erect; corolla with a distinct, usually spreading limb   23
23a. Limb of corolla purple, with teeth alternating with the lobes; style terminal
                                                            2. Heliotropium
b. Limb of corolla yellow, without additional teeth; style arising basally,
   between the 4 lobes of the ovary   **11. Arnebia**
24a. Low, dense, cushion-forming plants; corolla with yellow scales in the
   throat   **10. Eritrichium**
b. Taller, open plants, not forming cushions; corolla with folds, hairs or
   whitish scales in the throat   25
25a. Throat of corolla hairy, without scales or folds   26
b. Throat of corolla with scales or folds   27
26a. Throat of corolla with 5 longitudinal bands of hairs   **15. Buglossoides**

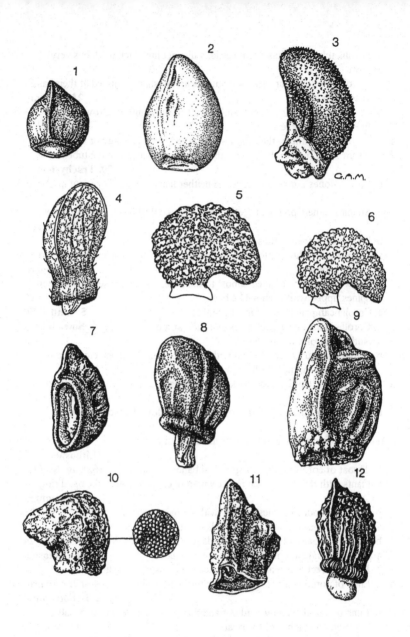

Figure 105. Boraginaceae: nutlets. 1, *Arnebia*. 2, *Lithospermum*. 3, *Lithodora*. 4, *Nonea*. 5, *Alkanna*. 6, *Alkanna* (another species). 7, *Trachystemon*. 8, *Symphytum*. 9, *Anchusa*. 10, *Moltkia*. 11, *Echium*. 12, *Borago*.

450

b. Throat of corolla with 5 tufts of hairs meeting to form a ring
17. Pulmonaria
27a. Nutlets each with a smooth or spiny incurved marginal wing
6. Omphalodes
   b. Nutlets not as above   28
28a. Corolla tubular, pendent   **22. Symphytum**
   b. Corolla with a distinct, spreading, lobed limb   29
29a. Nutlets covered with spines tipped with anchor-like hooks
9. Cynoglossum
   b. Nutlets variously ornamented, not as above   30
30a. Each nutlet with a distinct, oblique or recurved beak   **19. Alkanna**
   b. Nutlets not as above   31
31a. Nutlets flattened, 3-angled, fleshy   **16. Mertensia**
   b. Nutlets not flattened, rounded in outline, dry   32
32a. Calyx not enlarged in fruit; nutlets usually more or less tuberculate
25. Anchusa
   b. Calyx enlarged in fruit; nutlets various, not tuberculate 33
33a. Corolla yellow; scales in throat short, semicircular, cut into narrow, hairy
   lobes, joined by a ring of hairs   **18. Nonea**
   b. Corolla blue; scales in throat oblong-rectangular, not as above 34
34a. Corolla-throat open, not closed by the scales, with anthers included
   in the mouth of the tube   **26. Cynoglottis**
   b. Corolla-throat closed by conspicuous hairy scales, with anthers inserted
   in the upper half of the corolla-tube   **27. Pentaglottis**

1. Ehretia Linnaeus. 50/few. Evergreen or deciduous trees and shrubs. Leaves alternate, with or without hairs, simple, obovate, elliptic-ovate to lanceolate or oblong, stalked, entire or toothed. Flowers in terminal or axillary cymes or panicles. Calyx 5-lobed. Corolla-lobes spreading or reflexed, white, yellow or rarely blue. Stamens 5, usually protruding, anthers oblong to ellipsoid. Ovary with 2--4 cells; style terminal, with 2 stigmatic branches. Fruit a drupe, ovoid to almost spherical, seeds 2--4. *Tropical & subtropical Asia, Africa & America.*

2. Heliotropium Linnaeus. 250/2. Annual or perennial herbs, sometimes woody at base. Leaves generally hairy. Flowers strongly scented, in terminal or apparently axillary, branched cymes; bracts absent. Corolla with a long slender tube without scales and an abruptly expanded limb usually with teeth between the lobes. Stamens included in the corolla-tube. Style terminal, included, usually very short; stigma large and disc-like, or cone-shaped tapering to a very fine point, entire or with 2--4 lobes. Fruit of 2 or 4 nutlets, rarely only 1 nutlet persisting, variously ornamented. *Tropical & temperate areas.*

3. Myosotidium J.D. Hooker. 1/1. Stout perennial herbs. Stem 30--45 cm, hairless below, sparsely downy above, terminating in a large, dense corymb of spirally coiled cymes. Basal leaves numerous, with long, stout stalks, to 32 × 13 cm (length includes stalk), blade ovate with cordate base, hairless, fleshy; stem-leaves alternate, upper stalkless, uppermost hairy beneath. Calyx-lobes ovate, *c.* 4 mm, obtuse, stiffly hairy. Corolla blue, with short tube and spreading limb, *c.* 1.2 cm across; lobes white-margined, throat-scales yellow. Stamens equal. Nutlets 4, coherent at their apices, broadly ovate, c. 6 mm, their margin sinuous. *New Zealand (Chatham Islands)*.

4. Caccinia Savi. 5/1. Bushy, sprawling, biennial or perennial herbs. Stems hairless. Leaves glaucous, hairless except for numerous white tubercles. Inflorescence paniculate; bracts present, leaf-like. Calyx almost spherical, with 10 tuberculate keels in flower, lobes clasping the apex of the corolla-tube; fruiting calyx-lobes separating to near their middle to form a 5-angled, 10-keeled structure. Corolla-tube narrow, equalling the calyx; lobes spreading, linear-oblong; throat-scales pyramidal, protruding. Stamens protruding; anthers unequal, 1 large, 4 small, each with 2 forked basal appendages. Style protruding. Nutlets 1 or 2, rarely 3, depressed- ovoid or circular, with narrow, toothed margins. *Southwest & central Asia*. Figure 104, p. 448.

5. Lindelofia Lehmann. 12/2. Perennial herbs with stout rootstocks. Stems erect or ascending, usually softly hairy. Leaves more or less long-stalked, at least the basal tapered gradually at the base. Inflorescence a terminal panicle of cymes, without bracts. Calyx divided nearly to the base; lobes lanceolate to linear. Corolla various shades of blue or violet, cylindric, campanulate to funnel-shaped, the tube longer than the calyx and at least as long as, often much longer than the limb; throat-scales usually large, oblong, with or without appendages near the apex. Stamens protruding from the corolla-tube and sometimes slightly protruding from the corolla. Anthers extending at least partly above the base of the scales. Style protruding. Nutlets attached to the receptacle by apical scars, their sides and dorsal surfaces more or less densely covered with tiny barbed spines. *Southern centrla Asia (Pamir Alai and Tien Shan), Himalaya, Pakistan & Afghanistan, possibly also 1 species in Iran*.

6. Omphalodes Miller. 14/6. Perennial, biennial or annual herbs. Leaves alternate (ours) or lower opposite, hairy or hairless, sometimes glaucous. Inflorescence cymose, with or without bracts, or flowers solitary in leaf-axils. Calyx divided to at least the middle, more or less enlarged in fruit and then with lobes stellately spreading, sometimes reflexed. Corolla blue or white, with a short tube and spreading limb to shortly campanulate with short tube and flat or concave limb of spreading lobes; throat-scales present, papillose and more or less notched at apex. Anthers included in the corolla-tube. Receptacle usually flat and rudimentary, but conical in *O. linifolia* and *O. commutata*. Nutlets 4, equalling or shorter than calyx-lobes, often shortly hairy, rarely with minute bristles, wing present, either as a narrow thickened rim or broader and incurving, the wing-margin entire or toothed. *Portugal, northwest Spain, Mediterranean area, central Europe & southwest Asia*. Figure 104, p. 448.

7. Rindera Pallas. 27/1. Erect perennial herbs. Stems usually unbranched and overtopping the basal leaves, usually softly hairy. Basal leaves in a rosette, long-stalked; stem-leaves gradually becoming stalkless upwards. Inflorescence corymbose, of scorpioid cymes. Calyx lobed, but not almost to the base, lobes oblong, ovate or lanceolate, enlarged in fruit, woolly. Corolla cylindric-campanulate, 1.5--2 times the calyx; tube shorter (ours) or longer than the limb; throat-scales present. Stamens and style usually protruding from corolla. Nutlets flat, almost circular, each with a broad membranous wing, the central area of the nutlet dorsal surface with or without barbed appendages. *Balkans, southwest and central Asia eastwards to China (Xinjiang), with its major centre of diversity in central Asia*.

452

8. Solenanthus Ledebour. 17/1. Erect perennial herbs. Basal leaves large, long-stalked, hairy. Inflorescence a panicle of several spreading coiled cymes. Calyx divided nearly to base into oblong-elliptic lobes. Corolla more or less cylindric, with very indistinctly defined limb and tube, at most to 7 mm, the lobes small; throat-scales present. Stamens projecting from corolla. Style much longer than calyx and usually projecting from corolla. Nutlets ovoid or almost spherical, wingless, each with many spines tipped with anchor-like barbs which attach to fur or clothing. *South Europe, southwest & central Asia, 1 species in central China (Hubei).* Figure 104, p. 448.

9. Cynoglossum Linnaeus. 50/9. Perennial, biennial or rarely annual herbs. Leaves alternate, hairy; basal leaves long-stalked. Inflorescence an elongating 1-sided cyme; bracts usually absent. Calyx lobed to more than halfway, enlarged in fruit; lobes spreading or reflexed. Corolla cylindric or funnel-shaped; lobes broad, spreading, overlapping; throat with pyramidal, oblong or almost crescent-shaped appendages. Stamens included in corolla, inserted in tube. Style short, entire. Nutlets 4, depressed ovoid or spherical, covered with spines with anchor-like tips, ascending or widely divergent, back flat or convex, often with raised margins. *Cosmopolitan.*

10. Eritrichium Schrader. 4--10/3. Low, dense, often cushion-forming perennial herbs (ours). Stems (when present) and leaves hairy. Flowers in simple or branched, terminal 1-sided cymes. Calyx divided into 5 lobes. Corolla funnel-shaped or with very short tube and spreading limb; throat-scales present, often yellow. Stamens included in the corolla-tube. Fruit of 4 small nutlets, hairy or not. *Circumboreal and in mountains of central Europe, Himalaya and western North America.*

11. Arnebia Forsskål. 25/2. Perennial to annual herbs. Root yielding purple or deep red dye. Stems erect, branched or unbranched, hairy. Leaves with a more or less dense covering of hairs and bristles. Inflorescence (ours) a dense terminal head; cymes bracteate. Flowers often heterostylous, erect. Calyx deeply divided into 5 linear or lanceolate lobes. Corolla yellow (ours), more or less funnel-shaped, with a distinct limb; throat without scales. Stamens included, inserted at 1 or 2 levels or irregularly in the corolla-tube. Style included, 2-lobed or bifid, the branches entire (stigmas 2) or forked (stigmas 4). Nutlets ovoid or somewhat spherical or 3-sided, each beaked and with a ventral keel, brown. *Mediterranean area, northeast tropical Africa, southwest Asia, Caucasus, Himalaya.* Figures 104, p. 448 & 105, p. 450.

12. Lithospermum Linnaeus. 60/3. Annual or perennial herbs. Stems erect or spreading, bristly or shaggy hairy. Leaves hairy. Flowers in terminal 1-sided cymes or single in the leaf-axils; bracts usually numerous, leaf-like. Calyx 5-lobed more or less to the base, enlarged in fruit. Corolla cylindric to funnel-shaped, white, yellow, orange or blue, hairy on the outside; tube often with downy or glandular appendages at the throat. Stamens included. Style included; stigma 2-lobed. Nutlets 4, ovoid or ellipsoid, stony, white or pale yellowish brown, smooth (ours), rough or warty. *Almost cosmopolitan though absent from Australia.* Figures 104, p. 448 & 105, p. 450.

13. Lithodora Grisebach. 7/5. Evergreen or semi-evergreen dwarf shrubs. Leaves alternate, entire, linear to elliptic or obovate, often covered with sharp bristly hairs, margins curved in or rolled under. Cymes terminal, leafy, loose with 1--10 flowers. Calyx 5-lobed, enlarged with age; lobes narrow, unequal. Corolla blue, purple or white, funnel-shaped or with a long slim tube and abruptly flattened limb, throat lacking appendages, exterior hairless or hairy, interior hairless, lacking a basal ring; lobes ovate to almost circular, ascending or spreading. Stamens usually included; filaments thread-like; anthers oblong with obtuse or notched apex. Style simple or branched above usually included; stigmas 2. Nutlets usually solitary, ovoid or ovoid-cylindric, constricted above the base. *Southwest Europe to southern Greece, Turkey and Algeria.* Figures 104, p. 448 & 105, p. 450.

Formerly included in *Lithospermum*.

14. Lobostemon Lehmann. 28/3. Shrubs (ours) or small shrubs, rarely almost herbaceous, roughly bristly or hairy, rarely hairless. Leaves alternate, stalkless, sometimes armed with spiny hairs. Calyx-lobes free or almost so, rarely joined for part of their length. Corolla blue, red, pink, white or greenish yellow, tubular or funnel-shaped, lobes equal or 2 larger than the rest, erect or spreading. Stamens 5, borne on the corolla-tube. Nutlets 4, distinct, erect, tuberculate or roughly tuberculate, rarely almost smooth, occasionally with tufts of glass-like spikes on the tubercles. *South Africa.*

15. Buglossoides Moench. 10/3. Annual or perennial herbs, bristly hairy. Stems erect or prostrate. Leaves entire, with a more or less dense covering of hairs and bristles. Inflorescence a terminal cyme; bracts present. Calyx deeply 5-lobed, enlarged in fruit. Corolla cylindric to funnel-shaped, with 5 longitudinal bands of hairs inside, 5-lobed, lobes overlapping. Stamens included, inserted below the middle of the corolla-tube; style included. Nutlets smooth to roughly tuberculate. *South Europe, southwest Asia.*

Often included in *Lithospermum*.

16. Mertensia Roth. 50/9. Perennial herbs, leaves hairy or hairless but not fleshy or glaucous. Leaves entire, linear to cordate, alternate. Inflorescence a loose or dense modified coiled cyme; bracts absent. Calyx 5-parted, sometimes campanulate, often enlarged in fruit. Corolla tubular, funnel-shaped or campanulate, tube as long as or longer than calyx; throat with (ours) or without scales; lobes 5, spreading or almost erect. Nutlets flattened, strongly 3-angled, fleshy, smooth. *North temperate areas, most native to North America.*

17. Pulmonaria Linnaeus. 18/9 and several hybrids. Perennial hairy and/or glandular herbs with compact creeping rhizomes. Basal leaves hairy, ovate to narrowly lanceolate, entire, stalked, cordate, truncate or narrowed gradually at the base, obtuse to acute, plain green or with conspicuous green, silvery or whitish mottling or blotches. Inflorescence of bracteate cymes. Flowers funnel-shaped; calyx tubular to funnel-shaped, 5-angled, with 5 triangular lobes up to one-third as long as the tube. Corolla red, purple, violet or blue with 5 rounded lobes and a cylindric tube; throat with 5 tufts of hairs meeting to form a ring. Stamens included. Stigma capitate or slightly bifid. Nutlets to 5 × 4 cm, ovoid, usually hairy, at least at first. *Europe, southwest Asia.*

18. Nonea Medikus. 35/1. Annual or perennial herbs, often with glandular hairs. Basal and stem-leaves present, hairy; stem-leaves stalkless. Calyx divided to not more than halfway, enlarged and becoming ovoid-spherical in fruit. Corolla yellow (ours), white, pink, blue, violet, purple or brown, with long, slender tube and abruptly expanded limb

454

(ours) or funnel-shaped; throat with 5 small semicircular, hairy scales cut into lobes, and sometimes a ring of hairs. Stamens included in the corolla. Nutlets downy or hairless, either erect and ovoid with an erect beak and basal attachment scar, or transversely ovoid with a somewhat horizontal beak and nearly lateral scar. *Mediterranean area, southwest & central Asia with one species extending to eastern Siberia.* Figures 104, p. 448 & 105, p. 450.

19. Alkanna Tausch. 31/2. Perennial herbs. Leaves alternate, usually entire, hairy or bristly. Inflorescence of 1 or more terminal, bracteate cymes. Corolla blue or yellow, funnel-shaped, with a ring of hairs in the throat. Stamens and style included in the corolla-tube. Nutlets usually 2, almost kidney-shaped to obliquely ovoid, granular to tuberculate, stalked, each with a distinct recurved or oblique beak. *South Europe to Iran.* Figures 104, p. 448 & 105, p. 450.

20. Trachystemon D. Don. 2/1. Perennial herbs. Leaves hairy. Inflorescence a loose panicle. Calyx 5-lobed, cup-shaped, enlarged in fruit. Corolla 5-lobed, funnel-shaped, blue-purple; lobes linear, with the margins curved downwards and inwards; throat-scales in 2 series of 5, one at the throat, one deeper in the tube, with shaggy hairs or papillose. Stamens 5, protruding, filaments pink. Fruit of 4 nutlets, each with a thickened collar-like ring at the base. *Eastern Europe & northern Turkey.* Figures 104, p. 448 & 105, p. 450.

21. Brunnera Steven. 3/1. Rhizomatous perennial herbs. Leaves hairy. Inflorescence a terminal panicle; bracts absent. Calyx lobed almost to the base, enlarged in fruit, lobes linear-lanceolate. Corolla purple or blue, very small, with a short, campanulate tube, and spreading, ovate-circular lobes. Stamens included in the corolla, inserted at the middle of the tube. Style included; stigma capitate. Nutlets oblong-obovoid, erect, rough, warty, each with a thickened collar-like basal ring. *Western Europe to Siberia.*

22. Symphytum Linnaeus. 35/10 and several hybrids. Stiffly hairy perennial herbs. Leaves generally hairy, ovate-elliptic, cordate to wedge-shaped at the base, the basal long-stalked, stem-leaves shortly stalked, stalkless or decurrent. Flowers nodding, in ebracteate, terminal, spirally coiled cymes. Calyx campanulate or tubular, lobed from one-fifth to nine-tenths, enlarged in fruit. Corolla funnel-shaped or campanulate, widest above the middle, variously coloured, with 5 short, triangular to semicircular lobes erect or recurved at the apex; throat with 5 linear or lanceolate scales converging like a cone, usually with marginal papillae and occasionally protruding, alternating with 5 equal, included stamens inserted at about the middle of the corolla-tube at the level of the base of the scales; filaments not more than 5 times as long as anthers. Style simple, protruding; stigma small, capitate. Nutlets 4, ovoid, smooth or granulate and often wrinkled, with a collar-like caruncle protruding at the base. *Europe, Asia.* Figures 105, p. 450 & 106, p. 456

23. Onosma Linnaeus. 150/10. Biennial or perennial, bristly herbs. Flowers pendent, with bracts, in terminal, usually branched cymes. Calyx lobed almost to base. Corolla yellow, whitish or purplish, tubular to tubular-campanulate, with 5 short lobes, lacking throat-scales but with a basal ring of hairs. Stamens included or projecting from corolla-tube, anthers arrowhead-shaped, united at their bases or to their middles.

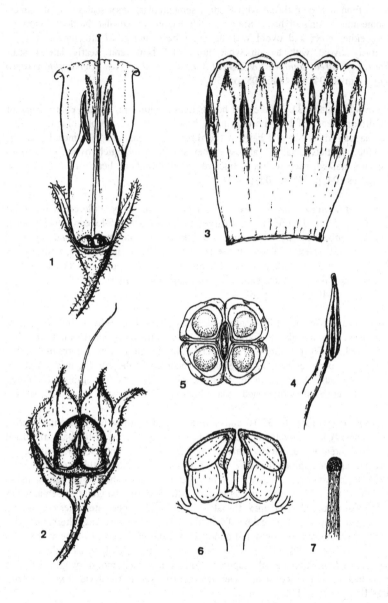

Figure 106. Boraginaceae. *Symphytum officinale.* 1, Longitudinal section of flower. 2, Calyx with two lobes removed to expose ovary. 3, Corolla opened out. 4, A single stamen. 5, Transverse section of ovary. 6, Longitudinal section of ovary through two developing nutlets. 7, Stigma.

456

Nutlets often 4, ovoid or triangular, smooth or with tubercles, flat-based. *Mediterranean area, southwest & central Asia.*

24. Cerinthe Linnaeus. 10/2. Annual (ours), biennial or perennial herbs; leaves alternate, hairless or almost so, usually glaucous. Inflorescence usually branched, with showy reddish to purple bracts. Calyx deeply lobed. Corolla yellow or dark red, tubular, lobes 5, shorter than to as long as tube, strongly recurved. Nutlets fused in 2 separate pairs. *Mediterranean area, Asia.*

25. Anchusa Linnaeus. 35/6. Annual, biennial or perennial herbs, occasionally woody at base. Leaves alternate, simple, entire to toothed. Flowers in terminal and axillary, bracteate cymes, usually elongated in fruit. Calyx 5-lobed. Corolla 5-lobed, blue, violet, white or yellow, funnel-shaped, tube straight or curved with an abruptly expanding limb of equal or unequal lobes, with 5 papillose or hairy scales in the throat. Stamens 5, included in corolla-tube or slightly protruding, variously inserted. Style included, stigma capitate. Nutlets 4, erect, oblique to horizontal, wrinkled or with net-like markings, usually more or less tuberculate, each with a thickened collar at the base. *Europe, northern and southern Africa & western Asia.* Figures 104, p. 448 & 105, p. 450.

26. Cynoglottis (Gusuleac) Vural & Kit Tan. 2/1. Perennial herbs; stems ascending to almost erect (ours), hairy. Leaves hairy. Inflorescence terminal, cymes numerous. Bracts present. Calyx divided to base. Corolla blue or blue-violet, radially symmetric, with a spreading limb and oblong-rectangular, papillose scales in the open throat of the tube. Anthers included in corolla-throat. Style included; stigma head-like. Nutlets erect, ornamented, each with a basal attachment-scar. *Europe.*

27. Pentaglottis Tausch. 1/1. Perennial evergreen herbs, bristly; stems ascending or erect, branched. Basal leaves ovate-oblong to ovate, to 40 cm, narrowed to long stalks. Inflorescence bracteate, of dense cymes. Calyx-lobes linear-lanceolate, cut almost to base, enlarged in fruit. Corolla-tube funnel-shaped, to 6 mm; limb to 1 cm wide, spreading, bright blue, with 5 hairy scales closing the mouth of the tube. Stamens 5, included, inserted above middle of the corolla-tube; style included, stigma head-like. Nutlets to 2 mm, blackish, ornamented. *Southern & western Europe.*

28. Moltkia Lehmann. 6/6 and some hybrids. Perennial herbs or small shrubs. Leaves alternate. Flowers in terminal bracteate cymes. Calyx deeply 5-lobed. Corolla blue, purple or yellow, funnel-shaped, without scales or annulus. Stamens projecting from or included in throat of corolla, filaments inserted at or above middle of tube; anthers not or only slightly sagittate at base, without projecting connectives at the apices but sometimes apiculate. Style protruding; stigma entire or notched. Nutlets usually smooth, with an upper keel. *Northern Italy to northern Greece & southwest Asia.* Figures 104, p. 448 & 105, p. 450.

29. Echium Linnaeus. 40/10. Annual, biennial or perennial herbs or shrubs, with tubercle-based bristles and an underlayer of usually short hairs. Inflorescence spike-like or paniculate, with bracteate cymes, often much enlarged in fruit. Calyx 5-lobed, divided almost to base. Corolla bilaterally symmetric, red, blue, purple, yellow or white, broadly to narrowly funnel-shaped, with a tapering tube and usually oblique, open limb. An annulus of 10 minute scales or tufts of hairs, or sometimes a collar-like membrane, is present at the base of the corolla. Stamens 5, unequal, inserted below the middle of the corolla. Style protruding; stigma head-like or bifid. Fruit 4, ovoid to 3-angled, erect, wrinkled. *Europe, northern & southern Africa, western Asia, Macaronesia.* Figures 104, p. 448 & 105, p. 450.

30. Myosotis Linnaeus. 100/11. Hairy annual, biennial or perennial herbs. Leaves alternate, entire and softly hairy. Flowers usually in paired cymes, bracts usually absent. Calyx 5-lobed, often enlarged in fruit. Corolla radially symmetric, 5-lobed, lobes flat or slightly concave, mostly blue, purple or white, sometimes yellow, corolla-throat bearing 5 white or yellow papillose scales, tube usually short. Stamens 5, usually included, filaments inserted near middle of corolla-tube. Style included, stigma capitate. Nutlets usually 4, brown to black, hairless, smooth and shiny, erect, ovoid and usually compressed. *Temperate areas of Europe, Asia, South America & Australasia.*

31. Borago Linnaeus. 3/2. Annual or perennial herbs. Leaves alternate, hairy. Flowers in branched, loose cymes; bracts usually present. Calyx lobed almost to the base, enlarging in fruit. Corolla blue, pink or white, flat to campanulate, tube short or absent, lobes widely spreading, lanceolate. Stamens protruding and forming a cone-shaped beak; each filament with a long narrow appendage at the apex. Style included; stigma capitate. Nutlets 4, obovoid, erect, rough. *Europe.* Figures 104, p. 448 & 105, p. 450.

172. VERBENACEAE

Shrubs or trees, sometimes herbs, commonly with 4-striped or -angled branches. Leaves opposite (usually at right-angles to the previous pair), rarely whorled or alternate, simple or compound; stipules absent. Inflorescences axillary or terminal, racemes, spikes, heads or variously cymose and often paniculate. Flowers bisexual or unisexual, bilaterally symmetric or almost radially symmetric. Calyx hypogynous, stamens borne on the corolla-tube. Calyx 4- or 5-lobed or -toothed, sometimes almost entire, persistent. Corolla united, tubular or enlarged above, often curved, 4- or 5-lobed, the lobes overlapping in bud. Stamens 4 (rarely 2 or 5), usually with one long pair and one short pair; staminodes sometimes present; anthers 2-celled, the cells often divergent, splitting lengthwise. Ovary superior, 2- or 4-celled. Style entire or shortly bifid, occasionally arising from between ovary-lobes. Ovules 1 or 2 in each cell, erect or pendent, placentation axile. Fruit a 2- or 4-celled drupe, capsule, or schizocarp. Seeds with straight embryo and thin endosperm. See figure 107, p. 460.

A family of about 76 genera and 3000 species from the tropics and subtropics, extending into temperate areas of the New World and a few in the Old World. An economically important family providing timbers, essential oils, teas, herbal medicines, fruits, gums, tannins and ornamentals.

1a.	Leaves palmately compound	**1. Vitex**
b.	Leaves simple, entire to 2-pinnatisect	2
2a.	Inflorescence a spike (rarely a raceme or corymb), often almost head-like; flowers stalkless or almost so	3
b.	Inflorescence a raceme, cyme or panicle; flowers stalked	6
3a.	Fruit with 4 stones	**2. Verbena**
b.	Fruit with 2 stones or mericarps	4

4a. Flowers in long, loose spikes **3. Aloysia**
  b. Flowers densely arranged in spikes or heads, sometimes solitary, or
    paired 5
5a. Fruit a fleshy drupe **4. Diostea**
  b. Fruit dry **5. Phyla**
6a. Calyx obconical to campanulate, entire or indistinctly 5-lobed
    6. Holmskioldia
  b. Calyx not as above 7
7a. Branches with spines **7. Rhaphithamnus**
  b. Branches without spines 8
8a. Fruit a capsule **8. Caryopteris**
  b. Fruit a drupe 9
9a. Corolla more or less radially symmetric; stamens equal **9. Callicarpa**
  b. Corolla more or less bilaterally symmetric; stamens of 2 lengths
    10. Clerodendrum

1. Vitex Linnaeus. 250/3. Trees or shrubs. Leaves opposite, palmately compound, with 3--7 leaflets, the 3 upper leaflets larger than the 2 basal leaflets. Flowers in axillary or terminal cymes or panicles. Calyx 5-toothed. Corolla 5-lobed, obliquely 2-lipped. Stamens 4, usually protruding. Style thread-like, 2-branched. Drupe spherical, succulent with bony endocarp. *Tropics & subtropics, Mediterranean area.*

2. Verbena Linnaeus. 250/10 and some hybrids. Herbaceous biennials or perennials, shrubs or woody-based herbs, some semi-evergreen, mostly hairy, erect to prostrate. Stems often 4-angled. Leaves opposite, simple, alternate or in whorls of 3, toothed, lobed or dissected, rarely entire. Flowers stalkless or almost so, usually in terminal narrow spikes whose branches elongate during fruiting. Calyx tubular, 5-angled, sometimes oblique, unequally 5-toothed, persistent in fruit. Corolla funnel-shaped or a long straight or curved tube with an abruptly expanded, weakly 2-lipped, 5-lobed limb. Stamens 4, paired, or rarely 2, included in corolla-tube; anther-connectives of longer stamens often each with a glandular appendage. Style terminal, 2-lobed, one lobe attached to stigma. Ovary spherical, 4-celled, often with 4 shallow lobes. Fruit of four 1-seeded pyrenes, often ridged or net-veined. *Cosmopolitan but largely centred in tropical & temperate America.* Figure 107, p. 460.
  Modern American taxonomists tend to divide the genus into two: *Verbena* in the strict sense (anther-connectives of longer stamens unappendaged; spikes elongate at flowering), and *Glandularia* Gmelin (anther-connectives of longer stamens append-aged; spikes flat-topped at first, later elongating). Species of both are cultivated..

3. Aloysia Jussieu. 37/2. Erect, branched, often lemon-scented shrubs. Leaves simple, opposite or in whorls of 3, entire or toothed. Flowers in loose terminal spikes. Calyx tubular-campanulate, 4-lobed. Corolla weakly 2-lipped. Stamens slightly protruding, unequal, inserted near middle of corolla-tube. Fruit with 2 stones. *North & South America.*

4. Diostea Miers. 3/1. Deciduous shrubs or small trees. Leaves opposite or whorled, simple, stalkless. Flowers stalkless or almost so, in short spikes. Calyx tubular or campanulate, 4- or 5-toothed. Corolla tubular, 5-lobed, with a lip; tube cylindric or curved. Stamens 4, in 2 pairs of unequal length, usually included in the corolla-tube.

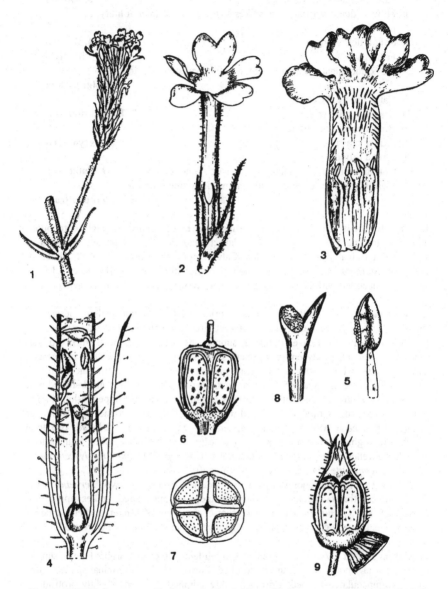

Figure 107. Verbenaceae. *Verbena rigida*. 1, Part of inflorescence (a lateral spike). 2, A single flower with bract. 3, Corolla opened out. 4, Longitudinal section of lower part of flower. 5, Stamen. 6, Longitudinal section of ovary. 7, Transverse section of ovary. 8, Stigma with unequal lobes. 9, Fruit.

Ovary 2-celled, ovules 2. Fruit a drupe, with two 1-seeded stones, enclosed by the persistent calyx. *South America.*

5. Phyla Loureiro. 15/1. Creeping or ascending herbaceous perennials. Non-flowering stems prostrate, rooting at the nodes; flowering stems 5--30 cm, ascending. Leaves simple, 1--2.5 cm, obovate or oblong to lanceolate, wedge-shaped at base, remotely toothed towards apex. Flowers in short, stout axillary spikes 5--7 mm across, enlarging considerably in fruit. Bracts obovate, usually in 2 or 3 series. Calyx deeply lobed. Corolla 2 mm across, weakly 2-lipped, slightly hairy, white to lilac. Stamens 4, included. Fruit dry, of 2 stones. *Tropics & warm temperate areas.*

6. Holmskioldia Retzius. 10/1. Climbing shrubs. Leaves opposite, simple, entire or toothed. Inflorescence a congested terminal panicle or an axillary raceme. Calyx usually red, obconical to campanulate, membranous; tube short, limb spreading, entire or indistinctly 5-lobed. Corolla-tube elongate, slightly curved; limb oblique, upper lip 2-lobed, lower lip 3-lobed, lobes reflexed. Stamens 4, in 1 long and 1 short pair, inserted below the middle of the corolla-tube, protruding; anthers with parallel cells. Style protruding, shortly bifid at apex. Ovary 4-celled, with 1 ovule per cell. Fruit an obovoid drupe with 1--4 stones included in the enlarged calyx. *Tropical Africa, Asia.*

7. Rhaphithamnus Miers. 2/1. Evergreen trees or shrubs with finely hairy shoots; branches armed with spines. Leaves opposite, simple, leathery, entire, almost stalkless. Inflorescence a raceme of up to 5 flowers. Calyx 5-lobed, campanulate, persistent. Corolla lilac, tubular, bilaterally symmetric, with 4 or 5 unequal lobes. Stamens 4, inserted at middle of corolla-tube; staminode present, lacking an anther. Ovary 4-celled, each cell with 1 ovule. Stigma bilobed. Fruit a fleshy, blue drupe with 2 stones. *Argentina, Chile.*

8. Caryopteris Bunge. 6--10/3 and a few hybrids. Small shrubs or herbs. Leaves opposite with short stalks, simple, almost entire or with forwardly pointing teeth, ovate to linear, dotted with minute yellow translucent glands. Flowers small, in panicles or axillary cymes. Calyx campanulate, deeply 5-lobed, lobes triangular to lanceolate. Corolla with short cylindric tube and 5 spreading, sometimes fringed lobes. Stamens 4, in 2 pairs of unequal length, protruding. Ovary imperfectly 4-celled with 4 ovules. Style thread-like, with 2 tapering pointed lobes at apex. Fruit a capsule, separating into 4 concave flaps with incurved margins retaining seeds. *Central & eastern Asia, from Mongolia to Himalaya (China, India) and Japan.*

9. Callicarpa Linnaeus. 140/5. Deciduous small trees and shrubs, lacking spines. Leaves opposite, simple, without stipules, entire, minutely toothed or minutely scalloped. Flowers 4- or 5-parted, usually both unisexual and bisexual in axillary cymes or in cymes above axils, the middle flowers in the cluster opening first. Calyx toothed almost to middle, or entire. Corolla more or less radially symmetric, funnel-shaped or a long slim tube with a flattened limb; lobes 5, narrow, spreading, equal. Stamens attached to base of the corolla-tube, equal, usually protruding. Stigma depressed-capitate or peltate. Fruit a spherical drupe, with fleshy exocarp and hard endocarp, the latter usually separating into 4 stones. *Warm areas.*

10. Clerodendrum Linnaeus. 450/few. Trees, shrubs or woody vines. Leaves opposite or whorled, simple, entire, indistinctly toothed or with shallow lobes, smelling unpleasant if crushed. Flowers fragrant, usually in complex panicles, condensed or elongate. Calyx tubular, campanulate, funnel-shaped or inflated, becoming fleshy, reflexed, the 5 sepals almost completely free or united for much of their length, often enlarging in fruit, green or coloured. Corolla strongly bilaterally symmetric with a slender tube, the 5 lobes obliquely set on the tube, unequal, spreading or reflexed. Stamens 4, protruding, slightly unequal. Fruit a fleshy, coloured drupe with 4 hard stones containing the seeds, surrounded by persistent calyx. *Mainly tropics & subtropics.*

173. CALLITRICHACEAE

Annual or perennial, terrestrial or aquatic herbs, usually monoecious. Stems slender, usually hairless. Leaves opposite or in flat rosettes at tips of floating stems, linear, elliptic, oblong to spathulate, entire or notched at apex, more or less stalkless; stipules absent. Flowers small, unisexual, solitary in the leaf-axils or sometimes 1 male and 1 female in the same leaf-axil. Bracts absent or 2. Perianth absent. Male flower: stamen 1, filament slender, anther 2-celled opening lengthwise, slits joining at apex. Female flower: ovary superior, of 2 carpels, 4-celled by the development of secondary septa; ovules 1 per cell; styles 2, thread-like. Fruit obovoid or more or less spherical, 4-lobed, each lobe winged or keeled, splitting into four 1-seeded nutlets. Seeds with fleshy endosperm.

A family of a single genus with an almost cosmopolitan distribution. The plants grow in still or slow to moderately fast-moving water or wet mud and are useful in ponds and aquariums as oxygenators.

1. Callitriche Linnaeus. 17/2. Description as for family. *Cosmopolitan.*

174. LABIATAE (*Lamiaceae*)

Shrubs or perennial, biennial or annual herbs, often glandular and aromatic; stems often 4-angled, at least before maturity. Leaves opposite, sometimes whorled, entire or toothed, usually simple, stalked or stalkless. Stipules absent. Flowers bisexual, male-sterile or rarely unisexual, solitary or arranged in cymes in the axils of upper leaves or bracts, often condensed into false whorls (verticillasters) which may form distinct clusters along the inflorescence-axis, or dense spike-like heads or racemes. Bracts similar to leaves or distinct; bracteoles present or absent, sometimes early deciduous. Calyx funnel-shaped or campanulate, 2-lipped, the upper lip with 2 and the lower with 3 teeth, but sometimes teeth fused, or with 5 more or less equal teeth, usually persistent and sometimes enlarging in fruit. Corolla strongly bilaterally symmetric, tubular at base with the 5-lobed limb forming 2 lips; upper lip hooded, straight or concave, sometimes indistinctly 2-lobed; lower lip 3-lobed; rarely corolla regular or upper lip reduced or rarely absent. Stamens rarely 2, more commonly 4 in 2 pairs of unequal length, all or one pair fertile, joined to corolla tube, arranged under or directed towards upper lip, spreading or protruding; rarely lying on or enclosed within lower lip; filaments sometimes joined. Ovary superior, of 2 united carpels, each divided longitudinally into 2 parts by a secondary septum producing an ovary with 4 apparently separate lobes. Style arising from base of lobes in all but a few genera; stigma bifid. Fruit of 4, occasionally 1, dry one-seeded nutlets, rarely fleshy drupes. See figures 108, p. 464 & 109, p. 472.

462

A family of about 200 genera and over 4000 species found throughout the world but especially in the Mediterranean area. The family is not easily grouped into natural units and has been variously classified by a number of authors. The following account is not therefore ordered according to any one system.

1a. Fertile stamens 2   2
 b. Fertile stamens 4 (absent in female flowers)   10
2a. Evergreen shrub; leaves linear with entire margins   **1. Rosmarinus**
 b. Plant not usually an evergreen shrub; if so, then leaves not as above   3
3a. Calyx distinctly 2-lipped or with the 3 lobes or teeth of the upper lip
     different from the 2 lobes or teeth of the lower   4
 b. Calyx with 5 more or less equal lobes, rarely the upper lobe larger and the
     remaining 4 equal   7
4a. Corolla divided into 5 lobes, the lower much larger than the others
                                                                  22. Collinsonia
 b. Corolla 2-lipped   5
5a. Upper corolla-lip divided into 4 lobes   **21. Perovskia**
 b. Upper corolla-lip not divided into 4 lobes   6
6a. Upper lip of corolla hooded and longer than lower   **8. Salvia**
 b. Upper lip of corolla erect, more or less flat and shorter than lower lip
                                                                     23. Hedeoma
7a. Corolla with 4 more or less equal lobes   **24. Lycopus**
 b. Corolla 2-lipped   8
8a. Flowers to 5 mm   **25. Cunila**
 b.Flowers 8 mm or more   9
9a. Flowers solitary or in few-flowered cymes   **26. Poliomintha**
 b. Flowers in dense whorls subtended by invlucral bracts   **9. Monarda**
10a. Upper corolla-lip absent or minute and represented by 2 small teeth; style
     arising from top of ovary   11
 b. Corolla 2-lipped with upper lip present or corolla of more or less equal
     lobes; style arising from base of ovary   12
11a. Lower lip of corolla 3-lobed; upper lip of 2 short teeth; tube with a ring
     of hairs inside   **4. Ajuga**
 b. Corolla without an upper lip, 5-lobed; tube without a ring of hairs inside
                                                                     3. Teucrium
12a. Calyx 2-lipped, enlarging in fruit, upper lip bearing a scale; upper
     corolla-lip hooded   **5. Scutellaria**
 b. Calyx not bearing a scale on the upper lip or if so, then upper corolla-
     lip not hooded; calyx enlarging or not in fruit   13
13a. Calyx large and campanulate, expanding in fruit, with spiny deflexed
     bracteoles   **10. Moluccella**
 b. Calyx not as above and without spiny deflexed bracteoles   14
14a. Calyx teeth 5 or 10, spiny, often hooked   **6. Marrubium**
 b. Calyx teeth or lobes 4 or 5, if 10, not spiny or hooked, or calyx 2-
     lipped   15
15a. Corolla not or weakly 2-lipped   16
 b. Corolla distinctly 2-lipped   18
16a. Corolla with 4 more or less equal lobes   **28. Mentha**
 b. Corolla with 5 more or less equal lobes or very weakly 2-lipped   17

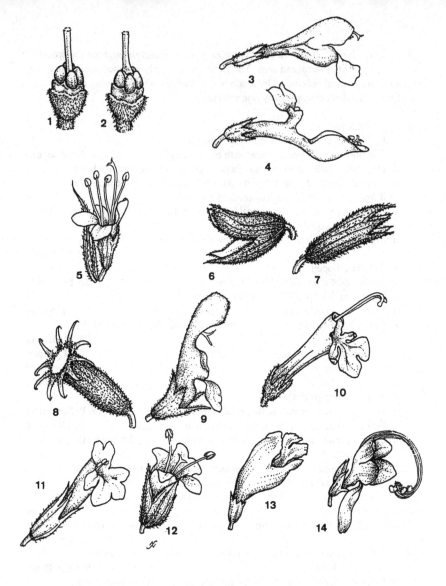

Figure 108. Labiatae: diagnostic features. 1, Style arising between nutlets. 2, Style arising from top of ovary. 3, Stamens under upper lip of corolla (e.g. *Nepeta*). 4, Stamens on lower lip of corolla (e.g. *Plectranthus*). 5, Stamens projecting (e.g. *Mentha*). 6, Calyx two-lipped (e.g. *Salvia*). 7, Calyx five-toothed (e.g. *Nepeta*). 8, Calyx ten-toothed (e.g. some *Marrubium*). 9, Upper lip of corolla hooded (e.g. *Lamium*). 10, Upper lip of corolla greatly reduced (e.g. *Teucrium*). 11, Lower lip of corolla expanded (e.g. *Clinopodium*). 12, Corolla-lobes more or less equal (e.g. *Lycopus*). 13, Corolla-tube swollen (e.g. *Physostegia*). 14, Atypical corolla-shape (*Trichostema*).

464

17a. Corolla more than 8 mm   **27. Monardella**
   b. Corolla 5 mm or less   **49. Perilla**
18a. Upper lip of corolla distinctly hooded or strongly concave, often longer
     than lower lip   19
   b. Upper lip of corolla flat or convex, sometimes slightly concave, usually
     equal to or shorter than lower lip   36
19a. Aromatic evergreen shrub or tree; upper corolla-lip very broad,
     concave or almost flat   **2. Prostanthera**
   b. Plants without above combination of characters   20
20a. Shrubs or herbs exceeding 1 m in height, with flowers very bright orange
     or red, over 2 cm   **13. Colquhounia**
   b. Herbaceous plants, if shrubs, flowers not bright red or orange usually less than
     2 cm   21
21a. Shrubs   22
   b. Herbaceous plants, sometimes woody at base   23
22a. Calyx with 5 spreading leaf-like lobes; nutlets fleshy   **48. Prasium**
   b. Calyx-lobes not leaf-like; nutlets dry   **14. Phlomis**
23a. Annuals; calyx with spiny teeth   **11. Galeopsis**
   b. Annuals or perennials; calyx usually without spiny teeth; if teeth spiny,
     plant not an annual   24
24a. Low-growing herbaceous plants rooting at nodes   25
   b. Plants erect, not rooting at nodes   26
25a. Flowers usually axillary; if inflorescence terminal, flowers not in dense
     spikes   **12. Lamium**
   b. Flowers in dense terminal cylindric spikes   **19. Prunella**
26a. Plant strongly aromatic, sometimes unpleasantly so   27
   b. Plant not or slightly aromatic   31
27a. Corolla tubular at base, inflated beyond calyx   **42. Dracocephalum**
   b. Corolla not as above   28
28a. Calyx tubular to conical, 10-veined, more or less expanded in upper half,
     often at right-angles, with 5--10 teeth   **17. Ballota**
   b. Calyx without above combination of characters   29
29a. Calyx net-veined   **20. Melittis**
   b. Calyx with 5, 10 or 15 veins   30
30a. Calyx with 15 veins   **40. Agastache**
   b. Calyx with 5 or 10 veins   **16. Stachys**
31a. Inflorescence a 1-sided loose, leafy raceme   **41. Meehania**
   b. Inflorescence not 1-sided and leafy   32
32a. Corolla tubular at base, inflated beyond calyx   33
   b. Corolla not inflated beyond calyx   34
33a. Calyx with 5 more or less equal teeth, faintly 10-veined   **15. Physostegia**
   b. Calyx 2-lipped, 15-veined   **42. Dracocephalum**
34a. Leaves often palmately lobed; calyx teeth spine-tipped; lower 2 deflexed
            18. Leonurus
   b. Leaves not palmately lobed; calyx teeth not as above   35
35a. Style-branches unequal   **14. Phlomis**
   b. Style-branches more or less equal   **12. Lamium**

36a. Leaves with 3, sometimes with 5 leaflets; herbaceous aromatic perennials
43. Cedronella
 b. Leaves simple; rarely pinnatisect; plants herbaceous or woody 37
37a. Calyx inflated in fruit; flowers 3 cm or more **44. Chelonopsis**
 b. Calyx not inflated in fruit, or if slightly inflated, then flowers less than
 3 cm 38
38a. Plant an annual; corolla not exceeding 5 mm, shaggily hairy outside
50. Elsholtzia
 b. Plant a herbaceous perennial or shrub; if annual, flowers exceeding 5
 mm or not shaggily hairy outside 39
39a. Plant a shrub (occasionally small) 40
 b Plant herbaceous, sometimes woody at base 52
40a. Clusters of flowers aggregated into short dense terminal or axillary spikes
 which are often arranged in panicles or false corymbs **38. Origanum**
 b. Flowers not aggregated into small spikes as above forming a compound
 inflorescence 41
41a. Creeping shrub often rooting at nodes 42
 b. Upright shrub or subshrub 44
42a. Leaves needle-like, more than 1.5 cm **29. Conradina**
 b. Leaves not needle-like, or if so, not more than 1 cm 43
43a. Bracteoles minute; bracts leaf-like, enlarged and coloured **35. Thymus**
 b. Bracteoles 3 mm or more; bracts not leaf-like, enlarged and coloured
33. Satureja
44a. Upper middle lobe of calyx more or less equalling other lobes, broadly
 triangular or modified into an appendage **52. Lavandula**
 b. Calyx not modified as above 45
45a. Bracteoles spiny **17. Ballota**
 b. Bracteoles not spiny or absent 46
46a. Calyx with long slender tube and abruptly expanded flattened limb
17. Ballota
 b. Calyx not as above 47
47a. Plant covered with dense white woolly hairs; corolla white or creamy
 yellow tipped with red or brown **7. Sideritis**
 b. Plant not covered with dense white woolly hairs; if so, corolla not white
 or creamy yellow tipped with red or brown 48
48a. Hairs stellate **51. Rostrinucula**
 b. Hairs simple 49
49a. Plants 80 cm or more **50. Elsholtzia**
 b. Plants less than 80 cm 50
50a. Bracts leaf-like, enlarged and coloured **35. Thymus**
 b. Bracts not leaf-like, enlarged and coloured 51
51a. Corolla mauve or purple **33. Satureja**
 b. Corolla white with violet markings **34. Micromeria**
52a. Calyx distinctly 2-lipped or lower teeth longer than upper 53
 b. Calyx with 5 more or less equal lobes 60
53a. Inflorescence terminal; leaves in a rosette; flowers dark blue
47. Horminum
 b. Inflorescence axillary; if terminal, leaves not in a rosette and flowers
 not dark blue 54

54a. Calyx swollen at 1 side at base, the tube constricted	**30. Acinos**
b. Calyx not swollen or tube constricted	55
55a. Calyx tube curved	**31. Clinopodium**
b. Calyx tube straight	56
56a. Plant with dense woolly hairs; flowers usually yellow	**7. Sideritis**
b. Plant without dense woolly hairs; flowers rarely yellow	57
57a. Flowers pale yellow becoming pinkish or white; plant strongly lemon-scented	**39. Melissa**
b. Flowers white, pink, blue or purple; plants not lemon-scented	58
58a. Plant an annual	**33. Satureja**
b. Plant a herbaceous perennial	59
59a. Bracts similar in shape and size to leaves; stems soft and fleshy	**45. Glechoma**
b. Bracts usually smaller and not similar in shape to leaves; stems not fleshy	**32. Calamintha**
60a. Flowers small in dense terminal clusters at the ends of branches, each subtended by 2 leaf-like bracts	**37. Pycnanthemum**
b. Flowers not arranged as above	61
61a. Leaves linear to linear-lanceolate, margins entire, stalks absent	**36. Hyssopus**
b. Leaves not linear or linear-lanceolate, or if so then margins toothed and stalks present	**46. Nepeta**

1. Rosmarinus Linnaeus. 3/2. Aromatic evergreen shrubs. Leaves opposite, stalkless, linear to semi-cylindric, margins entire and curved under. Inflorescence a short axillary raceme, 15--60 mm, few-flowered, bracteate, flowers bisexual, stalked. Calyx campanulate, 2-lipped, the upper lip entire, the lower 2-lobed; corolla protruding from calyx, 2-lipped; upper lip concave, bifid, lower prominent, concave. Stamens 2. Style single, usually distinctly protruding, not arising from between ovary-lobes. Nutlets 4, brown. *Mediterranean area (except the far eastern part).*

2. Prostanthera R. Brown. 100/2. Evergreen shrubs or undershrubs, rarely small trees, usually with strongly scented resinous glands. Leaves opposite, simple, rarely in 3-leaved whorls. Inflorescence a terminal dichasium, often appearing racemose or paniculate with leaves more or less absent from inflorescence, or leafy (flowers appearing axillary, borne singly in distal axils of branchlets). Calyx campanulate, 2-lipped, the lips entire or the lower slightly notched, usually persistent and greatly expanding in fruit, subtended by a pair of bracteoles. Corolla 2-lipped with a short tube and broad campanulate limb, upper lip erect or extended forward, 2-lobed and concave, lower usually spreading, 3-lobed, usually longer than the upper. Stamens 4, fertile, in pairs; anthers borne on filaments with connective often elongated into a tufted appendage joined to back of anther-cell, appendage sometimes short, or with one of the pair projecting as a spur. Style shortly divided, terminal, arising between the lobes of the ovary. Nutlets 4, enclosed by the persistent and usually much-enlarged calyx. *Australia.*

3. Teucrium Linnaeus. 100/17. Pleasantly or unpleasantly aromatic shrubs, subshrubs, perennial or annual herbs, generally hairy (hairs sometimes branched); stems 4-angled in section. Leaves opposite, entire, toothed, scalloped, lobed or pinnately divided, stalked or not. Flowers bisexual, generally in loose or dense, complex, often leafy inflorescences, sometimes condensed and head-like, sometimes long and spike-like, often the ultimate branches apparently raceme-like and bearing the flowers on 1 side only, occasionally in pairs or solitary in the leaf-axils. Calyx radially symmetric or 2-lipped, tube campanulate or tubular-campanulate, often swollen at the base, teeth 5, equal or unequal, some of them often reflexed or bent upwards. Corolla often hairy, tubular at the base, the limb without an upper lip, 5-lobed (2 outer lateral lobes, 2 median lateral lobes and a central, sometimes called the lip, which is generally convex and bent down from the rest). Stamens 4. Style arising from top of ovary. Nutlets with a raised surface network. *Mostly from the Mediterranean area & adjacent Europe & southwest Asia, but also occurring in America & in South Africa.* Figure 108, p. 464.

A species of **Trichostema** Linnaeus (see figure 108, p. 464) may occasionally be seen; it has the inflorescence covered with purple hairs.

4. Ajuga Linnaeus. 45/5. Herbaceous, evergreen annuals or perennials, usually with rhizomes. Leaves entire, toothed or lobed to deeply divided. Inflorescence of whorls of 2--6 bisexual flowers in a terminal spike. Bracts large, leaf-like. Bracteoles small or absent. Calyx campanulate, regularly 5-toothed, 10-veined. Corolla 2-lipped, the upper lip very short, the lower 3-lobed, tube longer than the calyx and with ring of hairs inside, blue, white or rose-pink. Stamens 4, in 2 pairs, usually protruding. Style arising from top of ovary. Nutlets 4, net-veined or wrinkled. *Temperate Eurasia; naturalised in South Africa & Australia.*

5. Scutellaria Linnaeus. 400/12. Perennial herbs, usually with rhizomes. Leaves entire, scalloped or toothed or sometimes pinnatifid, opposite. Flowers bisexual, in spikes which may be 1-sided with the flowers opposite or not, spiral, or 4-sided, with the flowers in pairs, serial pairs being at right angles to each other; bracts present, like the leaves, smaller than the leaves and with entire margins, or (reversed-) hood-like and enveloping the calyx. Calyx 2-lipped, the lips rounded and entire, more or less closing after flowering, the upper lip bearing a scale (scutellum) of varying form, the whole considerably enlarging in fruit. Corolla with a long tube which is more or less curved at the base so that most of the tube points upwards, 2-lipped, the lips scarcely spreading, blue, purplish, pink, whitish or yellow. Stamens 4, parallel, included in corolla and held under upper lip. Nutlets hairy or hairless, papillose or not, sometimes bearing minute hooked processes. *Almost cosmopolitan, though not found in the coldest areas.*

6. Marrubium Linnaeus. 40/6. Herbaceous perennials branching at base, usually silky or woolly. Leaves wrinkled, scalloped or toothed. Inflorescence many-flowered, in remote axillary dense-flowered, spherical whorls. Flowers bisexual. Bracteoles usually present. Calyx tubular, 10-veined, sometimes 5-veined, densely hairy within, teeth 5 or 10, spiny, equal or nearly so. Corolla white or purplish, tube usually shorter than calyx, with ring of hairs or hairless within, 2-lipped, upper lip erect, notched or entire, lower lip shorter, 3-lobed, spreading, middle lobe broad. Stamens 4, in 2 pairs, outer pair longer, not protruding but directed towards upper lip. Style divided at apex. Nutlets hairless. *Europe, north Africa & temperate Asia.* Figure 108, p. 464.

7. Sideritis Linnaeus. 100/7. Annuals, herbaceous perennials, subshrubs and shrubs; stems hairy, often densely so. Leaves entire or toothed, stalked. Inflorescence of 2--many-flowered whorls in dense or loose terminal spikes. Flowers bisexual. Bracts often leaf-like; bracteoles absent. Calyx 2-lipped, tubular to campanulate, 5-toothed, 10-veined. Corolla yellow, white or red, tube shorter than calyx; 2-lipped, upper lip erect or spreading, flattened, entire or 2-lobed, lower lip 3-lobed, lateral lobes reduced, spreading. Stamens 4, upper pair shorter, not protruding but directed towards upper lip. Style unequally bifid, included. Nutlets ovoid, smooth, hairless. *Mediterranean area, Atlantic Islands.*

8. Salvia Linnaeus. 1000/100. Shrubs, herbaceous perennials or annuals. Hairs simple, branched, stellate or glandular. Leaves stalkless to long-stalked, opposite, simple to pinnate. Flowering stems of few- to many-flowered interrupted whorls. Bracts deciduous or persistent. Calyx 2-lipped, funnel-shaped, campanulate or tubular, sometimes expanding in fruit, often covered in glandular and simple hairs; upper lip entire or with 3 teeth; lower lip with 2 teeth. Corolla 2-lipped, upper lip straight or sickle-shaped, longer than lower; tube with a ring of hairs within or hairless, sometimes folded at base, or with a pair of small papillae within; lower lip 3 lobed, central lobe frequently larger. Stamens 2, frequently enclosed in upper lip or protruding; anther sacs separated by a more or less well-developed connective which is often hinged on the filament; lower pollen-sacs sometimes sterile and reduced to a tooth-like or club-like appendage. Style smooth or hairy, frequently protruding, splitting at apex to form 2 branches or rarely entire. Nutlets usually 4, smooth. *Throughout the northern hemisphere, Africa & South America.* Figure 108, p. 464.

9. Monarda Linnaeus. 16/7. Erect, usually tall, herbaceous annuals or perennials with aromatic, mint-like foliage. Leaves toothed or entire, stalked or stalkless. Flowers bisexual, mostly more or less erect, in dense terminal or sometimes axillary whorls, mostly subtended by bracts which are frequently coloured. Calyx tubular, slender, with 5 more or less equal teeth, 15-veined, often somewhat hairy at the mouth. Corolla strongly 2-lipped, tube exceeding calyx, upper lip erect, arched or hooded, 2-lobed, lower lip 3-lobed, spreading, middle lobe broad. Fertile stamens 2, under upper corolla lip, protruding, with a second rudimentary pair sometimes present. Style unequally divided, protruding. Nutlets oblong, smooth. *North America.*

10. Molucella Linnaeus. 4/1. Annual hairless herbs. Leaves opposite, stalked, coarsely scalloped, often deeply cut, floral bracts similar. Inflorescence of 6--8-flowered whorls in a terminal spike, bracteoles spiny, deflexed; flowers bisexual. Calyx narrowly campanulate and curved at base, expanded into a wide, membranous limb, obscurely 2-lipped, with or without mucronate points to the main veins. Corolla white to purplish, 2-lipped, the upper hooded, the lower 3-lobed. Stamens 4, included within upper lip. Nutlets sharply 3-angled. *South Europe & Mediterranean area.*

One species of **Plectranthus** L'Héritier may occasionally be found; it will key out to *Molucella*, but the stamens lie on or within the lower corolla-lip (figure 108, p. 464).

11. Galeopsis Linnaeus. 10/2 and a hybrid. Annual herbs. Leaves opposite, simple. Flowers bisexual, in dense axillary and terminal whorls; bracts similar to stem-leaves; bracteoles subulate. Calyx tubular or campanulate, with 5 somewhat unequal, spine-pointed teeth. Corolla 2-lipped, upper lip hooded and laterally compressed, lower lip 3-lobed with two conical projections at the base; tube straight, exceeding calyx. Stamens 4, under upper corolla-lip. Nutlets 3-angled, rounded at apex. *Europe, temperate Asia.*

469

12. Lamium Linnaeus. 16/6. Herbaceous perennials and annuals, often stoloniferous or rhizomatous, sparsely to densely hairy. Leaves opposite, scalloped or toothed. Flowers bisexual, in 2--12-flowered whorls, the upper often crowded. Bracts leaf-like. Calyx tubular to campanulate with 5 more or less equal lobes. Corolla 2-lipped; upper lip arched or hooded and usually hairy, the lower 2- or 3-lobed, often streaked with contrasting colour, usually with reduced lateral lobes bearing teeth. A ring or partial ring of hairs (annulus) may or may not be present near base of corolla-tube. Stamens 4, lower pair longer than upper, both pairs included under upper corolla-lip. Anther cells widely divergent. Style bifid. Nutlets brown, acutely 3-angled. *Europe, north Africa & Asia.* Figures 108, p. 464 & 109, p. 472.

13. Colquhounia Wallich. 3/1. Semi-evergreen or deciduous shrubs, erect or somewhat straggly. Shoots 4-angled at first. Leaves opposite, stalked, with simple or stellate hairs, margins toothed. Flowers bisexual, in axillary clusters and whorls at the tips of the current season's growth. Calyx funnel-shaped or tubular, with 5 lobes and 10 veins. Corolla 2-lipped, yellow, red, orange or purple, sometimes bicoloured; tube longer than calyx, curved, wider at throat; upper lip simple, erect, lower 3-lobed, the middle lobe deflexed. Stamens 4, in 2 pairs, under upper lip, upper pair longer; protruding, filaments somewhat hairy. Ovary hairless; style thread-like, stigma 2-lobed. Nutlets oblong or oblanceolate, flattened on one side, tip slightly winged. *Areas north & south of the eastern Himalaya.*

14. Phlomis Linnaeus. 250/20. Perennial herbs and evergreen shrubs with extensive root systems, sometimes thickened or with tubers. Stems square in section, appearing round when covered in a dense layer of hairs. Leaves entire, opposite. Flowers bisexual, arranged in whorls round stems. Calyx tubular, bell- or barrel-shaped with 5 or 10 visible veins, 5 teeth held at various angles, equal or the outer 2 longer. Corolla yellow, pink, purple to white, often variable within a species; upper lip hood-shaped and usually laterally compressed; lower lip 3-lobed, the central lobe larger than laterals. Stamens 4, ascending under upper lip; anthers with forked ends, lower fork longer than upper. Style branches unequal. Nutlets 4, 3-sided, topped with hair or hairless. *Mediterranean area, through Iran and Afghanistan to the Himalayas, central Asia to Siberia, China & Korea.*

15. Physostegia Bentham. 12/2. Perennial herbs, generally hairless, with stiff, upright stems. Leaves opposite, more or less stalkless, linear to lanceolate or oblong, mostly toothed. Inflorescence a terminal spike of paired flowers, often crowded, each flower bisexual, subtended by usually a single bract. Calyx with 5 more or less equal teeth, tubular to campanulate, faintly 10-veined, enlarging in fruit. Corolla much longer than calyx, tubular at base, inflated beyond calyx, pink, purplish or white, 2-lipped, the upper lip hooded, straight, the lower 3-lobed, somewhat deflexed, the middle lobe the largest, rounded and notched. Stamens 4, enclosed in upper lip of corolla. Style slender, as long as stamens, stigma 2-lobed. Nutlets 3-angled. *Temperate North America.* Figure 108, p. 464.

16. Stachys Linnaeus. 275/15. Perennial herbs sometimes woody at the base (ours) or rarely annuals. Hairs generally present and forming a conspicuous covering. Leaves generally scalloped or scalloped-toothed, sometimes very faintly so. Flowers bisexual, in whorls of 3--20, bracts and bracteoles present, the whorls often congested into

470

raceme- or head-like inflorescences. Calyx tubular to campanulate, 5--10-veined, radially symmetric or slightly 2-lipped, with 5 teeth. Corolla with the tube included within or projecting from the calyx, 2-lipped, the upper lip concave, entire, notched or rarely bifid, the lower 3-lobed with large central lobe. Stamens 4, projecting from corolla-tube. Nutlets obovate to oblong or pyramidal. *Mainly in the Mediterranean area, but also in America and southern Africa.*

The genus is now generally interpreted to include the former Linnaean genus *Betonica.*

17. Ballota Linnaeus. 35/5. Perennial herbs or subshrubs with simple, clustered, branched or stellate hairs, often also with more or less stalkless glands. Leaves opposite, stalked, the base of the blade usually cordate, the margins toothed. Inflorescence of several to many whorls in a terminal, usually dense spike, each whorl subtended by leaf-like bracts and narrow bracteoles; flowers bisexual. Calyx tubular to conical, hairy, 10-veined, with a more or less expanded upper half, often at right-angles, 5--10-toothed, rarely more. Corolla 2-lipped, hairy outside; pink, purple or white, upper lip often crested, straight and 2-lobed or slightly hooded and notched, lower lip 3-lobed, sometimes blotched. Stamens 4, in 2 unequal pairs, protruding under the upper lip. Style slender, stigma 2-lobed. Nutlets ovoid, hairless. *Europe, Mediterranean area & western Asia.*

18. Leonurus Linnaeus. 14/5. Annual, biennial or perennial aromatic herbs. Stems more or less simple or loosely branched, square, often grooved, erect. Leaves opposite, palmately lobed or simple, the margins coarsely toothed or incised. Inflorescence of many-flowered whorls in usually crowded terminal spikes; flowers bisexual, bracts large, more or less leaf-like, bracteoles needle-like. Calyx conical to campanulate, not expanding in fruit, 5-veined, more or less 2-lipped with 3 straight upper teeth and 2 deflexed lower teeth, all spine-tipped. Corolla 2-lipped, tube usually as long as calyx, upper lip entire, hooded (in ours), lower lip 3-lobed, often spotted, the middle lobe the largest, white to reddish or purplish. Stamens 4, in two unequal pairs, under upper corolla-lip. Style slender, stigma 2-lobed. Nutlets 3-angled, flattened, sometimes hairy at apex. *Temperate Eurasia.*

19. Prunella Linnaeus. 7/3 and some hybrids. Semi-evergreen perennial herbs with spreading or prostrate vegetative shoots, usually rooting at nodes. Leaves opposite, simple, stalked. Flowers bisexual on erect stems in short, compact terminal spikes. Bracts distinct from leaves, usually ovate to rounded. Calyx tubular to campanulate, 10-veined, 2-lipped, upper lip broad, shortly 3-toothed, lower lip narrow, with 2 longer, subulate teeth. Corolla 2-lipped, tube straight, obconical, with ring of hairs inside; upper lip hooded; lower lip shorter, finely toothed. Stamens 4, the upper pair shorter than lower, included under upper corolla-lip; filaments with tooth or claw below anther. Nutlets oblong or ovoid, hairless, keeled. *Europe, Africa, Asia, & North America.*

20. Melittis Linnaeus. 1/1. Erect perennial herb 20--70 cm, very aromatic, even after drying, densely glandular or not. Stems square, with long hairs. Leaves opposite, stalks to 1 cm, hairy; blades 2--15 × 1--8 cm, oblong to ovate, sparsely hairy, more so on veins beneath, base broadly wedge-shaped to heart-shaped, margins coarsely toothed. Inflorescence leafy, with whorls of 2--6 bisexual flowers in the axils on stalks 6--8 mm.

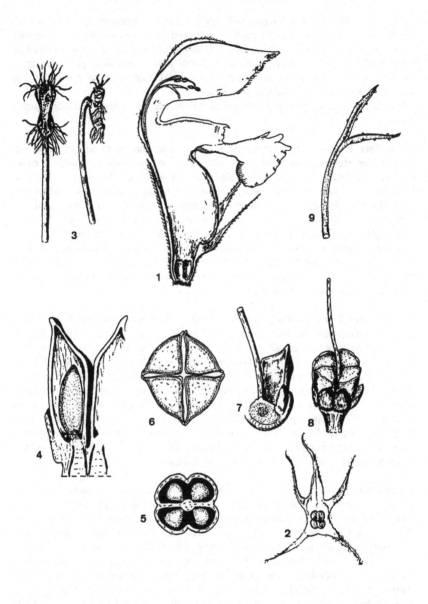

Figure 109. Labiatae. *Lamium album.* 1, Longitudinal section of flower. 2, Calyx from above, showing the four-lobed ovary in its base. 3, Stamen, front and side views. 4, Longitudinal section of one of the lobes of the ovary. 5, Transverse section of young ovary. 6, Older ovary, showing the development of 4 nutlets (lobes). 7, Part of the ovary showing one nutlet, the gynobasic style and the nectary. 8, The whole ovary after fertilisation. 9, Stigma and two-lobed style.

Calyx 1.2--2.5 cm, campanulate, net-veined, 2-lipped, the upper lip toothed or not, the lower 2-lobed, shorter, the lobes broadly triangular. Corolla 2.5--4 cm, tubular, 2-lipped, white, usually with a pink or purplish blotch on the lower lip, more rarely all pink or purple; the upper lip slightly hooded, shorter than lower lip, lower lip 3-lobed, the middle lobe the largest. Stamens 4, in 2 slightly unequal pairs, protruding into the hooded upper lip. Style slender, as long as stamens, stigma 2-lobed. Nutlets ovoid, 3-angled. *Europe.*

21. Perovskia Karelin. 7/3 and some hybrids. Deciduous, aromatic perennials, woody at base. Stems erect, hairy. Leaves opposite, scalloped to bipinnatisect, stalks to 1 cm. Flowers bisexual, in interrupted whorls of 4--6 in tall, branched, terminal panicles; bracts minute. Calyx tubular to campanulate, covered in hairs and white or yellow glands, upper lip entire to 3-toothed, lower lip 2-toothed. Corolla 2-lipped, twice as long as calyx, upper lip 4-lobed, lower lip entire, blue, violet or pink. Stamens 2 fertile, staminodes 2. Nutlets obovoid, brown, smooth. Flowers of two different types have been observed in some species: stamens may be protruding and style included or stamens may be included with style protruding. *Arid regions of Asia.*

22. Collinsonia Linnaeus. 5/1. Strongly aromatic perennial herbs with erect or spreading stems. Leaves opposite, simple, toothed and stalked. Flowers bisexual, in whorls arranged in a terminal panicle, bracts very small. Calyx campanulate, 15-veined, weakly 2-lipped (in ours) in flower, clearly 2-lipped in fruit; upper lip of 3 short spreading lobes, lower of 2 longer narrower lobes. Corolla 5-lobed, 4 lobes nearly equal, lower lobe much larger and fringed. Stamens 2 (in ours), protruding; filaments joined by a woolly ring at base. Nutlets spherical, more or less smooth. *Eastern North America.*

23. Hedeoma 38/1. Persoon. Annual or perennial herbs, aromatic. Leaves opposite, entire or remotely toothed. Flowers bisexual, in axillary whorls. Calyx tubular, swollen at base, 13-veined, 5-toothed, the lower 2 teeth longer than upper 3, ring of hairs (annulus) present within. Corolla 2-lipped, upper lip erect, shorter than lower lip, entire or 2-lobed, lower lip 3-lobed and spreading. Stamens 2, staminodes 2. Nutlets ovoid, smooth. *North & South America.*

24. Lycopus Linnaeus. 14/2. Perennial rhizomatous herbs. Leaves opposite, coarsely toothed. Flowers bisexual, in widely separated, dense, spherical whorls. Bracts leaf-like. Calyx with 5 more or less equal teeth (in ours). Corolla with 4 more or less equal lobes (in ours), tube shorter than calyx. Stamens 2, protruding, staminodes 2 or absent. Nutlets with thickened margins. *Northern temperate zones & Australia.* Figure 108, p. 464.

25. Cunila Linnaeus. 15/1. Perennial herbs or low-growing shrubs with aromatic stems, 4-angled in section. Leaves opposite, simple, entire or toothed, aromatic. Flowers bisexual, small, in clusters terminating the stem and also arising from upper axils of leaf-like bracts; bracteoles minute, linear. Calyx narrow-campanulate, with 5 equal lobes, hairy within throat. Corolla almost radially symmetric to weakly 2-lipped, upper lip erect, flat and notched, lower lip 3-lobed, purple to white. Fertile stamens 2, protruding. Nutlets ellipsoid, smooth. *Eastern USA to Uruguay.*

26. Poliomintha A. Gray. 7/2. Shrubs or shrubby perennials, much-branched. Leaves opposite, to 2 cm × 6 mm. Flowers bisexual, solitary or in few-flowered cymes in upper leaf-axils. Calyx cylindric with 5 more or less equal teeth. Corolla 2-lipped and hairy; upper lip 2-lobed, often indistinctly; lower lip 3-lobed. Stamens 2, staminodes 2. Style protruding and bifid. *South USA, north Mexico.*

27. Monardella Bentham. 19/3. Herbaceous, usually hairy, aromatic annuals or perennials. Leaves entire or toothed. Inflorescence a dense, rounded, terminal head subtended by broad bracts, which are frequently coloured. Flowers bisexual, many. Calyx tubular, slender, nearly equally 5-toothed, 10--15-veined, tube hairless within, teeth usually erect, triangular. Corolla red, pink, pale purple or white, weakly 2-lipped, lips almost equal, upper lip erect, 2-lobed, lower 3-lobed, horizontal or downwardly curved. Stamens 4, slightly protruding, all fertile. Style unequally divided at apex. Nutlets oblong-ovoid, smooth. *Western North America, mainly California.*

28. Mentha Linnaeus. 25/15 and several hybrids. Perennial or rarely annual herbs, rhizomatous or stoloniferous with characteristic scents, at least when fresh. Leaves opposite, entire to toothed. Flowers bisexual, small, purple, pink or white in axillary whorls. Bracteoles small or lacking. Calyx tubular or campanulate, with 4 or 5 equal or rather unequal teeth. Corolla almost radially symmetric, with 4 more or less equal lobes. Stamens 4, usually protruding. Nutlets ovoid, rounded at apex, smooth. *North & south areas of the Old World.* Figure 108, p. 464.

29. Conradina A. Gray. 7/1. Low shrubs. Leaves simple, opposite, in stiff clusters, needle-like, margins entire, curved downwards and inwards. Flowers bisexual, in axillary, 2--7-flowered whorls. Calyx tubular, 13-veined, 2-lipped, upper lip broad with 3 short lobes, lower of 2 longer narrow lobes. Corolla 2-lipped, upper lip erect, slightly notched, not hooded, lower lip spreading, 3-lobed, middle lobe slightly notched, purple, tube curved upwards and widening at throat. Stamens 4, included under upper corolla-lip. Nutlets 4, almost spherical, smooth. *Southeastern USA.*

30. Acinos Miller. 10/3. Annual or perennial herbs or low-growing shrubs. Leaves opposite, small, simple. Flowers bisexual in whorls often arranged in terminal inflorescences with bracts. Calyx tubular, 13-veined, 2-lipped, lower teeth longer than upper; tube usually curved, constricted near centre and swollen towards base. Corolla tubular, 2-lipped, upper lip not hooded. Stamens 4, curved under upper corolla-lip, not exceeding corolla. Nutlets 4, ovoid, smooth. *Europe, western Asia.*

31. Clinopodium Linnaeus. 10/5. Aromatic perennial herbs with straight hairs. Leaves short-stalked, margins toothed. Flowers bisexual in inflorescences of several, dense, many-flowered whorls in the axils of bracts. Calyx tubular, 13-veined, curved, hairy, 2-lipped, the lower teeth long-ciliate. Corolla 2-lipped, upper lip notched, lower lip 3-lobed, pink or purple or sometimes white. Stamens 4, lower pair shorter, not protruding, held under upper corolla lip. Style unequally 2-lobed, the lower lobe much larger than the upper. Nutlets more or less spherical, hairless. *North temperate areas.* Figure 108, p. 464.

Some authorities treat species of *Calamintha* under *Clinopodium* while others unite both genera with *Satureja*.

474

32. Calamintha Miller. 70/5. Aromatic perennial herbs with simple hairs, these sometimes glandular. Leaves stalked, elliptic to ovate or circular, entire to coarsely toothed. Inflorescence of several whorls in the axils of bracts smaller than leaves comprising paired, usually stalked cymes, few- to many-flowered; bracteoles narrow. Calyx tubular, 2-lipped, mostly hairy and glandular, with 11--13 veins, throat hairy, lower teeth longer than upper, ciliate. Corolla 2-lipped, purplish to pink, reddish or white, upper lip notched, not hooded, lower 3-lobed. Stamens 4, upper pair shorter, held under upper corolla-lip. Style with 2 unequal branches. Nutlets ovoid to sub-spherical, hairless. *Western Europe to central Asia.*

33. Satureja Linnaeus. 30/5. Annuals, perennial herbs or dwarf shrubs with aromatic glands. Leaves more or less stalkless, entire. Inflorescence of several whorls of more or less stalkless bisexual flowers or paired cymes in the axils of bracts, the flowers subtended by bracteoles. Calyx tubular 10-, rarely 13-veined, radially symmetric with more or less equal teeth or 2-lipped with the 3 upper teeth shorter, throat often hairy. Corolla 2-lipped, white or purplish, upper lip not hooded, lower lip 3-lobed. Stamens 4, curved, protruding or not. Style with 2 more or less equal lobes. Nutlets rounded, minutely hairy. *North temperate areas.*

See note under genus **31**.

34. Micromeria Bentham. 90/1. Aromatic perennial, rarely annual, herbs or dwarf shrubs. Leaves stalked, sometimes with margins curled under or thickened. Inflorescence of several whorls of paired, stalked cymes in the axils of bracts. Flowers bisexual. Calyx tubular or obconical, 13-, rarely to 15-veined, radially symmetric to weakly 2-lipped, with triangular to subulate teeth. Corolla 2-lipped, tube straight, lower lip 3-lobed, purple, mauve or white. Stamens 4, usually not protruding, curved upwards. Style 2-lobed. Nutlets hairless or minutely hairy. *Northern hemisphere.*

Some authorities include the genus in *Satureja*.

35. Thymus Linnaeus. 350/22 and several hybrids. Aromatic shrubs or subshrubs, usually hairy. Leaves small, entire, usually gland-dotted, stalked or not. Inflorescence spike-like or head-like, sometimes interrupted, the lower whorls separated from the upper. Flowers often functionally unisexual, though with both stamens and ovary; bracts leaf-like or enlarged and coloured; bracteoles minute. Calyx with a cylindric or campanulate tube and 2 lips, tube usually hairy outside and with a ring of hairs inside, 10--13-veined; lips usually unequal, the upper 3-toothed, the lower 2-toothed. Corolla-tube included in the calyx or projecting from it, 2-lipped, the upper lip notched or 2-lobed, not hooded, the lower 3-lobed. Stamens 4, included in the corolla or projecting. Style with 2 short, equal or unequal stigmatic lobes. Nutlets ovoid or oblong, smooth. *Mostly from Europe and the Mediterranean area, but extending to western Asia & the Himalaya.*

36. Hyssopus Linnaeus. 1--5/1. Aromatic, woody-based perennial herbs or subshrubs. Leaves opposite, stalkless, linear to oblong, entire. Flowers bisexual in spike-like terminal inflorescence, often 1-sided, of short-stalked whorls in the axils of reduced bracts, crowded at shoot-tips. Calyx coloured, tubular, with 5 more or less equal teeth and 15 veins, hairy and glandular. Corolla with a curved tube, 2-lipped, the upper lip notched, erect, the lower 3-lobed deflexed, the central lobe much larger than the laterals, deeply notched. Stamens 4, more or less equal, protruding, spreading. Style protruding, stigma 2-lobed. Nutlets oblong to ovate, slightly angled, usually hairless. *Mediterranean area, central Asia.*

37. Pycnanthemum Michaux. 17/2. Hairy or hairless, aromatic, perennial herbs. Leaves opposite, undivided, entire or toothed, usually hairy, stalked or stalkless. Inflorescence simple or branched with almost stalkless, very small, bisexual flowers in small, stalked, terminal heads subtended by 2 leaf-like bracts. Calyx 10--13-veined, with 5 equal, acute to acuminate lobes or more or less 2-lipped. Corolla 2-lipped, white or pink; upper lip notched or entire, almost flat, longer than lower; lower lip with 3 almost equal lobes. Stamens 4, straight, usually protruding. Nutlets smooth or tipped with short hairs. *North America.*

38. Origanum Linnaeus. 38/12 and several hybrids. Herbaceous perennials or subshrubs. Stems several, ascending to erect, usually branched. Leaves stalkess or stalked, elliptic to ovate or almost circular, usually entire, occasionally finely toothed. Whorls of flowers aggregated into dense spikes which are often arranged in panicles or false corymbs. Bracts distinct from the leaves in shape and size, usually conspicuous and overlapping, either membranous and whitish or purple, or leaf-like in texture and colour; flowers usually bisexual, occasionally bisexual and female. Calyx variable, radially symmetric and 5-toothed or bilaterally symmetric and 1- or 2-lipped, occasionally without a distinct tube. Corolla white, pink or purple, more or less equally 2-lipped, upper lip not hooded, tube sometimes swollen (saccate). Stamens 4, the lower pair longer, included within the corolla or projecting from it, straight or divergent. Nutlets small, ovoid, brown. *Mediterrranean area.*

39. Melissa Linnaeus. 3/1. Herbaceous perennials. Leaves ovate, toothed, with stalks. Inflorescence of few- to many-flowered whorls; bracteoles present. Calyx campan-ulate, 13-veined, 2-lipped, upper lip flattened, indistinctly 3-lobed, lower 2-toothed. Corolla-tube longer than calyx, curved upwards, hairless within, 2-lipped, upper lip 2-lobed, shorter, erect, lower spreading, 3-lobed, usually white. Stamens 4, in 2 pairs held under upper corolla-lip. Style nearly equally divided. Nutlets obovoid, smooth. *Mediterranean Europe and Asia; naturalised in northern Europe and North America.*

40. Agastache Gronovius. 22/6. Perennial aromatic herbs with upright simple or branched stems arising from a rhizome or woody base. Leaves opposite, stalked, the blades ovate to lanceolate or linear, coarsely toothed to entire, hairy or not. Inflorescence of whorls of bisexual flowers subtended by bracts in loose or dense terminal spikes, either solitary or in paniculate clusters. Calyx 15-veined, tubular to obconical with 5 teeth, either more or less equal or the upper 3 more or less united, tube and teeth often coloured, sometimes differently. Corolla narrowly funnel-shaped or tubular, pink, purplish, yellowish or white, 2-lipped, the upper lip weakly concave or erect, more or less 2-lobed, the lower 3-lobed and deflexed. Stamens 4, in 2 unequal pairs, either all protruding or 1 pair enclosed in the upper lip of corolla. Nutlets more or less spherical, usually minutely hairy at one end. *Eastern Asia, North America.*

41. Meehania Britton. 7/3. Stoloniferous perennial herbs or unbranched annuals. Flowering stems upright, sometimes woody at the base, nodes hairy. Leaves opposite, heart-shaped to lanceolate, stalked, thin-textured, margins toothed. Inflorescence of few-flowered whorls in terminal and axillary 1-sided leafy racemes or of paired axillary

476

flowers; bracts lanceolate to elliptic; flowers bisexual. Calyx 2-lipped, funnel-shaped or tubular, 15-veined, enlarged in fruit; upper lip 3-toothed, lower 2-toothed, teeth rounded-triangular to lanceolate. Corolla 2-lipped, narrowly funnel-shaped, blue or purplish; upper lip hooded, notched or 2-lobed, lower larger, 3-lobed, the central lobe the largest. Stamens 4, in two pairs, the upper slightly longer, not protruding, held under upper corolla-lip. Style slender, as long or longer than stamens, stigma 2-lobed. Nutlets oblong to ovoid, hairless or shortly hairy. *North America, Asia.*

42. Dracocephalum Linnaeus. 40/11. Aromatic annual or perennial herbs, sometimes woody at base, occasionally rhizomatous; stems usually erect. Basal leaves entire, toothed or pinnatisect. Stem-leaves similar or reduced to bracts. Flowers bisexual, whorled in axillary or terminal spikes. Calyx tubular or tubular-campanulate, 2-lipped, 15-veined, upper calyx-lobe larger than the other 4. Corolla 2-lipped, the upper lip notched, hooded and arched, the lower 3-cleft, with the middle lobe larger than the others, and notched or cleft, violet, purple, blue, rarely white or pinkish. Stamens 4, arranged under upper lip, equalling or rarely exceeding the corolla and curved under the upper lip. Stigmatic lobes equal. Nutlets 4, smooth. *Europe, Asia, north Africa & North America.*

43. Cedronella Moench. 1/1. Aromatic perennial herb to 1.5 m; stems hairless except for a few hairs at nodes, woody at base, 4-angled. Leaves opposite, with 3 or occasionally 5 leaflets, stalked, central leaflets 4--7 × 1.4--2.4 cm, lanceolate, lateral leaflets 2.5--4.5 cm × 8--14 mm, hairless above, downy beneath, toothed. Flowers bisexual in terminal inflorescence 3--8 cm, dense, of many-flowered whorls, the lowest often separated from upper; bracts undivided. Calyx to 1.5 cm, tubular to campanulate, with 5 equal teeth, 13--15-veined. Corolla to 2 cm, funnel-shaped, somewhat downy outside, 2-lipped, upper lip 2-lobed, weakly concave, equal to lower; lower lip 3-lobed, middle lobe longest, pink, lilac or white. Stamens 4, curved, equal to or just protruding beyond corolla, held under upper corolla-lip. Nutlets ovoid, smooth. *Canary Islands.*

44. Chelonopsis Miquel. 16/1. Perennial herb, often woody at base. Leaves opposite, undivided, stalked, toothed. Inflorescence of axillary whorls of 2--10 bisexual flowers. Calyx campanulate, 2-lipped, membranous, 10-veined, inflated in fruit. Corolla 2-lipped, upper lip notched, not hooded, lower lip 3-lobed, the centre larger than the outer 2, corolla-tube somewhat swollen beyond calyx. Stamens 4, included under upper lip of corolla, of 2 lengths, upper, outer pair longer than inner; anthers bearded. Styles with 2 equal lobes. Nutlets winged on upper margin. *Asia (Himalaya to Japan).*

45. Glechoma Linnaeus. 12/2. Perennial creeping herbs. Stems ascending to prostrate. Leaves opposite, simple, toothed, stalked. Flowers bisexual, directed to one side, in 2--5-flowered whorls with leaf-like bracts. Calyx tubular to campanulate, 15-veined, 2-lipped; upper lip 3-toothed, lower 2-toothed. Corolla 2-lipped, tubular, tube straight, tapered to base; upper lip flat, erect, lower 3-lobed, the central lobe largest, notched, usually blue to purple. Stamens 4, included under upper lip, upper pair longer than lower. Nutlets dark brown, smooth. *Europe, Asia.*

46. Nepeta Linnaeus. 250/20 and some hybrids. Herbaceous, usually aromatic perennials, rarely annuals. Leaves entire, toothed, or scalloped, stalked or stalkless. Bracts and bracteoles present. Inflorescence of whorls, spike-like or paniculate, terminal or axillary. Flowers bisexual or sometimes unisexual. Calyx 15-veined, with 5 nearly equal teeth. Corolla tubular or campanulate, dilated, hairless within, 2-lipped, upper lip erect, 2-lobed, lower 3-lobed with the middle lobe larger and concave, tube sometimes protruding from the calyx, often blue or white. Stamens 4, upper pair longer than lower held under upper lip of corolla. Nutlets smooth or tuberculate, ovoid or ellipsoid. *Eurasia, north and tropical Africa*. Figure 108, p. 464.

47. Horminum Linnaeus. 1/1. Perennial spreading by more or less horizontal rhizomes. Stems 15--20 cm sometimes to 45 cm, erect, unbranched, downy. Leaves mostly basal, blades 3--7 × 2--5 cm, ovate to elliptic or circular, margins scalloped, ciliate; stalks as long as, or longer than blades, ciliate. Inflorescence a terminal, more or less 1-sided spike of up to 15, 4--6-flowered whorls, each with a pair of bracts 4--11 mm, ovate, entire, ciliate, acute or acuminate; flowers bisexual. Calyx 2-lipped, hairy, with 13 veins; tube 4--5 mm, campanulate; lobes 4--5 mm, acuminate, the upper 3 curved upwards, the lower 2 more or less straight. Corolla 2-lipped, dark purplish blue striped white in throat, tube to 1.2 cm slightly curved upwards, widening gradually to mouth; upper lip *c*. 2 mm, straight, not hooded, notched, lower lip 3-lobed, middle lobe the largest to 4 mm. Stamens 4, more or less equal, under upper corolla-lip, protruding, spreading. Style protruding, stigma 2-lobed. *Europe*.

48. Prasium Linnaeus. 1/1. Shrub to 1 m, often hairless. Leaves opposite, blades to 5 × 2 cm, oval to ovate, apex acute, base cordate to truncate, toothed or scalloped, usually shiny; stalks 1--2 cm. Flowers bisexual, born terminally in 1- or 2-flowered whorls, in axils of leaf-like bracts. Calyx campanulate with 5 spreading, leaf-like lobes. Corolla 1.5--2 cm in length, 2-lipped; white or lilac-tinged, upper lip entire, hooded, shorter than lower; lower lip 3-lobed, middle lobe largest. Stamens 4, one pair longer than the other, under upper corolla-lip and slightly protruding. Style 2-lobed. Nutlets 4, fleshy. *Mediterranean area*.

49. Perilla Linnaeus. ?7/1. Aromatic annual herbs 30--150 cm, or sometimes to 2 m. Stems erect, much-branched, square, with short or long hairs. Leaves opposite, stalked, the blades ovate to circular, green to deep purple, hairy and glandular, especially beneath, margins toothed, veins prominent. Inflorescence of axillary and terminal 1-sided racemes; whorls 2-flowered, subtended by pairs of diamond-shaped to circular bracts; flowers bisexual. Calyx campanulate, 10-veined, 2-lipped with 3 upper and 2 lower teeth, enlarging and swelling on 1 side in fruit. Corolla tubular, with 5 more or less equal lobes or very weakly 2-lipped. Stamens 4, spreading, more or less equal, about as long as corolla. Style as long as stamens; stigma bilobed. Nutlets more or less spherical with raised net pattern. *Warm temperate Asia*.

50. Elsholtzia Willdenow. 40/3. Aromatic annual and perennial herbs, subshrubs or shrubs with simple hairs. Leaves simple, opposite, often toothed. Inflorescence of 2--many-flowered whorls in terminal and axillary spikes or heads, often very dense, cylindric or occasionally 1-sided, sometimes grouped in panicles; bracts small; flowers bisexual. Calyx campanulate or cylindric with 5 teeth, more or less equal or lower 2 longer. Corolla hairy and glandular outside, purplish, white or yellow; tube sometimes slightly curved, longer than calyx; lips 2, upper straight, flat or convex, entire or notched, lower 3-lobed, deflexed, middle lobe largest, the tip notched or finely toothed. Stamens 4, in 2 pairs of unequal length, protruding. Style usually longer than stamens, stigma 2-lobed. Nutlets oblong or ovoid, hairless or slightly hairy, tuberculate or smooth. *Temperate Old World.*

51. Rostrinucula Kudo. 2/1. Shrubs with stellate hairs. Leaves opposite, shortly stalked. Inflorescence of 6--10-flowered whorls in dense, terminal, cylindric racemes, elongate, nodding or pendulous; bracts broadly rounded-triangular, bracteoles narrowly elliptic, both falling quickly; flowers bisexual. Calyx campanulate, 10-veined, with 5 more or less equal teeth. Corolla tubular, 2-lipped, usually longer than calyx, glandular; upper lip erect, not hooded, red or purplish red, entire, lower lip 3-lobed, the middle lobe larger and concave. Stamens 4, more or less equal, protruding. Ovary stellate-hairy and glandular. Style protruding, stigma 2-lobed. Nutlets ellipsoid, 3-angled, beaked, stellate-hairy. *China.*

52. Lavandula Linnaeus. 32/12 and some hybrids. Aromatic evergreen shrubs, subshrubs or perennials with woody base. Leaves opposite, entire or dissected, margins often turned under. Inflorescence a compact terminal spike, with bracts borne opposite and alternating in all cultivated species, subtending a 5--7- (occasionally 3- or 9-) flowered cyme with minute bracteoles or a single-flowered cyme without bracteoles. Flowers bisexual, both calyx and corolla of 5 fused parts, usually strongly bilaterally symmetric. Upper middle lobe of calyx either more or less equal to other lobes, broadly triangular or modified into an appendage. Corolla 2-lipped, upper lip 2-lobed, not hooded, lower lip 3-lobed. Stamens 4, spreading. Nutlets hairless, smooth. *Canary Islands, Cape Verde, Madeira, the Mediterranean area, southwest Asia, north & northeast tropical Africa, Arabia & India.*

175. NOLANACEAE

Woody-based shrubs or annual or perennial herbs, usually decumbent, often glandular-hairy. Leaves usually of 2 kinds, basal and on the stems, simple, alternate or whorled, shortly stalked or almost stalkless, sometimes succculent; stipules absent. Flowers solitary or in clusters in leaf-axils, bisexual, usually radially symmetric, shortly stalked or almost stalkless. Calyx tubular to campanulate or urceolate, 4- or 5-lobed, lobes overlapping, sometimes unequal. Corolla campanulate or funnel-shaped, 5-lobed, sometimes 2-lipped, folded between lobes, blue to purple-blue, pink or white. Stamens 5, unequal, inserted at base of corolla-tube alternating with the corolla-lobes; filaments slender or enlarged towards base, often hairy. Anthers opening by longitudinal slits. Nectary-disc well-developed, fleshy, often lobed. Style single, terminal. Stigma with 2--5 lobes. Ovary superior, carpels more or less united, each with 1--several ovules; placentation axile. Fruit a schizocarp or cluster of nutlets, seeds 1--several.

A family of a single genus from desert, semi-desert and coastal areas of Chile and Peru, 1 from the Galapagos Islands (Ecuador). The family is similar to Convolvulaceae and Boraginaceae, but is usually considered closest to Solanaceae, in which it is sometimes included.

1. Nolana Linnaeus. 18/2. Description as for family. *Chile, Peru.*

176. SOLANACEAE

Small trees, shrubs or herbs, erect or climbing, annual or perennial, usually with glandular, simple or branched hairs, occasionally hairless; sometimes with prickles or thorns. Leaves spirally arranged, entire or variously lobed, pinnatisect or pinnate, often with a characteristic odour. Stipules absent, pseudostipules (basal leaflets or small axillary leaves) occasionally present. Flowers in nodal or internodal, leaf-opposed or terminal, cymes, racemes or panicles, rarely solitary, usually bisexual, when unisexual sometimes male and female flowers on separate plants, radially or bilaterally symmetric. Calyx of 5 (rarely 3--8) shallow or deep lobes, usually persistent and often enlarging, sometimes enclosing fruit. Corolla with a tube and 5 (rarely 4--10) lobes, rarely of free segments, wheel-shaped, star-shaped, campanulate, funnel-shaped, tubular, or 2-lipped. Stamens usually 5, sometimes 4 or 2 and then often with 1 or more staminodes, inserted on the corolla-tube, equal or unequal. Anthers with 1 or 2 cells, often forming a central cone or group around the style, opening by apical pores or inward longitudinal slits, rarely each with an apical appendage. Ovary superior, of 2 fused carpels, sometimes appearing 4-celled or irregularly 3-, 5- or more-celled, with numerous axile ovules; when cells 2, then the septum oblique. Fruit a berry or capsule, smooth or spiny. Seeds numerous, testa pitted or hairy. Embryo straight or curved, with endosperm. See figure 110, p. 484.

A widely distributed family of about 90 genera and 2600 species, with the greatest concentration in tropical and South America. A family containing many edible plants as well as poisonous plants of narcotic or pharmaceutical importance. Green parts of most species, including those grown for food, are poisonous if ingested.

1a. Anthers opening by apical pores **1. Solanum**
  b. Anthers opening by longitudinal slits 2
2a. Leaves pinnate to bipinnate 3
  b. Leaves simple 4
3a. Flowers inverted, corolla bilaterally symmetric; fertile stamens 2 with 3
    staminodes **22. Schizanthus**
  b. Flowers not inverted; corolla radially symmetric; stamens 5 all fertile
    12. Jaborosa
4a. Corolla urceolate **13. Salpichroa**
  b. Corolla not urceolate 5
5a. Fertile stamens 5 or 6 6
  b. Fertile stamens 4 or fewer, staminodes often present 18
6a. Corolla more than 13 cm **11. Datura**
  b. Corolla less than 13 cm 7
7a. Fruit a capsule 8
  b. Fruit a berry 20
8a. Stamens unequal in length 9
  b. Stamens equal in length 11
9a. Filaments united into a tube around the style for part of their length
    20. Nierembergia
  b. Filaments not united 10
10a. All leaves alternate; calyx-teeth mostly less than two-thirds of the total
     length of calyx; flowers opposed to or in axils of small bracts
     17. Nicotiana
  b. Upper leaves appearing opposite; calyx-teeth more than three-quarters
     of the total length of calyx; flowers solitary at leaf-nodes **18. Petunia**
11a. Calyx detaching near base of capsule by a line around its circumference,
     often reflexed; capsule usually spiny or tuberculate **11. Datura**

b. Calyx persistent; capsule smooth 12
12a. Capsule opening by flaps 13
b. Capsule opening by a lid, splitting around its circumference 15
13a. Leaves minute, overlapping, linear or spathulate, less than 7 mm wide
19. Fabiana
b. Leaves often large, basal or along stem, not as above, more than 6 mm
wide 14
14a. Plant completely hairless, not glaucous; seeds *c*. 3 mm **16. Vestia**
b. Plants glandular-hairy or glaucous; seeds 0.5--1 mm **17. Nicotiana**
15a. Flowers solitary or 2 or 3 together 16
b. Flowers more than 3 in panicles, umbels or 1-sided racemes 17
16a. Corolla more than twice as long as calyx, lobes short and inconspicuous;
fruiting-calyx tightly fitting fruit **7. Scopolia**
b. Corolla less than twice as long as calyx, lobes conspicuous, free, with
auricles at base; fruiting-calyx loosely fitting fruit **8. Anisodus**
17a. Flowers in panicles or umbels, noticeably stalked; lobes of fruiting-calyx
without spines **9. Physochlaina**
b. Flowers in 1-sided racemes, stalks very short or absent; fruiting-calyx-
lobes each with an apical spine **10. Hyoscyamus**
18a. Woody shrubs or climbers 19
b. Annuals, herbaceous or shrubby perennials **21. Salpiglossis**
19a. Anthers 2-celled **23. Anthocercis**
b. Anthers 1-celled **24. Cyphanthera**
20a. Calyx hardly expanding in fruit or, if expanding, not enveloping berry
12. Jaborosa
b. Calyx expanding in fruit, often enveloping or partly enveloping berry 21
21a. Plant herbaceous or a shrubby herb, not woody 22
b. Plant a woody shrub or small tree 25
22a. Leaves in a basal rosette at flowering time **6. Mandragora**
b. Leaves along well-developed stems at flowering time 23
23a. Calyx not papery and inflated **5. Atropa**
b. Calyx papery and inflated, enclosing fruit 24
24a. Calyx-lobes separate; ovary 3--5-celled **15. Nicandra**
b. Calyx-lobes fused along margins; ovary 2-celled **3. Physalis**
25a. Shrubs with spines **14. Lycium**
b. Shrubs without spines 26
26a. Corolla campanulate, to 1.5 cm **2. Withania**
b. Corolla tubular or funnel-shaped, at least 2 cm **4. Iochroma**

1. Solanum Linnaeus. 1700/27. Small trees (rarely exceeding 20 m), shrubs, herbs or
climbers, sometimes with tubers or rhizomes, hairless or with simple, branched, stellate
or glandular hairs, often with prickles on stems, leaves and calyces. Leaves alternate or
falsely paired, pairs often unequal, simple to pinnate, entire or lobed, rarely with
pseudostipules. Flowers solitary or in cymes, sometimes umbellate or racemose; at first
terminal, later lateral, leaf-opposed or internodal. Calyx campanulate with 4--6 (usually
5) teeth borne on the margin, rarely inflated. Corolla usually with 5 lobes, star- to
bowl-shaped or campanulate, white, yellow, blue, violet or purple. Stamens 5 or as
many as corolla-lobes, inserted near base of corolla-tube, often forming a conical

column around the style; filaments very short, slender or broad; anthers oblong or tapered, opening by apical pores. Ovary usually 2-celled with numerous ovules; stigma capitate, entire or bilobed, rarely bifid. Fruit a spherical, ovoid or ellipsoid, usually fleshy berry, rarely woody or dehiscent; sclerotic granules sometimes present. Seeds numerous, flattened, circular or kidney-shaped, sometimes winged; surface wrinkled, pitted, netted or falsely hairy. *Cosmopolitan but especially numerous in the American tropics.* Figure 110, p. 484.

2. Withania Pauquy. 11/1. Small shrubs to 2 m, with fine branched hairs. Leaves alternate or almost opposite on flowering shoots, simple, entire. Flowers solitary or in cymes, at leaf-nodes or terminal on lateral spurs, bisexual or unisexual. Calyx campanulate, 5- or 6-lobed, expanding in fruit. Corolla campanulate, 3--6-lobed. Stamens 5, attached near the base of the corolla; anthers 2-celled, opening by longitudinal slits. Ovary 2-celled; style thread-like, stigma capitate, bilobed. Fruit a spherical berry surrounded by the inflated calyx; seeds compressed, netted. *South Europe, western Asia to China, Canary & Cape Verde Islands, Africa.*

3. Physalis Linnaeus. 100/2. Annual or rhizomatous perennial herbs, often woody at the base, rarely shrubby, hairless or with simple, branched, stellate or glandular hairs. Leaves alternate, sometimes paired, linear to broadly ovate, stalked, with wavy, lobed or shallowly toothed margins. Flowers at branch-forks or leaf-nodes, solitary in cultivated species or in small clusters, with stalks that elongate in fruit. Calyx campanulate or tubular, 5-lobed, papery and inflated, expanding considerably and enveloping fruit. Corolla off-white, yellow, or with purple spots at the mouth of the tube, wheel-shaped, funnel-shaped or campanulate, often hairy at the mouth and with nectary pouches in the tube. Stamens 5, equal, inserted near base of tube; filaments flattened or narrow; anthers oblong-ovoid, opening by longitudinal slits. Ovary 2-celled, placentation axile; style protruding or included; stigma capitate or club-shaped. Fruit a fleshy berry, sometimes sticky, included in the inflated, 5- or 10-angled, ribbed or smooth calyx, which is often depressed around the stalk. Seeds few to many, wrinkled. *Widespread.*

4. Iochroma Bentham. 20/1. Small trees or shrubs, thorns absent (ours), usually woolly or downy on young branches and leaves. Leaves ovate to lanceolate, entire, stalked. Flowers with long stalks, hanging, in clusters, either in primary inflorescences of many (to 120) flowers near stem-apex, or secondary inflorescences of few (1--15) flowers from the nodes of the lower leaves or all along the branch. Calyx cup-shaped, urceolate, campanulate or inflated, with small teeth or large lobes or 2-cleft, expanding in fruit. Corolla funnel-shaped, campanulate or narrowly tubular, often slightly curved and inflated towards the middle, with 5 or 10 teeth or short lobes, blue to purple, red to orange, yellow, greenish yellow or white. Stamens 5, inserted near base of tube, sometimes fully or partly protruding. Anthers 2-celled, splitting lengthwise. Fruit a berry, spherical or conical-ellipsoid, partly or completely enclosed by the expanded calyx. Seeds many, pitted, circular to kidney-shaped; small sclerotic granules often present in fruit. *Colombia, Ecuador, Galapagos, Peru, Bolivia & northwestern Argentina.*

5. Atropa Linnaeus. 6/1. Upright perennial herbs with large swollen roots and creeping stolons. Leaves alternate or opposite, simple, entire. Flowers solitary, rarely

paired, in branch-forks or at leaf-nodes, stalked. Calyx campanulate, 5-lobed, expanded and star-shaped in fruit. Corolla funnel-shaped, cylindric or campanulate with 5 short lobes, yellow or purple. Stamens 5, almost equal, inserted at base of corolla, hairy towards base; anthers opening by longitudinal slits. Ovary 2-celled, with ring-shaped disc at base; stigma bilobed, peltate. Fruit a black-purple or yellow, many-seeded, 2-celled berry. Seeds with pitted surface. *Europe to China.*
All parts of the cultivated plant, including the berries, are very poisonous.

6. Mandragora Linnaeus. 5/2. Perennial herbs with large fleshy taproots. Leaves in rosettes at flowering time, rarely producing taller stems as the fruit develops, broadly ovate, lanceolate, oblanceolate or spathulate, with wavy, sometimes hairy margins. Flowers solitary at leaf-nodes, on short stalks. Calyx campanulate, 5-lobed, expanding in fruit. Corolla campanulate or funnel-shaped, with 5 or 6 lobes. Stamens 5 or 6, inserted below the middle of the corolla-tube, with filaments hairy at the base; anthers opening by longitudinal slits. Ovary 2-celled, with numerous ovules, surrounded at the base by a glandular disc; stigma capitate. Fruit a spherical or ovoid berry; seeds compressed with a pitted surface. *Europe, Mediterranean area, central Asia to China.*

7. Scopolia Jacquin. 2/1. Perennial herbs with fleshy, erect, branching stems and horizontal rhizomes. Leaves alternate, entire, membranous, stalked. Flowers solitary at leaf-nodes or branch-forks, pendent, green or tinged purple; stalks thread-like. Calyx campanulate, with 5 straight lobes, enlarging and enclosing fruit. Corolla at least twice as long as calyx, radially symmetric, cylindric to campanulate; limb 5-lobed, lobes short and inconspicuous. Stamens 5, equal, inserted at base of corolla-tube, included, opening by longitudinal slits. Stigma club-shaped. Fruit a capsule opening by a line around its circumference. *Central & southeastern Europe, Korea & Japan.*

8. Anisodus Link. 5/1. Herbaceous perennials with a swollen overwintering rootstock. Stems erect, branched, hairless or with simple or branched hairs. Flowers solitary at leaf-nodes, hanging. Calyx funnel-shaped or campanulate to urceolate, 10-veined, sometimes appearing 2-lipped; lobes 4 or 5, unequal. Corolla less than twice as long as the calyx, radially symmetric, campanulate to cylindric; lobes free, with overlapping margins. Stamens 5, equal, inserted near base of corolla-tube, included, anthers opening by longitudinal slits. Fruit a spherical or ovoid capsule, opening by splitting at apex or by a line around its circumference above the middle, included in the strongly ribbed expanded calyx, with enlarged stalk. *North India to western China.*
Often included in *Scopolia.*

9. Physochlaina G. Don. 11/1. Herbaceous perennials with swollen roots and short, creeping rhizomes. Stems erect, branched. Leaves alternate, entire or sinuous, rarely toothed. Flowers nodal or terminal, in panicles or racemes, usually with bracts, stalked. Calyx tubular to campanulate, funnel-shaped or urceolate, expanded in fruit with 10 prominent veins; lobes 5, equal or almost so. Corolla campanulate or funnel-shaped, cylindric towards base, white, yellow, purple to violet or green; lobes 5, slightly unequal, overlapping in bud. Stamens 5, equal, inserted towards middle of corolla-tube, protruding; anthers ovate, opening by longitudinal slits. Ovary 2-celled; style protruding, stigma wide, 2-lobed; disc conspicuous. Fruit a capsule, oblong or spherical, opening by a line around its circumference slightly above middle. Seeds kidney-shaped, pitted, brownish yellow. *Turkey to China.*

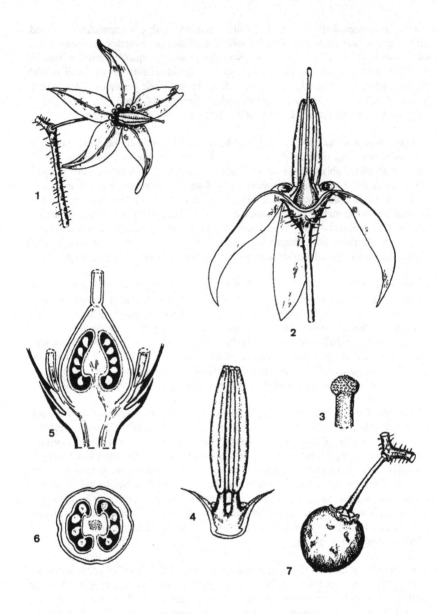

Figure 110. Solanaceae. *Solanum dulcamara*. 1, Single flower. 2, Longitudinal section of flower. 3, Stigma. 4, Two stamens. 5, Longitudinal section of ovary. 6, Transverse section of ovary. 7, Fruit.

484

10. Hyoscyamus Linnaeus. 30/2 and their hybrid. Annual, biennial or perennial herbs to 2 m (ours), most parts with sticky glandular hairs. Leaves alternate or in rosettes, simple, usually toothed or pinnatifid. Flowers in 1-sided racemes, stalks very short or absent. Calyx tubular to campanulate or urn-shaped, 5-lobed, expanded and becoming rigid and enveloping capsule after flowering, lobes usually with apical mucros. Corolla broadly funnel-shaped to campanulate, bilaterally symmetric, 5-lobed, often net-veined; lobes unequal, rounded. Stamens 5, equal, inserted at base or in lower half of corolla-tube, weakly or strongly protruding; anthers opening by longitudinal slits. Ovary 2-celled, often with basal annular nectary; style often curved upwards, stigma capitate. Fruit a capsule, opening by a line around the circumference at the top, the lid falling off. Seeds many, oblong to kidney-shaped, *c.* 2 mm, netted-pitted. *Europe, Mediterranean area, Middle East to central Asia.*

The plants are very poisonous, and several species are cultivated comercially as a source of tropane alkaloids.

11. Datura Linnaeus. 10/6. Annual or weakly perennial herbs, apparently dichotomously branched, with simple and/or glandular hairs. Leaves alternate, simple and entire or lobed, stalked. Flowers solitary in branch-forks, shortly stalked. Calyx tubular, 3--6-lobed, detaching near the base along a line around its circumference, base enlarging in fruit to form a reflexed skirt. Corolla single or double, funnel- or trumpet-shaped, white to pale purple or yellowish, tube 5-angled, green, throat white or coloured; limb 5-lobed with acuminate teeth, appearing 10-toothed in some species with extra acuminate projections between teeth. Stamens 5, equal, inserted at or below middle of corolla-tube; anthers elliptic, opening by longitudinal slits. Ovary 2-celled, often with 4 cells in the lower half, with small spines; nectary-ring present at base; stigma saddle-shaped. Fruit a capsule, ovoid to globular, with 2--4 cells, exterior with numerous stout or thin spines or tubercles, rarely smooth, opening by 4 valves. Seeds 100--300 per capsule, D-shaped, pitted, brown or black. *South-western USA & Mexico; 1 species naturalised in Australia.*

Species of **Brugmansia** Persoon are similar but tree-like, with generally larger flowers and the fruit a fleshy berry; they are in cultivation but are not really hardy in northern Europe.

12. Jaborosa Jussieu. 23/2. Perennial herbs with decumbent or procumbent stems and fleshy roots or creeping rhizomes. Leaves clustered, entire or deeply pinnately divided. Flowers at leaf-nodes, solitary, or up to 100 together in cymes. Calyx tubular or campanulate, 5-lobed. Corolla tubular to campanulate with 5 spreading lobes, radially symmetric, white, yellow to brown-purple, tube usually longer than lobes. Stamens 5, inserted on corolla-tube near apex with very short filaments or inserted towards base with longer filaments, protruding or included; anthers free, attached by back to filament, opening by longitudinal slits. Ovary 2--5-celled; stigma 2--5-lobed. Fruit a firm berry with many seeds, spherical; calyx persisting and enlarging at base of berry; stalk curved. Seeds compressed. *South America (mostly Argentina).*

13. Salpichroa Miers. 22/1. Scrambling, soft-wooded shrubs or climbers, with leaf-opposed branches. Leaves hairy, alternate or opposite in pairs of unequal size, entire, ovate. Flowers solitary or paired, in branch-forks and leaf-opposed, on short stalks. Calyx with 5 deep lobes, hairy. Corolla tubular or urceolate with 5 spreading or reflexed lobes, hairy. Stamens 5, inserted in or above the middle of the corolla; anthers opening by longitudinal slits. Ovary 2-celled, conical, with a fleshy disc at the base; stigma capitate. Fruit a fleshy, red or white berry, with persistent calyx; seeds flattened with a finely wrinkled surface. *South America, centred on Andean Peru, extending from Venezuela to northern Argentina.*

14. Lycium Linnaeus. 100/5. Usually thorny shrubs with shortly stalked, simple leaves, alternate or clustered. Flowers shortly stalked, solitary or in small clusters. Calyx cup-shaped with 5 teeth or 2-lipped with 2 or 3 teeth, expanding in fruit. Corolla funnel-shaped or cylindric with stamens attached in throat of corolla-tube. Stamens 5; anthers opening by longitudinal slits. Ovary 2-celled. Stigma shallowly 2-lobed. Fruit a berry. *Warm temperate areas; 3 species introduced in Europe.*

15. Nicandra Adanson. 1/1. Annual herb, with grooved, hairless stems to 2 m. Leaves alternate, ovate, 4--21 × 2--10 cm, apex acute, base gradually tapering; margins with wide shallow teeth, stalk 1.5--6.5 cm with narrow wings. Flowers solitary, nodal on stalks 6--24 mm that elongate in fruit. Calyx hairless, 5-lobed; lobes 9--20 mm, ovate with acute, mucronate apices and sagittate bases; adjacent lobes pressed against each other along the margins to form a wing, fused along lower third. Corolla campanulate, pale blue to mauve or white, with 5 lobes 1.2--2.5 cm × 5--15 mm, margin entire or notched. Stamens 5, inserted near the corolla-base, filaments 3--5.5 mm with dense down on the expanded bases; anthers yellow, ovate, 2--4 mm, opening by longitudinal slits. Ovary with 3--5 unequal cells and numerous ovules; disc present; style 3--6 mm with a capitate 3--5-lobed stigma. Fruit a spherical, rather dry berry, 1.1--2.2 cm across, pale yellow or purple, enclosed in the enlarged papery, netted calyx. Seeds disc-shaped to kidney-shaped, brown, 1--2 mm with a pitted surface. *Peru; widely naturalised in tropical & temperate areas.*

16. Vestia Willdenow. 1/1. Shrubs to 3.5 m, with erect branches, evergreen. Leaves alternate, 1.2--5 cm × 6--18 mm, obovate to oblong-elliptic, gradually tapered to short winged stalks, bright glossy green often tinged purple, strongly smelling, margins entire, hairless. Flowers 1--4, from the upper nodes, pendent. Calyx campanulate with 5 short lobes, expanding in fruit. Corolla tubular to funnel-shaped, 2.5--3.5 cm, with 5 spreading ovate to triangular lobes 6--9 mm, interior with short hairs, pale greenish yellow or purplish yellow. Stamens 5, equal, inserted in the lower quarter of the corolla, hairy at the base, strongly protruding; anthers *c.* 2 mm, opening by longitudinal slits. Ovary 2-celled, partly enveloped by the persistent corolla-base, style and stigma protruding. Fruit an ovoid capsule, opening by 4 lobes that become reflexed, the lower half enveloped by the tightly fitting expanded calyx. Seeds numerous, *c.* 3 × 1.5 mm, wingless, surface wrinkled. *Southern South America.*

17. Nicotiana Linnaeus. 70/11. Soft-wooded shrubs, biennials or annual herbs, usually sticky with glandular hairs. Leaves alternate, entire, stalked or stalkless. Flowers opposed to or subtended by small bracts, in terminal panicles or racemes or rarely clustered, shortly stalked. Calyx 5-toothed, shorter than corolla, persistent, enlarging in fruit, teeth mostly less than two-thirds of the whole length. Corolla radially or slightly bilaterally symmetric, tubular, funnel-shaped or with a long, thin tube that abruptly expands into a wide limb; lobes 5, erect, spreading or recurved, almost entire. Stamens 5, filaments equal or not, usually inserted where corolla-tube narrows, usually included; anthers less than 2 mm, with or without connective, opening by longitudinal slits. Ovary 2-celled, with a thickened disc at base; ovules numerous; style slightly bilobed. Fruit a capsule, opening by flaps; seeds minute with variable specific shape and surface. *America, Australia & Pacific Islands, 1 species in Africa (Namibia).*

18. Petunia Jussieu. 3/3 and some hybrids. Annual or perennial herbs, often trailing, usually with sticky, glandular hairs. Lower leaves alternate, the upper opposite, entire, simple. Flowers solitary, stalked, at leaf-nodes. Calyx campanulate, with 5 deep, narrow lobes. Corolla with a narrow tube circular in section and an abruptly expanded, flattened limb, bilaterally symmetric, with 5 lobes, white, pink to red to purple, often netted in the throat; lower 2 corolla-lobes not enclosing the upper 3 in bud. Stamens 5, inserted at or below the middle of the tube, with 2 pairs of different lengths and the fifth shorter; anthers less than 3 mm, opening by longitudinal slits. Ovary 2-celled, with a fleshy nectary-disc below; ovules numerous; stigma capitate, grooved. Fruit a capsule, shorter than calyx, with numerous small seeds. *South Brazil to northeastern Argentina & Bolivia.*

About 37 species formerly included in *Petunia* are now mostly transferred to the tropical genus *Calibrachoa* Llave & Lexarza, in which the lower 2 corolla-lobes do enclose the upper 3 in bud.

19. Fabiana Ruiz & Pavón. 25/1. Small evergreen heath-like shrubs to 3 m. Leaves small, heather-like, linear to cylindric, alternate, often with glandular hairs, sometimes sticky. Flowers solitary, from leaf-nodes, minutely stalked. Calyx tubular with 5 small lobes, glandular-hairy. Corolla cylindric, constricted at base, with 5 spreading lobes, usually creamy white, yellowish or pale bluish mauve. Stamens 5, included, inserted at the base of corolla-tube; filaments hairless; anthers with confluent or separated cells, opening by longitudinal slits. Ovary 2-celled; style equalling stamens; stigma capitate, bilobed. Fruit a capsule with 4 teeth, opening by valves, enveloped by persistent calyx. Seeds small, numerous. *Western South America from Peru to Chile & Argentina.*

20. Nierembergia Ruiz & Pavón. 30/3. Small shrubs or perennial herbs, decumbent, erect or creeping. Leaves alternate, simple, entire, narrow, hairless or hairy. Flowers solitary, terminal or nodal. Calyx tubular or campanulate, 5-lobed. Corolla-tube very narrow, elongate, abruptly expanded into a shallow, campanulate or funnel-shaped limb with 5 broad lobes, white to purple, usually with minute, stalked oil-glands towards the centre. Stamens 5, with 4 in 2 pairs and 1 shorter, inserted on or near the apex of the corolla-tube, often at different levels but with filaments united in part to form a tube around the style, protruding; anthers opening by longitudinal slits. Pollen released in monads or tetrads. Nectary-disc absent. Ovary 2-celled; stigma usually enlarged into a 2-armed or crescent-shaped structure between or above the anthers, protruding. Fruit a 2-celled capsule with deeply bifid valves. Seeds angular, with netted-pitted surface. *North America, South America (from the Andes of Colombia to Patagonia).*

21. Salpiglossis Ruiz & Pavón. 2/1. Annuals, covered with sticky, glandular hairs, producing a rosette of deeply sinuous, long-lanceolate leaves from a thickened taproot. Flowering-stem to 1 m, with linear-lanceolate leaves and bracts. Calyx campanulate with 5 long-triangular lobes. Corolla 2.3--4.5 cm, campanulate or tubular to funnel-shaped, colour and venation-pattern variable. Anthers attached to the filament by their backs, opening by longitudinal slits; pollen released as single grains, or in groups of 4 (tetrads). Ovary 2-celled, conical, with a circular disc at the base; style flattened towards top, with a broad, trumpet-shaped stigma. Fruit a capsule, opening by 4 valves. Seeds numerous, small. *Andes of Chile & Argentina.*

22. Schizanthus Ruiz & Pavón. 12/3 and some hybrids. Annual herbs to 90 cm; stems hairless or densely covered with adpressed glandular bristles. Leaves alternate, stalkless or stalked, pinnate to bipinnate in most cultivated species, margins deeply lobed to bipinnatisect. Inflorescence a terminal panicle; flowers bisexual, inverted, stalked, upright or spreading, each subtended by 2 leaf-like bracts. Calyx 5-lobed, almost radially symmetric, lobes linear-oblong to obovate to spoon-shaped. Corolla bilaterally symmetric, weakly or strongly 2-lipped, upper lip 3- to almost 5-lobed, lower lip 3-lobed, white or pink to violet to deep purple, often with a yellow area spotted with purple at the base of the upper lip. Stamens 5, 2 fertile with filaments shorter or longer than the corolla throat, 3 sterile, 2 upper stamens with filaments reduced; anthers opening by longitudinal slits. Ovary 2-celled. Fruit a papery capsule, spherical to ellipsoid. Seeds kidney-shaped, rough. *Chile & western Argentina.*

22. Anthocercis Labillardière. Few/1. Evergreen shrubs, hairless or with simple or glandular hairs. Leaves alternate or clustered, entire. Flowers rarely solitary, usually in terminal or nodal cymes, racemes or panicles, each subtended by a pair of bracts. Calyx cup-shaped or campanulate, 5-lobed. Corolla radially symmetric to slightly bilaterally symmetric, tubular to funnel-shaped; limb spreading, 5-lobed. Stamens 4, equal or in 2 pairs, inserted at base of corolla-tube, sometimes with a fifth staminode; anthers 2-celled, attached by the back to the filament, opening by longitudinal slits. Ovary 2-celled; stigmas capitate, 2-lobed. Fruit a smooth 2-valved capsule; valves bifid. Seeds oblong to kidney-shaped. *Southern Australia.*

23. Cyphanthera Miers. 9/2. Evergreen shrubs, usually densely hairy with branched hairs, simple and glandular hairs often also present. Leaves alternate or clustered, entire with glandular and/or stellate hairs. Flowers terminal or at leaf-nodes, 1--many. Calyx cup-shaped or campanulate, 5-lobed, hairy. Corolla more than 5 mm, tube broadly funnel-shaped to campanulate with purple lines; lobes spreading. Stamens 4, in 2 pairs; anthers 1-celled, kidney-shaped, opening by a slit across the top. Ovary 2-celled with nectary-disc. Fruit a capsule with few seeds. *Australia.*

177. BUDDLEJACEAE
Erect shrubs or small trees with stellate and/or glandular hairs. Leaves simple, opposite or rarely alternate, with stipules usually forming a line joining the leaf-bases (often reduced, sometimes absent). Calyx 4-lobed. Corolla radially symmetric, 4-lobed; lobes overlapping or edge-to-edge in bud. Stamens 4; anthers opening by longitudinal slits. Ovary superior, carpels usually 2, ovules numerous, axile. Fruit a capsule, rarely a berry. See figure 111, p. 490.

The family contains the single, widely-cultivated genus, *Buddleja.*

1. Buddleja Linnaeus. 90/25 and several hybrids. Shrubs, trees or woody-based herbs. Twigs usually 4-sided. Leaves usually opposite, usually woolly or hairy; stalks short. Inflorescence a long panicle, cluster or false spike. Calyx campanulate, 4-lobed. Corolla short, tubular, campanulate to almost disc-like or long-cylindric, 4-lobed. Stamens 4, filaments as long as, or shorter than anthers. Fruit a capsule or berry. Seeds numerous, very small. *Tropics & subtropics of America, Africa & Asia.* Figure 111, p. 490.

One species and its hybrids are widely introduced in northwest Europe.

178. SCROPHULARIACEAE

Mostly herbs (some hemi-parisitic), with a few evergreen or deciduous shrubs, rarely climbers. Leaves alternate or opposite, simple, pinnately lobed or toothed, stipules absent. Inflorescence racemose or cymose, with bracts and bracteoles. Flowers bisexual, usually bilaterally, rarely almost radially symmetric. Calyx 5-, rarely 4-lobed. Corolla 5-, rarely 4-lobed, mostly 2-lipped, sometimes with 1 or 2 spurs, very rarely absent. Stamens usually 4, sometimes 2, 3, or 5, borne on the corolla-tube, 2 longer than the others; 1 or more staminodes sometimes present. Anthers 2-celled, cells sometimes unequal, splitting longitudinally by slits. Ovary superior, of 2 united carpels with 2 cells, cross-wall horizontal; nectar-secreting disc present at base. Ovules usually numerous, placentation axile. Style simple or bilobed. Fruit usually a dry capsule, rarely indehiscent. See figures 112, p. 492, 113, p. 496 & 114, p. 500.

A family of about 3000 species and 220 genera, widely distributed throughout the world.

1a. Plant parasitic, vegetative parts without chlorophyll, white, underground	**50. Lathraea**
b. Plant free-living or hemi-parasitic, vegetative parts green, above ground	2
2a. Leaves all alternate or all basal	3
b. At least the lower leaves opposite or whorled	21
3a. Fertile stamens 2, staminodes 0--3	4
b. Fertile stamens 4, staminodes 0 or 1	7
4a. Corolla more or less equally 4-lobed, not at all 2-lipped	**39. Synthyris**
b. Corolla unequally 4- or 5-lobed, 2-lipped	5
5a. Calyx spathe-like, split to the base on the lower side, 2-lobed; fruit indehiscent, 1-seeded	**42. Lagotis**
b. Calyx not spathe-like or split as above, 2--5-lobed	6
6a. Corolla less than 2 cm, bluish purple	**41. Wulfenia**
b. Corolla 5 cm or more, yellow, or totally absent	**40. Besseya**
7a. Corolla more or less equally 4- or 5-lobed, not 2-lipped	8
b. Corolla unequally lobed, distinctly 2-lipped	13
8a. Flowers solitary in the axils of bracts which are similar to the foliage leaves	9
b. Flowers in spikes, racemes or panicles, in the axils of bracts which are clearly differentiated from the foliage leaves	10
9a. Corolla with the upper lobes white, the lower pink, or all yellow; leaves kidney-shaped to almost circular, toothed	**36. Sibthorpia**
b. Corolla white, pink or scarlet; leaves not as above	**34. Ourisia**
10a. Stamens protruding from the corolla-tube	11
b. Stamens included within the corolla-tube	12
11a. Corolla open, lobes much longer than the tube, almost radially symmetric; filaments conspicuously hairy with white, yellow or purple hairs	**5. Verbascum**
b. Corolla not open, lobes shorter than to as long as tube, strongly bilaterally symmetric; filaments all hairless	**38. Picrorhiza**

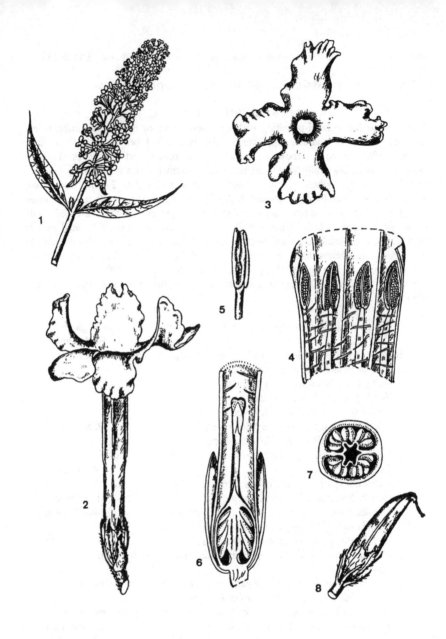

Figure 111. Buddlejaceae. *Buddleja davidii.* 1, Inflorescence. 2, A single flower. 3, Corolla from above. 4, Part of corolla-base opened out to show stamens. 5, Anther. 6, Longitudinal section of lower part of flower. 7, Transverse section of ovary. 8, Fruit before dehiscence.

12a. Leaves all basal; corolla-tube oblique or curved **34. Ourisia**
  b. Leaves alternate on the stems; corolla-tube straight **37. Erinus**
13a. Plants twining by stems or leaf-stalks 14
  b. Plants not twining 18
14a. Flowers hanging; calyx divided to about halfway, fleshy, pink, spreading
    almost at right-angles to the dark red corolla-tube **32. Rhodochiton**
  b. Combination of characters not as above 15
15a. Base of corolla-tube bulging and swollen at one side **30. Epixiphium**
  b. Base of corolla-tube not bulging and swollen at one side 16
16a. Anthers free; leaves palmately veined **31. Lophospermum**
  b. Anthers joined to each other by their margins; leaves not palmately veined
    17
17a. Corolla without a palate; capsule with equal cells **28. Maurandya**
  b. Corolla with a conspicuous, yellow palate; capsule with unequal cells
    29. Maurandella
18a. Leaves alternate on the stems **33. Digitalis**
  b. Leaves all in basal rosettes 19
19a. Corolla-tube without a bulge or swelling at one side at the base
    33. Digitalis
  b. Corolla-tube with a bulge or swelling at one side at the base 20
20a. Leaves stalked; stem-leaves deeply divided with 3--7 segments; anthers
    joined to form a ring-like structure **24. Anarrhinum**
  b. Leaves stalkless, undivided; anthers joined in pairs by their margins
    26. Nuttallanthus
21a. Fertile stamens 2, staminodes 0--3 22
  b. Fertile stamens 4, staminodes 0 or 1 32
22a. Corolla more or less equally 4- or 5-lobed, not 2-lipped 23
  b. Corolla very unequally lobed, clearly 2-lipped 29
23a. Prostrate herb forming mats; leaves stalkless; flowers solitary; corolla
    campanulate with 4 overlapping lobes; capsule 1-celled **4. Micranthemum**
  b. Combination of characters not as above 24
24a. Corolla-lobes always 5; calyx-lobes 5 or 6 **46. Chionohebe**
  b. Corolla-lobes usually 4, rarely a few corollas with 5 lobes; calyx-lobes
    3--5 25
25a. Leaves at least finely toothed or scalloped or rarely just wavy, and/or
    stalked 26
  b. Leaves entire, usually not stalked or stalks extremely short 28
26a. Shrubs with woody aerial branches, sometimes small; leaves evergreen
    45. Parahebe
  b. Herbs; leaves not evergreen 27
27a. Leaves in whorls of 3 or more, stalkless; tall erect herbs
    48. Veronicastrum
  b. Leaves opposite, very rarely some in whorls of 3, usually stalked; creeping
    to erect herbs **47. Veronica**
28a. Shrubs with persistent, woody, aerial branches **44. Hebe**
  b. Thin, creeping, mat-forming herb **45. Parahebe**
29a. Corolla with the lower lip inflated and much larger than the upper 30
  b. Corolla with the lips variously shaped and sized, but not as above 31

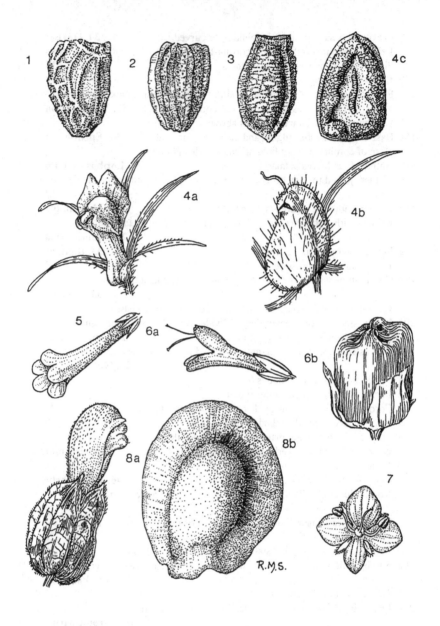

Figure 112. Scrophulariaceae: diagnostic features. 1, Seed of *Antirrhinum*. 2, Seed of *Chaenorhinum*. 3, Seed of *Linaria*. 4, *Misopates* (a, flower; b, fruit; c, seed). 5, Flower of *Wulfenia*. 6, *Lagotis* (a, flower; b, fruit). 7, Flower of *Veronica*. 8, *Rhinanthus* (a, flower, b, seed).

492

30a. Corolla white to pale lilac or flushed with yellow, spotted with purple; leaves irregularly toothed; capsule opening by 4 valves **14. Jovellana**
  b. Corolla yellow or rarely purple, the throat often purple-spotted; leaves entire to deeply lobed to pinnatifid; capsule opening by 2 valves
  15. Calceolaria
31a. Fowers solitary in the axils of foliage leaves **3. Gratiola**
  b. Flowers in terminal, spike-like racemes **43. Paederota**
32a. Corolla unequally to more or less equally 4- or 5-lobed, not 2-lipped 33
  b. Corolla very unequally lobed, clearly 2-lipped 39
33a. Lower leaves opposite, upper leaves alternate 34
  b. All leaves opposite 36
34a. Flowers usually more than 1.2 cm, not borne in dense, many-flowered spikes **12. Zaluzianskya**
  b. Flowers not more than 1.2 cm, in dense many-flowered spikes 35
35a. Corolla deeply 4-lobed, yellow or white **51. Hebenstretia**
  b. Corolla usually 5-lobed, blue **52. Selago**
36a. Flower inverted by a 180° twist in the flower-stalk **16. Alonsoa**
  b. Flower not inverted 37
37a. Flowers stalkless; capsule fleshy, berry-like **35. Hemiphragma**
  b. Flowers stalked; capsule not fleshy and berry-like 38
38a. Scrambling shrubs; corolla tubular, tapering to a swollen base
  6. Phygelius
  b. Perennial herbs, not scrambling; corolla with an oblique tube not swollen at the base, and spreading lobes **34. Ourisia**
39a. Corolla-tube with a bulge, swelling or spur at the lower side (rarely with 2 bulges, swellings or spurs) 40
  b. Corolla-tube without a bulge, swelling or spur 53
40a. Corolla with usually 1 (rarely 2) distinct, conical or parallel-sided spurs at the base 41
  b. Corolla with a rounded or elongate swelling or bulge (rarely 2 such swellings) at the base at 1 side 46
41a. Corolla with 2 spurs 42
  b. Corolla with a single spur 43
42a. Upper lip of corolla with a translucent 'window' **17. Diascia**
  b. Upper lip of corolla without a translucent 'window' **18. Nemesia**
43a. Mouth of spur with 2 rounded projections in the the middle, usually with bright orange hairs **18. Nemesia**
  b. Spur not as above 44
44a. Stoloniferous creeping perennial; leaves long-stalked, palmately veined; flowers solitary in the leaf-axils **23. Cymbalaria**
  b. Erect or ascending annual to perennial; leaves not as above; flowers in racemes 45
45a. Capsule opening by longitudinal slits; palate usually closing the mouth of the corolla **25. Linaria**
  b. Capsule opening by valves or pores; palate not closing the mouth of the corolla **20. Chaenorhinum**
46a. Corolla without a swollen palate partially or totally closing its mouth 47
  b. Corolla with a conspicuous palate 50

47a. Perennial herb, often woody below   **9. Penstemon**
  b. Annual herbs   48
48a. Leaves all opposite, cordate at base   **13. Collinsia**
  b. Leaves opposite or whorled below, alternate above, not cordate at base   49
49a. Calyx as long as corolla or almost so   **21. Misopates**
  b. Calyx much shorter than corolla   **26. Nuttallanthus**
50a. Capsule with 2 equal cells; leaves all opposite   51
  b. Capsule with 2 unequal cells; upper leaves alternate   52
51a. Upper lip of corolla 4-lobed; palate small not 2-lobed   **18. Nemesia**
  b. Upper lip of corolla 2-lobed; palate large, 2-lobed   **22. Asarina**
52a. Annual or short-lived perennial; style obliquely inserted on ovary
     27. Sairocarpus
  b. Woody-based perennials; style erect, not inserted obliquely on the ovary
     19. Antirrhinum
53a. Hemi-parasitic annuals; stems black-spotted; calyx 4-lobed, inflated;
     capsule winged   **49. Rhinanthus**
  b. Combination of characters not as above   54
54a. Leaves all basal or lower opposite, upper alternate   **2. Lindenbergia**
  b. Leaves all opposite   55
55a. Calyx with a distinct tube forming half the length or more   **1. Mimulus**
  b. Calyx with a very short tube or tube almost absent   56
56a. Stems square in section above; flowers more or less spherical, yellow,
     in panicles   **7. Scrophularia**
  b. Combination of characters not as above   57
57a. Shrubs or woody climbers, or if plants herbaceous, then leaves falling early
     and branches several from each node   **10. Keckiella**
  b. Herbs, sometimes woody at the base, not as above   58
58a. Staminode as long as or longer than the fertile stamens, usually hairy
     9. Penstemon
  b. Staminode shorter than the fertile stamens, not hairy   59
59a. Corolla with lower lip greatly exceeding upper; bases of filaments of
     fertile stamens densely hairy   **11. Nothochelone**
  b. Corolla with lower lip not greatly exceeding upper; bases of filaments of
     fertile stamens not hairy   **8. Chelone**

1. Mimulus Linnaeus. 100/20 and some hybrids. Hairless or hairy annuals, herbaceous perennials or shrubs. Leaves opposite, stalked or stalkless, pinnately veined or with 3--7 veins from near the base of the leaf; margins toothed or entire. Flowers solitary in leaf-axils or in leafy or bracteate racemes. Calyx tubular to campanulate, strongly 5-angled to folded, 5-toothed; fruiting calyces sometimes enlarged and inflated. Corolla 5-lobed, usually strongly 2-lipped with 2 upper and 3 lower lobes, upper lobes erect to reflexed, lower lobes usually spreading and held angled downwards. Mouth of corolla with 2 prominent, hairy ridges forming a palate which may sometimes more or less close the throat. Stamens 4, usually included; stigma 2-lobed. Capsule 2-valved, usually oblong, dehiscent to the base along one or both sutures. Seeds small, numerous, smooth to tuberculate or finely furrowed. *Predominantly in America but also in southern Africa, Asia & Australasia.*

2. Lindenbergia Lehmann. 15/1. Annual or perennial herbs, or more or less climbing small shrubs. Stem erect or ascending, sometimes pendent, branched or unbranched, softly short-hairy or hairless. Leaves simple, opposite or uppermost alternate, stalked, toothed. Flowers in terminal spike-like racemes or axillary, or solitary. Bracts like the leaves but usually smaller. Calyx campanulate or saucer-shaped, divided to middle or less into 5 acutely triangular lobes. Corolla 2-lipped, yellow, sometimes tinged purple or red at least on upper lip, throat or tube with orange or brown markings; upper lip erect, hairy in middle, lower longer than upper, hooded, with 3-lobed tip. Stamens 4, in 2 pairs. Style thread-like, often expanded towards apex, hairy at least near base (in ours). Fruit a 2-celled, ellipsoid to ovoid capsule, splitting along the cell wall; seeds numerous, elongate-ovoid or ellipsoid; seed-coat with net-like markings. *Old World tropics and subtropics.*

3. Gratiola Linnaeus. 25/2. Erect or creeping perennial herbs, hairless or with gland-like hairs. Leaves opposite, entire or toothed, usually stalkless. Flowers solitary, borne in the leaf-axils, subtended by 2 bracteoles. Calyx deeply 5-lobed, lobes unequal. Corolla-tube 2-lipped, upper limb flat, 2-lobed, the lower slightly larger, 3-lobed. Stamens 5 included in corolla, upper 2 fertile, lower 3 forming sterile staminodes. Stigma 2-lobed. Capsule 4-valved, seeds numerous. *Europe, America, Africa, temperate Asia & Australia.*

4. Micranthemum Michaux. 3/1. Short-lived, prostrate herb, forming mats and rooting at the nodes, hairless or almost so. Leaves opposite, mostly rounded, sometimes succulent, stalkless, margins entire. Flowers axillary, solitary, stalkless or with very short stalks. Calyx of 4 free segments. Corolla *c.* 2 mm, red, purple or white, campanulate, with 4 overlapping lobes. Stamens 2, protruding. Stigma with free, almost reflexed lobes, hairy. Capsules spherical, shorter than calyx, 1-celled. *North, Central & South America.*

5. Verbascum Linnaeus. 350/23 and some hybrids. Annual, biennial, perennial herbs or rarely small shrubs, occasionally spiny, often hairy, sometimes very densely so, with simple and/or branched hairs which are sometimes glandular. Leaves alternate, the basal usually in a distinct rosette, simple or divided. Flowers in terminal spikes, racemes or panicles, each usually subtended by a bract, sometimes also with bracteoles. Calyx 5-lobed, usually radially symmetric. Corolla with a short tube and usually 5 widely spreading lobes, radially symmetric or somewhat bilaterally symmetric, usually yellow, more rarely orange, purple, brownish or white, often hairy outside. Stamens 4 or 5, sometimes 4 plus 1 staminode; filaments mostly with long hairs which are white, yellow or purplish. Anthers of the 2 or 3 upper stamens always kidney-shaped, those of the 2 lower stamens sometimes obliquely inserted or decurrent on the filament. Capsule opening along the cross-walls; seeds numerous, small. *Eurasia & North Africa, centred in Turkey & adjacent Iran & Iraq.* Figure 113, p. 496.

Some species are sometimes included in the genus *Celsia* Linnaeus.

6. Phygelius Bentham. 2/2 and their hybrid. Evergreen shrubs with 4-angled shoots, woody at the base, more or less herbaceous above. Leaves opposite, toothed. Flowers pendent, in terminal cyme-like panicles, branches mainly opposite, alternate at the tips

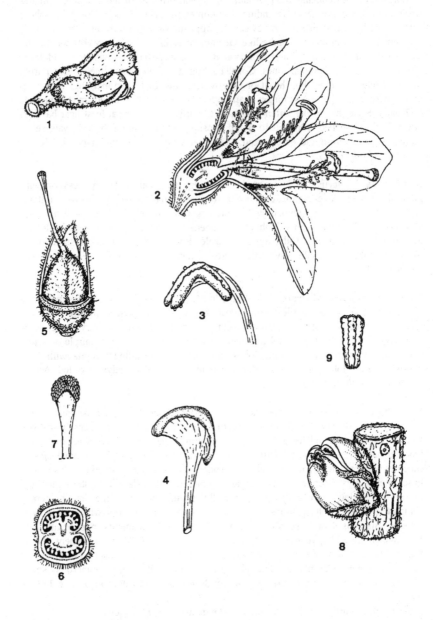

Figure 113. Scrophulariaceae. *Verbascum thapsus.* 1, Flower-bud. 2, Longitudinal section of flower. 3, Anther of lower stamen. 4, Anther of upper stamen. 5, Ovary. 6, Transverse section of ovary. 7, Stigma. 8, Fruit. 9, Seed.

496

of the inflorescence. Calyx 5-lobed. Corolla tubular, tapering to a swollen base, 5-lobed at the mouth. Stamens 4, protruding. *South Africa.*

7. Scrophularia Linnaeus. 200/4. Biennial or perennial herbs. Stems often 4-angled, angles sometimes winged. Leaves opposite (ours) or alternate. Flowers in cymes in bract-axils, forming terminal panicles or racemes. Calyx 5-lobed, equal, margin translucent or not. Corolla 2-lipped or more or less radially symmetric; tube short, spherical. Stamens 5, 4 fertile and 1 present or absent, sterile, scale-like staminode. Seeds many. *North temperate areas.*

8. Chelone Linnaeus. 4/4. Perennial herb. Stems erect, simple or sometimes branched, hairless. Leaves large, in pairs, ovate to lanceolate with marginal teeth, hairless. Flowers in simple terminal spikes, each subtended by 2 or 3 large sepal-like bracts which hide the short flower-stalks. Calyx deeply 5-lobed, ovate. Corolla-tube inflated, triangular in section, strongly 2-lipped, the upper lip notched and arched, the lower lip 3-lobed, all densely hairy. Fertile stamens 4, paired, filaments hairy; staminode 1, sterile, filament usually not hairy, shorter than the fertile stamens. Capsule broadly ovoid, *c*. 1 cm. Seeds flat, winged. *Eastern North America.*

9. Penstemon Schmidel. 250/60 and several hybrids. Shrubs or perennial herbs. Leaves opposite or occasionally in whorls of 3 or the upper alternate, lower usually stalked, upper stalkless. Inflorescence a raceme-like or cyme-like panicle. Calyx 5-lobed. Corolla tubular, almost radially symmetric to strongly 2-lipped, upper lip 2-lobed, lower 3-cleft. Stamens 4, paired, filaments arching. Staminode present, elongate, often dorsally hairy. Fruit a capsule, splitting through the cross-wall. Seeds numerous. *Mostly from North America but with 1 species from northeast Asia.*

10. Keckiella Straw. 7/4. Shrubs, stems woody at least at base. Leaves opposite or sometimes the upper ones alternate, leathery. Inflorescence a panicle, sometimes reduced to a spike-like raceme; bracts 5. Calyx of 5 essentially distinct, more or less equal segments. Corolla yellowish, whitish or red, glandular-hairy on the outside, strongly 2-lipped, upper lip divided into 2 lobes near the tip, lower lip with 3 reflexed lobes. Staminode well-developed, thread-like. Fertile stamens 4, in unequal pairs, filaments strongly hairy at base, anthers small and hairless. Stigmas united and head-like. Capsule brown, ovoid. *Western USA & Mexico (Baja California).*

11. Nothochelone Straw. 1/1. Perennial herb often woody at base, taproot short. Leaves opposite, all borne along the stem with the lower leaves much reduced, lanceolate to ovate, margins conspicuously toothed. Inflorescence with gland-like hairs, flowers borne in whorls. Calyx with 5 free lobes. Corolla rosy purple to light maroon, with gland-like hairs outside, 2-lipped, the lower lip greatly exceeding the upper. Stamens 4, anthers with long woolly hairs. *Northwest North America.*

12. Zaluzianskya Schmidt. 35/4. Sticky annuals and erect or clump-forming perennials. Leaves simple, entire or toothed, lower leaves opposite, upper leaves often alternate. Flowers with long slender tubes, produced in terminal spikes, some opening during the day, others at dusk, to display 5, deeply cleft, spreading petals. Calyx 5-toothed with deep clefts. Corolla 5-toothed, tube long. Stamens 4, 2 long and 2 short. Capsule 2-valved. *Southern Africa.*

13. Collinsia Nuttall. 25/1. Erect to spreading annuals, stems slender, hairless or hairy, often glandular. Leaves opposite, or in whorls of 3--5, ovate, oblong or lanceolate, sometimes softly hairy, lower shortly stalked, upper stalkless. Inflorescence a raceme of whorls of 1--8 axillary flowers; bracts present. Calyx 5-lobed, campanulate, lobes almost equal. Corolla tubular, white, pink, bicoloured or blue, 2-lipped, pouched at base, usually hairless outside, upper lip 2-lobed, double ridged at base, more or less reflexed, lower lip deflexed, 3-lobed, middle lobe shorter than laterals and folded longitudinally forming a pouch to enclose stamens and style. Stamens 4, in unequal pairs, attached unequally near throat base; staminode gland-like. Style elongate. Fruit a capsule, ovoid or spherical, valves 2-lobed. *North America, Mexico.*

14. Jovellana Ruiz & Pavón. 4/3. Perennial herbs or low shrubs; stems erect or ascending. Leaves opposite, simple, margin irregularly toothed. Inflorescence cyme-like, occasionally flowers solitary. Calyx 4-lobed, joined to the corolla at base. Corolla 2-lipped, lower lip inflated, without infolded lower lobes. Stamens 2. Capsule opening by 4 valves. *Chile, New Zealand.*

15. Calceolaria Linnaeus. 300/12. Annual or perennial herbs or shrubs. Leaves mostly opposite, sometimes whorled, basal or on the stems, usually simple, entire to deeply lobed or pinnatifid, stalks sometimes winged. Inflorescence cyme-like or flower solitary. Calyx-lobes 4. Corolla 2-lipped, the upper lip arched or hooded (rarely inflated), the lower lip inflated, with infolded lower lobes; yellow, rarely purple, the throat often red-spotted. Stamens 2. Fruit a dry capsule opening by 2 valves. *Mexico to southern South America.*

16. Alonsoa Ruiz & Pavón. 18/5 and some hybrids. Dwarf shrubs or branched perennial herbs in the wild, but often treated as annuals in cultivation. Stems 4-angled. Leaves opposite or whorled. Inflorescence terminal. Each flower inverted by a 180° twist in the stalk. Sepals 5. Corolla bilaterally symmetric, fused into a tube at the base, 5-lobed at the apex. Stamens 4. Fruit a capsule. *Mainly in the Peruvian Andes, but ranging from Mexico to Bolivia & Chile; 2 species in South Africa.*

17. Diascia Link & Otto. 50/16 and some hybrids. Annual herbs with slender rootstocks or perennial herbs with erect or decumbent stems, sometimes forming mats and rooting at the nodes. Leaves opposite, often diminishing in size up the stem, with or without stalks. Inflorescence a crowded or loose panicle. Flowers sometimes nodding; corolla with a short tube, usually with 2 lateral spurs with dark, stalkless glands within, limb 2-lipped, lower lip 3-lobed, upper 2-lobed with a translucent 'window' (occasionally the 'window' is split into 2 small patches). Stamens 4, some of them sometimes sterile. Ovary 2-celled, ovules 2--many in each cell. Capsule opening through the cross-wall; seeds usually strongly curved, ribbed, with raised net-veining or winged. *Southern Africa.*

18. Nemesia Ventenat. 65/4 and some hybrids. Annual or perennial herbs, woody at base. Leaves opposite. Flowers axillary or arranged in terminal racemes. Calyx 5-lobed, lobes scarcely overlapping. Corolla membranous, tube short, spurred or with a pocket or pouch, limb 2-lipped, mouth of spur with 2 rounded projections in the middle, usually with bright orange hairs; upper lip 4-lobed, lower entire or 2-lobed, with a palate almost enclosing throat. Stamens 4, filaments of lower pair sometimes curved around the upper pair. Ovary 2-celled. Capsule laterally compressed; seeds flattened, conspicuously winged. *South Africa, a few species from tropical Africa.*

19. Antirrhinum Linnaeus. 20/8. Dwarf shrubs or woody-based perennial herbs; stems erect or ascending, the lateral branches rarely twining. Lower leaves opposite, upper alternate, simple, circular to linear, pinnately veined, margin entire, stalked or stalkless. Flowers bilaterally symmetric, borne in racemes or solitary in leaf-axils. Flower-stalks not twining. Calyx deeply divided, lobes equal, margins entire. Corolla-tube broad, more or less cylindric, inconspicuously bulged at base, limb 2-lobed, basal palate prominent and closing mouth. Stamens 4, fertile, equal, not protruding from corolla, adjacent pairs of anthers joined at margin, not forming a ring-like structure. Capsule ovoid to almost spherical, cells unequal, many-seeded. *South Europe.* Figures 112, p. 492 & 114, p. 500.

20. Chaenorhinum (de Candolle) Reichenbach. 21/3. Annual herbs or woody-based perennials; stems erect or ascending, branches not twining. Leaves simple, pinnately veined, opposite and shortly stalked below, alternate and stalkless above. Flowers bilaterally symmetric, borne in loose terminal racemes; bracts like the leaves but smaller and narrower. Calyx deeply lobed, margins of the lobes entire. Corolla-tube cylindric, base with a straight spur, limb 2-lipped, the upper 2-lobed, the lower 3-lobed and with a low palate which does not close the mouth of the tube. Stamens 4, fertile, equal, anthers of adjacent pairs joined at margin and not forming a ring-like structure. Capsule with unequal cells, opening by flaps or a toothed pore. *Southwest Europe, North Africa & southwest Asia; widely naturalised in temperate countries.* Figure 112, p. 492.

21. Misopates Rafinesque. 7/1. Annual herbs with erect or ascending stems. Leaves opposite below, alternate above, simple, linear, elliptic or lanceolate, stalkless or with short stalks. Flowers borne in leafy racemes with bracts. Calyx deeply divided, lobes entire, unequal, usually exceeding corolla-tube. Corolla-tube cylindric or urn-shaped, limb 2-lipped. Fertile stamens 4, equal. Style simple, stigma capitate. Capsule ovoid to oblong, many-seeded. *Europe, north Africa, Atlantic Islands & southwest to south-central Asia.* Figure 112, p. 492.

22. Asarina Miller. 1/1. Perennial or woody-based herb; stems 9--50 cm, procumbent, trailing or ascending, simple or branched at base, with underground runners. Leaves 2.5--5 × 3--6.5 cm, opposite, ovate to heart-shaped or kidney-shaped, margin scalloped-toothed or lobed, stalks 3--7 mm. Flowers bilaterally symmetric, solitary in the leaf-axil, stalks 8--20 mm, spreading or almost erect. Calyx 1--1.4 cm, deeply divided, lobes linear-lanceolate or lanceolate, apex pointed, margin entire. Corolla 2-lipped, 3--4 cm, base of tube white with purple veins, lip yellow, palate a double bulge closing mouth of tube, pale or deep yellow, stalks becoming bent backwards and thickened in fruit. Stamens 4, fertile, equal, adjacent pairs of anthers joined at margin but not forming a ring-like structure. Capsule 8--12 mm, hairless. *Europe.*

23. Cymbalaria Hill. 10/5. Stoloniferous creeping perennials. Leaves almost circular, kidney-shaped, almost entire or bluntly lobed, and with palmate main veins. Flowers violet, white or purple, solitary, axillary. Calyx 5-lobed. Corolla 2-lipped, with narrow cylindric or conical spur at base, throat closed by a white or yellow boss-like swelling

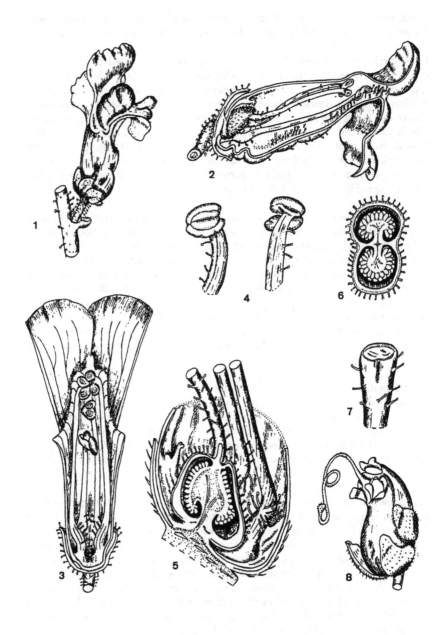

Figure 114. Scrophulariaceae. *Antirrhinum majus.* 1. Single flower. 2, Longitudinal section of flower (through vertical plane). 3, Longitudinal section of flower (through horizontal plane). 4, Stamen, front (L) and back (R) views. 5, Longitudinal section of lower part of flower. 6, Transverse section of ovary. 7, Stigma. 8, Fruit.

500

on the lower lip. Fruit a spherical capsule opening by apical slits. *Mediterranean area, southern Alps; some species much naturalised elsewhere.*

24. Anarrhinum Desfontaines. 8/1. Biennial or perennial herbs or dwarf shrubs; stems ascending to erect. Basal leaves often forming a rosette, stalked, pinnately veined or often with several veins from the base, margin entire, lobed or toothed; stem-leaves alternate, usually deeply dissected with 3--7 segments, sometimes undivided, margin of divisions entire or toothed. Flowers bilaterally symmetric, borne in racemes or panicles, stalked or almost stalkless; bracts simple or 3-fid, linear or linear-lanceolate. Calyx deeply or shallowly divided. Corolla-tube cylindric, spurred or bulging (pouched), limb 2-lipped, entire. Stamens 4, equal, included in corolla, anthers joined at edges and forming a ring-like structure. Capsule almost globe-like, splitting from apex more or less to base, many-seeded. *Mediterranean area.*

25. Linaria Miller. 150/14. Annual, biennial or perennial herbs, sometimes woody at base; stems prostrate, ascending or erect. Leaves simple, usually narrow, in whorls below, opposite above, margins entire. Flowers in terminal racemes or spikes, subtended by bracts. Calyx deeply 5-lobed. Corolla hairless except for the palate; tube cylindric with a basal spur; limb 2-lipped, the upper lip 2-lobed, the lower 3-lobed with a hairy palate which usually closes the mouth of the tube. Stamens 4, equal, included in corolla, opposing pairs of anther joined at margin and not forming a ring-like structure. Capsule more or less globe-shaped, with equal cells; seeds winged or wingless. *Europe, western Asia & North Africa.* Figure 112, p. 492.

26. Nuttallanthus Sutton. 4/1. Annual or biennial herbs, branches not twining; flowering stems erect, non-flowering stems prostrate or decumbent. Leaves simple, stalkless, pinnately veined, whorled or alternate, margin entire. Flowers bilaterally symmetric, borne in terminal racemes with bracts. Calyx deeply divided, lobes equal, shorter than corolla-tube, margin entire. Corolla-tube narrow, cylindric, with a very narrow spur or bulge at base, limb very unequally 2-lipped, the lower lip greatly exceeding the upper, the palate scarcely developed. Stamens 4, fertile, equal, included in corolla, adjacent pairs connected at the margins. Capsule with equal cells, each dehiscing by 4 or 5 flaps; seeds many, each with 4--7 ridges. *North America.*

27. Sairocarpus Sutton. 13/2. Annual or perennial herbs, stems erect or ascending, sometimes with bending or twining lateral branches in the inflorescence. Leaves opposite, alternate above, simple, margin entire, stalked or stalkless, pinnately veined, midrib ending in a small gland. Flowers solitary, borne in the leaf-axils or in terminal leafy racemes or panicles; flower-stalks straight, rarely twining; bracts leaf-like. Calyx deeply divided, margin entire. Corolla-tube cylindric, usually with pouch, 2-lipped, palate prominent; stamens 4, fertile, equal, adjacent pairs of anthers joined at margin. Capsule ovoid to oblong, cells unequal. *North America.*

28. Maurandya Ortega. 2/2. Twining perennial, woody at base. Leaves alternate, broadly triangular, rarely 5-lobed, margin entire, stalks twining. Flowers more or less bilaterally symmetric, borne in leaf-axils, stalks twining, becoming deflexed in fruit. Calyx-tube broad, compressed with 2 longitudinal folds, limb unequally 2-lipped, without basal palate. Stamens 4, fertile, opposing pairs of anthers joined at margin, not forming a ring-like structure, equal. Capsule oblong-spherical to oblong-ovoid, cells equal, many-seeded. *USA, central America.*

29. Maurandella (A. Gray) Rothmaler. 1/1. Perennial, woody-based herb; stems to 3 m, hollow, slender, climbing by twining leaf-stalks. Leaves 6--25 × 4--35 mm, alternate, broadly triangular to heart-shaped, 5-lobed, pointed at apex, hastate-sagittate at base, margin entire except basal lobes often mucronate, often with a lateral tooth near base; stalks 9--22 mm, narrowly winged at base. Flowers bilaterally symmetric, solitary in leaf-axils, stalks 1--2.5 cm, narrowly winged. Calyx deeply lobed, lobes 9--13 × 1.5--2.5 mm, larger in fruit, linear-lanceolate, pointed at apex. Corolla 1.6--2.3 cm, funnel-shaped, palate conspicuous almost closing mouth of corolla, spotted and veined yellow or white; basal spur not present; limb 2-lipped, dark blue to violet, rarely pink. Stamens 4, fertile, equal, the anthers joined at margin but not forming a ring-like structure. Capsule 7--9 mm, ovoid to oblong to spherical, cells unequal, hairless. *North America, Mexico.*

30. Epixiphium (A. Gray) Munz. 1/1. Hairless, annual herb, climbing by means of twining leaf-stalks; stems to 3 m, hollow. Leaves 1--4 × 1--3 cm, broadly triangular to heart-shaped, hastate or clasping stem at base, basal lobes with sharp points, upper leaves alternate, stalks 1--5.5 cm. Flowers bilaterally symmetric, solitary in leaf-axils. Calyx-lobes 1.2--1.8 cm × 1.5--4 mm, linear-lanceolate, deeply divided, almost equal, pointed at tip, bulging, margin entire, green in flower. Corolla 2.3--3.5 cm, funnel-shaped, bulging at base, without spur, curved below, pale bluish violet, basal palate partially closing mouth of corolla. Stamens 4, equal, anthers with adjacent pairs joined at margin and not forming a distinct ring-like structure. Capsule 1.2--2 mm, each cell opening by a broad solitary valve. *North America.*

31. Lophospermum D. Don. Few/3. Woody-based perennial herb, sometimes with tuberous roots; stems climbing by means of twining leaf- and flower-stalks, branches not twining. Leaves alternate, broadly triangular, heart-shaped, almost circular or almost kidney-shaped, margin toothed or scalloped to almost entire, palmately veined, the midrib not terminating with a gland. Flowers bilaterally symmetric, held horizontally, solitary or rarely in pairs in leaf-axils. Calyx narrowly funnel-shaped to urn-shaped, not inflated, deeply divided, green or sometimes tinged with red or purple. Corolla-tube broad, without a pouch, limb 2-lobed. Stamens 4, fertile, equal; anthers free. Capsule globe-shaped. *USA, Mexico & central America.*

32. Rhodochiton Otto & Dietrich. 3/1. Perennial, semi-woody herbs climbing by means of twining leaf-stalks. Leaves alternate, heart-shaped to broadly triangular, palmately veined, margin obscurely 5-lobed or with very small teeth or entire; stalks twining. Flowers almost radially symmetric, solitary in the leaf-axils, stalks pendent, not twining. Calyx broadly funnel-shaped or campanulate, somewhat inflated, shallowly divided with entire margins, shorter than the corolla-tube, brightly coloured with pink or purple. Corolla-tube broad, almost circular or somewhat 5-sided in section, with 4 shallow longitudinal folds, limb slightly 2-lipped, lobes entire. Fertile stamens 4, almost equal, free, protruding or included within the corolla-tube, connective not dilated; staminode small, with remains of anther. Style undivided, erect. Fruit a capsule, oblong to globe-shaped, many-seeded, opening longitudinally. Seeds with wart-like protuberance and longitudinal wings. *Mexico, Guatemala.*

33. Digitalis Linnaeus. 20/15 and some hybrids. Perennial (often short-lived in cultivation) or biennial herbs, rarely small shrubs, hairless or with glandular and non-glandular hairs. Stems often rising from a thickened rootstock. Leaves simple, alternate, sometimes mostly basal, margin entire or toothed. Flowers borne in terminal, often 1-sided racemes, bracts present. Calyx with 5 equal lobes, shorter than the corolla-tube. Corolla-tube inflated or campanulate, often constricted below, nodding; limb more or less 2-lipped, erect and spreading, the upper lip shorter than the lower, which is 3-lobed. Stamens 4, usually included. Fruit a capsule, ovoid to conical, opening through the cross-wall. Seeds numerous. *Mediterranean area to central Asia.*

The genus **Isoplexis** (Lindley) Loudon, is a small evergreen shrub, but is otherwise very similar. It is from the Atlantic Islands, and none of its species are really hardy in northwest Europe.

34. Ourisia Commerson. 20/10. Low-growing perennials, sometimes woody at base, often mat-forming with creeping rhizomatous shoots. Leaves opposite, mostly basal, margin entire or scalloped. Flowers axillary, solitary or in racemes which are sometimes umbel-like. Calyx usually deeply 5-lobed. Corolla-tube oblique or curved, lobes 5, spreading, overlapping in bud. Stamens 4, not protruding from corolla. Ovary 2-celled. Stigma capitate. Fruit a capsule, grooved on each side. *New Zealand, Australia (Tasmania), southern Andes of South America.*

35. Hemiphragma Wallich. 1/1. Slender creeping perennial, stems to 60 cm, sparsely downy. Leaves of 2 different kinds according to season: wet-season leaves opposite, shortly stalked, more or less circular, 5--20 ×6--15 mm, apex acute, base truncate, sparsely downy above, less so and pale green beneath; dry-season leaves in small, dense, ovoid-spherical clusters, rigid and needle-like, greyish green with marginal hairs. Flowers stalked, axillary, usually subtended by dry-season leaves. Calyx-lobes *c.* 2.5 mm, narrowly elliptic. Corolla shortly campanulate, with 5 spreading, almost equal lobes *c.* 2.5 mm, mostly pink, white towards the base. Stamens 4, inserted at the corolla-base but anthers reaching the mouth of the corolla-tube. Style shorter than corolla-tube. Fruit a fleshy capsule, green at first, later scarlet, ultimately blackish and shining. *Himalaya & mountainous parts of subtropical East Asia.*

36. Sibthorpia Linnaeus. 5/2. Slender, shortly hairy perennials with prostrate, creeping stems that root at the nodes. Leaves simple, stalked, kidney-shaped to almost circular, toothed. Flowers solitary in the leaf-axils. Calyx campanulate, 4- or 5-lobed; corolla more or less wheel-shaped, with short tube, 5-lobed. Stamens 3--5. Fruit a 2-celled capsule. *West Europe, Macaronesia, tropical Africa, Andes.*

37. Erinus Linnaeus. 2/1. Perennial herbs. Leaves alternate, simple. Inflorescence a terminal, bracteate raceme. Calyx divided almost to the base into 5 equal lobes. Corolla-tube cylindric; lobes 5, spreading, nearly equal. Stamens 4, in 2 equal pairs with the upper stamen the smallest of all, included in the corolla-tube. Stigma capitate with 2 lateral lobes. Fruit a 4-valved capsule, splitting through the cross-wall and the cells. Seeds numerous. *South Spain, Pyrenees & the Alps, North Africa.*

38. Picrorhiza Bentham. 1/1. Rhizomatous perennials, almost stemless, with stout woody rootstocks. Leaves in a basal rosette, elliptic, oblanceolate or spoon-shaped, conspicuously toothed, blackening when dry. Inflorescence a dense, spike-like raceme, with somewhat curled brown hairs; 1--4 small, alternate bract-like leaves present just below inflorescence; bracts oblong-lanceolate, brown-downy. Calyx-lobes overlapping in bud, narrowly lanceolate or lanceolate-elliptic, acute, brown-hairy. Corolla whitish, blue or violet, lobes 5, almost equal, slightly longer than tube, ovate or ovate-lanceolate, acuminate to gradually tapered, hairless outside, glandular and with marginal hairs within. Stamens not in unequal pairs; filaments protruding, all hairless; anther cells widely spreading greyish brown. Capsule ovoid, brown or blackish without glaucous bloom, more or less deeply grooved, opening through the cross-wall. Seeds *c.* 1.5 mm, with brown net-like markings and moderately glistening honeycomb-like pattern. *Himalaya.*

39. Synthyris Bentham. 14/3. Low-growing, tufted perennial herbs with a rhizomatous rootstock. Leaves radical, heart-shaped, kidney-shaped or deeply dissected, stalked. Inflorescence a narrow raceme, erect, unbranched. Calyx 5-lobed. Corolla lilac-blue or violet-blue, campanulate to wheel-shaped, 4-lobed, the upper 2 lobes united further than the lower 2. Stamens 2, protruding from corolla. Fruit a capsule, flattened, opening through the cells. Seeds disc-like. *Western North America.*

40. Besseya Rydberg. 9/2. Perennial herbs. Basal leaves heart-shaped to ovate-oblong, stalked, margin scalloped or toothed. Inflorescence a spike-like raceme, stalks with bract-like, alternate, stalkless leaves below. Calyx 2- or 4-lobed. Corolla 2-lipped, sometimes reduced or absent; limb notched almost to the base, violet-purple, yellow or white. Stamens 2, protruding from corolla. Fruit a flattened capsule which splits longitudinally along the cells. *Temperate North America.*
    Sometimes included in the genus *Synthyris.*

41. Wulfenia Jacquin. 5/4 and some hybrids. Rhizomatous perennial herbs, hairless or with simple or gland-like hairs. Leaves mainly basal, forming a rosette; stem-leaves alternate, undivided, stalked, margin scalloped, toothed or lobed. Flowers somewhat nodding, borne in terminal, spike-like racemes which are mostly one-sided; bracts present. Calyx with 5 equal lobes, linear to lanceolate. Corolla-tube cone-like with an obscurely 2-lipped limb; upper lip entire or notched, lower shortly 3-lobed. Stamens 2, inserted at apex of corolla-tube; filaments very short. Stigma capitate, notched. Capsule 4-valved, seed numerous. *Southeast Europe, west Asia, Himalaya.* Figure 112, p. 492.

42. Lagotis Gaertner. 30/1. Hairless perennial fleshy herbs with stout rhizomes. Stems absent or present, inflorescence on a scape. Leaves mainly basal, with winged leaf-stalks often dilated at the base. Inflorescence a terminal, bracteate, spike-like raceme. Calyx spathe-like, split to base on one side, 2-lobed, membranous. Corolla 2-lipped, purplish or white; upper lip entire or 2-lobed, lower 2- or 3-lobed. Stamens 2, more or less longer than corolla; anthers 1-celled, hairless, without mucros. Ovary with 2 ovules. Fruit hard, not opening; seeds 1 or 2. *Centred in China but extending westwards to eastern Turkey.* Figure 112, p. 492.

43. Paederota Linnaeus. 2/2. Perennials with simple stems to 30 cm, hairless or with white hairs. Leaves opposite, simple, with 3--10 pairs of teeth. Flowers borne in terminal spike-like racemes, bracts present. Calyx unequally 5-lobed. Corolla yellow or violet-blue (rarely pink), tube cylindric, limb 2-lipped, upper lip entire or rarely

2-lobed, the lower 3-lobed. Stamens 2, protruding or as long as corolla-tube. Stigma capitate. Capsule with many cells; seeds numerous. *European Alps.*

44. Hebe Jussieu. 100/45 and some hybrids. Evergreen shrubs, often of compact habit. Branchlets hairless or more often hairy, the hairs frequently confined to 2 lines. Leaves opposite, stalkless or short-stalked, often 4-ranked, entire, toothed or scalloped and, in some species (the 'whipcord' Hebes), small and scale-like. Juvenile leaves toothed or pinnatifid. Leaf-margins usually closely adpressed in bud, but in species with stalked leaves the leaf-bases are separated, leaving a variably shaped sinus. Flowers few to many, in stalked axillary or terminal racemes, spikes or panicles. Calyx 3--5-lobed, the lobes sometimes partly united. Corolla 4-lobed, bilaterally symmetric, white, pale or deep violet or pink. Stamens 2. Ovary 2-celled. Capsule ovoid or nearly spherical, usually rather flattened, with a cross-wall in most species across the wider dimension, dehiscing at the apex and partially along the cross-wall. Seeds numerous, more or less flattened. *Mostly endemic to New Zealand, a few in southeastern Australia & South America.*

45. Parahebe Oliver. 30/7. Prostrate or decumbent shrublets or herbaceous perennials. Branches often with dense brown, grey or white hairs, on older internodes hairs sometimes in 2 longitudinal lines, becoming hairless and covered with scars. Leaves simple, opposite with each pair at right-angles to the last, leathery, usually toothed or scalloped, stalked, sometimes shortly so, stalk clasping stem. Flowers in racemes, mainly in the upper leaf-axils, occasionally flowers solitary. Bracteoles present. Calyx with a short tube and 4 lobes, rarely 5. Corolla usually with 4 (rarely 5) lobes, the upper larger than the others, tube short. Stamens 2, inserted on either side at base of upper lobe. Stigma capitate. Ovary 2-celled. Capsules with the cross-wall at the narrowest diameter, more or less 2-lobed, occasionally separating into 4 valves. Seeds numerous, flattened circular. *New Zealand, New Guinea & Australia.*

46. Chionohebe Briggs & Ehrendorfer. 6/4. Compact, evergreen, cushion-forming perennials, small prostrate herbs or dwarf shrubs with woody stems. Leaves dense, leathery, overlapping, hairy or with marginal hairs, opposite, margin entire, stalkless. Flowers solitary, terminal, subtended by a pair of united bracts. Calyx 5-, rarely 6-lobed. Corolla with 5 spreading lobes, tube salver- or funnel-shaped, one-third to two-thirds of corolla length. Stamens 2, attached near to throat of corolla. Fruit a capsule, slightly laterally compressed, broadest at its apex, deeply 2-lobed, splitting at cell walls and septum into 4 valves. Seeds many. *New Zealand, Australia.*

A genus better known to gardeners as *Pygmaea* J.D. Hooker.

47. Veronica Linnaeus. 250/31. Perennials (ours), usually herbaceous, but sometimes woody at the base or dwarf shrubs. Leaves opposite at least below, more rarely in whorls of 3. Flowers usually in terminal or axillary racemes, more rarely borne singly in the leaf-axils. Calyx 4- or 5-lobed. Corolla flat, disc-shaped, unequally 4-lobed, bilaterally symmetric, blue, white or pink. Stamens 2, diverging. Ovary 2-celled with axile placentation. Capsule 2-celled, with a single persistent style. Seeds usually numerous. *Mainly Eurasia.* Figure 112, p. 492.

48. Veronicastrum Moench. 2/2. Like *Veronica*, but leaves in whorls of 3--6; calyx usually 5-lobed (rarely 4-lobed); corolla-limb not flat, the 4 more or less equal lobes forwardly directed. *Eastern North America, eastern Asia.*

49. Rhinanthus Linnaeus. 45/2. Annual hemi-parasitic herbs on grasses, with opposite, stalkless, toothed or entire leaves. Flowers in leafy-bracted terminal spikes. Calyx flattened and inflated in flower, enlarging in fruit, with 4 entire teeth. Corolla-tube yellow, upper lip laterally compressed, with 2 teeth, lower lip with 3 lobes. Stamens 4, included in the upper lip. Capsule compressed with a few large winged seeds. *North temperate areas, especially Europe.* Figure 112, p. 492.

50. Lathraea Linnaeus. 7/2. Herbaceous root-parasites of trees, without chlorophyll. Rhizomes creeping, covered with overlapping scales and bearing rootlets which become swollen at the point of attachment with the host. The flowers are borne singly in the axils of scales. Calyx campanulate, equally 4-lobed above. Corolla 2-lipped, the upper lip strongly concave, entire, the lower smaller and 3-lobed. Stamens slightly protruding from the corolla. Fruit a capsule. *Temperate Europe & Asia.*

Plants may be grown by planting seed close to a susceptible host-plant (generally *Salix* or *Populus*). The genus is often placed in the family *Orobanchaceae.*

51. Hebenstretia Linnaeus. 40/2. Shrubs, annual or perennial herbs, sometimes woody based. Leaves mostly alternate, the lower opposite. Flowers in dense, terminal spikes with bracts. Calyx with membraneous margins, entire or notched. Corolla divided for half its length into 4 lobes, the tube slender. Stamens 4, equal, inserted at base of corolla-limb. Ovary 2-celled. Fruit oblong or ovate. *Africa, most in South Africa.*

52. Selago Linnaeus. 150/1. Shrubs, perennials (sometimes woody-based), or annual herbs. Leaves solitary or grouped together, alternate or the lower opposite. Flowers borne in spikes, corymbs or panicles, bracts present. Calyx shortly or deeply 5-lobed, sometimes with membranous margins, sometimes ciliate. Corolla 5-, rarely 4-lobed, tube cylindric or funnel-shaped. Stamens equal. Ovary 2-celled, with a single ovule in each cell. Seeds flat on one face and concave on the other. *Africa, most in South Africa.*

179. GLOBULARIACEAE

Perennial herbs and small shrubs; usually evergreen. Indumentum of glandular hairs consisting of a single-celled stalk and a 2- or 4-celled tip. Leaves alternate, simple, entire, lacking stipules, often forming basal rosettes. Flowers hypogynous, bilaterally symmetric, closely grouped in a spherical (or rarely elongate) head surrounded by an involucre. Calyx 2-lipped, united into a tube with 5 acute or acuminate teeth. Corolla blue (rarely pink or white), 2-lipped, upper lip divided into 2 segments, lower lip longer, more or less deeply bilobed. Stamens 4, protruding, borne on the corolla-tube and alternating with lobes. Ovary superior, with 1 pendent ovule in a single cell. Style protruding, with a 2-lobed stigma. Fruit borne within persistent calyx.

A family of 10 genera and about 250 species centred in the Mediterranean area, western Asia and southern Africa. It is similar to the *Scrophulariaceae*, and now sometimes includes the family *Selaginaceae.*

1. Globularia Linnaeus. 20/13. Small hummock forming shrubs or low-growing perennial herbs. Leaves persistent, evergreen; usually steel-blue flowers. *Mediterranean area.*

180. BIGNONIACEAE

Trees, shrubs or woody climbers, rarely herbs, the climbers frequently twining. Leaves opposite or whorled, rarely alternate, usually pinnate or ternate, rarely simple or palmate; the terminal leaflet sometimes replaced by a tendril. Leaf-stalks often with basal glands; stipules absent, pseudostipules sometimes present. Flowers usually large and showy, solitary or in racemes or cymes, terminal or lateral, bisexual, bilaterally symmetric. Bracts and bracteoles inconspicuous. Calyx and corolla tubular below, usually 5-lobed above, sometimes 2-lipped. Corolla larger than calyx, funnel-shaped or campanulate. Stamens 4 (often 1 pair long and 1 pair short), rarely 2 or 5, borne on corolla-tube, arched under upper lip, anthers 2-celled; 1 or 3 staminodes often present. Ovary superior, usually 2-celled, with nectar-secreting disc round base, sometimes shortly stalked; ovules numerous, axile. Style long; stigma 2-lobed, sometimes touch-sensitive. Fruit usually a dry dehiscent capsule with flat, papery-winged seeds, less often a dryish or fibrous berry with unwinged seeds. See figures 115, p. 508 & 116, p. 510

About 120 genera, mostly from the humid tropical forests of the New World, also represented in Africa, Asia and Australia.

1a. Herbs, trees or shrubs, not climbing	2
b. Climbers, often twining or with tendrils	7
2a. Herbs, stems not woody below	3
b. Trees or shrubs; stems woody at least near base	4
3a. Leaves palmately divided (leaflets pinnately lobed)	**9. Argylia**
b. Leaves pinnately divided or lobed	**10. Incarvillea**
4a. Leaves pinnately divided	**3. Tecoma**
b. Leaves simple or palmately divided	5
5a. Shrubs to 5 m	**3. Tecoma**
b. Trees	6
6a. Fertile stamens 2	**2. Catalpa**
b. Fertile stamens 4	**1. Paulownia**
7a. Leaves bipinnate or bipinnately lobed	**8. Eccremocarpus**
b. Leaves once-pinnate	8
8a. Leaves usually with 2 leaflets and a terminal tendril, sometimes with 3 leaflets	9
b. At least some leaves on each plant with 5 or more leaflets (early leaves sometimes with 3 leaflets); tendrils absent	10
9a. Flowers pale to bright yellow	**4. Macfadyena**
b. Flowers orange, scarlet, red or purple	**5. Bignonia**
10a. Stamens not projecting	**7. Campsis**
b. At least 1 pair of stamens projecting from flower	11
11a. Corolla-lobes almost circular, spreading; 1 pair of stamens projecting, the other pair not projecting	**6. Campsidium**
b. Corolla-lobes ovate to oblong, the lower strongly recurved; all stamens projecting	**3. Tecoma**

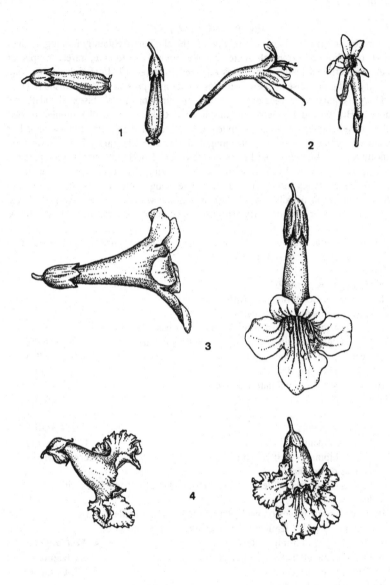

Figure 115. Bignoniaceae: corolla-shapes, from side (L) and front (R). 1, *Eccremo-carpus*. 2, *Tecoma*. 3, *Campsis*. 4, *Catalpa*.

1. **Paulownia** Siebold & Zuccarini. 7/4. Trees, deciduous in temperate areas but evergreen in tropics. Young bark smooth, with conspicuous lenticels. Branches opposite. Terminal buds absent. Leaves opposite, occasionally in whorls of 3, long-stalked; blade entire or with 3--5 shallow lobes. Inflorescence a large compound panicle of cymes. Cymes stalked or stalkless, with 1--8 (usually 3--5) flowers. Calyx campanulate or conical, hairy; lobes 5, almost equal but uppermost enlarged. Corolla purple, lilac or white, tubular to funnel-shaped or campanulate, 2-lipped, sometimes with conspicuous folds; lower lip 3-lobed, upper lip yellow and 2-lobed; Stamens 4, in 2 unequal pairs; anthers diverging. Ovary 2-celled. Fruit a capsule with numerous, small, winged seeds. *Laos, Vietnam & especially China, extending as far north-east as Manchuria.*

2. **Catalpa** Scopoli. 11/5 and a few hybrids. Deciduous trees. Leaves opposite or in whorls of 3, usually very large, entire or 3-lobed, often with red glandular patches in the vein-axils beneath, 3--5-veined at base, long-stalked. Flowers in terminal panicles, racemes or corymbs. Calyx irregularly splitting or 2-lipped. Corolla campanulate, 2-lobed, with 2 smaller lobes above and 3 smaller lobes beneath. Fertile stamens 2, curved. Ovary 2-celled. Style slightly longer than stamens. Fruit a long, tubular capsule with numerous flat seeds. Seeds *c.* 2.5 cm, each with a tuft of hairs at each end. *North America, the West Indies and eastern Asia.* Figures 115, p. 508 & 116, p. 510.

3. **Tecoma** Jussieu. 14/5. Small trees, shrubs or woody-based herbs, rarely climbers. Leaves simple or pinnate with a terminal leaflet, margin with forward-pointing teeth. Inflorescence a terminal raceme or panicle. Calyx cup-shaped, with 5 triangular, often apiculate lobes. Corolla tubular to campanulate or narrowly funnel-shaped, red, orange or yellow, usually hairless outside. Stamens 4, protruding or included, anthers hairless or hairy. Ovary narrowly cylindric, with 2 chambers in each cell; disc cup-shaped to cylindric. Fruit a flattened, linear capsule, hairless. Seeds numerous, 2-winged. *Africa, New World tropics.* Figure 115, p. 508.

4. **Macfadyena** de Candolle. 4/1. Climbers; branches with glandular patches between nodes. Leaves with 2 leaflets and a terminal, 3-fid, claw-like tendril. Inflorescence an axillary cyme or panicle, often reduced to 1--3 flowers. Calyx truncate or irregularly lobed to 2-lipped or spathe-like. Corolla tubular-campanulate, yellow, hairless outside. Anthers hairless. Fruit a narrow, linear, flattened capsule. Seeds 2-winged. *Mexico, West Indies to Uruguay.*

5. **Bignonia** Linnaeus. 1/1. Evergreen to deciduous, hairless climber to 20 m, with tendrils; stems ribbed. Leaves opposite, pinnate, with 3-fid terminal tendril clinging by discs; leaflets usually 2, 5--15 cm, oblong to lanceolate, stiff, entire; apex obtuse, acuminate; base cordate. Flowers axillary, solitary or 2--5 per cyme. Calyx campanulate, shallowly lobed. Corolla 4--5 cm, orange to scarlet, 2-lipped, widely trumpet-shaped; lobes rounded, *c.* 1--1.5 cm. Stamens not projecting, 1 pair long and 1 pair short, hairy below; anthers hairless. Ovary stalkless or almost so, with 1--3 rows of ovules. Capsule to 17 cm, linear, flattened. Seeds elliptic, winged. *North America.*

6. **Campsidium** Seemann. 1/1. Evergreen climber to 15 m. Leaves opposite, 10--15 cm, with 5--17 leaflets; leaf-stalks 1--3 cm; leaflets oblong-elliptic, entire or slightly toothed towards apex, stalkless, 8--4 × 5--15 mm, apices rounded to almost acute,

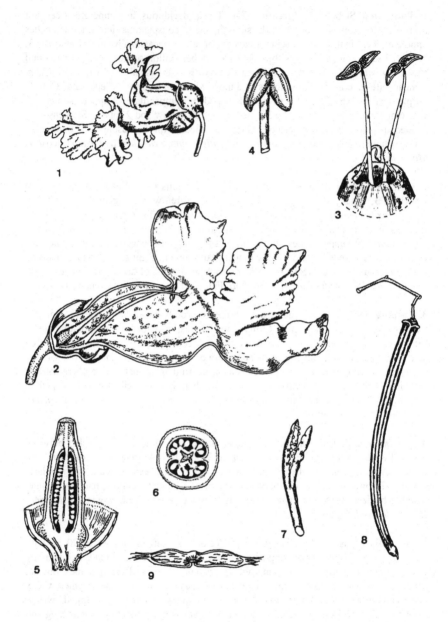

Figure 116. Bignoniaceae. *Catalpa bignonioides.* 1, Single flower. 2, Longitudinal section of flower. 3, Part of corolla-tube folded back to show two stamens and two staminodes. 4, Stamen. 5, Longitudinal section of ovary. 6, Transverse section of ovary. 7, Style with two-lobed stigma. 8, Fruit before opening. 9, Seed.

510

bases rounded to tapering, hairless apart from scattered reddish scales. Inflorescence 6--10 cm, a few-flowered terminal raceme, hairless or very slightly scaly; bracts linear to awl-shaped. Calyx 6--12 × 7--9 mm, green, campanulate, with 5 mucronate teeth, persistent, hairless apart from a few scales. Corolla 5-lobed, 3--3.5 cm × 5--8 mm, red, tubular, slightly constricted at apex, hairless except for some soft hairs inside and a few hairs round the stamens; lobes slightly hairy inside. Stamens 4, 2 protruding, 2 reaching corolla-mouth or just below it. Style stout; stigma 2-lobed. Ovary ovoid, slightly glandular-downy. Fruit 5.5--10 × 2.3--3 cm, a narrow elliptic-oblong capsule. Seeds thin, elliptic, surrounded by a brownish membranous wing. *Chile, Argentina.*

7. Campsis Loureiro. 2/2 and their hybrid. Deciduous climbers, often with clinging aerial roots at the nodes. Leaves opposite; leaflets 7--11, toothed, acuminate, hairless or hairy on veins beneath; tendrils absent. Flowers in terminal cymes or panicles. Calyx tubular to campanulate, with 5 unequal teeth. Corolla funnel-shaped, curved, wider above calyx, slightly 2-lipped; lobes rounded, widely spreading. Stamens 4 (2 long and 2 short), not projecting. Disc large. Ovary 2-celled. Capsule stalked, long and narrow. Seeds numerous. *North America, China, Japan.* Figure 115, p. 508.

8. Eccremocarpus Ruiz & Pavón. 6/1. Evergreen or deciduous climbers. Leaves bipinnate or bipinnately lobed, with terminal tendrils. Inflorescence a terminal raceme. Calyx campanulate, 5-lobed. Corolla tubular, entire or 2-lipped, scarlet, orange or yellow; tube long and narrow; widening above, throat swollen but abruptly contracted at mouth. Fruit an ovate to elliptic capsule, valves joined at top after splitting. *South Colombia & Peru to south Argentina.* Figure 115, p. 508.

9. Argylia D. Don. 12/1. Perennial herbs with erect viscid stems arising from a swollen stem base. All parts hairy. Leaves alternate, palmately divided into 5--9 incised, pinnatifid leaflets. Inflorescence-stalks axillary, ending in a raceme subtended by bracts. Calyx 5-parted. Corolla tubular, swollen on one side, bilaterally symmetric, 2-lipped, with 5 rounded lobes. Stamens 4, not projecting, staminode sometimes present. Style thread-like, stigma 2-lobed. Fruit a capsule, cells 2; seeds with or without membranous margins. *South America.*

10. Incarvillea Jussieu. 16/12. Annual or perennial herbs with tuberous or woody roots. Leaves in rosettes or on leafy stems to 2 m, mostly alternate, pinnate or pinnatisect with up to 11 pairs of leaflets and a terminal leaflet; margin entire or with forward-pointing teeth; upper leaves often tapered to the bases, hairy at least on veins. Inflorescence a terminal raceme or panicle with 1--many flowers; bracts and bracteoles present. Calyx-tube campanulate, 5-toothed. Corolla tubular, 5-lobed, bilaterally symmetric. Stamens 4, not projecting from corolla-tube; anthers clasping the style. Style hairless; stigma 2-lobed. Ovary to 1 cm, cylindric, 2-celled. Fruit a cylindric, 4-angled or 6-winged capsule. Seeds winged or hairy at apices. *Himalaya, central and eastern Asia.*

181. ACANTHACEAE

Herbs, sometimes woody at base, or shrubs. Stems and leaves often dotted or streaked with cystoliths. Leaves usually opposite, rarely basal, mostly entire, sometimes spiny. Stipules absent. Flowers bisexual, bilaterally symmetric, usually in spikes or racemes. Bracts and bracteoles usually present, often coloured, large and enclosing flowers.

511

Calyx 4- or 5-lobed, rarely 2-lipped, mostly shorter than corolla. Corolla 1- or 2-lipped or 5-lobed, tube cylindric, funnel-shaped or campanulate; when 2-lipped, upper lip usually entire, notched or 2-toothed; lower lip 3-lobed or -toothed. Stamens usually 4, in 1 long and 1 short pair, alternating with corolla-lobes, anthers 1- or 2-celled; when 2-celled, the cells more or less equal, borne at the same level, or separated by a development of the connective when they are often of different sizes and borne at different levels. Ovary superior, 2-celled, usually with 2 ovules in each cell. Stigma simple or equally or unequally 2-lobed. Fruit an often explosively dehiscent capsule. Seeds usually 4, often flattened, in most genera borne on hook-like appendages from the funicles and placenta (retinacula, jaculators) which forcibly eject the seeds at dehiscence. See figure 117, p. 514.

A pantropical family of about 250 genera and 2600 species with four main centres of distribution: Indo-Malaysia, Africa, Brazil and Central America. Only 2 genera have hardy representatives.

1a. Bracts conspicuously spiny, each with a terminal spine and up to 10
    narrowly triangular sharply spine-like teeth on either side   **2. Acanthus**
  b. Bracts without spines   **1. Ruellia**

1. Ruellia Linnaeus. 105/9. Evergreen perennial herbs woody at base, or shrubs. Cystoliths present. Leaves opposite, usually entire, some shallowly toothed or wavy, stalkless or stalked, apex acuminate or obtuse. Flowers showy, often brightly coloured, solitary or in axillary clusters, cymes or terminal panicles. Bracts 1 or absent, narrow, often thread-like, without spines; bracteoles 2 or absent, threadlike. Calyx deeply 5-lobed, lobes mostly equal. Corolla radially symmetric or almost so, blue to violet, white or red, often conspicuously veined, tubular to funnel-shaped, sometimes pouched; lobes 5, obtuse, ovate or rounded, spreading, almost equal, twisted in bud; tube straight, curved or abruptly bent. Stamens 4, fused at base into 2 pairs, joined to corolla-tube; anthers 2-celled, oblong, symmetric. Stigma with 2 unequal lobes. Fruit a cylindric or ovoid capsule, not laterally compressed. Seeds 4--20 per capsule, compressed, discoid, with water-absorbing hairs on margins. *Tropical America & temperate North America, Africa & Asia.*

2. Acanthus Linnaeus. 22/5. Perennial herbs to shrubs or small trees; herbaceous species (ours) usually with short rhizomes. Basal rosette-leaves simple to deeply pinnatifid or pinnatisect, non-spiny to densely spiny, usually with long stalks; inflorescence-stem leaves smaller, these progressively reduced from lower to upper stem and opposite or sometimes (in larger species) alternate. Inflorescence a dense spike with conspicuous spiny bracts. Bracts broadly ovate-concave, with a terminal spine and up to 10 narrowly triangular, sharply spine-like teeth on each side. Calyx 4-parted, the upper lip large and about equalling the corolla and somewhat hooded over it (appearing like an upper lip to the corolla), lateral lobes linear and inconspicuous, lower lip large and slightly shorter than the corolla, usually 3-toothed at the tip. Corolla with a very short thick cartilaginous tube; upper lip absent, lower lip with an oblong claw and lower lip with 3--5 lobes. Stamens 4, filaments stout, cartilaginous, curved; anthers large, at the throat of the flower, each with a conspicuous brush of hairs on one side. Style elongate. Capsule 4-seeded, tardily and irregularly if at all dehiscent. *Himalaya, to Australia, tropical Africa.* Figure 117, p. 514.

512

182. PEDALIACEAE

Annual or perennial herbs, often sticky; hairs shortly stalked with a head of 4 or more mucilage-filled cells. Leaves opposite, occasionally spirally arranged above, mostly simple; stipules absent. Flowers bisexual, bilaterally symmetric, usually solitary, borne in the leaf-axils; stalks mostly with glands near their bases. Calyx 5-lobed; corolla 5-lobed, tubular, sometimes with a basal spur; stamens usually 4, in 2 pairs, borne on the corolla-tube; ovary superior, 4-celled, style terminal, solitary, placentation axile. Fruit a 2-valved capsule, drupe or nut, often armed with horns, hooks or prickles.

A family of about 16 genera, native to the Old World tropics and subtropics, although some are naturalised in the New World. Only 1 genus is hardy in our area.

1. Sesamum Linnaeus. 16/1. Annual or perennial herbs, stems erect or creeping. Leaves stalkless or stalked, entire, lobed or divided, often on the same plant. Flowers solitary, in the leaf-axils. Calyx persistent or deciduous. Corolla obliquely campanulate, obscurely 2-lipped, the lowest lobe longer, white, pink or mauve, sometimes spotted within. Ovary 2-celled, almost cylindric. Capsule oblong or almost conical, grooved, with a beak at the apex. Fruit a capsule; seeds numerous, obovate, compressed, winged or wingless. *Africa, India & Sri Lanka.*

183. MARTYNIACEAE

Annual or perennial herbs, sometimes with tuberous roots. Stems viscid, with glandular hairs. Leaves opposite below, alternate above, simple. Stipules absent. Inflorescence a terminal raceme. Bracts small, deciduous; bracteoles 1 or 2. Flowers bisexual, bilaterally symmetric. Calyx of 5 free or partly joined sepals. Corolla tubular, campanulate or spreading, 5-lobed. Disc annular. Ovary superior, 1-celled; placentation parietal with a false wall; ovules few--many. Fruit with a fleshy exocarp which soon disintegrates exposing a woody endocarp with woody appendages, including apical horns partly derived from the style. Seeds rough, black.

A family of 3 or 4 genera containing 17--20 species from tropical and subtropical America, widely naturalised elsewhere. Many authors now include this family in the *Pedaliaceae*. Only 1 genus is hardy in our area.

1. Proboscidea Keller. 10/2. Annual or perennial herbs, strongly scented and covered with glandular hairs. Leaves opposite with long stalks. Inflorescence terminal. Flowers bisexual. Calyx with 4 or 5 teeth, split to the base on lower side, deciduous. Corolla campanulate to funnel-shaped, limb oblique, lobes 5, unequal, spreading. Stamens 4, included. Fruit a 2-celled capsule, (apparently 4-celled), crested at least below with a prominent incurved apical beak, endocarp woody, sculptured, exocarp fleshy, eventually separating from endocarp; horns as long as or longer than main body of fruit. Seeds numerous tuberculate. *North & South America.*

184. GESNERIACEAE

Epiphytic or terrestrial herbs, shrubs, occasionally climbers, rarely trees; sometimes with tubers, rhizomes (sometimes scaly), or stolons. Leaves usually opposite, rarely whorled or spirally arranged, or rarely a single basal leaf per plant, equal to strongly unequal, usually simple, rarely pinnatifid, entire to variously toothed. Flowers solitary or in cymes (flowers often paired) or racemes, bisexual. Calyx 4- or 5-lobed, or of 4 or 5 free sepals. Corolla 5-lobed, cylindric, occasionally bulging or spurred at base, sometimes pouched at throat, sometimes laterally compressed, 2-lipped or almost

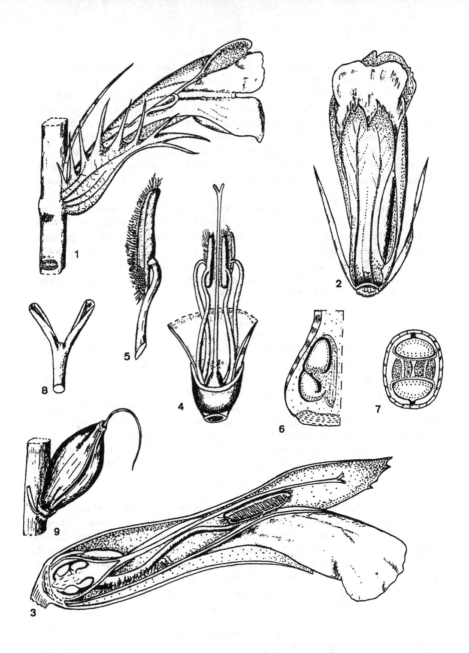

Figure 117. Acanthaceae. *Acanthus mollis.* 1, Side view of flower showing toothed bract. 2, Flower from above. 3, Longitudinal section of flower. 4, Calyx and corolla cut away to reveal stamens and ovary. 5, Anther. 6, Longitudinal section of half of the ovary. 7, Transverse section of ovary. 8, Apex of style with two stigmas. 9, Mature fruit.

radially symmetric; lobes 5, erect or spreading to reflexed, equal or unequal. Stamens borne on the corolla-tube, usually 4 in 2 unequal-sized pairs (upper stamen absent) or 2, or rarely 5, when alternate with corolla-lobes; staminodes 1--3, each replacing a missing stamen. Anthers variously shaped, joined at first, dehiscing by pores or longitudinal slits. Disc, if present, a ring of 1--5 separate or united glands. Ovary superior, half-inferior or inferior, with a simple terminal style and a simple or variously bilobed, sometimes mouth-shaped stigma, 1- or rarely 2-celled, ovules numerous, placentation usually parietal. Fruit a berry or a dry or fleshy capsule. Seeds numerous, small, sometimes with hair-like appendages at both ends.

A family of 150 genera and some 3700 species. Mostly they are tropical and sub-tropical, though there are a few genera in temperate Eurasia.

1a. Fertile stamens 2; staminodes 0 or 2 or 3   2
  b. Fertile stamens 4 or rarely 5, staminodes 0 or 1   4
2a. Small shrub; corolla urn-shaped and lop-sided, lobes crimson   **4. Sarmienta**
  b. Plant and corolla not as above   3
3a. Each filament with a prominent tooth at the apex; plant arising from a scaly rhizome   **12. Lysionotus**
  b. Filaments without teeth at the apex; scaly rhizome absent
      10. Didymocarpus
4a. Ovary distinctly 2-celled, a definite cross-wall present   **1. Asteranthera**
  b. Ovary 1-celled, sometimes appearing 2-celled by the meeting of the intrusive placentas in the middle (squeezing the ovary shows that these are not joined to each other)   5
5a. Well-developed, leaf-bearing stems present   6
  b. Stems absent, leaves all in basal rosettes, flowers or inflorescences on scapes   7
6a. Stems rough with stiff, conical hairs; flowers solitary, corolla-tube orange with red veins, lobes spreading, ciliate   **3. Rhabdothamnus**
  b. Stems, flowers and corolla-tubes not as above   **2. Mitraria**
7a. Flowers with very short tubes and flat, salver-shaped limbs   8
  b. Flowers with cylindric to almost bell-shaped tubes, the limb shorter, spreading or erect-spreading   9
8a. Anthers all distinct   **8. Ramonda**
  b. Anthers all united into a conical cap   **11. Conandron**
9a. Rhizomes present   10
  b. Rhizomes absent   11
10a. Rhizomes scaly; corolla orange to yellow-green; filaments not coiling after pollen is shed   **9. Briggsia**
  b. Rhizomes not scaly; corolla white to purplish or bluish; filaments coiling spirally after pollen is shed   **5. Corallodiscus**
11a. Anthers all distinct, neither joined nor touching at their tips   **7. Jancaea**
  b. Anthers united in pairs by their tips   **6. Haberlea**

1. Asteranthera Klotzsch & Hanstein. 1/1. Evergreen slender creeper or woody climber to 5 m (in the wild); young shoots pale- or whitish-hairy. Leaves opposite, those of each pair unequal, ovate-circular to ovate-elliptic, 1--4 cm × 6--20 mm, scalloped to toothed, hairy, dark green above, paler beneath. Flowers usually solitary, occasionally in pairs, axillary, stalks to 5 cm each with a very small leaf-like bract at the middle. Calyx 5-lobed, *c.* 1 cm, hairy, lobes toothed. Corolla tubular, to 6 cm, bright scarlet-red, sometimes blotched yellowish, limb 2-lipped, upper lip of 2 lobes, lower of 3, mouth to 3 cm across. Stamens 4, protruding, anthers united. Ovary superior, distinctly 2-celled, with a cross-wall. Nectary very poorly developed, ring-like. Fruit a fleshy capsule, green marked with red. *Southern Chile, Argentina.*

2. Mitraria Cavanilles. 1/1. Evergreen woody-based perennial herb or small shrub, stems climbing or straggling to 2 m, hairless or shaggy-hairy. Leaves 1--3 cm, opposite, those in each pair equal, ovate, acute, leathery and shiny, toothed, hairy, dark green. Flowers solitary, axillary on stalks 2--5 cm; bract large, 2-lobed. Calyx of 5 free sepals, 7--12 mm. Corolla tubular, much longer than calyx, to 5 cm, tube inflated about halfway, narrowing at the 5-lobed mouth. Stamens 4, inserted at the bottom of the corolla-tube, anthers united in pairs. Ovary superior, 1-celled. Nectary poorly developed, ring-like. Fruit a capsule, 8--15 mm, green flushed red. *Chile, Argentina.*

3. Rhabdothamnus Cunningham. 1/1. Much-branched shrub to 2 m, branches opposite, rough-hairy, with stiff, conical, several-celled hairs. Leaves opposite, those in each pair more or less equal, 1--5 × 1.5--5 cm, broadly ovate to almost circular, coarsely toothed, grey-green because of the hair-covering, stalks slender, to 1 cm. Flowers solitary from the leaf-axils, stalks to 4 cm. Calyx deeply 5-lobed, to *c.*1 cm, lobes ovate to triangular-acuminate, downy. Corolla with a more or less campanulate tube, orange with red veins, the limb 5-lobed with more or less equal, spreading, ciliate lobes. Stamens 4, slightly protruding from corolla, staminode 1. Ovary superior, 1-celled. Style curved at apex, to *c.* 1.5 cm; nectary ring-like. Fruit an ovoid capsule, 7--10 mm, tapering to the apex, enclosed in the persistent calyx. *New Zealand.*

4. Sarmienta Ruiz & Pavón. 1/1. Small, creeping, epiphytic shrub, semi-prostrate with stems to 15 cm which root at the nodes. Leaves opposite, those in each pair more or less equal, somewhat fleshy, 1--2.5 cm, with few teeth or small lobes near the apex. Flowers solitary from the leaf-axils. Calyx 5-lobed. Corolla 2--3 cm, pendent, tube urn-shaped, lop-sided, lobes short, all crimson. Stamens 2, projecting well beyond the corolla. Ovary superior, 1-celled. Fruit a capsule. *Chile, Argentina.*

5. Corallodiscus Batalin. 18/2. Perennial herbs with non-scaly rhizomes. Leaves in basal rosettes, hairless to densely woolly, veins deeply marked on the upper surface. Flowers solitary or many in cymes (sometimes umbel-like); bracts absent. Calyx more or less radially symmetric, divided to the base into 5 equal sepals. Corolla blue to purple (rarely yellow or white), tubular, with 2 lines of hairs runnning down the inside of the tube to the attachment of the filaments, limb 2-lipped, upper lip 2-lobed, shorter than the 3-lobed lower lip which is densely hairy inside; all the lobes more or less equal. Stamens 4, inserted on the corolla-tube above the base, filaments coiling after the pollen is shed; anthers joined in pairs; 1 staminode present. Ovary superior, oblong. Nectary ring-like. Stigma terminal, notched. Fruit a narrowly oblong to linear capsule (rarely ovoid), much longer than the calyx. *Himalaya to China and southeast Asia.*

6. Haberlea Frivaldsky. 2/2. Tufted perennial herbs, sometimes forming patches of loose leaf-rosettes. Leaves all basal, 3--8 × 2--4 cm, ovate or broadly oblong, shortly stalked, blunt, coarsely toothed, softly and densely hairy, especially beneath. Scapes

6--10 cm, with a few small, linear-lanceolate bracts. Flowers 3--5, slightly pendent. Calyx 5-lobed, the tube and the lobes of about the same length. Corolla 1.5--2.5 cm, with cylindric tube longer than the limb, which is unequally 2-lipped, all pale bluish violet, hairy and spotted with yellow inside. Stamens 4, not protruding, anthers united. Ovary superior. Stigma 2-lobed. Nectary forming a ring. Fruit a capsule. *Bulgaria, northeast Greece.*

7. Jancaea Boissier. 1/1. Tufted perennials, without rhizomes, forming clusters of neat rosettes. Leaves all in basal rosettes, 2--4 cm, broadly ovate, shortly stalked, blunt, entire, densely silver silky-hairy above, brown-hairy beneath. Scape 3--10 cm, ascending, 1--3-flowered. Calyx deeply 5-lobed. Corolla broadly campanulate, lobes 5 or occasionally 4, about as long as tube, bluish lilac. Stamens 4, not protruding, with 1 staminode; anthers free, bluish. Ovary superior. Stigma entire. Nectary ring-like. Capsule 6--7 mm, egg-shaped. *North Greece (Mt Olimbos).*

The genus name was originally spelled 'Jancaea' by Boissier in its first publication (1875); he corrected this to 'Jankaea' in volume **4** of his *Flora Orientalis* (1879), as the name was intended to commemorate a M. Janka. Though the corrected spelling was proposed for conservation in 1983, it was not accepted, and we must therefore use the silly variant arising from Boissier's mistake.

8. Ramonda Richard. 3/3. Tufted perennial herbs without rhizomes. Leaves all in basal rosettes, wrinkled, shortly white-hairy above, shaggily rusty-hairy beneath and on stalks. Scapes leafless, glandular-hairy, 1--6-flowered. Calyx 5-lobed, lobes longer than the tube. Corolla salver-shaped with a very short tube and a 5- or rarely 4-lobed limb which spreads widely and is almost radially symmetric. Stamens as many as the corolla-lobes, protruding, filaments stout, erect, anthers distinct, free. Ovary superior. Nectary obscure, ring-like. Fruit a capsule splitting longitudinally. *South Europe.*

9. Briggsia Craib. 23/2. Stemless, perennial, evergreen herbs arising from stout, scaly rhizomes. Leaves mostly in a basal rosette (rarely some opposite, when those of each pair equal), stalked or stalkless, entire or toothed, often downy. Flowers axillary, solitary or in umbel-like cymes. Calyx deeply 5-lobed. Corolla orange to yellow-green, tube cylindric below, inflated above, the limb 2-lipped, the upper lip 2-lobed, the lower 3-lobed, all the lobes more or less equal. Stamens 4, anthers united in pairs. Ovary superior. Nectary cup-like. Fruit an elongate capsule. *Himalaya, Burma & China.*

10. Didymocarpus Wallich. 150/2. Annual or perennial herbs, sometimes slightly woody at the base; stems fleshy, stout; vertical or horizontal rhizomes and stolons are present in perennial species. Leaves opposite, those in each pair unequal, rarely whorled, usually densely downy. Flowers solitary or many in axillary cymes; inflorescence-stalk fused to the short leaf-stalks. Calyx tubular or divided to the base, with 2, 3 or 5 lobes. Corolla with narrowly tubular to broadly funnel-shaped tube and 5-lobed, spreading limb. Fertile stamens 2, fused at the apex; staminodes 2 or 3. Ovary superior; tip of stigma unlobed. Nectary low, ring-like. Fruit 2-lobed or -forked at the apex. *India & China to the Malay Peninsula.*

11. Conandron Siebold & Zuccarini. 1/1. Perennial herbs, with shortly creeping rhizomes covered with long, dark brown hairs. Leaves few, all in a basal rosette, 10--30 × 3--8 cm, elliptic or ovate, fleshy, strongly wrinkled, finely toothed. Flowers in

517

a terminal cyme on a scape to 30 cm, which bears a few small bracts. Calyx 4--6 mm, with 5 narrow lobes. Corolla salver-shaped, *c.* 1.5 cm across, purple, with a short tube and 5 longer, spreading, triangular lobes. Stamens 5, attached to the base of the corolla, the anthers united into a cap around the style. Ovary superior, partially 2-celled. Nectary ring-like, low. Fruit a slender capsule *c.* 1 cm, containing numerous small seeds. *Japan, Ryukyu Islands.*

12. Lysionotus D. Don. 20/1. Evergreen, usually epiphytic shrubs or herbs with woody bases. Leaves opposite or in whorls of 3 (those in each pair or whorl more or less equal), toothed or almost entire. Flowers in terminal and axillary cymes or solitary in the leaf-axils. Calyx 5-lobed almost to the base, lobes narrow. Corolla tubular, limb 2-lipped, the upper lip 2-lobed, the lower 3-lobed. Stamens 2, attached near or above the middle of the corolla-tube, not projecting, each filament with a prominent tooth at the apex. Ovary superior, 1-celled. Stigma rounded, not projecting. Fruit a slender capsule containing many small, linear seeds. *Himalaya, northeast Asia.*

A genus of about 20 species, 1 occasionally grown.

185. OROBANCHACEAE

Perennial herbs lacking chlorophyll and parasitic on the roots of other plants, the radicle becoming an haustorium which penetrates the root of the host plant. Stems erect, usually simple, often fleshy. Leaves alternate, scale-like, often succulent at first. Flowers in a terminal spike, raceme or panicle, occasionally solitary. Calyx tubular, cup-shaped or 2-lipped. Corolla 5-lobed, 2-lipped or almost regular. Stamens 4. Ovary superior, 1-celled; placentation parietal; style single; stigma 2-lobed. Fruit a capsule; seeds small, numerous.

A family of about 230 species in 17 genera, very closely related to the *Scrophulariaceae* and thought by many to be indistinguishable from it. They are confined to the northern hemisphere, particularly the temperate and subtropical areas of Europe and Asia.

Although none is actively cultivated, plants may appear in the garden from time to time, especially members of the genus **Orobanche** Linnaeus (see also *Lathraea*, p. 505).

186. LENTIBULARIACEAE

Terrestrial or aquatic carnivorous perennials. Leaves simple, glandular in spirally-arranged rosettes (*Pinguicula*) or stems with alternate, opposite or whorled leaves, with trigger-release trap-mechanisms (*Utricularia*). Flowers bisexual. Inflorescence a bracteate raceme or flowers solitary (bracteate or not). Calyx 4- or 5-lobed or more or less 2-lobed, hypogynous. Corolla 5-lobed, radially or bilaterally symmetric, forming 2 lips, lower lip basally spurred. Stamens 2, borne on the corolla-tube. Ovary superior, of two 1-celled carpels, with free-central placentation. Stigma unequally 2-lobed, almost stalkless. Fruit usually a capsule opening by 2--4 valves or irregular, or splitting around the equator (sometimes indehiscent and 1-seeded, some *Utricularia* spp.).

1a. Leaves entire, glandular sticky, forming a basal rosette   **1. Pinguicula**
  b. Leaves alternate, whorled, or in a basal tuft, entire or with thread-like lobes
                                              2. Utricularia

1. Pinguicula Linnaeus. 40/20. Rosette-forming perennials. Leaves entire, glandular, sticky with digestive enzymes, spirally arranged in a basal rosette. Margins somewhat inrolled, becoming more so with stimuli. Species with seasonally different leaves, produce small-leaved winter rosettes; some species overwinter as a bud, a rootless resting stage, easily displaced. Some species produce vegetative gemmae (small, asexual, detachable buds) around the overwintering bud. Flowers solitary, violet, yellow or pink. Infloresence-stalk generally glandular-hairy. Calyx 2-lipped, upper 3-lobed, lower 2-lobed. Corolla 5-lobed, bilaterally symmetric, forming 2 lips, upper lip 2-lobed, lower lip longer, 3-lobed, sometimes almost radially symmetric; basally spurred. Anthers 2, borne on the corolla-tube. Style short, curved. Fruit a capsule with numerous seeds. *North temperate areas, Mexico.*

2. Utricularia Linnaeus. 250/17. Terrestrial or aquatic, carnivorous annuals or perennials. Stems merge with other vegetative parts, forming horizontal stolons, simple or branched of varying length, terete, hairless though sometimes glandular. Tubers sometimes forming at base of flower-stem or as part of the stem. Leaves variable in position and shape, alternate, opposite, in a whorl or in a tuft at base of flower-stem; terrestrial species: entire, stalked or stalkless, linear to circular, kidney-shaped, peltate, delicate or leathery; aquatic species: narrowly lobed or absent. Roots absent, rhizoids often present. Traps held underwater, spherical or ovoid, small, stalked or not; mouth close to or opposite stalk or lateral, subtended by various bracts. Bracts basifixed (attached at their base) or attached above the base. Calyx 2-lobed (ours). Stamens 2. *Almost cosmopolitan.*

Plants formerly included in the genus *Polypompholyx* Lehmann, with 4 calyx-lobes, are now treated as part of *Utricularia.*

187. MYOPORACEAE

Small trees and shrubs. Leaves simple, mostly alternate, usually gland-dotted. Flowers bisexual, solitary or in axillary cymes. Calyx of 5 fused sepals. Corolla bilaterally symmetric, with 5 fused overlapping petals, often 2-lipped. Stamens 4, sometimes with an upper posterior staminode, rarely 5, borne on the corolla-tube, alternating with the lobes. Ovary superior, 2-celled, placentation axile, ovules 4 or if more then in vertical rows in each cell. Fruit a drupe or separating into 1-seeded drupe-like segments.

A family of 5 genera (only 1 cultivated) from the southern hemisphere and south-east Asia.

1. Myoporum G. Forster. 32/6. Evergreen trees or shrubs, many heather-like, hairless or glutinous. Leaves alternate, occasionally opposite, entire or toothed, gland-dotted. Flowers axillary, usually in clusters, mostly white. Calyx 5-lobed. Corolla campanulate or funnel-shaped, 5-lobed. Stamens 4--6. Fruit a small, more or less fleshy drupe. *Australia, New Zealand, Pacific Islands, New Guinea, Mauritius & east Asia*

188. PLANTAGINACEAE

Annuals, biennials or perennials, sometimes dwarf shrubs. Leaves simple, rarely pinnatifid, alternate (usually in a basal rosette) or opposite, parallel-veined. Flowers small, wind-pollinated, usually bisexual, radially symmetric, in heads or spikes. Calyx of 3 or usually 4 free or united sepals. Corolla with a short tube and usually 4 rather papery lobes. Stamens 4, sometimes 1--3, inserted on the corolla-tube. Ovary superior, usually of 2 united carpels, ovules 1--many, placentation axile Fruit a capsule opening round its circumference, or rarely a nut. See figure 118, p. 522.

A cosmopolitan family of 3 genera and some 250 species, all but 4 of them in *Plantago*.

    1a. Plant terrestrial, without far-creeping stolons; flowers bisexual **1. Plantago**
      b. Plant aquatic, with far-creeping stolons; flowers male or female
                                                                      2. Littorella

1. Plantago Linnaeus. 250/20. Annuals, biennials, perennials or dwarf shrubs. Leaves in basal rosettes or opposite on branched stems. Flowers bisexual, parts in 4s, small, greenish and wind-pollinated (sometimes visited by insects for pollen), with small bracts, massed in stalked spikes. Corolla dish-shaped with a short tube, the lobes spreading or bent back. Stigmas maturing before the anthers (protogyny). Ovary with 2--4 cells. Fruit a papery capsule, splitting longitudinally. Seeds 1--many, often mucilaginous when wet. *Cosmopolitan.* Figure 118, p. 522.

2. Littorella Bergius. 3/1. Aquatic, hairless, tufted perennial, spreading by slender stolons, rooting to form clumps and underwater 'lawns'. Leaves all in basal rosettes, 2--15 cm, linear, half-cylindrical in section, spongy, with sheathing bases. Male flowers solitary, on stalks to 8 cm, with 4 minute greenish sepals and petals, and 4 stamens 1--2 cm; female flowers almost stalkless, 1--8 at the base of male flower-stalk, with style *c.* 1 cm. Fruit indehiscent, 1-seeded. *America, Eurasia.*

189. CAPRIFOLIACEAE

Herbs, shrubs, woody climbers or sprawling wintergreen perennials. Leaves opposite, entire or pinnate, stipules usually absent, if present never interpetiolar. Flowers usually bisexual (rarely some flowers sterile), usually in panicles, sometimes corymbose. Calyx 5- or rarely 4-lobed. Corolla with a long or short tube, 5- or rarely 4-lobed, radially or bilaterally symetric. Nectary present or absent. Stamens 4 or 5, attached to the corolla-tube between the lobes. Ovary inferior, rarely semi-inferior, usually 3-celled, sometimes 1 or 2 of the cells sterile; placentation axile. Fruit dry and 1-seeded or a capsule, berry or drupe. See figures 119, p. 524 & 120, p. 526.

About 15 genera, most of them cultivated, mainly from north temperate areas.

    1a. Plant a herb or trailing evergreen shrublet   2
      b. Plant a shrub   3
    2a. Plant a herb   **3. Triosteum**
      b. Plant a dwarf, trailing, evergreen shrublet   **5. Linnaea**
    3a. Inflorescence a corymb or panicle; corolla radially symmetric, without a
          nectary   4
      b. Inflorescence not as above; corolla radially to bilaterally symmetric, with a
          nectary   5
    4a. Leaves pinnate; fruit with 3--5 seeds   **1. Sambucus**
      b. Leaves entire, sometimes lobed; fruit with 1 seed   **2. Viburnum**
    5a. Ovary composed of 1 or 2 fertile cells each with a single ovule and 1 or 2
          sterile cells; fruit with fewer than 3 seeds   6
      b. All cells of the ovary fertile, with more than 3 ovules; fruit with 3 or more
          seeds   10

520

6a. Leaves markedly 3-veined from the base; inflorescence a panicle made up of 7-flowered dichasia; calyx-lobes enlarging and becoming rose to purple in fruit   **13. Heptacodium**
  b. Combination of characters not as above   7
7a. Corolla radially symmetric; fruit a berry   **4. Symphoricarpos**
  b. Corolla bilaterally symmetric; fruit not a berry   8
8a. Ovary 4-celled, enclosed between 2 large bracts which persist into fruit   **6. Dipelta**
  b. Ovary 3- or rarely 4-celled, bracts not as above   9
9a. Ovaries of adjacent flowers united, covered in long bristles   **7. Kolkwitzia**
  b. Ovaries of adjacent flowers free, without bristles   **8. Abelia**
10a. Fruit a capsule   11
  b. Fruit a berry   12
11a. Calyx deciduous; plant not stoloniferous; corolla rarely yellow   **9. Weigela**
  b. Calyx persistent; plant strongly stoloniferous; corolla yellow   **10. Diervilla**
12a. Ovary 4- or 5-celled, with a sterile neck; branches hollow   **11. Leycesteria**
  b. Ovary 2- or 3-celled, rarely 5-celled, without a sterile neck; branches usually not hollow   **12. Lonicera**

1. Sambucus Linnaeus. 9/6. Perennial herbs, shrubs or small trees. Pith soft, white to orange-brown. Leaves opposite, pinnate with a terminal leaflet; leaflets 3--15, with forwardly pointing teeth. Stipules usually absent, when present linear (sometimes glandular) or leaf-like, stipels present or absent. Inflorescence a terminal corymb. Flowers radially symmetric, bisexual (ours), stalked or not, often fragrant. Calyx very small. Corolla white or yellowish, sometimes reddish outside, wheel-shaped, tube small, lobes 5 (rarely 3 or 4), overlapping or edge-to-edge in bud. Anthers usually protruding, yellow or purple, opening to the outside of the flower or laterally. Stigma with 3--5 lobes. Ovary mostly inferior, with 3--5 united cells, each with 1 ovule. Fruit a spherical or ovoid, fleshy, berry-like drupe, with 3--5 hard-shelled stones. *Mostly northern hemisphere, a few species in the southern.*
 Sometimes placed in the *Adoxaceae* (p. 525), or a family of its own (*Sambucaceae*).

2. Viburnum Linnaeus. 210/30. Deciduous to evergreen shrubs, hairless or with simple or stellate hairs. Leaves opposite, sometimes in whorls of 3, sometimes leathery, stalked, stipules usually absent or joined to the leaf-stalks, glands absent or present, margin entire, toothed or 3--5-lobed. Inflorescence an axillary or terminal panicle or corymb, with many small, radially symmetric flowers, with or without sterile, enlarged marginal flowers which are sometimes bilaterally symmetric. Calyx 5-toothed, small. Corolla wheel-shaped, campanulate or tubular, 5-lobed, white, pinkish or greenish. Stamens 5, anthers versatile, alternating with corolla-lobes. Stigma 3-lobed, stalkless, ovary semi-inferior, 3-celled, 2 cells sterile; ovule solitary, pendent. Fruit a dry or juicy drupe, blue, black or red when mature. *Temperate and subtropical regions of Asia, Europe, North Africa & America.* Figure 119, p. 524.

3. Triosteum Linnaeus. 6/1. Perennial herbs. Leaves opposite, simple, sometimes united round the stem. Flowers solitary or in few-flowered clusters in the axils of leaves, sometimes terminal. Calyx 5-lobed, persistent. Corolla tubular, bilaterally symmetric, tube swollen at the base, lobes 5, unequal, the lower lobe directed downwards (ours), lobes overlapping in bud. Stamens 5. Ovary usually 4-lobed and

521

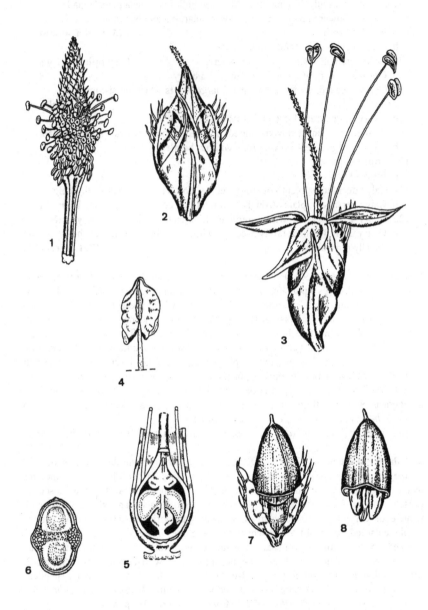

Figure 118. Plantaginaceae. *Plantago lanceolata.* 1, Inflorescence. 2, Single flower-bud subtended by four bracts. 3, Single flower. 4, Anther. 5, Longitudinal section of ovary. 6. Transverse section of ovary. 7, Young fruit. 8, Opened fruit.

4-celled (rarely 3- or 5-lobed and -celled), with 1 ovule per cell and 1 cell sterile. Fruit a dryish drupe, with 3 or fewer seeds. *Asia, North America.*

4. Symphoricarpos Duhamel. 18/8. Upright to prostrate deciduous shrubs; winter buds with 2 pairs of outer scales. Leaves opposite, shortly stalked, entire or variably lobed on vigorous shoots, without stipules. Flowers in axillary pairs or short spikes, small, bisexual, radially symmetric. Calyx 4- or 5-toothed. Corolla tubular to bell-shaped, radially symmetric, 4- or 5-lobed at the apex, sometimes hairy outside, often with long hairs in the throat. Stamens 4 or 5, borne on the corolla-tube, anthers included in the corolla-tube or projecting, occasionally projecting beyond the corolla-lobes. Ovary inferior, 4-celled, 2 cells each containing 1 fertile ovule, the other 2 cells with several sterile ovules. Fruit a 2-seeded rather dry berry, often long-persistent. *North America, China.*

5. Linnaea Linnaeus. 1/1. Dwarf, trailing, evergreen shrublet, stems rooting at the nodes. Leaves 5--20 × 3--15 mm, broadly ovate to elliptic, circular or obovate, toothed-scalloped in the upper half, tapering below to a stalk 2--3 mm, without stipules. Inflorescence-stalks terminating lateral branches 4.5--8 mm, glandular-hairy, with 2 lanceolate, membranous bracts 1.5--2 mm. Flowers in pairs, fragrant, stalks 5--20 mm, glandular-hairy, bracteoles 2, lanceolate, membranous. Calyx 2--5 mm, 5-lobed, lobes narrow-lanceolate. Corolla 5--12 mm, 5-lobed, funnel-campanulate, pinkish white, often marked with pinkish purple, hairy inside. Stamens 4, 2 long and 2 short, inserted near the base of the corolla-tube. Ovary 3-celled, 1 cell fertile, 2 sterile. Fruit dry, 1-seeded, *c.* 3 mm, densely glandular-hairy. *Scattered in mountainous & northern parts of the northern hemisphere.*

6. Dipelta Maximowicz. 4/3. Deciduous shrubs, generally many-stemmed. Leaves opposite, shortly stalked, lacking stipules. Flowers solitary or in few-flowered racemes, subtended by a variable number of bracts and bracteoles, 2 of which enlarge after flowering and surround the fruit. Calyx 5-lobed. Corolla tubular to campanulate, bilaterally symmetric, 5-lobed but 2-lipped. Stamens 4. Ovary inferior, 4-celled, all cells fertile with several ovules. Capsule enclosed by 2 large persistent bracts. *China.*

7. Kolkwitzia Graebner. 1/1. Deciduous shrub, to 2.5 m. Young shoots densely hairy, older shoots with peeling bark. Leaves broadly ovate, 2.5--8.5 × 1.5--5.5 cm, sparsely toothed, margin hairy. Inflorescence-stalks bristly hairy. Flowers paired, in terminal clusters on short lateral shoots. Calyx 5-lobed, narrow, persistent in fruit. Corolla campanulate, 5-lobed, to 2.5 cm, pink-white with a yellow throat. Stamens 4. Ovary 3-celled, 1 fertile, 2 sterile, covered with long, bristle hairs. Fruits paired, united, with persistent calyx. *Central to eastern China.*

8. Abelia R. Brown. 30/14 and some hybrids. Evergreen, semi-evergreen and deciduous shrubs, rarely small trees, with slender branches. Leaves opposite or occasionally in whorls of 3 or 4 on strong shoots, blades lanceolate or ovate, entire or shallowly to strongly toothed. Flowers solitary or clustered in leaf-axils, or clustered on a slender stalk. Sepals 2, 4 or 5 (variable in 1 hybrid), linear, oblanceolate or elliptic, persisting long after the corollas fall and usually pink or reddish. Corolla funnel-shaped or campanulate with a narrow lower tube, or salver-shaped with a cylindric tube and short spreading lobes, white, pink or yellowish, often marked with

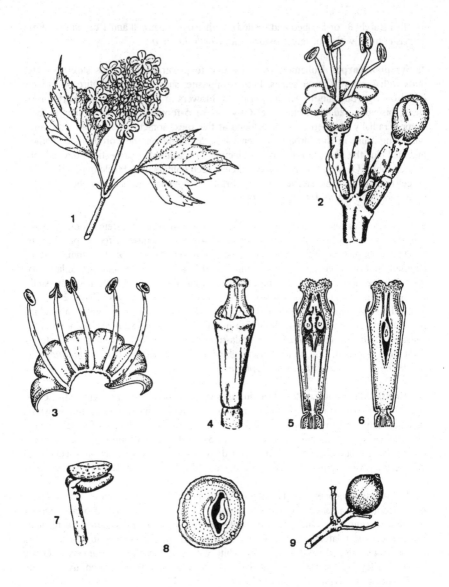

Figure 119. Caprifoliaceae. *Viburnum opulus.* 1, Inflorescence. 2, Part of inflorescence with one flower in bud and one open. 3, Corolla (opened out) and stamens. 4, Ovary. 5, Longitudinal section of young ovary showing two (of the three) ovules. 6, Longitudinal section of older ovary now containing only one developing ovule. 7, Stamen. 8, Transverse section of older ovary. 9, Fruit.

orange in the throat. Stamens 4, 2 short and 2 longer, the filaments attached above the base of the corolla. Style slender, longer than the stamens, stigma capitate. Ovary inferior, 3-celled, 1 or 2 cells fertile; ovules 5--9, but often only 1 maturing. Fruit an oblong leathery achene containing a single seed and topped by the conspicuous, persistent sepals. *Eastern Asia; 2 species in Mexico.*

9. Weigela Thunberg. 10/10 and some hybrids. Deciduous shrubs, usually rather upright, without stolons, branches with solid pith. Leaves opposite, stalked or not, toothed, usually hairy; stipules absent. Flowers rather large, solitary or in few-flowered cymes in the upper leaf-axils. Calyx with a tube and 5 lobes, the tube sometimes very short or absent, the whole deciduous, sometimes 2-lipped. Corolla weakly bilaterally symmetric, with a long tube which widens abruptly or gradually upwards, shallowly 5-lobed, often hairy outside, sometimes pale or greenish in bud, later white, white becoming red, red or yellow. Stamens 5, anthers usually free and hairless, sometimes united and hairy. Ovary 2-celled, with many ovules in each cell. Capsule opening down the cross-wall, many-seeded. *Eastern Asia.*

For many years the genus was included in *Diervilla*, and almost all the species have names in that genus.

10. Diervilla Miller. 3/3. Low shrubs similar to *Weigela*, but differing in being strongly stoloniferous and forming colonies, flowers much smaller, cymes numerous and forming clusters, calyx persistent on the fruit, corolla more or less 2-lipped, pale yellow to yellow. *North America.*

11. Leycesteria Wallich. 6/2. Deciduous shrubs with erect, hollow branches. Leaves opposite, simple, with or without stipules between the stalks. Flowers in terminal and axillary hanging spikes, with large colourful bracts subtending cymose whorls of 2--6 stalkless flowers. Calyx unequally 5-lobed. Corolla funnel-shaped, usually 5-lobed. Stamens 5. Ovaries 5-celled with many ovules per cell. Fruit a many-seeded berry. *Himalaya to western China.*

12. Lonicera Linnaeus. 200/100 and some hybrids. Deciduous, rarely semi-evergreen or evergreen shrubs or climbers; shoots usually not hollow. Leaves opposite, entire, rarely lobed, stalkless or with short stalks; stipules absent. Flowers in pairs on a common stalk or in whorls, subtended by 2 bracts and usually 4 bracteoles. Calyx 5-lobed. Corolla tubular or campanulate, 5-lobed, often swollen at the base, lobes usually forming 2 lips, the upper lip composed of 4 short lobes, the lower of a single strap-shaped lobe, sometimes lobes equal. Stamens 5. Ovary inferior, 2- or 3-celled, sometimes adjacent ovaries partially or wholly united. Fruit a many-seeded berry, rarely dry. *Throughout the northern hemisphere, mostly temperate, southwards in the Old World to North Africa, Java & Philippines, in the New World to Mexico.* Figure 120, p. 526.

13. Heptacodium Rehder. 1/1. Deciduous small shrubs or trees to 7.5 m. Bark peeling. Leaves 8--15 × 5--7 cm, ovate, entire or margins somewhat wavy, with 3 deeply impressed veins from the base; base rounded or cordate, apex acuminate. Flowers fragrant in wide terminal panicles made up of long-stalked, tiered, generally 7-flowered dichasia. Calyx 5-lobed. Corolla 5-lobed, bilaterally symmetric, creamy white, tube shorter than the lobes, downy outside. Stamens 5, protruding. Style and

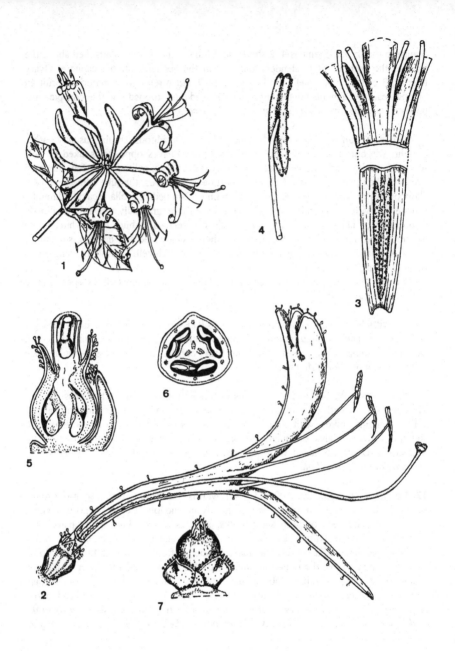

Figure 120. Caprifoliaceae. *Lonicera periclymenum*. 1, Inflorescence. 2, Longitudinal section of flower. 3, Longitudinal section of lower part of corolla-tube. 4, Stamen. 5, Longitudinal section of ovary. 6, Transverse section of ovary. 7, Developing fruit subtended by two bracts.

filament-bases hairy. Ovary inferior, 3-celled, 1 cell fertile, 2 sterile (with several ovule-rudiments). Fruit oblong, hairy, with persistent enlarged, rose to purple calyx-lobes at apex. Seeds terete. *Northeast China.*

190. ADOXACEAE

Hairless perennial herbs with short and erect or slender and far-creeping scaly rhizomes. Basal leaves with 3 almost equal leaflets or pinnate, the leaflets lobed. Stems solitary or 2--4, erect, slightly fleshy, with 2 opposite leaves similar to but smaller than the basal ones. Flowers small, stalkless, in a terminal cluster or in several clusters forming an interrupted spike. Calyx-lobes 2 or 3. Corolla with 3--5 spreading lobes. Stamens 3--5, but divided to the base and apparently 6--10. Ovary partially inferior, either 1-celled with a stalkless stigma or 3--5-celled with 3--5 styles. Fruit a green drupe, seldom formed.

A family of 3 genera, 1 occurring in temperate areas in Europe, North America and Asia and cultivated, 2 others recently described from northwest China.

1. Adoxa Linnaeus. 1/1. Perennial herb to 10 cm, with a slender, scaly, far-creeping rhizome. Basal leaves long-stalked, with 3 leaflets 1--3 cm, the leaflets with deep, rounded lobes. Stem erect, bearing 2 opposite, short-stalked leaves 8--15 mm. Flowers *c.* 6 mm wide, pale green, usually 5 in a terminal head. Calyx small, with 2 or 3 lobes. Corolla of terminal flower with 4 spreading, rounded lobes, of lateral flowers 5-lobed. Stamens 4 or 5, but divided to base and appearing to be 8 or 10. Ovary 3--5-celled, the apex tapering and bearing 3--5 spreading styles. Fruit a green drupe *c.* 6 mm wide, containing 3--5 seeds, but rarely formed. *Europe, Asia, North America.*

191. VALERIANACEAE

Annual or perennial herbs, sometimes woody at the base, the whole plant often with a distinct odour. Leaves opposite, whorled or all basal, without stipules, simple or pinnatisect and usually clasping at the base. Inflorescence cymose with many crowded flowers, bracts present. Flowers bisexual or unisexual, bilaterally symmetric. Calyx 0--5-lobed, very small in flower, often enlarging to a hairy pappus at the top of the fruit. Corolla funnel-shaped, the tube often bulging on 1 side or spurred below; lobes 5. Stamens 1--4, usually inserted near the base of the corolla-tube. Ovary inferior, 3-celled, 1 cell with a pendent ovule, the other 2 sterile. Fruit dry and indehiscent, the calyx usually persistent. See figure 121, p. 530.

A family of 10 genera and 300 species, mainly distributed throughout the northern hemisphere, South America and southeastern South Africa.

1a. Stamen 1; corolla-tube spurred near the base or prominently swollen near
    the middle   **1. Centranthus**
  b. Stamens 2--4; corolla-tube rarely spurred, sometimes slightly swollen   2
2a. Annual with forked branching   3
  b. Perennial, branching not forking   4
3a. Stamens 3, free; corolla-tube not more than twice as long as lobes
                                               2. Valerianella
  b. Stamens 2 or 3, 2 united; corolla-tube more than twice as long as lobes
                                                  3. Fedia
4a. Stamens 3; calyx-teeth feathery in fruit; corolla pink or white   **4. Valeriana**
  b. Stamens 4; calyx-teeth not feathery in fruit; corolla yellow   **5. Patrinia**

1. **Centranthus** de Candolle. 9/1. Annuals or rhizomatous perennials. Leaves simple or pinnate, often glaucous. Calyx of 10--25 linear, inrolled segments in flower, having the appearance of an epigynous ring and developing into a feathery pappus in fruit. Corolla red, pink or white, lobes 5, unequal, tube funnel-shaped and bulging on 1 side, or cylindric and spurred near the base, with an internal longitudinal membrane from the insertion of the spur to the mouth. Stamen 1. Stigma entire to 3-fid. Fruit compressed. *Mediterranean Europe.*

2. **Valerianella** Miller. 50/2. Annuals. Branching usually forking. Basal leaves spathulate, wedge-shaped at base or oblanceolate or oblong and almost stalkless, margins toothed or wavy; upper leaves with a few distinct teeth at the base. Inflorescence a contracted cyme. Calyx toothed or crown-like, regular or irregular, sometimes inflated, occasionally absent. Corolla bluish to pink; tube not more than twice as long as llimb, neither spurred nor bulging on 1 side. Stamens 3. Fruit with the 2 sterile cells variously developed. *Portugal, Mediterranean area, southwest Asia to Afghanistan.*

3. **Fedia** Gaertner. 3/1. Annuals with forked branching. Leaves opposite, simple. Flowers bisexual, borne in terminal, usually paired cymes. Calyx with 2--4 teeth, usually very small, not enlarging in fruit. Corolla purplish, lobes 5, unequal, the tube cylindric, more than twice as long as lip, obscurely swollen, not spurred. Stamens 2 or 3, 2 of them united. Stigma bifid. *Mediterranean area.*

4. **Valeriana** Linnaeus. 200/8. Rhizomatous or tuberous perennials. Stems usually unbranched. Leaves simple or pinnate, opposite. Flowers unisexual or bisexual, in compound, many-flowered cymes. Calyx-teeth 5--15, linear, inrolled in flower, enlarging and having the appearance of an epigynous ring, finally developing into a feathery pappus in fruit. Corolla pinkish or white, narrowly funnel-shaped, slightly swollen at base; lobes 5 (rarely 3), unequal, spreading. Stamens 3. Stigma shortly 3-lobed. Fruit compressed. *Temperate areas (absent from Australia).* Figure 121, p. 530.

5. **Patrinia** Jussieu. 15/3. Erect, rhizomatous perennials. Leaves stalked or not, mostly 1 or 2 times pinnately lobed. Cymes much branched. Calyx small, 5-toothed, crown-like, not feathery in fruit. Corolla yellow (white in non-cultivated species), lobes regular, spreading, tube spurred or swollen. Stamens 4. Fruit with the sterile cells often wing-like, with a wing-like bracteole at the base. *Temperate Asia.*

192. DIPSACACEAE

Annual, biennial or perennial herbs, rarely shrubs. Leaves usually opposite, sometimes in whorls, lacking stipules. Flowers borne in a dense head (capitulum) or occasionally in a spike of whorls. Heads often with the marginal flowers larger and 2-lipped (sometimes referred to as 'radiant'), subtended by 1--3 rows of usually herbaceous involucral bracts. Flowers usually bisexual, rarely female, usually bilaterally symmetric, each with a basal epicalyx (involucel) which may have a terminal corona or teeth; receptacle hairy or with chaffy scales. Calyx cup-shaped or 2-lobed, or divided into 4, 5 or more teeth. Corolla with a long tube and 4 or 5 lobes, almost equal or 2-lipped. Stamens 2 or 4, borne on corolla-tube, alternating with corolla-lobes, protruding. Ovary inferior, 1-celled, containing a single ovule. Stigma simple or 2-lobed. Fruit an achene, enclosed within the involucel and often crowned by the persistent calyx. See figure 122, p. 532.

A temperate family: the number of genera varies from 8 to 16 according to botanical opinion, but it is generally agreed that there are about 250 species native to Eurasia and Africa. *Morina* is sometimes put into its own family, *Morinaceae*.

1a. Flowers in whorls, which may sometimes be condensed into a more or less
      spherical head   **1. Morina**
  b. Flowers in heads   2
2a. Corolla 4-lobed   3
  b. Corolla 5-lobed   6
3a. Stems with prickles, if prickles absent then heads spherical and flowers cream
      with black anthers   **2. Dipsacus**
  b. Stems lacking prickles   4
4a. Receptacular scales absent, receptacle hairy   **4. Knautia**
  b. Receptacular scales chaffy or herbaceous   5
5a. Corollas white, cream or yellow   **3. Cephalaria**
  b. Corollas lilac to violet, rarely pink or white   **5. Succisa**
6a. Calyx-teeth 10 or more, feathery   **6. Pterocephalus**
  b. Calyx-teeth 5, not feathery   **7. Scabiosa**

1. Morina Linnaeus. 8/6. Perennial herbs. Leaves in whorls, mostly basal, usually spiny. Inflorescence a spike of many-flowered, bracteate whorls which may sometimes be condensed into a more or less spherical head. Calyx usually deeply 2-lobed, enclosed within a bristle-tipped involucel. Corolla with a curved tube, 2-lipped, 5-lobed. Fertile stamens 2. Fruit with an oblique apex, wrinkled. *Southeast Europe to Himalaya.*

2. Dipsacus Linnaeus. 15/5. Biennial or short-lived perennial herbs. Stems erect, usually prickly. Leaves opposite, often united into a cup at the base and collecting rain-water, simple to pinnatifid, margin toothed to lacerate. Flowers borne in terminal, usually long-stalked, ovoid, cylindric or spherical heads. Involucral bracts in 1 or 2 rows, linear or lanceolate, erect to spreading, spine-tipped. Receptacular scales spine-tipped or not. Calyx 4-ridged, cup-like, enclosed within the involucel, persistent in fruit. Corolla unequally 4-lobed. Stamens 4, more or less protruding. Fruit 4-angled, adpressed-hairy. Involucel 4- or 8-angled, with a short 4-lobed or 12-toothed limb at apex. *Europe, North Africa, Asia.*

3. Cephalaria Roemer & Schultes. 65/9. Annual, biennial or perennial herbs, rarely woody. Leaves opposite, toothed or pinnatifid. Involucral bracts chaffy. Receptacular scales similar to involucral bracts but larger. Calyx cup-shaped. Flowers borne in dense, spherical or ovoid, long-stalked heads, marginal flowers 2-lipped, larger than inner or not. Corolla tubular, 4-lobed. Stamens 4. Involucel 4- or 8-angled, 8-ridged, usually with 4 or more terminal teeth. Fruit a ribbed achene. *Europe, central Asia, northern and southern Africa.*

4. Knautia Linnaeus. 60/4. Annual and perennial herbs. Leaves opposite, simple to pinnatifid or pinnate. Flowers in long-stalked, flat-topped, heads with the marginal flowers 2-lipped, larger than the inner, sometimes female only, subtended by

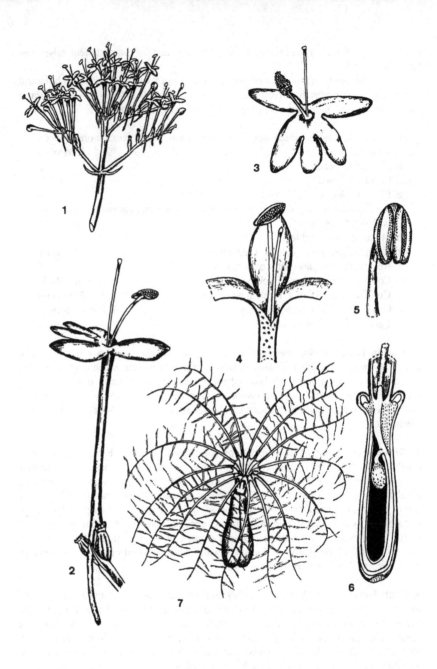

Figure 121. Valerianaceae. *Centranthus ruber.* 1, Inflorescence. 2, Single flower. 3, Corolla obliquely from above. 4, Longitudinal section of upper part of corolla-tube. 5, Stamen. 6, Longitudinal section of ovary. 7, Mature fruit crowned with a pappus.

530

herbaceous involucral bracts. Receptacle bearing bristles, not scales. Calyx shallowly cup-shaped with 8--16 bristles or teeth. Corolla-lobes 4, unequal. Stamens 4. Involucel 4-angled, cup-shaped, compressed, with or without tiny teeth. Fruit ovoid to cylindric, more or less hairy. *Europe, Mediterranean area, Caucasus & Siberia.* Figure 122, p. 532.

5. Succisa Haller. 4/1. Erect or decumbent, often downy, perennial herbs. Basal leaves in a rosette, oblong-oval to elliptic, sparsely downy, tapering to a short stalk. Stem-leaves somewhat narrower, occasionally toothed. Flowers bisexual or female, borne in long-stalked, hemispherical heads with all flowers similar. Involucral bracts herbaceous. Receptacular scales almost equalling corolla. Calyx shallowly cup-shaped, prolonged above into 4 or 5 bristle-tipped teeth, persistent in fruit. Corolla 4-lobed. Involucel 4-angled with 4 triangular herbaceous lobes. *Europe, western Asia, north & east Africa.*

6. Pterocephalus Adanson. 25/4. Annual or perennial herbs, subshrubs and shrubs. Leaves opposite, entire to pinnatisect. Heads hemispherical, marginal flowers 2-lipped, larger than inner, all subtended by narrow, herbaceous involucral bracts. Receptacular scales hairy or absent. Outer flowers conspicuously 2-lipped. Calyx stalked, with 10--24 long feathery teeth. Corolla-lobes 5. Stamens 5. Involucel grooved, with terminal bristles, minute teeth or a short corona. *Mediterranean area to west, central & eastern Asia.*

7. Scabiosa Linnaeus. 80/30. Annual, biennial or perennial herbs, rarely dwarf shrubs. Leaves opposite, simple to 2-pinnate. Flowers borne in long-stalked hemispherical or sometimes cylindric heads which have the marginal flowers somewhat 2-lobed and more or less larger than the inner, and subtended by 1--3 rows of herbaceous involucral bracts. Calyx cup-shaped, the upper part usually with 5 awn-like teeth, persistent in fruit. Corolla with a short tube and 5 unequal lobes. Stamens 4, rarely 2. Involucel usually cylindric, 8-grooved for entire length, or with 8 pits at top of tube and 8 grooves below, expanded at apex into a membranous corona with many teeth, sometimes projecting beyond the margin, veins and a flat or conical diaphragm which almost closes the top of the tube. *Europe, Asia, north & east Africa, South Africa.*

193. CAMPANULACEAE

Herbs, shrubs or vines, rarely arborescent herbs, with a clear or milky sap. Leaves mostly alternate, occasionally opposite or whorled, usually simple. Stipules absent. Flowers usually bisexual, radially or bilaterally symmetric, terminal and/or axillary, solitary or in various cymose or racemose inflorescences. Calyx with usually 3--5 (occasionally more) lobes united at the base, usually fused to ovary-wall (free in *Cyananthus*). Corolla-lobes 5, if radially symmetric, appearing almost free or united and campanulate, or if bilaterally symmetric, strongly 2-lipped, with 2 upper lobes and 3 lower. Stamens usually 5, free or united. Anthers free or united. Ovary inferior or semi-inferior, rarely superior, 2--5- (rarely to 10-) celled. Fruit a berry or capsule, of 2--5 cells. Seeds few to many. See figure 123, p. 536.

A cosmopolitan family of about 87 genera and approximately 2000 species.

1a. Flowers radially symmetric (Subfamily *Campanuloideae*)   2
  b. Flowers bilaterally symmetric (Subfamily *Lobelioideae*)   24

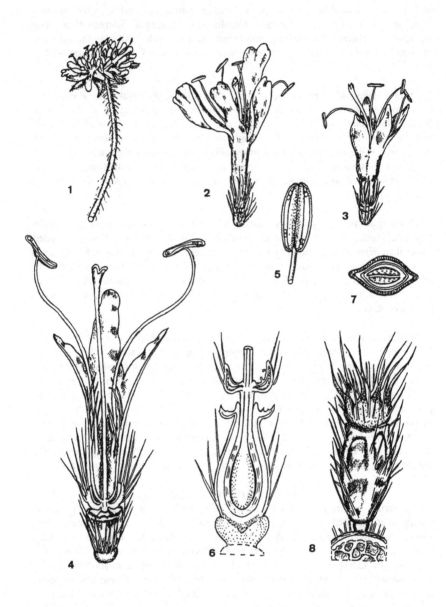

Figure 122. Dipsacaceae. *Knautia arvensis*. 1, Inflorescence. 2, Single marginal flower. 3, Single central flower. 4, Longitudinal section of calyx and corolla. 5, Stamen. 6, Longitudinal section of ovary. 7, Transverse section of young fruit. 8, Young fruit still attached to receptacle.

532

2a. Ovary superior; calyx inflated and not united to the ovary; anthers united at base   **16. Cyananthus**

  b. Ovary inferior or semi-inferior; calyx-tube partly or wholly united to ovary and not inflated; anthers united or not   3

3a. Capsule opening by apical valves above the calyx-lobes   4

  b. Capsule opening not as above   9

4a. Inflorescence a dense spherical head, subtended by an involucre of bracts; anthers united at base only   **19. Jasione**

  b. Inflorescence not a head, involucre absent; anthers free or entirely united 5

5a. Corolla tubular   **15. Leptocodon**

  b. Corolla star-shaped or campanulate   6

6a. Corolla star-shaped   7

  b. Corolla campanulate   8

7a. Plants twining or scrambling, strongly scented   **14. Codonopsis**

  b. Plants not twining or scrambling, not strongly scented   **21. Platycodon**

8a. Corolla greenish, yellowish or purplish, often with distinct veins, sometimes with irregular blotches of contrasting colour   **14. Codonopsis**

  b. Corolla blue or whitish   **17. Wahlenbergia**

9a. Capsule opening by irregular apical rupture above calyx-lobes; flowers in clusters, usually with an involucre or solitary and without an involucre of bracts; leaves linear, grass-like   **18. Edraianthus**

  b. Combination of characters not as above   10

10a. Capsule opening by lateral, longitudinal or transverse slits   **7. Ostrowskia**

  b. Capsule opening by lateral pores, irregular rupturing or indehiscent   11

11a. Capsule with lateral pores or clefts from middle to apex, usually fairly regular in shape   12

  b. Capsule with lateral pores usually below middle or at base, often irregular in shape, or with valve-like flaps   16

12a. Mature capsule strongly angular; corolla saucer-shaped   **20. Legousia**

  b. Mature capsule obovate or elliptic, round or slightly angular; corolla not as above   13

13a. Inflorescence a loose panicle or flowers solitary; corolla campanulate or star-shaped, rarely incised below middle, lobes not strap-shaped

      1. Campanula

  b. Inflorescence racemose, an interrupted spike, globular head or plants cushion-like with flowers solitary; corolla star-shaped and deeply incised to below middle, lobes strap-shaped, lanceolate or linear   14

14a. Anthers with very short filaments; nectar-dome strongly turgid, blue; stigma-lobes forming a spherical head; leaves pinnate   **12. Petromarula**

  b. Anthers with conspicuous, slender filaments; nectar-dome neither strongly turgid nor blue; stigma-lobes not forming a spherical head; leaves not pinnate   15

15a. Corolla-lobes tipped with distinct points; style strongly curved

      2. Campanulastrum

  b. Corolla-lobes not tipped with distinct points; style not strongly curved

      13. Asyneuma

16a. Corolla curved or claw-shaped, apices of lobes temporarily or permanently united; flowers usually aggregated into dense heads or spikes   17

b. Corolla tubular, star-shaped or campanulate, apices of lobes not united; flowers rarely aggregated into dense heads or spikes, more rarely in clusters   18
17a. Corolla-lobes separating at maturity and spreading   **10. Phyteuma**
  b. Corolla-lobes remaining distally united at maturity   **11. Physoplexis**
18a. Anthers permanently united in mature flowers   19
  b. Anthers not permanently united in mature flowers   20
19a. Leaves alternate, rarely whorled; top of ovary not bulging, not dark green; disc not inflated   **3. Symphyandra**
  b. Leaves occasionally whorled; top of ovary bulging, dark green; disc inflated   **4. Hanabusaya**
20a. Disc orange-yellow or whitish, fleshy, cylindric or a ring   **5. Adenophora**
  b. Disc colourless or greenish, not cylindric or ring-like   21
21a. Herbs or cushion plants; inflorescence a conspicuous corymb or flowers solitary and terminal; capsules spherical   22
  b. Herbs; inflorescence a panicle or flowers solitary or in clusters (not a corymb); capsules obovoid or elliptic   23
22a. Plant not cushion-like; inflorescence a flat-topped or hemispherical corymb to 30 cm, stalk with few leaves; corolla blue or mauve, rarely white   **8. Trachelium**
  b. Plant cushion-like, erect or trailing; inflorescence spherical, less than 15 cm, stalk leafy thoughout; corolla usually pale blue or white   **9. Diosphaera**
23a. Corolla usually campanulate, usually 5-lobed; capsule opening at base by 3--5 pores; stigma-lobes 3--5   **1. Campanula**
  b. Corolla star-shaped, lobes strap-shaped and spreading, 8--10-lobed; capsule opening at base by 3 or more pores; stigma-lobes 8--10   **6. Michauxia**
24a. Fruit a berry   **22. Pratia**
  b. Fruit a capsule   25
25a. Corolla-tube split to base on 1 side   **25. Lobelia**
  b. Corolla-tube not split   26
26a. Fruits splitting laterally   **26. Downingia**
  b. Fruits splitting apically   27
27a. Stamens free from the corolla-tube   **23. Laurentia**
  b. Stamens fused to the corolla-tube   **24. Isotoma**

1. Campanula Linnaeus. 300/55. Annual, biennial or perennial herbs, often woody at the base with thickened rootstocks. Leaves alternate, stalked or stalkless, simple, entire, toothed, sometimes lobed; basal leaves often in rosettes. Flowers erect or nodding, solitary or few to many, axillary or terminal, in racemes, panicles, spikes or clusters. Calyx 5-lobed, sometimes with reflexed appendages alternating with the lobes, persistent. Corolla 5-lobed, campanulate, tubular with spreading lobes or flat, divided usually to at least one-third but rarely below middle, blue, violet, lilac or white, rarely yellow. Stamens free, alternate with the corolla-lobes; filaments often over-arching the ovary to form a nectar-dome; anthers usually free. Style sometimes protruding; stigma-lobes 3--5 or rarely 2. Disc not prominent, usually colourless and concealed beneath nectar-dome. Ovary with 2--5 cells, inferior although upper part may bulge upwards. Capsule erect or nodding, obovoid or narrowly oblong, sometimes angular and ribbed, usually flat-topped, opening by basal, median or apical pores, rarely by irregular rupture or indehiscent. Seeds ellipsoid, numerous, small, brownish and lustrous. *Temperate areas of the northern hemisphere, especially the Mediterranean area & western Asia, but extending in tropical mountains of Africa to the equator.* Figure 123, p. 536.

2. Campanulastrum Small. 1/1. Annual or biennial herb to 1.8 m. Stems slender, erect, simple. Lower leaves forming a rosette, stalked, downy, ovate to heart-shaped, acuminate, scalloped; upper leaves lanceolate, toothed. Flowers erect or slightly nodding, in a long racemose inflorescence. Corolla flat, lobes lanceolate, tipped with distinct points, downy, spreading star-shaped, incised to below middle, pale blue (paler at centre) to white. Stamens with anthers free and slender filaments. Style curved and recurved. Ovary inferior. Capsule slightly angular, opening by irregular median to almost apical pores. *Southeastern USA.*

Formerly included in *Campanula*.

3. Symphyandra de Candolle. 10/4. Erect perennial herbs with woody stocks and resembling *Campanula* in habit. Leaves alternate, stalked (lower long-stalked), often heart-shaped. Flowers in spherical or partly paniculate inflorescences, axillary and terminal. Calyx deeply 5-lobed, with or without appendages. Corolla 5-lobed, campanulate or with spreading lobes, hairy or hairless, white, lilac or blue. Stamens 5; filaments free, hairy, dilated at base to form a nectar-dome; anthers permanently united in a tube around style. Style hairy; stigma-lobes 3. Ovary inferior, 3-celled, top not bulging or dark green. Disc not inflated. Capsule opening by 3 pores near base. Seeds ovate, flattened, shiny. *Balkans to Caucasus and northern Iran.*

4. Hanabusaya Nakai. 2/1. Hairless perennial herbs to 80 cm. Stems erect to slightly curved. Leaves *c.* 7 cm, alternate to slightly whorled, ovate or ovate-lanceolate, acuminate, tapered to cordate at base, coarsely toothed, stiff, sparsely hairy above, hairless or almost so beneath and sometimes bronze-purplish in colour, long-stalked. Inflorescence a loose, branched panicle, with *c.* 5 nodding flowers; bracts small, uppermost becoming linear. Calyx obconical, hairless, lobes linear, toothed, recurved in bud; appendages absent. Corolla tubular-campanulate, hairless, lobes ovate, obtuse, mucronate, slightly spreading, violet. Anthers permanently united in a tube around the style. Ovary inferior, top bulging, dark green. Disc short, inflated, concealed beneath expanded filament bases (nectar-dome). Style not protruding. Capsule nodding, opening by basal pores. *Korea.*

5. Adenophora Fischer. 60/16. Perennial herbs, often with fleshy thickened lower stems. Stems erect and straight, rarely decumbent, round or angular. Leaves alternate, opposite or whorled, hairy or hairless, broadly ovate or elliptic to lanceolate or linear, entire or toothed. Inflorescence loosely racemose, paniculate or flowers solitary. Flowers terminal or axillary, usually nodding. Calyx-lobes entire or toothed, usually spreading or recurved, appendages absent. Corolla more or less campanulate with short lobes, blue, rarely white. Stamens 5, free; filaments dilated and ciliate at the base. Style slender, equal to the corolla or often considerably protruding. Stigma 3-lobed. Ovary 3-celled, inferior. Disc cylindric or ring-like, fleshy, surrounding and concealing the base of the style, colourless or greenish. Capsule crowned by the persistent

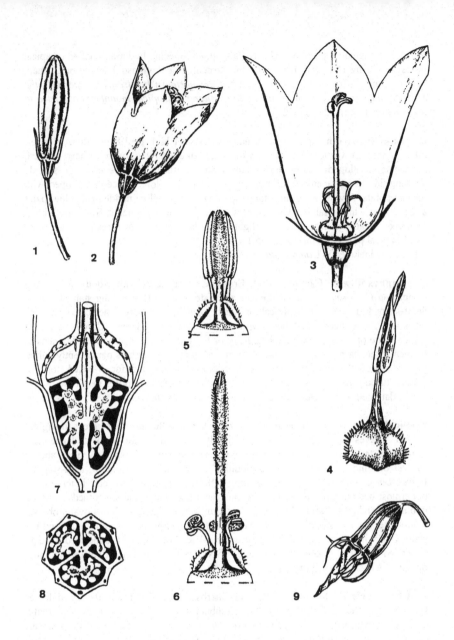

Figure 123. Campanulaceae. *Campanula rotundifolia.* 1, Flower-bud. 2, Single flower. 3, Longitudinal section of flower. 4, A single stamen. 5, Stamens and style in the male phase. 6, Stamens and style in the female phase. 7, Longitudinal section of ovary. 8, Transverse section of ovary. 9, Fruit.

536

calyx-lobes, opening by pores near base between veins. Seeds small and ovate, more or less flattened, numerous. *Eurasia, especially in the east.*

6. Michauxia L'Héritier. 7/2. Erect biennial herbs with thick stems to 2 m, usually robust, coarsely hairy. Leaves alternate, mostly ovate-oblong, lower leaves lanceolate or pinnatisect-lyrate, acute, tapered to base, irregularly toothed; upper stem-leaves stalkless and often clasping the stem. Inflorescence a dense interrupted spike, raceme or panicle; flowers axillary and terminal, almost stalkless. Calyx-tube broadly obconical with reflexed, acute appendages between lobes; lobes usually 8--10, coarsely hairy, acuminate, spreading to reflexed, much longer than ovary, enlarged and hardened in fruit. Corolla star-shaped, to 8 cm across, lobes usually 8--10, occasionally more, hairy, deeply lobed, narrow, linear-strap-shaped, spreading or reflexed, white or pale pink; almost stalkless. Stamens 8--10, free; filaments dilated at base to form a nectar-dome; anthers free, twisting after opening, pollen yellow. Style long and stout, stiffly hairy, long-protruding; stigma-lobes linear, 8--10. Ovary inferior, 8--10-celled, ovules numerous. Disc colourless. Capsule obconical, ribbed, cylindric, nodding, opening by 3, sometimes 8--10 pores at the base. Seeds small, ellipsoid, light brown, shiny. *Turkey, Caucasus to Iran & Israel.*

7. Ostrowskia Regel. 1/1. Erect, robust, hairless perennial herb with thick, hollow, unbranched stems to 1.8 m. Leaves in remote whorls, ovate, oblong or ovate-oblong, incised-toothed with mucronate or glandular teeth, glaucous; lower and middle leaves to 15 cm, similar in shape; uppermost reduced. Flowers in a spreading pyramidal raceme, stalks 5--10 cm, thickened at apices. Calyx short-cylindric or flattened-spherical, broadly conical at the base; lobes usually 7, sometimes 5--9, narrow linear-lanceolate, straight, acute, more or less leathery. Corolla to 8 cm, openly campanulate, with usually 7, sometimes 5--9, shallow, ovate lobes, hairless, milky blue or whitish with lilac veins. Stamens usually 7, sometimes 5--9, free. Style stout, cylindric; stigma 7-lobed. Nectar-dome absent. Ovary inferior, usually with 7, sometimes 5--9 cells, ovules numerous. Capsule obconical, apically constricted, opening by large, oblong, longitudinal, parallel slits around the middle. Seeds ovate-oblong, flattened, narrowly winged. *Central Asia.*

8. Trachelium Linnaeus. 2/1. Erect perennial herbs. Stems slender, woody towards base, smooth. Stock large, tuberous. Leaves ovate or lanceolate, elliptic or oblong, acute at each end, slender-stalked, sharply toothed, dark green or sometimes purplish. Inflorescence a dense or open, broad, flat-topped or hemispherical corymb with few leaves. Calyx with 5 linear lobes. Corolla 5-lobed, spreading, salver- or funnel-shaped, blue or mauve, rarely white. Anthers free. Style long-protruding; stigma capitate, 3-lobed, very slender. Ovary inferior, 3-celled. Disc inconspicuous. Nectar-dome absent. Capsule spherical or pear-shaped, opening by 3 basal pores. *West Mediterranean area from the Iberian Peninsula to Italy & north Africa; introduced in many areas, from Madeira to northwest Europe (British Isles).*

9. Diosphaera Buser. 5/3. Erect, cushion-like or tufted perennial herbs, fragile, often with woody bases. Stems numerous. Leaves small, with firm texture, hairless or hairy, overlapping, elliptic to broadly ovate, stalkless. Inflorescence a many-flowered, dense or loose corymb, or flowers solitary and terminal, on slender leafy stalks. Calyx-lobes linear-lanceolate; appendages absent. Corolla tubular, with 5 or 6 often spreading lobes, divided to a variable length, at most to middle, pale lavender-blue, pinkish or white. Filaments not or only slightly dilated at base; anthers free. Disc colourless, inconspicuous. Ovary inferior, with 2 or 3 cells, more or less circular, hairy. Style long-protruding or rarely included. Capsule spherical, opening by 2 or 3 basal pores. *South & east Greece to Turkey, Syria & Iran.*

10. Phyteuma Linnaeus. 15/10. Perennial herbs with thick, fleshy stocks and erect stems. Leaves alternate, lanceolate or ovate-lanceolate, the lower heart-shaped or obtuse, tapered at the base, toothed or double-toothed. Bracts linear or lanceolate to ovate, acuminate. Flowers small, almost stalkless, in short, more or less spherical spikes. Calyx with 4 or 5 deep lobes. Corolla with 4 or 5 lobes, strap-shaped, often curved and asymmetric in bud, united at their tips, separate only towards base at flowering, completely parted and reflexed later, dark blue to violet, lilac, white or pale yellow. Filaments linear, dilated at base; anthers free. Style long with 2 or 3 slender stigma-lobes. Nectar-dome present. Ovary inferior, with 2 or 3 cells, with free-central placentation. Capsule obconic or almost spherical, often asymmetric; opening medially or basally by 2 pores. Seeds many. *Europe.*

11. Physoplexis (Endlicher) Schur. 1/1. Tufted, hairless perennial herb to 15 cm. Leaves ovate-heart-shaped to oblong-elliptic, coarsely toothed, shiny dark green, stalked. Inflorescence a clustered umbel with up to 20 flowers; flower-stalks *c.* 5 mm. Corolla to 2 cm, claw-shaped, inflated at base, lobes permanently united at apices, pinkish violet, tipped blackish violet. Anthers free. Style long and very slender; stigma-lobes usually 2, thin, recurved. Ovary inferior. Capsule obconic or almost spherical, often asymmetric, opening medially or basally by 2 pores. *Alps.*

12. Petromarula Hedwig. 1/1. Erect, robust perennial to *c.* 1 m, minutely hairy above, hairless below. Leaves to 30 cm, in a large basal rosette, pinnate to pinnately lobed, slightly glossy, dark green; the lower long-stalked; lobes coarsely toothed, oval to oblong. Inflorescence a long, unbranched interrupted paniculate spike; flowers in clusters. Calyx deeply 5-lobed. Corolla 9--10 mm, 5-lobed, divided almost to base, lobes linear or strap-shaped, spreading to recurved, pale lilac-blue or whitish. Stamens appearing almost without filaments, though filaments swollen to form a turgid blue nectar-dome; anthers free; pollen orange-purple. Style relatively short, slightly curved; stigma-lobes short and rounded, capitate; anthers free. Ovary inferior. Capsule 3-celled, obconic or almost spherical, opening by 3 pores at the middle. *Greece (Crete).*

13. Asyneuma Grisebach & Schenck. 22/4. Biennial or perennial herbs with thin, fusiform stocks. Stems erect, simple or branched, with or without basal rosettes, rarely cushion-like. Basal leaves short-stalked, lanceolate to triangular-elliptic or ovate, toothed or entire; stem-leaves usually noticeably different from basal leaves, alternate, almost unstalked (stalks winged or not). Inflorescence a few-flowered, slender, often interrupted spike or a loose panicle, or flowers solitary. Flowers erect, short-stalked. Calyx 5-lobed, persistent. Corolla divided almost to base, lobes 5, linear or strap-shaped, not united at the apex, longer than calyx-lobes, and later recurved, blue or blue-violet, rarely whitish. Stamens 5, filaments broadening at the base to form a nectar-dome; anthers free; pollen yellow or purplish brown. Stigma-lobes 2 or 3. Ovary inferior, with 2 or 3 cells. Capsule cylindric, ovoid or more or less spherical, not constricted at apex, opening by 3 median, apical or almost apical pores. *Central and southern Europe to central Asia & India, extending to Japan.*

14. Codonopsis Wallich. 40/23. Erect or vine-like, hairless or downy perennial herbs, often strong-smelling. Stocks oblong, thick, spherical or spindle-shaped. Stems twining or climbing rarely straight, branching. Stem-leaves alternate, opposite or sometimes whorled or tufted, simple, entire, scalloped or toothed, stalked or stalkless. Flowers terminal or axillary in loose cymes or solitary, stalks hairless or downy. Calyx 5-fid, sometimes 4- or 6-fid, tube short, broad and united (or mostly so) to the ovary, hairless or downy; lobes ovate or triangular, acute or acuminate to obtuse. Corolla tubular, tubular-campanulate or star-shaped, lobes ovate or triangular, acute or obtuse, often with distinct veins and/or tessellated with irregular blotches, usually greenish to purplish, sometimes white, blue or yellow. Stamens with filaments dilated broadly or only slightly at the base, hairless or partly ciliate or densely hairy from base; anthers free. Nectar-dome present or absent, if absent then nectary often distinctly coloured and usually pentagonal. Style cylindric, frequently club-shaped, sometimes dilated and bulging upwards at base; stigma-lobes 3--5. Ovary inferior or semi-inferior, with 3--6 cells, obconic or almost spherical. Capsule conical with a flattened top, opening at the top by beaked valves, fleshy when young, becoming dry and hard, hairless or sparsely hairy, green, whitish, purple or bluish. Seeds numerous, elliptic, oblong or ovoid, wingless or sometimes slightly winged, brownish or whitish. *Turkestan to India, Malaysia, south China, Taiwan & Japan.*

15. Leptocodon (J.D. Hooker) Lemaire. 2/1. Twining perennial herb. Leaves alternate and opposite, stalked, blue-green. Flowers solitary, terminal and axillary. Calyx 5-lobed; tube short, united to ovary. Corolla shallowly 5-lobed, tubular, blue or bluish purple without prominent veins. Stamens 5, filaments long and very slender, broadening slightly at base; anthers free. Style very long and slender with 3 stigma-lobes. Nectar-dome absent. Ovary inferior. Disc with 5 prominent, blackish, club-shaped glands. Capsule 3-celled, opening by 3 valves. *Himalaya to western China.*

16. Cyananthus Bentham. 19/6. Low, spreading, trailing or ascending annual or perennial herbs with stout, occasionally branched stocks, usually more or less hairy. Leaves alternate, numerous, oblong-elliptic or obovate to wedge-shaped, entire, shallowly to deeply lobed and sometimes rather fleshy. Flowers terminal, solitary. Calyx 5-lobed, to about halfway or less, not united to ovary, often large and inflated. Corolla 5-lobed, funnel-shaped with a corona of white hairs at the throat; lobes spreading, ovate, rounded or acute, gentian-blue, blue-violet or dull yellow. Stamens 5, anthers united at base. Ovary superior, with 3--5 cells. Style with 3--5 stigmatic lobes. Capsule elliptic, opening by usually 5, rarely 3 or 4 valves at apex. *Himalaya (from Kashmir and Tibet) to the Hengduan Mountains of western China.*

17. Wahlenbergia Roth. 200/5. Erect or ascending, annual or perennial herbs. Stems simple or branched, often weak and widely divergent, hairless or hairy. Leaves opposite or alternate, often clustered on short lateral branchlets, rarely rosette-forming, often thickened, hairless above to softly hairly beneath, mostly along midrib, the margins curved inwards and downwards or upwards, entire, toothed, scalloped or wavy, rarely lobed. Inflorescence usually a cymose panicle, spike or flowers solitary, rarely an umbel; bracts narrow, bracteoles present. Flowers erect or nodding, terminal or axillary. Calyx 5-lobed, hemispheric to oblong-obconic; lobes erect, linear, persistent. Corolla campanulate, with 3--5 lobes and a distinct tube, divided usually to middle or almost star-shaped and divided almost to base; often hairy on outside and especially on mid-veins, blue, bluish purple or white, rarely with distinct veins. Stamens 3--5, free, not protruding, filaments often dilated at the base, hairy. Style sometimes with glands at top just below stigma-lobes; stigma-lobes 2--5, recurved, narrow. Nectar-dome present. Ovary with 2--5 cells, inferior or half-inferior with the apex bulging upwards, more or less spherical; placentation axile. Capsule usually erect, with 2--5 cells, opening by 2--5 valves (when 5, the cells opposite calyx-lobes) in the apex above calyx-lobes, obconic, hemispheric, narrow or cylindric. Seeds numerous, small, lenticular, ellipsoid or 3-angled, shining and brown. *Cosmopolitan but found mainly in the southern hemisphere.*

18. Edraianthus de Candolle. 10/7. Perennial tufted herbs. Stems numerous, often woody at base. Secondary rosettes produced from central stem which may then wither. Leaves alternate, linear. Inflorescence densely clustered or flowers solitary, terminal, bracteate, stalkless. Calyx 5-lobed, lobes hairy at margins, persistent. Corolla with 5 spreading lobes, divided one-third to one-half, tubular or almost campanulate, pale blue-violet, deep violet, pinkish or white. Stamens 5, free, filaments dilated at base; pollen yellow or pinkish purple. Style not protruding; stigma-lobes 2 or 3. Ovary inferior. Capsule with 2 or 3 cells, more or less spherical, ovoid or angular, flattened above, opening by irregular rupture of the apex and appearing like a deep funnel. Seeds numerous, ovate, flattened. *Mountains of southern Europe (from central Italy and Sicily to the Balkan Peninsula).*

19. Jasione Linnaeus. 13/5. Perennial, biennial or annual herbs, usually with a basal rosette of leaves and branched flowering stems, either erect and sometimes forming small clumps or spreading and forming small tufts, often woody at base. Leaves alternate, linear-oblong sometimes lanceolate to obovate, entire or remotely toothed, margin often thickened, papillose and ciliate, stalk absent, base often slightly decurrent. Flowering stems leafy at base and (in ours) leafless near the top. Inflorescence a head containing up to 300 small flowers; flower-maturation from margin towards centre. Head surrounded by 1 or more rows of involucral bracts which are shorter than flowers, usually green, in ours often purplish tinged. Flowers very shortly stalked. Calyx 5-lobed, tube ovoid or obconic, lobes erect. Corolla 5-lobed, divided almost to base into 5 linear-lanceolate, erect to spreading lobes, usually blue. Stamens 5; anthers united at base, around style. Style club-shaped, hairy in the upper-half and there initially coloured pink or purple by the pollen deposited on it; stigma-lobes 2 or 3, short, thick. Ovary inferior. Capsule hemispherical, flattened above and opening by 2 short, rounded, apical valves with wide openings. Seeds numerous, ovoid, smooth, shiny, usually brown. *Europe, Mediterranean area.*

20. Legousia Durande. 6/3. Erect, annual herbs. Stems simple or branched from base. Leaves alternate, entire, ovate to ovate-lanceolate, lower short-stalked, entire or almost waved, scalloped. Flowers solitary or in cymes, racemes or panicles, terminal and axillary. Calyx 5-lobed, divided almost to base. Corolla 5-lobed, saucer-shaped, lobes flat, spreading, blue, pinkish violet or white, each often with a darker base. Stamens 5, borne on base of corolla; filaments short, anthers free. Style downy, included in corolla; stigma-lobes 2 or 3, short, slender. Ovary inferior, cylindric, much longer than wide, usually 3-celled, rarely 1-celled. Capsule oblong, strongly angular, elongating rapidly after flowering, opening by 3 almost apical or median pores or slits. Seeds small, flattened, shiny, ovoid or more or less spherical to elliptic, brown. *Western Europe, Mediterranean area to Iran.*

21. Platycodon de Candolle. 1/1. Erect, blue-green and hairless perennial herbs, 40--70 cm. Rootstock fleshy, thick, white. Stems erect to ascending, simple, hairless or smooth. Leaves along lower half of stem or slightly above, then becoming reduced in size towards apex; alternate, almost opposite or sometimes whorled, ovate-lanceolate, tapered at base, rather coarsely-toothed, pale beneath, blue-green, stalked. Flowers terminal, solitary or sometimes 2 together, erect. Calyx blue-green, obconic, broadening above, 5-lobed, lobes straight or slightly recurved, triangular-acuminate, entire, much shorter than corolla; tube united to ovary. Corolla 5-lobed, 4--5 cm across, broadly star-shaped with spreading and/or recurved lobes; lobes divided one-third to one-half, ovate, acute, blue or purplish blue with dark blue-violet veins. Stamens 5, free; filaments broadening at the base forming an incomplete nectar-dome, hairy. Style broadened at base, club-shaped; stigma-lobes 5, short and thick, white. Ovary inferior, 5-celled. Capsule erect, ovoid, opening by 5 valves opposite calyx-lobes. Seeds ovate, large, shiny, blackish and winged. *East Siberia (Ussuriland), Korea, Japan, northeast China.*

22. Pratia Gaudichaud. 36/8. Perennial herbs with erect or creeping or prostrate stems, rooting at the nodes. Leaves linear, spathulate to circular, stalkless, shallowly toothed. Flowers in terminal racemes or, more commonly, solitary in the leaf-axils, inverted by a twisting of the flower-stalk. Calyx with 5 linear lobes. Corolla white, tube cylindric, entire and unsplit, or split to the base on the upper side, lobes more or less equal. Stamens 5 fused into a tube, the smaller anthers each with a terminal tuft of bristles. Ovary top-shaped, 2-celled; stigma-lobes about as wide as long, recurved at maturity. Fruit a spherical berry, deep red to purple-black. *Tropics & south temperate areas.*

23. Laurentia Adanson. 25/1. Annual herbs with erect stems (ours). Leaves spathulate to oblong, stalked, entire. Flowers solitary (ours), inverted by a twisting of the flower-stalks. Calyx with 5 linear lobes. Corolla-tube entire (not split as in *Lobelia*), limb strongly 2-lipped. Stamens 5 fused into a tube, the 2 shorter anthers each with a terminal tuft of bristles. Ovary top-shaped, 2-celled; stigma-lobes about as wide as long, recurved at maturity. Fruit an obconic capsule opening by apical slits. *Mediterranean area, South Africa.*

24. Isotoma Lindley. 8/3. Annual or perennial herbs with erect or prostrate, branched stems. Leaves lanceolate to circular, stalkless, unevenly toothed or pinnatifid. Flowers solitary in the leaf-axils, or in leafy racemes, inverted by a twisting of the flower-stalk. Calyx with 5 linear lobes. Corolla blue or white, tube cylindric, entire (not split as in *Lobelia*), lobes more or less equal. Stamens 5 fused into a tube, the 2 smaller anthers each with a terminal tuft of bristles. Ovary top-shaped, 2-celled; stigma-lobes about as wide as long, recurved at maturity. Fruit a spherical capsule opening by apical slits. *Australia, New Zealand.*

541

25. Lobelia Linnaeus. 350/21. Annual or perennial herbs with erect or decumbent, branched stems. Leaves usually simple, stalked or stalkless, unevenly toothed or rarely pinnatifid. Flowers solitary in the leaf-axils or in racemes, spirally arranged or 1-sided, inverted by twisting of the flower-stalks. Calyx with 5 linear lobes. Corolla red, yellowish, pink, lavender, blue or white, tube cylindric, split to the base on the upper side, lobes usually unequal, those of the lower lip larger than those of the upper. Stamens 5 fused into a tube, the 2 smaller anthers each with a terminal tuft of bristles. Ovary top-shaped, 2-celled; stigma-lobes about as wide as long, recurved at maturity. Fruit a spherical capsule opening by apical slits. *Tropical and warm temperate areas worldwide.*

26. Downingia Torrey. 13/1. Annual herbs with erect or decumbent, branched stems to 40 cm. Leaves 5--20 × 0.5--4 mm, linear to elliptic (in the cultivated species), stalkless, entire. Flowers solitary in the leaf-axils or in spikes, inverted by a twisting of the stalkless, long-cylindric ovary. Calyx-lobes 5, linear. Corolla-tube entire (not split as in *Lobelia*), limb strongly 2-lipped, the 2 lobes of the upper lip smaller than the 3 of the lower. Stamens 5 fused into a tube, the 2 smaller anthers each with a terminal tuft of bristles or a horn-like appendage. Ovary narrow, cylindric, stalk-like, 1-celled (in the cultivated species), stigma-lobes about as long as wide, recurved at maturity. Fruit a cylindric capsule opening by lateral slits. *Western North America, 1 species in Chile.*

A genus of 13 species, 1 occasionally cultivated.

194. GOODENIACEAE

Shrubs, herbs or, more rarely, trees. Leaves alternate, rarely opposite or whorled, simple. Stipules absent. Flowers bisexual, solitary or in cymes, racemes or heads. Sepals 3--5, free or united, sometimes reduced or minute, lobed. Corolla of 5 fused petals, lobes usually each with a membranous wing, bilaterally symmetric. Stamens 5, opposite sepals, anthers free or united. Ovary usually inferior, more rarely half-inferior or almost superior. Ovary with 2 cells (usually incomplete), rarely 4 or 1 with 1--2 ovules per cell, placentation axile or basal. Style 1, stigma sheathed. Fruit a capsule, drupe or nut. Seeds with endosperm, flat or winged.

A family with 11 genera and about 400 species, mostly found in Australasia, but with a few species from the Pacific, Atlantic and Indian Ocean islands, southern China, Africa and Chile. All are marginally hardy in northern Europe, but several are grown as half-hardy annuals, or are planted out in summer.

1a. Ovary with 1 or 2 ovules		2
b. Ovary with more than 2 ovules		3
2a. Anthers united to form a tube; corolla 2-lipped		**1. Dampiera**
b. Anthers free, corolla fan-like		**2. Scaevola**
3a. Anthers united to form a tube		**3. Lechenaultia**
b. Anthers free		4
4a. Corolla-lobes wingless; fruit fleshy		**4. Selliera**
b. Corolla-lobes winged; fruit usually a capsule		**5. Goodenia**

1. Dampiera R. Brown. 66/2. Subshrubs or herbs, sometimes suckering. Leaves variable, usually alternate, simple, hairless or densely hairy, margins entire, toothed or lobed; stalks present or absent. Flowers in racemes, cymes or solitary. Sepals 5, more or less united to the ovary, sometimes absent or replaced by long hairs. Bracteoles usually present, occasionally absent. Corolla bilaterally symmetric, blue, red, pink, white or yellow, united into a tube which is split down the upper sides and with its exterior hairy, rarely hairless; upper lobes 2, with auricles, winged; lower lobes 3, winged. Stamens 5, anthers united around the style. Style simple, straight, shorter than the corolla, hairless. Ovary with 1 or rarely 2 cells with 1 ovule in each cell. Fruit a nut. Seeds ovoid or curved with no wing. *Australia.*

2. Scaevola Linnaeus. 96/12. Shrubs, subshrubs or herbs. Leaves alternate, rarely opposite, entire or toothed; stalks present or absent. Flowers solitary or in racemes, spikes or cymes. Bracteoles usually present, in pairs. Sepals 5, usually united to form a tube which is united to the ovary. Corolla bilaterally symmetric, white, blue, purple or rarely yellow, hairy or hairless, united to form a tube which is split open down the lower side; lobes 5, usually equal, with wings. Stamens 5, free. Style simple, unbranched, bent at the apex, hairy or hairless, shorter than the corolla. Ovary with 1--4 cells, inferior; cells usually each with 1 ovule. Fruit indehiscent. Seeds ovoid, wingless. *Australia, Pacific Islands, Socotra, Hainan Island, 2 species widespread in the tropics.*

3. Lechenaultia R. Brown. 26/5. Perennial herbs, subshrubs or heathlike shrubs with several to many stems, usually hairless. Leaves linear to ovate, entire, usually hairless; stalks absent. Flowers stalkless, solitary or in terminal cymes. Bracts and bracteoles leaf-like, rarely reduced. Calyx 5-lobed, lobes linear to lanceolate, usually hairless. Corolla bilaterally symmetric, red, pink, blue, yellow, white or in combination, united into a tube which is often split down to the base; tube length variable, interior variously hairy, exterior usually hairless; upper lobes 2, sometimes winged; lower lobes, 3, sometimes winged. Stamens 5, filaments free or united, anthers usually united. Style simple, straight. Ovary with 2 cells, inferior. Ovules in 2 ascending rows, numerous. Fruit a capsule, linear, with 4 cells, longitudinal. *Australia, 1 species extending to New Guinea.*

The name is sometimes misspelled 'Leschenaultia'.

4. Selliera Cavanilles. 1/1. Perennial herbs. Stems prostrate to 50 cm, hairless, rooting at the nodes. Leaf-blades 1--11 cm × 2--35 mm, oblanceolate to obovate, simple, entire, alternate or usually clustered at the nodes. Flowers usually solitary, rarely in dense racemes, axillary; inflorescence-stalk to 5 cm. Bracts present. Bracteoles to 3 mm, linear. Sepals 5, 4--5 mm, ovate to oblong. Corolla 5--12 mm, bilaterally symmetric, reddish brown outside, interior usually white, united to form a hairless tube which is split down one side; corolla-lobes 5, more or less equal, acutely ovate, wings absent. Stamens 5, free. Style unbranched, sparsely hairy. Ovary with 2 cells, inferior. Ovules in 2 rows in each cell. Fruit 4--6 mm, ovoid to oblong, fleshy, dehiscent or indehiscent. Seeds *c.* 2 mm wide, circular to elliptic, numerous, each with a wing *c.* 0.5 mm wide. *Australia, New Zealand, Chile.*

5. Goodenia Smith. 179/4. Shrubs and herbs. Leaves variable, usually alternate, simple. Flowers in racemes or cymes, terminal or axillary. Sepals more or less united to the ovary or free, 5-lobed. Corolla bilaterally symmetric, yellow, blue or white, hairy or hairless, united into a tube which is split down one or two sides. Upper lobes 2, winged; lower lobes 3, winged. Stamens 5, free. Style unbranched or 2--4-fid. Ovary with 2, often incomplete, cells, inferior or half-inferior. Ovules in 2 rows, scattered or

543

rarely solitary. Fruit usually a capsule with 2--4 cells, rarely a nut or a drupe. Seeds flat, usually winged. *Australia, New Guinea, Philippines, China.*

195. BRUNONIACEAE
Densely hairy, herbaceous perennial to 1 m. Leaves basal, to 12 cm, obovate, entire. Flowers small in dense heads to *c*. 3 cm across, blue. Calyx 5-lobed, tube to 2 mm. Corolla 5-lobed, radially symmetric, lobes edge-to-edge in bud, tube to 4 mm. Stamens 5, inserted at base of corolla-tube; anthers adhering around style. Ovary superior, 1-celled, containing a single basal ovule. Fruit a nut to 2 mm, enclosed in the hardened calyx-tube.

A family of a single genus and species from Australia.

1. Brunonia R. Brown. 1/1. Description as for family. *Australia.*

196. STYLIDIACEAE
Annual or perennial herbs, rarely woody at base; latex absent. Leaves in basal rosettes, spirally arranged, linear, almost grass-like. Stipules absent. Inflorescence usually glandular-hairy. Flowers usually bisexual in terminal bracteate corymbs, racemes, cymes or panicles. Calyx tubular, 5-lobed, rarely 2- or 7-lobed, lobes free or sometimes 2 fused, or with free sepals. Corolla shortly tubular, 5-lobed, lobes overlapping, unequal, the lowermost lobe smaller and modified forming the lip (or labellum). Stamens 2, filaments free from corolla but united to style forming a sensitive column. Stigma entire or 2-lobed, concealed between the anthers or protruding from them. Nectary-disc or pair of glands above ovary. Ovary inferior, of 2 fused carpels, 2-celled or 1-celled at base; placentation axile to free-central, ovules numerous. Fruit a capsule, 1- or 2-celled, rarely indehiscent. Seeds small, usually numerous.

A family of 5 genera and about 150 species from Australia, New Zealand, south-east Asia and Chile.

1. Stylidium Willdenow. 136/3. Annual or perennial herbs. Leaves basal, tufted or arising from stem, alternate or in tufts, margins entire or toothed. Calyx 5-lobed more or less united into 2 lips. Corolla tubular, 5-lobed, lobes in 2 pairs usually arranged laterally, remaining lower lobe (the lip or labellum) usually smaller and reflexed or occasionally nearly as long and curved upwards; lobes overlapping in bud. Column elongate, bent down or folded, sensitive to stimuli. Stamens 2, anthers either side of stigma at top of column. Stigma undivided. Ovary incompletely 2-celled. Fruit a capsule, 2-valved, spherical, linear or lanceolate. Seeds usually numerous. *South-east Asia, New Zealand, Australia.*

The sensitive trigger-action of the column (a structure formed by the united stamens and style) gives the plants their common name (Triggerplants) and describes their unique method of insect pollination.

197. COMPOSITAE *(Asteraceae)*
Annual to perennial herbs, subshrubs, shrubs or small trees, often aromatic; sap sometimes milky. Leaves usually alternate or all basal, more rarely opposite or whorled, simple, entire to lobed, or pinnately or palmately compound; stipules usually absent, occasionally lowermost leaflets appearing superficially as stipules. Flowers small, grouped together in characteristic heads (capitula), the flowers borne on a flat, convex or conical receptacle and surrounded by 1 or several whorls of bracts forming

an involucre; the receptacle may be hairy or scaly (at least in part) or naked; the bracts of the involucre may be free or variously united, and sometimes bear appendages of various forms. Heads rarely 1-flowered, when clearly so by reduction. Flowers unisexual or bisexual (heads usually with flowers bisexual or of both sexes or female and bisexual, occasionally plants dioecious), of various forms, occurring in various combinations. The central flowers usually form a disc and are generally bisexual, with a campanulate, usually 5-lobed, radially symmetric corolla and a calyx represented by a pappus of hairs or scales, a corona or absent; sometimes the outermost of such disc-flowers are female only, with a thread-like, narrowly tubular corolla which is scarcely lobed, or occasionally the corolla is apparently 2-lipped, or the flowers are effectively male; the head may contain only such flowers, or there may be 1 or more whorls of flowers which may be female, bisexual or sterile in which the corolla has a strap-shaped extension (a ray or ligule), thus being bilaterally symmetric. In heads which contain both disc- and ray-flowers, the rays are generally 3-toothed at the apex. In heads which contain only ray-flowers, the rays are usually 5-toothed at the apex (generally such plants have milky latex), or are 3-toothed at the apex, when the heads are effectively 'double' variants of heads with ray- and disc-flowers. Very occasionally the ray-flowers with 3-toothed apices to the rays also have 2 other lobes, which are smaller, rendering the corolla 2-lipped. In all cases there are usually 5 (rarely 4) stamens, whose filaments are free, but whose anthers are united into a tube around the style, and an inferior ovary of 2 united carpels, containing a single basal ovule, and with a style branched above into 2 stigmas. The fruit, which is dry and indehiscent, is technically a cypsela (a 1-seeded, hard, indehiscent fruit formed from an inferior ovary), but is simply referred to as a fruit in this account; it is usually crowned by the pappus, if there is one; this may be of simple or rough bristles, plumose hairs, scales, teeth, awns, or a small corona. See figures 124--132, pp. 548, 550, 552, 554, 556, 558, 560, 576 & 598.

Probably the largest family of Flowering Plants, with almost 1500 genera and about 22,000 species, absolutely cosmopolitan.

The family itself is generally very easily recognised, but this is not the case with the genera within it, which are often distinguished on small technical characters of the anthers, styles, stigmas, pappus and fruits which are not easily seen or understood until quite a large amount of comparative material has been studied. In the present account, we have tried to avoid such difficult characters as far as possible, but this is not always the case. It seems likely that the genera have, in fact, been over-split and some regrouping may well occur over the next twenty years.

In this account the widely-used term 'florets' for the individual flowers (of whatever type) has been avoided, and 'flowers' is used instead. Flowers with rays (ligules) are referred to as 'ray-flowers', and flowers without rays are referred to as 'disc-flowers'; the elongate corolla-extension of ray-flowers is referred to as a 'ray'. The bracts forming the involucre are referred to as 'involucral bracts', the use of the term 'phyllaries' being eschewed.

The key to the genera provided here is based, ultimately, on the excellent keys written by Gunnar Wagenitz for *Parey's Blumengärtnerei*, edn 2, **2**: 688--695 (1960) and for Hegi, *Illustrierte Flora von Mitteleuropa*, edn 2, 4(3): XXXVI--XLIV (1979). The present key is more extensive than either of these, however, and any errors will be the responsibility of the present editor.

  1a. All flowers in the head with rays   2
   b. All flowers, or at least the inner, tubular or rarely 2-lipped   3
  2a. Rays 5-toothed at the apex; milky latex almost always present
    Group 9 (p. 567)
   b. Rays 3-toothed at the apex; milky latex absent (these are 'double'
      variants of plants which normally have disc-flowers in the head)
    Group 8 (p. 566)
  3a. Plants with spines on leaves or stems (i.e. plants thistle-like)
    Group 1
   b. Plants not thistle-like, not spiny on leaves or stems   4
  4a. Heads with tubular flower only (the outer sometimes enlarged or thin
      and thread-like)   5
   b. Heads with central tubular (rarely 2-lipped) flowers, and marginal
      ray-flowers   6
  5a. Pappus made up of numerous hairs or bristles   *Group 2* (p. 547)
   b. Pappus absent or a membranous ridge, a corona, or made up of scales,
      tubercles, or at most 5 bristles   *Group 3* (p. 553)
  6a. Pappus mostly of numerous hairs or bristles, scales present or
      absent   *Group 4* (p. 555)
   b. Pappus absent or a membranous ridge, a corona or made up of scales,
      tubercles, or at most 5 bristles   7
  7a. Leaves, at least in part, opposite, very rarely in whorls   *Group 7* (p. 565)
   b. Leaves alternate or all basal   8
  8a. Receptacle with chaffy scales subtending most of the flowers, or, at
      least in a ring between the disc- and ray-flowers   *Group 6* (p. 563)
   b. Receptacle with no chaffy scales, or with the occasional scale, but not
      as above   *Group 5* (p. 561)

Group 1

  1a. Climber; leaves with tendrils   **3. Mutisia**
   b. Plants not climbing; tendrils absent   2
  2a. Heads 1-flowered, aggregated into spherical secondary heads   **7. Echinops**
   b. Heads not as above   3
  3a. Corolla 2-lipped   **4. Nassauvia**
   b. Corolla not 2-lipped, more or less equally 5-lobed   4
  4a. Leaves white-veined or variegated above   5
   b. Leaves neither white-veined nor variegated above   6
  5a. Outer and middle involucral bracts each with a toothed appendage
      ending in a long, spreading spine   **20. Silybum**
   b. Outer and middle involucral bracts without toothed appendages, each
      ending in a slender spine   **21. Galactites**
  6a. Tall, grey-leaved plants, the leaves decurrent as wings on the stems;
      receptacular scales absent, but flowers borne in fringed pits on the
      receptacle   **17. Onopordum**
   b. Plants not as above   7
  7a. At least some fruits with the pappus-hairs plumose   8
   b. Pappus-hairs never plumose, or pappus of scales or absent   11
  8a. Ovary and fruit hairy   **8. Carlina**

b. Ovary and fruit hairless   9

9a. Small spines present on the leaf upper surface and on the margins between the larger spines   **19. Cirsium**

b. Spines absent from the leaf upper surface; all marginal spines more or less equal (large)   10

10a. Receptacle thick and fleshy   **14. Cynara**

b. Receptacle not thick and fleshy   **15. Ptilostemon**

11a. Outer fruits coarsely wrinkled, without pappus, inner fruits smooth, with pappus   **28. Carthamus**

b. All fruits similar, all with pappus   12

12a. All flowers yellow   **27. Cnicus**

b. Flowers blue or purple, rarely white   13

13a. Stems without spiny-margined wings   **29. Carduncellus**

b. Stems with spiny-margined wings   14

14a. Receptacular scales present, hair-like   **18. Carduus**

b. Receptacular scales absent, but flowers borne in fringed pits on the receptacle   **17. Onopordum**

Group 2

1a. At least some of the leaves opposite or rarely whorled   **186. Eupatorium**

b. Leaves all alternate or basal   2

2a. Outer flowers trumpet-shaped, larger than most of the inner, corolla 5- or more-lobed, sterile; bracts each with a membranous or spine-like appendage (rarely reduced to a rim)   3

b. Outer flowers not distinguished from the inner in this way; bracts not as above   4

3a. Receptacle not hairy; corollas yellow; spines of involucral bract-appendages reddish, pinnate   **27. Cnicus**

b. Receptacle hairy; combination of characters not as above   **26. Centaurea**

4a. Leaves all basal, inflorescences on scapes which may be covered with scale-leaves, but are without foliage leaves   5

b. Foliage leaves borne on the flowering stems or stems absent, heads borne directly on the rosette of leaves   8

5a. Foliage leaves large, but developing after the flowers; scapes usually bearing many heads; plants dioecious   **141. Petasites**

b. Foliage leaves with the flowers, small; scapes generally 1- or few-headed, or if heads several, then all aggregated together; plants not dioecious   6

6a. Heads several, but condensed into a complex 'head of heads'   **69. Craspedia**

b. Heads solitary   7

7a. Leaves silvery-white to yellow woolly, stalkless, lanceolate to oblong   **72. Leucogenes**

b. Leaves brownish hairy, long-stalked, blades almost circular, palmately lobed   **142. Homogyne**

8a. Involucral bracts in 1 row (occasionally very minute bracts at the base outside the others)   9

b. Involucral bracts in 2 or more rows   14

9a. Heads purple or white   **136. Cacalia**

b. Heads yellow, orange-yellow or red   10

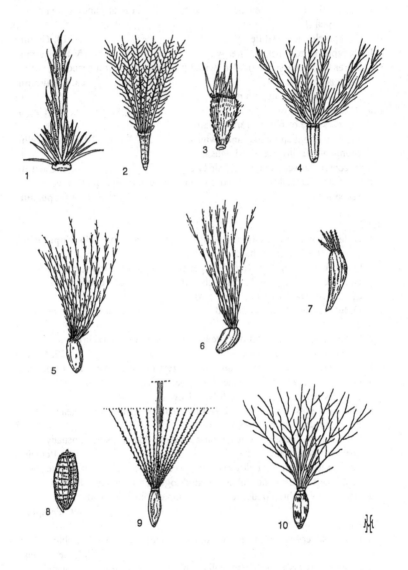

Figure 124. Compositae: fruits. 1, *Echinops*. 2, *Carlina*. 3, *Xeranthemum*. 4, *Sauss-urea*. 5, *Cynara*. 6, *Ptilostemon*. 7, *Arctium*. 8, *Onopordum*. 9, *Carduus*. 10, *Cirsium*.

10a. Plant annual **134. Emilia**
  b. Plant perennial or a shrub 11
11a. Plant a succulent shrub **132. Senecio**
  b. Plant a shrub or herb, not succulent 12
12a. Plant herbaceous **132. Senecio**
  b. Plant a shrub 13
13a. Hairs stellate, forming a white wool; leaves finely toothed or scalloped
                                        144. Bedfordia
  b. Hairs not stellate, sometimes forming white wool; leaves entire or with
    neatly wavy margins, appearing scolloped **145. Brachyglottis**
14a. Receptacle bearing hairs or scales subtending most of the flowers 15
  b. Receptacle without hairs or scales subtending the flowers or rarely
    with a few such scales 26
15a. Involucral bracts hooked at the apex **16. Arctium**
  b. Involucral bracts not hooked at the apex 16
16a. Heads stalkless; corollas pale cream or golden yellow, rarely pale pink;
    pappus-hairs in 2 series, those of the inner row longer than those of the
    outer, all twisted at the base **6. Berardia**
  b. Combination of characters not as above 17
17a. Involucral bracts acuminate, apices lying flat or recurved, without
    spine-like or membranous appendages 18
  b. Involucral bracts each with a spine-like or membranous appendage 24
18a. Evergreen shrub with heather-like (ericoid), often glandular leaves
                                          64. Cassinia
  b. Combination of characters not as above 19
19a. Pappus in 1 row; receptacular bracts narrow, multifid; corolla pink to
    purple; leaves often crowded in rosettes near the apices of the branches
                                          9. Staehelina
  b. Combination of characters not as above 20
20a. Pappus plumose, in 1 or 2 rows **11. Saussurea**
  b. Pappus not plumose, in several rows, sometimes scale-like 21
21a. Leaves not both pinnately divided and densely white- or grey-felted
    beneath **22. Serratula**
  b. Leaves pinnately divided and densely white- or grey-felted beneath 22
22a. A few pappus-hairs longer than all the others; pappus-hairs not united
    into a ring at the base **12. Jurinea**
  b. All pappus-hairs of more or less the same length (rarely a few outer
    shorter than the others), all united into a ring at the base 23
23a. Heads stalkless, 1 to each rosette **13. Jurinella**
  b. Heads stalked, not as above **23. Stemmacantha**
24a. Flowers blue-violet with a bluish style **14. Cynara**
  b. Flowers purple 25
25a. Pappus longer than the fruit **24. Leuzea**
  b. Pappus much shorter than the fruit **26. Centaurea**
26a. Shrubs 27
  b. Annual, biennial or perennial herbs 34
27a. Mat- or cushion-forming shrubs to 10 cm; involucral bracts in 4 rows
                                      99. Nardophyllum
  b. Plant not as above 28

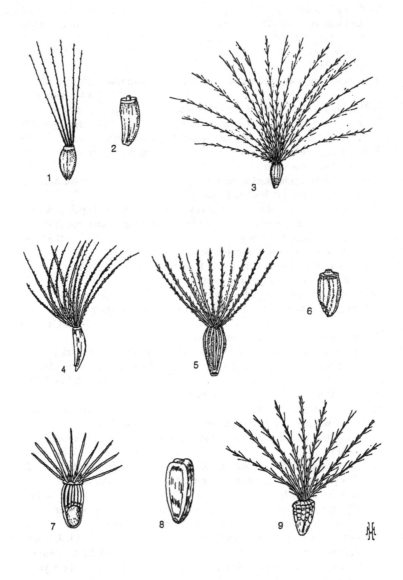

Figure 125. Compositae: fruits. 1, *Silybum*. 2, *Galactites*. 3, *Leuzea*. 4, *Serratula*. 5, *Serratula* (another species). 6, *Centaurea*. 7, *Cnicus*. 8, *Carthamus*. 9, *Carduncellus*.

550

28a. Heads 1-flowered   **101. Olearia**
    b. Heads with more than 1 flower   29
29a. Involucral bracts in 5 rows; flowers 5   **79. Chrysothamnus**
    b. Involucral bracts not as above; flowers usually more than 5   30
30a. Leaves heather-like (ericoid), linear, not more than 5 mm wide   31
    b. Leaves not heather-like, not linear, absolutely and relatively broader   32
31a. Heads numerous in corymbs; flowers white   **63. Ozothamnus**
    b. Heads solitary; flowers yellow   **91. Chrysocoma**
32a. Corollas white or purple; inner involucral bracts usually purplish; pappus
       of many rough or plumose bristles, flattened on their outer side
                                                   187. Brickellia
    b. Combination of characters not as above   33
33a. Leaves densely white-hairy, generally so on both surfaces
                                             73. Helichrysum
    b. Leaves not white-hairy on either surface   **90. Baccharis**
34a. Cushion-forming plants with leaves not more than 6 mm, entire, regularly
       arranged   35
    b. Plants not as above   36
35a. Heads with more female flowers than bisexual flowers   **62. Raoulia**
    b. Heads with more bisexual flowers than female   **73. Helichrysum**
36a. Involucral bracts membranous or each with a membranous, often
       coloured, petal-like appendage; leaves entire, white-woolly   37
    b. Involucral bracts mostly herbaceous, generally green or greenish; leaves
       generally not as above   44
37a. Heads congested at the ends of the stems, surrounded by leaves
       arranged in a star-shape   **71. Leontopodium**
    b. Heads not as above   38
38a. Pappus-hairs plumose from the base to the apex   39
    b. Pappus-hairs rough, if plumose then so only towards the apex   40
39a. Hairless or very slightly hairy annuals or rarely perennials   **56. Rhodanthe**
    b. Densely hairy or velvety woody-based perennials or subshrubs
                                               55. Syncarpha
40a. Flowers all unisexual; heads unisexual or with flowers of both sexes   41
    b. All heads containing bisexual flowers   42
41a. Plants strictly dioecious; pappus-hairs of male flowers with thickened,
       plumose apices; stems usually less than 20 cm   **66. Antennaria**
    b. Plants not strictly dioecious; pappus-hairs of all flowers without thickened,
       plumose apices; stems usually more than 20 cm   **67. Anaphalis**
42a. Heads generally brownish, not conspicuous, containing more female
       than bisexual flowers   **74. Gnaphalium**
    b. Heads conspicuous because of coloured bracts, containing more bisexual
       than female flowers   43
43a. Fruits 4-angled in section   **68. Bracteantha**
    b. Fruits terete or somewhat compressed in section   **73. Helichrysum**
44a. Flowers golden yellow   45
    b. Flowers white, greenish, red or blue, rarely yellowish white   46
45a. Pappus-hairs unequal   **86. Aster**
    b. Pappus-hairs equal   **60. Inula**

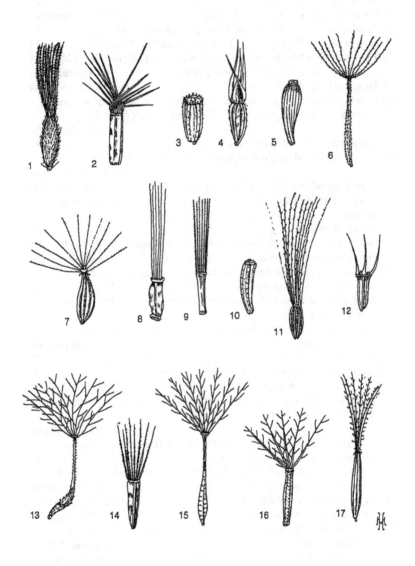

Figure 126. Compositae: fruits. 1, *Perezia*. 2, *Berardia*. 3, *Cichorium*. 4, *Catananche*. 5, *Lapsana*. 6, *Crepis*. 7, *Lactuca*. 8, *Cicerbita*. 9, *Prenanthes*. 10, *Sonchus*. 11, *Andryala*. 12, *Tolpis*. 13, *Urospermum*. 14, *Hieracium*. 15, *Hypochoeris*. 16, *Leontodon*. 17, *Scorzonera*.

46a. Heads arranged in strict spikes or racemes   **185. Liatris**
   b. Heads arranged in panicles   **50. Vernonia**

Group 3

1a. At least the lower leaves opposite   2
  b. All leaves alternate or all basal   6
2a. Leaves conspicuously 3-veined; heads each with 5 disc-flowers and 5 involucral bracts in a single series, in panicles; corolla with purple throat and white limb   **184. Stevia**
  b. Combination of characters not as above   3
3a. Flowers yellow   **172. Bidens**
  b. Flowers white, blue, purplish or pale pink   4
4a. Flowers 3--5 per head   **183. Piqueria**
  b. Flowers many per head   5
5a. Pappus absent; plants often grown as annuals   **182. Ageratum**
  b. Pappus present; shrubs   **181. Oxylobus**
6a. Small trees, shrubs or subshrubs   7
  b. Annual, biennial or perennial herbs   11
7a. Leaves narrowly obovate, with a few teeth in the upper half, hair-covering dense beneath and showing on the margins, containing forked and tree-like hairs   **109. Ajania**
  b. Combination of characters not as above   8
8a. Heads 2- or 3-flowered, aggregated into more or less spherical 2nd-order heads; leaves minute (to 4 mm)   **70. Leucophyta**
  b. Heads not as above; leaves generally larger   9
9a. Heads in spikes or racemes, small, inconspicuous   **110. Artemisia**
  b. Heads in panicles or solitary, larger, conspicuous   10
10a. Heads solitary, or few together, long-stalked; corollas persistent in fruit   **111. Santolina**
  b. Heads in dense panicles; corollas not persistent   **112. Otanthus**
11a. Involucral bracts with large, entire, membranous, often coloured appendages or bracts themselves pink or red-brown   12
  b. Involucral bracts without appendages or sometimes with membranous margins or each with a tiny, fringed appendage   14
12a. At least the upper leaves pinnatifid; receptacle covered with smooth bristles; fruits hairy, ribbed and wrinkled   **25. Amberboa**
  b. Combination of characters not as above   13
13a. Leaves decurrent; pappus cup-shaped, unequally 4-angled
    65. Ammobium
  b. Leaves not decurrent; pappus of scales   **10. Xeranthemum**
14a. Receptacle with chaffy scales over most of its surface   15
  b. Receptacle without chaffy scales over most of its surface, hairy, bristly, or hairless   16
15a. Flowers yellow   **112. Otanthus**
  b. Flowers white, blue or red   **155. Marshallia**
16a. Receptacle bristly; involucral bracts with small, fringed appendages
    26. Centaurea
  b. Receptacle not bristly, sometimes hairy; involucral bracts without appendages   17

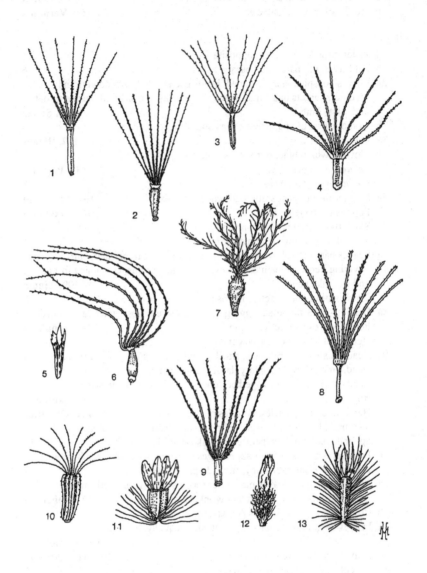

Figure 127. Compositae: fruits. 1, *Inula*. 2, *Pulicaria*. 3, *Ozothamnus*. 4, *Cassinia*. 5, *Ammobium*. 6, *Antennaria*. 7, *Craspedia*. 8, *Leontopodium*. 9, *Helichrysum*. 10, *Vernonia*. 11, *Arctotis*. 12, *Berkheya*. 13, *Gazania*.

554

17a. Outer flowers of the head enlarged, white, pink or yellow; stamens
    projecting beyond corolla   **146. Chaenactis**
   b. Combination of characters not as above   18
18a. Heads in spikes or racemes, small, inconspicuous   **110. Artemisia**
   b. Heads solitary or in panicles, usually conspicuous   19
19a. Flowers lavender to purple   **147. Palafoxia**
   b. Flowers yellow or brownish, sometimes greenish   20
20a. Upright herbs   21
   b. Low mat- or cushion-forming plants   22
21a. Pappus completely absent   **117. Chrysanthemum**
   b. Pappus a corona   **107. Tanacetum**
22a. Receptacle flat to convex; central flowers bisexual   **128. Cotula**
   b. Receptacle conical; central flowers female-sterile   **129. Leptinella**

Group 4

1a. Climber with tendrils   **3. Mutisia**
  b. Herbs or shrubs, not climbing, without tendrils   2
2a. Disc- and ray-flowers of contrasting colours, the disc-flowers usually
    yellow or brownish, the ray-flowers usually white, violet or blue   3
  b. Disc- and ray-flowers of essentially the same colour, generally yellow,
    the disc-flowers sometimes brownish   24
3a. Shrubs   4
  b. Annual, biennial or perennial herbs   7
4a. Leaves deciduous, herbaceous; rays usually blue   **92. Felicia**
  b. Leaves evergreen, leathery; rays generally white, rarely blue   5
5a. Receptacle with scales   **98. Chiliotrichum**
  b. Receptacle without scales   6
6a. Heads solitary on long, leafless stalks; leaves entire, very thick and fleshy,
    hairless and shining above, felted beneath, 6.5--16 × 3--7.5 cm; rays
    yellow   **103. Pachystegia**
  b. Heads usually in corymbs, if solitary then borne on leafy stalks; leaves
    not as above; rays usually not yellow   **101. Olearia**
7a. Heads borne on scapes (which may bear scale-leaves), the foliage leaves
    all in a basal rosette   8
  b. Heads borne on leafy stems   13
8a. Leaves at least beneath, densely white-felted   **102. Celmisia**
  b. Leaves variously hairy or hairless, not as above   9
9a. Heads solitary; fruits flattened, with a conspicuous, translucent glandular
    wing   **97. Brachycome**
  b. Combination of characters not as above   10
10a. Both ray- and disc-flowers 2-lipped; leaves pinnately lobed; rays white
    inside, reddish outside   **2. Leibnitzia**
  b. Combination of characters not as above   11
11a. Pappus-bristles somewhat expanded into scales at the base; involucral
    bracts with membranous margins   **96. Townsendia**
  b. Pappus-bristles not as above, fine and hair-like; involucral bracts
    herbaceous   12

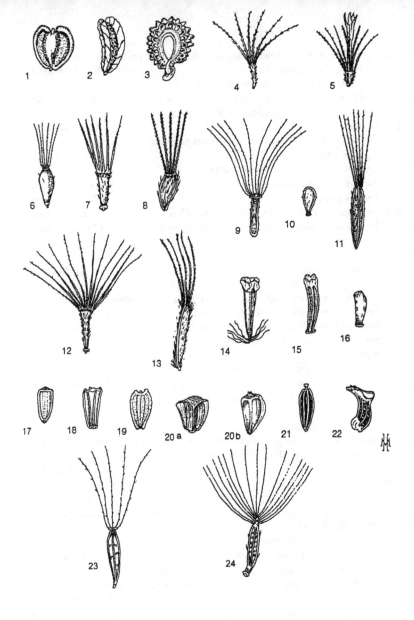

Figure 128. Compositae: fruits. 1, *Dimorphotheca*. 2, *Osteospermum*. 3, *Calendula*. 4, *Solidago*. 5, *Solidago* (another species). 6, *Erigeron*. 7, *Erigeron* (another species). 8, *Chrysocoma*. 9, *Felicia*. 10, *Bellis*. 11, *Celmisia*. 12, *Olearia*. 13, *Pachystegia*. 14, *Ursinia*. 15, *Tanacetum*. 16, *Artemisia*. 17, *Achillea*. 18, *Anthemis*. 19, *Chrysanthemum*. 20a & b, *Xanthophthalmum*. 21, *Leucanthemum*. 22, *Coleostephus*. 23, *Aster*. 24, *Aster* (another species).

556

12a. Pappus made up of about 5 bristles, between which there are short scales
          94. Bellium
  b. Pappus made up of numerous hairs, scales absent   **86. Aster**
13a. Outer involucral bracts leaf-like, larger than the others; pappus made up
    of bristles and a corona   **88. Callistephus**
  b. Outer involucral bracts not as above; pappus made up only of hairs   14
14a. Ray-flowers sterile; leaves entire or toothed, white-woolly at first, later
    often hairless; rays violet to pink   **82. Corethrogyne**
  b. Combination of characters not as above   15
15a. Leaves very finely pinnatisect and white-woolly; plant prostrate, forming
    loose mats   **106. Allardia**
  b. Combination of characters not as above   16
16a. Stem-leaves fleshy; pappus of 5 bristles and several smaller scales
          94. Bellium
  b. Combination of characters not as above   17
17a. Involucral bracts with translucent and/or membranous margins   18
  b. Involucral bracts entirely herbaceous   19
18a. Pappus of several short bristles and 2 or 4 longer bristles, none of them
    swollen and scale-like at the base   **95. Boltonia**
  b. Pappus of many more or less equal bristles which are somewhat swollen
    and scale-like at the base   **96. Townsendia**
19a. Involucral bracts in 1 series, sometimes with small or reflexed outer
    bracts at the base   20
  b. Involucral bracts in 2 or more series   21
20a. Plants herbaceous   **132. Senecio**
  b. Plants shrubby, evergreen   **145. Brachyglottis**
21a. Ray-flowers in several whorls, rays narrowly linear   **89. Erigeron**
  b. Ray-flowers in a single whorl, rays relatively broader   22
22a. Fruits flattened; pappus with bristles to 1 mm, much shorter than the fruit,
    united into a small disc below   **87. Kalimeris**
  b. Fruits not flattened; pappus with bristles free, more than 1 mm, almost
    as long to as long as the fruit   23
23a. Pappus-bristles in 2--3 rows   **86. Aster**
  b. Pappus-bristles in 1 row   **92. Felicia**
24a. Involucral bracts in 1--2 (rarely 3) rows   25
  b. Involucral bracts in 3 or more rows   36
25a. Involucral bracts united to each other for most of their length   26
  b. Involucral bracts totally free from each other, or united only at the
    extreme base   27
26a. Leaves deeply lobed or divided; disc-flowers usually fertile **135. Euryops**
  b. Leaves entire, fleshy; disc-flowers sterile   **131. Othonna**
27a. Leaves simple, in almost opposite pairs, often cordate at base, bases of
    stalks clasping the stems; stems brown-downy; ray-flowers 2--8
          139. Sinacalia
  b. Combination of charcaters not as above   28
28a. Shrubs   **145. Brachyglottis**
  b. Herbs   29
29a. Ray-flowers with 2-lipped corollas   **5. Perezia**

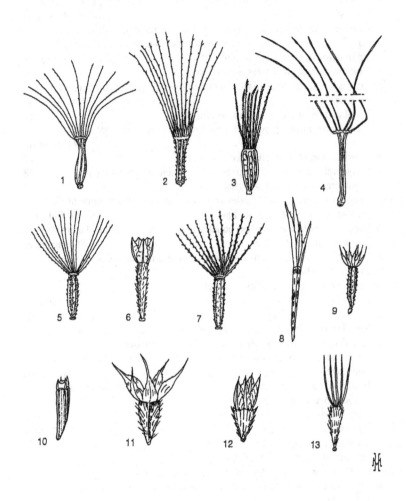

Figure 129. Compositae: fruits. 1, *Senecio*. 2, *Senecio* (another species). 3, *Cremanth- odium*. 4, *Tussilago*. 5, *Brachyglottis*. 6, *Chaenactis*. 7, *Arnica*. 8, *Tagetes*. 9, *Lasthenia*. 10, *Eriophyllum*. 11, *Gaillardia*. 12, *Marshallia*. 13, *Layia*.

b. Ray-flowers with corollas which are not 2-lipped 30
30a. At least some of the leaves opposite **148. Arnica**
 b. Leaves all basal or all alternate 31
31a. Inflorescences produced on a scaly scape before the leaves **140. Tussilago**
 b. Inflorescences produced after the leaves, on leafy shoots or on shoots arising from existing basal rosettes 32
32a. Receptacle convex **143. Doronicum**
 b. Receptacle flat 33
33a. Heads solitary (if several together, then all pendent) 34
 b. Heads usually several together, not as above 35
34a. Heads pendent on arching scape-like stems **138. Cremanthodium**
 b. Heads erect on upright stems **83. Stenotus**
35a. Leaves usually widened and cordate at the base; disc-corolla-tubes each with a very narrow basal part, widened above, the whole usually more than 6 mm **137. Ligularia**
 b. Leaves not widened and cordate at the base; disc-corolla-tubes not as above, rarely as long as 6 mm **132. Senecio**
36a. Shrubs **81. Haplopappus**
 b. Annual, biennial or perennial herbs 37
37a. Involucral bracts with fringed or spine-like, membranous appendages **26. Centaurea**
 b. Involucral bracts without such appendages 38
38a. Both disc- and ray-flowers 2-lipped 39
 b. Disc-flowers tubular, ray-flowers with tongue-like rays 40
39a. Leaves in a basal rosette **1. Gerbera**
 b. Leaves borne on the stem **5. Perezia**
40a. Heads very small, in spikes or narrow racemes 41
 b. Heads medium to large, in panicles or cymes 42
41a. Rays pale yellow or white, conspicuous; leaves oblanceolate **85. × Solidaster**
 b. Rays yellow, inconspicuous, if pale or white then leaves elliptic or diamond-shaped **84. Solidago**
42a. Involucral bracts very narrow, with rough, spreading tips, usually sticky **80. Grindelia**
 b. Involucral bracts not as above, usually relatively broader and generally adpressed 43
43a. Fruits of ray-flowers more or less 3-angled, those of the disc-flowers compressed with an outer pappus of narrow scales, an inner of 30--45 bristles **78. Heterotheca**
 b. Combination of characters not as above 44
44a. Pappus of an outer row of united scales surrounding a ring of 14--20 hairs **61. Pulicaria**
 b. Pappus of more than 20 hairs, scales absent 45
45a. Anthers narrowed to the base; pappus-hairs barbellate **60. Inula**
 b. Anthers rounded at the base; pappus-hairs not barbellate **81. Haplopappus**

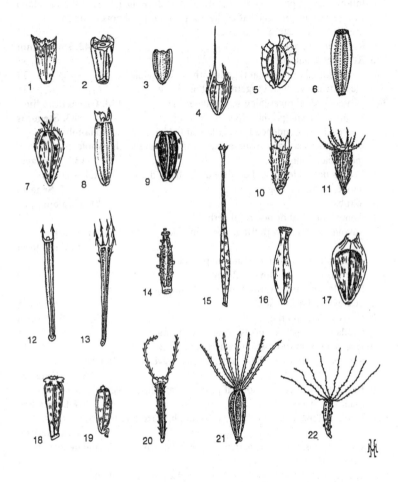

Figure 130. Compositae: fruits. 1, *Echinacea*. 2, *Rudbeckia*. 3, *Ratibida*. 4, *Zinnia*. 5, *Verbesina*. 6, *Balsamorhiza*. 7, *Helianthella*. 8, *Wyethia*. 9, *Helianthus*. 10, *Tithonia*. 11, *Tridax*. 12, *Bidens*. 13, *Bidens* (another species). 14, *Coreopsis*. 15, *Cosmos*. 16, *Dahlia*. 17, *Silphium*. 18, *Ageratum*. 19, *Piqueria*. 20, *Stevia*. 21, *Liatris*. 22, *Eupatorium*.

560

1a. Rays with 5-toothed apex; heads large, 5--10 cm across; outer bracts with large, leaf-like appendages with margins with spine-like bristles
49. Stokesia

  b. Combination of characters not as above   2

2a. Receptacle covered with bristles or dense hairs   3

  b. Receptacle with no bristles, hairs usually absent, rarely a few present   4

3a. Bracts herbaceous; receptacle swollen   **154. Gaillardia**

  b. Bracts with a membranous margin or appendage; receptacle flat
26. Centaurea

4a. Involucral bracts united for some distance above the base   5

  b. Involucral bracts entirely free from each other or united at the extreme base   7

5a. Involucral bracts with spiny margins   **53. Berkheya**

  b. Involucral bracts with margins which are not spiny   6

6a. Sap milky   **54. Gazania**

  b. Sap not milky   **151. Eriophyllum**

7a. Rays yellow or reddish, often dark or white at the base, often 3-coloured with pale base, darker just above, the rest paler or a different colour; annual herb; disc-flowers red or purple; ray-fruits triangular in section, disc-fruits laterally flattened   **119. Ismelia**

  b. Combination of characters not as above   8

8a. Ray-flowers yellow or orange to brownish, or mainly yellow, reddish or white towards the base   9

  b. Ray-flowers white, blue, purple or reddish   26

9a. Fruits hairy   10

  b. Fruits hairless   13

10a. Outer involucral bracts with appendages   **51. Arctotis**

  b. Outer involucral bracts without appendages   11

11a. Tall perennial herbs, over 40 cm; bracts strongly reflexed, at least by fruiting, often before   **153. Helenium**

  b. Lower plants, stems not more than 30 cm or, if taller, then annual; bracts remaining upright   12

12a. Bracts in a single row; rays inconspicuously 3-toothed   **151. Eriophyllum**

  b. Bracts in 2--3 rows; rays conspicuously 3-lobed   **152. Hymenoxys**

13a. Fruits in the one head varying in size, form, degree of curving and wartiness   14

  b. All fruits in the one head of the same form   17

14a. Shrubs with obovate, pinnately lobed, ultimately toothed leaves
120. Argyranthemum

  b. Combination of characters not as above   15

15a. Disc-flowers sterile; achenes (of ray-flowers) variable   **77. Calendula**

  b. Disc-flowers fertile, their achenes differing in form from those of the ray-flowers   16

16a. Fruits of disc-flowers obovoid to ellipsoid with thickened margins; achenes of ray-flowers 3-angled and incurved   **75. Dimorphotheca**

  b. Fruits of disc-flowers cylindric to cylindric-3-angled, those of ray-flowers 3-angled with the ribs often winged; corolla-tube of disc-flowers 2-winged   **118. Xanthophthalmum**

17a. Involucral bracts in 1 row **151. Eriophyllum**
  b. Involucral bracts in 2 or more rows 18
18a. Involucral bracts herbaceous, with spreading, sharp tips, the whole
  involucre and young head resinous-sticky **80. Grindelia**
  b. Involucre and young head not as above 19
19a. Heads solitary, or rarely 2--10 together in a loose raceme or corymb 20
  b. Heads more than 5 together, in dense corymbs 25
20a. Scapes or flowering stems hollow; leaves often mostly basal; fruit with
  3 wings on the back, the lateral wings curved inwards at the margins, the
  central straight; pappus of 4 scales or absent **51. Arctotis**
  b. Combination of characters not as above 21
21a. Annual; involucral bracts wide, fan-shaped; fruit curved, with 8--10 ribs
  which coalesce into a callus at the base; leaves finely toothed
  126. Coleostephus
  b. Combination of characters not as above 22
22a. Leaves entire to lobed, woolly beneath **52. Haplocarpha**
  b. Leaves at least 1-pinnatisect or pinnate, not woolly beneath 23
23a. Mat-forming plants; rays at first yellowish, becoming purple-tinged and
  ultimately blackish purple **125. Rhodanthemum**
  b. Combination of characters not as above 24
24a. Leaves 1-pinnatisect, not gland-dotted **123. Leucanthemopsis**
  b. At least some leaves 2-pinnatifid to pinnatisect, gland-dotted
  117. Chrysanthemum
25a. Subshrubs or shrubs **120. Argyranthemum**
  b. Perennial herbs **108. Tanacetum**
26a. Bracts completely herbaceous, without membranous, dark brown or
  translucent margins 27
  b. At least some of the bracts with membranous, dark brown or translucent
  margins 29
27a. Heads in racemes or corymbs borne on leafy stems **153. Helenium**
  b. Heads solitary on leafless scapes 28
28a. Pappus absent **93. Bellis**
  b. Pappus of small scales (sometimes with a few bristles) **94. Bellium**
29a. Fruits dimorphic, those of the disc-flowers different from those of the ray-
  flowers 30
  b. Fruits not dimorphic, all achenes in the one head similar 31
30a. Herbs **75. Dimorphotheca**
  b. Shrubs **120. Argyranthemum**
31a. Fruits hairy 32
  b. Fruits not hairy, sometimes wrinkled or warty 35
32a. Leaves palmately veined, clearly stalked **133. Pericallis**
  b. Leaves not as above 33
33a. Leaves entire **96. Townsendia**
  b. Leaves toothed, lobed or pinnately divided 34
34a. Leaves pinnately divided into thread-like segments; flowering stems not
  hollow **97. Brachycome**
  b. Leaves lobed or toothed; flowering stems hollow **51. Arctotis**

35a. Leaves simple, entire or toothed or lobed, not normally pinnatifid,
    pinnatisect or pinnate   36
  b. Leaves at least 1-pinnatifid or pinnatisect, often 2--3-pinnatifid or
    pinnatisect   43
36a. Rays white above, violet-blue beneath   **76. Osteospermum**
  b. Rays not as above   37
37a. Leaves glandular-punctate   **122. Leucanthemella**
  b. Leaves not glandular-punctate   38
38a. Fruits flattened and winged; pappus of some very short bristles and 2
    longer bristles   **95. Boltonia**
  b. Fruits not as above; pappus a small corona or entirely absent   39
39a. Fruits not ribbed; involucral bracts ultimately reflexed   **100. Lagenophora**
  b. Fruits 3--10-ribbed; involucral bracts not reflexed   40
40a. Fruits with 5--8 ribs   41
  b. Fruits with 10 ribs   42
41a. Heads solitary, leaves mostly in rosettes   **108. Arctanthemum**
  b. Heads usually in panicles or corymbs; leaves not mostly in rosettes
                                                      107. Tanacetum
42a. Involucral bracts in 2 or 3 rows   **124. Leucanthemum**
  b. Involucral bracts in 4 rows   **121. Nipponanthemum**
43a. Leaves 1-pinnatifid to pinnatisect (rarely almost entire)   44
  b. Leaves 2--3-pinnatifid to pinnatisect   48
44a. Pericarp of fruit with resin canals   45
  b. Pericarp of fruit without resin canals   46
45a. Leaves stalkless; ribs of fruit rounded; root-tips red   **124. Leucanthemum**
  b. Leaves stalked; ribs of fruit wing-like; root-tips not red
                                                      125. Rhodanthemum
46a. Leaves pinnatisect, the segments with sharp points; rays blue
                                                      97. Brachycome
  b. Combination of characters not as above   47
47a. Tufted perennial herbs   **123. Leucanthemopsis**
  b. Perennial herbs, not tufted   **124. Leucanthemum**
48a. Leaf-lobes thread-like or awl-shaped   **127. Matricaria**
  b. Leaf-lobes flat, not as above   49
49a. Pappus a corona; heads usually in corymbs; rays up to 10 mm
                                                      107. Tanacetum
  b. Pappus absent; heads usually solitary; rays more than 10 mm
                                                      117. Chrysanthemum

Group 6
  1a. Scales on the receptacle only between the ray-flowers and the disc-
    flowers, otherwise the receptacle not scaly   2
  b. Scales all over the surface of the receptacle or scattered   3
  2a. Perennials or monocarpic herbs with a spherical rosette of silvery,
    sword-shaped leaves and brownish flowers in panicles
                                                      156. Argyroxiphium
  b. Annual herbs; other characters not as above   **157. Layia**
  3a. Leaves entire, toothed, scalloped or slightly lobed   4
  b. Leaves pinnatifid to pinnatisect or pinnate   20

563

4a. Rays white   5
  b. Rays yellow, orange, reddish or pink   6
5a. Plant a shrub   **105. Eumorphia**
  b. Plant a herb   **113. Achillea**
6a. Shrubs or subshrubs   7
  b. Annual, biennial or perennial herbs   8
7a. Subshrubs; ray-flowers in 2 rows, female, style present   **57. Asteriscus**
  b. Shrubs; ray-flowers in 1 row, sterile, style absent   **165. Encelia**
8a. Leaves with broadly reversed-triangular, widely hastate or sagittate blades;
    receptacular scales wrapped around flowers   **166. Balsamorhiza**
  b. Leaves not as above; receptacular scales generally not as above   9
9a. Ray-flowers sterile   10
  b. Ray-flowers female   15
10a. Receptacle conspicuously conical or cylindric   11
  b. Receptacle flat or slightly raised in the centre   12
11a. Ray-flowers yellow, orange or brownish   **159. Rudbeckia**
  b. Ray-flowers pink   **158. Echinacea**
12a. Fruits strongly compressed from side to side   13
  b. Fruits terete or angled, rarely very weakly compressed   14
13a. Fruits unwinged   **167. Helianthella**
  b. Fruits winged   **164. Verbesina**
14a. Pappus of numerous scales (together, sometimes, with a few bristles)
     170. Tithonia
  b. Pappus of 2 tubercles which fall early   **169. Helianthus**
15a. Fruits strongly compressed and winged   16
  b. Fruits terete or 3- or 5-angled, or very slightly compressed   17
16a. Fruits compressed from front to back   **180. Silphium**
  b. Fruits compressed from side to side   **164. Verbesina**
17a. Pappus absent or forming a small corona   18
  b. Pappus of tubercles or scales   19
18a. Fruits of ray-flowers 3-angled, more or less compressed, those of disc-
    flowers several-angled; anthers not bearded at the base
     58. Buphthalmum
  b. Fruits all similar, several-angled, slightly compressed; anthers bearded
    at the base   **59. Telekia**
19a. Pappus of scales   **57. Asteriscus**
  b. Pappus of 1--3 bristles with scales between them   **168. Wyethia**
20a. Involucre of 2 series of bracts, sharply distinguished, the outer much
    shorter than the others, and spreading   **173. Coreopsis**
  b. Involucre not as above   21
21a. Ray-fruits joined at the base to the subtending receptacular scale and
    to parts of 2 adjacent sterile flowers   22
  b. Combination of characters not as above   23
22a. Resinous dots present on stems and leaves; outer bracts very broad,
    more or less circular   **177. Berlandiera**
  b. Resinous dots absent from stems and leaves; outer bracts less
    conspicuously broad   **179. Engelmannia**

23a. Involucral bracts with membranous or translucent margins 24
  b. Involucral bracts entirely herbaceous 28
24a. Pappus made up of 5 well-developed scales, with or without 5 additional bristle-like scales **104. Ursinia**
  b. Pappus absent or formed by a corona, rarely made up of a few small scales 25
25a. Fruits angled or cylindric, not strongly compressed 26
  b. Fruits strongly flattened 27
26a. Corollas of disc-flowers with the tube saccate at base, enclosing the top of the fruit **115. Chamaemelum**
  b. Corollas of disc-flowers not saccate as above **116. Anthemis**
27a. Fruits unwinged; heads usually small, in panicles **113. Achillea**
  b. Fruits, at least those of the marginal flowers, winged; heads larger, solitary **114. Anacyclus**
28a. Ray-flowers female; fruit strongly compressed, 2-winged **180. Silphium**
  b. Ray-flowers sterile; fruits generally not as above 29
29a. Fruits flattened, not clearly angled **160. Ratibida**
  b. Fruits clearly 4-angled, not flattened 30
30a. Receptacle highly conical or cylindric; pappus absent or forming a small corona **159. Rudbeckia**
  b. Receptacle somewhat raised in the centre; pappus made up of small scales and a few bristles **170. Tithonia**

Group 7

1a. Receptacle without scales 2
  b. Receptacle with scales subtending most flowers 5
2a. Prostrate shrubs **105. Eumorphia**
  b. Herbs, usually erect 3
3a. Involucral bracts united for most of their length **149. Tagetes**
  b. Involucral bracts free or united at extreme base 4
4a. Leaves all opposite; rays yellow **150. Lasthenia**
  b. Lower leaves opposite, upper alternate; rays white or purple **147. Palafoxia**
5a. Shrubs or subshrubs 6
  b. Annual, biennial or perennial herbs 8
6a. Leaves heather-like, crowded, simple or 3-fid at the apex **105. Eumorphia**
  b. Leaves not as above 7
7a. Leaves silky-hairy **130. Eriocephalus**
  b. Leaves hairless or rough-hairy **175. Dahlia**
8a. Pappus made up of rough, backwardly projecting tubercles 9
  b. Pappus various or absent, not as above 11
9a. Involucral bracts united to the middle or above **176. Thelesperma**
  b. Involucral bracts free or united only at the extreme base 10
10a. Rays yellow **172. Bidens**
  b. Rays not yellow **174. Cosmos**
11a. Fruit beaked **174. Cosmos**
  b. Fruit not beaked 12

12a. Leaves pinnately divided   13
  b. Leaves simple   16
13a. Involucral bracts united to each other to the middle   **176. Thelesperma**
  b. Involucral bracts free or united only at the extreme base   14
14a. Involucral bracts in 1 row, no distinct outer row present   **171. Tridax**
  b. Involucral bracts in 2 clearly distinct rows   15

15a. Outer involucral bracts large, reflexed, narrowed to the base   **175. Dahlia**
  b. Outer involucral bracts adpressed or erect, not narrowed towards the
    base   **173. Coreopsis**
16a. Corollas of ray-flowers persistent on the ripe fruits   17
  b. Corollas of ray-flowers not persistent on the ripe fruits   19
17a. Leaves toothed   **162. Heliopsis**
  b. Leaves entire   18
18a. Leaves stalkless; outer involucral bracts shorter than the inner   **163. Zinnia**
  b. Leaves distinctly stalked; outer involucral bracts as long as the inner
     161. Sanvitalia
19a. Heads with only 4 or 5 ray-flowers   **178. Chrysogonon**
  b. Heads with more than 5 ray-flowers   20
20a. Fruits compressed from side to side   **167. Helianthella**
  b. Fruits not compressed, or compressed from front to back   21
21a. Ray-flowers sterile, not setting fruit   **169. Helianthus**
  b. Ray-flowers female, setting fruit   22
22a. Fruits not obviously compressed, top-shaped, silky-hairy   **171. Tridax**
  b. Fruits compressed from front to back, not as above   23
23a. Rays numerous, in 2 or 3 rows   **180. Silphium**
  b. Rays less numerous, in a single row   **173. Coreopsis**

Group 8

1a. Leaves all basal, heads on scapes   **93. Bellis**
  b. Leaves borne on the flowering stems   2
2a. Involucral bracts united for more than half their length   **149. Tagetes**
  b. Involucral bracts free or united only at the extreme base   3
3a. Receptacle hairy or bristly between the flowers   4
  b. Receptacle either scaly bweeen the flowers, or scales and hairs absent   5
4a. Involucral bracts herbaceous, without appendages   **154. Gaillardia**
  b. Involucral bracts with a membranous or fringed appendage **26. Centaurea**
5a. At least the lower leaves opposite   6
  b. Leaves all alternate   10
6a. Involucre of 1 or 2 rows inner rows of adpressed bracts and a clearly
    distinguished outer row   7
  b. Involucre of several rows but these not clearly distinguished as above   8
7a. Outer involucral bracts reflexed, narrowed to their bases; roots with
    thickened tubers   **175. Dahlia**
  b. Outer involucral bracts adpressed or erect, broad to the base; roots
    without tubers   **173. Coreopsis**
8a. Leaves entire, stalkless   **163. Zinnia**

   b. Leaves toothed, usually stalked   9

9a. Receptacle strongly convex; pappus absent   **162. Heliopsis**

   b. Receptacle flat or slightly raised in the centre; pappus formed from 2
     early-deciduous tubercles   **169. Helianthus**

10a. Pappus of hairs or bristles   11

   b. Pappus of scales or tubercles, or corona-like, or absent   12

11a. Heads large; outer involucral bracts leaf-like   **88. Callistephus**

   b. Heads generally rather small; outer involucral bracts not leaf-like

                                                86. Aster

12a. Involucral bracts entirely herbaceous   13

   b. Involucral bracts with membranous or translucent margins   16

13a. Receptacle with scales subtending all or most flowers   14

   b. Receptacle without scales or scales few and scattered   15

14a. Receptacle flat or slightly raised in the centre   **169. Helianthus**

   b. Receptacle conical or cylindric   **159. Rudbeckia**

15a. Heads solitary   **77. Calendula**

   b. Heads on corymbs or panicles   **153. Helenium**

16a. Receptacle without scales or scales few and scattered

                             117. Chrysanthemum

   b. Receptacle with scales subtending all or most flowers   17

17a. Heads solitary; leaves 2-pinnatisect   **116. Anthemis**

   b. Heads in corymbs or panicles; leaves toothed   **113. Achillea**

Group 9

1a. Latex absent   2

   b. Latex present, generally milky   3

2a. All flowers with 2-lipped corollas, the upper lip 2-lobed, the lower
    3-lobed or -toothed, ray-like; involucral bracts without appendages

                                            5. Perezia

   b. Outer flowers with large, 5-lobed rays, inner flowers almost tubular; outer
    involucral bracts with large, leaf-like appendages which are edged with
    small spines   **49. Stokesia**

3a. Plant scapose, all leaves basal (sometimes a few scale-leaves borne
    on the scapes)   4

   b. At least 1 leaf borne on the flowering stems   11

4a. Receptacle with scales subtending all or most flowers   **45. Hypochoeris**

   b. Receptacle without scales, though sometimes hairy   5

5a. At least some of the pappus-hairs plumose   6

   b. All pappus-hairs simple, not plumose   7

6a. Pappus-hairs in 1 or 2 rows   **46. Leontodon**

   b. Pappus-hairs in 3 or more rows   **47. Scorzonera**

7a. Scapes unbranched, without bracts   **32. Taraxacum**

   b. Scapes branched or unbranched, usually bearing bracts   8

8a. Involucral bracts in several rows   9

   b. Involucral bracts in 2 rows, those of the outer row much smaller than
    those of the inner   10

9a. Plants usually stoloniferous; leaves never distinctly stalked; rays often
    with a reddish stripe outside, sometimes red; pappus-hairs mainly in 1
    row, a few shorter than the rest   **43. Pilosella**
 b. Plants not stoloniferous; leaves often distinctly stalked; rays yellow or
    greenish, without red stripes; pappus-hairs in 2 rows   **42. Hieracium**
10a. At least some pappus-hairs thickened at the base; rays usually turning
    greenish on drying   **40. Tolpis**
 b. Pappus-hairs not thickened at the base; rays not turning green on
    drying   **34. Crepis**
11a. Leaves spiny   **39. Sonchus**
 b. Leaves not spiny   12
12a. Fruits without pappus   **33. Lapsana**
 b. Fruits with a pappus of hairs or scales   13
13a. Fruits compressed   14
 b. Fruits not compressed   18
14a. Fruits beaked   **35. Lactuca**
 b. Fruits not beaked   15
15a. Rays yellow   16
 b. Rays blue-purple   17
16a. Pappus of uniform persistent or deciduous hairs; at least the outer
    involucral bracts with membranous or translucent margins; stems spiny
        38. Launaea
 b. Pappus of a few rough deciduous hairs and persistent, softer hairs borne
    in groups; involucral bracts without membranous or translucent margins;
    stems not spiny   **39. Sonchus**
17a. Heads with up to 5 flowers; involucre 3--5 mm wide   **37. Prenanthes**
 b. Heads with about 10 flowers; involucre more than 7 mm wide
        36. Cicerbita
18a. At least some of the fruits with a pappus of scales   19
 b. All fruits with a pappus of hairs   21
19a. Receptacle with scales, at least near the margin   **31. Catananche**
 b. Receptacle entirely without scales   20
20a. Rays purple or blue or rarely white   **30. Cichorium**
 b. Rays yellow   **40. Tolpis**
21a. Receptacle with scales subtending most flowers   22
 b. Receptacle without scales   24
22a. Pappus-hairs plumose   **45. Hypochoeris**
 b. No pappus-hairs plumose   23
23a. Receptacular scales enclosing the bases of the flowers   **41. Andryala**
 b. Receptacular scales not enclosing the bases of the flowers   **34. Crepis**
24a. At least some pappus-hairs plumose   25
 b. No pappus-hairs plumose   27
25a. Involucral bracts in 2 or more rows   **47. Scorzonera**
 b. Involucral bracts in 1 row   26
26a. Leaves deeply toothed to pinnatifid   **44. Urospermum**
 b. Leaves entire, neither toothed nor pinnatifid   **48. Tragopogon**
27a. Fruits densely velvety   **47. Scorzonera**
 b. Fruits not velvety, usually not hairy   28

28a. Pappus of rigid hairs which are somewhat expanded at their bases

40. Tolpis

b. Pappus of usually soft hairs which are not expanded at their bases   29

29a. Receptacle with long silky hairs some of which exceed the flowers

41. Andryala

b. Receptacle hairless or with short hairs, much shorter than the flowers   30

30a. Plants usually stoloniferous; leaves never distinctly stalked; rays often with a reddish stripe outside, sometimes red; pappus-hairs mainly in 1 row, a few shorter than the rest   **43. Pilosella**

b. Plants not stoloniferous; leaves often distinctly stalked; rays yellow or greenish, without red stripes; pappus-hairs in 2 rows   **42. Hieracium**

1. Gerbera Linnaeus. 30/few. Perennial herb. Leaves basal, upper surface with adpressed, lower with shaggy hairs. Heads bisexual, solitary, erect; outer bracts smaller than inner. Ray-flowers in 1 or 2 rows, female, corolla 2-lipped, outer lip an elongate ray, inner lip short and usually bifid; disc-flowers bisexual, fertile, corolla 2-lipped. Bases of anthers sagittate. Style-branches of disc-flowers broadly lanceolate. Fruits cylindric, ribbed. Pappus a single row of rough bristles. *Africa to Indonesia.*

2. Leibnitzia Cassini. 5/1. Perennials. Leaves in a basal rosette, pinnately lobed. Heads solitary on scapes, of 2 forms: in spring with ray-flowers and in autumn cleistogamous with tubular flowers; bracts linear, overlapping in a few series. Ray-flowers in spring fertile or sterile, corolla 2-lipped, the outer lip elongate, 3-toothed; tubular corolla slightly 2-lipped; autumn flowers fertile; anthers sagittate, the tails joined. Fruits fusiform, somewhat flattened, hairy; pappus bristles many, persistent. *Himalaya, south & east Asia.*

3. Mutisia Linnaeus filius. 60/9. Perennial shrubs or climbers, hairless or with felted hairs; stems sometimes with wings. Leaves alternate, simple or pinnate, stalkless, often each terminating in a tendril; leaflets entire or toothed, sometimes with dense felted hairs beneath. Heads always solitary, erect or pendulous, involucre cylindric or oblong-campanulate; bracts overlapping in several rows, broadly lanceolate, each often with an apical appendage. Ray-flowers female, disc-flowers many, bisexual, usually yellow. Fruit cylindric or spindle-shaped to top-shaped, hairless; pappus of long, stiff, plumose hairs, tawny or white. *South America (mostly Andes).*

4. Nassauvia Jussieu. 37/3. Perennial, rhizomatous herbs, sometimes woody at base or dwarf shrubs. Leaves alternate, often overlapping, stalkless, usually rigid and spiny, margin entire or toothed, apex often recurved. Heads very short or stalkless, 5-flowered, solitary or grouped into a dense, spherical to ovoid inflorescence, axillary. Receptacle flat, without bracts, involucre cylindric, bracts in 2 rows. Flowers all bisexual, tubular, 2-lipped, outer lip larger, 3-toothed, inner lip bifid; lobes white or yellowish. Fruit hairless or velvety, with pappus of 3--15 scales which fall early or many plumose bristles. *South America (Andes).*

5. Perezia Lagasca. 30/2. Perennials without milky latex, rootstocks rhizomatous, tuberous or a simple taproot; stems circular in cross-section, smooth, hairless or downy. Basal leaves simple, clasping stem at base; stem-leaves present. Heads solitary or up to 3 together, terminal; involucral bracts in 3--9 series. Flowers bisexual, almost spherical or obconical, 2-lipped; inner lip bifid, outer lip 3-toothed, ray-like; involucral bracts in more than 3 series. Pappus in several series, bristle-like, united at base. *South America (Andes)*. Figure 126, p. 552.

6. Berardia Villars. 1/1. Perennial herbs. Stem absent or to *c.* 12 cm, with dense cobwebby hairs, with papery sheaths towards base. Leaves alternate, 8--14 cm, rarely to 20 cm, ovate, obovate or almost rounded, entire or slightly toothed, leathery, densely covered with greyish white cobwebby hairs; stalks 1--8 cm, decurrent on stems. Heads solitary, 5--7 cm, hemispherical, stalkless. Receptacle without scales or with a few, scattered scales. Involucral bracts in 3 or 4 rows, herbaceous, entire, without appendages, wedge-shaped, the inner almost as long as the flowers. All flowers bisexual and tubular. Corolla cream, pale golden yellow or occasionally pale pink. Filaments winged, without bristles at the base. Fruit almost cylindric, brown or yellow; pappus-hairs plumose, 1.2--2.1 cm, twisted at base, yellowish; innermost row longer than the outer. *Alps*. Figure 126, p. 552.

7. Echinops Linnaeus. 120/7. Generally tall, herbaceous perennials, occasionally annuals (not ours). Stems grooved, generally hairy, often cobwebby. Leaves alternate, pinnatisect, spiny. The inflorescence is a terminal, spherical, umbel, composed of many heads each containing a single bisexual disc-flower. Fruit cylindric, angled, hairy. Pappus of free to united, scale-like bristles. *Mediterranean area to central Asia, mountains of Africa*. Figure 124, p. 548.

8. Carlina Linnaeus. 28/3. Annual to perennial herbs, sometimes woody at the base. Leaves alternate or basal, entire to deeply divided, usually with spine-toothed margins. Heads stalkless to shortly stalked, solitary or in cymose, often corymbose inflorescences. Receptacle scaly or bristly; scales divided at the apex or almost to the base, into linear segments. Involucral bracts in many rows, the outer leaf-like, spine-toothed, the inner entire, shiny, radiating when dry. Ray-flowers absent; disc-flowers bisexual, corolla 5-lobed. Ovary hairy; fruits oblong, hairy; pappus a ring of usually basally fused, feathery hairs. *Europe, Asia, the Mediterranean area & Macaronesia*. Figure 124, p. 548.

9. Staehelina Linnaeus. 8/1. Glandular-hairy shrub. Leaves alternate, simple, leathery, white-hairy, often crowded in rosettes at the ends of branches, margins toothed. Heads few to many in terminal corymbs, rarely solitary. Involucre cylindric, bracts in 2 or more series, overlapping and lying flat, unequal, oblong to ovate, pointed at apex, margins entire. Flowers all without rays, bisexual; corolla tubular, deeply 5-lobed, pink. Receptacular scales narrow, many-lobed. Fruits oblong with dense silky hairs. Pappus a single row of white, branched hairs. *South Europe*.

10. Xeranthemum Linnaeus. 5/3. Annuals with erect, rather slender stems. Leaves alternate, narrow, entire, white-hairy, not decurrent on the stems. Heads solitary at the end of long stalks. Involucral bracts persistent, papery, the outer shorter than the inner, which are petal-like and coloured. Receptacle with narrow, pointed, entire scales. Ray-flowers absent, outermost disc-flowers sterile, the rest bisexual. Fruits narrowly egg-shaped, silky-hairy; pappus of 5--15 straw-like scales. *Mediterranean area to Iran*. Figure 124, p. 548.

11. Saussurea de Candolle. 130/4. Perennial herbs with alternate, non-spiny, entire to pinnatisect leaves. Heads in corymbs or panicles. Involucre with 2 rows of overlapping non-spiny bracts. Receptacle flat, with dense chaffy bristles. Ray-flowers absent, disc-flowers bisexual, tubular, anthers with long acute terminal appendages and feathery tails. Fruits cylindric, 4-ribbed, hairless. Pappus of rough, plumose bristles in 1 or 2 rows. *Europe, Asia, North America, 1 species in Australia*. Figure 124, p. 548.

High-alpine species are grown by enthusiasts, but are very difficult to maintain.

12. Jurinea Linnaeus. 250/4. Herbaceous perennials. Leaves alternate, without spines, simple, lyrate or pinnatisect, sparsely hairy above, with dense, short, white hairs beneath. Heads solitary or in a corymb, cylindric or globular; receptacle scaly. Ray-flowers absent; disc-flowers bisexual, lilac to purplish. Involucral bracts in 5 or 6 or more series, linear to lanceolate, straight or recurved. Fruit 4--5 mm, 4-sided, narrowly top-shaped; pappus of free, stiff hairs, a few of them longer than the rest, in several rows. *Eurasia, especially central Asia*.

13. Jurinella Jaubert & Spach. 2/1. Stemless perennial. Leaves all basal, simple, lyrate or pinnatifid, grey-hairy beneath. Heads solitary, stalkless, borne on the rosette, almost spherical, receptacle scaly. Ray-flowers absent, disc-flowers lilac, vanilla-scented. Involucral bracts in 4 or 5 series, lanceolate, the inner almost hairless, the outer somewhat hairy and recurved or reflexed. Fruit rounded, smooth; pappus of more or less equal, stiff hairs. *Southwest Asia*.

14. Cynara Linnaeus. 10/3. Perennial herbs. Stems ridged, to 2 m, rarely absent. Both basal and stem-leaves usually present, deeply dissected and spiny. Flowers in large cup-sized and -shaped heads with mauve coloured flowers. Receptacle thick and fleshy. Ovary and fruit hairless. Pappus-hairs plumose. *Mediterranean area, Canary Islands*. Figure 124, p. 548.

15. Ptilostemon Cassini. 14/2. Spiny biennials or perennials. Leaves alternate, entire, sinuately lobed or pinnatifid, more or less hairless above, densely white-hairy beneath, spiny. Heads disc-shaped; receptacle hairy, not becoming thick and fleshy. Involucral bracts overlapping, leathery, spine-tipped. Ray-flowers absent, disc-flowers white to pink or purple. Fruit scarcely compressed, hairless; pappus-hairs in several rows, plumose. *Mediterranean & adjacent areas*. Figure 124, p. 548.

16. Arctium Linnaeus. 11/3. Erect biennial herbs with long, stout taproots. Leaves alternate, woolly-hairy, entire or with few marginal teeth. Heads solitary or in corymb- or raceme-like clusters, ovoid-conical to spherical or hemispherical. Involucre hairless or with cobwebby hairs; bracts numerous, overlapping in several rows, subulate with adpressed bases, the outer long, rigid, spreading, with hooked apices. Receptacle scaly. Flowers all without rays, tubular, bisexual, purple or white. Anthers acuminate above, sagittate below. Style swollen at base, the branches wedge-shaped. Fruit oblong, compressed, surface rough; pappus hairs rough, golden-yellow, free to base. *Europe, Asia*. Figure 124, p. 548.

17. Onopordum Linnaeus. 40/5. Robust biennial to short-lived perennial herbs. Stems erect, often stout, branched above, broadly winged, spiny. Leaves alternate, sinuately toothed to pinnatifid, stalkless and decurrent. Heads large, solitary or few on thick, rigid stalks, with flat receptacles covered with fringed pits in which the flowers are borne. Involucral bracts spine-tipped, leathery, in 2--4 densely overlapping rows. Ray-flowers absent, disc-flowers 2--3 cm, usually purple. Ovary hairless; pappus rough-hairy or plumose. *Europe, North Africa & central Asia.* Figure 124, p. 548.

18. Carduus Linnaeus. 90/few. Annuals, biennials or perennials. Stems and leaves usually armed with spiny teeth, teeth on the stems borne on wings decurrent from the leaf-bases. Leaves alternate. Involucre of numerous overlapping, appressed, rigid, spiny-tipped bracts. Receptacle scaly, scales hair-like. Ray-flowers absent; disc-flowers bisexual. Fruits all similar, ovoid, compressed, hairless. Pappus-hairs rough. *Europe, the Mediterranean area, western Asia, extending to mountains of east Africa.* Figure 124, p. 548.

19. Cirsium Miller. 250/5. Spiny perennial or biennial herbs, rarely annuals. Leaves alternate, entire to pinnately divided, with spiny margins on upper surfaces and spiny teeth or lobes, small spines alternating with larger spines. Involucral bracts overlapping, often with conspicuous oil-glands and an apical spine. Receptacular scales numerous. Flowers all tubular, bisexual, purple, pink, white or yellow. Fruits hairless, oblong, swollen or compressed, the apex with a distinct ring, surrounding a central projection. Pappus of several rows of feathery hairs, the inner longer then the outer, that of the outermost flowers with fewer simple hairs. *Europe, Asia, North America.* Figure 124, p. 548.

20. Silybum Adanson. 2/1. Annual or biennial, robust spiny herbs. Stems unwinged, hairless or somewhat cobwebby-hairy. Leaves alternate, often pinnately lobed, hairless, shiny, with spiny margins, white-veined or -variegated above; upper leaves smaller, clasping the stem. Heads almost spherical. Outer and middle involucral bracts each with a toothed appendage ending in a long spreading or recurved spine. Ray-flowers absent. Pappus a ring of hairs. *Mediterranean area.* Figure 125, p. 550.

21. Galactites Moench. 3/1. Erect annuals or biennials. Stems erect, branched above. Leaves alternate, lanceolate, pinnatifid with spine-tipped segments, veined and spotted with white above, white-cottony beneath. Heads solitary or several together. Ray-flowers absent; central disc-flowers bisexual, the outermost larger, spreading and sterile. Involucral bracts cottony, without toothed appendages, at least the outer ending in a slender, sharp, greenish spine. Fruits cylindric, hairless; pappus of plumose, deciduous bristles, which are united in a ring at the base. *Mediterranean area, Canary Islands.* Figure 125, p. 550.

22. Serratula Linnaeus. 70/1. Spineless perennial herbs; stems erect. Leaves alternate, ovate-lanceolate, finely to coarsely toothed to very deeply pinnatifid, not white- or greyish-felted beneath. Heads borne in loose panicles or almost stalkless in a compact cluster; involucral bracts in several rows, overlapping, margins entire, usually without appendages; innermost always exceeding middle bracts. Receptacle scaly. Flowers mostly bisexual, tubular, ray-flowers absent. Fruit hairless. Pappus of several rows of narrow scales, rarely absent. *Europe to North Africa & Japan.* Figure 125, p. 550.

23. Stemmacantha Cassini. 20/1. Spineless perennial herbs; stems erect. Leaves ovate-lanceolate, finely to coarsely toothed to very deeply pinnatifid, densely white-felted beneath. Heads borne in loose panicles or in a compact cluster; receptacle scaly. Involucral bracts overlapping, without distinct appendages; innermost always exceeding middle bracts. Flowers bisexual, fertile, tubular, ray-flowers absent. Fruit hairless. Pappus of several rows of barbed hairs united into a ring at the base, the barbs not longer than the width of the hairs. *South Europe, North Africa, Asia & Australia.*

The 1 cultivated species is generally found under the name *Leuzea centaurioides* (Linnaeus) Holub.

24. Leuzea de Candolle. 3/1. Spineless perennial herbs; stems erect. Leaves ovate-lanceolate, finely to coarsely toothed to very deeply pinnatifid. Heads borne in loose panicles or almost stalkless in a compact cluster; receptacle scaly. Involucral bracts overlapping, with entire or lacerate appendages; innermost always exceeding middle bracts. Flowers bisexual, tubular, purple. Anthers with obtuse basal appendages. Fruit hairless. Pappus of several rows of feathery hairs longer than the body of the fruit, united into a ring at the base. *Southwest Europe.* Figure 125, p. 550.

25. Amberboa (Lessing) de Candolle. 7/1. Annual or biennial herbs, mostly hairless, spineless. Stems erect, branched. Leaves entire to pinnatifid, alternate. Heads solitary, broadly ovoid, involucral bracts in several series, overlapping, obtuse, green, the inner with brownish, translucent appendages. Receptacle bearing smooth bristles. Flowers white, yellowish, pink or red to lilac, the outer sterile, larger and radiating, their corollas divided into several lobes. Fruits hairy, ribbed and wrinkled, laterally compressed, attachment conspicuous, bordered. Pappus of several series of narrow scales, rarely absent. *Southwest Asia.*

26. Centaurea Linnaeus. 450/30. Annual to perennial herbs or rarely shrubs, rarely spiny, usually variously hairy, often with glands. Leaves very variable, entire to twice pinnatisect or bipinnate, borne on the stems or all basal, sometimes the bases decurrent on the stems as wings. Heads ovoid to spherical, or rarely flattened. Involucral bracts in several series, overlapping and more or less rigid, each usually with an apical appendage, these very variable, sometimes very large, sometimes each terminating in a spine. Receptacle with bristles. Outer flowers sometimes longer than inner and radiating, sometimes sterile or female with staminodes, funnel-shaped with 5--8 corolla-lobes or thread-like and inconspicuous with 5 corolla-lobes; central flowers bisexual. Fruits laterally compressed, rounded at the apex, each often with an oily appendage. Pappus variable, generally of several series of unequal bristles, rarely absent. *Mediterranean area, Europe & southwest Asia, North America, Australia* (1 species). Figure 125, p. 550.

27. Cnicus Linnaeus. 1/1. Annual herbs with stems 10--60 cm, cobwebby-hairy. Leaves alternate, 20--30 × 5--8 cm, oblong in outline, deeply pinnately lobed to almost entire; lower leaves stalked, upper stalkless, clasping the stem, overlapping each other; all leathery, with spiny teeth, light green with prominent white veins beneath. Heads 2.5--4 × 2--3 cm, solitary, surrounded by the upper leaves, involucral bracts brown. Receptacle-scales numerous, bristle-like. Ray-flowers absent; inner flowers bisexual; outer flowers minute, sterile. Corolla yellow. Fruits 6--8 × 2--2.5 mm, ribbed, brown; pappus yellow. *South Europe east to central Asia.* Figure 125, p. 550.

Some authors now include this genus in *Centaurea.*

28. Carthamus Linnaeus. 14/1. Spiny annuals, hairless. Leaves alternate, pinnatifid or entire, margin spiny, basal more divided than stem-leaves. Involucral bracts in many rows, overlapping, spiny, the outer leaf-like, the inner sometimes each with an apical appendage. Heads solitary; flowers all without rays, all bisexual, corolla vermilion to yellow. Fruits 4-angled, those from the outer flowers coarsely wrinkled, without a pappus, those from the inner smooth, sometimes with a persistent pappus of of linear scales. *Mediterranean area.* Figure 125, p. 550.

29. Carduncellus Adanson. 25/2. Perennial herbs, usually spiny, stemless or with a simple or branched stem, with cobweb-like hairs. Basal leaves pinnate to lyrate, stem-leaves toothed or with wavy margins. Involucral bracts in many rows, overlapping, spiny, the outer leaf-like, the inner almost circular to ovate, with fringed appendages. Heads with tubular flowers only, bisexual; corolla blue or purple; filaments bearded. Fruits 4-angled, surface rough, hairless. Pappus of many rows of narrow ciliate scales to plumose bristles, joined at the base and deciduous. *Mediterranean area.* Figure 125, p. 550.

30. Cichorium Linnaeus. 8/2. Annual to perennial herbs with latex. Stems mostly solitary, branched. Leaves alternate, pinnatifid with downward-pointing lobes, or toothed. Heads numerous, axillary; ray-flowers present, bisexual, rays 5-lobed, usually blue, rarely pink or white; disc-flowers absent. Involucre with bracts in 2 rows, the outer shorter. Receptacle almost flat, without scales. Fruits obovoid, slightly 5-angled, truncate at apex; all with pappus of 1--3 rows of short scales. *Mediterranean area, extending to western & central Europe and Ethiopia.* Figure 126, p. 552.

31. Catananche Linnaeus. 5/3. Stems 1--few, often adpressed-hairy, with latex. Leaves mostly basal, linear to linear-oblanceolate, entire or remotely toothed. Heads 1--5 on long stalks, all flowers with rays. Receptacle flat, with long, thread-like scales, at least near the margins. Involucral bracts in several series, overlapping, membranous with dark mid-veins. Rays somewhat papery, blue, white or yellow. Fruit oblong, with 5 angles and 5--10 ribs; pappus with 1 row of 5--7 ovate, long-awned scales. *Mediterranean area.* Figure 126, p. 552.

32. Taraxacum Weber. ?/few. Biennial and perennial herbs with simple or branched, often very deep taproots, milky latex usually copious. Leaves in a basal rosette, simple, toothed or lobed, the lobes or teeth often pointing towards the bases of the leaves. Heads usually solitary on a hollow scape; flowers all with 5-toothed rays. Receptacle without scales or bristles, pitted. Involucre campanulate to cylindric with bracts in 2 series, the inner more or less linear, equal, erect, the outer shorter and often wider, usually spreading or reflexed. Corollas usually yellow, more rarely white or pink, often the rays each with a darker stripe beneath. Fruit obovoid to fusiform, slightly compressed, ribbed, often with small spines near the apex, and usually with a long, often warty beak. Pappus of many rows of simple, rough white hairs. *North temperate areas, temperate South America.* Figure 131, p. 576.

A genus of an uncertain number of species, including many apomictic microspecies, few cultivated; many are troublesome weeds.

33. Lapsana Linnaeus. 9/1. Annual to short-lived perennial herbs with latex. Leaves alternate, basal and lower stem leaves oblong to lanceolate or ovate, usually deeply lobed, with a large terminal lobe; upper stem-leaves linear to narrowly lanceolate, slightly toothed or entire. Heads many in open panicles. Ray-flowers present, bisexual, rays 5-toothed, yellow. Disc-flowers absent. Involucral bracts in 2 rows, those of the outer few and small. Receptacular scales absent. Fruit flattened slightly, ribbed, without a beak, more than half as long as the involucral bracts. Pappus absent. *Temperate Eurasia.* Figure 126, p. 552.

34. Crepis Linnaeus. *c.* 200/2. Perennial herbs. Leaves toothed to pinnatisect. Involucral bracts in 2 rows, those of the outer row much shorter than those of the inner. Flowers all with rays, yellow or orange with reddish purple outer faces, to purplish pink. Fruits fusiform. Pappus of many rows of soft, white, simple hairs. *Northern hemisphere.* Figure 126, p. 552.

35. Lactuca Linnaeus. 100/4. Annual, biennial or perennial herbs with latex. Stems usually solitary, erect, branched in the upper part. Leaves alternate, entire to pinnatifid, sometimes prickly. Heads in dense panicles. Involucral bracts in several rows. Receptacle without scales. Ray-flowers present, bisexual, rays 5-toothed, yellow, bluish or lilac; disc-flowers absent. Fruits compressed, beaked, with pappus of 2 equal rows of simple, white or yellowish hairs. *Mostly north temperate areas.* Figure 126, p. 552.

36. Cicerbita Wallroth. 18/4. Perennial rhizomatous or tap-rooted herbs with latex. Stems usually solitary, branched. Leaves alternate, lobed, clasping the stem. Heads numerous, each with *c.* 10 flowers. Involucre more than 7 mm wide, bracts in 3 or more series. Receptacle without scales. Ray-flowers bisexual, with blue, lilac or violet, rarely yellow (not ours), 5-toothed rays. Disc-flowers absent. Fruits flattened, not beaked, all with pappus of 2 rows of simple hairs, the outer shorter than the inner. *Mountains of Europe, Asia, North Africa & North America.* Figure 126, p. 552.

37. Prenanthes Linnaeus. 30/1. Perennial herbs with latex and usually single, much-branched stems. Leaves basal and on the stems, entire to lobed, the stem-leaves with auricles clasping the stems. Heads many in panicles, of up to 5 purplish ray-flowers with 5-toothed rays, disc-flowers absent. Involucral bracts in 2 or 3 rows. Receptacle without scales. Fruits compressed, not beaked. Pappus of 2 or 3 rows of equal, simple hairs, those of the outer row not thickened near their bases. *North temperate areas, mountains of Africa.* Figure 126, p. 552.

38. Launaea Cassini. 30/1. Spiny shrubs or herbs with latex. Leaves few, alternate or mostly basal. Heads in corymbose panicles. Involucral bracts in several rows, at least the outer with membranous or translucent margins. Receptacle lacking scales. Ray-flowers present, bisexual, rays 5-toothed, yellow. Disc-flowers absent. Fruit cylindric, slightly compressed, not beaked, ribbed. Pappus uniform, of several rows of simple hairs. *Europe to eastern Asia, Canary Islands, South Africa.*

39. Sonchus Linnaeus. 60/2. Annual, biennial or perennial herbs, sometimes shrubby, with latex. Leaves alternate, often deeply lobed and sometimes softly spiny. Heads usually in corymbose panicles. Involucral bracts without translucent margins, in 3 overlapping rows. Receptacle without scales. Ray-flowers present, bisexual, rays 5-toothed, yellow. Disc-flowers absent. Fruits compressed, with 1--4 ribs on each face,

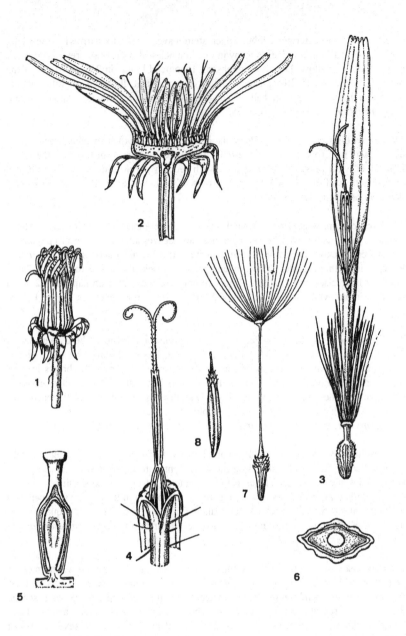

Figure 131. Compositae. *Taraxacum officinale*. 1, Young head showing involucre, the inner bracts erect, the outer reflexed. 2, Longitudinal section of head. 3, Single flower. 4, Corolla rolled back to show the stamens with free filaments and united anthers. 5, Longitudinal section of ovary. 6, Transverse section of ovary. 7, Mature fruit with long beak bearing the pappus. 8, Body of fruit showing ribs.

576

yellow or brownish. Pappus of rough deciduous bristles and persistent fine hairs. *Eurasia, Africa.* Figure 126, p. 552.

40. Tolpis Adanson. 20/1. Annual to perennial herbs with milky latex. Stems solitary or several, usually branched. Leaves mostly basal, entire to toothed or lobed. Heads solitary or several in a cymose panicle, with ray-flowers only, rays 5-toothed at the apex. Involucre campanulate with narrow bracts in 2 or 3 rows; receptacle flat, without scales or hairs. Corollas mostly yellow, though the inner may be purple-brown. Fruit ribbed, truncate at the apex, pappus of long or short bristles. *Mediterranean area, Atlantic Islands.*

41. Andryala Linnaeus. 30/1. Annual or perennial herbs (ours), sometimes woody below, with latex. Leaves entire to pinnatisect, often more or less stem-clasping. Heads solitary (ours) or numerous in panicles, of ray-flowers with 5-toothed yellow rays only, disc-flowers absent. Receptacle scaly, the scales enclosing the bases of the flowers, also pitted with the pit-margins ciliate-lacerate. Fruits oblong, not compressed, truncate at the apex, with 8--10 conspicuous ribs. Pappus of greyish hairs, falling as a complete unit. *Mediterranean area, Atlantic Islands.* Figure 126, p. 552.

42. Hieracium Linnaeus. ?/10. Perennial herbs with branched stocks but without stolons. Stems usually solitary, without or with numerous leaves, usually branched. Leaves entire to deeply incised-toothed, distinctly stalked. Heads 1--many, involucral bracts in many rows. Corollas all with rays, yellow. Receptacle flat, pitted, without scales. Style yellow or discoloured (appearing dirty, stained). Fruits 2.5--5 mm, cylindric, the ribs apically confluent into an obscure ring; pappus-hairs in 2 rows, both with long and short intermixed. *Europe, northern Asia, North America.* Figure 126, p. 552.

Perhaps about 10,000 or more apomictic segregates.

43. Pilosella Hill. 18/3 (and 1 hybrid). Perennial herbs with stolons. Stems (scapes) 1--many, without, or with few leaves, sometimes branched. Leaves entire or slightly finely toothed, gradually tapered at the base and not distinctly stalked. Heads 1--many, involucral bracts in many rows. Corollas all with rays, yellow, often with a red stripe on the outer face of each, or orange to reddish. Receptacle flat, pitted, without scales. Fruits to 2.5 mm, each rib stiffly projecting above to form a finely scalloped apex; pappus-hairs in 1 row with a few shorter than the rest. *Europe, western Asia; introduced in North America, Australia & new Zealand.*

44. Urospermum Scopoli. 2/1. Hairy perennial with latex. Stems 20--40 cm, little-branched, hairy. Leaves alternate, deeply toothed or pinnatifid with backward-pointing lobes, sometimes entire, softly hairy; stalks winged. Upper leaves narrower, clasping. Heads solitary or 2 or 3 on long stalks, 3.5--5 cm across. Receptacle without scales. Ray-flowers present, bisexual, rays 5-toothed, sulphur-yellow, the outer each with a purplish red stripe beneath. Disc-flowers absent. Involucral bracts 7 or 8 in 1 series, lanceolate, velvety. Fruits 4--5.5 mm, irregularly papillose, each with a conspicuous beak to 1.4 cm, dark brown; pappus stiffly plumose, reddish brown. *Mediterranean area.* Figure 126, p. 552.

577

45. Hypochoeris Scopoli. 50/1--2. Hairy perennials with latex. Leaves mostly or all basal, entire or sinuately or pinnately lobed. Heads solitary or few; receptacle scaly (not scaly in *Leontodon*). Ray-flowers present, bisexual, rays 5-toothed, yellow, the outer often each with a greenish or reddish stripe beneath; disc-flowers absent. Involucral bracts in 1--several rows, overlapping, blackish green. Fruits cylindric, beaked; pappus of 1--2 rows of plumose hairs. *Eurasia, North Africa & South America.* Figure 126, p. 552.

46. Leontodon Linnaeus. 40/2. Perennials or sometimes annuals with latex. Leaves all basal, 5--30 cm, toothed to pinnatifid; scapes with scales above. Heads solitary on the scape; receptacle without scales (scaly in *Hypochoeris*). Ray-flowers present, bisexual, rays 5-toothed, yellow or orange. Disc-flowers absent. Involucral bracts in 2--several rows, overlapping. Fruits cylindric; pappus of 1 row of plumose hairs, with or without an outer row of finely toothed hairs. *Eurasia.* Figure 126, p. 552.

47. Scorzonera Linnaeus. 150/7. Annual to perennial herbs, rarely shrubby, with latex. Stems solitary to several, from a thick rootstock. Leaves alternate, entire to pinnatisect. Heads solitary to many together, all flowers with rays; receptacle naked, almost flat. Involucre ovate to cylindric, bracts overlapping, in several rows. Rays yellow, purple or white. Fruits cylindric; pappus of several rows of hairs, usually plumose at least at base, or outer ones rough. *Mediterranean area to central Asia.* Figure 126, p. 552.

48. Tragopogon Linnaeus. 50/5. Annual to perennial herbs with taproots and milky latex. Leaves alternate, with parallel veins, linear to linear-lanceolate, entire, stem-leaves clasping. Heads solitary or few, all flowers with rays. Receptacle naked. Involucre narrow, campanulate or cylindric, bracts in 1 row, the apex of each reflexed at maturity. Rays yellow or purple. Fruit ribbed, often beaked, beak separated from pappus by a groove; pappus of plumose hairs. *Temperate Eurasia, Mediterranean area.*

49. Stokesia L'Héritier. 1/1. Erect, perennial herb to 1 m, without latex. Leaves alternate, *c.* 20 × 7 cm, elliptic to oblong-lanceolate, margins entire or spinose toward base. Heads to 10 cm across, solitary and terminal or few to many in a corymb. Involucral bracts in many series, oblong to lanceolate, the outer each with a leaf-like bristly appendage at the tip. Receptacle flat. Ray-flowers present, rays white, yellow to pale lavender to deep indigo, 5-toothed. Disc-flowers bisexual. Fruit 3- or 4-angled, hairless. Pappus of awns or scales. *Southeast USA.*

50. Vernonia Schreber. 1000/3--4. Annual to perennial herbs, shrubs or trees. Leaves alternate, usually simple, margins entire or toothed. Head hemispherical to campanulate, cylindric or top-shaped, in terminal, corymbose or paniculate clusters. Receptacle flat, naked. Involucral bracts overlapping, in several series, green or coloured, not membranous. Flowers all tubular, bisexual, purple to rose, rarely white. Fruit ribbed. Pappus of purple, brown or white bristles or scales. *Globally distributed in tropical & warm regions, temperate in North America.* Figure 127, p. 554.

51. Arctotis Linnaeus. 50/few. Short-stemmed or rosette-forming annuals or perennials usually grown as annuals. Scapes or flowering stems usually hollow. Leaves alternate or all basal, entire to lyrate-pinnatifid, often white-woolly. Heads large to medium-sized, solitary and long-stalked. Involucral bracts in several overlapping series, the outer each with an appendage. Ray-flowers female, corolla commonly yellow or white, fertile or sometimes neuter; disc-flowers bisexual, corolla yellow, or purple-black, fertile, radially symmetric, tubular, deeply 5-lobed. Fruits oblong, hairy, with 3 wings on the back, the lateral wings incurved, the central straight. Pappus of small delicate scales or wanting. *South Africa to Angola.* Figure 127, p. 554.

As interpreted here, the genus includes species formerly placed in *Venidium* Lessing.

52. Haplocarpha Lessing. 10/1. Mat- or rosette-forming perennial herbs with leaves in rosettes, linear to almost circular, entire or irregularly toothed to pinnatifid, woolly beneath. Heads solitary, shortly stalked or not; involucre campanulate, bracts in 2 or 3 ranks, receptacle flat, naked; ray-flowers yellow, female, corolla linear to oblong, 3-toothed; disc-flowers tubular, bisexual, corolla 5-lobed. Fruit ribbed, rough. Pappus of scales or absent. *Africa.*

53. Berkheya Ehrhart. 150/2. Perennial herbs, subshrubs or shrubs, often spiny. Leaves pinnate, pinnatisect or pinnatifid, usually stalked and often densely white-felted hairy beneath. Heads with or without rays, usually terminal, few to several in spreading clusters. Involucral bracts basally united, in several ranks, spreading and usually spiny. Ray-flowers sterile, yellow, rarely white, purple- or brown-tinged, sometimes absent; disc-flowers yellow, rarely purple to brown, corolla tubular, 5-toothed. Pappus usually of scales. *Southern Africa.* Figure 127, p. 554.

54. Gazania Gaertner. 16/6. Variable perennial or annual herbs often with milky sap. Leaves mainly crowded at the base or on short stems, entire or pinnatisect, usually white-felted beneath. Heads large, solitary, on long, usually leafless stalks. Involucral bracts united into a lobed cup. Rays yellow, white, orange or reddish, each often with a greenish or blackish spot at the base. Disc-flowers of the same colour as the rays. Fruits ovoid to obconical, silky. Pappus of scales in 2 series. *Tanzania to South Africa.* Figure 127, p. 554.

55. Syncarpha de Candolle. 25/2. Perennial herbs or shrubs with stalkless, densely hairy to woolly leaves which are often overlapping and held erect. Heads solitary or in corymbs, stalked or stalkless. Involucres hemispherical, bracts yellow, pink, white or brown, opaque, persistent. Ray-flowers absent; disc-flowers bisexual or unisexual, corollas yellow. Fruits with spherical hairs. Pappus of many plumose bristles, united below. *South Africa.*

A genus formerly included in the heterogeneous and illegitimately named *Helipterum* de Candolle.

56. Rhodanthe Lindley. 25/5. Annual or perennial herbs, hairless to densely white-woolly. Leaves alternate, variable in shape. Heads hemispherical to top-shaped or cylindric, solitary or in few-flowered corymbs, involucral bracts many, overlapping in several series, at least the inner with narrow, translucent-margined claws and expanded, petal-like upper portion, all long-persistent. Ray-flowers absent; disc-flowers few to many, usually bisexual and fertile, occasionally the central flowers male only. Anthers appendaged. Fruits mostly hairy, generally silky-velvety. Pappus of several bristles plumose from the base, in some species each bristle tipped with a dense tuft of lateral cilia. *Australia*.

Formerly called *Acroclinium* A. Gray, or included in *Helipterum* (see note under previous genus).

57. Asteriscus Miller. 3/2. Annual, biennnial or perennial herbs. Stems leafy. Leaves stalked. Heads terminal. Involucral bracts in 2 rows. Receptacle convex, with scales. Ray-flowers in 1--3 rows, female, yellow; disc-flowers numerous, the tube compressed and sometimes with 1 or 2 narrow wings. Outer fruits flat, winged; inner fruits slightly compressed, not or scarcely winged. Pappus of free, short, translucent scales. *S Europe, particularly the Mediterranean part*.

58. Buphthalmum Linnaeus. 2/1. Perennial herbs. Leaves alternate, simple. Heads conspicuous on long leafy stalks. Involucre with bracts in several rows. Receptacle convex with numerous scales round the fruits. Ray-flowers present, female, rays yellow. Disc-flowers bisexual, yellow. Anthers not bearded at base. Fruits of ray-flowers 3-angled, more or less compressed; fruits of disc flowers several-angled. Pappus a papery rim, toothed or with few longer teeth. *Central & south Europe, west Asia*.

59. Telekia Baumgarten. 2/1. Tall perennial herbs. Leaves alternate, entire or toothed. Heads solitary or few, terminal. Receptacle convex, with numerous scales which are folded round the fruits. Ray-flowers female, rays very long and narrow, yellow. Disc-flowers bisexual, yellow. Anthers bearded at the base. Fruits with several angles, terete or slightly compressed. Pappus of scales united into a short crown. *Central Europe to Caucasus*.

60. Inula Linnaeus. 90/22. Biennial or perennial herbs, rarely small shrubs, with erect stems, often robust and leafy. Leaves alternate, simple. Heads solitary or in panicles. Receptacle flat. Ray-flowers present or absent; if present, female, rays 3-toothed, yellow. Disc-flowers bisexual, yellow. Involucral bracts in several rows, often green and leaf-like. Anthers narrowed to the base. Fruits usually cylindric, sometimes 4- or 5-angled or spindle-shaped. Pappus of unequal, simple, roughened hairs, usually free. *Eurasia, Africa*. Figure 127, p. 554.

61. Pulicaria Gaertner. 40/1. Rather hairy perennials with fleshy, scaly, whitish stolons. Stems 20--60 cm, erect, branched. Leaves alternate, simple, entire, oblong or more or less lanceolate, up to 8 cm, semi-clasping at the base, paler and hairier beneath. Heads 1.5--3 cm in diameter, numerous in loose, flat-topped panicles; ray-flowers present, female, rays yellow. Disc-flowers bisexual, yellow. Involucral bracts linear, downy, in 4 or 5 overlapping rows. Fruits 1--1.5 mm, hairy. Pappus whitish, an outer row of whitish scales united at the base, and an inner row of hairs. *Widely distributed in the Old World*. Figure 127, p. 554.

62. Raoulia J.D. Hooker. 20/7. Cushion-forming or creeping, evergreen spineless herbs, sometimes woody at the base, often forming cushions or mats. Leaves usually crowded, alternate, not more than 6 mm. Heads terminal, solitary, stalkless or almost so, with only disc-flowers (sometimes with the inner involucral bracts resembling ray-flowers). Involucral bracts in 2 or more rows. Receptacle without hairs or scales subtending the flowers or rarely with a few such scales. Outer flowers female, corolla thread-like, 2--5-toothed; inner flowers bisexual, corolla tubular, 5-toothed. Anthers sagittate, tails thread-like. Fruit more or less oblong, hairless to hairy; pappus-hairs 15--25 in 1 series or 50--150 in several series. *New Zealand, New Guinea.*

Natural hybrids occur between *Raoulia* and *Leucogenes* in New Zealand (× **Leucoraoulia** Allan). One occasionally cultivated is × **Leucoraoulia loganii** (Buchanan) Cockayne & Allan (*R. × loganii* Anon.).

63. Ozothamnus R. Brown. 50/4. Evergreen shrubs and woody-based perennial herbs. Leaves alternate, simple, often with margins rolled under, or else clasping the stems. Heads with disc-flowers only, borne in dense corymbs; ray-flowers absent. Involucre oblong-ovoid to campanulate. Involucral bracts overlapping, with papery appendages, often conspicuously radiating, white and simulating ray-flowers. Disc-flowers bisexual, or the outer sometimes female. Fruit hairless or with few silky hairs; pappus of bristles. *Australia, New Zealand, New Caledonia.* Figure 127, p. 554.

64. Cassinia R. Brown. 20/4. Evergreen shrubs or undershrubs, rarely herbs. Leaves alternate, entire, usually small and scale-like. Heads small, numerous in terminal corymbose panicles, possessing disc-flowers only. Receptacle usually with membranous, white-tipped scales. Involucral bracts translucent and membranous or coloured, in several overlapping series, inner series often spreading. Disc-flowers few, bisexual or a few outer ones female, corolla pale brown-green, tubular and 5-toothed. Fruit angled, compressed or almost terete, with a pappus of slender bristles, free or fused at base. *Australia, New Zealand, South Africa.* Figure 127, p. 554.

65. Ammobium R. Brown. 2/1. Erect, branched or simple, woolly perennial herbs. Stems with herbaceous wings decurrent from the leaf-bases. Leaves to 30 × 5 cm, in a basal rosette and alternate on the stems, lanceolate, white, shaggy-hairy, stem-leaves smaller upwards. Heads to 2.5 cm across, solitary or few in a terminal corymb, with disc-flowers only. Involucre hemispherical, bracts petal-like or thin and membranous, ovate, in several series. Receptacle shortly conical. Disc-flowers numerous, bisexual, orange-yellow. Fruit slightly compressed, unequally 4-angled, hairless; pappus a blunt or shortly 2--4-toothed membranous cup. *Eastern Australia.* Figure 127, p. 554.

66. Antennaria Gaertner. Up to 75/6. Dioecious perennials, usually with white-woolly hairs. Leaves alternate, usually entire. Heads many-flowered, solitary to many in a congested inflorescence; ray-flowers absent. Involucral bracts in several rows, overlapping, often coloured, not spreading in a star-shaped in fruit.. Male flowers purplish with tailed anthers, style simple, pappus-hairs conspicuously thickened above; female-flowers with a branched style and conspicuous pappus, its bristles united at base. Fruits 3-angled or slightly flattened. *North temperate areas & South America.* Figure 127, p. 554.

67. Anaphalis de Candolle. 100/5. Perennial herbs, rarely shrubs, erect, white-felted. Leaves alternate, linear to lanceolate, entire. Heads in flat-topped corymbs, bisexual, of 2 kinds: most flowers female, a few of the central flowers male, or most flowers male, a few of the outer female, all yellow, ray-flowers absent. Receptacle without scales or bristles. Involucral bracts in several rows, overlapping, the outer white, dry, the inner petal-like. Fruits small, thin; pappus of slender bristles, none of which are expanded above. *North temperate regions, tops of tropical mountains.*

68. Bracteantha Everett. 5/1. Usually annual herbs (ours). Leaves alternate, glandular-hairy, entire, stalkless. Heads terminal, solitary or up to 8 in an open inflorescence, hemispherical to almost spherical. Involucral bracts in 8--12 series, entire, dry and membranous, white, yellow or pink. Receptacle flat, lacking scales. Ray-flowers absent, disc-flowers mostly bisexual, outermost sometimes female, yellow. Fruit oblong, 4-angled, hairless; pappus of free or shortly fused, barbed bristles. *Australia; naturalised in many countries.*

69. Craspedia G. Forster. 8/2. Annual to perennial, usually evergreen herbs. Leaves in basal rosettes and on stems, entire. Heads in spherical or cylindrical clusters, with disc-flowers only; stalks stiff, unbranched. Receptacle with scales. Involucral bracts ovate to oblong, early deciduous. Disc-flowers 3--8, white or yellow, rarely pink to purple, tubular. Fruit silky-hairy; pappus of feathery bristles. *Australia, New Zealand.* Figure 127, p. 554.

70. Leucophyta R. Brown. 1/1. Small shrub to 40 cm. Leaves alternate, to 5 cm, linear-lanceolate, erect, stalkless, triangular, adpressed to the stems. Heads aggregated into dense secondary spherical clusters to 1.2 cm across, subtended by a few silvery leaves. Involucral bracts *c.* 10, oblong, membranous. Receptacle conical, without scales. Ray-flowers absent. Disc-flowers bisexual, 2 or 3 in each head. Fruit a pappus of 8--12 plumose bristles with flattened axes, united at the base. *Temperate Australia.*

71. Leontopodium Cassini. 35/2. Clump-forming perennial herbs. Stems erect or ascending. Leaves alternate, mostly crowded at base, simple, entire. Flower-heads with disc-flowers only, arranged in a cluster surrounded by ray-like, lanceolate, white-felted leaves forming a star-like shape. Receptacle concave, lacking bracts. Disc-flowers bisexual, tubular. Fruit hairy or hairless, cylindric; pappus of minutely toothed hairs. *Mountains of Europe & Asia.* Figure 127, p. 554.

72. Leucogenes Beauverd. 2--3/2. Clump-forming, woolly perennial herbs, woody at base. Leaves overlapping, stalkless. Heads with disc-flowers only, arranged in a dense cluster subtended by prominent leaves. Involucral bracts overlapping in many series. Receptacle slightly convex, lacking bracts. Inner disc-flowers bisexual, numerous; outer disc-flowers female, few in 1 or 2 series; all flowers tubular, corollas 4- or 5-toothed. Fruit hairy; pappus with 20--25 hairs, thickened towards apex. *New Zealand.*

73. Helichrysum Miller. 500/15. Annual to perennial, usually erect herbs (sometimes woody at base) or shrubs. Leaves alternate or opposite, entire, flat or with margins rolled under. Heads with both disc- and ray-flowers or disc-flowers only, solitary or several in a corymb.Involucral bracts in few to many overlapping series, rigid, membranous, white or otherwise, hairless or hairy. Receptacle flat or convex, usually lacking scales. Flowers few to numerous, yellow; outer female, often lacking rays, but sometimes with well-developed rays. Inner flowers bisexual, funnel-shaped or tubular. Fruit almost terete, angled or slightly compressed, hairless; pappus of few to many, smooth, bristly or feathery, deciduous bristles, occasionally absent. *Europe, central Asia, South Africa, Australia.* Figure 127, p. 554.

582

74. Gnaphalium Linnaeus. 150/2. Annual, biennial or perennial herbs, sometimes woody. Leaves alternate, simple, entire. Heads solitary or in small to large clusters, arranged in corymbs, spikes or racemes. Ray-flowers absent, disc-flowers yellow or whitish, the central flowers bisexual, the outer female. Involucral bracts in 2--several rows, membranous, spreading in a star-shape in fruit. Fruits oblong or obovoid, not ribbed; pappus-hairs in 1 row, thickened at apex. Fruit small, oblong. *Cosmopolitan.*

75. Dimorphotheca Moench. 7/4. Aromatic annual or perennial herbs or small shrubs. Leaves usually membranous, entire to deeply toothed or lobed. Heads solitary on long stalks above the leaves. Involucral bracts in a single series. Ray-flowers female, setting fruit, sometimes with 4 staminodes. Disc-flowers bisexual and setting fruit. Fruits of 2 kinds, those from the ray-flowers more or less terete, straight or curved, smooth or wrinkled and warty, those from the disc-flowers flattened and with thickened margins, smooth. Pappus absent. *Southern Africa.* Figure 128, p. 556.

76. Osteospermum Linnaeus. 70/3. Perennial herbs usually woody below, or small shrubs. Leaves alternate, membranous to somewhat fleshy, entire to toothed. Heads solitary on long stalks above the leaves. Involucral bracts in a single series or obscurely in 2 series. Ray-flowers female, without stamens or with 4 staminodes. Disc-flowers apparently fully fertile, with stamens and ovary, but not setting fruit, often the outer distinguished from the inner by their hooded corolla-lobes. Fruits all of one kind (from the ray- flowers). Pappus absent. *Southern Africa.* Figure 128, p. 556.

77. Calendula Linnaeus. 20/1. Annual to perennial herbs, usually glandular and aromatic. Leaves alternate, simple, slightly fleshy. Heads solitary, terminal, conspic-uous; receptacle flat, without hairs or scales. Involucral bracts in 1 or 2 ranks, more or less equal, not conspicuously reflexed in fruit. Disc-flowers male; ray-flowers female. Fruits hairless, beaked, without pappus, varying in size, curvature and degree of wrinkling within the same head. *Mediterranean area, southwest Asia, Atlantic Islands.* Figure 128, p. 556.

78. Heterotheca Cassini. 30/1. Annual to perennial herbs with erect, simple or branched stems. Leaves alternate, usually simple, entire or toothed. Heads solitary to many together, when often in corymbose clusters. Involucre more or less bell-shaped, with 3--5 rows of bracts. Both ray- and disc-flowers present, yellow. Fruit obconical, hairless or downy, that of the ray-flowers more or less 3-angled, pappus absent or of bristles, that of disc-flowers compressed, the outer pappus of narrow scales, the inner of 30--45 bristles. *North America, Mexico.*

79. Chrysothamnus Nuttall. 16/1. Perennial aromatic herbs or shrubs, usually much-branched and with erect stems. Leaves alternate, stalkless. Heads numerous, in cymes or panicles, usually with 5 disc-flowers, ray-flowers absent; involucre cylindric, bracts in 3--5 series, often clearly disposed in 5 vertical ranks, free, overlapping. Corollas bright yellow, lobes spreading. Style-branches long, slender, projecting. Fruits narrow-ly cylindric, 5-ridged, pappus of numerous soft bristles. *Western North America.*

80. Grindelia Willdenow. 55/6. Annual, biennial or perennial herbs, sometimes woody at base. Leaves simple, entire, toothed or pinnatifid, often with glands producing a sticky resin. Heads solitary or in corymb-like panicles, many-flowered. Ray-flowers usually present. Involucre resinous-sticky, of 4--8 series of bracts, their bases more or less hardened and adpressed, the apices broadly thread-like, ascending-erect, spreading or reflexed. Receptacle flat, surface deeply pitted but without scales subtending the flowers. Ray-flowers yellow, 14--45, female, fertile, the rays linear-oblong to linear-spathulate. Disc-flowers yellow, bisexual. Fruits more or less cylindric, sometimes ribbed or with 1--3 knob-like processes at the apex. Pappus of 2--15 bristles or scales. *America.*

81. Haplopappus Cassini. 150/2. Annual or perennial herbs and shrubs, usually resinous and glandular. Leaves usually alternate, entire to lobed or dissected. Heads solitary to many, involucral bracts almost equal or overlapping; ray-flowers female, usually fertile, yellow; disc-flowers bisexual, corolla 5-lobed, yellow. Anthers rounded at the base. Fruit angled, hairless to densely hairy; pappus of 1 or more rows of unequal roughened bristles. *Western North America, South America.*

82. Corethrogyne de Candolle. 3/1. Perennial herbs, sometimes woody at the base, with a covering of soft, white woolly hairs at first, later sometimes becoming hairless above. Leaves alternate, entire or toothed. Heads solitary or clustered, each with many flowers. Involucre obconical to hemispherical, bracts overlapping in several rows, narrow, with erect or spreading green tips. Central disc-flowers bisexual, yellow, style-branches linear with appendages, each bearing a tuft of rigid yellow hairs. Ray-flowers sterile, rays linear, violet to pink. Fruit narrowly obconical with a pappus of numerous tawny hairs. *Western USA (California).*

Sometimes included in the otherwise uncultivated larger genus *Lessingia* Chamisso.

83. Stenotus Nuttall. 5/1. Mat-forming evergreen perennial, woody at base. Leaves crowded towards apex of branch, rigid, margin entire, stalkless. Heads solitary, on leafless stalks. Involucral bracts, linear to ovate, in 2 or 3 rows. Ray-flowers 6--15, female, yellow; disc-flowers with deeply 5-toothed corollas, bisexual. Fruit hairless or with dense silky hairs; pappus of many soft white bristles. *Western North America.*

84. Solidago Linnaeus. 100/up to 40. Perennials with sometimes woody stems, usually more or less hairy, often with foliage at ground-level in winter. Leaves alternate, simple, entire or toothed. Flower-heads usually many, in our species usually less than 2 cm wide, with disc- and ray-flowers. Receptacle flat or nearly so, without scales. Involucral bracts in several rows, free, chaffy, with a green central zone towards the tip; green zones usually not forming a regular pattern. Flowers deep yellow, rarely pale yellow or white. Ray-flowers female, usually with a small ray; disc-flowers bisexual, with anthers entire at base. Ovary and fruit ribbed or almost unribbed, usually hairy. Pappus of rough bristles. *Mostly America, 1 species in Eurasia.* Figure 128, p. 556.

There are many known wild hybrids, often between strongly dissimilar species. Deliberate hybridisation in cultivation has produced some garden-worthy cultivars.

85. × Solidaster Wehrhahn. 1/1. Like *Solidago* but with the rays pale yellow or white, more conspicuous than in small-headed species of *Solidago. Garden Origin.*

Known in only cultivation and sometimes found under the name × *Asterago* Everett.

86. Aster Linnaeus. 250/32. Cultivated species perennial, usually herbaceous, often with foliage at ground level in winter. Stems usually eglandular-hairy, more so above than below. Leaves alternate, simple in culltivated species, toothed or entire, increasingly bract-like above (the uppermost here referred to as sterile bracts). Heads 1--many, with disc-flowers and, usually, ray-flowers. Receptacle without scales. Involucral bracts in 2 or more rows, free, at least the inner chaffy at the base, at least at the sides. Ray-flowers in 1 row, female or sterile, with white to blue, purple or pink corollas longer than the involucre. Disc-flowers bisexual, their corollas usually yellow, becoming reddish purple. Anthers without basal appendages. Pappus-hairs usually rough, in several rows, approximately equal in length or sometimes some of the outermost much shorter than the rest, at least nearly as long as the fruits. Ovary and fruit not or only slightly compressed, usually more or less ribbed, often hairy. *Northern hemisphere.* Figure 128, p. 556.

87. Kalimeris (Cassini) Cassini. Several/4. Like *Aster* but always rhizomatous, with the leaves sometimes pinnatifid, most of the involucral bracts about the same length, and pappus-bristles not more than 1 mm, not more than half as long as the achene. Plants more or less hairy. Heads in a loose corymb. *Eastern Asia.*

Often known as *Calimeris* Nees.

88. Callistephus Cassini. 1/1. Erect, branched, annual herb to 80 cm. Leaves ovate to triangular-ovate, coarsely toothed, stalked, the upper more spathulate, less toothed and more shortly stalked. Heads to 12 cm across, solitary and terminal; involucral bracts in many series, those of the outer series green and leaf-like, often reflexed. Ray-flowers white to pale mauve to violet or reddish purple; disc-flowers many, tubular, yellow, but in many cultivars largely replaced by ray-flowers (heads double). Fruits compressed. Pappus of 2 rows of bristles. *China.*

There are numerous cultivars.

89. Erigeron Linnaeus. 150/20 (and several hybrids). Annual, biennial or perennial herbs, occasionally somewhat woody at the base. Leaves alternate, entire, toothed or somewhat divided. Heads solitary or in racemes which are often corymbose. Involucral bracts narrow, in several rows, all more or less similar in size, green at least in the centre, hairy. Receptacle flat, without scales. Flowers numerous; disc- and ray-flowers present, occasionally the outermost disc-flowers with thread-like, unlobed corollas, Style-appendages short. Pappus single or double, always with central bristles, sometimes with outer fine hairs or scales. Fruits usually 2-veined, hairy. *North temperate areas, most in North America.* Figure 128, p. 556.

90. Baccharis Linnaeus. 400/3. Dioecious shrubs. Leaves alternate, simple. Heads solitary or in false spikes; receptacle flat, smooth; involucral bracts in 3--4 rows, overlapping. Flowers unisexual, corollas of female thread-like, those of male tubular, 5-lobed, with non-functional stigma. Fruits flattened. Pappus of 1 or more rows of bristles, distinctly straw-coloured. *America.*

91. Chrysocoma Linnaeus. 20/1. Annual to perennial herbs or small shrubs. Leaves alternate, linear, entire or with a few lobes. Involucre hemispherical or broadly bell-shaped, bracts in 3 or 4 rows, each with membranous margins. Heads solitary, terminal, with disc-flowers only. Receptacle flat or convex, without scales. Corollas yellow. Fruit with a pappus of 1 row of fine bristles. *South Africa.* Figure 128, p. 554.

92. Felicia Cassini. 83/7. Annual, biennial, perennial herbs or small shrubs, sometimes with rhizomes, often bristly and/or glandular. Leaves opposite or alternate, occasionally the lower opposite, the upper alternate, entire or finely toothed. Heads usually on long stalks, solitary or not. Involucral bracts in 2 or more rows, entirely herbaceous. Ray-flowers in a single whorl, female, rays usually blue, more rarely white; disc-flowers yellow, bisexual. Fruits flattened, often each with a defined marginal area or wing, often hairy. Pappus of a single row of equal or unequal, simple, toothed or ciliate bristles, deciduous or persistent. *Africa (mostly Cape Province of South Africa)*. Figure 128, p. 556.

93. Bellis Linnaeus. 8/4. Short annual or perennial herbs. Leaves alternate or basal, entire to scalloped. Heads solitary, stalked. Involucral bracts in 2 rows. Receptacle cone-shaped to almost flat; scales absent. Ray-flowers female, rays entire or almost so, white, often tinged purplish crimson; disc-flowers with 4- or 5-lobed, yellow corollas. Fruits compressed; pappus absent or short and rudimentary. *Europe, Mediterranean area*. Figure 128, p. 556.

One species is a very common lawn-weed.

94. Bellium Linnaeus. 4/2. Small perennial herbs. Leaves all basal or alternate, more rarely almost whorled, long-stalked, entire. Heads solitary. Involucral bracts green or translucent and shining, in a single row. Receptacle hemispherical to conical, without scales. Ray-flowers female, rays white or tinged red beneath; disc-flowers each with a 4- or 5-lobed corolla with a campanulate tube. Fruits slightly compressed, downy; pappus of an outer ring of 4--6 (rarely to 10) translucent scales which are up to half as long as the fruit, and an inner ring of bristles which are as long as or longer than the fruit. *Mediterranean area*.

95. Boltonia L'Héritier. 4/1. Hairless perennial herbs, often stoloniferous. Leaves more or less entire, often turned edgewise. Heads many-flowered, in open, loose panicles, receptacle conical or hemispherical. Involucral bracts overlapping, in approximately 2 series, each with narrow, translucent margins. Ray-flowers female, disc-flowers bisexual. Fruits very flattened, with winged margins. Pappus of several very short bristles and usually 2 or 4 longer bristles (awns). *North America*.

96. Townsendia J.D. Hooker. 20/7. Biennial or perennial herbs, often with the leaves in basal rosettes. Leaves generally spathulate to oblanceolate, hairy or hairless. Heads borne on long or very short stalks, sometimes borne within the rosette of leaves, large or small, receptacle conical or almost flat. Involucral bracts in 2--7 series, each usually hairy and with translucent margins. Ray-flowers female, 10--100 but usually 20--40, blue, purplish, white or pink. Disc-flowers bisexual, usually yellow, sometimes tinged with pink or purple. Fruits compressed, 2- or rarely 3-ribbed, usually hairy with hairs with forked, separated tips. Pappus of specialised bristles with apparently lacerated margins, these sometimes very short, sometimes small scales present as well. *Western North America*.

97. Brachycome Cassini. *c.* 25/3 (and some hybrids). Annual to perennial herbs, usually grown as annuals. Leaves basal or borne on the stems, entire to pinnatisect, alternate. Heads solitary or numerous. Involucral bracts numerous in 2 (rarely 3) rows, green, entire or with translucent, ragged, ciliate margins. Receptacle flat to conical, pitted or not. Ray-flowers female, rays white, blue, violet or pink (very rarely yellow), in a single row. Disc-flowers bisexual, yellow or dark brown to almost black. Anthers each often with a prolonged connective. Fruit sometimes flattened, often winged, hairy or hairless. Pappus of microscopic or somewhat larger bristles, more rarely absent. *Australia.*

98. Chiliotrichum Cassini. 7/1. Evergreen shrubs. Leaves alternate, entire. Heads solitary, borne at the ends of the branches, ray-flowers female, disc-flowers bisexual, corolla with 5 teeth. Involucral bracts in 2--3 or more rows. Pappus of 2 or 3 rows of rigid hairs. Fruit cylindric, glandular. *Temperate South America.*

99. Nardophyllum Hooker & Arnott. 7/1. Low-growing or dwarf shrubs, densely branched, often forming mats or cushions, hairy or hairless, Leaves alternate, small, entire. Heads solitary at the ends of the branches. Involucre bell-shaped. Receptacle convex with few or no scales. Involucral bracts overlapping in several rows, the outer smaller than the inner. Flowers all similar, bisexual, corolla tubular, 5-lobed. Fruit cylindric or obconic, with 4 or 5 ribs. Pappus of numerous unequal bristles. *Southern South America.*

100. Lagenophora Cassini. 25/1. Perennial herbs usually with rhizomes. Leaves basal and borne on the stems. Heads on leafy stems or leafless scapes. Involucral bracts with membranous, translucent margins, ultimately reflexed. Receptacle without scales. Ray-flowers female, rays white or purplish; disc-flowers tubular, bisexual, corolla 5-lobed, style-arms flattened with lanceolate-triangular, non-receptive tips. Anthers obtuse at their bases. Fruits compressed. Pappus absent. *New Zealand, Australia, temperate South America.*

101. Olearia Moench. 180/35 (plus some hybrids). Evergreen shrubs or small trees. Leaves alternate or opposite, simple, scattered or crowded or clustered, leathery or membranous, with densely felted hairs beneath; margins entire or toothed. Heads axillary or terminal, stalked or stalkless, solitary or in corymbs or panicles, with or without ray-flowers, occasionally containing only a single (disc-) flower. Receptacle without scales or bristles. Disc-flowers bisexual, tubular, yellow, white, violet, cream or purple; ray-flowers female, white, yellowish, purple, lilac, mauve or pale blue, or ray-flowers absent. Involucral bracts overlapping, in several series. Fruits ribbed or striped, terete or compressed, downy, hairy or hairless; pappus of unequal short stiff hairs, often slightly thickened at tips. *Australia, New Zealand, Lord Howe Island & New Guinea.* Figure 128, p. 556.

102. Celmisia Cassini. 60/30 (and some hybrids). Short, clump-forming evergreen herbs with leaf-bases forming a pseudostem or woody-based perennials with leaves tufted or along branches, either decumbent or cushion-forming. Leaves usually lanceolate, ovate or obovate occasionally needle-like, often with a deciduous membranous skin (pellicle) above and densely felted-hairy beneath. Scape with 1 head (rarely more) and various bracts along its length. Disc-flowers yellow (1 species purple, not ours), bisexual; ray-flowers white (occasionally pinkish or purplish in 2 species), female. Receptacle pitted, more or less convex. Fruit ribbed, hairless to hairy; pappus of hairs. *New Zealand, southeast Australia including Tasmania.* Figure 128, p. 556.

103. Pachystegia Cheeseman. 1--3/1. Spreading shrubs with densely felted-hairy branchlets. Leaves crowded in rosettes at ends of branches, very leathery, entire; upper surface hairless, shining, lower with densely felted hairs. Heads solitary on long stout stalks. Receptacle without scales or bristles. Ray-flowers in 2 or 3 series, female, white, disc-flowers bisexual, yellow, tubular. Receptacle slightly convex, pitted. Involucre ovoid with numerous involucral bracts, densely overlapped in many series; innner with slender, recurved tips. *New Zealand.*

Sometimes placed in *Olearia.*

104. Ursinia Gaertner. 37/3. Annual to perennial herbs or small shrubs. Leaves alternate, usually pinnatisect to bipinnatisect. Heads solitary on long stalks. Involucral bracts in several series, at least the inner with membranous margins. Ray-flowers present, rays usually yellow, sometimes each with a dark spot at the base; disc-flowers bisexual, usually yellow, sometimes purplish towards the apex. Receptacular scales present, each often enfolding a disc-flower, sometimes with an appendage at the apex. Fruits curved. Pappus a single series of 5 scales, or of 2 series, the outer of 5 scales, the inner of 5 bristles, in fruit sometimes appearing to mimic a membranous perianth. *Southern Africa.* Figure 128, p. 556.

105. Eumorphia de Candolle. 6/2. Erect or prostrate heather-like shrubs. Leaves mostly opposite, crowded, simple; translucent glands can be seen in the leaves of some species after they have been boiled. Heads terminal, solitary or in loose corymbs, on short or very short stalks. Receptacle flat or slightly convex, almost always with some scales subtending flowers but these sometimes very few or perhaps totally absent. Involucre spherical to cylindric, bracts in 3--5 series, papery and translucent, hairy, hairless or ciliate, the outermost sometimes glandular on their backs. Ray-flowers female, fertile, rays white or pinkish on the backs; disc-flowers with 5-lobed corolla with a distinct tube and limb. Fruits 10--12-ribbed (rarely with up to 18 ribs), minutely papillose, pappus absent. *South Africa.*

106. Allardia Decaisne. 8/1--2. Tufted or somewhat spreading alpine herbs forming mats. Leaves alternate, usually in rosettes, pinnatifid or rarely entire. Heads large, solitary, terminal. Involucre hemispherical, bracts in 2 or 3 series, with dark brown or purplish margins. Receptacle convex, without scales. Ray-flowers female, fertile or sterile, large, rays white, pink or bluish violet; disc-flowers tubular, yellow or the 5-lobed corolla-limb bluish violet. Fruits 5-angled, faintly 5--10-ribbed, generally covered with stalkless glands, sometimes hairy. Pappus of many red or brown bristles which are as long as or longer than the corollas. *Afghanistan, Himalaya, China.*

107. Tanacetum Linnaeus. 50/12. Perennial herbs, often with rhizomes, sometimes somewhat woody at the base; often densely hairy with simple or bifid hairs mixed with stalkless glands, rarely hairless. Leaves entire, toothed, pinnatifid or 1--3-pinnatisect, primary segments usually distant but occasionally close and contiguous, apparently joined by their hair-covering. Heads usually in sparse or dense corymbs, rarely solitary. Involucral bracts in 3 or 4 series, overlapping, often with translucent margins. Receptacle flat, without scales. Ray-flowers present or absent, when present female, rays sometimes small, white, yellow or pink; disc-flowers bisexual, yellow. Fruits cylindric or club-shaped, 5--10-ribbed, often glandular, hairless. Pappus a short corona, usually unevenly toothed or lobed, sometimes developed on one side of the

588

fruit only, occasionally completely absent. *Europe, Mediterranean area.* Figure 128, p. 556.

108. Arctanthemum (Tzvelev) Tzvelev. 4/1. Perennial herbs with somewhat woody rootstocks, bearing rosettes of leaves. Leaves alternate, lobed, toothed or rarely entire. Heads solitary. Involucral bracts in 3 series, with dark brown, translucent margins; receptacle conical to convex, without scales. Ray-flowers female, rays white; disc-flowers with 5-lobed yellow corollas with tubes generally bearing stalkless glands. Fruits oblong, usually somewhat 5--8-ribbed; pappus absent. *Arctic areas.*

Originally included in *Chrysanthemum* or *Dendranthema.*

109. Ajania Poljakov. 34/1. Subshrubs or shrubs. Leaves alternate, evergreen or half-evergreen, simple, with *c.* 5--7 lobes in the upper half, dark green and hairy above, densely white-hairy beneath, the hair-covering visible round the margins, containing forked and tree-like hairs. Heads small, depressed-spherical, in corymbs or rarely solitary. Receptacle convex to conical, without scales. Involucral bracts overlapping in 3 series, hairy. Ray-flowers absent, but the outer series of disc-flowers are female with tube-like, unlobed corollas; central disc-flowers bisexual, fertile, tubular to bell-shaped, corolla 4- or 5-lobed, yellow to purplish. Fruit obovoid, thin-walled, faintly 4--6-ribbed. Pappus absent. *Eastern Asia.*

110. Artemisia Linnaeus. 300/28. Annual, biennial or perennial herbs or shrubs, usually aromatic. Leaves alternate, usually compound, often with very narrow segments. Heads small, in racemes or panicles, cylindric to spherical or hemispherical. Involucre with bracts in a few series, the bracts themselves often with translucent margins. Receptacle hairy or not. Flowers all similar, bisexual and fertile, or the outer flowers with thread-like, unlobed corollas and female, the inner bisexual or apparently so (though functionally male), with 5-lobed, white, yellowish, reddish, purplish or brownish corollas, the tube sometimes hairy or glandular. Fruits usually smooth, occasionally hairy. Pappus absent. *North temperate areas.* Figure 128, p. 556.

Some grown as herbs or flavourings. Some authors recognise the genus *Seriphidium* (Besser) Poljakov, which includes those species in which all the flowers in the head are of the same type.

111. Santolina Linnaeus. 12/6. Aromatic dwarf shrubs, often white due to dense close covering of fine white hairs but sometimes almost hairless and green. Leaves alternate, narrow, linear, toothed to pinnate or bipinnate, segments often indistinct due to their inrolling and the dense covering of felty hairs, the leaves therefore often appearing cylindric. Heads stalked, solitary or several together on a branched stalk. Involucral bracts in several rows, closely appressed. Receptacular scales conspicuous, subtending each flower. Flowers all tubular, usually bisexual, yellow-orange to creamy white, persistent in fruit. Fruit oblong, weakly angled; pappus absent. *West Mediterranean area, especially Spain.*

112. Otanthus Hoffmannsegg & Link. 1/1. Rhizomatous perennial herbs, somewhat woody at base, covered with conspicuous, white woolly hairs. Stems to 50 cm, ascending. Leaves alternate, fleshy, oblong to oblong-lanceolate, entire or obscurely toothed, stalkless, stem-clasping at the base. Heads in corymbs. Receptacle convex, with scales. Involucre hemispherical, 7.5--12 mm across; bracts in 2 or 3 series, densely woolly. Ray-flowers absent. Disc-flowers all bisexual, tubular, yellow, the corolla 5-lobed, swollen and spongy at base. Fruits oblong with 3 or 4 ribs, thin-walled, glandular. Pappus absent. *West Europe, Mediterranean area.*

113. Achillea Linnaeus. 100/40. Perennial herbs, often with woody bases. Leaves alternate, entire to deeply pinnatisect. Heads with both female and bisexual flowers, stalked or almost stalkless, usually arranged in terminal corymbs or rarely solitary and terminal. Involucral bracts in many overlapping series, with translucent margins. Flowers white or yellow, rarely pink; ray-flowers female, rays with 3 teeth; disc-flowers bisexual, radially symmetric, with 5 equal teeth. Receptacle flat or convex, with transparent scales. Achenes hairless, smooth, compressed, not winged; pappus absent. *Northern hemisphere.* Figure 128, p. 556.

114. Anacyclus Linnaeus. 9/2. Annual or perennial herbs. Leaves alternate, pinnatisect. Heads with both bisexual tubular and usually female ray-flowers, stalked, solitary. Receptacle with chaffy scales. Involucral bracts in 3 overlapping series, translucent. Ray-flowers female, white or yellow, rays with 3 teeth. Disc-flowers yellow, with 5 sometimes irregular teeth. Fruits compressed, each surrounded by a translucent wing. Pappus of a corona. *Mediterranean area.*

115. Chamaemelum Miller. 6/1. Aromatic annual or perennial herbs, sometimes woody below. Leaves 1–3-pinnatisect, alternate. Heads generally solitary, rarely in loose corymbs. Involucral bracts in 2 or more series, with translucent margins, gradually decreasing in size outwards. Receptacle hemispherical, with scales. Ray-flowers usually present, female or sterile; disc-flowers fertile, with long-tubed, 5-lobed corollas, the base of each corolla-tube saccate around the top of the ovary. Fruits obovoid with 3 very thin ribs, slightly compressed. Pappus absent. *Europe, North Africa, southwest Asia.*

116. Anthemis Linnaeus. *c.* 100/7. Annual to perennial herbs, sometimes woody at the base, rarely small shrubs, often densely hairy, the hairs often forked or T-shaped. Leaves 1–3-pinnatisect or rarely simple, primary segments usually 3, each divided pinnately or palmately. Heads solitary on long stalks; ray-flowers usually present in cultivated species. Receptacle convex or conical, with scales which are often as long as the disc-flowers or longer, linear-lanceolate to needle-like or oblanceolate, papery or parchment-like, acute or 3-toothed at their apices. Involucre with bracts in usually 3 series, at least the inner with conspicuous, translucent margins. Ray-flowers female, usually fertile, rays white or yellow, rarely purplish, corolla-tubes somewhat persistent on the fruits. Disc-flowers with tubular, 5-lobed corollas which are sometimes inflated towards the base, yellow. Fruits obconical, rounded or diamond-shaped in section, pappus absent or the rim at the apex elongated to form a rounded or strap-shaped auricle. *Mainly Mediterranean area, southwest Asia.* Figure 128, p. 556.

117. Chrysanthemum Linnaeus. 20/2 (with some hybrids). Perennial herbs, often somewhat woody at the base. Leaves alternate, 1- or 2-pinnatisect or lobed, often gland-dotted. Heads in loose corymbs or solitary with ray- and disc-flowers or 'double' with all the flowers with rays which are 3-toothed at the apex. Involucre with bracts in 3 rows, bract-margins dark brown. Receptacle convex to conical, without scales. Ray-flowers (when distinguishable from disc-flowers) female, fertile, rays white, pink, yellowish, coppery or brownish. Disc-flowers bisexual, with tubular, 5-lobed corollas which are usually glandular on the tubes. Fruits (when produced) all similar, cylindric-obconical to obovoid, faintly 5--8-ribbed. Pappus absent. *Eurasia*. Figure 128, p. 556.

Many cultivars are grown. The name 'Chrysanthemum' has now been conserved to apply to these plants (the florists' Chrysanthemums), which were formerly known as *Dendranthema* (de Candolle) Desmoulins.

118. Xanthophthalmum Schultz Bipontinus. 2/2. Annual herbs. Leaves alternate, simple but pinnately divided, often deeply so, the upper sometimes not divided, sometimes greyish. Heads more or less solitary at the ends of the branches. Involucre hemispherical, bracts overlapping in 3 or 4 series, each bract with broad, translucent margins. Receptacle convex, without scales. Rays yellow, cream or white. Disc-flowers numerous, yellow, tubular, tube 2-winged. Fruits of 2 forms, those of the disc-flowers cylindric to cylindric and 3-angled, those of the ray-flowers 3-angled, with the angles often winged. Pappus absent. *Europe*. Figure 128, p. 556.

This genus was previously the central core of the Linnaean genus *Chrysanthemum*; however, this latter name has now been conserved for the florist's Chrysanthemum and its allies (see above), so the unfamiliar, almost unacceptably long and scarcely pronounceable name given above has to be used for the present genus.

119. Ismelia Cassini. 1/1. Upright annual herb. Leaves alternate, pinnatisect, with linear lobes, somewhat fleshy. Heads solitary or few together in loose corymbs. Involucre with keeled bracts which are wide and many-veined, with resin-canals. Receptacle without scales. Ray-flowers female, large, the ray basically yellow, often reddish or white at the base or occasionally all tinged with red. Disc-corollas red to purple. Ray-fruits triangular in section, 3-winged; disc-fruits laterally flattened; pappus absent. *Morocco; widely escaped from cultivation in Europe*.

120. Argyranthemum Schultz Bipontinus. 22/6 (and some hybrids). Shrubs. Leaves alternate, lobed or 1--2 times pinnately divided, green or glaucous. Heads on corymbs or rarely solitary. Involucral bracts in 3 series, with translucent margins. Receptacle without scales. Ray-flowers present, female, rays linear or linear-lanceolate, usually white, rarely pink or yellowish. Disc-flowers bisexual, tubular-campanulate, corolla-lobes usually yellow. Fruits variously curved, flattened and winged, those of the ray-flowers often differing from those of the disc-flowers. Pappus usually a small corona. *Canary Islands*.

121. Nipponanthemum Kitamura. 1/1. Shrubs to 1 m, shortly downy above. Leaves alternate, close, obovate, obtusely toothed towards the apex, tapered but not stalked, fleshy, shining. Heads borne on long leafless stalks, solitary, to 6 cm across. Involucre to *c.* 1 cm long, with bracts in 4 distinct series, downy, brownish on the margins. Receptacle without scales. Ray-flowers several, female, fertile, rays white, 2.2--3 cm; disc-flowers yellow, corolla 5-lobed. Fruits 10-ribbed, thin-walled; pappus a corona of small scales. *Japan*.

122. Leucanthemella Tzvelev. 2/1. Perennial herbs. Leaves simple, alternate, gland-dotted, oblong-lanceolate, stalkless, 2--4-lobed at the base, the central stem-leaves with forwardly-directed teeth. Heads few in loose corymbs. Receptacle strongly

591

convex, without scales. Involucral bracts in 2 or 3 rows, with translucent or dark brown margins. Ray-flowers apparently female, sterile, in a single row, the tube strongly compressed but not winged. Disc-flowers with 5-lobed corollas bearing stalkless glands. Fruits all similar, 7--12-ribbed; pappus absent, though a small apical rim is present on each fruit. *Eastern Europe to China & Japan.*

123. Leucanthemopsis (Giroux) Heywood. 10/2. Creeping, tufted perennials with woody bases. Leaves alternate or in rosettes, toothed to pinnatifid, often comb-like and spathulate in outline, not gland-dotted. Heads solitary, on long, leafless stalks. Receptacle convex, without scales. Involucral bracts in 2 or 3 series, margins translucent to dark brown. Ray-flowers female, fertile, rays yellow, white or rarely pinkish; corollas of the disc-flowers 5-lobed, the tube somewhat swollen at base in fruit. Fruits all similar, 3--10-ribbed, without resin-canals. Pappus a delicate, papery and translucent crown. *Mediterranean area, southeast Europe.*

124. Leucanthemum Miller. 35/2 (and some hybrids). Perennial herbs with red-tipped roots. Leaves alternate, entire to pinnately divided. Heads usually large, solitary, long-stalked or rarely 2--6 together in a loose corymb. Receptacle convex or rarely conical, without scales. Involucral bracts in 2 or 3 rows. Ray-flowers well developed, female, fertile; disc-flowers yellow, corolla 5-lobed, basally swollen and spongy, especially after fertilisation. Fruits obconical-cylindric, 10-ribbed; pappus a corona or a posterior auricle. *Temperate Eurasia.* Figure 128, p. 556.

125. Rhodanthemum Wilcox et al. 12/3. Mat-forming perennials with stolons or rhizomes which are often woody and irregularly twisted. Leaves mostly in rosettes, generally long-stalked, 1- or 2-pinnatisect into 3 leaflets/segments. Heads solitary, mostly on scapes, rarely these with small, bract-like leaves, or foliage leaves in the upper part. Receptacle convex, without scales. Involucral bracts overlapping in 2 or 3 series, rounded at their apices, yellowish, papery and translucent with tan or dark brown margins. Ray-flowers female, fertile or sterile, rays white, pink, reddish or creamy orange to yellow; disc flowers with 5-toothed corollas, yellow or occasionally purplish on the lobes, the tube basally swollen and spongy at maturity. Fruits 5--12-ribbed or -winged, with resin-canals. Pappus a papery corona, sometimes rather irregular. *North Africa, Spain.*

The authors first gave this genus the name *Pyrethropsis*, but this was not validly published.

126. Coleostephus Cassini. 3/1. Annual herbs. Leaves alternate, finely to deeply toothed. Heads solitary or few in loose corymbs, long-stalked. Involucral bracts in 2 or 3 rows, brownish with translucent papery margins, wide and somewhat fan-shaped. Receptacle flattened-convex to conical, without scales. Ray-flowers female, fertile or sterile, rays golden yellow, many-veined. Disc-flowers with 5-lobed yellow corollas which are swollen, spongy and compressed at their bases going into fruit; lobes with more or less developed appendages. Fruits curved, 8--10-ribbed, ribs whitish and converging into a basal callus. Pappus a whitish, translucent, oblique corona or auricle, as long as the fruit. *Mediterranean area.* Figure 128, p. 556.

127. Matricaria Linnaeus. 7/1. Aromatic or sweetly-scented annual herbs. Stems erect or ascending, branched. Leaves 2- or 3-pinnatisect, ultimate segments narrow and thread-like. Heads solitary at ends of branches or somewhat corymbose. Receptacle conical, hollow, without scales. Involucral bracts overlapping in 2--3 series, with translucent, papery margins. Ray-flowers usually present, female, fertile; disc-flowers bisexual, fertile, corollas 4- or 5-lobed, the tube somewhat swollen. Fruits mucilaginous when wetted, slightly compressed, weakly 3--10-ribbed, the ribs more strongly developed on the posterior surface. Pappus absent or present as a small corona, rarely a single posterior scale. *Europe, Mediterranean area, southwest Asia, North America; introduced elsewhere.*

128. Cotula Linnaeus. 90/3. Annual or perennial herbs, if perennial then rhizomatous or stoloniferous. Leaves alternate, usually 1- or 2-pinnate. Heads small, solitary, generally on long stalks, the stalks sometimes swollen beneath the heads. Receptacle flat to conical, without scales. Involucral bracts in 1 or 2 series, usually papery and translucent, at least on the margins. Ray-flowers absent. Outer flowers female, fertile, in 1--several rows, generally stalked, corollas yellowish, brownish, greenish or sometimes absent; inner flowers with usually 4-lobed corollas. Fruits flattened, often laterally winged, sometimes hairy. Pappus absent. *More or less cosmopolitan.*

Many species formerly placed in this genus are now included in the next.

129. Leptinella Cassini. 40/6. Plants very like *Cotula* but differing in having heads mainly on axillary scapes, receptacle conical, outer flowers of the head tubular (corollas not or scarcely lobed) and female, in 1--several rows, the corolla-tube inflated and with a hollow space between the outer and inner surfaces; central flowers female, sterile, with 2--4-lobed yellowish, brownish or greenish corollas; in some species the heads are strictly unisexual. The plants are generally mat-forming and the corollas white, yellow or purplish to almost black. *New Zealand, Australia, South America.*

A genus formerly included in *Cotula.*

130. Eriocephalus Linnaeus. 30/1. Much-branched aromatic shrubs. Leaves alternate or opposite, simple or lobed (on the same plant), usually narrow and needle-like. Heads many, in terminal, umbel-like corymbs. Involucral bracts in 2 unequal series, those of the outer series 4 or 5, translucent and papery, hairy or not; those of the inner series generally densely hairy and variously united. Receptacle with hairy scales. Ray-flowers female, fertile, rays white or lilac, short and wide (absent in some non-cultivated species); disc-flowers functionally male, with the stigma-lobes fused together, corolla 5-lobed, yellow or mauve. Fruits flattened, with 2 lateral ribs. Pappus absent. *Southern Africa.*

131. Othonna Linnaeus. 100/7. Perennial, woody or more or less succulent shrubs or herbs, some with more or less subterranean tubers and crown of deciduous, fleshy leaves, the creeping herbs evergreen. Leaves alternate, usually with felt in their axils, otherwise hairless, often glaucous. Heads solitary or few together on slender, sparingly forked stalks, cup-shaped, with a single whorl of more or less united involucral bracts. Disc-flowers yellow, central male, outer female or rarely bisexual; rays yellow or purple. Style-branches truncate, each with a conical brush of hairs. Pappus of bristles. *South Africa.*

132. Senecio Linnaeus. 1500/50. Annual or perennial herbs, stem- or leaf-succulents or climbers, sometimes becoming woody below, hairless or sparsely hairy to densely woolly; hairs sometimes glandular. Leaves alternate, deciduous or semi-deciduous, stalked or stalkless, entire, toothed, scalloped, pinnately lobed or lyrate, often merging into inflorescence-bracts above; basal rosettes sometimes present. Heads solitary or few to many in loose or dense corymbs. Involucral bracts in a single row, sometimes with a few small extra bracts (calyculus) at the base. Receptacle flat, without scales. Ray-flowers usually present, sterile or female, rays minutely 3-toothed. Disc-flowers bisexual. Fruits ribbed, hairless or with short hairs. Pappus of fine, toothed hairs, rarely absent. *Cosmopolitan.* Figure 129, p. 558.

133. Pericallis D. Don. 14/3 (and a hybrid). Shrubs or perennial herbs. Leaves alternate or in a basal rosette, palmately veined, stalked, margin simple or lobed. Heads solitary or in a corymb or panicle, bisexual; involucre bell-shaped, of many bracts in a single row; receptacle without scales or bristles. Ray-flowers white, pink, mauve or purple (never yellow). Fruit compressed with 10 ribs, disc-fruit densely hairy, falling early; ray-fruit sparsely hairy, pappus absent. *Azores, Canary Islands.*

134. Emilia Cassini. 100/2. Annual herbs. Leaves alternate, often predominantly basal and greatly reduced above, margin almost entire to toothed or unevenly pinnately lobed, clasping stem at base. Inflorescence corymbose with 1--few heads. Heads of disc-flowers only, bisexual, ray-flowers absent; involucral bracts 8--10 in 1 row; receptacle without scales. Flowers 10--100, corollas red or purple. Pappus of many white hairs. Fruits 5--ribbed, finely downy, outer fertile, inner sterile. *Old World tropics.*

135. Euryops (Cassini) Cassini. 97/10. Evergreen shrubs. Leaves alternate, stalkless, loosely to closely arranged and often overlapping, margins entire to variously divided. Inflorescence-stalks either axillary or terminal, simple, leafless. Involucral bracts in 1 row. Ray-flowers female, fertile with a cylindric tube, ray yellow or orange; style bifid, branches linear; pappus of a few to many bristles, sometimes barbed or absent; fruits hairy or hairless. Disc-flowers bisexual, fertile or sterile; corolla tubular, lobes with distinct mid-veins; style bifid or sometimes simple; stamens 5; pappus and fruits as with ray-flowers. *South Africa to Arabia.*

136. Cacalia Linnaeus. 50/1. Perennial herb. Leaves alternate, kidney-shaped, when juvenile convoluted and folded down to the stalk like a closed umbrella. Inflorescence a panicle; involucral bracts in a single row, of 2 types, alternating wider with broad membranous margins, and narrower with very narrow membranous margins; receptacle flat or indistinctly raised. Flowers bisexual, 1--20, fertile, all tubular, corolla white, narrowed below into a slender tube, upper portion bell-shaped, anthers each with a narrow triangular apical appendage, arrow-shaped at base, united in pairs in neighbouring anthers. Style-branches elongate with short hairs. Fruit cylindric in section, truncate, slightly narrowed at tip, gradually narrowed at base, hairless, many-ribbed; pappus bristles many, snow-white. *Asia, North America.*

137. Ligularia Cassini. 130/13 (and some hybrids). Perennial herbs 30--300 cm, hairless or bearing simple, crisped or cobwebby hairs; stems erect, unbranched. Leaves basal and alternate on stems; blades triangular or kidney-shaped to narrowly elliptic, simple and entire or toothed to deeply palmately divided with 1- or 2-pinnatifid segments; basal and lower leaves stalked, stalks usually flattened or channelled above,

594

rarely hollow or with entire to deeply toothed wings, sheathing at base; upper leaves much reduced, usually stalkless. Heads solitary to very many in terminal racemes, panicles or corymbose heads. Involucres cylindric to broadly bell-shaped, rarely hemispheric; bracts free, more or less equal, apparently in 1 row. Receptacle flattened, without scales. Corollas yellow to orange. Ray-flowers to *c.* 24 (absent in some non-cultivated species), female; rays entire or 3-toothed. Disc-flowers bisexual, corollas with very slender bases and slightly to distinctly broadened limbs with 5 usually recurved teeth; anther-bases without tails; styles truncate. Fruits oblong-oblanceolate or sometimes obovoid, hairless; pappus capillary, pure white to reddish, brownish or purplish. *Southwest China, Himalaya.*

138. Cremanthodium Bentham. 70/4. Perennial herbs 10--90 cm, hairless or with simple, crisped or cobwebby hairs. Stems simple, often scape-like. Leaves mostly basal, triangular or kidney-shaped to narrowly elliptic, entire, toothed or pinnatifid, on stalks with sheathing bases; stem-leaves alternate, similar to basal and spreading or oblong and pressed to stem. Heads usually solitary or rarely 2--20 in racemes, nodding. Involucre broadly bell-shaped or hemispheric; involucral bracts more or less equal, often blackish, apparently in 1 whorl or rarely with outer, larger, spreading bracts present. Receptacle flattened, without scales. Ray-flowers female; rays yellow, rarely white, pink or purple, entire or 3-toothed. Disc yellow or brownish, rarely pink or purple; flowers bisexual; corollas with narrow tubular bases usually broadening to bell-shaped limbs, with 5 often erect teeth. Anther-bases not tailed. Style-branches truncate. Fruits oblong-oblanceolate, hairless; pappus of white or brownish hairs. *South-west China, Himalaya.* Figure 129, p. 558.

139. Sinacalia Robinson & Brettell. 4/1. Perennial herbs. Stems simple. Leaves simple, in almost opposite pairs, ovate to almost rounded, cordate at base, stalks somwhat expanded, clasping the stem. Heads solitary to numerous in loose terminal compound clusters; involucre narrowly cylindric to bell-shaped; bracts 4--8 in a single row, margins papery. Ray-flowers 2--8, yellow; disc-flowers 2--numerous, yellow; style truncate, covered with blunt protuberances. Fruit cylindric, hairless. Pappus of bristles. *China.*

140. Tussilago Linnaeus. 1/1. Rhizomatous perennial, far-creeping and producing clusters of shoots. Leaves all basal, to 25 cm, long-stalked, heart-shaped, rounded, shallowly and irregularly toothed, cobwebby beneath, becoming much larger and with more angular margins in summer. Heads produced before the leaves appear, 1.5--3.5 cm across, solitary on erect, unbranched stems up to 20 cm, covered with lanceolate, fleshy scales. Flowers golden-yellow, the disc-flowers male, few, the ray-flowers female, numerous, narrow, reddish or orange beneath. Receptacle hairless. Involucral bracts numerous, purplish. Flower-stems lengthening in fruit; fruits *c.*3 mm, slender, brown, with a pappus of toothed, white hairs. *Eurasia; naturalised in North America.* Figure 129, p. 558.

141. Petasites Miller. 15/7. Dioecious perennial herbs with stout rhizomes, forming patches. Leaves mostly basal, long-stalked, often large and appearing after flowering. Flowering stems stout, with scale-leaves and usually numerous heads. Male heads with central, tubular male flowers and with or without an outer ring of female flowers; female heads with tubular to 2-lipped corollas, usually with a few central, sterile flowers. Involucral bracts in 1--3 rows. Receptacle flat, without scales. Fruits cylindrical, hairless; pappus of simple rough hairs. *Eurasia, North America.* The genus includes species formerly separated as *Nardosmia* Cassini.

142. Homogyne Cassini. 3/2. Perennial rhizomatous herbs with kidney-shaped basal leaves and a few small leaves on stem leading to an inflorescence of 1--3 heads Involucral bracts in a single row. The central flowers are bisexual with a 5-lobed corolla but the outer flowers are all female with obliquely truncate corollas. The fruits are cylindric and slightly ribbed; pappus of several rows of simple hairs. *Mountainous parts of Europe.*

143. Doronicum Linnaeus. 30/6 (and some hybrids). Stout perennial herbs with flaccid, somewhat decurrent leaves and a hair-covering variously consisting of glandular and eglandular hairs. Rootstock tuberous or rhizomatous, fleshy or fibrous, simple, clothed with long-stalked basal leaves. Stem simple or branched, hollow, ridged; stem-leaves stalked to fiddle-shaped and stem-clasping, somewhat bristly. Heads 1--many in a corymb; receptacle convex, without scales or bristles. Involucral bracts in 1 or 2 (rarely 3) series, ray- and disc-flowers yellow; disc-flowers with multiseriate pappus; ray-flowers usually lacking pappus. Fruit slightly grooved, often bearing adpressed hairs. *Europe & temperate Asia to Himalaya.* Figure 132, p. 598.

144. Bedfordia de Candolle. 3/1. Tall shrubs or small trees. Young branches covered with stellate woolly hairs. Leaves evergreen, alternate, soft and spreading with entire or scalloped margins, the undersurface covered with flattened, papery, scale-like stellate hairs. Inflorescences axillary. Heads with only disc-flowers, *c.* 9 mm across, yellow, solitary or in clusters or dense panicles. Involucral bracts 4--5 mm, *c.* 8 in a single row, obtuse or nearly acuminate, overlapping, with dry, translucent margins. Fruit cylindric and striped, attached to a pappus of bristles. *Eastern Australia.*

145. Brachyglottis Forster & Forster. 29/8 (and some hybrids). Shrubs or rarely tree-like, usually rather conspicuously white-, yellowish- or buff-hairy. Leaves evergreen, alternate, entire, margins sometimes regularly wavy. Heads with both ray- and disc-flowers, or ray-flowers absent, in corymbose panicles, rarely solitary, fragrant. Involucral bracts in 1 series, with scattered, recurved bracts outside them, leathery, usually white-hairy or glandular. Ray-flowers female, if absent then outer disc-flowers tubular and female; central disc-flowers bisexual. Rays yellow, cream or white. Fruits hairy or hairless, often with resin-ducts, ribbed. Pappus of numerous, persistent, finely barbed hairs. *New Zealand, Chatham Island, Australia (Tasmania).* Figure 129, p. 558.

146. Chaenactis de Candolle. 25/2. Small to robust annual, biennial or low-growing perennial herbs. Leaves hairless, downy or glandular, alternate or in a basal rosette, simple to irregularly pinnatisect. Heads solitary or in corymbs, often with long stalks. Involucre hemispherical to bell-shaped, 10--50-flowered; involucral bracts more or less equal, in 1, 2 or occasionally 4 rows, slightly overlapping, linear and acuminate to oblong and obtuse, herbaceous, usually densely downy when young. Receptacle flat, with a few bristles or, more usually, naked. Corolla 5-toothed, tubular or funnel-shaped, the outer marginal flowers sometimes enlarged. Corollas white, yellow or pinkish, hairless or downy, often glandular. Stamens projecting beyond corollas.

Fruit slender club-shaped, striped, laterally compressed or terete, usually downy, brownish to black. Pappus in 1 or 2 rows of translucent scales with toothed margins, the scales of the marginal flowers often much shorter than those of the inner flowers, or absent. *Western North America, Mexico.* Figure 129, p. 558.

147. Palafoxia Lagasca. 12/1. Annual or perennial herbs or dwarf shrubs. Leaves alternate or the lower opposite, narrowly linear or lanceolate to oblong, entire, stiffly hairy, sometimes densely glandular. Heads disc-like or sometimes radiate in corymbose clusters or panicles. Involucre cylindric to obconical; bracts more than 6, in 2 or 3 rows, leaf-like in texture, or with margins and tip papery, somewhat unequal; receptacles flat, naked. Ray-flowers absent (ours); disc-flowers with radially symmetric, 5-lobed corollas, white to violet. Fruit linear or narrowly obpyramidal, 4-angled, softly hairy. Pappus of 4--12 scales. *South USA, Mexico.*

148. Arnica Linnaeus. 32/4. Rhizomatous herbaceous perennials. Stems erect, simple or branched from the base. Leaves opposite, stalked or not. Heads up to 30, in cymes or rarely solitary, hemispheric to obconic, 1--3 × 1--3 cm (rays may extend beyond). Involucral bracts more or less in 2 series, herbaceous. Receptacle convex, fringed or with stiff hairs, frequently margin of achene scar raised, more or less radiate. Ray-flowers female, occasionally with 5 staminodes, rays pale yellow to deep orange, 1--4 cm × 4--12 mm, with achenes similar to those of disc-flowers but shorter, broader, and with shorter pappus. Disc-flowers bisexual, yellow to orange; corolla tubular to goblet-shaped, 5-lobed; anthers yellow or purple, base minutely auriculate. Fruits cylindric or tapered at both ends, 3--10 × 0.7--1.5 mm, more or less 5--10-veined, hairless or with stiff hairs or glandular, with a white ring at base. Pappus-bristles slender, white to tawny, beaded to plumose, more or less as long as disc-corolla. Leaves opposite (occasionally the upper leaves alternate), stalked or not. *Northern hemisphere.* Figure 129, p. 558.

149. Tagetes Linnaeus. 50/6. Annual or perennial herbs, occasionally with woody stocks; stems solitary or frequently much-branched to form a small bush, to 1.5 m; strongly scented. Leaves generally opposite, upper leaves sometimes alternate, simple or, more usually, pinnatifid or pinnate, often conspicuously dotted with oil-glands, margins frequently conspicuously toothed. Heads of various sizes, solitary or in loose, leafy panicles or dense clusters at the end of the branches, with or without ray-flowers. Involucral bracts in 1 or 2 rows, outer minute or absent, inner united into a toothed, cylindric, bell-shaped or fusiform cup, dotted with oil-glands. Ray-flowers usually present, few, female, fertile, reddish brown, orange, yellow or white. Disc-flowers bisexual, fertile, corolla-lobes triangular to lanceolate, coloured as the ray-flowers. Fruit slender, club-shaped or linear, angular or terete. Pappus of 2--10, entire, mostly unequal, more or less united scales and/or bristles. *South USA, Central America, extending south to Argentina.* Figure 129, p. 558.

150. Lasthenia Cassini. 16/2. Annual or perennial herbs to 60 cms. Leaves opposite, usually slightly united at their bases, narrow, simple or pinnatisect. Heads often showy, few together or solitary on long stalks with ray-flowers. Involucre cylindric to hemispherical, bracts few to many in 1 or 2 rows, free or united at the base, sometimes with prominent midribs. Receptacle conical or hemispherical. Ray-flowers yellow, whitish or greenish, female; disc-flowers with corolla 4- or 5-lobed, yellow, bisexual,

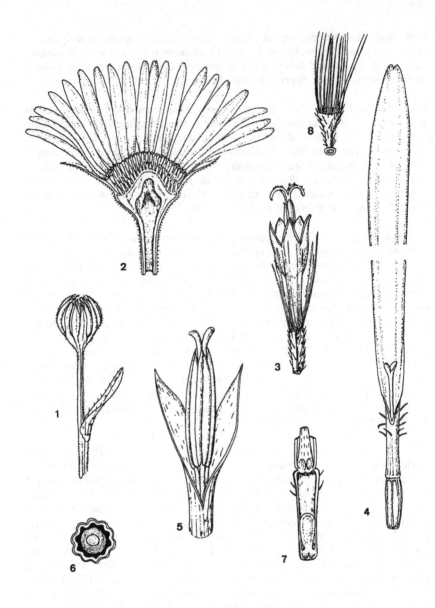

Figure 132. Compositae. *Doronicum plantagineum.* 1, Inflorescence-bud. 2, Longitudinal section of head. 3, Single disc-flower. 4, Single ray-flower. 5, Upper part of disc-flower, corolla partially removed to show united anthers. 6, Transverse section of ovary. 7, Longitudinal section of ovary. 8, Fruit.

598

widely bell-shaped with spreading lobes. Fruit narrow, terete or compressed, hairless or downy. Pappus absent or, when present, of awned or frayed scales. *Pacific North America, Chile.*

151. Eriophyllum Lagasca. 12/1. Annual or perennial herbs or dwarf shrubs. Leaves alternate (ours), lobed, toothed or pinnatifid or, in the annual species, entire, covered in densely felted hair. Heads solitary or in leafy cymose clusters or panicles; ray-flowers usually present. Involucre hemispheric to oblong, bracts in 1 (or 2) rows, erect and rigid, more or less united at the base, often reflexed at the apex. Receptacle pitted, convex, conical or flat, usually naked. Ray-flowers usually 4--15 female, fertile, rays oval to oblong, yellow, occasionally absent. Disc-flowers bisexual, usually yellow, sometimes downy or glandular. Fruit slender, 4- or 5-angled or -ribbed, often laterally compressed. Pappus usually of papery scales frequently with frayed margins, or absent. *Western North America, northwest Mexico.* Figure 129, p. 558.

152. Hymenoxys Cassini. 25/2. Hairy perennial herbs. Leaves alternate, basal or scattered along stems, or both, narrow, entire or 1- or 2-pinnatisect. Receptacle convex or conical. Involucral bracts in 1--3 series. Ray-flowers 10--20, female, rays broadly oblong, yellow, very distinctly and conspicuously 3-lobed at the apex. Disc-flowers yellow, numerous, bisexual, fertile. Fruit obovoid, 5--10-ribbed or -angled, densely hairy; pappus of 5--12 scales. *Temperate America.*

153. Helenium Linnaeus. 40/3. Annual, biennial or perennial, bitter and resinous herbs, with erect, simple or branched, sometimes densely downy stems. Leaves alternate, stalkless or shortly stalked, usually narrow, simple, toothed or pinnatisect, dotted with glands, sometimes densely downy, often decurrent as wings along stem. Heads stalked, solitary or loosely corymbose, usually with ray-flowers. Involucres saucer-shaped, bracts more or less equal, narrow, linear to subulate, in 1--3 rows, spreading or deflexed at maturity, dotted with glands. Receptacle oblong, hemispherical or convex. Ray-flowers female, fertile or sterile, with yellow or purplish brown rays; disc-flowers bisexual, fertile, corolla 4- or 5-toothed, yellow or purplish brown with glandular lobes. Fruit obpyramidal, 4--8-angled or -ribbed, usually hairy. Pappus of 5--10 entire, toothed or frayed acuminate or awn-tipped scales. *Southern USA, Mexico.*

154. Gaillardia Fougeroux. 28/3. Annual, biennial and short-lived perennial herbs or rarely subshrubs. Leaves alternate or in basal rosettes, simple to pinnatisect, hairy, sometimes with resin-glands. Heads solitary or in a loose corymbose panicle, on long stalks, ray-flowers usually present. Involucral bracts leaf-like, in 2 or 3 rows, slender, linear or lanceolate, overlapping, usually becoming reflexed. Receptacle convex or almost spherical, pitted, the pit-margins sometimes extended, or bristly. Ray-flowers female or sterile, rays bright and showy, yellow to crimson with yellow to purple tips, sometimes absent; disc-flowers yellow to dark brown or purple, fertile, corolla-lobes with jointed hairs. Fruit 5-ribbed, more or less woolly, at least at the base. Pappus of 5--12 thin scales, longer than the fruit, each with the midrib excurrent as an awn. *Western & southern USA, Central & South America.* Figure 129, p. 558.

155. Marshallia Schreber. 7/1. Erect perennial herbs. Leaves alternate, simple, margins entire, stalkless; stem-leaves sometimes clasping at the base. Heads with

599

tubular, bisexual, fertile disc-flowers only, ray-flowers absent. Receptacle convex to cone-shaped, usually hollow, scaly. Involucre hemispherical or campanulate; bracts in 1 or 2 series, often overlapping. Corollas tubular, creamy white, pale lavender, pink or purple. Fruit club-shaped; pappus of scales. *North America*. Figure 129, p. 558.

156. Argyroxiphium de Candolle. 5/1. Monocarpic, erect, rosette-forming shrubs. Leaves linear to lanceolate, in whorls of 5--7, sometimes fused at base, green and hairless to densely silver-hairy. Heads usually with both ray- and disc-flowers, few to numerous, in elongated racemes·or small panicles. Receptacle with scales in a ring between the ray- and disc-flowers. Involucral bracts forming a cup-shaped false involucre enclosing the usually numerous disc-flowers. Ray-flowers few to many, sometimes absent, strap-shaped, white or yellow to purple-pink. Fruit 3--10-ribbed; pappus of 2--10 scales. *Hawaii*.

157. Layia Hooker & Arnott. 15/1. Annual herbs, often covered with black glandular hairs. Leaves alternate, toothed or pinnately lobed, narrowly lanceolate. Heads solitary on long stalks; receptacle with a ring of scales between the ray- and disc-flowers. Involucral bracts in 2 rows. Ray-flowers bisexual, ray 3-lobed; disc-flowers bisexual. Anthers black or yellow. Styles 2-lobed. Fruit slightly flattened; pappus of bristles, awns or scales. *Western North America*. Figure 129, p. 558.

158. Echinacea Moench. 9/3. Rhizomatous or deeply rooted perennial herbs. Stems unbranched, erect, rather stout. Leaves alternate, simple. Heads usually solitary on thick, rigid stalks, each with a tall conical receptacle covered with stiff, prickly scales. Involucral bracts in 2--4 densely overlapping rows. Rays 2--8 cm, sterile, usually purple; disc-flowers brownish purple. Fruits 4-sided; pappus a short, toothed, papery crown. *Eastern USA*. Figure 130, p. 560.

159. Rudbeckia Linnaeus. 15/7. Perennial herbs, sometimes woody at base, occasionally annuals or biennials. Leaves alternate. Heads terminating the stems or branches, many-flowered. Involucral bracts leaf-like, in 2 series, spreading. Receptacle conical or often more or less elongated and spike-like, scaly with the scales concave. Ray-flowers sterile, in a single series; disc-flowers tubular, bisexual, corolla cylindric, teeth erect or spreading. Rays yellow (rarely bicoloured), usually elongated, sometimes spreading or drooping; disc-corollas, and styles purple or brownish purple, sometimes greenish yellow. Anthers reddish brown. Fruits 4-sided, obpyramidal, hairless, flat at the apex. Pappus absent, or minute and crown-like, rarely conspicuous. *North America*. Figure 130, p. 560.

160. Ratibida Rafinesque. 6/2. Erect, roughly hairy biennial or perennial herbs to 1.5 m. Leaves alternate, pinnate to pinnatifid. Heads solitary, showy, on long stalks. Receptacle spherical to cylindric or cone-shaped; receptacular scales thin, subtending both ray- and disc-flowers, more or less clasping fruit. Involucral bracts in 1 series, linear or linear-lanceolate, leaf-like, outer   longer and reflexed or spreading. Ray-flowers yellow or partly purple, rays toothed. Disc-flowers bisexual; corolla short, yellow-brown; anthers with 2 short triangular lobes at base; style-branches slightly flattened with ovate to subulate, hairy appendages. Fruit compressed, angled or winged; pappus of 1 or 2 teeth, bristles or small scales, or absent. *North America, northern Mexico*. Figure 130, p. 560.

600

161. Sanvitalia Lamarck. 7/1. Annual or perennial, low, simple or branched, hairy herbs. Leaves opposite, joined at bases, simple, entire or lobed in some perennial species. Heads few, often stalkless. Receptacle hemispheric to narrowly conical; receptacular scales folded around disc-flowers. Involucral bracts overlapping, in 2 or 3 series, dry or leaf-like at tips. Ray-flowers female, fertile, persistent on fruit, white or yellow. Disc-flowers bisexual, fertile, brownish purple or greenish white, often shorter than receptacular scales. Anthers dark, each with an acute pale appendage, almost entire to slightly lobed at base. Style-branches red, linear, tips hairy and truncate. Ray-fruit irregularly 3-angled with pappus of 3 awns, ray persistent, white or usually yellow; central disc-fruit smooth, biconvex, 2-winged, outer disc-fruit compressed, often 4-angled, wingless with pappus of awns or absent. *Southwest USA, Central America (Mexico to Guatemala).*

162. Heliopsis Persoon. 12/1. Loosely branched, erect perennial herbs. Leaves opposite, coarsely toothed, 3-veined, stalked. Receptacle conical, sometimes narrowly so, (especially in fruit); receptacular scales concave and clasping. Involucral bracts overlapping, in 1 or 2 series. Ray-flowers female, rarely sterile, yellow, becoming papery, persistent on fruit. Disc-flowers partly enclosed in receptacular scales, bisexual; anthers entire or nearly so at base; style-branches flattened, with short hairy appendages. Fruit 4-angled (those of ray-flowers 3-angled); pappus absent or a short irregular crown. *North America.*

163. Zinnia Linnaeus. 20/5. Annual or perennial herbs or shrublets. Leaves opposite, entire, usually stalkless or shortly stalked, bases sheathing stem. Heads borne on hollow stalks; receptacle scaly, scales overlapping, oblong to obovate, often darker near apex. Involucre with bracts in 2 or 3 rows, the outer shorter than the inner. Disc-flowers tubular, usually yellow, ray-flowers with rays 1--3-lobed, white, yellow, orange, red or purple, persistent in fruit. Fruit angled; pappus of awns present or absent. *South-central USA to Argentina.* Figure 130, p. 560.

164. Verbesina Linnaeus. 150/3. Annual to perennial herbs, shrubs and trees, usually hairy. Stems often winged. Leaves opposite or rarely alternate, margins commonly toothed or roughly pinnately lobed. Heads with both ray- and disc-flowers (ours), solitary or several in clusters. Receptacle convex or conical; receptacular scales folded, partly enclosing fruit. Involucral bracts overlapping in 2--several series. Ray-flowers female, rays entire or 2- or 3-toothed, white or yellow, sometimes absent. Disc-flowers bisexual, usually fertile, yellow or white; anthers obtuse or with obscure triangular lobes at base; style branches flattened, with short or lengthened acute hairy appendages. Fruit flattened, 2-winged, rarely wingless; pappus of 1--3 (usually 2) awns, or rarely absent. *Central & eastern USA.* Figure 130, p. 560.

165. Encelia Adanson. 15/1. Hairy or almost hairless shrubs. Leaves evergreen, alternate, simple, stalked, margin entire. Inflorescence of 1--5, bisexual, hemispherical heads; involucral bracts in 2 or 3 rows, free, slightly overlapping. Ray-flowers sterile: style absent, yellow; disc-flowers many, corolla-tube slender, throat abruptly expanded, lobes triangular; style-branches flattened. Fruit strongly flattened, edged with long hairs, faces hairless; pappus absent. *California to Chile.*

166. Balsamorhiza Nuttall. 14/1. Erect or ascending perennial herbs. Stems sparsely leafy, branched from base. Leaves opposite and forming basal rosettes, lanceolate to almost sagittate, entire or toothed to 2-pinnatifid. Heads large, usually solitary. Receptacle convex or flat with scales. Involucral bracts in 2--4 series, leathery or leaf-like, linear-lanceolate or oblong. Ray-flowers female, fertile, yellow, rays with 2 or 3 teeth at apex. Disc-flowers bisexual, fertile, yellow. Fruit oblong, 3- or 4-angled, hairless; pappus usually absent. *Western North America.* Figure 130, p. 560.

167. Helianthella Torrey & Gray. 8/1. Perennial herbs. Stems simple or with a few branches, leafy. Leaves opposite (ours) or alternate, oblong-lanceolate to linear, entire, 3-veined; basal leaves stalked, upper leaves usually stalkless. Heads large, solitary or few in flat-topped panicles. Receptacle flat or convex; receptacular scales clasping fruits. Involucral bracts in 2--4 series, lanceolate, loose, more or less leaf-like. Ray-flowers sterile, rays oblong, 3-toothed, yellow. Disc-flowers numerous, bisexual, fertile, yellow to purple- or brown-tinged, anthers with minute triangular lobes at base; style-branches flattened, with short, blunt, triangular, hairy appendages. Fruit strongly compressed at right-angles to the involucral bracts, often slightly 2-winged; pappus a crown of scales and 2 persistent awns, or absent. *Western North America, Mexico.* Figure 130, p. 560.

168. Wyethia Nuttall. 14/4. Erect or ascending, simple or branched, coarse, pleasantly scented, leafy perennial herbs. Stem leaves alternate, linear to almost circular or ovate-triangular, entire or toothed, usually stalked; basal leaves often large. Heads with both ray- and disc-flowers, solitary or few together, terminal or axillary. Receptacle flat to broadly convex; receptacular scales firm, persistent, clasping the fruit. Involucral bracts in 2--4 series, usually lanceolate to obovate, the outer sometimes leaf-like. Flowers yellow. Ray-flowers female, fertile, rays 2- or 3-toothed, usually yellow. Disc-flowers numerous, bisexual, fertile, light yellow; anthers almost entire or with minute triangular lobes at base; style-branches slender, hairy almost to base. Fruit oblong to linear-oblong, more or less 4-angled, those of rays 3-angled; pappus of scales and often 1--4 awns or teeth. *Western North America.* Figure 130, p. 560.

169. Helianthus Linnaeus. 70/14 (and some hybrids). Annual to perennial herbs, often with fibrous or tuberous roots and rhizomes. Leaves simple, alternate or opposite below, becoming alternate above, usually lanceolate to ovate, veins usually branching from base of midrib, margins toothed or rarely entire. Heads terminal, solitary or in corymbs, loose racemes or panicles, usually both ray- and disc-flowers present, rarely only ray- or disc-flowers. Involucral bracts in 2 or more overlapping series, usually roughly hairy. Receptacle flat to low-convex, usually with dry membranous scales. Ray-flowers usually in a single series, sterile, yellow or red. Disc-flowers bisexual, yellow, brown, red or purple, arranged in concentric circles, with floral maturity proceeding inwardly as the head matures. Fruit laterally compressed, angular; pappus of 2 dry membranous awns. *Mostly North America, a few species in South America.* Figure 130, p. 560.

170. Tithonia Jussieu. 10/2. Tall perennial herbs, often woody at base. Leaves usually alternate, stalked and noticeably 3-veined. Heads large, borne on long and conspicuously swollen inflorescence stalks, with yellow or orange flowers. Involucral bracts in 2--5 series, almost equal or graduated. Receptacle convex, with stiff, striate scales. Ray-flowers large and showy, usually infertile; disc-flowers bisexual and fertile. Fruit slightly flattened or weakly 4-sided; hairy; pappus of awns or scales, persistent. *Mexico, Central America.* Figure 130, p. 560.

171. Tridax Linnaeus. 26/1. Annual to perennial herbs. Leaves opposite, entire to toothed, pinnatifid or pinnatisect. Heads small, solitary or few in a cymose panicle. Receptacle flat to convex. Involucral bracts in 2 or 3 series, apices often tinged purple. Ray-flowers female, white, yellow or pink-red, 2-lipped, outer lip with 2 or 3 teeth, inner lip smaller with 1 or 2 lobes. Disc-flowers yellow. Fruit obconical, compressed at apex, hairless or densely hairy; pappus of scales and occasionally bristles. *North America, Mexico.* Figure 130, p. 560.

172. Bidens Linnaeus. 240/2. Annual or perennial herbs. Leaves opposite, entire or pinnately divided. Heads solitary, usually stalked. Involucral bracts in 2 rows, the outer usually herbaceous and often leaf-like, the inner membranous, often with papery margins. Receptacle flat or slightly convex, with scales. Ray-flowers 3--12, in 1 row, yellow, sterile or occasionally absent. Disc-flowers bisexual, tubular. Fruit obovoid-oblong, flattened, or 4-angled, usually with hairy margins; pappus of 2--5 stiff, bristle-like hairs or awns, barbed with backward-pointing spines. *America, also in Africa, Australia & tropical Asia.* Figure 130, p. 560.

173. Coreopsis Linnaeus. 50/c. 2. Annual or perennial herbs (rarely shrubs). Leaves usually opposite, entire or variously lobed or dissected, 1--3-pinnate. Heads stalked, involucral bracts in 2--4 rows, more or less united at the extreme base, outer bracts leaf-like, or membranous, adpressed or spreading, inner usually larger, brown or yellow, membranous. Receptacle with scales. Ray-flowers female, fertile or sterile, rays usually yellow or apices red or brown. Disc-flowers bisexual, mostly fertile. Fruit often with wings and 2 awn-like teeth at apex. *America.* Figure 130, p. 560.

174. Cosmos Cavanilles. 25/3. Annual or perennial herbs, hairless or hairy. Stems often furrowed, erect or spreading and freely branched. Leaves opposite, undivided or lobed or 1--3-pinnatisect, green above, often paler beneath, hairless or hairy, stalked or stalkless. Heads 1--many, borne terminally on long stalks in loose corymbs, with both ray- and disc-flowers. Involucre hemispherical, receptacle flat and scaly; bracts in 2 series, reflexed or erect, membranous, striped, apices sometimes coloured. Outer ray-flowers sterile, in 1 series, entire or slightly toothed at the apex, pink or violet to dark-purple or blood-red, more rarely deep orange, yellow or white. Disc-flowers fertile, bisexual, tubular, purple, blood-red or yellow. Fruit beaked, linear to spindle-shaped, pappus of 2--8 often barbed awns, or absent. *South USA, Central America.* Figure 130, p. 560.

175. Dahlia Cavanilles. 28/4. Herbaceous or suffruticose perennials with tuberous rootstocks. Stems erect, usually arising singly but occasionally 3--5, branched or unbranched, 40--1000 cm. Leaves opposite or whorled, pinnately compound. Heads with ray- and disc-flowers, borne on slender stalks; involucral bracts in 2 rows, the outer spreading or reflexed, rather fleshy, the inner erect and membranous. Ray-flowers sterile, rays ovate to elliptic, acute, whitish violet, deep purple, pink, red, orange or yellowish, usually with conspicuous parallel veins; disc-flowers yellow

603

throughout or with red or purple tips, fertile flowers tubular (all or some replaced by ray-flowers in many cultivars), with corolla 5-toothed; filaments hairless, style-branches bearded dorsally, spreading or slightly reflexed at maturity. Fruits brown or greyish black, linear or linear-oblanceolate in outline, pappus absent or of 2 rudimentary teeth. *Central America, mainly Mexico.* Figure 130, p. 560.

An enormous range of cultivars (in excess of 20,000), is registered and classified into horticultural groups, largely on the basis of inflorescence structure.

176. Thelesperma Lessing. 12/1. Annual to perennial herbs, sometimes woody at base, hairless or becoming so. Leaves opposite or upper leaves alternate, pinnate with linear to thread-like segments. Heads small, solitary on long stalks. Receptacle flat; receptacular scales flat or concave, somewhat clasping fruit. Involucral bracts in 2 series; inner series united for at least the lower half, margins membranous; outer series smaller, spreading, leaf-like. Ray-flowers 8--10, yellow to brown-red, sterile, sometimes absent; disc-flowers red to purple, bisexual, fertile; style-branches flattened with short, hairy appendages; anthers obtuse at base. Fruit linear to linear-oblong, sometimes with small protuberances, hairless; pappus of 2 retrorsely barbed awns, or absent. *Western North America, southern South America.*

177. Berlandiera de Candolle. 12/1. Hairy perennial herbs. Leaves alternate, scalloped to pinnatifid with a large terminal lobe and smaller lateral lobes. Receptacle flat or almost so, with scales. Ray-flowers 5--12, female, yellow. Disc-flowers sterile, with undivided styles; anthers entire or with minute triangular lobes at base. Fruit wingless, usually hairy on inner surface, attached to adjacent receptacular bracts or to the subtending involucral bract, or both, the whole falling together; pappus a tiny incomplete crown or 2 short awns, or absent. *South USA, Mexico.*

178. Chrysogonum Linnaeus. 1/1. Rhizomatous perennial herbs to 40 cm. Leaves to 10 × 7 cm, opposite and basal, triangular-ovate, acute to obtuse, entire or toothed, tapering gradually or almost cordate at base, hairy; stalks long. Heads to 3.5 cm across, solitary on long stalks. Receptacle flat, with thin membranous receptacular scales. Involucral bracts in 2 series; inner series lanceolate to ovate, papery, subtending ray-flowers; outer series larger, elliptic, leaf-like. Ray-flowers few, female, yellow. Disc-flowers yellow, male, with undivided styles, and anthers almost entire at base. Fruit flattened-obovoid, 4-angled, to 4 mm; pappus a short crown. *North America.*

179. Engelmannia Torrey & Gray. 1/1. Erect, hairy, perennial herbs, 30--70 cm. Leaves 5--15 cm, alternate, oblong, stalked, deeply pinnatisect with pinnatifid to toothed lobes. Heads in cymose panicles; stalks long, slender. Receptacle flat; outer receptacular scales in pairs adhering to the base of each involucral bract. Involucral bracts loose, leathery, in several series. Ray-flowers 8--10, female, golden yellow; style bifid. Disc-flowers apparently bisexual but female-sterile; style entire, long, hairy. Ray-fruits broadly obovoid, compressed; pappus of 2 awns. *South & central USA.*

180. Silphium Linnaeus. 23/3. Perennial herbs. Leaves opposite, alternate, whorled or all basal, entire, toothed or pinnate. Heads stalked, in corymbs or panicles. Receptacle flat to hemispherical with thin receptacular scales. Involucral bracts usually overlapping, in 2 or more series, leaf-like to dry and membranous; inner series subtending ray-flowers. Ray-flowers in 2 or 3 series, female, yellow, rarely white. Disc-flowers yellow, female-sterile, with undivided styles; anthers entire or with minute triangular lobes at base. Ray-fruit hairless, strongly compressed, winged on margins; pappus of 2 awns sometimes present. *Eastern North America.* Figure 130, p. 560.

181. Oxylobus A. Gray. 4--5/1. Decumbent herbs to low shrubs. Leaves opposite, short-stalked; blades small, ovate, elliptic to obovate, margins scalloped to almost entire, surface usually with glandular dots, sparsely to densely glandular-downy beneath, rarely hairless. Inflorescence loosely to densely corymbose to almost cymose; stalks in the cultivated species sticky, glandular-downy or rarely hairless. Involucre glandular-downy, broadly obconical. Involucral bracts 10--16, not overlapping, in 2 or 3 series, mostly unequal, spreading at maturity. Receptacle slightly convex, hairless. Flowers 20--75 per head, all tubular. Corollas white or pink, each with a long narrow basal tube and a narrowly bell-shaped limb; lobes triangular, longer than wide, inner surface densely papillose, outer surface smooth. Fruit fusiform, 5-ribbed, bearing short, sharply pointed bristles; pappus of a few short, persistent, irregular and finely cut scales. *South Mexico, Central America.*

182. Ageratum Linnaeus. 40/3. Annual and perennial herbs, sometimes woody at the base. Stems terete, striped with sparse hairs. Leaves opposite, margins entire to toothed, underside usually with large, sometimes partially sunken glandular dots, commonly 3-veined from or near base. Involucral bracts 30--40, lanceolate, tough. Receptacle cone-shaped or columnar, smooth or bearing small, dry, membranous bracts or scales; each series not overlapping. Heads in cymes; flowers 20--125 per head, all tubular. Corollas white, blue or lavender, funnel-shaped or with distinct basal tubes; lobes triangular, about as long as wide, papillose on inner surface, sometimes with minute, stiff bristles on outer surface. Fruits prismatic, 4--5-ribbed, hairless or with short, sharp bristles on ribs. *Central & South America.* Figure 130, p. 560.

183. Piqueria Cavanilles. 7/1. Erect, annual to perennial herbs, sparsely branched. Stems terete to slightly quadrangular. Leaves mostly opposite, stalks short to absent, blades ovate to lanceolate, margins scalloped to toothed, 3-veined from the base. Inflorescence loosely paniculate with loose to dense, almost cymose branches, stalks of heads moderately long. Involucral bracts 3--5, not overlapping, in one series, equal. Receptacles flat, hairless. Flowers 3--5 per head, as many as involucral bracts. Corolla with short, narrow, densely hairy basal tube, throat short, bell-shaped, nearly smooth on inner surface or with hair-like papillae, especially in lower part of throat; lobes oblong-triangular, densely papillose on inner surface. Fruits ribbed; pappus absent; fruit attachment usually asymmetric and large, with a sinuous vascular trace. *Mexico.* Figure 130, p. 560.

184. Stevia Cavanilles. 230/1. Mostly erect, annual or perennial herbs or shrubs. Stems terete to slightly hexagonal, partially striped. Leaves of the cultivated species opposite, stalkless, blades lanceolate to spathulate with the broadest part above the middle, membranous or leathery in texture, conspicuously 3-veined, surfaces usually with soft hairs, sometimes glandular, often with proliferous buds in the axils; toothed, teeth pointing forwards above the middle. Inflorescence loosely paniculate, heads on stalks as long as the involucres, or in dense corymbose clusters at branch-tips. All species have heads with 5 disc-flowers and 5 involucral bracts in 1 series. *South- west USA to central Argentina, absent from the Amazonian areas.* Figure 130, p. 560.

185. Liatris Gaertner. 43/2--3. Erect perennial herbs. Rootstock an almost spherical corm. Leaves alternate, and in a basal rosette, linear, entire, surfaces with glandular spots. Heads arranged in a spike or raceme, opening from the uppermost downwards, campanulate; bracts 20--25, overlapping in 3--5 unequal series, mostly persistent, broadly rounded to oblong-lanceolate and sometimes coloured and petal-like, almost flat, hairless. Receptacle without hairs or scales. Flowers 3--80 per head, all similar and tubular. Corolla usually purple, sometimes lavender or white, lobes linear-lanceolate to linear-oblong, usually hairy within, though not so in most cultivated species. Fruits 10-ribbed, bristly. Pappus of many plumose or beaded (ours) bristles. *North America.* Figure 130, p. 560.

186. Eupatorium Linnaeus. 1200/*c.* 20. Annual to perennial herbs, erect. Stems terete, striped, downy or hairless. Leaves mostly opposite or whorled, upper leaves almost opposite to alternate; blades linear to ovate, broadly triangular, or 3-lobed, margins with teeth pointing forwards to almost entire. Inflorescence a corymbose or pyramidal panicle; heads on short stalks. Involucral bracts 10--22, in 2--5 series, usually persistent; inner bracts deciduous in some species. Receptacle flat or weakly convex, hairless. Flowers all without rays, 3--30 per head. Corollas white to purple, lavender, or pink, narrowly funnel-shaped or with constricted basal tube and a narrow to broadly campanulate limb; outer surface with glands often concentrated at base of throat and on outer surfaces of lobes, rarely with a few hairs; lobes triangular to oblong-ovate, usually cylindric. Pappus of 25--40 rough persistent bristles. *Temperate North America to temperate South America, Eurasia.* Figure 130, p. 560.

187. Brickellia Elliot. 98/1. Erect shrubs, or annual to perennial herbs. Stems terete, striped. Leaves alternate and stalked in the cultivated species and broadly triangular-ovate, margins usually toothed, commonly with 3 prominent veins. Inflorescence usually with heads stalkless to long-stalked, clustered in a leafy, ovoid panicle, sometimes corymbose or cymose, rarely heads solitary. Involucral bracts 14--45, almost overlapping, in 3--5 series in our species, persistent and spreading with age, lanceolate to oblanceolate, rarely with expanded herbaceous tips. Receptacle flat to slightly convex, without scales. Flowers 8--18 in a head, all tubular (ours). Corolla white to cream, with lobes as long as wide to twice as long as wide, ovate-oblong to triangular, smooth. Fruits prismatic, with 10 ribs, bristly. Pappus of 10--80 persistent bristles in 1 series, flattened on outer surfaces, rough with very short stiff hairs to having a densely feathered fringe on lateral margins. *Western USA, Mexico & Central America.*

abscission zone. A predetermined layer at which leaves or other organs break off.

achene. A small, dry, indehiscent, 1-seeded fruit in which the fruit-wall is of membranous consistency and free from the seed.

aciculus. See p. 24 (*Nomocharis*).

acuminate. With a long, slender point.

adpressed. Closely applied to a leaf or stem and lying parallel to its surface but not adherent to it.

adventitious. (1) Of roots: arising from a stem or leaf, not from the primary root derived from the radicle of the seedling. (2) Of buds: arising somewhere other than in the axil of a leaf.

aggregate fruit. A collection of small fruits, each derived from a single carpel, closely associated on a common receptacle, but not united. *Ranunculus* and *Rubus* provide familiar examples.

alternate. Arising singly, 1 at each node; not opposite or whorled (figure 133.2, p. 610).

anastomosing. Describes veins of leaves which rejoin after branching from each other or from the main vein or midrib.

anatropous. Describes an ovule which turns through 180° in the course of development, so that the micropyle is near the base of the funicle (figure 136.2, p. 616).

androgynophore. A stalk which raises the stamens and ovary some distance above the petals in some flowers.

annual. A plant which completes its life-cycle from seed to seed in less than 1 year.

anther. The uppermost part of a stamen, containing the pollen (figure 135.3 & 4, p. 614).

apetalous. Describes a flower without a corolla (petals).

apical. Describes the attachment of an ovule to the apex of a 1-celled ovary (figure 136.8, p. 616).

apiculate. With a small point.

apomictic. Reproducing by asexual means, though often by the agency of seeds, which are produced without the usual sexual nuclear fusion.

arachnoid. Describes hairs which are soft, long and entangled, suggestive of cobwebs.

areole. A cushion-like pad, usually bearing spines: see p. 162 (*Cactaceae*).

aril. An outgrowth from the region of the hilum, which partly or wholly envelops the seed; it is usually fleshy.

arilloid. An outgrowth from some region of a seed other than the hilum, which resembles an aril.

ascending. Prostrate for a short distance at the base but then curving upwards so that the remainder is more or less erect; sometimes used less precisely to mean pointing obliquely upwards.

attenuate. Drawn out to a fine point.

auricle. A lobe, normally 1 of a pair, at the base of the leaf-blade, bract, sepal or petal.

awn. A slender but stiff bristle on a sepal or fruit.

axil. The upper angle between a leaf-base or leaf-stalk and the stem that bears it (figure 133.1, p. 610).

axile. A form of placentation in which the cavity of the ovary is divided by septa into 2 or more cells, the placentas being situated on the central axis (figure 136.10, p. 616).

axillary. Situated in or arising from an axil (figure 133.1, p. 610).

back-cross. A cross between a hybrid and a plant similar to one of its parents.

basal. (1) Of leaves: arising from the stem at or very close to its base. (2) Of placentation: describes the attachment of an ovule to the base of a 1-celled ovary (figure 135.6 & 7, p. 614).

basifixed. Attached by its stalk or supporting organ by its base, not by it back (figure 135.3, p. 614).

berry. A fleshy fruit containing 1 or more seeds embedded in pulp, as in the genera *Berberis, Ribes* and *Phoenix.* Many fruits (such as those of *Ilex*) which look like berries and are usually so-called in popular speech, are in fact drupes.

biennial. A plant which completes its life-cycle from seed to seed in a period of more than 1 year but less than 2.

bifid. Forked; divided into 2 lobes or points at the tip.

bilaterally symmetric. Capable of division into 2 similar halves along 1 plane and 1 only (figure 135.8 & 9, p. 614).

bipinnate. Of a leaf: with the blade divided pinnately into separate leaflets which are themselves pinnately divided (figure 133.18, p. 610).

blade. A broadened part, furthest from the base, usually called the limb (see below), of a petal, corolla or similar organ, which has a relatively narrow basal part -- the claw or tube (figure 135.5 & 6, p. 614).

bract. A leaf-like or chaffy organ bearing a flower in its axil or forming part of an inflorescence, differing from a foliage leaf in size, shape, consistency or colour (figure 134.2 & 3, p. 612).

bracteole. A small, bract-like organ which occurs on the flower-stalk, above the bract, in some plants.

bractlet. A very small bract.

bulb. A seasonally dormant underground bud, usually fairly large, consisting mainly of a number of fleshy leaves or leaf-bases.

bulbil. A small bulb, especially one borne in a leaf-axil or in an inflorescence.

bulblet. A small bulb developing from a larger one.

caespitose. Tufted.

callus. A hardened swelling.

calyptra. Applied to any cap-like covering.

calyx. The sepal; the outer whorl of a perianth (figure 135.1, p. 614).

campanulate. Bell-shaped.

campylotropous. Describes an ovule which becomes curved during development and lies with its long axis at right-angles to the funicle (figure 136.4, p. 616).

capitate. Compact and approximately spherical, head-like.

capitellate. Grouped into small, dense clusters.

capitulum. An inflorescence consisting of small flowers (florets), usually numerous, closely grouped together so as to form a 'head', and often provided with an involucre.

capsule. A dry, dehiscent fruit derived from 2 or more united carpels and usually containing numerous seeds.

carpel. One of the units (sometimes interpreted as modified leaves) situated in the centre of a flower and together constituting the gynaecium or female part of the flower (ovary). If more than 1, they may be free or united. They contain ovules and bear a stigma (figure 135.1 & 2, p. 614).

carpophore. See p. 155 (*Caryophyllaceae, Silene*), and p. 382 (*Umbelliferae*).

caruncle. A soft, usually oil-rich appendage attached to the seed near the hilum.

catkin. An inflorescence of unisexual flowers, made up of relatively conspicuous, usually overlapping bracts, each of which subtends a small apetalous flower or a group of such flowers; catkins are generally pendent, but some are erect.

chromosome. One of the small, thread-like or rod-like bodies consisting of nucleic acid and containing the genes, which appear in a cell nucleus shortly before cell division.

ciliate. Fringed on the margin with usually fine hairs.

cincinnus. A coiled cyme. See p. 59 (*Commelinaceae*) and figure 134.7, p. 612.

circinate. Coiled at the tip, so as to resemble a crozier.

cladode. A branch which takes on the function of a leaf (the leaves being usually vestigial).

clavate. Club-shaped.

claw. The narrow base of a petal or sepal, which widens above into the limb or blade (figure 135.6, p. 614).

cleistogamous. Describes a flower with a reduced corolla which does not open but sets seed by self-pollination.

clone. The sum total of the plants derived from the vegetative reproduction of an individual, all having the same genetic constitution.

cluster. An imprecisely defined inflorescence in which a number of flowers arise from the same leaf- or bract-axil.

column. (1) In the *Orchidaceae*, a solid structure in the centre of the flower consisting of the style and stigma united to the stamen or stamens (figure 19.1, p. 98). (2) In the *Gramineae*, see p. 61.

compound. (1) Of a leaf: divided into separate leaflets. (2) Of an inflorescence: bearing secondary inflorescences in place of single flowers. (3) Of a fruit: derived from more than 1 flower.

compressed. Flattened from side to side.

comus. A tuft of infertile flowers or bracts sometimes found at the top of an inflorescence.

connective. The tissue which separates the 2 lobes of an anther, and to which the filament is attached.

contorted. Of calyx or corolla: folded so that the margins of adjacent organs overlap so that each overlaps, and is overlapped by, one other.

cordate. Describes the base of a leaf-blade which has a rounded lobe on either side of the central sinus (figure 133.26, p. 610).

corm. An underground, thickened stem-base, often surrounded by papery leaf-bases and superficially resembling a bulb.

cormlet. A small corm produced from an already existing, larger one.

corolla. The petals: the inner whorl of a perianth.

corona. (1) A tubular or ring-like structure attached to the inside of the perianth (or perigynous or epigynous zone), either external to the stamens or united with their

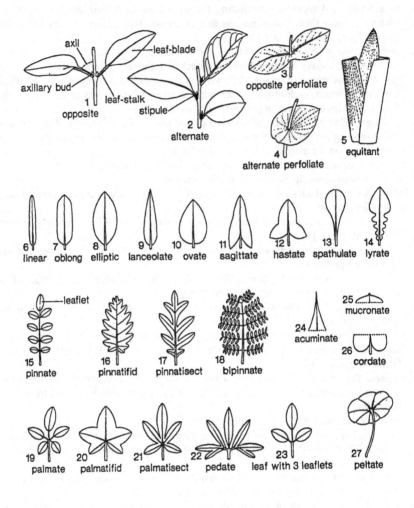

Figure 133. Leaves. 1--5, Leaf-insertion types. 6--14, Leaf-blade outlines. 15--23, Leaf-dissection types. 24--26, Leaf-apex and -base shapes. 27, Attachment of leaf-stalk to blade.

610

filaments; it is usually lobed or dissected. (2) A ring-like structure or circlet of appendages on the outside of a tube formed by united filaments.

coronal scale. See p. 155. (*Caryophyllaceae, Silene*).

corymb. A broad, flat-topped inflorescence. In the strict sense the term indicates a raceme in which the lowest flowers have stalks long enough to bring them to the level of the upper ones (figure 134.6, p. 612).

costapalmate. See p. 82 (*Palmae*).

cotyledon. One of the leaves preformed in the seed.

crisped. (1) Of hairs: strongly curved so that the tip lies near the point of attachment. (2) Of leaves, leaflets or petals: finely and complexly wavy.

cristate. With elevated, irregular ridges.

crownshaft. See p. 82 (*Palmae*).

culm. See p. 61 (*Gramineae*).

cultivar. A more or less uniform assemblage of plants (usually selected in cultivation) which is clearly distinguished by 1 or more characters (morphological, physiological, cytological, chemical or others) and which, when reproduced, retains its distinguishing characters. This term is derived from *culti*vated *vari*ety. Its name can be Latin in form, e.g. 'Alba', but is more usually in a modern language, e.g. 'Madame Lemoine', 'Frühlingsgold', 'Beauty of Bath'.

cupule. A group of bracts, united at least at the base, surrounding the base of a fruit or a group of fruits.

cyathium. See p. 297 (*Euphorbia*).

cyme. An inflorescence in which the terminal flower opens first, other flowers being borne on branches which arise below it (figure 134.4, 5 & 7, p. 612).

cymule. See p. 293 (*Geranium*).

cystolith. A concretion of calcium carbonate found within the cells of the leaf in some plants; they can sometimes be seen when the leaf is viewed against the light, or felt as tiny hard lumps when the leaf is drawn between finger and thumb.

declinate. Of stamens or styles: arching gradually downwards and then gradually upwards towards the apex.

decumbent. More or less horizontal for most of its length but erect or semi-erect near the tip.

decurrent. Continued down the stem below the point of attachment as a ridge or ridges.

dehiscent. Splitting, when ripe, along 1 or more predetermined lines of weakness.

dendroid. Of hairs: branched, tree-like.

dichasial. Resembling a dichasium.

dichasium. A form of cyme in which each node bears 2 equal lateral branches (figure 134.4, p. 612).

dichotomous. Divided into 2 equal branches; regularly forked.

diffuse parietal. A type of placentation in which the ovules are borne all over the carpel walls.

dioecious. With male and female flowers on separate plants.

diploid. Possessing in its normal vegetative cells 2 similar sets of chromosomes.

disc. A variously contoured, ring-shaped or circular area (sometimes lobed) within a flower, from which nectar is secreted.

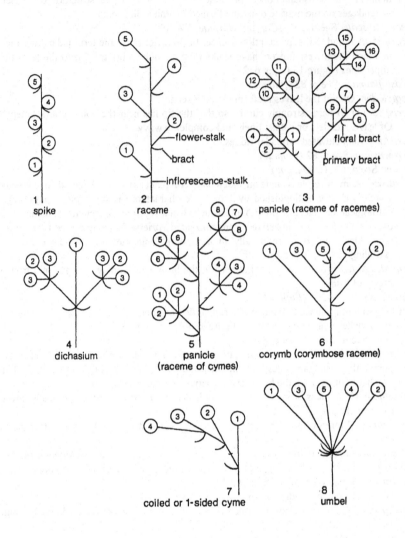

Figure 134. Inflorescence-types.

612

dissected. Deeply divided into lobes or segments.

distylic. Having flowers of different plants either with long styles and shorter stamens or with long stamens and shorter styles.

domatia. Modified projections for shelter-parasites.

dorsifixed. Attached to its stalk or supporting organ by its back, usually near the middle (figure 135.4, p. 614).

double. Of flowers: with petals much more numerous than in the normal wild state.

drupe. An indehiscent fruit in which the outer part of the wall is soft and usually fleshy but the inner part is stony. A drupe may be 1-seeded as in *Prunus* or *Juglans*, or may contain several seeds, as in *Ilex*. In the latter case, each seed is enclosed in a separate stony endocarp and constitutes a pyrene.

drupelet. A miniature drupe forming part of an aggregate fruit.

elaiosome. An oily appendage on a seed.

ellipsoid. As elliptic, but applied to a solid body.

elliptic. About twice as long as broad, tapering equally both to the tip and the base (figure 133.8, p. 610).

embryo. The part of a seed from which the new plant develops; it is distinct from the endosperm and seed-coat.

endocarp. The inner, often stony layer of a fruit-wall in those fruits in which the wall is distinctly 3-layered.

endosperm. A food-storage tissue found in many seeds, but not in all, distinct from the embryo and serving to nourish it and the young seedling during germination and establishment.

entire. With a smooth, uninterrupted margin; not lobed or toothed.

epicalyx. A group of bracts attached to the flower-stalk immediately below the calyx and sometimes partly united with it.

epigeal. The mode of germination in which the cotyledons appear above ground and carry on photosynthesis during the early stages of establishment.

epigynous. Describes a flower, or preferably the petals, sepals and stamens (or perianth and stamens) of a flower in which the ovary is inferior (figure 135.13 & 14, p. 614).

epigynous zone. A rim or cup of tissue on which the sepals, petals and stamens are borne in some flowers with inferior ovaries (figure 135.14, p. 614).

epiphyte. A plant which grows on another plant but does not derive any nutriment from it.

equitant. Of leaves: folded so that they are V-shaped in section at the base, the bases overlapping regularly, as in many *Iridaceae* (figure 133.5, p. 610).

exocarp. The outer, skin-like layer of a fruit-wall in those fruits in which the wall is distinctly 3-layered.

fall. See *Iridaceae*, p. 49 and especially figure 7, p. 50.

farina. The flour-like wax present on the stem and leaves of many species of *Primula* and of a few other plants.

fascicle. A bunch of leaves or flowers, often enclosed at the base by a sheath.

fasciculate. Clustered, with the constituent units appearing to arise from a common point.

fastigiate. With all the branches more or less erect, giving the plant a narrow tower-like outline.

filament. The stalk of a stamen, bearing the anther at its tip (figure 135.3 & 4, p. 614).

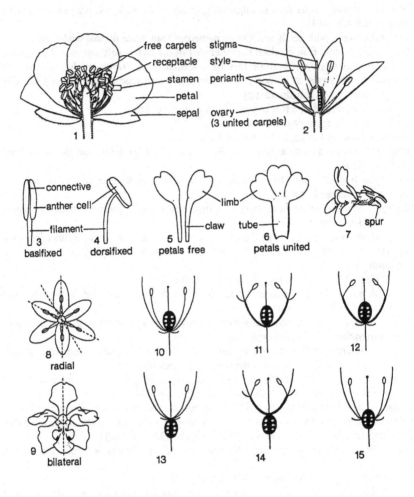

Figure 135. Flowers. 1 & 2, Two different flowers in longitudinal section, illustrating floral parts. 3 & 4, Two stamens showing alternative types of anther-attachment. 5--7, Some terms relating to petals. 8 & 9, Floral symmetry, planes of symmetry shown by broken lines. 10--15, Position of ovary: 10--12, Superior ovaries; 13--14, Inferior ovaries; 15, Half-inferior ovary; 11, Perigynous zone bearing sepals, petals and stamens. 12, Perigynous zone bearing petals and stamens; 14, Epigynous zone bearing sepals, petals and stamens.

614

filius. Used with authority names to distinguish between parent and offspring when both have given names to species, e.g. Linnaeus (C. Linnaeus, 1707--1788), Linnaeus filius (C. Linnaeus, 1741--1783), son of the former).

flagellate. Bearing whip-like structures.

floret. A small flower, aggregated with others into a compact inflorescence.

folioliferous. (Hairs) branch-like, bearing leaf-like structures.

follicle. A dry dehiscent fruit derived from a single free carpel and opening along one suture.

free. Not united to any other organ except by its basal attachment.

free-central. A form of placentation in which the ovules are attached to the central axis of a 1-celled ovary (figure 136.5, p. 616).

fruit. The structure into which the gynaecium is transformed during the ripening of the seeds; a *compound* fruit is derived from the gynaecia of more than one flower. The term 'fruit' is often extended to include structures which are derived in part from the receptacle (*Fragaria*), epigynous zone (*Malus*) or inflorescence-stalk (*Ficus*) as well as from the gynaecium.

funicle. The stalk of an ovule.

fusiform. Spindle-shaped; cylincric but tapered gradually at both ends.

gamete. A single sex-cell which fuses with one of the opposite sex during sexual reproduction.

gland-dotted. With minute patches of secretory tissue usually appearing as pits in the surface, as translucent dots when viewed against the light, or both.

glandular. (1) Of a hair: bearing at the tip a usually spherical knob of secretory tissue. (2) Of a tooth: similarly knobbed or swollen at the tip.

glaucous. Green strongly tinged with bluish grey; with a greyish waxy bloom.

glochid. See p. 162 (*Cactaceae*).

glume. A small bract in the inflorescence of a grass (*Gramineae*) or sedge (*Cyperaceae*); also used in a narrower sense to denote the (usually) 2 small bracts at the base of a grass spikelet (figure 12, p. 62).

graft-hybrid. A plant which, as a consequence of grafting, contains a mixture of tissues from 2 different species. Normally the tissues of 1 species are enclosed in a 'skin' of tissue from the other species.

gynaecium. The female organs (carpels) of a single flower, considered collectively, whether they are or united.

gynophore. The stalk which is present at the base of some ovaries (and the fruits developed from them).

half-inferior. Of an ovary: with its lower part inferior and its upper part superior (figure 135.15, p. 614).

haploid. Possessing in its normal vegetative cells only a single set of chromosomes.

hastate. With 2 acute, divergent lobes at the base, as in a mediaeval halberd (figure 133.12, p. 610).

hastula. See p. 82 (*Palmae*).

haustorium. The organ with which a parasitic plant penetrates its host and draws nutriment from it.

head. See *capitulum*.

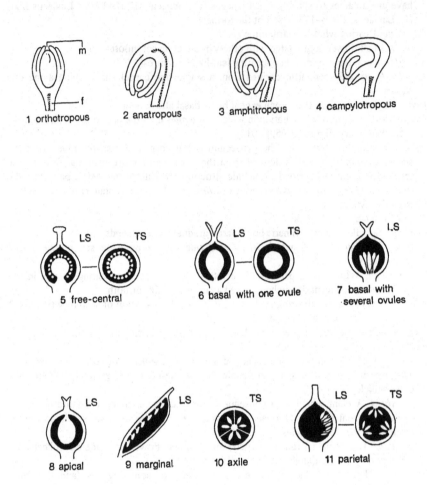

1 orthotropous   2 anatropous   3 amphitropous   4 campylotropous

LS   TS   5 free-central

LS   TS   6 basal with one ovule

I.S   7 basal with several ovules

LS   8 apical

LS   9 marginal

TS   10 axile

LS   TS   11 parietal

Figure 136. Ovules and placentation. 1--4, Ovule forms (f, funicle; m, micropyle). 5--11, Placentation types (LS, longitudinal section; TS, transverse section).

616

herb. A plant in which the stems do not become woody, or, if somewhat woody at the base, do not persist from year to year.

herbaceous. Of a plant: possessing the qualities of a herb, as defined above.

heterostylous. Having flowers in which the length of the style relative to that of the stamens varies from one plant to another.

hilum. The scar-like mark on a seed indicating the point at which it was attached to the funicle.

hip. A fruit formed from a tubular, cup-like or urceolate fleshy perigynous zone, containing several free carpels (see p. 255, *Rosa*).

hybrid. A plant produced by the crossing of parents belonging to two different named groups (e.g. genera, species, subspecies, etc.). An F1 hybrid is the primary product of such a cross. An F2 hybrid is a plant arising from a cross between two F1 hybrids (or from the self-pollination of an F1 hybrid).

hydathode. A water-secreting gland immersed in the tissue of a leaf near its margin.

hypocotyl. That part of the stem of a seedling which lies between the top of the radicle and the attachment of the cotyledon(s).

hypogeal. The mode of germination in which the cotyledons remain in the seed-coat and play no part in photosynthesis.

hypogynous. Describes a flower, or, preferably the petals, sepals and stamens (or perianth and stamens) of a flower in which the ovary is superior and the petals, sepals and stamens (or perianth and stamens) arise as individual whorls on the receptacle (figure 135.10, p. 614).

imbricate. With margins of adjacent organs of the same kind overlapping.

incised. With deep, narrow spaces between the teeth or lobes.

included. Not projecting beyond the organs which enclose it.

indefinite. More than 12 and possibly variable in number.

indehiscent. Without preformed lines of splitting, opening, if at all, irregularly by decay.

indumentum. Any hair-covering.

induplicate. See p. 82 (*Palmae*).

inferior. Of an ovary: borne beneath the sepals, petals and stamens (or perianth and stamens) so that these appear to arise from its top (figure 135.13 & 14, p. 614).

inflorescence. A number of flowers which are sufficiently closely grouped together to form a structured unit (figure 134, p. 612).

infraspecific. Denotes any category below species level, such as subspecies, variety and form. To be distinguished from *subspecific*, which means relating to subspecies only.

integument. The covering of an ovule, later developing into the seed-coat. Some ovules have a single integument, others 2.

internode. The part of a stem between 2 successive nodes.

interpetiolar (stipules). Borne between the leaf-bases.

interprimary vein. See p. 85 (*Araceae*).

intrapetiolar (stipules). Borne between the leaf-base and the stem.

involucel. A whorl of united bracteoles borne below the flower in some *Dipsacaceae*.

involute. Leaf-margins which are rolled upwards and inwards in bud.

involucre. A compact cluster or whorl of bracts around the stalk at or near the base of some flowers or inflorescences or around the base of a capitulum; sometimes reduced to a ring of hairs.

keel. A narrow ridge, suggestive of the keel of a boat, developed along the midrib (or rarely other veins) of a leaf, sepal or petal; in the particular case of the flowers of *Leguminosae*, a structure formed by the uniting along at least parts of their lower margins of the 2 lower petals.

laciniate. With the margin deeply and irregularly divided into narrow and unequal teeth.

lamellate. Made up of thin plates.

lanceolate. 3--4 times as long as wide and tapering more or less gradually towards the tip (figure 133.9, p. 610).

layer. To propagate by pegging down on the ground a branch from near the base of a shrub or tree, so as to induce the formation of adventitious roots.

leaflet. One of the leaf-like components of a compound leaf.

legume. The typical fruit of the *Leguminosae*, formed from a single carpel and opening down both sutures; legumes are, however, variable, some indehiscent, others breaking into 1-seeded segments, etc.

lemma. The lower and usually stouter of the 2 horny or membranous bracts which enclose the flower of a grass (*Gramineae*) -- figure 12, p. 62.

lenticel. A small, slightly raised interruption of the surface of the bark (or the corky outer layers of a fruit) through which air can penetrate to the inner tissues.

liana. A woody climber.

lignotuber. A woody, underground tuber.

ligule. A small herbaceous, petaloid or membranous flap (more rarely a line of hairs) at the base of a leaf-blade or petal (or perianth-segment).

limb. A broadened part, furthest from the base, of a petal, corolla or similar organ which has a narrow basal part (the claw or tube). Effectively synonymous with *blade*.

linear. Parallel-sided and many times longer than broad (figure 133.6, p. 610).

lip. (1) A major division of the apical part of a bilaterally symmetric calyx or corolla in which the petals or sepals are united; there is normally an upper and lower lip, but either may be missing. (2) That petal in the flower of an orchid (*Orchidaceae*) which differs from the other 2; it usually occupies the lowest position in the flower and serves as an alighting place for insects -- see figure 19, p. 98. (3) A staminode or petal, suggestive of the lip of an orchid, in the flower of some other groups (e.g. *Zingiberaceae*).

lodicule. Small, scale-like structures in the flowers of most grasses (*Gramineae*) which occur between the lemma and the reproductive organs; by their swelling, they cause the lemma and palea to diverge at flowering time (figure 12, p. 62).

lyrate. Pinnatifid or pinnatisect, with a large terminal and small lateral lobes.

marginal. Of placentation: describing the placentation found in a free carpel which contains more than 1 ovule.

medifixed. Of a hair: lying parallel to the surface on which it is borne and attached to it by a stalk (usually short) near its mid-point.

mentum. See figure 19, p. 98 (*Orchidaceae*).

mericarp. A carpel, usually 1-seeded, released by the break-up at maturity of a fruit formed from 2 or more joined carpels.

mesocarp. The central, often fleshy layer of a fruit-wall in those fruits in which the wall is distinctly 3-layered.

micropyle. A pore in the integument(s) of an ovule and later in the coat of a seed (figure 136.1, p. 616).

monocarpic. Flowering and fruiting once, then dying.

monoecious. With separate male and female flowers on the same plant; male flowers may contain non-functional carpels (and vice versa).

monopodial. A type of growth pattern in which the terminal bud continues growth from year to year.

mucronate. Provided with a short, narrow point at the apex (figure 133.25, p. 610).

mycorrhiza. A symbiotic association between the roots of a green plant and a fungus.

nectary. A nectar-secreting gland.

neuter. Without either functional male or female parts.

node. The point at which 1 or more leaves or flower parts are attached to an axis.

nut. A 1-seeded indehiscent fruit with a woody or bony wall.

nutlet. A small nut, usually a component of an aggregate fruit.

obconical. Shaped like a cone, but attached at the narrow end.

obcordate. Inversely heart-shaped, the notch being apical.

oblanceolate. As lanceolate, but attached at the more gradually tapered end.

oblong. With more or less parallel sides and about 2--5 times as long as broad (figure 133.7, p. 610).

obovate. As ovate, but attached at the narrower end.

obovoid. As ovoid, but attached at the narrower end.

obsolete. Rudimentary; scarcely visible, reduced to insignificance.

ochrea. A sheath, made up of the stipules of each leaf-base, which surrounds the stem in many species of *Polygonaceae*.

opposite. Describes 2 leaves, branches or flowers attached on opposite sides of the axis at the same node (figure 133.1, p. 610).

orthotropous. Describes an ovule which stands erect and straight (figure 136.1, p. 616).

ovary. The lower part of a carpel, containing the ovules(s) (i.e. excluding style and stigma); the lower, ovule-containing part of a gynaecium in which the carpels are united (figure 135.1 & 2, p. 614).

ovate. With approximately the outline of a hen's egg (though not necessarily blunt-tipped) and attached at the broader end (figure 133.10, p. 610).

ovoid. As ovate, but applied to a solid body.

ovule. The small body from which a seed develops after pollination (figure 136, p. 616).

palea. The upper and usually smaller and thinner of the 2 bracts enclosing the flower of a grass (*Gramineae*) -- see figure 12, p. 62.

palmate. Describes a compound leaf composed of more than 3 leaflets, all arising from the same point, as in the leaf of *Aesculus*; also used to described similar venation in simple leaves (figure 133.19, p. 612).

palmatifid. Lobed in a palmate manner, with the incisions pointing to the place of attachment, but not reaching much more than halfway to it (figure 133.20, p. 610).

palmatisect. Deeply lobed in a palmate manner, with the incisions almost reaching the base (figure 133.21, p. 610).

panicle. A compound raceme, or any freely branched inflorescence of similar appearance (figure 134.3, p. 612).

papillose. Covered with small blunt protuberances (papillae).

parietal. A form of placentation in which the placentas are borne on the inner surface of the walls of a 1-celled ovary, or, rarely in a similar manner in a septate ovary (figure 136.11, p. 616).

pectinate. With leaves, leaflets, or hairs in regular, eyelash-like rows.

pedate. With a terminal lobe or leaflet, and on either side of it an axis curving outwards and backwards, bearing lobes or leaflets on the outer side of the curve (figure 133.22, p. 610).

peltate. Describes a leaf or other structure with the stalk attached other than at the margin (figure 133.27, p. 610).

pepo. A gourd fruit; a 1-celled, many-seeded, inferior fruit, with parietal placentas and pulpy interior.

perennial. Persisting for more than 2 years.

perfoliate. Describes a pair of stalkless opposite leaves of which the bases are united, or a single leaf in which the auricles are united so that the stem appears to pass through the leaf or leaves (figure 133.4, p. 610).

perianth. The calyx and corolla considered collectively, used especially when there is no clear differentiation between calyx and corolla; also used to denote a calyx or corolla when the other is absent (figure 135.1 & 2, p. 616).

pericarp. The fruit wall.

pericarpel. See p. 82 (*Cactaceae*).

perigynous. Describing a flower, or preferably the petals, sepals and stamens (or perianth and stamens) of a flower in which the ovary is superior and the petals, sepals and stamens (or perianth and stamens) are borne on the margins of a rim or cup which itself is borne on the receptacle below the ovary -- it often appears as though the sepal, petals and stamens (or perianth and stamens) are united at their bases (figure 135.11 & 12, p. 614).

perigynous zone. The rim or cup of tissue on which the sepals, petals and stamens are borne in a perigynous flower (figure 135.11 & 12, p. 614).

petal. A member of the inner perianth-whorl (corolla) used mainly when this is clearly differentiated from the calyx. The petals usually function in display and often provide an alighting place for pollinators (figure 135.1, p. 614).

petaloid. Like a petal in texture and colour.

phyllode. A leaf-stalk taking on the function and, to a variable extent, the form of a leaf-blade.

pinnate. Describes a compound leaf in which distinct leaflets are arranged on either side of the axis (figure 133.15, p. 610). If these leaflets are themselves of a similar compound structure, the leaf is termed *bipinnate* (similarly *tripinnate*, etc.).

pinnatifid. Lobed in a pinnate manner, with the incisions reaching not much more than halfway to the axis (figure 133.16, p. 610).

pinnatisect. Deeply lobed in a pinnate manner, with the incisions almost reaching the axis (figure 133.17 p. 610).

pistillode. A sterile ovary in a male flower.

placenta. A part of the ovary, often in the form of a cushion or ridge, to which the ovules are attached.

placentation. The manner of arrangement of the placentas (figure 136, p. 616).

plicae. See p. 427 (*Gentiana*).

pollen-sac. One of the cavities in an anther in which pollen is formed; each anther normally contains 4 pollen-sacs, 2 on either side of the connective, those of each pair separated by a partition which shrivels at maturity.

pollinium. Regularly-shaped masses of pollen formed by a large number of pollen-grains cohering (*Orchidaceae, Asclepiadaceae*).

polycarpic. See p. 197 (*Papaveraceae*).

polyploid. Possessing in the normal vegetative cells more than 2 sets of chromosomes.

pome. A fruit made up of a fleshy receptacle, surrounding and attached to 1 or more free or united carpels, whose walls are often parchment-like.

procumbent. Creeping along the substrate.

proliferous. Giving rise to plantlets or additional flowers on stems, leaves or in the inflorescence.

prophyll. See p. 53 (*Crocus*).

protandrous. With anthers beginning to shed their pollen before the stigmas of the same flower are receptive.

protogynous. With stigmas becoming receptive before the anthers in the same flower shed their pollen.

pseudobulb. A swollen, above-ground internode or group of internodes, green (at least when young), characteristic of epiphytic *Orchidaceae*; see figure 18, p. 96.

pulvinus. A swollen region at the base of a leaflet, leaf-blade or leaf-stalk.

pustulate. With low swellings.

pyrene. A small nut-like body enclosing a seed, 1 or more of which, surrounded by fleshy tissue, make up the fruit of, for example, *Ilex*.

raceme. An inflorescence consisting of stalked flowers arranged on a single axis, the lower opening first (figure 134.2, p. 612).

rachilla See p. 61 (*Gramineae*).

rachis. The axis of a pinnate leaf.

radially symmetric. Capable of division into 2 similar halves along 2 or more planes of symmetry (figure 135.8, p. 614).

radiate. (1) Spreading outwards from a common centre. (2) Possessing ray-flowers, as in the *Compositae*.

radicle. The root pre-formed in the seed and normally the first visible root of a seedling.

ramiform. Of hairs: branched.

raphe. A perceptible ridge or stripe, at one end of which is the hilum, on some seeds.

receptacle. The tip of an axis to which the floral parts, or perigynous zone (when present), are attached (figure 135.1, p. 614).

reduplicate. See p. 82 (*Palmae*).

reflexed. Bent sharply backwards from the base.

resupinate. See p. 97 (*Orchidaceae*).

revolute. Of leaf-margins: rolled under and inwards.

rhizome. A horizontal stem, situated underground or on the surface, serving the purpose of food-storage or vegetative reproduction or both; roots or stems arise from some or all of its nodes.

rootstock. The compact mass of tissue from which arise the new shoots of a herbaceous perennial. It usually consists mainly of stem tissue, but is more compact than is generally understood by rhizome.

rostellum. See p. 95 (*Orchidaceae*).

ruminate. With tissues of different layers interdigitating, waved and overlapping.

runner. A slender, above-ground stolon with very long internodes.

sagittate. With a backwardly directed basal lobe on each side, like an arrow-head (figure 133.11, p. 610).

samara. A winged, dry, indehiscent fruit or mericarp.

saprophytic. Dependent for its nutrition on soluble organic compounds in the soil. Saprophytic plants do not photosynthesise and lack chlorophyll; some plants, however, are *partially saprophytic* and combine the two modes of nutrition.

sarcostesta. See p. 82 (*Palmae*).

scale-leaf. A reduced leaf, usually not photosynthetic.

scape. A leafless flower-stalk or inflorescence-stalk arising usually from near ground level.

scarious. Dry and papery, often translucent.

schizocarp. A fruit which, at maturity, splits into its constituent mericarps.

scion. A branch cut from one plant to be grafted on the rooted stock of another.

seed. A reproductive body adapted for dispersal, developed from an ovule and consisting of a protective covering (the seed-coat) and embryo, and, usually, a food-reserve.

semi-parasite. A plant which obtains only part of its nutrition by parasitism.

sepal. A member of the outer perianth-whorl (calyx) when 2 whorls are clearly differentiated as calyx and corolla, or when comparison with related plants shows that a corolla is absent. The sepals most often function in protection and support of other floral parts (figure 135.1, p. 614).

septum. An internal partition.

sheath. The part of a leaf or leaf-stalk which surrounds the stem, being either tubular or with free but overlapping margins.

shrub. A woody plant with several stems or branches arising from near the base, and of smaller stature than a tree.

silicula. See p. 207 (*Cruciferae*).

siliqua. See p. 207 (*Cruciferae*).

simple. Not divided into separate parts.

sinus. The gap or indentation between 2 lobes, auricle or teeth.

spadix. A spike with numerous small flowers borne on a usually fleshy axis which may project beyond the topmost flower, often wholly or partly enclosed by a large bract (*spathe*) -- see p. 85.

spathe. A large bract at the base of flower or inflorescence and wholly or partly enclosing it; the term is used in some families to denote collectively 2 or 3 such bracts.

spathulate. With a narrow basal part, which towards the apex is gradually expanded into a broad, blunt blade.

spicate. Similar to a spike.

spike. An inflorescence or subdivision of an inflorescence, consisting of stalkless flowers arranged on a single axis (figure 134.1, p. 612).

spikelet. A small spike forming one of the units of a complex inflorescence in the grasses -- *Gramineae*; figure 12, p. 62.

spiral (arrangement of leaves or other organs). Borne 1 at each node, the nodes arranged along the stem in a spiral.

spur. An appendage or prolongation, more or less cylindric, often at the base of an organ. The spur of a corolla or single petal or sepal is usually hollow and often contains nectar (figure 135.7, p. 614).

stamen. The male organ, producing pollen, generally consisting of an anther borne on a filament (figure 135.1, p. 614).

staminode. An infertile stamen, often reduced or rudimentary or with a changed function.

standard. See *Iridaceae* (p. 46) and *Leguminosae* (p. 267), in which the term is applied to the uppermost petal, which is often larger than the rest and reflexed upwards or backwards.

stellate. Star-like, particularly of branched hairs.

stigma. The part of a style to which the pollen adheres, normally differing in texture from the rest of the style (figure 135.2, p. 614).

stipe. (1) See p. 95 (*Orchidaceae*). (2) See p. 84 (*Araceae*).

stipel. An organ equivalent to a stipule, but borne at the base of a leaflet rather than a leaf.

stipule. An appendage, usually 1 pair, found beside the base of the leaf-stalk in many flowering plants, sometimes falling early, leaving a scar. In some cases the 2 stipules are united; in others they are partly united to the leaf-stalk.

stock. A rooted plant, often with the upper parts removed, on to which a scion may be grafted.

stolon. A far-creeping, more or less slender, above-ground or underground rhizome giving rise to a new plant at its tip and sometimes at intermediate nodes.

stoma. A microscopic ventilating pore in the surface of a leaf or other herbaceous part.

strophiole. See p. 157 (*Caryophyllaceae*).

style. The usually slender, upper part of a carpel or gynaecium, bearing the stigma (figure 135.2, p. 614).

stylopodium. See p. 382 (*Umbelliferae*).

subtend(ed). Used of any structure (e.g. a flower) which occurs in the axil of another organ (e.g. a bract); in this case the bract subtends the flower.

subulate. Narrowly cylindric, and somewhat tapered to the tip.

sucker. An erect shoot originating from a bud on a root or rhizome, sometimes at some distance from the parent plant.

superior. Of an ovary: borne at the morphological apex of the flower so that the petals, sepals and stamens (or perianth and stamens) arise on the receptacle below the ovary (figure 135.10--12, p. 614).

suture. A line marking an apparent junction of neighbouring parts.

sympodial. A type of growth pattern in which the terminal bud ceases growth, further growth being carried on by a lateral bud; of some importance in *Orchidaceae*.

syncarp. A compound fruit formed from the growing together of several ripening ovaries.

tendril. A thread-like structure which by its coiling growth can attach a shoot to something else for support.

terete. Approximately circular in cross-section; not necessarily perfectly cylindric, but without grooves or ridges.

terminal. Borne at the apex of a shoot or branch.

tessellated. With a chequered pattern of light and dark squares.

tetraploid. Possessing in its normal vegetative cells 4 similar sets of chromosomes.

throat. The part of a calyx or corolla transitional between the tube and limb or lobes.

trichome. See p. 162 (*Cactaceae*).

triploid. Possessing in its normal vegetative cells 3 similar sets of chromosomes.

tristylic. Having flowers of different plants with long, short or intermediate-length styles; the stamens of each flower are of 2 lengths which are not the same as the style-length of that flower.

truncate. As though with the tip or base cut off at right angles.

tuber. A swollen underground stem or root used for food-storage.

tubercle. A small, blunt, wart-like protuberance.

tunic. The dead covering of a bulb or corm.

turion. A specialised perennating bud in some aquatic plants. Consists of a short shoot covered in closely packed leaves, which persists through the winter at the bottom of the water.

umbel. An inflorescence in which the flower-stalks arise together from the top of an inflorescence-stalk; this is a *simple umbel* (figure 134.8, p. 612). In a *compound umbel* the several stalks arising from the top of the inflorescence-stalk terminate not in flowers but in secondary umbels.

undivided. Without major divisions or incisions, though not necessarily entire.

urceolate. Shaped like a pitcher or urn, hollow and contracted at or just below the mouth.

utricle. See p. 91 (*Carex*).

valvate. Of sepals or petals: edge-to-edge in bud, not overlapping, though the edges may sometimes be folded inwards (e.g. perianth-segments of *Clematis*).

valve. (1) That part of the fruit, covering the seeds, that falls during dehiscence. (2) a partially detached flap of an anther which opens to release the pollen.

vascular bundle. A strand of conducting tissue, usually surrounded by softer tissue.

vein. A vascular strand, usually in leaves or floral parts and visible externally.

venation. The pattern formed by the veins in a leaf.

ventricose. Swollen on 1 side.

versatile. Of an anther: flexibly attached to the filament by its approximate mid-point so that a rocking motion is possible.

vessel. A microscopic water-conducting tube formed by a sequence of cells not separated by end-walls.

vesicle. A small bladder-like sac or cavity filled with fuid or air.

vesicular. Possessing vesicles.

viscidium. See p. 95 (*Orchidaceae*).

viviparous. Bearing young plants, bulbils or leafy buds which can take root; they can occur anywhere on the plant and may be interspersed with, or wholly replace, the flowers in an inflorescence.

whorl. A group of more than 2 leaves or floral organs inserted at the same node.

wing. A thin, flat extension of a fruit, seed, sepal or other organ; or, for *Leguminosae*, the 2 lateral petals (each of which is known as a wing).

xerophytic. Drought-tolerant. Can also describe the environment in which drought-tolerant plants live.

Ehrharta, 66
Elaeagnaceae, 342
Elaeagnus, 342
Elaeocarpaceae, 328
Elatinaceae, 351
Elatine, 352
Eleocharis, 91
Eleutherococcus, 378
Elliottia, 402
Elmera, 236
Elodea, 7
Elsholtzia, 478
Elymus, 70
Embothrium, 141
Emilia, 594
Emmenanthe, 446
Emmenopterys, 436
Empetraceae, 408
Empetrum, 408
Encelia, 601
Engelmannia, 604
Enkianthus, 405
Entelea, 329
Eomecon, 203
Epacridaceae, 409
Epacris, 409
Epigaea, 402
Epilobium, 369
Epimedium, 183
Epipactis, 101
Epixiphium, 502
Eragrostis, 67
Eranthis, 174
Ercilla, 147
Eremurus, 19
Erica, 404
Ericaceae, 399
Erigeron, 585
Erinacea, 289
Erinus, 503
Eriobotrya, 263
Eriocephalus, 593
Eriochloa, 69
Eriogonum, 144
Eriolobus, 261
Eriophorum, 90
Eriophyllum, 599
Eritrichium, 453

Erodium, 293
Eruca, 222
Eryngium, 390
Erysimum, 215
Erythrina, 279
Erythronium, 23
Escallonia, 243
Escalloniaceae, 231
Eschscholzia, 204
Eschscholzioideae, 197
Eucalyptus, 363
Eucnide, 353
Eucommia, 137
Eucommiaceae, 137
Eucryphia, 191
Eucryphiaceae, 190
Eumorphia, 588
Euodia, 303
Euonymus, 320
Eupatorium, 606
Euphorbia, 298
Euphorbieaea, 294
Euphorbiaceae, 297
Euptelea, 171
Eupteleaceae, 170
Euryops, 594
Eustoma, 427
Evolvulus, 443
Exochorda, 250
Fabaceae, 267
Fabiana, 487
Fagaceae, 131
Fagopyrum, 144
Fagus, 133
Fallopia, 144
Fascicularia, 59
× Fatshedera, 379
Fatsia, 381
Fedia, 528
Felicia, 586
Fendlera, 241
Ferraria, 51
Ferula, 395
Festuca, 77
Fibigia, 217
Ficus, 139
Filipendula, 253
Flacourtiaceae, 343

Foeniculum, 392
Fontanesia, 422
Forestiera, 422
Forsythia, 423
Fortunearia, 225
Fothergilla, 225
Fragaria, 260
Francoa, 240
Francoaceae, 231
Frangula, 325
Frankenia, 351
Frankeniaceae, 351
Franklinia, 193
Fraxinus, 422
Fremontodendron, 338
Fritillaria, 23
Fuchsia, 368
Fumana, 349
Fumaria, 206
Fumariaceae, 197
Fumarioideae, 196
Funkia, 21
Gagea, 22
Gaillardia, 599
Galactites, 572
Galanthus, 42
Galax, 397
Galaxia, 53
Galega, 282
Galeopsis, 469
Galium, 438
Galtonia, 29
Garrya, 375
Garryaceae, 374,
Gaudinia, 73
Gaultheria, 408
Gaura, 369
Gaylussacia, 408
Gazania, 579
Geissorhiza, 55
Gelasine, 53
Gelsemium, 425
Genista, 290
Gentiana, 427
Gentianaceae, 425
Gentianella, 427
Gentianopsis, 427
Geraniaceae, 291

Printed in the United States
By Bookmasters